# Advances in Intelligent Systems and Computing

Volume 826

**Series editor**

Janusz Kacprzyk, Polish Academy of Sciences, Warsaw, Poland
e-mail: kacprzyk@ibspan.waw.pl

The series "Advances in Intelligent Systems and Computing" contains publications on theory, applications, and design methods of Intelligent Systems and Intelligent Computing. Virtually all disciplines such as engineering, natural sciences, computer and information science, ICT, economics, business, e-commerce, environment, healthcare, life science are covered. The list of topics spans all the areas of modern intelligent systems and computing such as: computational intelligence, soft computing including neural networks, fuzzy systems, evolutionary computing and the fusion of these paradigms, social intelligence, ambient intelligence, computational neuroscience, artificial life, virtual worlds and society, cognitive science and systems, Perception and Vision, DNA and immune based systems, self-organizing and adaptive systems, e-Learning and teaching, human-centered and human-centric computing, recommender systems, intelligent control, robotics and mechatronics including human-machine teaming, knowledge-based paradigms, learning paradigms, machine ethics, intelligent data analysis, knowledge management, intelligent agents, intelligent decision making and support, intelligent network security, trust management, interactive entertainment, Web intelligence and multimedia.

The publications within "Advances in Intelligent Systems and Computing" are primarily proceedings of important conferences, symposia and congresses. They cover significant recent developments in the field, both of a foundational and applicable character. An important characteristic feature of the series is the short publication time and world-wide distribution. This permits a rapid and broad dissemination of research results.



More information about this series at http://www.springer.com/series/11156

Sebastiano Bagnara · Riccardo Tartaglia
Sara Albolino · Thomas Alexander
Yushi Fujita
Editors

# Proceedings of the 20th Congress of the International Ergonomics Association (IEA 2018)

## Volume IX: Aging, Gender and Work, Anthropometry, Ergonomics for Children and Educational Environments

Springer

*Editors*
Sebastiano Bagnara
University of the Republic of San Marino
San Marino, San Marino

Thomas Alexander
Fraunhofer FKIE
Bonn, Nordrhein-Westfalen
Germany

Riccardo Tartaglia
Centre for Clinical Risk Management and Patient Safety, Tuscany Region
Florence, Italy

Yushi Fujita
International Ergonomics Association
Tokyo, Japan

Sara Albolino
Centre for Clinical Risk Management and Patient Safety, Tuscany Region
Florence, Italy

ISSN 2194-5357 ISSN 2194-5365 (electronic)
Advances in Intelligent Systems and Computing
ISBN 978-3-319-96064-7 ISBN 978-3-319-96065-4 (eBook)
https://doi.org/10.1007/978-3-319-96065-4

Library of Congress Control Number: 2018950646



This Springer imprint is published by the registered company Springer Nature Switzerland AG
The registered company address is: Gewerbestrasse 11, 6330 Cham, Switzerland

# Preface

The Triennial Congress of the International Ergonomics Association is where and when a large community of scientists and practitioners interested in the fields of ergonomics/human factors meet to exchange research results and good practices, discuss them, raise questions about the state and the future of the community, and about the context where the community lives: the planet. The ergonomics/human factors community is concerned not only about its own conditions and perspectives, but also with those of people at large and the place we all live, as Neville Moray (Tatcher et al. 2018) taught us in a memorable address at the IEA Congress in Toronto more than twenty years, in 1994.

The Proceedings of an IEA Congress describes, then, the actual state of the art of the field of ergonomics/human factors and its context every three years.

In Florence, where the XX IEA Congress is taking place, there have been more than sixteen hundred (1643) abstract proposals from eighty countries from all the five continents. The accepted proposal has been about one thousand (1010), roughly, half from Europe and half from the other continents, being Asia the most numerous, followed by South America, North America, Oceania, and Africa. This Proceedings is indeed a very detailed and complete state of the art of human factors/ergonomics research and practice in about every place in the world.

All the accepted contributions are collected in the Congress Proceedings, distributed in ten volumes along with the themes in which ergonomics/human factors field is traditionally articulated and IEA Technical Committees are named:

I. Healthcare Ergonomics (ISBN 978-3-319-96097-5).
II. Safety and Health and Slips, Trips and Falls (ISBN 978-3-319-96088-3).
III. Musculoskeletal Disorders (ISBN 978-3-319-96082-1).
IV. Organizational Design and Management (ODAM), Professional Affairs, Forensic (ISBN 978-3-319-96079-1).
V. Human Simulation and Virtual Environments, Work with Computing Systems (WWCS), Process control (ISBN 978-3-319-96076-0).

VI. Transport Ergonomics and Human Factors (TEHF), Aerospace Human Factors and Ergonomics (ISBN 978-3-319-96073-9).
VII. Ergonomics in Design, Design for All, Activity Theories for Work Analysis and Design, Affective Design (ISBN 978-3-319-96070-8).
VIII. Ergonomics and Human Factors in Manufacturing, Agriculture, Building and Construction, Sustainable Development and Mining (ISBN 978-3-319-96067-8).
IX. Aging, Gender and Work, Anthropometry, Ergonomics for Children and Educational Environments (ISBN 978-3-319-96064-7).
X. Auditory and Vocal Ergonomics, Visual Ergonomics, Psychophysiology in Ergonomics, Ergonomics in Advanced Imaging (ISBN 978-3-319-96058-6).

Altogether, the contributions make apparent the diversities in culture and in the socioeconomic conditions the authors belong to. The notion of well-being, which the reference value for ergonomics/human factors is not monolithic, instead varies along with the cultural and societal differences each contributor share. Diversity is a necessary condition for a fruitful discussion and exchange of experiences, not to say for creativity, which is the "theme" of the congress.

In an era of profound transformation, called either digital (Zisman & Kenney, 2018) or the second machine age (Bnynjolfsson & McAfee, 2014), when the very notions of work, fatigue, and well-being are changing in depth, ergonomics/human factors need to be creative in order to meet the new, ever-encountered challenges. Not every contribution in the ten volumes of the Proceedings explicitly faces the problem: the need for creativity to be able to confront the new challenges. However, even the more traditional, classical papers are influenced by the new conditions.

The reader of whichever volume enters an atmosphere where there are not many well-established certainties, but instead an abundance of doubts and open questions: again, the conditions for creativity and innovative solutions.

We hope that, notwithstanding the titles of the volumes that mimic the IEA Technical Committees, some of them created about half a century ago, the XX Triennial IEA Congress Proceedings may bring readers into an atmosphere where doubts are more common than certainties, challenge to answer ever-heard questions is continuously present, and creative solutions can be often encountered.

## Acknowledgment

A heartfelt thanks to Elena Beleffi, in charge of the organization committee. Her technical and scientific contribution to the organization of the conference was crucial to its success.

## References

Brynjolfsson E., A, McAfee A. (2014) The second machine age. New York: Norton.
Tatcher A., Waterson P., Todd A., and Moray N. (2018) State of science: Ergonomics and global issues. Ergonomics, 61 (2), 197–213.
Zisman J., Kenney M. (2018) The next phase in digital revolution: Intelligent tools, platforms, growth, employment. Communications of ACM, 61 (2), 54–63.

Sebastiano Bagnara
Chair of the Scientific Committee, XX IEA Triennial World Congress
Riccardo Tartaglia
Chair XX IEA Triennial World Congress
Sara Albolino
Co-chair XX IEA Triennial World Congress

# Organization

## Organizing Committee

| | |
|---|---|
| Riccardo Tartaglia (Chair IEA 2018) | Tuscany Region |
| Sara Albolino (Co-chair IEA 2018) | Tuscany Region |
| Giulio Arcangeli | University of Florence |
| Elena Beleffi | Tuscany Region |
| Tommaso Bellandi | Tuscany Region |
| Michele Bellani | Humanfactor[x] |
| Giuliano Benelli | University of Siena |
| Lina Bonapace | Macadamian Technologies, Canada |
| Sergio Bovenga | FNOMCeO |
| Antonio Chialastri | Alitalia |
| Vasco Giannotti | Fondazione Sicurezza in Sanità |
| Nicola Mucci | University of Florence |
| Enrico Occhipinti | University of Milan |
| Simone Pozzi | Deep Blue |
| Stavros Prineas | ErrorMed |
| Francesco Ranzani | Tuscany Region |
| Alessandra Rinaldi | University of Florence |
| Isabella Steffan | Design for all |
| Fabio Strambi | Etui Advisor for Ergonomics |
| Michela Tanzini | Tuscany Region |
| Giulio Toccafondi | Tuscany Region |
| Antonella Toffetti | CRF, Italy |
| Francesca Tosi | University of Florence |
| Andrea Vannucci | Agenzia Regionale di Sanità Toscana |
| Francesco Venneri | Azienda Sanitaria Centro Firenze |

## Scientific Committee

| | |
|---|---|
| Sebastiano Bagnara (President of IEA2018 Scientific Committee) | University of San Marino, San Marino |
| Thomas Alexander (IEA STPC Chair) | Fraunhofer-FKIE, Germany |
| Walter Amado | Asociación de Ergonomía Argentina (ADEA), Argentina |
| Massimo Bergamasco | Scuola Superiore Sant'Anna di Pisa, Italy |
| Nancy Black | Association of Canadian Ergonomics (ACE), Canada |
| Guy André Boy | Human Systems Integration Working Group (INCOSE), France |
| Emilio Cadavid Guzmán | Sociedad Colombiana de Ergonomia (SCE), Colombia |
| Pascale Carayon | University of Wisconsin-Madison, USA |
| Daniela Colombini | EPM, Italy |
| Giovanni Costa | Clinica del Lavoro "L. Devoto," University of Milan, Italy |
| Teresa Cotrim | Associação Portuguesa de Ergonomia (APERGO), University of Lisbon, Portugal |
| Marco Depolo | University of Bologna, Italy |
| Takeshi Ebara | Japan Ergonomics Society (JES)/Nagoya City University Graduate School of Medical Sciences, Japan |
| Pierre Falzon | CNAM, France |
| Daniel Gopher | Israel Institute of Technology, Israel |
| Paulina Hernandez | ULAERGO, Chile/Sud America |
| Sue Hignett | Loughborough University, Design School, UK |
| Erik Hollnagel | University of Southern Denmark and Chief Consultant at the Centre for Quality Improvement, Denmark |
| Sergio Iavicoli | INAIL, Italy |
| Chiu-Siang Joe Lin | Ergonomics Society of Taiwan (EST), Taiwan |
| Waldemar Karwowski | University of Central Florida, USA |
| Peter Lachman | CEO ISQUA, UK |
| Javier Llaneza Álvarez | Asociación Española de Ergonomia (AEE), Spain |
| Francisco Octavio Lopez Millán | Sociedad de Ergonomistas de México, Mexico |

# Contents

## Anthropometry

## Ergonomics for Children and Educational Environments

# Aging

# Impact of Exercise and Ergonomics on the Perception of Fatigue in Workers: A Pilot Study

A. C. H. Pinetti[1(✉)], N. C. H. Mercer[2], Y. A. Zorzi[2], F. Poli[3], E. Nogiri[3], A. C. Lima[3], and M. R. Oliveira[4]

[1] SESI, Arapongas, Parana, Brazil
aline.pinetti@sistemafiep.org.br
[2] SESI, Curitiba, Parana, Brazil
[3] Caemmun Industria e Comercio de Moveis Ltda, Arapongas, Parana, Brazil
[4] University Pitagoras Unopar, Londrina, Parana, Brazil

**Abstract.** The supervised implementation of psychophysiological recovery breaks and physical exercise in the workplace can be motivational factors for employees to be productive and remain for a longer time at work2. The aim of this study was to evaluate whether ergonomic conditions and exercise programs can reduce fatigue before, during and after work hours. In general, participants that conducted psychophysiological recovery breaks and an exercise program showed less fatigue compared with those that did not perform the program, principally in the times during and after work hours. However, the results showed no differences between groups that exercised either with or without psychophysiological recovery breaks, suggesting that the practice of exercise can be as important as rest. In conclusion, psychophysiological recovery breaks, ergonomic conditions and exercise programs may help to reduce fatigue during and after work hours.

**Keywords:** Ergonomics · Exercise · Occupational Health

## 1 Introduction

Sedentary life style and work demands can lead to musculoskeletal disorders and reduced work ability [1]. Work ability is a person's ability to do work in his working life associated with the specific demands of work tasks [2]. A long workday connected with forced work positions and reduced lack of recovery present high risk for work disability [3]. Exercise programs are indicated as an intervention approach for the prevention and management of disability in the work place [4].

The potential beneficial effects of exercise programs on work ability can be understood from a combination of various physical and psychological mechanisms [5]. Michishita et al. evaluated workers that performed passive rest passive and active rest (10 min of exercise), demonstrating that the practice of active breaks in workplace units is important for improving physical-functional variables [6]. In fact, the assessment of recovery breaks and exercise is suitable, principally when both variables are evaluated together. Few studies have evaluated the impact of exercise programs and recovery breaks on worker fatigue in the workplace.


S. Bagnara et al. (Eds.): IEA 2018, AISC 826, pp. 3–7, 2019.
https://doi.org/10.1007/978-3-319-96065-4_1

The aim of this study was to evaluate whether ergonomic conditions and exercise programs can reduce fatigue before, during and after work hours and improve the work ability of participants.

# 2 Methods

## 2.1 Study Design Subsection Sample

This cross-sectional study was conducted at a furniture company in Arapongas, Brazil.

## 2.2 Participants

A total of 20 participants were invited to participate. The participants were divided into 4 groups (G1 - Exercise with rest, n = 5, age = 37, and body mass index - BMI = 25; G2 - Exercise without rest, n = 5, age = 35, BMI = 26; G3 - No exercise with rest, n = 5, age = 37, BMI = 25; G4 - No exercise without rest, n = 5, age = 38, BMI = 26). Ethics approval was obtained from the Local Ethics Committee (CEP/UNOPAR: 2.531.274).

## 2.3 Physical and Individual Measures

The methodology used in this ergonomic action follows the Ergonomic Work Analysis. This method consists of several steps. The first step is to identify the initial demand and the respective reformulation, global analysis and choice of critical work conditions. The hypotheses are formulated to be validated or rejected through a systematic analysis. The validation of the hypothesis is then performed to obtain an accurate diagnosis of the studied situation. From the diagnosis it is possible to make suggestions for improvement in work conditions.

The exercise program was performed twice per week, on alternate days, with approximately 40 min of exercise in each session, totaling 8 sessions. The program included both resistance exercises (eccentric and concentric)/stretching and aerobic exercises. Exercises such as squats, unilateral knee flexion, extension, adduction, unilateral hip abduction, plantar flexion while standing, adduction and extension of the arms, flexion and extension of the elbow were all performed (inexpensive materials such as mats, sticks, balls, elastic bands, dumbbells and ankle weights were used).

To quantify individual fatigue and body-part discomfort, surveys with a 10-point Borg scale were used. The scale shows values from 0 = nothing at all to 10 = extremely strong. The survey was administered to each participant before, during and after the workday.

### 2.4 Analysis

Means and standard deviations are presented. First, the Shapiro-Wilk test was used to evaluate the normality of the variables and determine which tests would be used. Kruskal-Wallis test was used to assess differences between the results of the fatigue for each group (Exercise with rest, Exercise without rest, No exercise with rest and No exercise without rest). All statistical analyses were performed with SPSS 20.0 for Windows (SPSS Inc., Chicago, IL, USA) with a level of significance of 0.05.

## 3 Results

According to the Ergonomic Work Analysis, there was a high risk for the upper limbs, mainly due to the repetitiveness/frequency factor and the inadequate amount of recovery time. Concomitant with this factor, workers reported dissatisfaction with the absence of breaks during the working day.

In general, no differences were found between characteristics of the participants (age, $P = 0.322$ and BMI, $P = 0.453$). The participants that conducted the psychophysiological recovery breaks and exercise program showed less fatigue when compared with the groups that did not perform the program, principally in the moments during and after work hours (G1 < G3/G4, P = <0.04; and G1 < G3/G4, P = <0.05, Table 1).

**Table 1.** Comparison between groups (1) Exercise with rest, (2) Exercise without rest, (3) No exercise with rest and (4) No exercise without rest, for fatigue before, during and after work. N = 20

| | Exercise with rest | Exercise without rest | No exercise with rest | No exercise without rest | Kruskal-Wallis (*P*) | (*Direction*) |
|---|---|---|---|---|---|---|
| FTG (before) | 0 [0–0] | 0 [0–0] | 0 [0–1] | 0 [0–1] | (0.283) | – |
| FTG (during) | 2 [2–3] | 3 [3–3] | 5 [3–6] | 5 [3–6] | (0.005)* | **<0.025** *(1 < 3,4)* |
| FTG (after) | 2 [2–3] | 2 [0–2] | 5 [4–7] | 4 [4–6] | (0.007)* | **<0.015** *(1 < 3,4)* |

Values are presented in median and [interval interquartile range 25–75]. Fatigue – FTG. * Significant differences between groups.

No differences among the groups were found before the work hours. Figure 1 shows that the legs were the part of the body that had more fatigue.

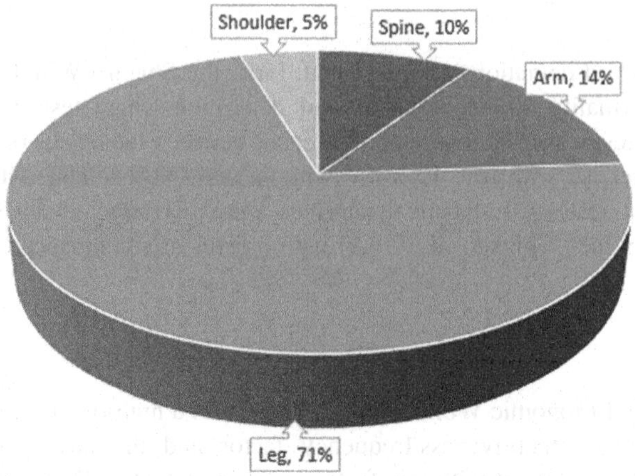


**Fig. 1.** Local fatigue reported by participants

## 4 Discussion

Ergonomic intervention associated with psychophysiological recovery breaks and exercise programs may help to reduce fatigue of employees, demonstrating gain in work efficiency and day to day tasks. However, the results showed no differences between groups with exercise either with or without the psychophysiological recovery breaks, indicating that the practice of exercise can be as important as rest.

Based on these results, we believe that it is important to perform exercise programs in the workplace. Evidence to supports the use of exercise to reduce work-related fatigue [7]. The physiological and motivational factors associated with adherence to a program are conditions important for performing exercise in the workplace. In fact, the practice of exercise may help for a positive association between physical activity, health and maintenance of functional capacity.

In conclusion, the findings suggest that the ergonomic aspects associated with exercise programs may help in reduction of fatigue principally during and after work hours. These results can contribute to health promotion and prevent deterioration of work ability.

## References

1. Bugajska J, Sagan A (2014) Chronic musculoskeletal disorders as risk factors for reduced work ability in younger and ageing workers. Int J Occup Saf Ergon 20(4):607–615
2. Chiu MC, Wang MJ, Lu CW, Pan SM, Kumashiro M, Ilmarinen J (2007) Evaluating work ability and quality of life for clinical nurses in Taiwan. Nurs Outlook 55(6):318–326
3. Jakobsen MD, Sundstrup E, Brandt M, Jay K, Aagaard P, Andersen LL (2015) Physical exercise at the workplace prevents deterioration of work ability among healthcare workers: cluster randomized controlled trial. BMC Public Health 15(1):1174

4. Cullen K, Irvin E, Collie A, Clay F, Gensby U, Jennings P et al (2017) Effectiveness of workplace interventions in return-to-work for musculoskeletal, pain-related and mental health conditions: an update of the evidence and messages for practitioners. J Occup Rehabil 1–15
5. de Vries JD, van Hooff ML, Geurts SA, Kompier MA (2015) Efficacy of an exercise intervention for employees with work-related fatigue: study protocol of a two-arm randomized controlled trial. BMC Public Health 15(1):1117
6. Michishita R, Jiang Y, Ariyoshi D, Yoshida M, Moriyama H, Yamato H (2017) The practice of active rest by workplace units improves personal relationships, mental health, and physical activity among workers. J Occup Health 59(2):122–130
7. de Vries JD, van Hooff ML, Guerts SA, Kompier MA (2017) Exercise to reduce work-related fatigue among employees: a randomized controlled trial. Scand J Work Environ Health 43 (4):337–349

# The Use of Technology as a Creative Means of Ergonomics to Support the Realization of Activities in an Aging Population: A Review of the Literature

Pierre-Yves Therriault[1(✉)], Galaad Lefay[2(✉)], Marie-Michèle Lord[1(✉)], and Alexe Desaulnier[1(✉)]

[1] Université du Québec à Trois-Rivières, Trois-Rivières, QC G9A 5H7, Canada
{pierre-yves.therriault,marie-michele.lord, alexe.desaulniers}@uqtr.ca
[2] Université Paris Descartes, 75006 Paris, France
galaad.lefay@uqtr.ca

**Abstract.** Given the aging of the population in most Western countries, the challenges to support the daily activities of the aging population are of interest to ergonomists. The use of innovative technologies to support older people when they encounter difficulties in carrying out their daily activities is already possible. The purpose of this presentation is to provide insights into what is known to date in the use of assistive technologies for an aging population. To do this, a synthesis of knowledge has been realized. Peer reviewed papers of good scientific quality was reviewed (n = 39). Considering the development of ergonomically-based technology products, a noticeable difference between the design and usage objectives has been noted. This implies that stakeholders, such as ergonomists, therapists do not use technology in the same direction for which it was designed. In addition, it is possible to note that the least explored objectives are those related to the concept of development and capacity optimization of the elderly. Thus, the proposal of an integrative concept to document both the development, use, introduction and evaluation of assistive technologies for the elderly can be very interesting for the discipline and will be discussed in this article.

**Keywords:** Assistive technology · Rehabilitation · Technology continuum
Enabling environment

## 1 Introduction

The aging population is a topical issue in the Western countries. The field of ergonomics for aging people with physical, mental and cognitive difficulties involves the participation of different health workers. Their objectives are to promote the support and adaptation, the development or recovery of the autonomy of the clients in the perspective of a greater engagement, comfort and security. In this context, the challenges to support the daily activities of the aging population are of interest to ergonomists.


S. Bagnara et al. (Eds.): IEA 2018, AISC 826, pp. 8–15, 2019.
https://doi.org/10.1007/978-3-319-96065-4_2

According to the World Health Organization (WHO) [1], the 20$^{th}$ coming years, the proportion of elderly people will become increasingly. Be old doesn't mean having a disability but the risk to be with disabilities increase with age [2]. Moreover, aging for handicapped, aging generates news challenges complicating his daily life [3]. Thus, in future years, the social and community services offered to this group will need to be better adapted [4].

Several strategies are used to support persons with disabilities, the use of assistive technologies is one of these [5]. Barlow, Singh, Bayer and Curry [6] demonstrated that technologies are associated with reduced consumption of health services, Bouchard [7] pointed out that the use of technologies leads to an improvement in the daily functioning of the elderly. Khosravi and Ghapanchi [8] indicate that positives impacts related to the use of technology affect not only individuals, but also those around them, especially the caregivers and relatives. In doing so, the use of technologies now appears to be an effective strategy for reducing social costs while at the same time leading to improvements in the quality of life of users [9]. Technology could be make consider environments like a way to compensate for disabilities, support the person in is emancipation and increase his sense of security. Assistive technologies are defined as any commercially acquired, modified, manufactured, or personalized object, device or system that is used to support a person with functional deficits in daily life [10]. In accordance with this proposal, the term technology used in this article refers to "any technical object intended to prevent, compensate, alleviate or neutralize difficulties in an individual".

The purpose of this paper is to provide an overview of current knowledge about assistive technologies in rehabilitation. More specifically, several questions are addressed: (1) What are the main actors encountered when it comes to the use of assistive technologies with the aging population? (2) What are the main families of assistive technologies considered in the literature? (3) Why are these types of technologies developed? (4) Why are these devices used? (5) Are technologies fit with the three enabling environment criteria proposed by Falzon [11]?

## 2 Method

A review of the literature was carried out according to the approach proposed by Fink [12] in several stages: (a) selection of the databases to be explored; (b) inventorying of keywords and descriptors, (c) use of purification strategies to refine the search, (d) use of selection criteria to choose the relevant texts, (e) knowledge of the content of the texts and finally, (f) analysis and classification of texts.

The corpus was first compiled using the periodicals available on the databases: Academic Search Complete, CAIRN, CINAHL, Cochrane Library, Google Scholar, MEDLINE, OT Seeker, PsycInfo, Sciences Direct and Web of Science. In order to identify the relevant documentation in the different databases, the use of several descriptors or keywords was necessary. The following terms have been used: "assistive technology" AND "accessibility" or "rehabilitat" or "autonomy" or "health and wellness" or "cognitive assistance" or "health" or "social participation".

In order to control the number of texts identified from the key words and descriptors and to define which ones to retain, three inclusion criteria were applied. The first consists of selecting texts from peer-reviewed journals, the second, only the writings published between December 2000 and December 2015, the third was to keep only French and English language articles. Subsequently, after removing the duplicates, the identified articles were screened by three evaluators. The titles and summaries of the articles were examined in order to identify the texts judged most relevant to the subject of the research. Once the screening was completed, the full texts of the selected articles were obtained and read in order to verify more in depth, using an analysis grid, their eligibility and relevance to the research questions guiding this study.

## 3 Results

Results will be presented in five separate sections. A total of 59 articles were selected. The articles are from several countries and 39 are specific the population concerned. The results show that a wide range of categorization and many vocabularies are used to describe the domain of assistive technologies.

### 3.1 The Technological Continuum and Its Actors

Given that technology is from the outset a human creation, an object conceived, manufactured and used, it implies a process of use in which different actors are involved. In terms of literature, these actors interact together in a more or less flexible continuum. As part of this study, this continuum is referred to as the "technological continuum" and three main types of actors contribute to it, designers, facilitators and consumers. First, the category of designers refers to people who design technological objects (ex.: computer scientists, engineers, ergonomists, mathematicians): their goal is to create devices that meet the needs of various stakeholders and consumers. The participants in this group are mostly in academic circles, in research and development departments, or work independently (inventor). The second category refers to facilitators. They are intermediaries between designers and consumers (ex.: sellers and distributors, health workers). They integrate technology into their actions with people with varying needs to carry out their occupations. Finally, the third category is that of consumers of technology. These are the people who are confronted with obstacles in carrying out their daily activities. In doing so, consumers mobilize a variety of assistive technologies to realize their daily business patterns and live in different living spaces, private or public, in the community. Consumers want access to easy-to-use, comfortable and safe technology.

### 3.2 Categories of Assistive Technology

Despite the presence of a very varied nomenclature when it comes to assistive technologies, three large families comprise the vast majority of the devices presented in the literature. These main families are **domotics, adapted housing** and **intelligent habitat**. In each of these large families, different technologies are introduced (see Table 1).

**Table 1.** Families of assistive technology

| Domotic | Adapted housing | Intelligent habitat |
|---|---|---|
| • Approach that automatically manages various functions of a living space in order to provide comfort and safety to its occupants or enable them to communicate with external helpers [13, 20] | • Spaces designed to overcome certain difficulties of a consumer with physical or cognitive difficulties<br>• Includes some temporary or permanent support technologies [5] | • Spaces developed from the perspective of maintaining at home. It can include various robotic equipment, home automation or teleservice [29] |

### 3.3 Design Objectives

Four main goals guide the work of designers: **ensuring safety and surveillance, develop autonomy**, **providing comfort** and **supporting communication**. The designer seeks as a first objective to ensure consumer safety and surveillance [6, 16, 21, 27, 56]. Technological devices thus provide a safe environment for consumers in a given space, or in the expression of their behaviors. It is also expected that technological devices will provide a sense of security. The second goal is to develop the autonomy of consumers given their singular characteristics [8, 9, 12, 17, 21]. In this case, the assistive technology aims to promote or compensate for a lack of autonomy for the performance of a particular task. The third goal is to provide maximum comfort to people with disabilities [5, 18]. The goal is that the consumer feels a physical and psychic well-being in a living space. Finally, the final goal of designers is to support communication [8, 9, 18, 28]. In this case, the technological devices are intended to enable the consumer to communicate with their relatives, or other community members.

### 3.4 Objectives of Facilitators

Four main objectives guide the use of technologies by facilitators: **improving functional abilities, assisting singular capacities**, **compensating for singular capacities** and **stimulating the functions of the person**.

The primary goal is to stimulate the functions [12, 26]. This goal involves technological devices that encourage the realization of an activity by encouraging the consumer to act, to continue his action or to maintain an effort. The second goal is to improve functional abilities [14, 24, 25]. Several technological devices are selected in order to enable a consumer to develop his capacities or his functions and his field of action. The third goal is to assist the singular abilities [13, 15, 19]. Different technological devices are used with the view to support and assist the consumer in the pursuit of the tasks he wishes to initiate and carry out and which he is not fully capable of performing independently. The fourth and last goal is the compensation of singular abilities [15, 23, 40, 56]. Some assistive technologies are identified to perform a task that the consumer cannot independently carry out.

### 3.5 Enabling Environment Objectives

Studying the aims of the different actors through the prism of the three goals of the enabling environment proposed by the field of ergonomic constitute a good idea. Actually, Falzon [11] affirm that the technology must perceive like a complex ensemble, grouping tools, equipment, technics but also knowledge and methods needed to doing of human activities. This theoretical posture suggests that always to take into account the environment where this technology is implemented. According to Falzon, an enabling environment needs to meet three criteria: (1) Prevention (preserved the human capabilities and secure the latter); (2) Universality (take into account the human diversity and privileged integration, inclusion and social recognition) and (3) Development (contributes to the development of new capabilities from the user).

Data analysis demonstrates the majority of the articles analyzed respect the principle of prevention and universality but not necessarily the developmental criteria (see Table 2).

**Table 2.** Number of articles dealing technology with according to the enable environment criteria (n = 59)

| | Prevention | Universality | Development |
|---|---|---|---|
| Nb of articles | 58 | 56 | 40 |

## 4 Discussion

The aim of this article was to draw the picture of current knowledge about the use of assistive technologies in rehabilitation.

First, the synthesis of knowledge has brought to light an important categorization and nomenclature used to describe the technologies. It appears that when a single entity is treated in a set of domains as in the field of assistive technology, the associated evidence can take all sorts of directions. As Piau, and his collaborators [9] point out, this cause a great confusion.

Second, a divergence in the aims between designers and facilitators is found. Designers tend to create technologies with the main objectives: (1) to provide security and surveillance, (2) to develop autonomy, (3) to provide comfort, and (4) to support communication. While facilitators, acting as intermediaries between designers and technology users, integrate technology into their actions for the person with rehabilitation needs, while also pursuing four main objectives: (1) to improve functional abilities, (2) to assist singular capacities, (3) to compensate for singular abilities, and (4) to stimulate the person's functions. For both groups of actors, the objectives differ and do not always seem to correspond. However, the analysis of scientific literature shows that very little, if any, model seems to support the use of assistive technologies across the technological continuum. Indeed, the various actors of this continuum do not seem to mobilize a transversal model that would not only standardize the terms used but also guide the design and use objectives. Perhaps the absence of a model guiding the technological continuum could explain the divergence between the objectives currently pursued? This may constitute a barrier to collaboration between technology creators and rehabilitation practitioners and may indicate that stakeholders do not use assistive technology in the same way as the function for which they are created. For example, there is a growing body of research on the potential drift in the use of social media between physicians and their patients, initially created for the general public and now increasingly used to reach clients in remote areas [30].

Third, in association with the ergonomic perspective, it seem that only two of enabling environment three criteria proposed by Falzon [11] are be considered. The conception of technologies not include systematicly the developmental aspect and not contributes to the potential development of capabilities of the user. In a context of the technology change, it's obligatory to consider the development of human capacities being.

Moreover, this review of the literature also highlighted that decision-makers are not currently actors involved in the technological continuum. This group of actors refers, for example, to governments and granting agencies that govern technological development. Yet, from a collaborative perspective, it is important to guide the actions of different actors, not only with conceptual frameworks and models, but also with clear policies. Mobilization of all, including decision-makers, seems essential to anchor best practices and foster collaboration.

## 5 Strengths and Limitations

The results of this synthesis of knowledge should be analyzed with caution due to methodological constraints, in particular because of inclusion criteria based on peer review in scientific journals, the number of articles studied which remains limited given the disciplinary fields concerned and the fact that the articles selected do not cover all the possible experiences. On the other hand, this article is a first step towards a better understanding of the vast field of assistive technology, a constantly expanding and innovative field of daily activities in private and public spaces public.

## 6 Conclusion

In view of the different objectives pursued by the actors involved in assistive technologies and the heterogeneity of the literature, it seems legitimate to question whether an integrative concept would effectively document the development, the use, and evaluation of these devices in a more homogeneous manner. Combining the knowledge of all through a unifying concept would possibly create a place of communication and encounter between the designer, the facilitators and the consumer as well as at the level of the decision-makers. Using the concept of enabling environment such as a common thread could be a solution. The search for such a concept thus seems an interesting avenue of research.

The enthusiasm for technological solutions is present in the literature in several fields. It is now necessary to ensure that the actors working in these fields have the necessary tools that will frame development in order to homogenize the process across the technological continuum.

**Conflict of Interest.** No conflict of interest to declare.

## References


1. World health Organization (2016): World report on disability
2. Laville A, Volkoff S (1993) Age, santé, travail: le déclin et la construction. In: Actes du XXVIIIème congrès de la SELF, pp 22–24. SELF, Genève
3. Azéma B, Martinez N (2005) Les personnes handicapées vieillissantes: espérances de vie et de santé; qualité de vie. Revue française des affaires sociales 2:295–333
4. Ministère de la santé et des services sociaux: Plan stratégique: 2015–2020. (2015)
5. Bobillier-Chaumon MÉ, Ciobanu RO (2009) Les nouvelles technologies au service des personnes âgées: entre promesses et interrogations–une revue de questions. Psychologie française 54(3):271–285
6. Barlow J, Singh D, Bayer S, Curry R (2007) A systematic review of the benefits of home telecare for frail elderly people and those with long-term conditions. J Telemed Telecare 13(4):172–179
7. Bouchard B (2013) Recherche sur les technologies d'assistance our le maintien domicile des ersonnes atteintes d'Alzheimer. In.: Journée de la recherche du FRQNT, Sherbrooke, Québec
8. Khosravi P, Ghapanchi AH (2015) Investigating the effectiveness of technologies applied to assist seniors: a systematic literature review. Int J Med Inform 85(1):17–26
9. Reeder B, Meyer E, Lazar A, Chaudhuri S, Thompson HJ, Demiris G (2013) Framing the evidence for health smart homes and home-based consumer health technologies as a public health intervention for independent aging: a systematic review. Int J Med Inform 82(7):565–579
10. Edyburn DL (2004) Assistive technology in K-12 schools: understanding the impact of the AT consideration mandate
11. Fink A (1998) Doing the review: a reader's guide chapter. In: Fink A (ed) Conducting research literature reviews: from the internet to paper. Sage, Los Angeles
12. Agree E (2014) The potential for technology to enhance independence for those aging with a disability. Disabil Health J 7(1 Suppl): S33–S39. MEDLINE with Full Text, Ipswich, MA. Accessed 2 Feb 2018
13. Bismuth S, Villars H, Durliat I, Boyer P, Oustric S (2012) Gerontotechnologies likely to enable patients with soft cognitive deficit and alzheimer's disease at the light stage to stay home. Les cahiers de l'année gérontologique 4(3):310–319
14. Bobillier-Chaumon MÉ, Cuvillier B, Durif-Bruckert C, Cros F, Vanhille M, Salima B (2014) Concevoir une technologie ambiante pour le maintien à domicile: une démarche prospective par la prise en compte des systèmes d'activité. Le travail humain 77(1):39–62
15. Boll S, Heuten W, Meyer EM, Meis M (2010) Development of a multimodal reminder system for older persons in their residential home. Inform Health Soc Care 35(3):104–124
16. Brummel-Smith K, Dangiolo M (2009) Assistive technologies in the home. Clin Geriatric Med 25(1):61–77

17. Cao J, Xie SQ, Das R, Zhu GL (2014) Control strategies for effective robot assisted gait rehabilitation: the state of art and future prospects. Med Eng Phys 36(12):1555–1566
18. Chan M, Estève D, Escriba C, Campo E (2008) A review of smart homes - present state and future challenges. Comput Methods Programs Biomed 91(1):55–81
19. Chen K, Chan AHS (2013) Use or non-use of gerontechnology - a qualitative study. Int J Environ Res Publ Health 10(10):4645–4666
20. Culnaert E, Galy S, Chotard A, Tomas J (2009) Séniors et dépendance: La domotique au service du maintien à domicile. L'Aquitaine numérique - La lettre d'information d'AEC 21:9–16
21. Eaton L, Gordon D, Doorenbos A et al (2014) Development and implementation of a telehealth-enhanced intervention for pain and symptom management. Contemp Clin Trials 38(2): 213–220. MEDLINE with Full Text, Ipswich, MA. Accessed 2 Feb 2018
22. Karmarkar AM, Dicianno BE, Graham JE, Cooper R, Kelleher A, Cooper RA (2012) Factors associated with provision of wheelchairs in older adults. Assistive Technol 24(3):155–167
23. Kristoffersson A, Coradeschi S, Loutfi A, Severinson-Eklundh K (2011) An exploratory study of health professionals' attitudes about robotic telepresence technology. J Technol Hum Serv 29(4):263–283
24. Laufer Y, Dar G, Kodesh E (2014) Does a Wii-based exercise program enhance balance control of independently functioning older adults? a systematic review. Clin Interv Aging 9:1803–1813
25. Lockey K, Jennings MB, Shaw L (2010) Exploring hearing aid use in older women through narratives. Int J Audiol 49(8):542–549
26. Melander-Wikman A, Fältholm Y, Gard G (2008) Safety vs privacy: elderly persons' experiences of a mobile safety alarm. Health Soc Care Commun 16(4):337–346
27. Melillo P, Castaldo R, Sannino G, Orrico A, de Pietro G, Pecchia L (2015) Wearable technology and ECG processing for fall risk assessment, prevention and detection. In: 37th Annual International Conference of the IEEE Engineering in Medicine and Biology Society (EMBC), pp 7740–7743. IEEE
28. Nef T, Urwyler P, Büchler M, Tarnanas I, Stucki R, Cazzoli D, Mosimann U (2015) Evaluation of three state-of-the-art classifiers for recognition of activities of daily living from smart home ambient data. Sensors 15(5):11725–11740
29. Roy P (2012) Modèle possibiliste pour la reconnaissance d'activités: habitat intelligent. Université de Sherbrooke, Sherbrooke
30. Dejean PH, Naël M (2004) Ergonomie du produit. Falzon DP (Ed) Ergonomie, Paris: PUF, pp 463–477

# How Age and Pace of Work Affect Movement Variability During Repetitive Assembly Tasks

Martine A. Gilles(✉), Clarisse Gaudez, Jonathan Savin, Aurélie Remy, Olivier Remy, and Pascal Wild

INRS Lorraine, 1, rue du Morvan, CS60027, 54 519 Vandœuvre Cedex, France
martine.gilles@inrs.fr

**Abstract.** During production, companies aim to ensure optimal productivity and quality. With this in mind, workstation designers tend to assume that operators will perform tasks in a uniform manner, and tend not to include movement variability parameters in their designs. The aim of this study was to characterise movement variability during repetitive assembly tasks performed at a defined pace. 62 right-handed men in three different age groups were asked to affix a handle on a base with two nuts at two different paces. Particular attention was paid to how two factors influenced movement variability: the operator's and the pace of work. Variability was observed in assembly way of doing when the procedure was not imposed. The variability observed during assembly, as performed for this study, was unaffected by operators' age or the pace of work. No effect of variability was observed on the duration of assembly cycles, nor on the adaptation to changes in pace. In contrast, variability allowed operators alternatives to repetitive movement which could potentially exert strain on the locomotor system. Allowing operators the possibility to spontaneously use variable movements during repetitive tasks appears to be an important element to consider when designing workstations.

**Keywords:** Movement variability · Repetitive task · Age

## 1 Introduction

Companies generally aim to achieve significant productivity with optimal quality. With this in mind, work station designers tend to assume that operators will perform tasks in a uniform manner, and tend not to include movement variability parameters in their designs. However, movement variability could be considered beneficial when seeking to limit musculo-skeletal disorders.

Movement variability is an essential characteristic of human movement [1, 2]. It appears to be closely linked to the process of control and regulation of movement [3, 4]. Movement variability provides adaptability and flexibility, both of which are essential to comply with operators' personal characteristics and the specificities of the task to be performed, while also taking environmental constraints into consideration [2]. Movement variability is observed in all actions controlled by the sensori-motor system; it has been observed between individuals and for the same individual [5–9]. It is generally highlighted by differences in segmented whole-body kinematics and/or in muscular


S. Bagnara et al. (Eds.): IEA 2018, AISC 826, pp. 16–22, 2019.
https://doi.org/10.1007/978-3-319-96065-4_3

activity when performing repetitive tasks [10]. Most frequently, it has been addressed in studies comparing two distinct populations such as women versus men [11] or novices versus experts [12].

The aim of this study was to identify variability in assembly tasks repeatedly performed at a fixed pace. Particular attention was paid to how two factors influenced this variability: the operator's age – an intrinsic factor to the operator – and the pace of work – an extrinsic factor.

## 2 Protocols and Methods

### 2.1 Task Analysed

The experiments were performed in a laboratory. Subjects were asked to complete a repetitive assembly task in conditions similar to those encountered at an assembly-line workstation. The prescribed task involved five successive actions to be performed cyclically and repetitively without interruption at an imposed work pace. The five actions were (1) collection of an assembly base from a base distributor/collector, (2) movement between the distributor and the workstation, (3) collection of a handle and two nuts stored at floor-level under the workstation, (4) assembly at the workstation (Fig. 1), (5) return to the distributor after completing the assembly to initiate a new assembly cycle. No recommendations were made on how the assembly task per se should be performed. The height of the workstation was adapted to each subject's size. In this paper, the analysis was exclusively focused on the assemblies performed at the workstation.

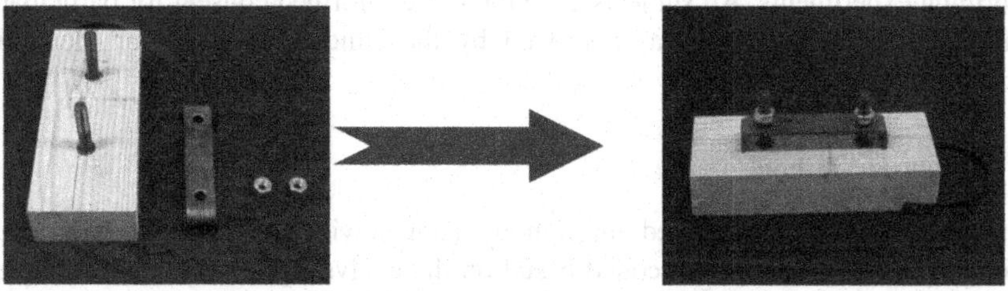

**Fig. 1.** Assembly task to be performed: affix the handle to the base using the two nuts.

Assembly was performed over two 20-min work sessions. For each session, a different work pace was defined. The pace was either comfortable - with a maximum of 49 assemblies to be performed - or rapid, with a maximum of 60 assemblies to be performed. The paces were applied in random order, with as many subjects from each age-group starting with each one. The pace of assembly was continuously monitored during the work session, and subjects were given a verbal reminder when they did not respect the required cadence.

### 2.2 Participants

Sixty-two right-handed men volunteered to participate in the experiment. Two selection criteria were used during the recruitment procedure. The subjects had to be in one of the following three age-groups: junior (J) 30 to 35 years-old, middle-aged (M) 45 to 50 years-old, or senior (S) 60 to 65 years-old. The mean physical characteristics for each age-group are presented in Table 1. In addition to the age requirements, all subjects had to be employed in or retired from a profession considered "physically demanding" so that functional capabilities could be considered homogeneous across all groups.

**Table 1.** Characteristics of the three subject groups.

| Characteristics | | Junior | Middle-aged | Senior |
|---|---|---|---|---|
| Number of subjects | | 19 | 21 | 22 |
| Age (years) | Mean | 32.5 | 47.0 | 61.8 |
| | Range | 30–35 | 45–50 | 60–65 |
| Weight (kg) | Mean | 75.4 | 78.7 | 81.7 |
| | Range | 53–103 | 60–110 | 62–106 |
| Height (m) | Mean | 1.77 | 1.74 | 1.73 |
| | Range | 1.55–1.87 | 1.63–1.91 | 1.65–1.88 |
| BMI (kg.m$^{-2}$) | Mean | 24.1 | 25.9 | 27.2 |
| | Range | 19.7–34.5 | 20.6–34.6 | 21.0–35.5 |

Volunteers were recruited either through a temping agency or through small ads published in local newspapers. Subjects' functional capacities were assessed before performing experiments. All subjects gave their free informed consent for participation in this study. The protocol was approved by the ethics committee for biomedical research.

### 2.3 Analysis Method

The work sessions were filmed throughout using a video camera. Each assembly performed by the subjects was coded based on the analysis grid presented in Table 2. Two main actions were distinguished:

- Positioning (P): engagement of a nut on one of the threaded shafts set into the base.
- Tightening (T): winding of the nut onto one of the threaded shafts until it abuts on the handle to be attached.

For the positioning and tightening actions, the hand performing the action as well as the order (right-hand, left-hand or simultaneous) in which the threaded shafts were loaded with nuts were coded. Finally, the sequence in which the two phases of these two actions was performed (assembly mode) was also a source of information for the analysis. A total of 6672 assemblies were coded.

**Table 2.** Coding used for actions identified from video recordings of assembly tasks.

| Actions | | Coding |
|---|---|---|
| *Mode of positioning (P) of nuts on the threaded shafts:* | | *P* |
| Positioning on the right-hand threaded shaft | with the right hand | 1 |
| | with the left hand | 2 |
| Positioning on the left-hand threaded shaft | with the right hand | 1 |
| | with the left hand | 2 |
| Order of positioning of the nuts | right first | 1 |
| | left first | 2 |
| | simultaneously | 3 |
| *Mode of tightening (T) of nuts on the threaded shafts:* | | *T* |
| Tightening on the right-hand threaded shaft | with the right hand | 1 |
| | with the left hand | 2 |
| Tightening on the left-hand threaded shaft | with the right hand | 1 |
| | with the left hand | 2 |
| Order of tightening of nuts | right first | 1 |
| | left first | 2 |
| | simultaneously | 3 |
| *Mode of assembly (A) of the nuts on the threaded shafts* | | *A* |
| Assembly | PP-TT | 1 |
| | PT-PT | 2 |

From this analysis, the different "ways of doing" used by operators during assembly were identified by crossing the variables in the grid. The number of ways of doing used by each subject and at each pace was analysed using a mixed linear model with the variable "subject" as random effect and the "age-group" and "work pace" variables as fixed effects.

## 3 Protocols and Methods

Coding of the actions revealed a total of 144 theoretically possible ways of doing to perform the task. This number omits impossible ways of doing, such as tightening the nuts before engaging them on the threaded shaft. From the code, 44 of the 144 possible ways of doing were observed. The order in which the nuts were placed on the threaded shafts was the main factor distinguishing between ways of doing. The three most frequently used ways of doing are presented in Fig. 2. They represent 78.8% of all the assemblies coded. The most frequently represented way of doing corresponds to 2266 of the 6672 assemblies, i.e., 34.0% of all assemblies. The second way of doing was used in 1894 assemblies (28.4%) and the third was used in 1099 assemblies (16.4%). Moving down the ranks of the ways of doing used, the number of assemblies where the less popular ways of doing were observed decreased rapidly, with the 4th way of doing used in only 295 assemblies (4.4%). The decrease continued progressively for the remaining 40 ways of doing.

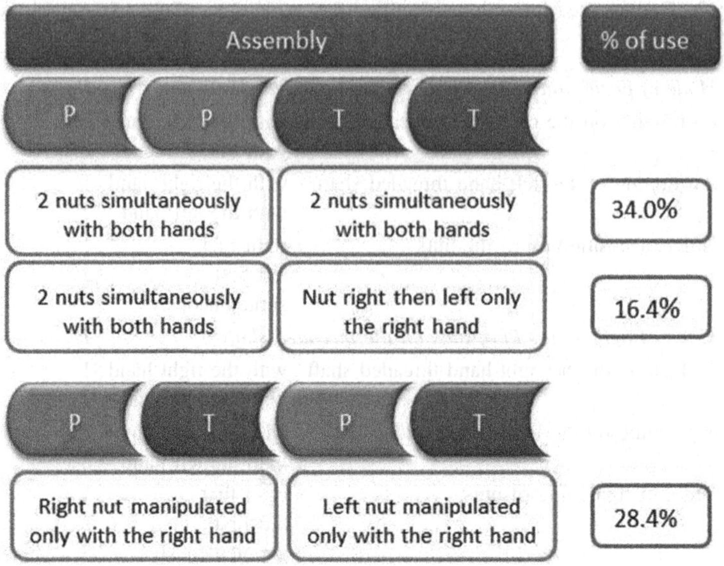


**Fig. 2.** Identification of the activity performed by each hand for the three most commonly used ways of doing. The percentages of use were calculated relative to all the assemblies performed during the experiment; P = positioning, T = tightening.

The description of the 44 ways of doing observed reveals that the right-hand nut was always placed with the right hand, whereas all other actions could potentially be performed by either hand. Likewise, some assemblies were carried out exclusively with the right hand whereas the equivalent with the left hand was never observed. The most frequently encountered way of doing was of a (P-P) (T-T) type. In this situation, the subjects used both hands simultaneously to place the two nuts and tighten them on each of the threaded shafts. The second most frequent way of doing was of a (P-T) (P-T) type. In this case, the nut was placed then tightened immediately, starting with the right-hand threaded shaft on the base. Subjects only used their right hands to manipulate the nuts in this way of doing, using their left hands to maintain the base on the table. The third most frequently observed way of doing was another (P-P) (T-T) type: both nuts were first placed simultaneously using both hands, and then only the right hand was used to tighten each of the nuts in turn, starting with the right-hand threaded shaft.

For each subject, the number of different ways of doing used when performing assemblies was counted for each of the paces set. The distribution of the number of ways of doing used as a function of age-group and pace is presented in Table 3. At the comfortable pace, 45% of subjects used only a single way of doing, whereas 40% used 2 to 5 ways of doing, and 15% used more than 6 ways of doing. At the rapid pace, 32% of subjects used only a single way of doing, whereas 48% used between 2 and 5 ways of doing, and 30% used more than 6 ways of doing. The differences between the number of ways of doing used by each subject and at each pace were not statistically significant. Similarly, no significant difference was observed for the number of ways of doing used by the different age-groups as the pace increased.

**Table 3.** Distribution of the number of ways of doing used by each age-group and at each pace; No WoD = Number of ways of doing, J = junior, M = middle-aged, S = senior.

| Assemblies performed at the comfortable pace | | | | | | | | | | | | | |
|---|---|---|---|---|---|---|---|---|---|---|---|---|---|
| No WoD used | 1 | 2 | 3 | 4 | 5 | 6 | 7 | 8 | 9 | 10 | 11 | 12 | 13 |
| J | 12 | 1 | 2 | 1 | 1 | 1 | – | – | – | 1 | – | – | – |
| M | 7 | 1 | 4 | 3 | 3 | – | 1 | – | – | – | 1 | 1 | – |
| S | 8 | 2 | 2 | 3 | 2 | 4 | – | – | 1 | – | – | – | – |
| Assemblies performed at the rapid pace | | | | | | | | | | | | | |
| No WoD used | 1 | 2 | 3 | 4 | 5 | 6 | 7 | 8 | 9 | 10 | 11 | 12 | 13 |
| J | 7 | 7 | – | 2 | 2 | – | 1 | – | – | – | – | – | – |
| M | 5 | 1 | 5 | 3 | 2 | 1 | 3 | – | – | – | 1 | – | – |
| S | 8 | 2 | 3 | 3 | – | 2 | 1 | 1 | – | 1 | – | – | 1 |

Moreover, the time required to complete an assembly presented no significant differences between the different age-groups for each pace. Although naturally, the time required to complete a cycle was significantly reduced at the increased pace (Table 4).

**Table 4.** Statistical analysis of duration of an assembly: mean values (sd) for each group of subjects at the two work paces; J = junior, M = middle-aged, S = senior, bw = between.

| | Comfortable pace | | | Rapid pace | | | p values | | |
|---|---|---|---|---|---|---|---|---|---|
| Age-group | J | M | S | J | M | S | bw age | bw pace | IA |
| Assembly duration (s) | 12.7 (2.5) | 12.0 (2.8) | 12.8 (2.8) | 11.4 (2.6) | 10.7 (2.8) | 11.3 (3.1) | 0.08 | <0.001 | 0.92 |

## 4 Discussion and Conclusion

Variability was observed in assembly ways of doing when the procedure was not imposed. The variability observed during completion of the assembly task performed for this study was not affected by operators' age or the pace of work. This variability had no effect on the adaptation to the duration of assembly when the pace was increased or decreased. Whatever the age-group considered, as expected, the duration of assembly decreased as the pace increased. However, variability allows operators to implement alternatives to repetitive movement potentially demanding for the locomotor system. Allowing operators the possibility of expressing the spontaneously existing movement variability during repetitive tasks appears to be an important element to consider when designing workstations.

## References

1. Berthoz A (1997) Le sens du mouvement. Odile Jacob (Eds), Paris
2. Glazier PS, Wheat JS, Pease DL, Bartlett R (2006) Dynamic system theory and the functional role of movement variability. In: Davids K, Bennett S, Newell K (eds) Movement system variability. Human Kinetics Publishers, Champaign (IL), pp 49–72
3. Diniz A, Wijnants ML, Torre K, Barreiros J, Crato N, Bosman AM, Hasselman F, Cox RF, Van Orden GC, Delignières D (2011) Contemporary theories of 1/f noise in motor control. Hum Mov Sci 30(5):889–905
4. Latash ML, Scholz JP, Schröner G (2002) Motor control strategies revealed in the structure of motor variability. Exerc Sport Sci Rev 30(1):26–31
5. Jackson JA, Mathiassen SE, Dempsey PG (2009) Methodological variance associated with normalization of occupational upper trapezius EMG using sub-maximal reference contractions. J Electromyogr Kinesiol 19(3):416–427
6. Madeleine P, Lundager B, Voigt M, Arendt-Nielsen L (2003) The effects of neck-shoulder pain development on sensory-motor interactions among female workers in the poultry and fish industries. a prospective study. Int Arch Occup Environ Health 76(1):39–49
7. Madeleine P, Lundager B, Voigt M, Arendt-Nielsen L (2003) Standardized low-load repetitive work: evidence of different motor control strategies between experienced workers and a reference group. Appl Ergon 34(6):533–542
8. Mathiassen SE, Burdorf A, Van der Beek AJ (2002) Statistical power and measurement allocation in ergonomic intervention studies assessing upper trapezius EMG amplitude. a case study of assembly work. J Electromyogr Kinesiol 12(1):45–57
9. Mathiassen SE, Möller T, Forsman M (2003) Variability in mechanical exposure within and between individuals performing a highly constrained industrial work task. Ergonomics 46(3):800–824
10. Terrier P, Schutz Y (2003) Variability of gait patterns during unconstrained walking assessed by satellite positioning (GPS). Eur J Appl Physiol 90(5–6):554–561
11. Svendsen JH, Madeleine P (2010) Amount and structure of force variability during short, ramp and sustained contractions in males and females. Hum Mov Sci 29(1):35–47
12. Madeleine P, Voigt M, Mathiassen SE (2008) The size of cycle-to-cycle variability in biomechanical exposure among butchers performing a standardised cutting task. Ergonomics 51(7):1078–1095

# Identifying Factors Related to the Estimation of Near-Crash Events of Elderly Drivers

Misako Yamagishi[1,2(✉)], Takashi Yonekawa[2], Makoto Inagami[2], Toshihisa Sato[3], Motoyuki Aakamatsu[3], and Hirofumi Aoki[2]

[1] Aichi Shukutoku University, Katahira, Nagakute, Aichi, Japan
yamagiko@asu.aasa.ac.jp
[2] Nagoya University, Furo-cho, Nagoya, Aichi, Japan
[3] National Institute of Advanced Industrial Science and Technology, Higashi, Tsukuba, Ibaraki, Japan

**Abstract.** This study attempted to identify factors associated with driving behavior of elderly drivers to assess their safety and estimate their risk during naturalistic driving. We performed binomial logistic regression using self-reported past crash involvement as a response variable to identify critical factors and provided an estimation model has 18 variables. However, applying driver category based on crash and near-crash events (CNCs) collected from naturalistic driving study employed on-dash cam instead of self-reported crash involvement to the previous model showed lower predictive performance (0.63 for sensitivity and 0.51 for specificity). This implies that the model based on self-reported crash experiences was difficult to detect for drivers with CNC during naturalistic driving. Then, we performed binomial logistic regression based on CNC involvement and indicated another model, where the predictive performance was improved, with 0.81 for sensitivity and 0.70 for specificity. To predict the number of CNCs as drivers' risk, this study adopted Poisson regression analysis using nine variables selected from the second model. The analyses showed a plausible model and significant variables for the estimation of CNCs. Mini-Mental State Examination (MMSE) was one of the better predictor putting in this model, and showed the probability that lower performance associated with higher number of CNCs. This model for CNC estimation would be helpful for the development of safety programs for elderly drivers with possible incidents.

**Keywords:** Elderly drivers · Poisson regression analysis · Naturalistic driving

## 1 Introduction

This study aims to reduce the risk while driving for elderly drivers. In Japan, increase of the number of elderly people with driving licenses as well as traffic accidents related to elderly drivers are recognized as critical issues. Although development of automated driving system is encouraged to correct this issue, elderly drivers will still have to drive a vehicle by themselves for the next several years. Therefore, to ensure elderly drivers' safe mobility and road safety of all road users, it is important to identify the factors associated with drivers in critical incidents.


S. Bagnara et al. (Eds.): IEA 2018, AISC 826, pp. 23–31, 2019.
https://doi.org/10.1007/978-3-319-96065-4_4

From the 100-car Naturalistic Driving Study [1], similar driving monitoring paradigm has been employed in many studies to investigate detailed driving behavior during naturalistic driving. Some of them collected crash and near-crash events (CNCs) while driving, and intend to identify factors related to observed CNCs. CNCs of elderly drivers have been included in several studies; for example, Guo and Fang [2] investigated risk factors such as age, gender, personality and the critical-incident events (CIE) associated with individual driving risk and built a logistic prediction model to predict high-risk drivers by using data of the 100-car Naturalistic Driving Study. They found that drivers' personality, age, and CIE had significant impacts on the CNC risk, and a logistic prediction model using the CIE rates could identify high- and moderate-risk drivers. Guo, Fang and Antin [3] evaluated the relationship between senior drivers' fitness profiles and driving risk represented by CNC rate. First, they adopted a principal component analysis to reduce the dimensionality of the fitness assessment metrics. Then through a negative binomial regression, they indicated that right-eye contrast sensitivity metrics are significantly related to CNC risk. Huishingh et al. [4] examined the association between visual sensory and visuo-cognitive functions and the rates of CNC involvement among elderly drivers by using a naturalistic study design as part of the Strategic Highway Research Program 2 (SHRP2), and Poisson regression suggested that elderly drivers with impaired contrast sensitivity and far peripheral vision have significantly higher rates of crash involvement than those without these impairments. The similarities between these studies is that they focused on CNC during naturalistic driving, adjusted CNC by time or distance driven, attempted to find the relationship between human characteristics and CNC risk, adopted some regression analysis to build up the model, and find risk factors or impact of the risk factors. For practical assessment and screening of drivers, CNC risk of elderly drivers should be considered according to different levels of human characteristics.

The purpose of this study is to identify factors associated with drivers with CNCs to estimate CNC risk of elderly drivers. The analyses were organized by following procedure: (1) we identified the critical factors related to CNCs from our driving-related human characteristics database described below, (2) we examined the relationship between the identified factors and CNCs, and (3) by applying plausible factors affecting CNCs, we predicted the number of CNCs during driving. The hypotheses are that critical factors in the estimation model comprise visual functioning as well as the factors from previous researches, and that the estimation model will provide predictive values varying with performance change of single critical factor under identical circumstances. This procedure will make it possible to employ to drivers' assessment and subsequent safe program for elderly drivers based on naturalistic driving.

## 2 Methods

### 2.1 Participant Drivers

This study used data from the Nagoya Center of Innovation (COI) project supported by the Japan Science and Technology Agency (JST). The project involved a sample of about 300 drivers, of which 68 drivers residing in and around Nagoya city between the ages of 50 and 83 (average age = 67.5 years old, SD = 7.87) participated in the naturalistic driving study. The drivers were licensed, drove regularly, and planned to keep the drive recorder on their private vehicles for a while (at least 6 months). However, data of 59 drivers of all were used in this study through exclusion criteria such as data loss and high score on the lie scale (maximum was 4 points). Most of them enrolled from February 2015, although some drivers stopped the enrollment and surrendered the drive recorder for various personal reasons. The analysis period of this study is up to around August 2016, while the recording period of the driving data was 6 months to 2 years (average period = 12 months, SD = 6.9). The total distance driven was 411,775 km and the total number of driven days was 14,162 days.

## 2.2 CNC Detection

The drive recorder (BU-DRHD421 from Yupiteru Corporation) was used for this study to record naturalistic driving. These devices recorded the front view at 10–15 fps using the front-side camera, the location information and speed from GPS, and the acceleration of the vehicles measured by triaxial accelerometer. The recordings were saved under three modes: continuous recording (GPS and acceleration with video), event recording (GPS and acceleration with video of the triggered events), and driving record (GPS and acceleration for entire period).

CNCs were extracted from the event recording, which was activated by triggers such as sudden acceleration changes and impact to the vehicles, and represented records for 10 s before and after the trigger. Although there was a total of 14,985 events in the recordings, these contained insignificant events due to circumstances of vehicles and roads such as irregular impact due to road bumps and aggressive door opening and closing action. After excluding events due to non-accident events, in order to extract moments that were believed to be risky events from the event recordings, two observers confirmed all of the records and selected based on the scene such as making contact with other vehicles, pedestrians, structures or traffic lights and great acceleration (–0.40 G on the deceleration side).

## 2.3 Human Characteristics Database

This study collected various human characteristics data related to driving. Cognitive functioning was measured using Mini-Mental State Examination (MMSE) [5], Trail Making Test (TMT) part A and B [5], the AIST-Cognitive Aging Test (AIST-CAT) [6], DrivingHealth Inventory® (DHI) [7], and driving aptitude test (DAT) consisting of reaction time (RT) tasks and attention tasks. Visual functioning was evaluated by ophthalmic tests according to static visual acuity, kinetic visual acuity, night vision, horizontal visual field and contrast sensitivity. Subjective assessments for beliefs and attitudes about driving were conducted using five questionnaires: driving habits and individual attributions questionnaire, questionnaire on self-awareness changes in cognitive and physical functions while driving [8], questionnaire on compensatory driving strategies [9], workload sensitivity questionnaire (WSQ), and driving style questionnaire (DSQ) [10]. In addition to the subjective assessments, driving habits such as distance driven were obtained from the driving record mode from the drive recorders. These measurements had provided around 100 performances and indices, and organized as Data Repository for Human Life-Driving Anatomy (Dahlia) in this project.

### 2.4 Ethical Considerations

This study was implemented under the approval of the Institute of Innovation for Future Society of Nagoya University Ethics Committee. When drivers enrolled in this study, the study's purpose, significance, implementation methods and acquired data and its usage were explained to them, and consent was obtained from them.

## 3 Results

### 3.1 Identifying Factor Related to Detection of Drivers with CNCs

We conducted a logistic regression to identify the factors to distinguish between risky and safe drivers and constructed an estimation model of self-reported past crash involvement ("never" and "once or more" in the past three years) as a response variable from driving habits questionnaire in the Dahlia. Of almost 100 performances and indices involved the Dahlia, explanatory variables in this analysis were selected based on the influence on self-reported crash involvement, and predictive performance of the model was evaluated in terms of sensitivity, specificity, accuracy, and positive predictive value (PPV).

Based on values of odds ratio and deviance derived from each single regression analysis, eighteen variables were selected as explanatory variables related to age, performances of cognitive functioning and response time, and subjective assessment, and a logistic regression yielded a model (first model) having better predictive performance for four values: 0.81 for sensitivity, 97.7 for specificity, 93.2 for accuracy, and 92.9 for PPV. This suggested that the model would yield a plausible estimation of positive judgment on past crash involvement, and the 18 variables would be significant indices for discriminating safe and risky drivers. The same analysis was conducted, replacing the response variable with driver category according to CNC (drivers with one or more CNCs and drivers without CNCs). Predictive performance was 0.63 for sensitivity, 0.51 for specificity, 0.54 for accuracy and 0.32 for PPV, implying that the performance was lower than the model for self-reported crash involvement, and the model for self-reported crash involvement was difficult to apply to CNC involvement during driving. Again, logistic regression was performed to compute a model for identifying drivers with CNCs by using 18 variables comprising by different variables from the previous model, which are related to age, performances of cognitive functioning, RT, visual functioning, attention, and subjective assessment. The analysis showed the second model, and the predictive performance was 0.81, 0.70, 0.76 and 0.77, respectively, which was higher than the previous model for driver category according to CNC involvement. The result indicated that the second model might be more plausible to detect elderly drivers with CNC probability. Table 1 shows the comparison of predictive performances among the three analyses.

**Table 1.** Predictive performances for logistic regression among the three analyses

| Model | Response variable | Sensitivity | Specificity | Accuracy | PPV |
|---|---|---|---|---|---|
| First model | Self-reported crash involvement | 0.81 | 0.98 | 0.93 | 0.93 |
| First model | Driver category based on CNC | 0.63 | 0.51 | 0.54 | 0.32 |
| Second model | Driver category based on CNC | 0.81 | 0.70 | 0.76 | 0.77 |

### 3.2 Examination of the Association Between Selected Factors and CNC

The relationship between critical factors and CNC probability was examined using the second model. Although self-reported crash involvement and driver category with CNCs as response variables were binomial, CNCs observed from naturalistic driving monitoring had more information and advantages compared with binomial variables. For example, CNC incorporated the details of incidents such as number and circumstance of CNC, and was more dynamic rather than self-reporting because of observed and objective data during naturalistic driving records. Thus, we adopted the Poisson regression analysis using the number of CNCs as response variable to calculate the estimation model for CNC probability, as follows.

Poisson regression analysis using the second model with 18 variables indicated significant seven variables according to Wald statistics: score of MMSE, performance of TMT part B (sec), performance of Maze 2 in DHI (sec), values of coefficient of variation (CV) of simple and discrimination RT in DAT, contrast sensitivity in right eye (Area Under Log Contrast Sensitivity Function; AULCSF), and "hesitation for driving" score in DSQ. Focusing on significant seven variables, a low-deviance and compact model was investigated. However, considering that the potential of encountering a CNC depend on the frequency of driving, and studies have adjusted CNCs by driving habits such as time or distance driven, this analysis should include distance driven as a driving habits factor in the estimation model. Four Poisson analyses were performed to examine the possibility, and indicated different models: (A) all model containing 18 variables, (B) the model with seven select variables, (C) model of selected seven variables and two driving habits (distance driven per day and total distance driven), and (D) model of 7 variables and distance driven per day with the log of total distance driven as an offset. Table 2 described the likelihood of these models. Comparison of likelihood as model among these four models indicate that the (D) was better for CNC involvement than the other models, and the results suggest that a few cognitive performances, visual functioning, hesitation for driving, and distance driven might be associated with the CNC probability.

**Table 2.** Likelihood for Poisson regression analysis of the four models.

| Model | AIC* | Deviance | Log likelihood |
|---|---|---|---|
| (A) | 245.5 | 111.1 | −103.8 |
| (B) | 243.2 | 132.8 | −114.6 |
| (C) | 223.4 | 106.9 | −101.7 |
| (D) | 205.9 | 91.4 | −93.9 |

*AIC = Akaike information criterion

### 3.3 Estimation of CNCs Using Poisson Regression Analysis

Poisson regression analysis used 8 critical factors and an offset (the model D) was performed, and the result was listed in Table 3.

**Table 3.** Poisson regression analysis results of the CNCs.

| | β | SE | Z value | Pr |
|---|---|---|---|---|
| Intercept | 5.872 | 2.190 | 2.681 | 0.007 |
| MMSE | −0.333 | 0.062 | −5.374 | **0.000** |
| TMT part B | −0.008 | 0.005 | −1.549 | 0.121 |
| MAZE 2 (DHI) | 0.011 | 0.015 | 0.717 | 0.473 |
| CV of simple RT (DAT) | 0.020 | 0.008 | 2.399 | **0.016** |
| CV of discrimination RT (DAT) | −0.046 | 0.015 | −3.080 | **0.002** |
| Contrast sensitivity (right) | −1.671 | 0.487 | −3.430 | **0.001** |
| Hesitation for diving (DSQ) | −0.367 | 0.174 | −2.106 | **0.035** |
| Distance driven per day | −0.027 | 0.008 | −3.485 | **0.000** |

Using the model, predictive values of the number of CNCs were computed by inputting given performance values of each factor. As a hypothesis, if the drivers are plotted randomly across the distribution of the number of CNCs and distance driven, the model delivers higher number of CNCs to lower levels of performance drivers as in Fig. 1, which illustrates the hypothesis on variation of the probability of CNCs corresponding to performance levels of a single factor.

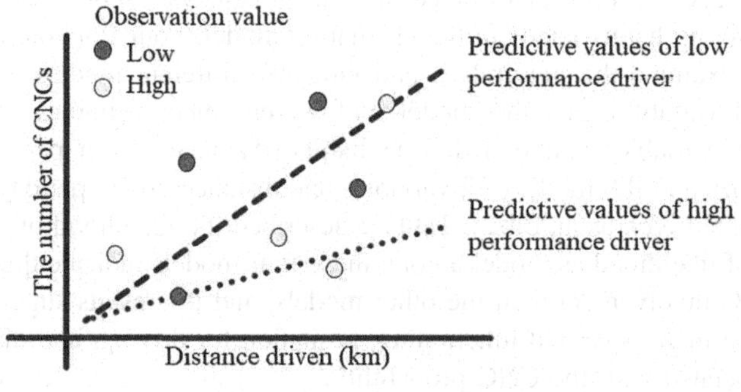


**Fig. 1.** The hypothesis on predictive values of low- and high-performance drivers and the association between the number of CNCs and distance driven.

Focusing on MMSE, consistent with the hypothesis, when performances of the other variables are determined in median constantly, predictive values of CNCs increase under lower MMSE score, which was represented by a decline in cognitive functions. The result is shown in Fig. 2. Similar results were obtained from the CV of simple RT task and contrast sensitivity. However, TMT part B and CV of discrimination RT task show different tendencies, that predictive values of CNCs increase under higher performance. Unlike MMSE, these factors did not have equal variances on distance driven between each variable of low and high-performance drivers, which might reflect a different distribution of performances to the predictive values.

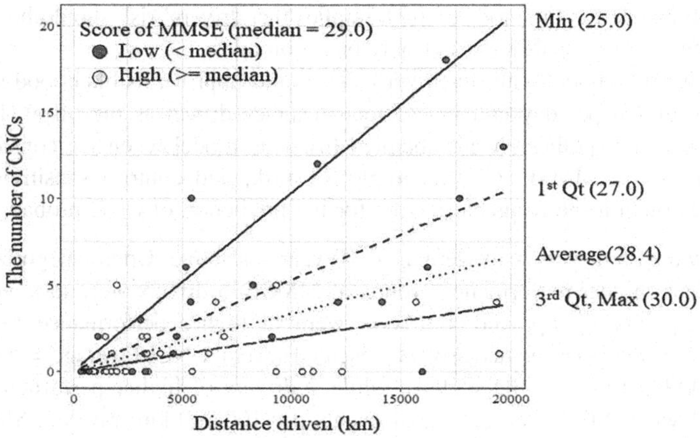


**Fig. 2.** Regression lines on the scatterplots of the distance driven and the number of CNCs for each MMSE performance level. The dots represent the number of CNCs of each driver. Min = minimum, 1st Qt = first quantile, 3rd Qt = third quantile, Max = maximum.

For "hesitation for driving," when inputting minimum score in the model, the predictive values of CNCs output higher frequency of CNCs, where drivers of lower hesitation (i.e., positive attitude to drive) tended to drive longer distance ($t$ value (28) = −1.82, $p$ = 0.08), and such drivers have a potential to encounter more CNCs.

# 4 Discussion

This study investigated the identifying factors related to estimation of CNCs, examined the association between the factors and CNCs, and estimated the probability of CNCs according to critical factors. The results are as follows:

- Binomial logistic regressions indicated that the plausible model comprising 18 variables (age, cognitive performance, RT, attention, visual function, subjective assessment) selected by driver category based on observed CNCs have higher predictive performance, suggesting that this model could be applied to the detection of possible drivers with CNCs. However, since self-reported crash involvement had different features from CNCs while driving, the first and second models could not interact with each other.

- The 18 variables in the second model examined the association with CNCs to find critical factors for estimation of the number of CNCs. Poisson regression analysis has indicated a better predictive model comprising seven significant variables, namely MMSE, TMT part B, CVs of RT performances, contrast sensitivity, and hesitation for driving. Selection of contrast sensitivity supports previous studies [3, 4] as well as our hypothesis and reveals certain factor for CNCs. However, a variable of age was not included in the model, which implies that drivers' risk should be evaluated based on various capabilities, not only based on age.
- The estimation model with seven variables would improve its likelihood by adding distance driven per day and log of total distance driven as the offset. Predictive values as our hypothesis were obtained from the model when we consider given performance of MMSE, CV of simple RT task, and contrast sensitivity. These variables seem to be plausible factors for the prediction of CNC probability.

There were 25 drivers who were aged 70 years or above. Upon categorizing them into lower and higher performance groups of MMSE, 5 drivers were in lower performance (25 points) group, and 4 drivers were in higher performance (30 points) group. All of lower performance drivers presented several CNCs (average = 55.8 events per 100,000 km traveled, SD = 29.4), while 2 drivers of higher performance group presented fewer CNCs (average = 20.4 events per 100,000 km traveled, SD = 24.8). Although most of CNCs have been encountered at intersections, their situations were different in two driver groups based on MMSE performance level, where the low-performance group tended to have CNCs in stop-sign-controlled intersections (15.2%), while the high-performance group tended to have CNCs in signalized intersections (37.5%). These observations suggest that drivers belonging to the low-performance group of MMSE had deficits in conditions required for processing a variety of information simultaneously, such as an awareness of peripheral vehicles and objects at stop-sign-controlled (unsignalized) intersections [11]. Therefore, drivers with lower MMSE performance will need assistance with their maneuver through systems on private vehicles, and we propose training programs for their attention and information processing capacities.

This study could identify the critical factors in relation to CNC risk, and estimate CNC probability using these factors. This estimation might be applied to the procedure of driving aptitude tests and driving observation for elderly drivers. Today, drivers aged 70 years or above receive a mandating lesson and driving aptitude tests as part of driving license renewal procedure in Japan. If they could pass the tests, they keep driving regularly, and if they fail the tests, they require further tests or medical assessments. However, considering the increase of elderly drivers in the future, a large number of medical practitioners will be required in this stage, and their temporal and mental workload would rise. Applying the present estimation model could be helpful to detect safe and risky drivers. Further, instead of medical assessments, we suggest a possibility of observation procedures after the tests by naturalistic driving monitoring of drivers who are predicted to be higher CNC probability based on their abilities and attributions.

**Acknowledgement.** This study was conducted with the support of the "Center of Innovation (COI) program," which is part of the research result institute of the Japan Science and Technology Agency (JST.) We have also listed the members who provided their cooperation with the collection and analysis of the DR data as a way to express our sincere gratitude.

## References

1. Neale LV, Dingus AT, Klauer GS, Sudweeks J, Goodman M (2005) An overview of the 100-car naturalistic study and findings. National highway traffic safety administration paper number 05-0400 (2005)
2. Guo F, Fang Y (2013) Individual driver risk assessment using naturalistic driving data. Accid Anal Prev 61:3–9
3. Guo F, Fang Y, Antin FJ (2015) Older driver fitness-to-drive evaluation using naturalistic driving data. J Saf Res 54:49–54
4. Huisingh C, Levitan BL, Marguerite RI, Maclennan P, Wadley V, Owsley C (2017) Visual sensory and visual-cognitive function and rate of crash and near-crash involvement among older drivers using naturalistic driving data. Invest Ophthalmol Vis Sci 58(7):2959–2967
5. Anstey JK, Wood J, Lord S, Walker GJ (2005) Cognitive, sensory and physical factors enabling driving safety in older adults. Clin Psychol Rev 25(1):45–65
6. Suto S, Kumada T (2010) Effects of age-related decline of visual attention, working memory and planning functions on use of IT-equipment. Jpn Psychol Res 52(3): 201–215
7. TransAnalytics Health & Safety Services. DrivingHealth.com. http://drivinghealth.com/. Accessed 18 Apr 2018
8. Akamatsu M, Hayama K, Takahashi J, Iwasaki A, Daigo H (2006) Cognitive and physical factors in changes to the automobile driving ability of elderly people and their mobility life: Questionnaire survey in various regions of Japan. IATSS Res 30(1):38–51
9. Sato T, Akamatsu M, Aoki H, Kanamori H, Yamagishi M (2016) Relations between elderly drivers cognitive functions and their compensatory driving behaviors. Humanist 5th Conference
10. Stanton AN, Landry S, Di Bucchianico G, Vallicelli, A (2014) Advances in human aspects of transportation. In: Proceedings of the AHFE 2014 International Conference on Human Factors in Transportation
11. Ball K, Owsley C (1991) Identifying correlated of accident involvement for the older driver. Hum Factor 33(5):583–595

# The Use of Auditory Presentations in Assisting Older Adults for Processing Healthcare Information

Dyi-Yih Michael Lin(✉) and Yuan-Ju Hung

Department of Industrial Management, I-Shou University,
Kaohsiung City 840, Taiwan
dlin@isu.edu.tw

**Abstract.** The present study examined the extent to which different auditory presentations could assist the older adult to better retain browsed healthcare information as a function of sound orientation. 18 students from a local university (serving as a control) and 20 older adults aged over 65 from a local lifelong learning Center were recruited to participate in a 2 × 3 × 2 slit-plot factorial experiment where age, auditory display and sound orientation were manipulated as independent variables. The auditory display was a within-subject factor consisting of narration, earcon, and narration plus earcon. Sound orientation was a between-subject factor comprising treatment levels of presence and non-presence. Perusal performance was evaluated by recall hit rate and the subject's preference. Preliminary results indicated that the older subject was significantly disadvantaged as compared to the young counterpart. Auditory information presented with sound orientation resulted in better recall performance. The narration plus earcon interface outperformed the other two auditory conditions in terms of recall hit rates and subjective preference. Both the older and young groups reported that inclusion of auditory display improved their memory and attention with earcon plus narration receiving the most favorable preference. Sound orientation, however, was not considered by both age groups as resulting in perceived assistance.

**Keywords:** Auditory interface · Cognitive aging · Human factors

## 1 Introduction

Considering the pervasive advance in digital technology, navigation over electronic contents has become a major interaction activity. Such perusal of digital contents involves mental resources including memory and spatial cognition. As literature has well documented that aging is normally associated with decline in cognitive abilities, design of appropriate interfaces for older adults to successfully interact with digital contents thus bears significance [1, 2]. This is particularly true when it comes to leaning healthcare knowledge for the older adult. Interface design has been visually oriented, which leaves many issues remaining unresolved concerning the role of the auditory modality in human-computer interaction [3]. For example, how does the verbalized auditory display (e.g., narration) perform differentially from the non-verbal (e.g., earcons) counterpart in


S. Bagnara et al. (Eds.): IEA 2018, AISC 826, pp. 32–35, 2019.
https://doi.org/10.1007/978-3-319-96065-4_5

achieving stronger retention of browsed contents? By addressing the issues and alike, the present study aims to investigate the effect of auditory design on cognitive performance as well as user experience in digital contents perusal with aging considerations.

# 2 Methods

## 2.1 Experimental Design

A 2 × 3 × 2 slit-plot factorial experiment was conducted where age, auditory display and sound orientation were manipulated as independent variables. Age was a quasi between-subject factor defined by older vs. young (control). Auditory display was a within-subject factor consisting of narration, earcon, and narration plus earcon. Sound orientation was a between-subject factor comprising treatment levels of presence vs. non-presence. Perusal performance was evaluated by recall hit rate. User experience was reported by the subject concerning how the two types of auditory manipulation were perceived to improve their retention of and attention to the perused healthcare information.

## 2.2 Subjects

18 students from a local university and 18 older adults aged over 65 from a local lifelong learning Center were recruited to participate in the experiment. All of the subjects were informed of how the experiment will be proceeded and signed participation consent. The young group consisted of 12 male and 6 female students whose age ranged from 22 to 24 years old. The older group consisted of 11 male and 7 female subjects whose age ranged from 65 to 75 years old. Every one of the 36 subjects was an experienced computer users and none of them reported having hearing problems.

## 2.3 Material/System

Healthcare information such as cardiovascular disease, diabetes and dementia, etc. was selected as the experimental material. Three types of auditory interface were developed to display the experimental information and were presented to the subject randomly via right or left channel of a headset.

**The Healthcare Information System.** The system was developed by Flash CS 4.0 where 12 major healthcare issues such as cardiovascular disease, diabetes and dementia, etc. were selected as the experimental contents and were displayed in a 3 (rows) × 4 (columns) tile-based layout. Selection of the 12 tiles was made on a touch screen (see Fig. 1).

**Narration Display.** Narrated information associated with the aforementioned 12 healthcare topics was recorded by GoldWave and the digitalized narrations were imbedded into the Flash-based 3 × 4 tiles matrix.

**Earcon Display.** C major piano note was chosen as the earcon and was developed by Simple Piano. The upper, middle and lower rows of the 3 × 4 tiles matrix were associated with high (523.3 Hz), medium (261.7 Hz), and low (130.8 Hz) frequency major C respectively. The rationale behind the design intends to provide the subject with a stronger cue as to the location of the topics in the 3 × 4 tiles matrix (see Fig. 2).

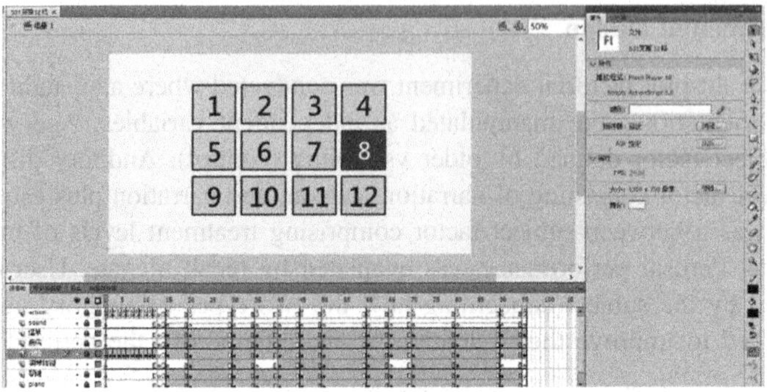


**Fig. 1.** Snapshot of the 3 × 4 tiles matrix developed in the Flash CS 4.0 system

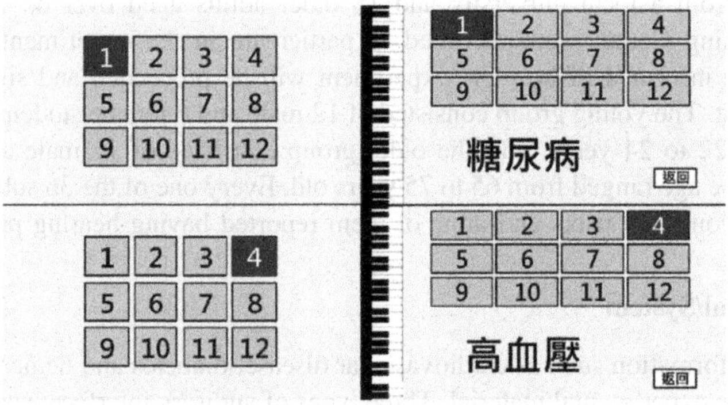


**Fig. 2.** The earcon design for diabetes (tile 1) and high blood pressure (tile 4) which are both upper row tiles associated with high frequency piano Major C note.

**Table 1.** Means and (standard deviations) of retention rate under the manipulation of auditory display and sound orientation as a function of age difference. (SO: Sound Orientation; N: Narration; E: Earcon; N+E: Narration plus Earcon)

| Older | | | | | | Young | | | | | |
|---|---|---|---|---|---|---|---|---|---|---|---|
| 0.39 (0.28) | | | | | | 0.56 (0.30) | | | | | |
| SO | | | Non-SO | | | SO | | | Non-SO | | |
| 0.44 (0.28) | | | 0.34 (0.28) | | | 0.63 (0.32) | | | 0.50 (0.27) | | |
| N | E | N+E | N | E | N+E | N | E | N+E | N | E | N+E |
| 0.42 (0.25) | 0.36 (0.31) | 0.53 (0.29) | 0.26 (0.22) | 0.25 (0.28) | 0.42 (0.33) | 0.69 (0.33) | 0.44 (0.33) | 0.75 (0.25) | 0.44 (0.27) | 0.39 (0.22) | 0.67 (0.25) |

## 3 Results and Conclusions

The descriptive data of the retention performance was shown in Table 1. The ANOVA results indicated that none of the three first-order interaction effects are significant. Neither is the second-order interaction. However, the three main effects were all significant ($F_{(1,72)} = 11.56$, $p < 0.05$ for age effect, $F_{(1,72)} = 4.16$, $p < 0.05$ for sound orientation and $F_{(2,16)} = 3.88$, $p < 0.05$ for auditory display. As auditory display consists of three treatment levels, we further conducted a post-hoc Tukey pair-comparison test with 5% as the significance level. The Tukey test result indicated that while the difference between narration plus earcon and narration, and between narration and earcon were only due to a random error, the narration plus earcon display significantly outperformed the earcon counterpart in terms of how well the subject is able to retain the healthcare information processed.

The self-reports for user experience revealed that the majority of the older subject are in favor of the presentation of auditory display, feeling that the presentation could assist them to improve retention of and attention to the information perused. The subjective preference of the older subject towards sound orientation, nevertheless, indicated otherwise. The majority of the older subject reported that inclusion of sound orientation improves neither retention of nor attention to the performance.

## References


1. Park D, Schwarz N (2000) Cognitive aging: a primer. Psychology Press, Philadelphia
2. Fisk AD, Roger WA, Charness N, Czaja SJ, Sharit J (2009) Designing for older adults: principles and creative human factors approaches, 2nd edn. CRC Press, Boca Raton
3. Baldwin CL (2012) Auditory cognition and human performance: research and applications. CRC Press, Boca Raton

# Human Factors for Dementia: Evidence-Based Design

Charlotte Jais, Sue Hignett(✉), and Eef Hogervorst

Loughborough Design School, Loughborough University, Loughborough, UK
{C.Jais3, S.M.Hignett}@lboro.ac.uk

**Abstract.** Designing care environments for people living with dementia is a complex challenge as the key stakeholder may have difficulty communicating their capabilities, limitations and preferences. This paper describes the use of evidence-based design personas in a multi-disciplinary team with architects and chartered human factors specialists. Four individual personas (Alison, Barry, Christine and David) and a couple persona (Chris and Sally) were used to bring the voices of the people living with different stages of dementia to the design process. Their changing/fluctuating symptoms were communicated in two formats (wheel and matrix) within an inclusive design process to adapt a Victorian semi-detached house. The demonstrator house presents evidence-based design, adaptation and support solutions to support people living with dementia to age well at home.

**Keywords:** Dementia · Human factors/ergonomics · Care · Environment

## 1 Introduction

Dementia is an increasing global issue, with a worldwide estimate of 46 million people living with dementia and expected to rise to around 131.5 million people by 2050 [1]. The diverse symptoms of dementia (cognitive, perceptual, functional and communicative changes) contribute to the challenge of designing appropriate and supportive care environments [2, 3]. A well-designed care environment can improve quality of life and enable those with dementia to maintain some level of independence [4–6], but a poorly designed care environment can have negative consequences for quality of life and wellbeing [5].

*'A person's home is not just the place where they live, but also a place of work for home care workers'* [7]. Home care aims to satisfy peoples' health and social needs in their homes by '*providing appropriate and high-quality home-based healthcare and social services, by formal and informal care-givers, with the use of technology when appropriate, within a balanced and affordable continuum of care'* [8]. However, the home setting presents challenges for the more established (acute) caregiver-patient interactions and requires adaptation of policies, protocols and routines [9].

People living with dementia (PWDem) may find it difficult to communicate their design needs due to cognitive difficulties which are a core feature of dementia. To address this challenge, a human factors/ergonomics (HFE; [10]) inclusive design approach was taken to develop dynamic personas for PWDem (changing/fluctuating limitations and capabilities).


S. Bagnara et al. (Eds.): IEA 2018, AISC 826, pp. 36–43, 2019.
https://doi.org/10.1007/978-3-319-96065-4_6

## 2 Design Personas

Personas have more commonly been used in fields, such as marketing and software design [11]. It was suggested that this concept could be used in dementia design to communicate the needs, symptoms, limitations and abilities of PWDem, and could be enhanced by an evidence-based approach. Design personas can be used to represent the needs of archetypal PWDem to provide designers (architects) with access to information which would be particularly useful for people in middle to later stages, where communication abilities deteriorate further.

A systematic literature review established the framework for a scoping study to explore activities of daily living (ADLs) and instrumental ADLs (iADLs) for PWDem [2]. The results suggested that eating, toileting, social interaction and physical activity were the most important activities which needed to be considered in the design of dementia care environments. The first version of the personas was based around design needs associated with eating and toileting; social interaction was promoted through supporting communal dining. Physical activity was supported by enabling independent mobilisation to the toilet. The subsequent versions were reviewed and revised iteratively in a series of focus groups and interviews with architects, care support workers, and care home developers. This included the addition of clinical assessment scores for the Mini Mental State Exam (MMSE; [12]), the Montreal Cognitive Assessment (MoCA; [13]), the Addenbrooke's Clinical Examination III (ACE-III; [14]) and the Abbreviated Mental Test (AMT; [15]). Icons were also added to illustrate content such as eyesight or hearing problems, and requiring assistance with activities of daily living, for ease of reference.

The final personas describe four stages of living with dementia as Alison, Barry, Christine and David [16]. They are available in two formats; firstly, as a dynamic wheel to indicate a good, average or bad day (Fig. 1) with green representing a good day, amber an average day and red a bad day. Secondly, as a matrix (Fig. 2) to show symptoms, care needs and design needs across good, average and bad days for each persona without having to scroll through the persona wheel.

Barry (second stage of living with dementia) is a 74 year old retired postman. He has been diagnosed with dementia and is currently still living in his own home. He has difficulty with word finding, planning and organising and has a small risk of falls (physical changes). On an average day he sometimes forgets recent events (including personal information about himself) and may need support for numerical tasks. On a bad day he may have communication difficulties and be confused, particularly in new environments (navigation information processing). These cognitive and physical symptoms and care needs are linked as a 'good day' in the next level, Christine (Fig. 2).

Christine (third stage of living with dementia) includes a generic 'ageing' characteristic of hearing loss since the age of 70 years (she is described at 82 years). She uses bilateral hearing aids which may present dementia-related challenges for her in remembering to wear them and for usability with physical dexterity in their operation and maintenance (for example battery changes). As she is a talented musician, this hearing challenge can contribute to frustration with her changing capabilities and limitations.

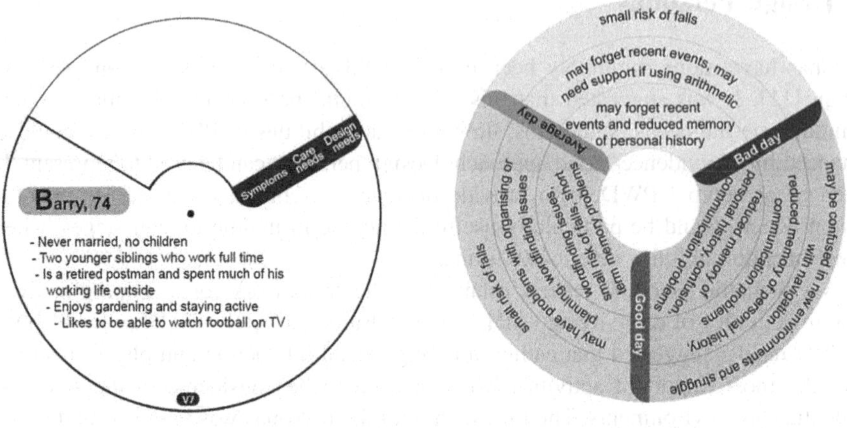


**Fig. 1.** Persona Barry as a dynamic wheel (color figure online)

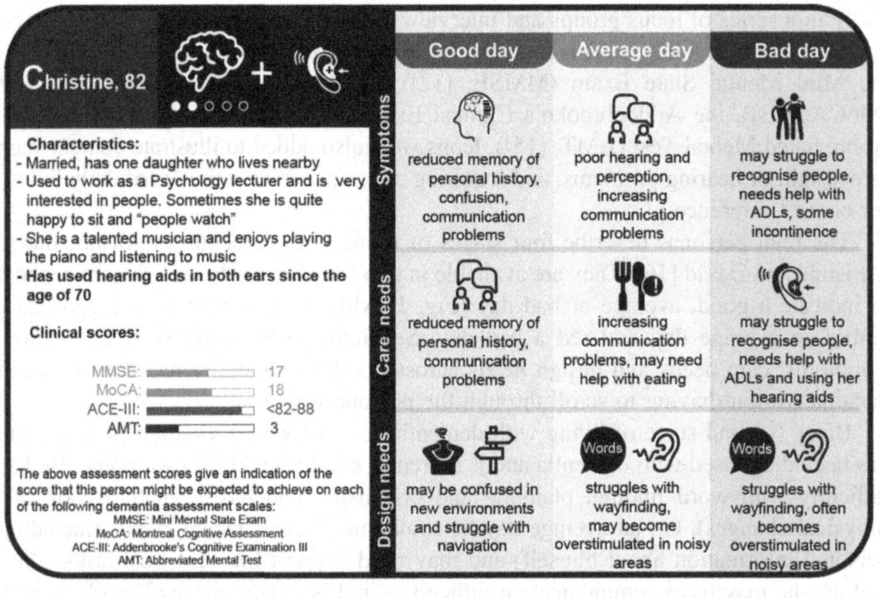


**Fig. 2.** Persona Christine as a matrix (more detailed information about Chris)

The final iteration was the development of a persona representing a couple to reflect the needs of a home care environment. This uses the same format as the individual personas, but has an inner wheel to represent the needs of Chris (PWDem. Figure 3), and his wife (carer), Sally. Chris has the same characteristics as Christine (third stage of living with dementia). They have been married for over 50 years but Sally is starting to find it harder to care for Chris as his dementia progresses. If their 'bad' days coincide,

Chris may struggle to recognise people, exhibit sundowning syndrome [17] and become frustrated (lash out physically). Sally needs considerable help and may awake all night to care for him.

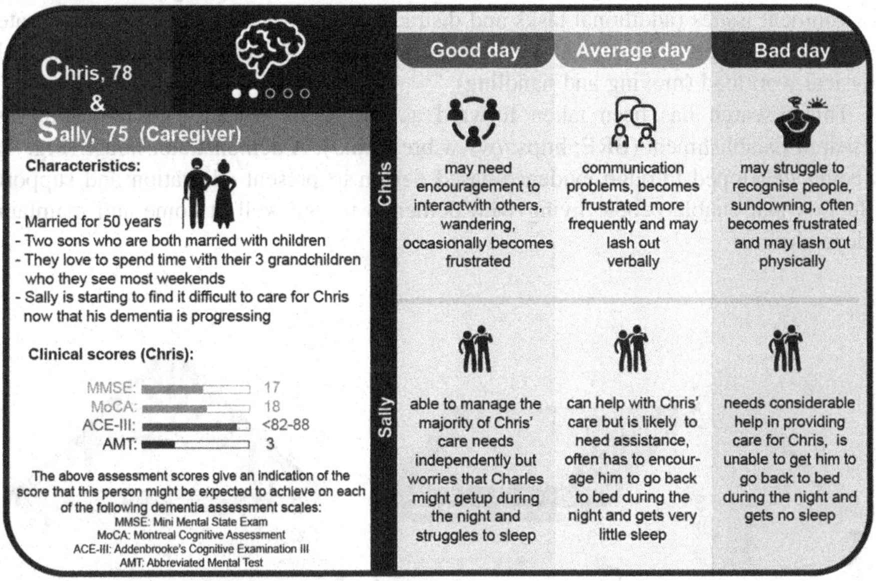


**Fig. 3.** Chris and Sally (couple persona)

The final personas were evaluated and validated in a third study by discussing the personas with professional caregivers (care managers, nurses and care assistants). The participants responded that the personas were a useful representation of PWDem; they '*recognised*' needs and symptoms displayed by their residents and could identify features of the care home that might cause difficulties for the personas (as experienced by their residents). Overall, the personas were confirmed as a useful design communication tool, with applications in both care home design and care delivery.

## 3 Chris and Sally's House

Providing care and treatment at home presents challenges for PWDem whether care is delivered from one or multiple organisation(s) or within different models of home care. These may include the hospital in the home, patient-centred medical home, home first policies and aging-in-place. Hignett, Edmunds-Otter and Keen [18] summarised the state of science for physical interactions (context for design) with the results grouped as environment (health policy, physical and social), artefacts (equipment and technology), tasks (procedures and work schedules) and care recipient/provider. This included permanent and temporary building design and access, communication and lone working, provision of equipment and consumables, and clinical tasks. The care tasks were categorised as basic care, including personal hygiene, mobilisation, nutrition and social company, and advanced care, including medication administration, tube feeding, and operating home care technology (e.g. ventilator, electric wheelchair). There was research evidence identifying risks associated with awkward working positions, social environment issues (additional tasks and distractions), abuse and violence, inadequate team (peer) support, problems with workload planning, needle stick injuries and physical workload (moving and handling).

This research has been taken forward as part of a project with the Building Research Establishment (BRE; https://www.bre.co.uk/). A demonstrator house (Fig. 4) is being developed through evidence-based design to present adaptation and support solutions that enable people living with dementia to age well at home and maintain independence.

**Fig. 4.** Victorian semi-detached house, with side view

Dementia design guidelines have been critically appraised [6, 19–25] and found to be mostly based on professional consensus and stakeholder opinions rather than robust research evidence. The personas address this gap by integrating a systematic literature review with empirical qualitative and quantitative data from stakeholders (scoping study, focus groups, interviews, and care home observations).

The Victorian house (re)design plans were reviewed and discussed by the multi-disciplinary team using the personas to explore usability within a theoretical framework of inclusive design [26]. The final layout has key features of dementia friendly toilets downstairs and upstairs, and a through-floor lift (rather than a stair lift) to achieve end of life care in the first floor bedroom with an adjacent kitchenette (for overnight support/respite carers). A ceiling track hoist is included to support moving and handling activities during full (end-of-life) care for David.

The research evidence incorporated within the personas includes the use of colour as a navigation aid (and cue) [5, 27–29]. For example, Namazi and Johnson [29] found that PWDem achieved better success with navigation tasks in environments using primary colours, improved contrast and concrete nomenclature (rather than abstract) with supporting diagrams, and indirect cues from signage placement.

## 4 Conclusion

Chris and Sally's house gives an example of how university-based research can be taken from empirical data through to real life application. The multi-disciplinary team has benefited by incorporating human factors expertise as a 'linking discipline' between clinical and safety science and design to produce an inclusive design. We hope that this offers both an exemplar for housing adaptation and also potential for incorporation into future building regulations to support ageing well at home.

**Acknowledgements.** Design Star Doctoral Consortium funded the development of the Design Personas (AHRC PhD studentship). Loughborough University Enterprise Project Group funded the design development work with Building Research Establishment.

## References

1. Prince M, Wimo A, Guerchet M, Ali G-C, Wu Y-T, Prina M et al (2015) World Alzheimer Report 2015 The Global Impact of Dementia An Analysis of Prevalence, Incidence, Cost and Trends. https://www.alz.co.uk/research/WorldAlzheimerReport2015.pdf
2. Jais C, Hignett S, Hogervorst E (2016) Developing personas for use in the design of dementia care environments. In: Proceedings of the healthcare ergonomics & patient safety (HEPS) conference, Toulouse, 5–7 October, pp 210–216
3. Jais C, Hignett S, Habell M, Brown A, Hogervorst E (2016) Defining activities of daily living for the design of dementia care environment. In: Sharples S, Shorrock S, Waterson P (eds) Contemporary ergonomics 2016. Proceedings of the annual conference of the chartered institute of ergonomics & human factors. Taylor & Francis, London
4. Cioffi JM, Fleming A, Wilkes L, Sinfield M, Le Miere J (2007) The effect of environmental change on residents with dementia: the perceptions of relatives and staff. Dementia 6 (2):215–231
5. Day K, Carreon D, Stump C (2000) The therapeutic design of environments for people with dementia: a review of the empirical research. Gerontologist 40(4):397–416
6. Fleming R, Purandare N (2010) Long-term care for people with dementia: environmental design guidelines. Int Psychogeriatr IPA 22(7):1084–1096
7. Taylor BJ, Donnelly M (2006) Risks to home care workers: professional perspectives. Health Risk Soc 8(3):239–256
8. WHO (2008) The solid facts: homecare in Europe. http://www.euro.who.int/__data/assets/pdf_file/0005/96467/E91884.pdf. Accessed 11 June 2015
9. Duke M, Botti M, Hunter S (2012) Effectiveness of pain management in hospital in the home programs. Clin J Pain 28(3):187–194

10. Dul J, Bruder R, Buckle P, Carayon P, Falzon P, Marras WS et al (2012) A strategy for human factors/ergonomics: developing the discipline and profession. Ergonomics 55 (4):377–395
11. Adlin T, Pruitt J (2010) The essential persona lifecycle - your guide to building and using personas. Morgan Kaugman, Burlington
12. Folstein MF, Folstein SE, McHugh PR (1975) Mini-mental state: a practical method for grading the cognitive state of the patient for the clinician. J Psychiatr Res 12(3):189–198
13. Nasreddine Z, Phillips N, Bédirian V, Charbonneau S, Whitehead V, Colllin I et al (2005) The montreal cognitive assessment, MoCA: a brief screening tool for mild cognitive impairment. J Am Geriatr Soc 53(4):695–699
14. Hsieh S, Schubert S, Hoon C, Mioshi E, Hodges JR (2013) Validation of the Addenbrooke's cognitive examination III in frontotemporal dementia and Alzheimer's disease. Dement Geriatr Cogn Disord 36:242–250
15. Hodkinson HM (1972) Evaluation of a mental test score for assessment of mental impairment in the elderly. Age Ageing 1(4):233–238
16. Jais C, Hignett S, Galindo Estupiñan Z, Hogervorst E (2017) Evidence based dementia personas: human factors design for people living with dementia. In: Proceedings of 18th research-technical international conference: ergonomics for people with disabilities, 21–22 November 2017, Lodz, Poland
17. Khachiyants N, Trinkle D, Son SJ, Kim KY (2011) Sundown syndrome in persons with dementia: an update. Psychiatry Investig 8:275–287
18. Hignett S, Edmunds Otter M, Keen C (2016) Safety risks associated with physical interactions between patients and caregivers during treatment and care delivery in home care settings: a systematic review. Int J Nurs Stud 59:1–14
19. Calkins M (2001) Creating successful dementia care settings. Health Professions Press, Baltimore
20. Moore KD, Geboy LD, Weisman GD (2006) Designing a better day: guidelines for adult and dementia day services centers. The Johns Hopkins University Press, Baltimore
21. Fisk AD, Roger WA, Charness N, Czaja S, Sharit J (2009) Designing for older adults: principles and creative human factors approaches, 2nd edn. CRC Press, Boca Raton
22. Department of Health (2015) Health building note 08-02 dementia friendly health and social care environments. The Stationary Office, London
23. Grey T, Pierce M, Cahill S, Dyer M (2015) Universal design guidelines dementia friendly dwellings for people with dementia, their families and carers. National Disability Authority, Dublin. http://universaldesign.ie/Web-Content-/UD_Guidelines-Dementia_Friendly_Dwellings-2015-Introduction.pdf
24. Halsall B, McDonald R (2016) Volume 1 - design for dementia - a guide with helpful guidance in the design of exterior and interior environments. Halsall Lloyd Partners, Liverpool. http://www.hlpdesign.com/images/case_studies/Vol1.pdf
25. Halsall B, McDonald R (2016) Volume 2 - design for dementia - research projects, outlines the research projects and describes the participatory approach. Halsall Lloyd Partners, Liverpool. http://www.hlpdesign.com/images/case_studies/Vol2.pdf
26. British Standards Institute (2005) BS 7000-6 design managements systems - part 6: managing inclusive design. British Standards Institute, London
27. Calkins MP (1988) Design for dementia. Planning environments for the elderly and the confused. National Health Publishing, Baltimore
28. Chaâbane F (2007) Falls prevention for older people with dementia. Nurs Stand 22(6):50–55

29. Namazi KH, Johnson BD (1991) Physical environmental cues to reduce the problems of incontinence in Alzheimer's disease units. Am J Alzheimer's Care Relat Disord Res 6 (6):22–28
30. Author F (2016) Article title. Journal 2(5), 99–110
31. Author F, Author S (2016) Title of a proceedings paper. In: Editor F, Editor S (eds) Conference 2016, LNCS, vol. 9999. Springer, Heidelberg, pp 1–13
32. Author F, Author S, Author T (1999) Book title. 2nd edn. Publisher, Location
33. Author F (2010) Contribution title. In: 9th International proceedings on proceedings. Publisher, Location, pp 1–2
34. LNCS Homepage. http://www.springer.com/lncs. Accessed 21 Dec 2016

# How Do Municipal Workers Perceive the Changes in Activity Demands, Based on Age?

C. A. Ribeiro[1(✉)], T. P. Cotrim[1,2], V. Reis[3], M. J. Guerreiro[3], S. M. Candeias[3], A. S. Janicas[3], and M. Costa[3]

[1] Ergonomics Laboratory, Universidade de Lisboa, Lisbon, Portugal
camiladriribeiro@gmail.com
[2] CIAUD, Faculdade de Arquitetura, Universidade de Lisboa, Lisbon, Portugal
[3] Health and Safety Department, Municipality of Sintra, Sintra, Portugal

**Abstract.** Ageing is a complex and natural process that has been extensively studied. Several studies indicate that high physical exposure and job strain in midlife were strongly associated with the severity of disability in later life. In order to characterize the activity demands of the work, according to age, a prospective study was designed. In 2015 the sample included 885 participants and, in 2017, the follow up comprised 1167 participants. A self-administered questionnaire was used. The questionnaire was composed of questions related to sociodemographic characteristics, as well as related to the physical activity demands, classified through a Likert scale of 5 points. The results found that the postural characteristics and manual handling of loads obtained more favourable results in the year 2017, with the categories never/seldom presenting higher percentages. However, when we evaluated the age distribution, older workers presented a higher percentage of physical demands in their work activity, in both moments of the study, such as manual handling loads between 1–4 kg, working with trunk flexion and rotation, and precision tasks. Also, older workers belong mainly to the operational assistant category and have fewer training opportunities. The results indicate that is necessary to implement strategies for age management, so that older workers can remain active and healthy.

**Keywords:** Age · Municipal workers · Activity demands · Ergonomics Occupational Health

## 1 Introduction

The effects of age are described as a trend, not yet fully measured in intensity, influenced by the characteristics of the physical and social environment and by genetic programming, with an impact on physical and cognitive abilities (Ilmarinen 2001; Laville and Volkoff 2004; Warr 2001).

However, despite the age-related declines, older people are able to adapt, compensating the limitations, with their residual abilities used more efficiently, using the experience gained over the years, allowing or not to establish a new balance between functional impairment and task performance. The work activity, in turn, can accentuate


S. Bagnara et al. (Eds.): IEA 2018, AISC 826, pp. 44–51, 2019.
https://doi.org/10.1007/978-3-319-96065-4_7

the process of decline or enrich the experience, within the limits imposed by the organization and the resources of work (Cotrim and Simões 2013; Ilmarinen 2001; Laville and Volkoff 2004).

Several studies indicated that there is a predictive pattern between age and risk of musculoskeletal injuries, which occur predominantly due to cumulative exposure to workplace hazards. The predictors of these injuries vary according to the age group, which points to the need for interventions directed at specific populations (Oakman et al. 2016).

The increase in costs associated with occupational diseases, absenteeism and early retirement, reflect this impact on the health of workers, alerting to the need for strategic changes in the labour reality (Ahola et al. 2013; Marchand and Blanc 2010).

In order to characterize the physical activity demands of the work, according to age, a prospective study with municipal workers was designed beginning in 2015, in the Portuguese municipality of Sintra. The present paper aims at characterizing the differences on the perception of physical activity demands of the municipal workers based on age, comparing the results from 2015 and 2017.

## 2 Methodology

### 2.1 Methods

The analysis was done using a questionnaire composed by questions related to sociodemographic characteristics and to the physical activity demands, regarding postural characteristics and manual handling of loads, evaluated through a 5-point Likert scale.

The population of this study included 1,667 workers of a Portuguese municipality. In 2015 the response rate was of 54.7%, with a total of 885 participants. In 2017, the follow up comprised 1167 participants, corresponding to a response rate of 70%.

The questionnaire was self-administered during the year 2015 and the follow up was done in 2017. Workers were asked about their interest in participating in this study and those who agreed to volunteer signed the Informed Consent Form. The confidence level assumed for the statistical analysis was 95%.

## 3 Results

### 3.1 Socio-Demographic Characteristics

The sample showed an average age of 46.9 years (sd = 8.3) in 2015, and 48.4 years (sd = 8.7) in 2017. The mean age was higher in the follow up and the differences on the mean values of age between the two moments of the research were statistically significant, according to a T-Student test ($P \leq 0.001$). The variable work seniority presented a lower average in 2017 (19.4 years) comparing with 2015 (20.4 years), and the difference was statistically significant (Table 1). Age and work seniority were correlated, according to the R-Pearson test (2015: $r = 0.615$ $p = 0.01$; 2017: $r = 0.617$ $p = 0.01$).

**Table 1.** Characterization of the variables age and work seniority in 2015 and 2017

| | 2015 | | | | | 2017 | | | | | P value |
|---|---|---|---|---|---|---|---|---|---|---|---|
| | N | Min | Max. | Mean | S.D | N | Min. | Max. | Mean | S.D | |
| Age | 885 | 25 | 69 | 46.9 | 8.3 | 1167 | 21 | 68 | 48.4 | 8.7 | $\leq 0.001$ |
| Work seniority | 885 | 1 | 46 | 20.4 | 8.7 | 1167 | 1 | 45 | 19.4 | 9.9 | 0.023 |

In both years there were a higher percentage of workers in the female gender, with a high school level of education, working as operational assistants and of workers who have had training in the last two years (Table 2).

**Table 2.** Socio-demographic characteristics

| | | 2015 | | 2017 | |
|---|---|---|---|---|---|
| | | N | % | N | % |
| Gender | Woman | 548 | 65.6 | 689 | 61.8 |
| | Man | 287 | 34.4 | 425 | 38.2 |
| Professional category | Operational assistant | 287 | 33.7 | 370 | 32.7 |
| | Technical assistant | 323 | 37.9 | 420 | 37.1 |
| | White collars | 229 | 26.9 | 330 | 29.2 |
| | Municipal police | 13 | 1.5 | 11 | 1.0 |
| Qualifications | Elementary to Junior high school | 242 | 28.2 | 314 | 27.9 |
| | High school | 324 | 37.8 | 411 | 36.5 |
| | Graduated/Post Graduated | 291 | 34.0 | 402 | 35.7 |
| Training in the last two years | Yes | 449 | 52.0 | 564 | 50.0 |
| | No | 415 | 48.0 | 564 | 50.0 |

When analysing the differences in the mean age between the categories of the socio-demographic variables, we have found significant differences in both moments of the research, according to an ANOVA. The mean age was higher among workers who work as Operational Assistants ($P \leq 0.001$), with a basic education level ($P \leq 0.001$) and who had not received training in the last two years ($P \leq 0.001$).

### 3.2 Characteristics of the Activity Demands

Regarding the type of demands of the activity, in 2015, 48.2% of the workers considered that the nature of their demands was both mental and physical and 46.1% consider them to be mental. In 2017, we observed a reversal of these values, where 52% of workers considered that the nature of their demands were mental and 42.7% consider them to be both mental and physical (Table 3). The differences found between the two moments of the research were statistically significant according to an ANOVA ($p = 0.026$).

**Table 3.** Activity demands characterization

| | 2015 | | 2017 | |
|---|---|---|---|---|
| | N | % | N | % |
| Mental | 397 | 45.8 | 598 | 52.0 |
| Physical | 51 | 5.9 | 62 | 5.4 |
| Both | 418 | 48.3 | 491 | 42.7 |

Also, were found statistically significant differences on the mean age between the categories of demands of the activity, in both moments of the research ($P \leq 0.001$ for both years), according to an ANOVA. The average age was higher in the category of physical demands in 2015 and 2017 (Table 4), indicating that older workers often perform work activities that require physical effort.

**Table 4.** Age characterization by activity demands, in 2015 and 2017.

| | 2015 | | | | | 2017 | | | | |
|---|---|---|---|---|---|---|---|---|---|---|
| | N | Min | Max | Mean | S.D | N | Min | Max | Mean | S.D |
| Mental | 397 | 26 | 67 | 46.2 | 7.1 | 598 | 25 | 66 | 47.6 | 8.2 |
| Physical | 51 | 35 | 66 | 51.1 | 8.9 | 62 | 32 | 67 | 52.7 | 9.0 |
| Both | 418 | 25 | 69 | 47.0 | 9.1 | 491 | 21 | 68 | 48.6 | 9.1 |

The variables "Seated work", "Working with trunk rotation", "Precision tasks done with the hands and fingers", and "Manual handling of loads between 1–4 kg" were the most frequent work activity characteristics in both moments. When comparing these variables between 2015 and 2017, it was found that the category "never/seldom" obtained higher percentages in the year 2017, indicating differences statistically significant in their frequency between the two years, this can be explained by an increase in the participation of white-collar workers in the 2017' sample (Table 5).

**Table 5.** Characteristics of work activity in 2015 and 2017

| | | 2015 | | 2017 | | P value* |
|---|---|---|---|---|---|---|
| | | N | % | N | % | |
| Working with trunk flexion | Very Frequent/Frequent | 156 | 20.4 | 198 | 19.7 | ≤0.001 |
| | Sometimes | 217 | 28.3 | 211 | 21.0 | |
| | Never/Seldom | 393 | 51.3 | 598 | 59.4 | |
| Working with trunk rotation | Very Frequent/Frequent | 203 | 26.1 | 238 | 23.5 | ≤0.001 |
| | Sometimes | 270 | 34.7 | 279 | 27.6 | |
| | Never/Seldom | 304 | 39.1 | 494 | 48.9 | |

(*continued*)

**Table 5.** (*continued*)

| | | 2015 | | 2017 | | P value* |
|---|---|---|---|---|---|---|
| | | N | % | N | % | |
| Seated work | Very Frequent/Frequent | 606 | 71.6 | 784 | 72.1 | 0.013 |
| | Sometimes | 128 | 15.1 | 123 | 11.3 | |
| | Never/Seldom | 112 | 13.2 | 180 | 16.6 | |
| Crouching or Kneeling | Very Frequent/Frequent | 67 | 8.8 | 85 | 8.5 | 0.034 |
| | Sometimes | 127 | 16.7 | 124 | 12.4 | |
| | Never/Seldom | 567 | 74.5 | 794 | 79.2 | |
| Work with arms above shoulders | Very Frequent/Frequent | 67 | 8.7 | 106 | 10.5 | ≤0.001 |
| | Sometimes | 160 | 20.7 | 137 | 13.5 | |
| | Never/Seldom | 545 | 70.6 | 769 | 76.0 | |
| Precision tasks with hands and fingers | Very Frequent/Frequent | 367 | 46.5 | 491 | 47.6 | 0.035 |
| | Sometimes | 165 | 20.9 | 168 | 16.3 | |
| | Never/Seldom | 258 | 32.7 | 372 | 36.1 | |
| Apply force manually | Very Frequent/Frequent | 157 | 20.5 | 172 | 17.1 | ≤0.001 |
| | Sometimes | 194 | 25.4 | 168 | 16.7 | |
| | Never/Seldom | 413 | 54.1 | 664 | 66.1 | |
| Manual handling of loads 1–4 kg | Very Frequent/Frequent | 216 | 27.9 | 256 | 25.2 | ≤0.001 |
| | Sometimes | 247 | 31.9 | 224 | 22.1 | |
| | Never/Seldom | 312 | 40.3 | 535 | 52.7 | |
| Manual handling of loads 5–9 kg | Very Frequent/Frequent | 127 | 16.5 | 161 | 16.0 | ≤0.001 |
| | Sometimes | 174 | 22.6 | 159 | 15.8 | |
| | Never/Seldom | 470 | 61.0 | 687 | 68.2 | |

*ANOVA with Welch adjustment.

When analyzing the age distribution by the characteristics of the work activity, it was found higher age averages in the "very frequent/frequent" categories, in 2015, for the variable "precision tasks with hands and fingers" and, in 2017, for the variables "working with trunk rotation" and "manual handling of loads between 1–4 kg". The mean age was also higher in the category "sometimes", in 2015, for the variables "working with trunk flexion", and, in the 2017, for "seated work". These differences were statically significant according to an ANOVA test (Table 6).

**Table 6.** Characteristics of work activity by age, in 2015 and 2017

| | | 2015 | | | | | P value | 2017 | | | | | P value |
|---|---|---|---|---|---|---|---|---|---|---|---|---|---|
| | | N | Min | Max | Mean | S.D | | N | Min. | Max | Mean | S. D | |
| Precision tasks with hands and fingers | Very Frequent/Frequent | 367 | 26 | 69 | 47.2 | 8.5 | 0.016 | 491 | 21 | 68 | 47.8 | 8.7 | |
| | Sometimes | 165 | 25 | 66 | 45.2 | 8.1 | | 168 | 25 | 66 | 48.4 | 9.4 | |
| | Never/Seldom | 258 | 29 | 65 | 45.9 | 7.9 | | 372 | 26 | 67 | 47.0 | 8.2 | |
| Working with trunk rotation | Very Frequent/Frequent | 203 | 25 | 69 | 46.2 | 8.2 | | 238 | 21 | 67 | 48.6 | 9.1 | 0.029 |
| | Sometimes | 270 | 26 | 67 | 46.9 | 8.6 | | 279 | 21 | 67 | 46.5 | 8.9 | |
| | Never/Seldom | 304 | 26 | 65 | 45.7 | 7.7 | | 494 | 26 | 68 | 47.6 | 8.1 | |
| Manual handling of loads 1–4 kg | Very Frequent/Frequent | 216 | 26 | 69 | 46.7 | 8.3 | | 256 | 21 | 66 | 48.7 | 8.7 | 0.039 |
| | Sometimes | 247 | 26 | 67 | 46.5 | 8.1 | | 224 | 21 | 67 | 47.2 | 8.5 | |
| | Never/Seldom | 312 | 25 | 67 | 45.6 | 8.1 | | 535 | 26 | 68 | 47.1 | 8.5 | |
| Working with trunk flexion | Very Frequent/Frequent | 156 | 26 | 69 | 46.1 | 8.4 | 0.044 | 198 | 21 | 66 | 48.5 | 9.5 | |
| | Sometimes | 217 | 25 | 66 | 47.4 | 8.2 | | 211 | 21 | 67 | 47.5 | 8.7 | |
| | Never/Seldom | 393 | 26 | 67 | 45.7 | 8.0 | | 598 | 25 | 68 | 47.1 | 8.3 | |
| Seated work | Very Frequent/Frequent | 606 | 25 | 67 | 46.6 | 7.5 | | 784 | 25 | 68 | 48.2 | 8.3 | 0.047 |
| | Sometimes | 128 | 26 | 67 | 46.6 | 10.2 | | 123 | 27 | 67 | 48.8 | 9.1 | |
| | Never/Seldom | 112 | 26 | 69 | 46.6 | 9.3 | | 180 | 21 | 66 | 46.4 | 9.9 | |

## 4 Discussion

The average age of our samples was higher in the follow up, what corresponds to a stable population with low level of workers' turnover and corresponds to the ageing process of the working population. These values were higher than the mean age found in other national studies (Cotrim et al. 2017), but correspond to the ageing trend in the portuguese public administration.

In this study we found that the mean age was higher among the group of workers with the category of operational assistant, basic education and who had no training opportunities in the last two years. These results point to a vulnerability of this group of older workers, especially associated with the physical demands of the work activity. Several studies found that physical demands at work are the main determinants of poor work ability among the older workers (Lindberg et al. 2006; Savinainen et al. 2004). Ilmarinen (2001) refers that with advancing age, although workers become physically less fit, they become mentally stronger.

Still, municipal workers have a diversity of tasks that may explain the higher perception by the older groups not only of precision tasks with hands and fingers and seated work (related with administrative and office work), but also of working with trunk rotation or flexion and manual handling of loads between 1–4 kg (concerning the blue collar tasks). Some studies showed that older workers showed dual results concerning their work ability, what can be explained by a healthy worker effect or a greater vulnerability of this group (Lindberg et al. 2006). These duality may, also, be present in our sample, due to the heterogeneity of professional categories and individual characteristics within a large municipality, where work seniority is high and correlated with age.

Regarding the mentally demanding tasks, that are referred very frequently by our sample, some studies showed that the stimulation to the use of specific cognitive abilities, has a great impact on the learning capacity, showing significant effects in the reduction of cognitive losses until the age of 70 years. It is observed that more time is needed for older workers to retain new knowledge, although they can achieve the same levels of performance as younger workers (Schaie and Willis 2010). What leads to the definition of strategies that must include the opportunity of these workers to mobilize the experience to match the variety of work situations, aiming at higher levels of job satisfaction and motivation.

## 5 Conclusion

Organizations must manage workers' resources by adapting workplaces and tasks aiming at maintaining the productivity and promoting active and healthy ageing.

According to Lindberg et al. (2006), promoting excellent work ability in older workers tends to be more dependent of physical factors, work content, namely, clear work tasks, and a positive feed back. These determinants must be integrated by the municipality in different and complementary interventions ir order to promote an healthy ageing.

## References

Ahola K, Salminen S, Toppinen-Tanner S, Koskinen A, Väänänen A (2013) Occupational burnout and severe injuries: an eight-year prospective cohort study among finnish forest industry workers. J Occup Health 55(6):450–457. https://doi.org/10.1539/joh.13-0021-OA

Cotrim T, Carvalhais J, Neto C, Teles J, Noriega P, Rebelo F (2017) Determinants of sleepiness at work among railway control workers. Appl Ergon 58:293–300. https://doi.org/10.1016/j.apergo.2016.07.006

Cotrim T, Simões A (2013) Idade e Trabalho. In: Varela M (ed) IBER international business and economics review, 1st edn. Revista de Gestão Economia e Comunicação, Lisboa, pp 303–317. https://doi.org/364596/13

Ilmarinen J (2001) Aging workers. Occup Environ Med 58(8):546–552. https://doi.org/10.1136/oem.58.8.546

Laville A, Volkoff S (2004) 9. Vieillissement et travail. Ergonomie 145. https://doi.org/10.3917/puf.falzo.2004.01.0145

Lindberg P, Josephson M, Alfredsson L, Vingard E (2006) Promoting excellent work ability and preventing poor work ability: the same determinants? Results from the Swedish HAKuL study. Occup Environ Med 63(2):113–120. https://doi.org/10.1136/oem.2005.022129

Marchand A, Blanc ME (2010) The contribution of work and non-work factors to the onset of psychological distress: an eight-year prospective study of a representative sample of employees in Canada. J Occup Health 52(3):176–185. https://doi.org/10.1539/joh.L9140

Oakman J, Neupane S, Nygård CH (2016) Does age matter in predicting musculoskeletal disorder risk? An analysis of workplace predictors over 4 years. Int Arch Occup Environ Health 89(7):1127–1136. https://doi.org/10.1007/s00420-016-1149-z

Savinainen M, Nygård C-H, Arola H (2004) Physical capacity and work ability among middle-aged women in physically demanding work – a 10-year follow-up study. Adv Physiother 6 (3):110–121. https://doi.org/10.1080/14038190310017309

Schaie KW, Willis SL (2010) The Seattle Longitudinal Study of adult cognitive development. ISSBD Bull 57(1):24–29. http://www.pubmedcentral.nih.gov/articlerender.fcgi?artid=3607395&tool=pmcentrez&rendertype=abstract

Warr P (2001) Age and work behaviour: physical attributes, cognitive abilities, knowledge, personality traits and motives. Int Rev Ind Organ Psychol 16:1–36. https://doi.org/10.1002/9780470696392.ch4

# User-Centered Design Process to Develop Motor Speech Disorder Treatment Assistive Tool

Hsin-Chang Lo[1(✉)], Bo-Kai Peng[1], and Chia-Chen Li[2]

[1] Department of Product Design, Ming Chuan University, Taoyuan, Taiwan
lohc@mail.mcu.edu.tw, kevinapon@gmail.com
[2] Division of Rehabilitation, Taipei City Hospital Zhongxing Branch, Taipei City, Taiwan
ceoliuli@gmail.com

**Abstract.** With the advent of an aging society, the number of people suffering from degenerative diseases, which often lead to communication disorders such as motor speech disorders (MSDs). Intensive and repeated practice is essential for individuals with MSDs. Because of work overload, speech and language therapists (SLTs) might cope with the excess load of treatment, which may lead to reduce of rehabilitation effects. Nowadays smart mobile devices such as the iPad and tablet PC have changed the way of therapy. Therefore, this study proposes the feasibility of assistive technology intervention through the user-centered design process in MSDs treatment. First, it uses Delphi expert assessment interviewing the SLTs to realize how they work about MSDs. Second, the SLTs define the weighting of requirements using the analytic hierarchy process. The most two important two criteria are: evaluate willing to practice, and easy to review the home exercises. Third, it transfers the requirements into design criteria and develop the assistive APP, MOUTH EXERCISE. SLTs can select and demonstrate correct video to the individuals instead of exercising their oral muscles. Finally, SLTs participate the usability evaluation to validate the APP. They suggest that MOUTH EXERCISE can not only as an assistive manner of speech therapy, but also reduce the workload of them. It concludes that the developed APP is a useful tool for STLs. In the future, it will develop home practice mode to enhance the rehabilitation motivation of individuals.

**Keywords:** User-centered design · Motor speech disorder
Speech and language therapist

## 1 Introduction

With the advent of an aging society, the number of people suffering from degenerative diseases is on the increase [1]. These diseases often lead to communication disorders such as motor speech disorders (MSDs) and apraxia of speech. MSDs are impairments in the systems and mechanisms that control the movements necessary for the production of speech. They are a group of disorders resulting from disturbances in muscular control, weakness, slowness, or incoordination of the speech mechanism due to damage to the


S. Bagnara et al. (Eds.): IEA 2018, AISC 826, pp. 52–58, 2019.
https://doi.org/10.1007/978-3-319-96065-4_8

central nervous system [2]. Individuals with MSDs occupy a considerable proportion with communication difficulties. According to the Mayo Clinic's investigation from 1987–2001, MSDs accounted for 41% with acquired communication disorders and 58% with neurogenic ones [3]. About 92% of individuals with MSDs is Dysarthria, and they must be treated through repeated vocal and oral exercises. Most of the individuals with communication disorder took clinical treatment once a week for about 30 to 60 min per treatment [4]. Clinically, the speech and language therapist (SLT) will ask individuals to imitate their mouth-like repetitions to produce correct voice. The repetitive oral exercise may accelerate mouth muscle fatigue of therapists.

It is suggested that ratio of 26.2 professional SLT per one hundred thousand (‱) people is the best according to Enderby and Davies' research. The current ratio of SLT in Taiwan is approximately 0.56 ‱, which is far behind the US 5.45 ‱ and Australia 3.52 ‱ [5]. According to the survey, the incidence of whispering and swallowing disorders in the elderly in Taiwan is more than 70%, far exceeding the 36% of children and 16% of adults [5]. Faced with the continuous increase in demand for speech therapy, SLTs can only be arranged for group therapy, which shorten the time for each treatment course or also lengthen the treatment cycle to cope with excessive workload. This may decrease the treatment quality, and lead to a reduction in the treatment effectiveness. Many assistive technologies have been used in the field of speech therapy. Mobile device such as iPad or tablet PCs combined with APPs are able to use as assistive tool, which has considerable potential for the practice of MSDs treatment. The study indicated that the use of APP is a win-win strategy, in which the effectiveness of treatment is better for some individuals and it is easier for SLTs to guide individuals. In addition, SLT who familiar with mobile devices and APPs can strengthen the connection between individuals and their families, and can also treat individuals in everyday life [6]. The application of APP can improve the effect of treatment, and can reduce the burden of care, therefore may be a new treatment channel. The aim of this study is to proposes the feasibility of assistive technology intervention through the user-centered design process in MSDs treatment.

## 2 Methods

The four processes of user-centered design were used in this study [7]. The Delphi method and analytic hierarchy process were used to specify context of use and specify requirements of treatment, respectively. In the first process, the 9 SLTs were asked to conduct in-depth interviews to explore clinical requirements in the MSDs treatment process. The SLTs was asked to screen relatively important items by majority decision to seek consensus on related issues, therefore confirm clinical requirements. In the second process, it constructs a hierarchical structure and a set of pairwise comparison matrices for clinical requirements. The 9 SLTs, has to indicate which of the two elements the respondent prefers. Then a 9-point scale is used to measure the strength of this preference by means of verbal judgments [8]. Use the priorities obtained from the comparisons to weigh the priorities in the level immediately below. Do this for every element. Then for each element in the level below add its weighed values and obtain its overall or global priority. The weights of criteria represent the importance of the clinical requirements. Then the requirements will be transferred into the design criteria. In the third process of design and prototyping, the 3 SLTs were asked to review the operation process developed by the design team to confirm whether each function meets the requirements. After confirmation, the design team proceeded with detailed design and prototyping of assistive tool. In the last process of evaluation, the 3 SLTs were invited to participate the usability test in the verification of prototype.

## 3 Results

### 3.1 Specify Context of Clinical Requirements

The interviews included clinical interactions with individuals and problems encountered. From the interview result, this study find that the inconvenience of current MSDs treatment comes from the insufficiency of teaching materials, which cause difficulty in clinical and home practice. Then the 9 STLs screening out identified items together by a majority in a total of 13 items. Among them, items 4 and 5 are integrated into convenient create and modify teaching materials, and 6 and 7 are integrated to enhance the individual's willing to practice (Table 1).

**Table 1.** User requirements of STLs on MSDs treatment

| Items identified together | Votes | Confirmed clinical requirements |
|---|---|---|
| 1. Diadochokinetic exercise is tedious | 9 | • Reduce Diadochokinetic exercise |
| 2. Hope the individual can practice at home | 9 | • Increase times of individual practice at home |
| 3. Unable to review the home exercises | 8 | • Easy to review the status of home exercises |
| 4. Unable to meet different levels of illness | 8 | • Convenient create and modify teaching materials |
| 5. Not easy to create teaching materials | 5 | |
| 6. Individuals lack the willing to practice themselves | 8 | • Enhance the individual's willing to practice |
| 7. Individuals hope to have the opportunity to practice by themselves | 7 | |
| 8. The textbook is too fragmented | 7 | • Teaching materials are systematic |
| 9. Evaluation process is time-consuming | 7 | • Easier evaluation process |
| 10 Unable to review rehabilitation results | 6 | • Easy to review long-term rehabilitation results |
| 11. Assigning homework is not convenient | 6 | • Easy to assign homework |
| 12. The content of teaching materials does not correspond to daily life | 6 | • Materials close to daily life |
| 13. Textbooks are too cartoonish | 5 | • Materialization of teaching materials |

### 3.2 Specify Requirements of MSDs Treatments

This study establishes a hierarchical structure. The goal is the clinical requirements; and the second level are the three criteria of teaching materials, processes, and practice. The third level is the sub-criteria, which are materialization of teaching materials, materials close to daily life, convenient production teaching materials, systemic teaching materials, easier evaluation process, easy to assign home practice, reduce Diadochokinetic exercise, easy to review the home exercises, easy to review rehabilitation results, increase frequency of home practice, evaluate willing to practice. The results of analytic hierarchical process are as follows. It concluded that the assistive tool design direction should focus on evaluate willing to practice, easy to review the home exercises, increase frequency of home practice, easy to assign home practice, easy to review rehabilitation results criteria (Table 2).

**Table 2.** The importance of design criteria

| Criteria | Sub-criteria | Weighting | Order |
|---|---|---|---|
| Teaching materials | Materialization of teaching materials | 0.015 | 11 |
| | Materials close to daily life | 0.046 | 6 |
| | Convenient production teaching materials | 0.040 | 7 |
| | Systemic teaching materials | 0.036 | 8 |
| Processes | Easier evaluation process | 0.021 | 10 |
| | Easy to review rehabilitation results | 0.059 | *5* |
| | Easy to review the home exercises | 0.092 | *2* |
| | Reduce Diadochokinetic exercise | 0.030 | 9 |
| | Easy to assign home practice | 0.069 | *4* |
| Practice | Increase frequency of home practice | 0.087 | *3* |
| | Evaluate willing to practice | 0.505 | *1* |

### 3.3 Create Design Solutions and Develop Assistive Tool

This study proposes an innovative assistive tool operate on iPad for MSDs treatment called "MOUTH EXERCISE" APP. This is essentially as a teaching material for the SLTs during the treatment, which includes four functions: oral exercise, step process communication, practice review, and practice assignment.

(1) Oral exercise: User can select the function of the exercise and then select the exercise item (Fig. 1(a)). After the selection is completed, the demonstration video with voice prompts will be played once before training starts (Fig. 1(b)). The upper half of the screen during training was a demonstration video, and the lower half was the individual's practice (Fig. 1(c)). Considering that user has its own customary exercises protocol and different levels of individuals requires a variety of exercises, user can add practice exercise items by recording its own model videos.

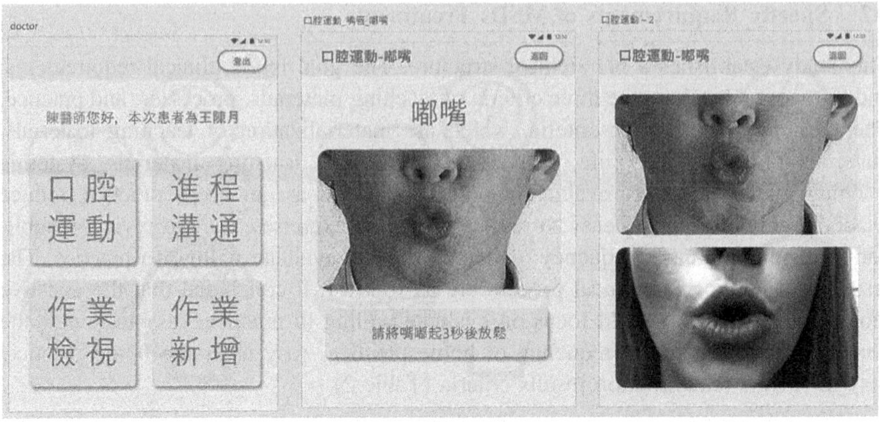


(a) (b) (c)

**Fig. 1.** The prototype of MOUTH EXERCISE.

(2) Practice review: This function recorded the individual's practice activities when individual at home, including the daily and the overall degree of completion. User can quickly view the individual's home practice records. If all of the tasks are completed, then the date would be showed as gray, whereas the date would be showed in orange color if there is a lack. Through the difference in color, user can quickly grasp the practice situation.
(3) Step process communication: This function is based on the eight-step process communication [9]. User enter the target word and record the demonstration video, then start practice. From the steps in the beginning, MOUTH EXERCISE provide a demonstration video and ask the individual to follow it, gradually reduce the number of guidance as the steps increase, and finally just give the words to the individual to read out.
(4) Practice assignment: This function is to allow the user to easily assign a home practice programs so that the individual can better understand the daily items and frequency of the exercise. User can also give advice on the exercise. After the individual reviewing the exercise advice, they can strengthen the project.

### 3.4 Evaluate Assistive Tool Prototype

After the functional prototype is fabricated, this study invites 3 SLTs to operate, then provide their advice on the MOUTH EXERCISE (Fig. 2). Their comments were summarized as follow: (1) The practicing videos can give some visual instruction, such as pulling the arrow to the side, or a circle in the middle of the project to help the individual know where the target is. (2) Pictures or color used in the selection page for distinguish are needed, because the individual and caregiver would obstacles in use if they can not understand the text. (3) In terms of step process communication, it is recommended that SLT can switch between different steps at will. Individuals do not have to follow each step, for whom with good level can do it from the next step or skip steps.

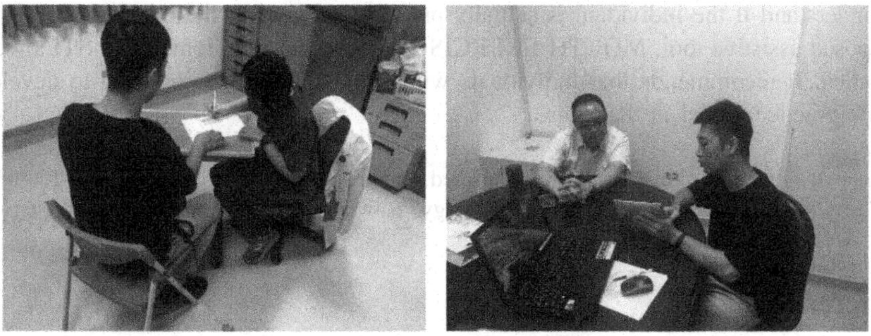

**Fig. 2.** Prototype usability test of STLs.

## 4 Discussion

The assistive tool, MOUTH EXERCISE, has the following three advantages: (1) The oral exercise function improves the individual's willingness to practice. Traditional paper-based education instructions are too boring. The MOUTH EXERCISE helps the individual understand the mouth movement through the demonstration video. The video guides are also provided during the training, so that the individuals can see the mouth shape by themselves. Through the visual and auditory instruction, the individuals are more willing to practice and exercise. (2) The practice review function facilitates SLT inspect home practice status. The MOUTH EXERCISE can record the individual's exercise process. In addition to the classification according to the practice date, the individual's practice completion degree is displayed. The completed/unfinished items are differentiated by color, which allows SLT to quickly understand how well the individual is performing at home. The degree of completion provide a function of reminding the individual, so that the individual knows what other practice items have not been executed. (3) The practice assignment function facilitates the assignment of the individual's home practice program. The traditional paper-based education instruction can only arrange the individual for homework or leave a note, which is hard to motivate the individual to practice. In order to achieve better home practice effect, SLT can assign the appropriate number of exercises and times for the specific individuals. Individuals only need to follow the instructions step by step to complete the home practice.

## 5 Conclusions and Suggestions

The conclusions of this study are as follows. Compared with the problems encountered in the clinics, SLTs care the practice of individuals after the they left the clinic. Therefore, evaluate willing to practice and easy to review the home exercises became the main clinical requirement. However, the current practice through paper-based instruction is impractical. Not only is it difficult for the individual to actually perform the exercise program, but the inability to record practice would cause SLTs is not easy to understand if the individual is actually practicing. Through the intervention of the proposed assistive tool, MOUTH EXERCISE, the clinical requirements of STLs were satisfied. It recommends that individuals with MSDs should be included to develop home practice model in the future.

**Acknowledgements.** This work was sponsored under grant of MOST 106-2410-H-130-035-MY2 by the Ministry of Science and Technology, Taiwan.

## References

1. Ministry of health and welfare Homepage. https://www.mohw.gov.tw/mp-2.html. Accessed 22 Dec 2017
2. Battle DE (2012) Communication disorders in multicultural and international populations, 4th edn. Mosby, Missouri
3. Duffy JR (2005) Motor speech disorders: Substrates, differential diagnosis, and management. Elsevier Mosby, St. Louis
4. Baker E, McLeod S (2011) Evidence-based practice for children with speech sound disorders: part 1 narrative review. Lang Speech Hear Serv Sch 42(2):102–139
5. Hwa S. The speech-language-hearing association, Taiwan. https://www.mohw.gov.tw/mp-2.html. Accessed 22 Dec 2017
6. Dunham G (2011) The future at hand: Mobile devices and apps in clinical practice. ASHA Lead 16:4–5
7. Norman DA, Draper SW (1986) User-dentered system design: new perspectives on human-computer interaction. Lawrence Earlbaum Associates, Hillsdale
8. Saaty TL (1980) The analytic hierarchy process. McGraw-Hill, New York

# Force Variability and Musculoskeletal Pain in Blue-Collar Workers

Kristoffer Larsen Norheim[1,2(✉)], Jakob Hjort Bønløkke[2], Øyvind Omland[2], Afshin Samani[1], and Pascal Madeleine[1]

[1] Sports Sciences, Department of Health Science and Technology, Faculty of Medicine, Aalborg University, Aalborg, Denmark
kln@hst.aau.dk

[2] Clinic of Occupational Medicine, Aalborg University Hospital, Aalborg, Denmark

**Abstract.** Blue-collar workers with physically demanding occupations have a high prevalence of musculoskeletal disorders accompanied by pain. Previous research has suggested that reduced motor variability may increase the risk of developing musculoskeletal disorders. Here we present preliminary data from an ongoing cross-sectional examination of physical performance in elderly manual workers. This paper includes data from 20 male workers (age 52–70 years). Handgrip force variability was measured using a digital hand dynamometer during an endurance trial where the workers exerted 30% of their maximal isometric contraction force until task failure. Absolute variability (standard deviation), relative variability (coefficient of variation), and the complexity of the force signal (sample entropy) were computed. The workers were dichotomized into two groups: no to mild pain/discomfort (No pain) and moderate to severe pain/discomfort (Pain) to investigate the effects of musculoskeletal pain on force variability. This dichotomy was done based on the rating of pain/discomfort within the last seven days, where workers reporting $\geq 3$ score in the upper extremities were allocated to the pain group (Pain, n = 9, No pain, n = 11). No significant between-group differences in force variability during the endurance trial were found. Both absolute and relative variability increased significantly over time. These preliminary data do not support a difference in force variability between blue-collar workers with or without musculoskeletal pain or discomfort in the upper extremities.

**Keywords:** Handgrip strength · Endurance · Motor control · Discomfort

## 1 Introduction

Labor force participation of elderly workers is expected to increase in the coming years [1]. Workers with physically demanding occupations, such as construction workers and electricians, have a high prevalence of musculoskeletal disorders [2] and pain [3, 4]. Consequently, blue-collar workers may not be able to keep up with the trend in prolonged labor market participation.

Work involving repetitive movements have been associated with pain and discomfort [5]. It is therefore of interest to study how 'variation' in motor output is


S. Bagnara et al. (Eds.): IEA 2018, AISC 826, pp. 59–67, 2019.
https://doi.org/10.1007/978-3-319-96065-4_9

associated with pain. Indeed, previous research has suggested that reduced motor variability may increase the risk of developing musculoskeletal disorders [6]. Moreover, low signal complexity (i.e. the structure of variability) has been related to musculoskeletal discomfort and pain [7]. Although healthy elderly may be able to continue working until normal retirement age, those with musculoskeletal discomfort or pain may be forced to withdraw from the labor market prematurely. This is especially problematic for manual workers, who are dependent of their strength and endurance capacity to adequately perform their work [2].

The present study aims to investigate the association between handgrip force variability and musculoskeletal pain or discomfort in blue-collar workers. We hypothesize that low force variability and signal complexity are associated with pain.

## 2 Methods

### 2.1 Participants

Here we present preliminary data from an ongoing cross-sectional examination of physical performance in elderly (age 50–70 years) blue-collar workers, which has been described in detail previously [8]. This conference paper includes the first 20 male workers that have been tested. Participants were informed about the purpose of the study and gave oral and written informed consent to participate. The study was carried out in accordance to the Helsinki declaration and is approved by the ethics committee of region North Jutland (N-20160023).

### 2.2 Musculoskeletal Pain

Musculoskeletal pain/discomfort was assessed using a modified Danish version of the Standardized Nordic Questionnaire [9]. It included questions about days with pain or discomfort within the last 12 months in the following regions: neck/shoulder, arms (elbow, wrist and hand), lower back, hips, and knees. Each region is graded on a 6-point scale (1 = 0 days, 2 = 1–7 days, 3 = 8–30 days, 4 = 31–90 days, 5 = >90 days, 6 = every day). In addition, the degree of pain in these areas within the last seven days was graded on an 11-point scale from 0 (no pain/discomfort) to 10 (most pain/discomfort imaginable). To investigate the effects of musculoskeletal pain on force variability, the tested workers were dichotomized into two groups: no to mild pain/discomfort (hereby referred to as "No pain") and moderate to severe pain/discomfort (hereby referred to as "Pain"). This was done based on the rating of pain/discomfort within the last seven days, where workers reporting $\geq 3$ score [10, 11] in the neck/shoulder or arm regions (upper extremities) were allocated to the Pain-group.

### 2.3 Force Variability

Handgrip force variability was measured using a digital hand dynamometer (Model G100, Biometrics Ltd, Gwent, UK) during an endurance trial, as described previously [8]. Briefly, the participants began by completing three trials (separated by 2 min) to measure their isometric maximal voluntary contraction (MVC) force. MVC was measured as a moving average over 100 ms. Then, in an endurance trial, the participants exerted 30% of their MVC force until task failure (Fig. 1). The test was terminated when the force dropped below 28% MVC for more than 5 consecutive seconds. To measure absolute and relative variability from the force signal, we calculated standard deviation (SD) and coefficient of variation (CV), respectively. Sample entropy (SaEn) was computed using a fixed embedding dimension $m = 2$ and a tolerance distance $r = 0.2 \times \mathrm{SD}$ to measure the complexity of the force signal [12]. After discarding the first and the last five seconds of each trial, total endurance time was normalized between workers by extracting seven epochs of 3 s in steps of 16.7% (i.e. from 0 to 100%). SD, CV and SaEn were then calculated over these seven epochs. All data processing was conducted using Matlab 2017b® (Mathworks, USA).

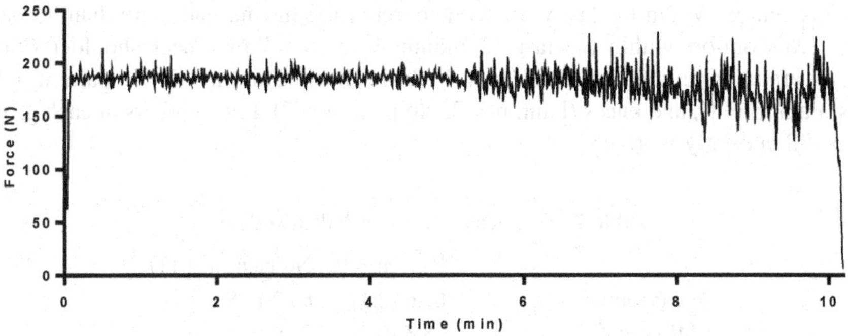


**Fig. 1.** Sample plot of an endurance trial from one worker from the No pain-group, for whom the target force of 30% maximal voluntary contraction was 189 N.

## 2.4 Statistics

A factorial repeated measures ANOVA was used to determine the effect of group (between-subjects factor) and time (within-subjects factor, 7 time epochs) on the dependent variables force variability and complexity during the endurance trial. To test the robustness of the group dichotomy (i.e. $\geq 3$ pain), we investigated one lower and two higher cut-off values ($\geq 2$, $\geq 4$ and $\geq 5$). It has previously been suggested that motor variability in the pre-fatigued state may predict the development of muscle fatigue [13, 14]. Therefore, a multiple regression was run to predict endurance time from SD, CV and SaEn measured from the first epoch, in addition to MVC, age, BMI and group (coded as Pain = 1 and No pain = 2). Variables were excluded from the model by stepwise elimination. Statistical analyses were carried out using SPSS 25.0 (SPSS Inc., Chicago, Illinois, USA). Mean (SEM) is reported and statistical significance set to $p < 0.05$.

# 3 Results

## 3.1 Participants

Out of the 20 included workers (age 52–70 years), half (i.e. 10) where still currently working full-time jobs. However, all workers had worked or were working in a manual job: mostly as construction workers (n = 12), but also electricians (n = 5) and other manual professions (n = 3).

## 3.2 Musculoskeletal Pain

Characteristics of the workers in the Pain and No pain-group are presented in Table 1. There were no significant between-group differences in age or BMI. Self-reported levels of pain/discomfort during the last seven days were significantly different between groups in the neck/shoulder and arm regions. Regarding musculoskeletal pain/discomfort within the last year, workers reporting having had more than 90 days of pain/discomfort within the last 12 month were as follows: neck/shoulder (Pain: n = 4), arms (Pain: n = 1, No pain: n = 1), lower back (Pain: n = 3, No pain: n = 1), hips (Pain: n = 1), and knees (Pain: n = 3, No pain: n = 2). Five workers in each group were still currently working.

**Table 1.** Characteristics of included workers.

| | Pain (n = 9) | No pain (n = 11) |
|---|---|---|
| Age (years) | 62.6 (2.1) | 62.2 (1.8) |
| BMI (kg/m$^2$) | 28.8 (0.7) | 27.9 (1.3) |
| Pain/discomfort last 7 days (degree) | | |
| Neck/shoulder | 4.6 (1.0)* | 0.7 (0.2) |
| Arms (elbow, wrist, hand) | 2.7 (0.9)* | 0.4 (0.2) |
| Lower back | 3.3 (0.5) | 1.9 (0.7) |
| Hips | 1.3 (0.6) | 0.7 (0.3) |
| Knees | 2.9 (1.1) | 2.8 (0.9) |

*Significant differences between groups, $p < 0.05$.

## 3.3 Force Variability

There was no significant difference in maximal voluntary contraction force between the Pain and No pain-group (461 (26) N vs 504 (23) N, $p = 0.228$, respectively), whereas endurance time was significantly shorter for the Pain-group compared to the No pain-group (228 (12) s vs 284 (20) s, $p = 0.035$, respectively). No between-group difference could be detected in force variability (SD and CV), nor signal complexity (SaEn) during the endurance trial (SD: $p = 0.879$, CV: $p = 0.906$, SaEn: $p = 0.224$, Fig. 2). However, a significant main-effect of time was found for all three outcomes (all $p < 0.01$). Pairwise comparisons revealed significant changes over time as illustrated in Fig. 2. The sensitivity analyses with two higher cut-off scores for the group dichotomy

($\geq 4$ and $\geq 5$) resulted in similar results as in the main analysis (data not shown). Contrary, using a lower cut-off ($\geq 2$), we found a significant group-effect for SaEn ($p = 0.033$), wherein the Pain-group (now n = 12) had lower signal complexity during the endurance trial.

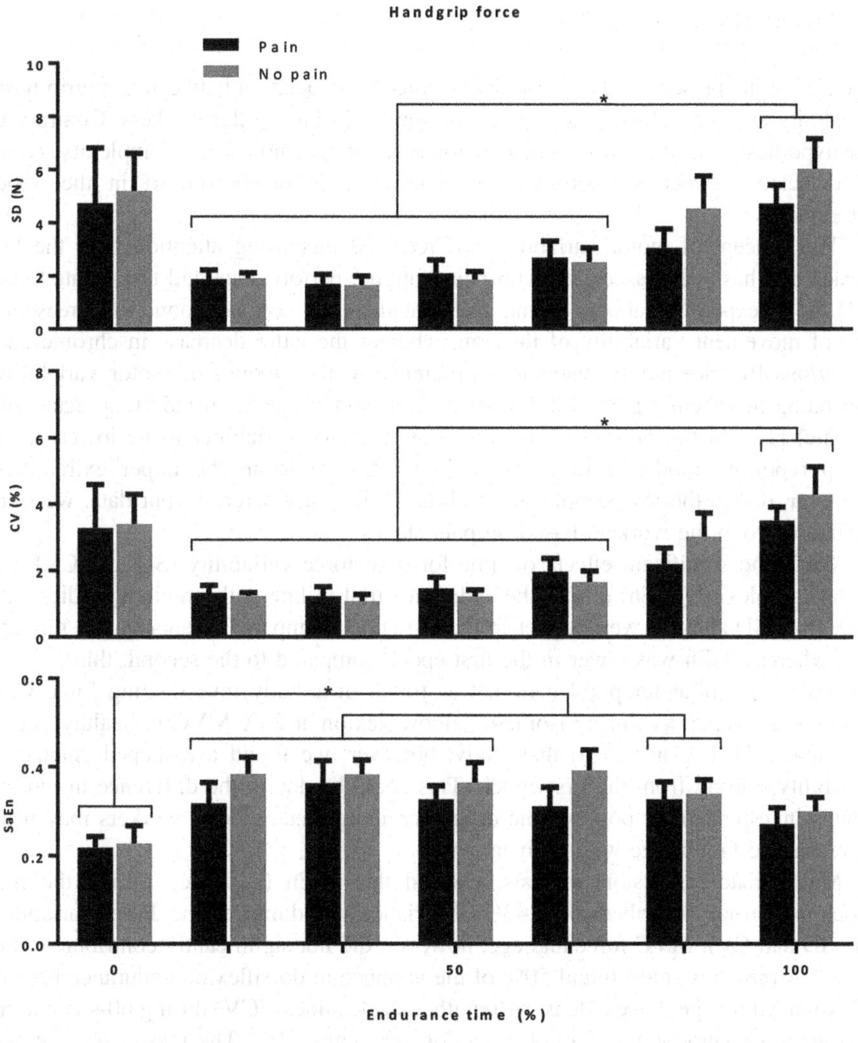


**Fig. 2.** Mean + SEM for standard deviation (SD, N), coefficient of variation (CV, %) and sample entropy (SaEn) during the endurance trial. *Significant differences between epochs ($p < .05$).

Regarding the multivariate regression analysis, group and SaEn was found to significantly predict endurance time, $F(2, 17) = 6.356, p = 0.009, R^2 = 0.428$, adjusted $R^2 = 0.361$ (endurance time [s] = 222.9 + 60.6 × group − 265.9 × SaEn). SD, CV, MVC and age did not significantly contribute to the regression model.

## 4 Discussion

The aim of the present study was to investigate the association between handgrip force variability and musculoskeletal pain or discomfort in blue-collar workers. Contrary to our hypothesis, no difference in neither force variability, nor signal complexity, could be detected between workers with or without pain or discomfort in the upper extremities.

The concept of motor variability has received increasing attention over the last decade and has been associated with pain, fatigue, performance and task-related skill [13]. Acute experimental pain during a simulated cutting task was found to increase the size of movement variability of the arm, whereas the latter decrease in chronic pain conditions [6]. Seemingly, there is a difference in the amount of motor variability, depending on whether pain is felt as discomfort and the stage of pain (e.g. acute and chronic) [15]. In the present study, we expected force variability to be lower in the group reporting moderate to severe pain or discomfort in the upper extremities. However, due to the low sample size and the preliminary nature of our data, we were unable to group the workers based on pain stages.

We found significant effects of time for both force variability (SD and CV) and signal complexity (SaEn) during the endurance trial in line with previous studies, e.g. [12]. Both SD and CV were higher in the last epoch compared to epochs two through five, whereas SaEn was lower in the first epoch compared to the second, third, fourth and sixth. A similar temporal pattern was found in a study investigating force variability and complexity during isometric elbow flexion at 20% MVC in healthy young participants [12]. Contrary to that study, however, we found a u-shaped change in variability starting from the first epoch. This could be due to the difference in muscle groups investigated, or possibly the difference in age, i.e., elderly workers may need more time to familiarize with a given task.

Multivariate regression analysis revealed that SaEn (measured during the first epoch) and group described about 43% of variance in endurance time. Force variability (i.e. SD and CV), MVC force and age, however, did not significantly contribute to the model. A previous study found 50% of the variance in dorsiflexion endurance time to be explained by age, knee extensor strength, and steadiness (CV) during 60-s isometric dorsiflexion contraction at 20% 1-repetition maximum [16]. The reason for age and MVC not being significant predictors in the present study could partly be explained by the homogeneity of both age and MVC. Alternatively, task differences i.e., pushing against a force transducer (as in the present study) versus supporting an inertial load (as in the aforementioned study), have been shown to affect endurance time [17]. Thus, the results cannot be readily compared.

Interestingly, the complexity of the force signal (SaEn) was a significant predictor of performance, so that a 0.1 increase in SaEn would reduce endurance time by 27 s. In other words, the more complexity and randomness in the beginning of the endurance trial, the poorer the performance. Sensitivity analysis of the group-dichotomization also revealed a significant group-difference for SaEn when the dichotomy was made at $\geq 2$ pain (on the 0–10 scale). This indicated, as hypothesized, that a lower signal complexity was associated with pain, and possibly that the null-finding in the main analyses was due to a type I error.

In previous studies, no difference was found in force steadiness (SD) during isometric wrist extension between patients with elbow pain and healthy controls [18], whereas force steadiness (CV) has been shown to decrease during experimental muscle pain (hypertonic saline injection) [19]. Contrary, a study investigating discomfort during 96 min of sitting on a force platform found that an increase in variability (SD) and a decrease in complexity (SaEn) of the center of pressure signal was associated with increased discomfort [7]. These findings, coupled with those of the present study, suggest a relationship between motor control patterns, pain/discomfort, and subsequently performance. However, whether one precedes the others is still uncertain [15], and its relevance in changes in pain stages warrants further research; specifically, whether motor variability and complexity can be manipulated to reduce the risk of musculoskeletal discomfort or pain. Although some studies have shown promise using biofeedback systems to manipulate motor variability [20, 21], it is a field still in its infancy.

The present study was limited by its small sample size. Moreover, fatigue is a multifaceted phenomenon which can be attributed to muscular or peripheral factors, but also psychological aspects such as motivation and mood [22]. Although identical strong verbal encouragement was provided during the endurance trial, these factors are difficult to account for and could potentially have affected our results. Lastly, the cross-sectional design limits the interpretation of our results to investigate associations without inferring causality. Nonetheless, this design still enables us to benchmark important effects of pain and discomfort on motor control.

In summary, no significant difference in force variability during the endurance trial could be found between the Pain and No pain-group in a subsample population. Thus, these preliminary data do not support a difference in force variability between blue-collar workers with or without musculoskeletal pain or discomfort in the upper extremities; however, a larger sample size would enable comparisons between different pain stages (e.g. acute and chronic) and thereby reveal important information about the association between pain and force variability.

**Acknowledgements.** The project is funded by Arbejdsmiljøforskningsfonden (project number 20140072843). Also, the authors would like to thank the rest of the ALFA-group, which includes Kirsten Fonager, Henrik Bøggild, Johan Hviid Andersen, and Claus Dalsgaard Hansen.

## References


1. Larsen M, Pedersen PJ (2017) Labour force activity after 65: what explain recent trends in Denmark, Germany and Sweden? J Labour Mark Res 50:15–27. https://doi.org/10.1007/s12651-017-0223-7
2. Tuomi K, Huuhtanen P, Nykyri E, Ilmarinen J (2001) Promotion of work ability, the quality of work and retirement. Occup Med Oxf Engl 51:318–324
3. Caban-Martinez AJ, Lowe KA, Herrick R, Kenwood C, Gagne JJ, Becker JF, Schneider SP, Dennerlein JT, Sorensen G (2014) Construction workers working in musculoskeletal pain and engaging in leisure-time physical activity: findings from a mixed-methods pilot study. Am J Ind Med 57:819–825. https://doi.org/10.1002/ajim.22332
4. de Zwart BC, Broersen JP, Frings-Dresen MH, van Dijk FJ (1997) Repeated survey on changes in musculoskeletal complaints relative to age and work demands. Occup Environ Med 54:793–799
5. Eurofound (2017) Sixth European working conditions survey – overview report (2017 update). Publications Office of the European Union, Luxembourg
6. Madeleine P, Mathiassen SE, Arendt-Nielsen L (2008) Changes in the degree of motor variability associated with experimental and chronic neck-shoulder pain during a standardised repetitive arm movement. Exp Brain Res 185:689–698. https://doi.org/10.1007/s00221-007-1199-2
7. Søndergaard KHE, Olesen CG, Søndergaard EK, de Zee M, Madeleine P (2010) The variability and complexity of sitting postural control are associated with discomfort. J Biomech 43:1997–2001. https://doi.org/10.1016/j.jbiomech.2010.03.009
8. Norheim KL, Bønløkke JH, Samani A, Omland Ø, Madeleine P (2017) The effect of aging on physical performance among elderly manual workers: protocol of a cross-sectional study. JMIR Res Protoc 6:e226. https://doi.org/10.2196/resprot.8196
9. Kuorinka I, Jonsson B, Kilbom A, Vinterberg H, Biering-Sørensen F, Andersson G, Jørgensen K (1987) Standardised Nordic questionnaires for the analysis of musculoskeletal symptoms. Appl Ergon 18:233–237
10. Boonstra AM, Stewart RE, Köke AJA, Oosterwijk RFA, Swaan JL, Schreurs KMG, Schiphorst Preuper HR (2016) Cut-off points for mild, moderate, and severe pain on the numeric rating scale for pain in patients with chronic musculoskeletal pain: variability and influence of sex and catastrophizing. Front Psychol 7:1466. https://doi.org/10.3389/fpsyg.2016.01466
11. Boonstra AM, Schiphorst Preuper HR, Balk GA, Stewart RE (2014) Cut-off points for mild, moderate, and severe pain on the visual analogue scale for pain in patients with chronic musculoskeletal pain. Pain 155:2545–2550. https://doi.org/10.1016/j.pain.2014.09.014
12. Svendsen JH, Madeleine P (2010) Amount and structure of force variability during short, ramp and sustained contractions in males and females. Hum Mov Sci 29:35–47. https://doi.org/10.1016/j.humov.2009.09.001
13. Srinivasan D, Mathiassen SE (2012) Motor variability in occupational health and performance. Clin Biomech 27:979–993. https://doi.org/10.1016/j.clinbiomech.2012.08.007
14. Skurvydas A, Brazaitis M, Kamandulis S, Sipaviciene S (2010) Muscle damaging exercise affects isometric force fluctuation as well as intraindividual variability of cognitive function. J Mot Behav 42:179–186. https://doi.org/10.1080/00222891003751835
15. Madeleine P (2010) On functional motor adaptations: from the quantification of motor strategies to the prevention of musculoskeletal disorders in the neck-shoulder region. Acta Physiol Oxf Engl 199(Suppl 679):1–46. https://doi.org/10.1111/j.1748-1716.2010.02145.x

16. Justice JN, Mani D, Pierpoint LA, Enoka RM (2014) Fatigability of the dorsiflexors and associations among multiple domains of motor function in young and old adults. Exp Gerontol 55:92–101. https://doi.org/10.1016/j.exger.2014.03.018
17. Hunter SK, Ryan DL, Ortega JD, Enoka RM (2002) Task differences with the same load torque alter the endurance time of submaximal fatiguing contractions in humans. J Neurophysiol 88:3087–3096. https://doi.org/10.1152/jn.00232.2002
18. Mista CA, Monterde S, Inglés M, Salvat I, Graven-Nielsen T (2018) Reorganised force control in elbow pain patients during isometric wrist extension. Clin J Pain. https://doi.org/10.1097/AJP.0000000000000596
19. Salomoni SE, Graven-Nielsen T (2012) Experimental muscle pain increases normalized variability of multidirectional forces during isometric contractions. Eur J Appl Physiol 112:3607–3617. https://doi.org/10.1007/s00421-012-2343-7
20. Holtermann A, Søgaard K, Christensen H, Dahl B, Blangsted AK (2008) The influence of biofeedback training on trapezius activity and rest during occupational computer work: a randomized controlled trial. Eur J Appl Physiol 104:983–989. https://doi.org/10.1007/s00421-008-0853-0
21. Samani A, Holtermann A, Søgaard K, Madeleine P (2010) Active biofeedback changes the spatial distribution of upper trapezius muscle activity during computer work. Eur J Appl Physiol 110:415–423. https://doi.org/10.1007/s00421-010-1515-6
22. Enoka RM, Duchateau J (2016) Translating fatigue to human performance. Med Sci Sports Exerc 48:2228–2238. https://doi.org/10.1249/MSS.0000000000000929

# Aging and Hand Functions Declining: Assistive Technology Devices for Assistance in Daily Life Activities Performance

Bianca Marina Giordani(✉) and Milton José Cinelli

Santa Catarina State University, Florianópolis, SC, Brazil
giordanibianca@gmail.com

**Abstract.** With the aging process several morphological changes occur in the human being. The upper limbs suffer a functional decline, beyond the possibility of developing diseases that occur as result of advancing age. All these factors contribute to the decay of functions that are important in the elderly's lives, depriving some seniors of performing manual actions autonomously. Therefore, assistive technology (AT) devices become an alternative to lessen the effects of aging, it provides means for a more active participation of these individuals in different contexts and everyday situations, giving them access to a more independent life. Thus, the present study intended to gather some devices of assistive technology found in the literature that seek to support manual functions, especially those compatible with the needs of the elderly, and, consequently, assist people to carry out common tasks. To this end, the method used was a literature review in the main databases. The purpose of the review was to analyze the qualified bibliography and to identify the AT devices as potential aid in the performance of activities of daily living, focusing on manual functions. The devices have been classified by type (what parts of the upper limbs they are intended), the functions that enhance and the manual tasks they attend. The investigation provided the knowledge of recent studies that address relevant devices with potential in assist the senile population to recover the ability to undertake activities with the upper limbs, especially regarding manual grip and pinch, as well as the limb movement.

**Keywords:** Assistive technology · Elderly · Activities of daily living

## 1 Introduction

Aging is relentless on the human body. Over the years, a series of changes, morphological and functional, fall under the individuals, reducing the corporal capacity in supplying the necessary demand to maintain a life with health [1]. With the aging process, the body loses its ability to adapt to different situations, being more susceptible to stress factors, physical and environmental [2].

Physical decline is one of the most noticeable dimensions of aging, it imposes a number of restrictions on the subjects' lives, which can lead to a loss of sense of control over their own lives, making the elderly even more vulnerable [2]. Studies have pointed


S. Bagnara et al. (Eds.): IEA 2018, AISC 826, pp. 68–77, 2019.
https://doi.org/10.1007/978-3-319-96065-4_10

out that senescent individuals perceive independence as the main parameter indicating their satisfaction with life [3].

A determining factor in the period that an elderly person is able to live on their own is their ability to deal autonomously with everyday products [4], in other words, it is related to the performance of daily life activities by the individual.

The manual function determines the quality of performance skills needed in day-to-day work-related functions and recreational activities. The upper limbs undergo many changes related to the aging process, in addition to being directly affected by diseases such as osteoporosis, osteoarthritis, rheumatoid arthritis and others. All these factors contribute to the decline of several functions that are important in the daily life of the elderly [5].

"Aging has a degenerative effect on hand function, including declines in hand and finger strength and ability to control submaximal pinch force and maintain a steady precision pinch posture, manual speed, and hand sensation" [6]. With affected manual muscle function elderly individuals have a lower ability to maintain stable submaximal forces, also time required to manipulate small objects increases and the pinch strength decays [7]. The decrease in manual force that results from aging is also related to lower manual dexterity in the elderly, a very important function for interaction with objects [8].

Therefore, many of the manual functions performed by the human being as a young person become more difficult with the advancing age. Some actions are nearly impossible to be fulfilled by the elderly, leading to a loss of independence, something that has direct effects on the quality of life [9]. Data collected by Incel et al. [7] from the hands of the elderly support the hypothesis that the muscular function of the hand correlates with functional dependence in the elderly.

To overcome difficulties in performing activities of daily living, the elderly need assistance or ancillary devices [10]. Thus, one way of alleviating the impact of the effects of aging on the daily activities of the senescent is using assistive technology devices. Assistive technology (AT) is the term used to designate every product or service directed to subjects with specific needs, especially people with disabilities, the elderly and people with chronic diseases. The ATs provide means for a more active participation of these individuals in the situations of daily life and, consequently, give access to a more independent life [11].

Understanding that assistive technology devices can provide support for greater autonomy for the elderly, the present study sought to gather some of the orthosis that are presented in the recent literature that are dedicated to enhancing the manual functions, especially those that can be used in order to assist the performance of activities of daily living. For this, a systematic review of literature was carried out, starting from the search of terms related to hand functions, assistive technologies and activities of daily living. With this, it was possible to identify some studies that are being carried out in this sense, indicating devices that could be used by the elderly in their daily practices.

## 2 Method

A systematic review consists in the search of data in the literature using a clear and organized research method, with the purpose of performing critical appreciation and synthesis of the information collected [12]. In the present study, the objective of the review was to analyze the literature on the assistive technology devices that are being described as potential assistance in the performance of activities of daily living, focusing on manual functions. Previous research showed that only few articles were targeting elderly, so the research was done with a more comprehensive focus, searching for devices that could potentially be used by that audience.

Therefore, a protocol containing information relevant to the conduction of the research was initially established. From the protocol, research was carried out on ProQuest, Scopus and Web of Science databases. With the initial portfolio, the first filter was executed, the works that seemed to deal with the established topic were searched by the reading of titles. After the first selection, the selected articles had their summaries read and the studies to be completely read have been chosen. After the complete reading, the final portfolio was obtained (Table 1).

**Table 1.** Systematic review final portfolio.

| | Article title | Authors |
|---|---|---|
| 1 | 3-D Printed Orthotic Hand with Wrist Mechanism Using Twisted and Coiled Polymeric Muscles | Sharma et al. [13] |
| 2 | EMG Pattern Classification to Control a Hand Orthosis for Functional Grasp Assistance after Stroke | Meeker et al. [14] |
| 3 | Case report on the use of a custom myoelectric elbow–wrist–hand orthosis for the remediation of upper extremity paresis and loss of function in chronic stroke | Dunaway et al. [15] |
| 4 | A soft robotic extra-finger and arm support to recover grasp capabilities in chronic stroke patients | Hussain et al. [16] |
| 5 | Development, design and validation of an assistive device for hand disabilities based on an innovative mechanism | Conti et al. [17] |
| 6 | Preliminary Findings of Feasibility of a Wearable Soft-robotic Glove Supporting Impaired Hand Function in Daily Life | Radder et al. [10] |
| 7 | Assisting drinking with an affordable BCI-controlled wearable robot and electrical stimulation: A preliminary investigation | Looned et al. [18] |
| 8 | MUNDUS project: Multimodal Neuroprosthesis for daily Upper Limb Support | Pedrocchi et al. [19] |

## 3 Results

The purpose of the review was to bring together devices that could help the elderly in the different activities of daily living, being products compatible with the capacities and limitations of these users, as well as being accessible and practical. Thus, only eight articles expressed the intentions of the promoted review. All of them presented AT devices developed to improve the performance of activities of the daily life of users with a decrease in manual functions. A summary of all the studies has been carried out and it can be seen below.

## 3.1 Analysis of References

The article by Sharma, Saharan and Tadesse [13] focuses on the design and performance of an orthosis to aid in the execution of basic hand movements in order to improve the quality of life of its users. The low-cost device, created by additive manufacture, facilitates the movement of all three finger joints. Tests with the device showed the range of movements it allows, especially pinch and manual grip movements, demonstrating the ability of the orthosis to flex and extend the fingers.

In Meeker et al. [14] is presented an electromyography-controlled exotendon that promotes the extension and flexion of the fingers, aiding in the grasping of objects. The glove-shaped device detects the user's intention and it is activated, assisting on a task. The study ran tests where users were encouraged to use the device to hold an object and move it around. Results showed that the device allows the performance of functional movements.

Dunaway et al. [15] report the use of a customized elbow-wrist-hand myoelectric orthosis (MEWHO) for a 62-year-old man that suffered a myocardial infarction. In this way, the orthosis was used with the purpose of promoting assistance in the manual functions, promoting assistance for elbow extension/flexion and manual gripping. Using the orthosis, the user can execute activities such as loading heavy objects, opening doors, preparing meals and other activities of daily living, making him more independent.

In Hussain et al. [16] the authors discuss the use of a sixth robotic finger to aid in holding objects for users who have this functionality reduced or non-existent. The device is triggered from the muscular response of the user's facial muscles by electromyography. In a pilot test, the sixth finger was used along with an arm support to compensate the arm weight and it proved to be an alternative to assisting in the performance of daily living activities.

The work of Conti et al. [17] presented a device designed for the care of people with hand disabilities, performed according requirement of portability, low cost and modularity. The product in question is a robotic orthosis, fabricated in additive manufacture, designed to be a portable exoskeleton able to assist people with physical disabilities in performing their daily tasks. The device was centered on the aid for opening the hand, starting from the geometry of the upper limb of the users. The prototype mechanism acts on the fingers, from the movements of the phalanges, being activated by cable.

The work developed by Radder et al. [10] deals with the testing of a robotic glove to assist people with weak hands, especially the elderly, in the performance of routine activities. The device, which was developed by the authors, provides support for the thumb, middle and ring fingers when the user actively initiates a prehension movement. In addition, it offers support for hand opening and manual gripping by touch sensors at the fingertips that sends a signal to the control unit that pulls the artificial tendons attached to the glove.

The study by Looned et al. [18] demonstrated the use of an exoskeleton arm in performing the task of drinking a glass of water. To achieve the objective of this study, the robotic arm orthosis was used combined with a system of functional electrical stimulation and electroencephalography, which translated into movements of the impaired upper limb. The results showed that the proposed system has the potential to help individuals with disabilities in arm and hand functions to perform daily activities independently.

Pedrocchi et al. [19] deal with an assistive device to recover the interaction capacity of severely disabled people. The device allows a greater range of arm movement and greater functionality in the hands. It exploits any residual user control and it is adapted to the level of severity or disease progression. The system can be driven by muscle activation, head or eye movement and brain signals. From this, the authors conducted tests with users, performing two tasks. The results showed the device as promising in the assistance of activities of daily living.

A summary of the articles is presented in Table 2, for a better visualization of assistive technology devices, objectives and activities of daily living that they proposed assist.

**Table 2.** Synthesis of AT devices, objectives and activities of daily living.

| Article | AT device | Objective | Activities of daily living |
|---|---|---|---|
| Sharma et al. [13] | Hand Orthosis | Facilitate the movement of the finger joints (flexion and extension) | Holding brushes, pliers, soda can and among others |
| Meeker et al. [14] | "Exotendon" | Assist in hand flexion/extension | Holding objects |
| Dunaway et al. [15] | Elbow-Wrist-Hand Myoelectric Orthosis | Promote assistance for elbow extension/flexion and manual grip | Carrying a basket of clothes, lifting heavy objects, preparing meals and opening doors |
| Hussain et al. [16] | Robotic Finger *(Soft-SixthFinger)* | Hand grip | Holding objects |
| Conti et al. [17] | Hand Orthosis | Hand grip | Holding objects and social interactions (shaking hands) |
| Radder et al. [10] | Robotic Glove (*ironHand*) | Assistance for manual opening and gripping | Opening a bottle, cutting food, twisting a cloth, turning a key and holding a book |
| Looned et al. [18] | Robotic Arm Orthosis | Assistance for elbow flexion/extension and forearm pronation/supination | Drinking a glass of water |
| Pedrocchi et al. [19] | MUNDUS – Hand and Arm Orthosis | Supports arm weight and reduces the muscle activation. Assists in manual gripping | Drinking something, scratching, changing a TV program, pushing hair away from the eyes and among others |

# 4 Discussion

The studies of the final portfolio presented assistive technology devices with potential to assist people with weaknesses in the upper limbs. However, each work has particularities, which will be dealt with below.

## 4.1 Target Public

Even though all the selected studies deal with ATs that may be used by the elderly in order to reduce functional losses and to assist in the performance of activities, only the study conducted by Radder et al. [10] had the elderly public as the central user of the product. Dunaway et al. [15], Sharma et al. [13], Meeker et al. [14] and Hussain et al. [16] pointed out that the devices they mentioned were intended for people who suffered from stroke. In Looned et al. [18] the focus were people who suffered trauma or neurological diseases.

Pedrocchi et al. [19] did not mention the target audience. Conti et al. also did not specify public but reported that the device in question is made for users with different weaknesses in the upper limb, especially those with difficulties in the hand opening gesture. However, the tests ran by the authors were done with a user with spinal muscular atrophy.

In general, even when directed at different users, the characteristics of the devices presented prove to be valid for several situations where aids are needed for the functionalities of the upper limbs, especially when the strength and range of movements are lacking in the limbs of the individuals in question, difficulty found in some elderly people. Therefore, although most of the devices gathered in this work are not specially developed for the senile public, they can be suitable for use by these individuals, and may be a way to aid in the day-to-day activities of such users, stimulating their autonomy.

## 4.2 Device Types

As a rule, all devices are orthotics, since all are devices that assist the members in question, they seek to potentiate the member that already exists. In neither case is the device used to replace the body segment. Although they are in the same class of assistive technology, the presented orthosis are different since they correspond to distinct parts of the superior member and they are destined to be used in different situations.

In Dunaway et al. [15] and Pedrocchi et al. [19] are presented elbow-hand-wrist orthosis, covering most of the upper limb. Such orthotics give the idea of providing greater assistance to the user, since they cover most of the functions that may be disabled in a user with more severe conditions.

The device of Looned et al. [18] was an orthosis of the elbow region, assisting arm and forearm. These devices are responsible for supporting this region in activities that need support for the limb weight and a greater range of movements. It can be used in various positions and activities performed by the user.

In the works of Sharma et al. [13], Meeker et al. [14], Hussain et al. [16], Conti et al. [17] and Radder et al. [10] the orthosis used correspond only to the region of the hand, focusing in assist activities that demand to grasp objects. Sharma et al. [13] presents a hand orthosis in additive manufacture consisting in rings to be embedded in the segments of the fingers and palm connected by "polymeric muscles", which aid in flexion/extension of the fingers. The exotendon of Meeker et al. [14] is a glove with external tendons, controlled by myoelectric signals that predict the intention of the user to perform extension/flexion of the fingers. In Hussain et al. [16] the device is an extra finger that holds objects in case of users who do not have enough grip strength to hold something. Conti et al. [17] present a hand exoskeleton, with a mechanism acting on the fingers, from the movements of the phalanges, aiding in the movements of opening the hands. The work of Radder et al. [10] proposes a robotic glove that captures the stimuli of the users' fingers, providing grip strength and greater hand opening during manipulation of objects.

Therefore, the type of device is related to the functions that it intends to assist. Elbow-wrist-hand devices encompass the functionalities of hand, arm and forearm orthosis. They are probably the choice for individuals with more severe conditions and weaknesses in the functions of the upper limb in general. The other types of orthosis are directed to specific regions of the upper limb, being used depending on the type of deficiency of the individual.

## 4.3 Activities of Daily Living

In the same way that the type of orthosis relates to the functions that are improved by it, they are linked with the activities of daily life that are intended to assist. Consequently, hand orthosis will be used in actions that require greater manual functionality, arm and forearm orthosis in those that depend on support to move the limb and those of the elbow-hand-wrist will provide assistance in both cases.

Hand orthosis are intended for activities that require manual and finger grip. The orthotic tests presented by Sharma et al. [13] shows the ability of the device to hold objects of various sizes, from tweezers to cans of soda. However, the tests performed were not with a real user, which can alter the results and limits the situations where the product could actually be used.

The orthosis of Meeker et al. [14] executed a pick and place test, demonstrating the device's ability to assist in tasks that require holding objects and positioning them. In the article this is demonstrated from the hold of a tennis ball and a minor object.

The tests performed with the sixth robotic finger of Hussain et al. [16] presented that the device is able to assist in tasks that require gripping force, holding objects and moving them in cases of deficient and even non-existent manual functions. However, it is limited to actions that do not require pinching movements.

The hand exoskeleton developed by Conti et al. [17] is intended for activities that require hand opening, important functionality for grasping objects. The authors argue that this functionality is not only relevant to the activities of everyday life, but also to social interactions, such as shaking hands with another individual. Performing specific tasks has not been described.

The robotic glove of Radder et al. [10] is an apparatus that provides assistance for actions that depend on the strength and movements of the fingers. In this case, the product is intended to aid in opening and closing a bottle, cutting food, twisting a cloth, turning a key and holding a book.

Arm and forearm support products act in activities that require movement of the upper limb. In the work of Looned et al. [18] the orthosis sought to assist and stimulate elbow flexion/extension and also pronation/supination of the forearm. In this case, the activity in which the device was tested was the act of drinking a glass of water, which requires the user to expend effort to raise the arm so that it reaches the mouth.

The elbow-wrist-hand orthosis act in activities in which both the gripping forces of the hand and the movement of the arm are necessary. The orthosis in Dunaway et al. [15] proposes to promote assistance for extension/flexion of the elbow and manual grip, which has an impact on the accomplishment of bilateral tasks. In this way, it provides everything from loading a laundry basket, lifting heavy objects, using tape measure, preparing meals, opening doors and among other things. Pedrocchi et al. [19] presents an orthosis with similar resources. According to the authors, the device supports arm weight and reduces the level of muscular activation required to perform some daily activities, as well as assist in manual gripping, which translates into support for simple tasks like drinking something, scratching, changing autonomously a television program, remove hair from the eyes and among others.

## 5 Conclusion

As the contingent of the elderly is increasing, actions to assist them in the performance of activities and, consequently, to stimulate their independence will be increasingly necessary. Therefore, investigating and knowing them is important in order to recognize the prospects for a reality that is established. The functionality of the hand, especially the grip strength, in the elderly, is of extreme importance in performing activities of daily living, such as holding objects, using a handrail or bus supports, performing housework, self-care activities, maintain the various daily activities with autonomy [20].

In this context, assistive technology devices appear as an important resource to assist the elderly population, since they provide support for a more independent life, allowing the elderly to perform a series of tasks that are often difficult depending on the health condition of the body.

Thus, the systematic review carried out in this article sought to show some devices that could help in the interaction of the elderly and objects that are part of daily activities, helping to achieve the necessary forces and movements when an elderly person is no longer able to do so.

The results showed that there are still few studies in the literature that address ATs to be used in the enhancement of manual functions in general, independent of the target audience. However, the devices found by the systematic review were interesting in several senses, besides being able to help users, especially the elderly, recover the capacity to perform activities with the upper limbs. In the cases presented, the apparatuses showed to be promising in the increase of grip and pinch forces and also in the movement of the upper limb.

In addition, the review presented demonstrates the lack of solutions for the elderly public, since only one of the studies focused on the senior user in fact. It is identified a knowledge gap and an opportunity for research that deals with specific solutions for the senile individual, given the expressiveness of such a population segment and the growing demand for products that provide a better quality of life for these people.

**Acknowledgments.** The authors gratefully acknowledge UDESC (*Santa Catarina State University*) and CAPES (*Coordenação de Aperfeiçoamento de Pessoal de Nível Superior* – Brazil) for the financial support.

## References

1. Perracini MR, Fló CM, Guerra RO (2009) Funcionalidade e Envelhecimento. In: Perracini MR, Fló CM (eds) Funcionalidade e Envelhecimento. Guanabara Koogan, Rio de Janeiro, p 557
2. Spirduso WW (2005) Dimensões Físicas do Envelhecimento. Manole, Barueri
3. Nowak E (2006) Anthropometry for the needs of the elderly. In: Karwowski W (ed) International encyclopedia of ergonomics and human factors, 2nd edn. CRC Press, Boca Raton, pp 258–265
4. Daams BJ (2006) Force exertion for (consumer) product design: problem definition. In: Karwowski W (ed) International encyclopedia of ergonomics and human factors, 2nd edn. CRC Press, Boca Raton, pp 347–349
5. Carmeli E, Patish H, Coleman R (2003) The aging hand. J Gerontol - Ser A Biol Sci Med Sci 58:146–152
6. Ranganathan VK, Siemionow V, Sahgal V, Yue GH (2001) Effects of aging on hand function. J Am Geriatr Soc 49:1478–1484. https://doi.org/10.1046/j.1532-5415.2001.4911240
7. Incel NA, Sezgin M, As I et al (2009) The geriatric hand: correlation of hand-muscle function and activity restriction in elderly. Int J Rehabil Res 32:213–218. https://doi.org/10.1097/MRR.0b013e3283298226
8. Martin JA, Ramsay J, Hughes C et al (2015) Age and grip strength predict hand dexterity in adults. PLoS ONE 10:1–18. https://doi.org/10.1371/journal.pone.0117598
9. Bagesteiro LB (2009) Função de Membro Superior e Envelhecimento. In: Perracini MR, Fló CM (eds) Funcionalidade e Envelhecimento. Guanabara Koogan, Rio de Janeiro, p 557
10. Radder B, Prange-Lasonder GB, Kottink AIR et al (2016) Preliminary findings of feasibility of a wearable soft-robotic glove supporting impaired hand function in daily life - a soft-robotic glove supporting ADL of elderly people. In: Rocker C, Ziefle M, Odonoghue J et al (eds) Proceedings of the international conference on information and communication technologies for ageing well and E-health (ICT4AWE 2016), pp 180–185. Scitepress, Setubal
11. Mello MAF de (2009) Tecnologia Assistiva. In: Perracini MR, Fló CM (eds) Funcionalidade e Envelhecimento. Guanabara Koogan, Rio de Janeiro, p 557
12. Sampaio RF, Mancini (2007) Systematic review studies: a guide for careful synthesis of scientific evidence. Rev bras fisioter 11: 77–82. https://doi.org/10.1590/s1413-35552007000100013

13. Sharma A, Saharan L, Tadesse Y (2017) 3-D printed orthotic hand with wrist mechanism using twisted and coiled polymeric muscles. In: Proceedings of the ASME 2017 international mechanical engineering congress and exposition (IMECE2017). ASME, Tampa, USA, pp 1–6
14. Meeker C, Park S, Bishop L, et al (2017) EMG pattern classification to control a hand orthosis for functional grasp assistance after stroke. In: 2017 international conference on rehabilitation robotics (ICORR), pp 1203–1210. IEEE, London
15. Dunaway S, Brianna Dezsi D, Perkins J et al (2017) Case report on the use of a custom myoelectric elbow–wrist–hand orthosis for the remediation of upper extremity paresis and loss of function in chronic stroke. Mil Med 182:e1963–e1968. https://doi.org/10.7205/MILMED-D-16-00399
16. Hussain I, Salvietti G, Spagnoletti G et al (2017) A soft robotic extra-finger and arm support to recover grasp capabilities in chronic stroke patients. Biosyst BioRobot 16:57–61. https://doi.org/10.1007/978-3-319-46532-6_10
17. Conti R, Allotta B, Meli E, Ridolfi A (2017) Development, design and validation of an assistive device for hand disabilities based on an innovative mechanism. Robotica 35:892–906. https://doi.org/10.1017/S0263574715000879
18. Looned R, Webb J, Xiao ZG, Menon C (2014) Assisting drinking with an affordable BCI-controlled wearable robot and electrical stimulation: a preliminary investigation. J Neuroeng Rehabil 11. https://doi.org/10.1186/1743-0003-11-51
19. Pedrocchi A, Ferrante S, Ambrosini E et al (2013) MUNDUS project: multimodal neuroprosthesis for daily upper limb support. J Neuroeng Rehabil 10. https://doi.org/10.1186/1743-0003-10-66
20. Wagner PR, Ascenço S, Wibelinger LM (2014) Hand grip strength in the elderly with upper limbs pain. Rev Dor 15:182–185. https://doi.org/10.5935/1806-0013.20140040

# Falls from Tractors in Older Age: Risky Behaviors in a Group of Swedish and Italian Farmers Over 65

Federica Caffaro[1], Peter Lundqvist[2], Margherita Micheletti Cremasco[3(✉)], Eva Göransson[2], Stefan Pinzke[2], Kerstin Nilsson[4], and Eugenio Cavallo[1]

[1] Institute for Agricultural and Earthmoving Machines (IMAMOTER), National Research Council of Italy (CNR), Turin, Italy

[2] Department of Work Science, Business Economics and Environmental Psychology, Swedish University of Agricultural Sciences, Alnarp, Sweden

[3] Department of Life Sciences and Systems Biology, University of Torino, Turin, Italy
margherita.micheletti@unito.it

[4] Division of Occupational and Environmental Medicine, Lund University, Lund, Sweden

**Abstract.** The frequent mounting and dismounting the tractor required by many farming operations increases the risk of falls, particularly for older farmers. The present study explored the risk factors related to tractor ingress and egress in older farmers from two countries with a different tradition in terms of safety culture: Sweden and Italy. Eighteen male farmers aged 65 + (8 from Skåne region, southern Sweden, and 10 from Piedmont region, northwestern Italy) were observed while mounting and dismounting their most used tractor, to investigate the routine behaviors adopted and to identify possible sources of risk of fall. The presence of three critical behaviors was recorded: the maintenance of three-point contact with the machine when entering and exiting the cab; facing the cab and the use of the last step when exiting. Farmers were also interviewed about their health status, attitudes toward safety, and perceived risks while performing the task. The results showed that similar unsafe behaviors were adopted by most of both Swedish and Italian participants; in particular, none of the farmers got off the tractor by facing the cab. Older farmers from both countries referred to age and previous experience as the major protective factors against falls, without acknowledging that new risks can rise from the age-related changes in their motor skills. The results raised some considerations about the need to develop targeted elderly-centered solutions to support the correct mounting/dismounting behaviors, both in the design of the machines and in information campaigns and training courses, which may have a cross-cultural validity.

**Keywords:** Agricultural machinery · Aging · Slips and falls


S. Bagnara et al. (Eds.): IEA 2018, AISC 826, pp. 78–86, 2019.
https://doi.org/10.1007/978-3-319-96065-4_11

# 1 Introduction

Agriculture is one of the most dangerous occupational sectors in both developing and industrialized countries and machinery is one of the main source of injuries [1]. Many people continue working in the agricultural sector well beyond their retirement age [2], making the risks even worse, because of the functional limitations related to the aging process [3]. Farmers older than age 65 are at higher risk for serious injuries and particularly susceptible to musculoskeletal disorders (MSDs) [4]. As for the general agricultural population, tractor operation is the leading cause of fatal and non-fatal accidents, also among older farmers, mainly in crop production [5]. The frequent mounting and dismounting the tractor required by many farming operations increases the risk of falls [6], because of the decline in postural balance with age [7], and it also enhances the exposure to biomechanical overload for lower-back and knee, because of the movements performed on the access path [8], with a cumulative effect, increasing with years of exposure and aging [9].

In the chain of events leading to a fall accident, important risk factors are represented by the attitudes toward risk of farmers and their behaviors [10]. The few available studies showed that older farmers often deny their declining abilities and consider the risk as an ineluctable part of their job, thus underestimating their risk exposure [11]. In addition, the familiarity with the tasks, the machinery and the equipment often leads to routine behaviors, lowering the attention rate and risk awareness when working [12].

## 1.1 Context and Aims of the Study

The participants selected for the present investigation came from two European countries with a high rate of older farmers and a very different tradition in terms of regulations for occupational safety and spreading of safety culture: Sweden and Italy.

In Sweden, approximately 22% of the population will be aged 65 years or older by 2020 [13], and, as regards farm labor force, the latest available statistics report that 31% of entrepreneurs and 30% of farm managers are aged over 65 [14]. In Italy, 22.6% of the agricultural workforce exceeds 55 years of age: of these, about 7% is made up of individuals aged over 65 [15]. With regard to company management, the farm managers under 40 years of age are 10% and those under the age of 30 only 2% [16].

Sweden is frequently considered as a benchmark in safety studies for its widespread safety culture, the effectiveness of its laws in reducing work-related risks [17], and the specific interventions designed in the agricultural sector to favor workers' safety [18]. As concerns Italy, up to 2008, Italy had only piecemeal and weak provisions for occupational health, combined with poor implementation capacities [19]. In 2008, in Italy, the adoption of the Legislative Decree 81/2008 led to an overall reorganization of the safety and health at the workplace, which tidied up and simplified the previous existing legislation. The aim of the Legislative Decree 81/2008 –which includes, inter alia, and implements the European Directive 89/391/EEC- is not only to modify or to repeal the previous laws, but also to focus on a new vision of "risk" and on a new concept of "danger", in compliance with the European directives.

Despite these differences in occupational safety regulations, agriculture is still a hazardous sector both in Italy and in Sweden, especially for older farmers. In Sweden, more than half of all occupational fatalities in agriculture and forestry involve workers aged over 55 [20] and agricultural machinery-related injuries are particularly common among farmers aged over 60 [21]. In particular, Pinzke and Lundqvist [22] found that 29% of injuries in agriculture were due to falls, involving especially farmers over 50 [20]. A similar picture emerged in Italy: in 2010, 105 out of 338 accidents involving agricultural vehicles (over 30%), occurred to operators aged over 65 [23]. Falls from tractor represent the second most frequent cause of death, after tractor rollover [24]. Also the occupational diseases are particularly widespread: out of 7,748 occupational diseases reported in Italy in 2012, two thirds concerned musculoskeletal diseases and among these, almost 61% of the cases involved individuals aged over 50 [25].

Based on the previous considerations, this study aimed at analyzing risk factors for falls from tractors in a group of Swedish and Italian farmers aged 65 and over. Consistent with the aims of the ergonomic discipline [26], the study investigated farmers' capabilities and their limitations in relation to the tasks performed and machinery used, considering both physical and cognitive components and both observable and non-observable aspects. The final aim was to identify the targets of possible user-oriented interventions, based on both a (re)design of farm equipment and machinery, and on training actions, to reduce risks and to lead to a better fitness between the person and the environment. Investigating these issues in two countries with a very different tradition in terms of safety culture and regulations allowed to explore whether some attitudes and behaviors are culturally-nested or they inform older farmers regardless their nationality.

## 2 Materials and Methods

### 2.1 Participants

Participants were selected according to the following criteria: age (65 and over) and being active in crop farming. A list of possible participants was provided by the LRF (*Lantbrukarnas Riksförbund*, the Federation of Swedish Farmers) in Sweden and by *Coldiretti,* one of the Italian biggest farmers' union, in Italy.

Respondents were contacted by telephone and, if willing to participate, were met at their own farm. The participation was voluntary, and all the farmers gave their written informed consent prior to their inclusion in the study. Eight male farmers over 65 in Skåne region, southern Sweden, and ten Italian farmers in Piedmont region, north-western Italy, took part in the study. Table 1 reports their main socio-demographic characteristics.

### 2.2 Instruments

Both subject-based and observation-based techniques, typical of an ergonomic approach devoted to integrate and compare information coming from targeted samples [27], were adopted to collect the data:

**Table 1.** Demographic description of the sample.

| | | Sweden | Italy |
|---|---|---|---|
| | | *n* | *n* |
| Education | Primary | 2 | – |
| | Lower secondary | – | 6 |
| | Upper secondary | 2 | 2 |
| | University | 4 | 2 |
| Tenure | Owner | 7 | 3 |
| | Not owner | 1 | 7 |
| Work alone on farm | Yes | 3 | 5 |
| | No | 5 | 5 |
| | | *M (SD)* | *M (SD)* |
| Age | | 69.9 (5.4) | 71.0 (4.3) |
| Experience in farming (years) | | 47.5 (11.6) | 53.9 (14.8) |
| Farm size (ha) | | 130.9 (93.4) | 30.5 (17.7) |

1. A questionnaire about the perceived health status and reported musculoskeletal disorders (MSDs). Two items of the SF-36 Health Survey [28] were used to investigate the perceived general health status and its difference compared to the past year. MSDs were assessed by means of the Swedish version of the Standardized Nordic Questionnaire-SNQ [29]. A standard socio-demographic and work experience form followed.
2. A semi-structured interview, to investigate farmers' attitudes about working life and safety in farming, and previous history of fall accidents. Diagnosed diseases and the role attributed by farmers to their occupation in the development of the diseases was also explored.
3. An observation of the older farmers' interaction with their most used tractor, to investigate the automatic behaviors adopted when dismounting the tractor cab. All the tractors involved into the investigation were equipped with a cab to protect the driver in case of rollover, in accordance with the Swedish and Italian regulations which require all newly manufactured tractors in the country to have ROPS installed since 1959 and 1974 respectively [30]. The video-recording was taken from a lateral view on the participant, to optimize the view of the targeted behaviors. Since observations may be supplemented by a verbal description from the operator of the decision processes taking place [27], the participants were also asked to report any perceived difficulty or risk related to the tasks, adopting the thinking aloud technique [31].

The questionnaire and the observations were driven by the first author. Meetings lasted about one hour for each participant.

### 2.3 Data Analysis

Two independent judges analyzed the videos using an observational grid (accordance rate = 85%). The grid assessed how the participants mounted and dismounted the tractor and the compliance with safety guidelines [32] with regard to: maintaining a three-point contact while entering and exiting the cab; facing the cab while exiting; use of the last step while exiting. Descriptive statistics were computed for data regarding health status and musculoskeletal discomfort, while the verbal protocols were analyzed by a content analysis.

## 3 Results

Participants rated their health as good, very good, or even excellent in both countries. However, four of the Swedish participants declared to suffer from hearing loss, one from diabetes, and another from respiratory diseases. Four Italian farmers reported impairments in visual acuity. All the participants rated their health status as about the same as 1 year before. One Italian farmer declared to feel much better compared to the past year and another one to feel somewhat worse. With regard to musculoskeletal disorders, the body areas in which the Swedish and the Italian participants reported to have experienced pain or discomfort during the last 12 months are represented in Fig. 1. The farm work was cited as the cause of the musculoskeletal disorders by 5 Swedish and 8 Italian participants. However, farming was perceived as a way to keep active and a good proxy of the health status.

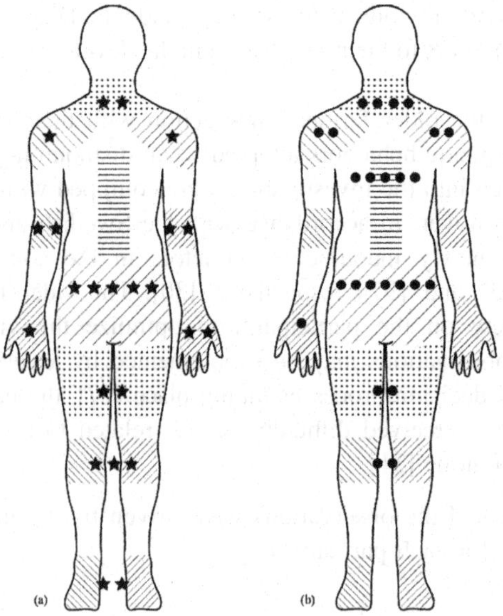


**Fig. 1.** Body areas in which the Swedish (a) and the Italian (b) participants reported feeling discomfort/pain (one symbol each time the participants indicated that area).

With regard to previous history of falls from agricultural machinery, two Swedish and three Italian participants reported having experienced some falls from machinery or equipment within the past 12 months. Many of both Swedish and Italian participants cited working experience as a protective factor against accidents. Overall, the favored prevention strategies adopted by the participants were performing activities without being in a hurry, counting on experience, and using common sense.

Data from the observations are reported in Table 2. The results showed many unsafe behaviors performed by both Swedish and Italian participants. In particular, the most recurrent unsafe behavior was represented by exiting the cab by facing forward, thus increasing the risks of slips and falls [32]. The recommended three-point contact with the machine while mounting and dismounting the tractor was maintained by a minority of farmers, especially among the Swedish participants. The majority of the participants from both countries did not perceive any effort or risk while mounting and dismounting the tractor. However, the height of the cab steps was reported as critical when mounting the machine: for five Swedish farmers and three Italian farmers, one additional step would be useful to reduce the effort in getting on and off the tractor many times a day (Table 3).

**Table 2.** Behaviors adopted by Swedish participants while mounting and dismounting the tractor.

| Participant | Mount | Dismount | | |
|---|---|---|---|---|
| | 3-point contact | 3-point contact | Face cab | Jump from last step |
| 1 | | | | X |
| 2 | | | | X |
| 3 | | X | | |
| 4 | X | X | | X |
| 5 | X | | | |
| 6 | | | | |
| 7 | | | | |
| 8 | X | | | |

**Table 3.** Behaviors adopted by Italian participants while mounting and dismounting the tractor.

| Participant | Mount | Dismount | | |
|---|---|---|---|---|
| | 3-point contact | 3-point contact | Face cab | Jump from last step |
| 1 | | | | |
| 2 | X | X | | |
| 3 | X | X | | |
| 4 | | X | | |
| 5 | X | | | X |
| 6 | X | | | X |
| 7 | | | | |
| 8 | X | X | | |
| 9 | | X | | |
| 10 | X | X | | |

## 4 Discussion

This study performed an ergonomic analysis of fall risks in a group of Swedish and Italian older farmers, considering both cognitive and physical, non-observable (attitudes and perceptions) and observable (behaviors) aspects.

Attitudes toward risks and routine behaviors are a fundamental part of farming culture, and they play an important role in determining risk exposure. Similar attitudes and behaviors emerged in the two groups of participants, even though they came from two countries with a very different tradition in terms of safety culture and promotion of occupational safety.

Farming activities and health status appeared to be tightly interwoven: many injuries and disorders were reported as having been caused by farm-work, but both Swedish and Italian farmers still enjoyed their work. This multifaceted relation between health and farming is consistent with the results from previous studies [11], and it highlights the importance of taking into account farmers' personal beliefs, to develop tailored interventions promoting older farmers' health and safety.

As regards risk perception and attitudes toward safety, all the participants shared the idea that the best strategy to avoid accidents and injuries was to use the 'common sense' and previous experience when accomplishing a task. The role played by this latter variable should be carefully considered, since previous studies have shown that familiarity and experience may actually lead to an overconfidence in the use of the devices, reducing the attention rate and decreasing risk awareness [12]. Specific and periodic training programs on safety issues may be designed, addressing farmers at their younger ages, at the start of their career: in this way, the correct safety practices would become an integral part of farmers' experience and common sense, thus making these variables effective protective factors.

Two main issues arose as interwoven in defining risk exposure when interacting with machinery: farmers' behaviors and machinery design. There was a high prevalence of unsafe behaviors while mounting and dismounting the tractor, which have documented effects on the risk of falls [5, 10] and biomechanical overload [4]. Moreover, the design of the access steps was reported as hindering these movements and many participants declared that an additional step could be useful to facilitate the entering. However, technical improvement may not be sufficient, given the importance of routine behaviors developed in many years of interaction with the machinery. If these behaviors may be reduced by a redesign of some technical elements of the tractors, more information campaigns should stress the importance of always performing the appropriate safe behavior to avoid accidents.

## 5 Conclusions

Falls from the tractor represent an important source of fatal and non-fatal accidents, and it increases the risk of musculoskeletal disorders, especially among older farmers. Older farmers involved in the present study showed many unsafe behaviors while entering and exiting the tractor cab, and mainly referred to common sense and their expertise as the best practices to avoid accidents. However, common sense usually disappears in stressful conditions, thus it is necessary to have solid safety habits to decrease risky situations when the attentional resources are limited. Unsafe behaviors performed while interacting with the machinery could be addressed both by targeted campaigns, which consider the role played by routine behaviors, and by directly involving the older farmers in the testing of new machinery design.

The present study pointed out similar behaviors and attitudes in both Swedish and Italian participants, even though the Swedish farmers came from a country typically considered a benchmark as regards occupational health and safety issues. These results raise some considerations about the need to design elderly-centered interventions to promote occupational safety and health, which may have a cross-cultural validity.

## References

1. ILO. http://www.ilo.org/global/industries-and-sectors/agriculture-plantations-other-rural-sectors/lang–en/index.htm. Accessed 12 Nov 2017
2. Andrews GR (2001) Demographic and health issues in rural aging: a global perspective. J Rural Health 4:323–327
3. Koolhaas W, Van Der Klink JJ, Groothoff JW, Brouwner S (2012) Towards a sustainable healthy working life: association between chronological age, functional age and work outcomes. Eur J Pub Health 22(3):424–429
4. Mitchell L, Hawranik P, Strain L (2002) Age-related physiological changes: considerations for older farmers' performance of agricultural tasks. University of Manitoba, Centre of Aging, Winnipeg
5. Myers JR, Layne LA, Marsh SM (2009) Injuries and fatalities to U.S. farmers and farm workers 55 years and older. Am J Ind Med 52(3):185–194
6. NIOSH. http://www.cdc.gov/niosh/docs/2000-116/pdfs/2000-116.pdf. Accessed 12 Oct 2017
7. Piirtola M, Era P (2006) Force platform measurements as predictors of falls among older people - a review. Gerontology 52(1):1–16
8. Leskinen T, Suutarinen J, Väänänen J, Lehtela J, Haapala H, Plaketti P (2002) A pilot study on safety of movement practices on access paths of mobile machinery. Saf Sci 40(7–8):675–687
9. Morgan LJ, Mansfield NJ (2014) A survey of expert opinion on the effects of occupational exposures to trunk rotation and whole-body vibration. Ergonomics 57(4):563–574
10. Caffaro F, Roccato M, Micheletti Cremasco M, Cavallo E (2017) Falls from agricultural machinery: risk factors related to work experience, worked hours, and operators' behavior. Hum Factors J Hum Factors Ergon Soc 60(1):20–30
11. Amshoff SK, Reed DB (2005) Health, work, and safety of farmers ages 50 and older. Geriatr Nurs 26(5):304–308
12. McLaughlin AC, Fletcher LM, Sprufera JF (2009) The aging farmer: human factors research needs in agricultural work. In: Proceedings of the human factors and ergonomics society annual meeting, vol 53(18), pp 1230–1234. Human Factors and Ergonomic Society, Santa Monica, CA
13. Bornefalk A, Yndeheim O (2004) Can we rely on the elderly? Report SOU 2004:44. Government Offices of Sweden, Stockholm

14. Statistics Sweden. http://www.jordbruksverket.se/webdav/files/SJV/Amnesomraden/Statistik,%20fakta/Sysselsattning/JO30/JO30SM1401/JO30SM1401.pdf. Accessed 21 Sep 2017
15. ISTAT. http://www.istat.it/it/files/2013/07/Italia_agricola.pdf. Accessed 20 Nov 2017
16. ISTAT. https://www.istat.it/it/files/2011/03/1425-12_Vol_VI_Cens_Agricoltura_INT_CD_1_Trimboxes_ipp.pdf. Accessed 15 Jan 2018
17. Spangenberg S, Baarts C, Dyreborg J, Jensen L, Kines P, Mikkelsen KL (2003) Factors contributing to the differences in work related injury rates between Danish and Swedish construction workers. Saf Sci 41(6):517–530
18. Lundqvist P, Svennefelt CA (2014) Swedish strategies for health and safety in agriculture: a coordinated multiagency approach. Work 49(1):33–37
19. Eichener V (1997) Effective European problem-solving: lessons from the regulation of occupational safety and environmental protection. J Eur Public Policy 4(4):591–608
20. Nilsson K, Pinzke S, Lundqvist P (2010) Occupational injuries to senior farmers in Sweden. J Agric Saf Health 16(1):19–29
21. Pinzke S, Nilsson K, Lundqvist P (2014) Farm tractors on Swedish public roads–age-related perspectives on police reported incidents and injuries. Work 49(1):39–49
22. Pinzke S, Lundqvist P (2007) Occupational accidents in Swedish agriculture. Agric Eng Res 13:159–165
23. ASAPS. http://dati.asaps.it/articoli/Art_2010/annuario_10/cap_13.pdf. Accessed 14 Dec 2017
24. Regione Piemonte. http://www.regione.piemonte.it/sanita/cms2/images/allegati/piano_nazionale_agricoltura_2014_2018.pdf. Accessed 06 Oct 2017
25. INAIL. https://www.inail.it/cs/internet/docs/ucm_107754.pdf. Accessed 10 Nov 2017
26. Kroemer KH (2005) 'Extra-ordinary' ergonomics: how to accommodate small and big persons, the disabled and elderly, expectant mothers, and children. Taylor & Francis, London
27. Kirwan B, Ainsworth LK (eds) (1992) A guide to task analysis: the task analysis working group. Taylor & Francis, London
28. Ware JE Jr (1992) Sherbourne CD (1992), The MOS 36-item short-form health survey (SF-36): I. Conceptual framework and item selection. Med Care 30:473–483
29. Kuorinka I, Jonsson B, Kilbom A, Vinterberg H, Biering-Sørensen F, Andersson G, Jørgensen K (1987) Standardised Nordic questionnaires for the analysis of musculoskeletal symptoms. Appl Ergon 18(3):233–237
30. Biddle EA, Keane PR (2012) Action learning: a new method to increase Tractor Rollover Protective Structure (ROPS) adoption. J Agromedicine 17(4):398–409
31. Lewis CH (1982) Using the "Thinking Aloud" method in cognitive interface design (Technical report). RC-9265. IBM Thomas J. Watson Research Center, Yorktown Heights NY
32. HSE. http://www.hse.gov.uk/pubns/indg185.pdf. Accessed 05 Oct 2015

# Virtual Aging – Implementation of Age-Related Human Performance Factors in Ergonomic Vehicle Design Using the Digital Human Model RAMSIS

Hans-Joachim Wirsching[1(✉)] and Michael Spitzhirn[2]

[1] Human Solutions, Europaallee 10, 67685 Kaiserslautern, Germany
Hans-Joachim.Wirsching@human-solutions.com
[2] Chair of Ergonomics and Innovation Management, Chemnitz University of Technology, Erfenschlager Straße 73, 09126 Chemnitz, Germany

**Abstract.** For user centered product design the application of digital human models has been widely established in automotive industry. These models are used to optimize the vehicle ergonomics in an early phase of the CAD design.

The integrated human behavior models are based on experimental data of healthy average age people in general. Hence these models cannot be used to develop ergonomic human machine interfaces that are also suitable for ageing customers. Applying the current digital human engineering process to elderly occupants, however, is a requirement gaining importance with the demographic change in industrial countries.

In order to fulfil this requirement, the most important human engineering applications of the digital human model RAMSIS were extended to take age-related human performance changes into account. A comprehensive literature review the research was conducted about age-related vision limits, acuity, glance eversion time, joint angle limits as well as joint strength over age.

The results were transferred to the RAMSIS human model to extend the simulation database. The enlarged simulation database facilitates applying the current human engineering applications of RAMSIS to occupants with age-related restrictions.

The differences in simulation results of different age groups were analyzed with respect to relevance and significance in the ergonomic vehicle development process. The automotive test use cases were set up for the criteria posture, reachability, force effort and visibility for occupant models in a vehicle model. The simulation results differ significantly between the age groups.

Hence, the extended simulation methods of RAMSIS support human engineering applications for age-based automotive product planning.

**Keywords:** Aging · Digital human model · Vehicle ergonomics
Virtual aging


S. Bagnara et al. (Eds.): IEA 2018, AISC 826, pp. 87–97, 2019.
https://doi.org/10.1007/978-3-319-96065-4_12

## 1 Motivation and Objective

For the user centered product design within virtual planning processes, the application of digital human models has been widely established in automotive industry since the middle of the nineties [4]. These models are used by engineers to optimize the vehicle ergonomics in an early phase of the CAD design, thus decreasing expensive physical tests with real subjects in hardware prototypes [11].

While anthropometric data for sizing digital human models are usually provided by international standard specifications for different age groups, the human behavior models are currently based on experimental data of healthy mean age people in general. Hence these models cannot be used for the difficult challenge of vehicle manufacturers to develop ergonomic human machine interfaces that are also suitable for ageing customers. Applying the current digital human engineering process to elderly occupants, however, is a requirement gaining importance with the demographic change in industrial countries and with a large palette of vehicles sold to the steadily growing population segment of elderly and wealthy people [2, 4].

In order to fulfil this industrial requirement, the most important human engineering applications of the digital human model RAMSIS should be extended to take age-related human performance changes into account as RAMSIS is mainly used in the digital product development of vehicle manufactures [4, 22].

## 2 Ageing the Digital Human Model RAMSIS

RAMSIS was extended by age-related human factors in the context of the research project “VirtualAging”. This project was funded by the German Federal Ministry of Education and Research to develop digital tools for the age-based production and product design. The main focus was the extension of the digital human models ema (production design) and RAMSIS (product design) with age-specific performance factors.

After collecting detailed user requirements and a comprehensive literature review on available age-related human performance data, the most user relevant parameters of anthropometrics, flexibility, force and vision were identified for an integration in the digital human model RAMSIS (Fig. 1) and analyzed with respect to the technical feasibility in digital human models [4, 18].

Hence, the research focused on age-related changes of the anthropometric data and limits of gaze and peripheral vision fields, acuity, glance eversion time, limits of joint motion range as well as joint strength distributions over age.

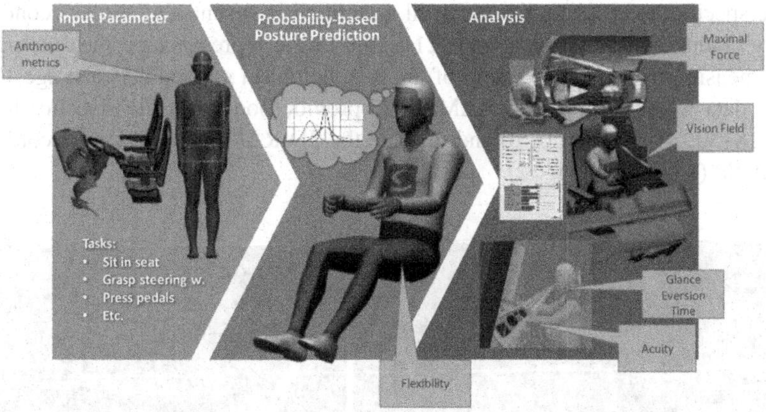


**Fig. 1.** Age-related human factors in human engineering application process of RAMSIS

# 3 Age-Specific Performance Factors in RAMSIS and Impact for Applications

## 3.1 General Process

The results of identified age-related changes of human performance [15–19] were processed and transferred to the digital human model RAMSIS to extend its existing simulation database. The enlarged simulation database facilitates applying the utilization of the current human engineering standard applications of RAMSIS to occupants with age-related human performance restrictions. These applications include e.g. the analysis of visibility conditions of displays, task-specific posture prediction, reachability and operation force analysis of devices in virtual vehicle environments.

The differences in simulation results of different age groups were analyzed with respect to relevance and significance in the ergonomic vehicle development from a user's point of view. A design change outside a threshold of 10 mm is significant and may require actions in general. For the testing of significance of the age-related changes, typical automotive use cases were set up for the criteria posture, reachability, force effort and visibility for the manikin in a vehicle model. The quantitative simulation results such as body point coordinates, accessibility and vision limit surfaces as well as force values significantly differ between the age groups. The age-specific factors and their corresponding impact in applications are given in the following subsections. The investigation is carried out by using a standard P50 anthropometric manikin in order to remove the influence of other age-specific factors on the results of each human performance factor. More details can be found in the respective sections.

## 3.2 Anthropometrics

Anthropometry is important for designing products and work processes and has a significant effect on planning of functional spaces such as reach envelops. Essential measures are body measurements and mass distributions [3, 4, 11].

Age-specific anthropometrics are well quantified in recent size surveys conducted in many countries around the world [21]. Due to the representative measurement of body dimension over a large section of the population (in general between age 18 and 70), the data were integrated in RAMSIS to provide body measure statistics for age groups 18–30, 30–50 and 50–70 and to scale the 3D-RAMSIS Manikins according to user criteria (left side of Fig. 2).

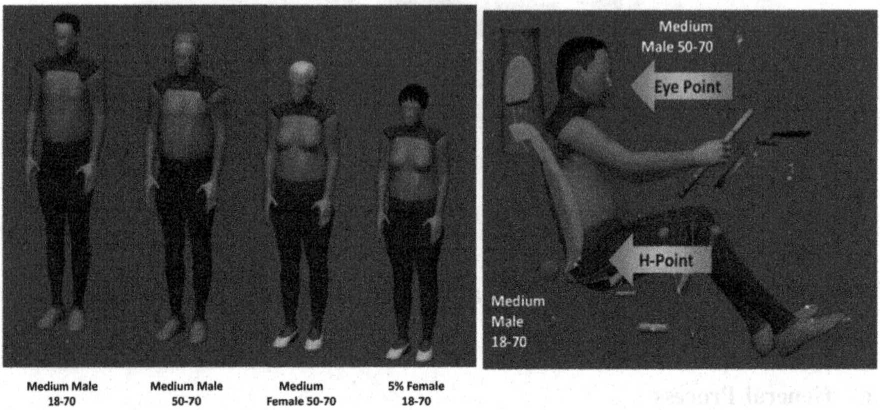


**Fig. 2.** Anthropometrics and its impact on product design; left: RAMSIS Manikins created from the anthropometrical database SizeGERMANY; right: Differences in posture simulation for manikins of different age groups (18–70, 50–70)

Age-specific body measures also have a crucial impact on the task-specific posture simulation and lead to different postures and positions [3]. This effect can be observed in the standard application of calculating a driving posture for medium male manikins of age groups 18–70 and 50–70 in a car. The differences are measured for the important eye and H-point positions (right side of Fig. 2).

The differences in positions amount 18 mm (eye point) as well as 26 mm (H-point). This is more than the design significance threshold of 10 mm (see Sect. 3.1) and may require actions in general. Nevertheless, the consideration of age-specific anthropometrics can be ignored, when the design is tested with the well accepted lower boundary female manikin of 5 percentile body height, as her body size is in most of the cases smaller than the size of a medium elderly male (Fig. 2).

### 3.3 Flexibility

Flexibility describes the functional capacity of a joint to move within a complete range of articulation, the range of motion (ROM). The human body consists of about 100 joints, each of which has function-dependent joint mobility. So it is possible that a person has a high mobility in one joint and a low mobility in another joint compared to other people [1, 7].

In order to move efficiently and with little effort, an adequate range of motion in the joints is important. Inadequate agility results in reduced performance in work-related activities and can contribute to back pain, which is costly [7]. Age-specific flexibility as joint angle limits have been studied in many surveys worldwide [15–17]. In order to find a consistent database a meta-analysis was performed on 14 international databases published in German or English literature [15, 17]. Finally, the results of 125 surveys were processed to build up age specific joint angle limits for all relevant joints in the digital human model RAMSIS. Due to the large individual range of age-related flexibility the results were provided for the percentiles 5, 50 and 95 [1, 6, 15, 17].

For consistency purposes the joint angle limits of the age group 50–70 were determined relatively with respect to the age group 18–30 (Fig. 3). These ratios were used to calculate joint angle limits for different percentiles (P05, P50, P95) in RAMSIS based on the available standard limits [8].

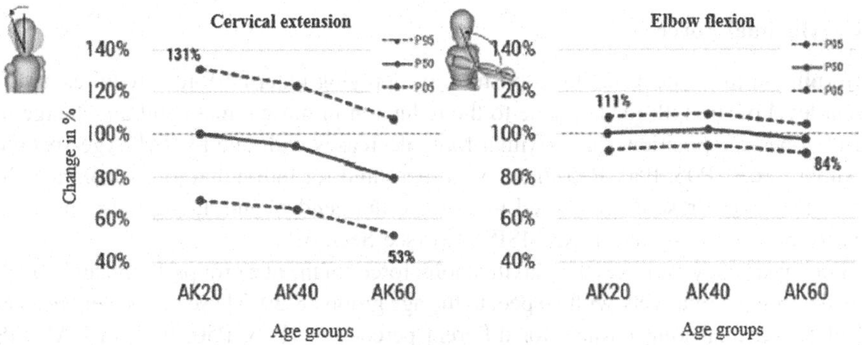


**Fig. 3.** Comparison of different age-related changes in joint movements for cervical extension and elbow flexion according to the percentiles 5, 50, 95

Age-specific flexibility has also an impact on the task-specific posture simulation and leads to different postures and positions [14]. This effect can be observed in the standard application to test the reachability of devices in a driving posture for a medium male manikin subjected to standard and age-specific joint angle limits (P05, P50, P95). We consider just one manikin size in order to remove the influence of age-specific anthropometrics (see Sect. 3.1). The differences are measured for the reachability envelope of the right arm in the neighborhood of a multi-media display (Fig. 4).

The reachability differences amount 7 mm (standard to 50. percentile group 50–70 year) and 28 mm (standard to 5. percentile group 50–70 year). The amplitude for the 5. Percentile flexibility is significant in vehicle design, because it is larger than the design significance threshold of 10 mm (see Sect. 3.1). In contrast, the reduced flexibility of an average elderly does not play a major role in this case.

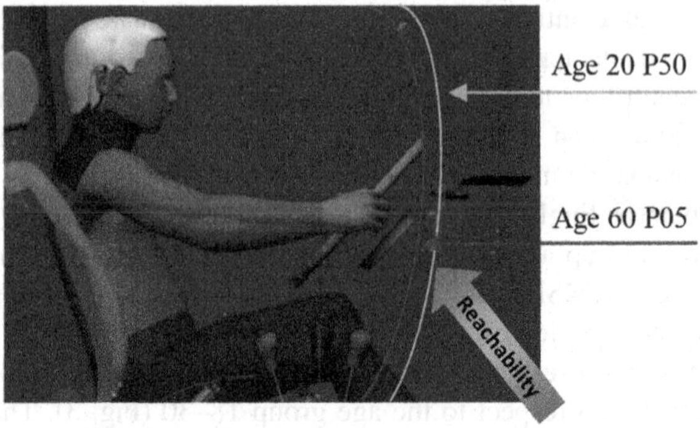


**Fig. 4.** Differences of reachability envelopes of right arm for manikins of different flexibilities (standard and age group 50–70, P05, P50, P95)

### 3.4 Maximal Force

Performing a movement, taking a posture, or applying external forces requires a sufficient level of strength [3, 11]. Due to the reduction in muscle mass and the change in muscle fiber composition, the maximal force decreases with age [3, 14]. Age-specific maximal forces (P05, P50, P95) have been measured for individual joints from age 20 to 80 [18]. These results were used to validate the gender- and age-specific maximal joint torque data available in RAMSIS [13] (see Sect. 3.1).

For consistency purposes the maximal joint forces of the age groups 40–50 and 50–70 were determined relatively with respect to the age group 18-30. These ratios were used to calculate maximal joint torques for different percentiles (P05, P50, P95) in RAMSIS based on the available standard maximal joint torques. The resulting maximal joint torques for the age groups 30–50 and 50–70 were compared with the existing data in RAMSIS through comparing the corresponding maximal force calculations in different applications. Since the differences are about 10%, the available maximal torque database in RAMSIS is comparable to the maximal joint force data of [18].

Age-specific joint forces have also an impact on the simulation of operational forces in product design. This effect can be observed in the standard application to test the necessary forces for releasing the hand brake for a medium male manikin subjected to age-specific maximal joint forces (age groups 18–70 and 50–70). The differences are determined for the maximal force of the right arm on the hand brake lever (pushing and pulling) (Fig. 5).

The maximal force differences between age group 50–70 and 18–70 are 50 N (approx. 14% reduction compared to the younger age group) for pulling task and 53 N (approx. 14% reduction) for pushing task. Differences more than 10% are significant in vehicle design, especially in case force values are close to the maximum feasible force of the operator or close to the minimal force of the operational device.

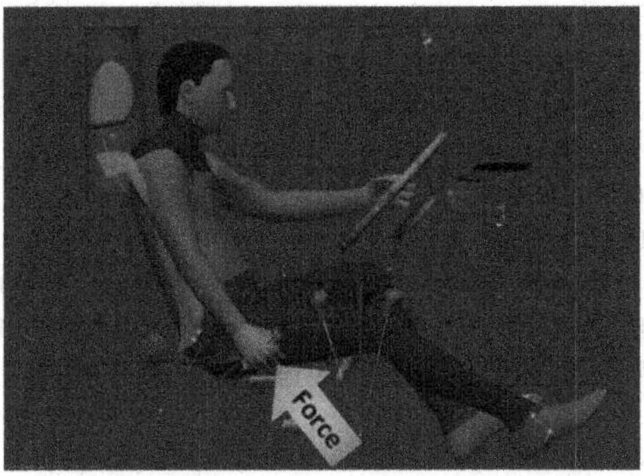


**Fig. 5.** Differences of maximal force of right arm for manikins of different age groups (comparison of age group 18–70 and 50–70)

### 3.5 Vision

80 to 90% of the information is perceived by vision [3]. For the integration of human vision the factors gaze and vision field, Glance eversion time, Acuity and age specific visus values were considered.

Gaze and vision field represent the size of the vision field depending on movement of the head or eyes [3]. To be able to perceive objects, these have to be in the field of view, sight or field of vision. The visual field decreases due to the reduction of the pupil diameter as well as the opacity of the eye medium from about 50–60 years [6]. Age-specific data for the gaze field and for the peripheral vision field from [5, 6]. The ratio between view limitations of young and elderly people were calculated from these data to reduce the existing standard vision limits in RAMSIS for corresponding limits of the age group 50–70.

Age-specific fields of view have an impact on the size of visibility areas. This effect can be observed in an application for providing optimal position areas for primary instruments (e.g. speedometer) for a medium male manikin subjected to age-specific gaze limits (standard age group 18–70 and age group 50–70). The differences are measured between the gaze field boundaries around the dash board (Fig. 6).

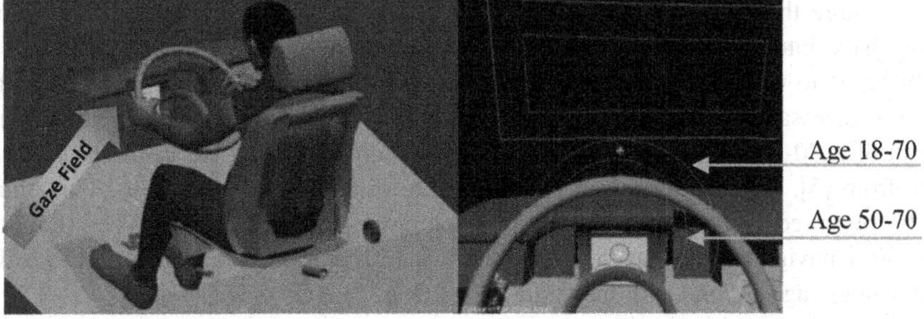


**Fig. 6.** Differences of gaze fields for manikins of different age groups (standard, 50–70)

The difference between both gaze fields amounts 34 mm. This amplitude is significant in vehicle design, as it is larger than the design significance threshold of 10 mm (see Sect. 3.1). Moreover, elderly people need an additional head movement to watch on the instruments, which also enlarges the glance eversion time.

The glance eversion time defines the required fixation time to change the point of view [3]. Age-specific data for the glance eversion time estimation is split up into glance moving time [10] and accommodation time [20]. The ratio between glance eversion time of young (age 25) and elderly (age 60) people was calculated from these data [10, 20] to increase the existing standard glance eversion time in RAMSIS for the age group 50–70.

Age-specific glance eversion times lead to different operation times. This effect can be observed in an application for evaluating the glance eversion time from the driving front view on the road to the navigation display for a medium male manikin (Sect. 3.1) subjected to age-specific glance eversion times (standard age group 18–70 and age group 50–70; Fig. 7)

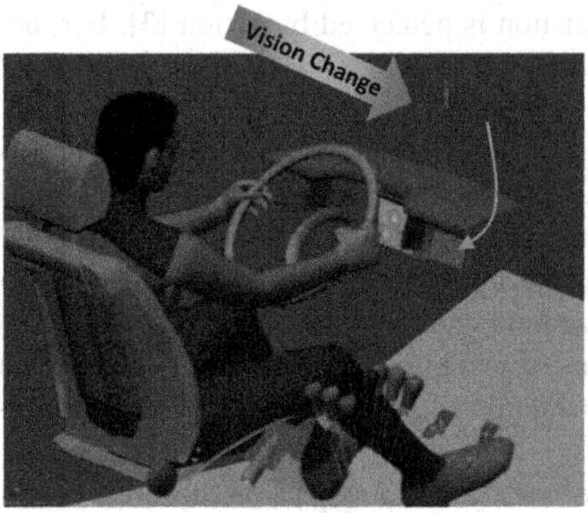


**Fig. 7.** Differences of glance eversion times for manikins of different age groups (standard age group 18–70, 50–70)

The difference between both glance eversion times amounts 292 ms. This amplitude is significant in vehicle design, as the entire vision prevention from the front road is up to more than 2 s [22] which also takes additional time to understand the information have into account.

In order to distinguish perceived objects from the environment, sufficient visual acuity is necessary [3]. At the age of 80, only half of the original normal vision (visual acuity) of a 20-year-old is present [3, 5]. Age-specific data for the visual acuity was taken from [5]. Age-specific visus values lead to different character sizes. This effect can be observed in an application for evaluating necessary character sizes (Landolt ring) on a navigation display for a medium male manikin subjected to age-specific visus values (age 30 and 60) during day-time (Fig. 8).

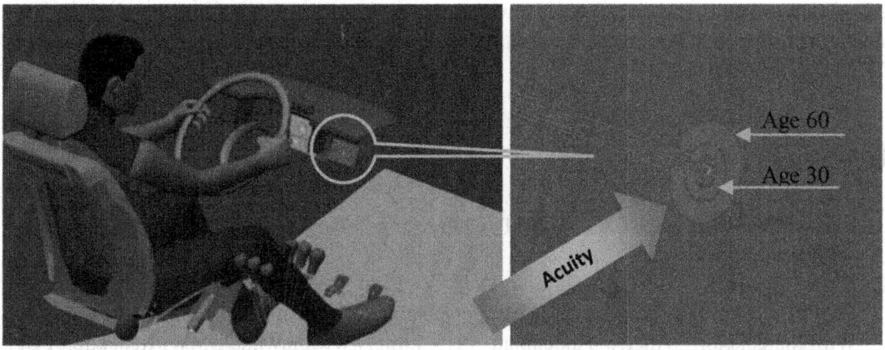


**Fig. 8.** Differences of necessary character sizes for manikins of different age (30, 60)

The difference in acuity (gap size of Landolt ring) is 0.16 mm. Although this assessment does not consider glasses of the occupant, the evaluation may be helpful, since many drivers do not use their varifocals due to the view restrictions.

## 4 Conclusion

Digital human models are used to optimize the vehicle ergonomics in an early phase of the CAD design. Due to a lack of data these models cannot be used to develop ergonomic human machine interfaces that are also suitable for ageing customers.

Based on a comprehensive literature review, the digital human model RAMSIS was extended by the most user relevant age-related human factors regarding anthropometrics, flexibility, force and vision. This extension facilitates applying the current human engineering standard applications of RAMSIS to occupants with age-related performance restrictions.

Several use cases were set up (analysis of visibility conditions of displays, task-specific posture prediction, reachability and operation force analysis of devices) for demonstrating the impact of age-related factors in ergonomic design.

The differences in simulation results of different age groups were analyzed with respect to relevance and significance in the ergonomic vehicle development from a user's point of view. In most of the cases the results differ significantly between age groups. Since the tests were just conducted on a single manikin with isolated age-related factors, the differences will increase when enlarging the test sample and consider combinations of age-related factors (e.g. anthropometrics plus flexibility).

Hence, the extended simulation methods of RAMSIS reflect the consequences of age-related performance changes of humans and support important human engineering applications for age-based automotive product planning.

The presented approach follows a pragmatic process for adapting single human performance factors. Currently the response of elderly people to their restrictions as compensating behavior strategies are not considered and should be an objective of future expensive studies.

**Acknowledgments.** The article would have been impossible without the support of the German Federal Ministry of Education and Research (Project: VirtualAging, FKZ: 01IS15002C) from 01.05.2015 to 31.10.2017.

## References

1. Bandy WD, Berryman Reese N (2010) Joint range of motion and muscle length testing. Saunders
2. BMWi (2010) Wirtschaftsfaktor Alter - Eine Initiative des BMWi und des BMFSFJ. Demographischer Wandel: Perspektiven für Anbieter und Märkte ausloten. Wirtschaftsfaktor Alter, Faktenblatt 1. http://www.bmwi.de/BMWi/Redaktion/PDF/Publikationen/wirtschaftsfaktor-alter-faktenblatt-1-marktpotenzial,property=pdf,bereich=bmwi2012,sprache=de,rwb=true.pdf
3. Bubb H, Bengler K, Grünen RE, Vollrath M (2015) Automobilergonomie. Springer Fachmedien, Wiesbaden
4. Bullinger-Hoffmann AC, Mühlstedt J (2017) Homo sapiens digitalis - virtuelle ergonomie und digitale menschmodelle. Springer, Heidelberg
5. ISO 22411:2008 - Ergonomics data and guidelines for the application of ISO/IEC Guide 71 to products and services to address the needs of older persons and persons with disabilities
6. Huaman AG, Sharpe JA Vertical saccades in senescence. Invest Ophthalmol Vis Sci 34: 2588–2595
7. Kaminsky LA (2014) ACSM's health-related physical fitness assessment manual. 4th edn American College of Sports Medicine, Baltimore
8. Kapandji IA (1970) The physiology of the joints. Churchill Livingstone, New York
9. NHTSA (2013) Visual-Manual NHTSA Driver Distraction Guidelines for In-Vehicle Electronic Devices. FR 04232013
10. Purucker C, Naujoks F, Prill A, Krause T, Neukum A (2014) Vorhersage von Blickabwendungszeiten mit Keystroke-Level-Modeling. Mensch und Computer
11. Schlick C, Bruder R, Luczak H (2010) Arbeitswissenschaft. Springer, Heidelberg
12. Seidl A, Bubb H (2006) Standards in anthropometry, handbook on standards and guidelines in ergonomics and human factors. Lawrence Erlbaum Associates
13. Seitz T et al (2005) FOCOPP - an approach for a human posture prediction model using internal/external forces and discomfort. SAE digital human modelling conference 2005, Paper 2005-01-2694
14. Spirduso W, Francis K, MacRae PG (2005) Physical dimensions of aging. 2nd edn Human kinetics, Champaign Illinois
15. Spitzhirn M (2017) Integration altersbedingter Veränderung der Beweglichkeit zur altersgerechten Arbeitsprozessgestaltung in digitalen Menschmodellen. In: Gesellschaft für Arbeitswissenschaft EV (eds) Soziotechnische Gestaltung des digitalen Wandels, 63. Kongress der Gesellschaft für Arbeitswissenschaft, pp 1–6. GfA-Press, Dortmund
16. Spitzhirn M, Bullinger AC (2016) Entwicklung eines Konzepts zur altersgerechten Arbeitsgestaltung mittels digitaler Menschmodelle. In: Müller E (eds) Smarte fabrik & smarte arbeit - industrie 4.0 gewinnt Kontur. VPP 2016 Vernetzt planen und produzieren. Wissenschaftliche Schriftenreihe des IBF. Sonderheft 22
17. Spitzhirn M, Bullinger AC (2018) Kritische Reflexion zu verschiedenen Datenquellen zu altersbedingten Veränderungen der Beweglichkeit. In: Gesellschaft für Arbeitswissenschaft EV (eds) ARBEIT(S).WISSEN.SCHAF(F)T - Grundlage für Management & Kompetenzentwicklung, 64. Kongress der GfA, pp 1–6. GfA-Press, Dortmund

18. Spitzhirn M, Kaiser A, Bullinger AC (2016) Virtual Aging - Nutzerbefragung zu Anforderungen und Bedarf zur Integration altersspezifischer Veränderungen des Menschen in digitale Menschmodelle. In: Bullinger (eds) 3D SENSATION, innteract conference, pp 238–252, aw&I-Verlag, Chemnitz
19. Stoll T, Huber E, Seifert B, Michel BA, Stucki G (2000) Maximal isometric muscle strength: normative values and gender-specific relation to age. Clin Rheumatol 19:105–113
20. Temme L, Morris A (1989) Speed of accommodation and age. Optom Vis Sci 66(2):106–112
21. Trieb R, Seidl A, Rissiek A, Wirsching H-J (2008) Weltweite anthropometrische Reihenmessungen mit BodyScanning – Überblick über die wichtigsten Projekte, Verfahren und Ergebnisse. Konferenz der Gesellschaft für Arbeitswissenschaften (GfA), München
22. Van der Meulen P, Seidl A (2007) Ramsis – the leading cad tool for ergonomic analysis of vehicles. digital human modeling. Lecture Notes in Computer Science, vol 4561, pp 1008–1017

# Aging Ebook: CIIP Silver Book on Aging and Work

Olga Menoni[1(✉)], Rinaldo Ghersi[2], and Susanna Cantoni[3]

[1] Department of Services and Preventive Medicine, IRCCS Cà GRANDA, Milan, Italy
olga.menoni@policlinico.mi.it

[2] CIIP, Italian Inter-Associative Prevention Council, Milan, Italy
ghersir@libero.it

[3] CIIP, Milan, Italy
susanna.cantoni@live.it

**Abstract.** Italy, like many other European countries, is experiencing a rapid aging of its workforce. Such phenomenon is due to demographic and economic factors as well as to the increase in retirement age. The *partial* replacement of retiring workers by younger ones cause two main problems: on the one hand, employers ignore which tasks to assign to aged workers; on the other hand, the workload of younger workers increase significantly.

CIIP, i.e. the Italian Inter-Associative Prevention Council, involves professional and scientific societies whose members are either public or private occupational physicians, ergonomists or professionals in the field of occupational and environmental safety and health. These societies, according to their different functions and peculiarities, deal with ergonomic issues and approaches such as work organization, biomechanics, toxicology, also focusing on the relation between human beings, machines and the environment. CIIP set up a working group on the issue of "Aging and work" with the aim of analyzing the topic and proposing management methods for aging in workplace.

We published a free E-book, available online, including a general part and a more specific one devoted to work in the healthcare sector.

The CIIP Group actively promotes public meetings on the aging topic in the Italian territory.

We provided physio-pathological information about aging in the workplace, taking into account the multiple approach defined by Illmarinen J and Others and highlighting resources and critical aspects of contemporary Italian contexts. Inspired by an interesting proposal and checklist by French INRS and by local researches and professionals of Clinica del Lavoro in Milan, the Group proposes risk assessment methods considering employees' age together with their physical, mental and organizational ergonomic tasks.

However, the Group underlines that workplace health promotion is necessary but not sufficient if a specific prevention plan to reduce risks is not implemented. It is therefore necessary to update the actual welfare support, allowing gradual retirement.

The text is written in Italian. English version can be found on CIIP's website [14].


S. Bagnara et al. (Eds.): IEA 2018, AISC 826, pp. 98–103, 2019.
https://doi.org/10.1007/978-3-319-96065-4_13

**Keywords:** Work · Aging · Ageing · Occupational health · Ergonomics
Healthcare

## 1 CIIP Commitment

CIIP, i.e. the Italian Inter-Associative Prevention Council, involves professional and scientific societies whose members are either public or private occupational physicians, ergonomists or professionals in the field of occupational and environmental safety and health. These societies, according to their different functions and peculiarities, deal with ergonomic issues and approaches such as work organization, biomechanics, toxicology, also focusing on the relation between human beings, machines and the environment. CIIP set up a working group on the issue of "Aging and work" with the aim of analyzing the topic and proposing management methods for aging in workplace.

## 2 The Aim of the Aging E-Book

Italy, like many other European countries, is experiencing a rapid aging of its workforce. Such phenomenon is due to demographic and economic factors as well as to the increase in retirement age. The *partial* replacement of retiring workers by younger ones cause two main problems: on the one hand, employers ignore which tasks to assign to aged workers; on the other hand, the workload of younger workers increase significantly.

The main aim of the E-book is to provide a synthesis of numerous studies and experiences, along with some proposals on the multiple approach towards active aging, addressed to:

- Human Resource Departments
- Employers
- Health and Safety managers and officers
- Workers' safety representatives
- Workers and trade unions
- Health and safety technicians
- Occupational physicians

The E-book wants to contribute to the promotion of discussions and interventions.

Given the Italian welfare system (that lead to a general aging of the workforce) and the inadequate employee turnover, particular attention has been devoted in the E-book to healthcare professionals.

## 3 The Aging E-Book: Methods and Contents

The working group gathered and analyzed the extensive literature and experiences on the subject, discussed and elaborated evaluations and different proposals publishing a free E-book [14], available online and addressed to professionals, employers and employees. The Group actively promotes public meetings on the aging topic in the Italian territory.

Each chapter of the aging e-book presents boxes highlighting fundamental concepts and key-points. Starting form epidemiological data and analysis, the text provides physio-pathological information about aging in the workplace, taking into account the multiple approach defined by Illmarinen J and Others [7] and highlighting resources and critical aspects of contemporary Italian contexts. Inspired by an interesting proposal and checklist by French INRS [15] and by local researches and professionals of Clinica del Lavoro in Milan, the Group proposes risk assessment methods considering employees' age together with their physical, mental and organizational ergonomic tasks. However, the Group underlines that workplace health promotion is necessary but not sufficient if a specific prevention plan to reduce risks is not implemented. It is therefore necessary to update the actual welfare support e.g. allowing gradual retirement.

With the aim of encouraging critical and concrete approaches, the E-book includes general and specific checklists addressed to employers, advisors and workers. Further assessment and management tools for subjects over the age of 50 and exposed to harsh working conditions are also available online for free, as attachments of the E-book.

The present edition includes a general part and a more specific one devoted to work in the healthcare sector.

Bibliography, links to websites and to panels of discussion will allow the reader to verify and elaborate the topic of aging in the workplace.

## 4 Ergonomic Approach

The ergonomic approach (micro and macro ergonomic approach) seems to hint at a discipline that shares several goals with Occupational Medicine:

- Prevention of work-related distress, disorders or diseases (with an eye to health protection as defined by the World Health Organization (WHO) in Art. 2.1 or by Italian Legislation).
- Adaptation of work conditions for "all ages", in line with the slogan of the European campaign, hence a better redeployment of aged workers, both "healthy" and "with issues".
- Maintenance of productivity in terms of both quality and quantity despite aging workers.

Physical, cognitive and organizational ergonomics could help us reach the desired goals towards the prevention of work-related disorders and may contribute to healthy aging by safeguarding employability.

Although a few cases have been faced in the past, ergonomic experts are yet to express their opinion about a real adjustment of working conditions to aged workers.

The problem of shiftwork and night work organization in workers over the age of 50 has also been tackled in the ebook.

Specific ergonomic proposals [1–6] in order to avoid damages due to the alteration of circadian rhythms have been formulated.

In particular:

(a) Reduce and avoid, if possible, night work past the age of 50
(b) Allow workers to work shifts that best suit their lifestyle (e.g. morning shifts or afternoon shifts)
(c) Reduce workload
(d) Reduce working hours and/or increase rest days
(e) Implement regular health checks
(f) Provide training and advice on the best coping strategies for sleep, diet, stress management and regular physical exercise.

## 5 Aging in the Healthcare Sector

Currently available methodologies for Manual Patient Handling risk assessment do not provide any classifications by age, as observed in the review issued by 2012 Technical Report ISO 12296, specifically dedicated to patient handling and entitled "Ergonomics - Manual handling of people in the healthcare sector".

We analyzed the aging phenomenon in Italy from 1999 to the present [8–11], providing a proposal for risk assessment based on the reduced tolerance of the musculoskeletal system in subjects aged more than 50 [12, 13].

In case of heavily overloaded work environments, screening workers' health condition is recommended. The aging E-book, in particular, provides a questionnaire for the most frequent musculoskeletal disorders by age and job sector. The data collected can be confronted with unexposed groups of workers.

**Table 1.** Mean age of healthcare professionals in wards

| Year | Sector | Mean age | N° exposed subjects | % Herniated disk |
|---|---|---|---|---|
| 96–99 | Hospital wards | **36** | 1566 | 6 |
| | Nursing homes | | 1535 | 12 |
| 2003 | Hospital wards | **36,5** | 2603 | 14 |
| 2006–8 | Hospital wards in Liguria region | **42** | 1994 | 14 |
| 2008–9 | Nursing homes in Veneto region | **41** | 178 | 8 |
| 2015 | Hospital wards in Apulia region | **48** | 2717 | 17 |

The topic of Organizational ergonomics has also been explored in the E-book, with an eye on possible professional errors, lower quality care and work-related stress factors.

A tailored version of the risk assessment tool has been provided for the healthcare sector. This tool, available on the CIIP website for free, has been clearly presented in the general part of the E-book.

The research unit "Ergonomics of Posture and Movement" (EPM) together with local researches and professionals of Clinica del Lavoro in Milan conducted numerous multi-centre studies in different hospitals in northern and central Italy (1999). Two further studies (2003–2012) have also been undertaken to estimate the risk of manual patient handling and the prevalence of WMSDs among nurses [8, 9].

Table 1 shows the extent of the problem reporting the mean age of healthcare professionals in wards: changes occur in 2004–2006.

The phenomenon of the aging workforce in the healthcare sector has particularly emerged in Liguria, Piemonte and Apulia regions.

A detailed analysis of the distribution by 10-year age groups shows an aging trend towards higher age groups (Table 2), this leading to a difficult management of the so called "active aging": an actual management and monitoring of preventive strategies, as indicated in ISO TR 12296 technical report, is the only solution.

**Table 2.** Distribution by age group of studies mentioned in Table 2

| Year of the study | <= 25% | 26–35% | 36–45% | >45% |
|---|---|---|---|---|
| 96–99 hospitals | 13% | 48% | 23% | 16% |
| 96–99 nursing homes | 8% | 40% | 35% | 17% |
| 2003 hospitals | 4,7% | 45,3% | 30% | 20% |
| 2006–8 hospitals | 1% | 16% | 49% | 33% |
| 2008–9 nursing homes | 5% | 20% | 34% | 41% |
| 2015 hospitals | 0,1% | 6,9% | 29% | 64% |

## 6 Age-Sensitive Approach to Risk Management

The main risk factors for workers in the healthcare sectors are:

- Physical ergonomic factors: patient handling and lifting; awkward or painful postures;
- Psychosocial factors: fast work pace or excessive workload; emotional demands; threats and physical violence; shiftwork; work-life balance;
- Biological factors: risk of exposure to potentially infectious body fluids/airborne agents;
- Injury risk: accidental falls; needlestick injuries.

As far as aging is concerned, physical ergonomic factors and psychosocial factors are particularly important.

Proposals for specific assessment of physical ergonomic factors have been formulated, providing a specific checklist to assess important risk factors associated with age (Excel file) based on the content of INRS checklist previously mentioned. Each important factor in terms of age is associated with preventive strategies that need to be implemented.

Proposals for age-sensitive manual handling risk assessment have also been included in the E-book. Useful tools can be found on the CIIP website for free [14].

## References


1. Costa G, Goedhard W, Ilmarinen J (eds) (2005) Assessment and promotion of work ability, health and well-being of ageing worker. Elsevier, Amsterdam, p 435
2. Costa G (1998) Guidelines for the medical surveillance of shiftworkers. Scand J Work Environ Health 24:151–155
3. Hakola T, Härmä M (2001) Evaluation of a fast forward rotating shift schedule in the steel industry with a special focus on ageing and sleep. J Hum Ergol 30:35–40
4. Härmä M, Ilmarinen J (1999) Towards the 24-hour society – new approaches for aging shift workers? Scand J Work Environ Health 25:610–615
5. Härmä M, Kandolin I (2001) Shiftwork, age and well-being: recent developments and future perspectives. J. Human Ergol 30:287–293
6. Knauth P, Hornberger S (1998) Changes from weekly backward to quicker forward rotating shift systems in the steel industry. Int J Industr Ergon 21:267–273
7. Ilmarinen J, Rantanen J (1999) Promotion of work ability during ageing. Am J Industr Med 1:21–23
8. Battevi N, Consonni D, Menoni O, Ricci MG, Occhipinti E, Colombini D (1999) Application of the synthetic exposure index in manual lifting of patients: preliminary validation experience. Med Lav 90(2):256–275 Italian version
9. Battevi N, Menoni O, Ricci MG, Cairoli S (2006) MAPO index for risk assessment of patient manual handling in hospital wards: a validation study. Ergonomics 49(7):671–687
10. Battevi N, Menoni O, Alvarez-Casado E (2012) Screening of patient manual handling risk using the MAPO method. Med Lav 103(1):37–48 Italian version
11. Menoni O, Ricci MG, Panciera D, Battevi N, Colombini D, Occhipinti E, Greco A (1999) Assessment of exposure to manual patient handling in hospital wards: methods and exposure indices (MAPO) and classification criteria. Med Lav 90(2):152–172 Italian version
12. Jager M, Jordan C, Theilmeier A, Luttmann A, Dolly Group (2007) Spinal-load analysis of patient-transfer activities. In: Buzug TM, Holz D, Weber S, Bongartz J, Kohl-Bareis M (eds) Advances in medical engineering. Springer, Berlin, pp 273–278
13. Marras WS (2008) The working back. A systems view. Wiley-Interscience, Wiley, Hoboken
14. CIIP website. https://www.ciip-consulta.it/index.php?option=com_phocadownload&view=category&id=8:aging-ebook-en&Itemid=609
15. INRS Website. http://www.inrs.fr/accueil/produits/mediatheque/doc/publications.html?refINRS=ED%206097

# Early Detection of Fatigue Based on Heart Rate in Sedentary Computer Work in Young and Old Adults

Ramtin Zargari Marandi[1,2(✉)], Pascal Madeleine[1], Nicolas Vuillerme[1,2,3], Øyvind Omland[4], and Afshin Samani[1]

[1] Department of Health Science and Technology, Sport Sciences, Aalborg University, 9220 Aalborg, Denmark
rzm@hst.aau.dk
[2] University Grenoble Alpes, AGEIS, 38000 Grenoble, France
[3] Institut Universitaire de France, 75231 Paris, France
[4] Aalborg University Hospital, Clinic of Occupational Medicine, Danish Ramazzini Center, 9000 Aalborg, Denmark

**Abstract.** Given the growing number of working elderly, monitoring fatigue developing at work is of utmost importance. In this study, 38 participants (18 elderly and 20 young adults) were recruited to perform a prolonged computer task including 240 cycles while their heart rate was measured. In each cycle, the participants memorized a random pattern of connected points then replicated the pattern by clicking on a sequence of points to complete an incomplete version of the pattern. Task performance in each cycle was calculated based on the accuracy and speed in clicking. After each 20 cycles (one segment), participant rated their perceived fatigue on Karolinska Sleepiness Likert scale (KSS). The mean and range of heart rate, HRM and HRR respectively, in each cycle were calculated and together with the performance were averaged across each segment. Statistical analysis revealed that HRR followed an increasing trend in both young and elderly groups as time on task (TOT) increased, $p < 0.001$. The HRM exhibited a tendency to increase with TOT in both groups, $p = 0.063$. The performance increased in the elderly group and fluctuated in the young group, $p < 0.001$. The KSS increased in both groups with TOT, $p < 0.001$. No interactions between TOT segments and groups were found in any of the measures except in case of performance indicating a higher performance for the young group with a fluctuating temporal pattern. The results provide insights on the feasibility of using heart rate as an index to monitor fatigue in both young and elderly computer users.

**Keywords:** Fatigue · Aging · Heart rate variability

## 1 Introduction

Aging workforce underlines the need for human centered design, specifically for mentally demanding tasks such as computer work. This requires acquiring relevant biological data in an unobtrusive manner. One of the accessible choices is heart rate


S. Bagnara et al. (Eds.): IEA 2018, AISC 826, pp. 104–111, 2019.
https://doi.org/10.1007/978-3-319-96065-4_14

monitoring which can accurately and reliably be measured with affordable and unobtrusive wearable heart rate sensors or fitness trackers [1, 2].

One important problem during prolonged sedentary computer work is the development of fatigue. However, in tasks with low level of muscle contraction like computer work, fatigue development often remains undiscovered if conventional metrics of fatigue (e.g. reduced maximal force output) are utilized [3]. If fatigue development is detected at early stages, an informed pausing regime can impede the process [4]. Previous studies lend support to the usability of heart rate in the detection of fatigue [5–7]. Since heart rate is regulated by the autonomic nervous system [8], this association is seemingly plausible. However, further studies have been suggested to explore the association between heart rate and fatigue development [6], especially amongst individuals from different age groups.

Fitness trackers usually provide limited access to cardiac responses which may only include instantaneous heart rate. This constrains the computation of cardiac features to statistical characteristics of instantaneous heart rate (e.g. mean and range in current study) across a task-relevant time window.

In this study, we aimed to analyze the changes of mean and range of heart rate during a prolonged computer work both in young and old adults to determine if those metrics could be used as biomarkers for the detection of fatigue development. We employed a fitness tracker to record heart rate while individuals performed the sedentary computer work with mental demands. We hypothesized that the mentioned heart rate features would reflect the progression of time-on-task (TOT).

## 2 Materials and Methods

### 2.1 Participants

Twenty participants as a young group, nine females, aged 23 (*SD* 3) years, and 18 participants, 11 females, aged 58 (*SD* 7) as an elderly group voluntarily participated in this study. They were all right-handed, had normal or corrected to normal vision, and reported no background of mental or psychological disorders, and no history of chronic fatigue. The participants were asked to obstinate alcohol for 24 h, and caffeine, smoking and drugs for 12 h prior to experimental days. Two participants (one from each group) were dropped out for missing data. The study was approved by The North Denmark Region Committee on Health Research Ethics, project number N-20160023, and conducted in accordance with the Declaration of Helsinki.

### 2.2 Experimental Procedure

A task (WAME 1.0, [9]) was developed in a graphical user interface (MATLAB R2015b) based on standard models of computer work [10, 11]. The task was displayed on a computer screen, 19 in. LCD monitor (1280 $\times$ 1024 pixels), while the participants were sitting behind a desk to perform it using a computer mouse. The task has been described in our previous work [11]. Briefly, each cycle began by memorizing a random pattern of connected points displayed on a computer screen following by disappearance of the pattern and then replicating the pattern while only the points constructing the pattern were shown, Fig. 1. The replication was performed by clicking using a computer mouse on the points to redraw the connecting lines in a sequential manner. After each 20 cycles, i.e. one segment, participants rated their perceived fatigue on Karolinska Sleepiness Likert scale (KSS), [12] ranging from 1 up to 9 respectively corresponding to "Very alert" and "Very sleepy, fighting sleep", while the task execution was paused for 5 s. An extra point (distracting point) was also shown during the replication, which must not have been clicked. A new cycle with a new pattern appeared after the offset of each cycle with no pause in between. The patterns were generated randomly subject to some constraints. No loop or crossings were allowed for the lines connecting the points. The length of the lines and the angles made in the connection of two lines were limited to ensure central location of the connected pattern in the center of computer screen. The experiments were performed in two time slots (10–12 a.m. or 1–3 p.m.) to lower the possible effects of variation in circadian rhythms. The first two sections of each cycle took 2.34 s, and the third section of a cycle took 5.06 s. Five short episodes (5-min) of the task were performed by the participants before the prolonged (40-min) task to avoid learning effect.

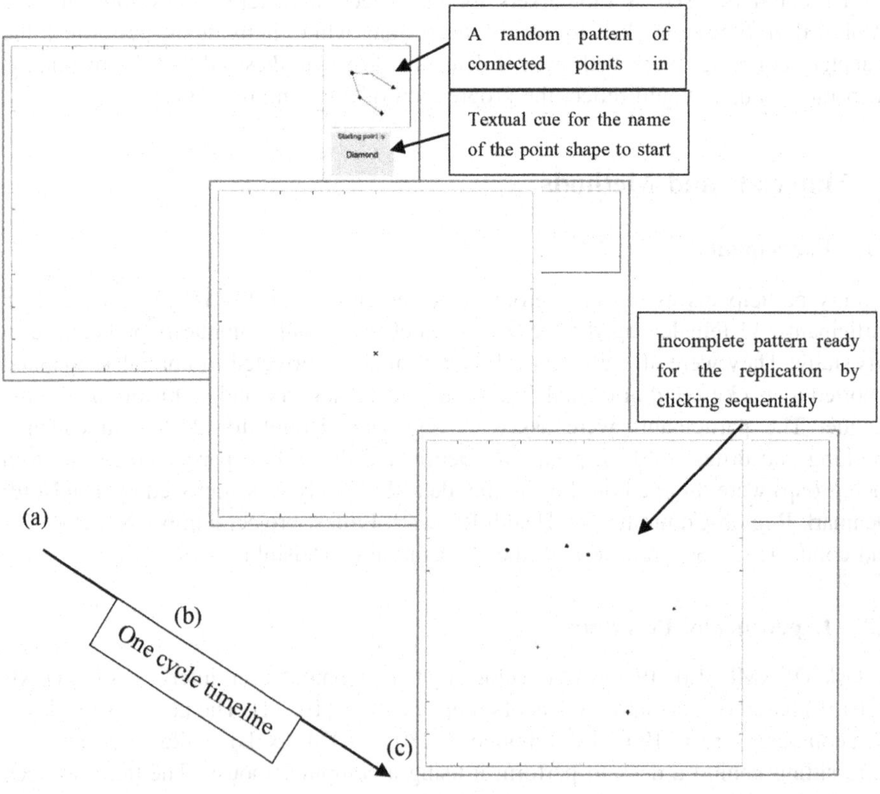


**Fig. 1.** A cycle timeline of the computer task. (a) Memorization of a pattern, (b) Disappearance of the pattern and fixation on a cross in the center, (c) Replication of the recently viewed pattern.

### 2.3 Data Recording

Heart rate was recorded during the task using A300 fitness tracker (Polar Electro Oy, Finland). This device provides instantaneous heart rate in one second intervals. The asserted accuracy of the heart rate measurement is about $\pm 1\%$ or 1 bpm in stable conditions which may also apply in our experimental setting with sedentary computer work. The validity and usability of this device have been approved in an independent study [13]. We extracted the mean and range of heart rate (respectively indicated by HRM and HRR) during each cycle and averaged them across cycles for each segment.

In addition, we measured the performance of doing the task based on the clicking accuracy and speed. As described previously [9], the accuracy was computed based on how complete the patterns were replicated considering all clicks. The speed referred to how fast the pattern replication was drawn by the participants. The performance measure was monotonically related to how well the participants performed the task.

### 2.4 Statistics

Repeated-measures analysis of variance used to examine the effects of change in TOT segments (1–12) on cardiac features (HRM and HRR), performance, and KSS. If the assumption of sphericity was not met, a Greenhouse-Geisser correction was applied. Bonferroni adjustment was used for pairwise comparison across the mental load levels. To account for a probable mediating effect of performance on the effect of TOT segments, we introduced averaged performance as the covariate to the statistical model following the recommendations for applying the analysis of covariance approach to a within-subjects design [14]. Level of significance was set at 0.05.

## 3 Results

As illustrated in Fig. 2(a), there was an increasing trend in HRR in both groups with the main effect of TOT, Fig. 2(a), $F(5.6, 189.3) = 4.8$, $p < .001$, $\eta_p^2 = .2$. No interaction between TOT segments and groups was found in HRR. However, there was a significant difference between groups in HRR, $F(1, 34) = 20.3$, $p < .001$, $\eta_p^2 = .4$. Pairwise comparisons revealed that HRR in the first TOT segment was significantly lower than the 7$^{th}$, and 9$^{th}$ to the 12$^{th}$ segment in the young group. There was a tendency to increase in HRM with increase in TOT segments in both young and elderly groups, $F(3.0, 102.7) = 2.6$, $p = .063$, $\eta_p^2 = .1$, Fig. 2(b). No significant difference between groups was found in HRM.

The performance fluctuated in the young group and increased in the elderly group with the increasing TOT, Fig. 2(c), $F(11, 374) = 4.0$, $p < .001$, $\eta_p^2 = .1$. An interaction between TOT segments and groups on the performance was observed, $F(11, 374) = 2.4$, $p = .007$, $\eta_p^2 = .1$. There was a significant difference in the performance between groups, $F(1, 34) = 32.8$, $p < .001$, $\eta_p^2 = .5$, with the higher performance in the young group compared with the elderly group. Pairwise comparisons revealed that the performance was significantly higher in the 11$^{th}$ segments compared

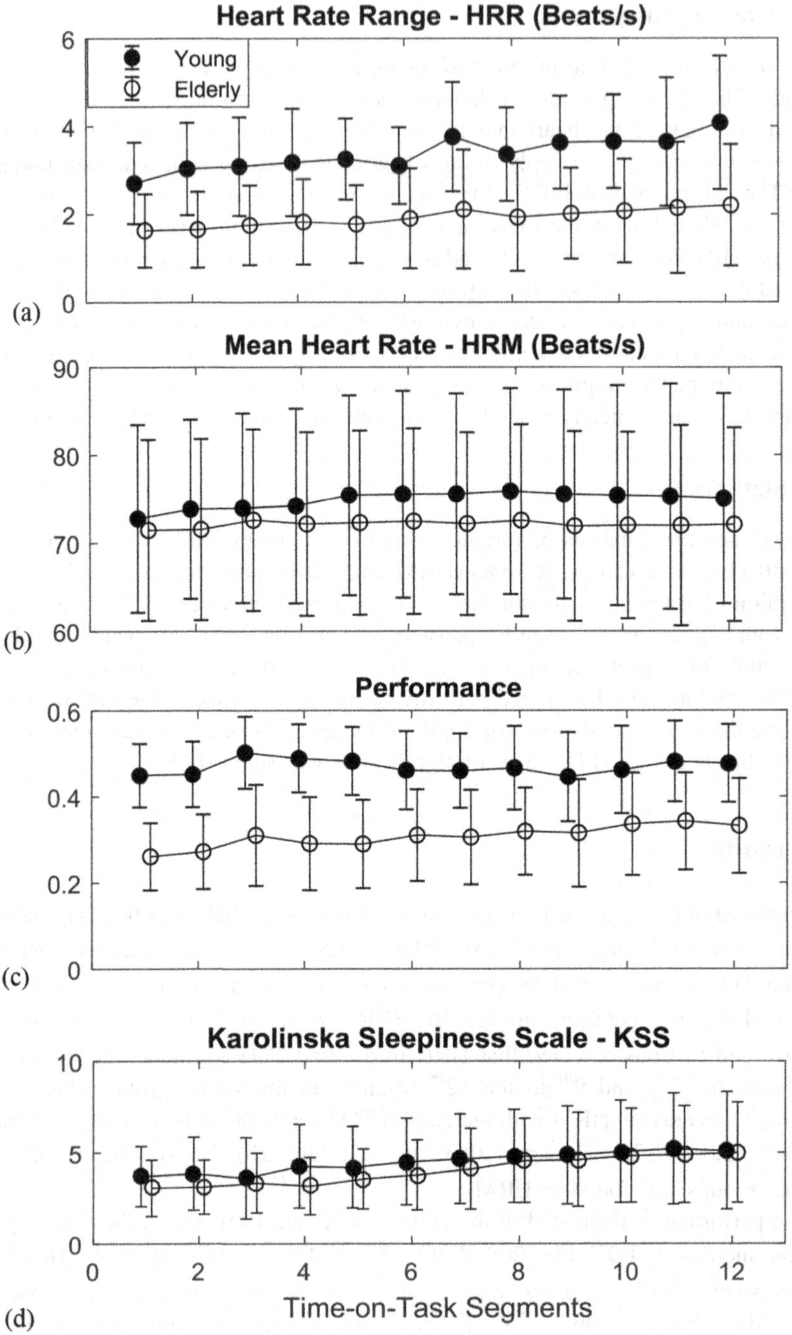


**Fig. 2.** The illustration of mean (circles) and SD (error bars) of (a) HRR, (b) HRM, (c) Performance, and (d) KSS in the young and elderly groups for each time-on-task (TOT) segment.

with the first TOT segment in elderly group. In the young group, it was the third segment with significantly higher performance compared with the second TOT segment. The effect of TOT segments on HRR remained significant, $F(5.5, 181.6) = 7.1$, $p < .001$, $\eta_p^2 = .2$, after introducing the averaged performance as a covariate to the statistical model.

As illustrated in Fig. 2 (d), KSS significantly increased with the increase in TOT segments in both groups, $F(1.8, 59.9) = 12.8$, $p < .001$, $\eta_p^2 = .3$. There was no interaction between TOT segments and groups in KSS. No significant difference between groups was observed in KSS. The KSS in first TOT segment was significantly lower than those of segments $5^{th}$–$12^{th}$.

## 4 Conclusion

The results showed that HRR was sensitive to fatigue development. The performance and perceived mental fatigue changed in concordance with increasing TOT. This, in sum, lends support to the possibility of using HRR as an index to detect fatigue development in sedentary computer work among young and elderly individuals.

HRM reflects both the sympathetic and parasympathetic activation levels which respectively contribute on increasing and slowing the heart beats [15]. Insensitivity of HRM to the development of fatigue supports that the heart rate by itself may not reflect fatigue progression, but its variability as measured by HRR contained the relevant information to capture these changes. This observation is in agreement with previous studies showing that the heart rate variability (HRV)-based metrics usually perform superior to the mean of heart rate [16, 17]. As such, the mean of heart rate corresponded to very light work [18].

The higher HRR observed in the young group compared with the elderly group is in accordance with previous evidence that the variability of heart rate decreases with age [19]. Further, the overall level of performance was higher for the young compared with the elderly computer workers. This may contribute to the reported higher HRR. After adjusting the statistical model for the averaged performance, the main effect of TOT on HRR remained significant suggesting that the effect was not mediated by the averaged performance.

The data (e.g., RR intervals) required to compute HRV components such as standard deviation of RR intervals a.k.a. SDNN may not be accessible in some fitness trackers, however, statistical features of variation in instantaneous heart rate such as HRR would partially provide this information. This cardiac feature together with relevant features from other unobtrusive modalities such as eye-tracking [9] is suggested to be incorporated into psychophysiological models of fatigue for healthy young and elderly individuals. The present study offers new perspectives for ergonomics on monitoring mental load at work. Monitoring fatigue level using cardiac features could be further studied to avoid occupational burnout [20] since cardiac features also reflect various emotional states and stress levels at work [21].

**Acknowledgments.** This project was funded by the Veluxfonden (project number: 00010912).

## References


1. Mukhopadhyay SC (2015) Wearable sensors for human activity monitoring: a review. IEEE Sens J 15(3):1321–1330
2. Phan D, Siong LY, Pathirana PN, Seneviratne A (2015) Smartwatch: performance evaluation for long-term heart rate monitoring. In: 2015 international symposium on bioelectronics and bioinformatics (ISBB). IEEE, Beijing, pp 144–147
3. Yung M, Wells RP (2016) Responsive upper limb and cognitive fatigue measures during light precision work: an 8-hour simulated micro-pipetting study. Ergonomics 60(7):940–956
4. Samani A, Holtermann A, Søgaard K, Madeleine P (2010) Active biofeedback changes the spatial distribution of upper trapezius muscle activity during computer work. Eur J Appl Physiol 110(2):415–423
5. Wright RA, Stewart CC, Barnett BR (2008) Mental fatigue influence on effort-related cardiovascular response: extension across the regulatory (inhibitory)/non-regulatory performance dimension. Int J Psychophysiol 69(2):127–133
6. Lal SKL, Craig A (2001) A critical review of the psychophysiology of driver fatigue. Biol Psychol 55(3):173–194
7. Patel M, Lal SKL, Kavanagh D, Rossiter P (2011) Applying neural network analysis on heart rate variability data to assess driver fatigue. Expert Syst Appl 38(6):7235–7242
8. Behar J, Ganesan A, Zhang J, Yaniv Y (2016) The autonomic nervous system regulates the heart rate through cAMP-PKA dependent and independent coupled-clock pacemaker cell mechanisms. Front Physiol 7:419
9. Marandi RZ, Samani A, Madeleine P (2017) The level of mental load during a functional task is reflected in oculometrics. In: IFMBE Proceedings EMBEC & NBC 2017. Springer, Singapore, pp 57–60
10. Samani A, Holtermann A, Søgaard K, Madeleine P (2009) Active pauses induce more variable electromyographic pattern of the trapezius muscle activity during computer work. J Electromyogr Kinesiol 19(6):e430–e437
11. Marandi RZ, Madeleine P, Omland Ø, Vuillerme N, Samani A (2018) Reliability of oculometrics during a mentally demanding task in young and old adults. IEEE Access 6:17500–17517
12. Åkerstedt T, Gillberg M (1990) Subjective and objective sleepiness in the active individual. Int J Neurosci 52(1–2):29–37
13. Vooijs M, Alpay LL, Snoeck-Stroband JB, Beerthuizen T, Siemonsma PC, Abbink JJ, Sont JK, Rövekamp TA (2014) Validity and usability of low-cost accelerometers for internet-based self-monitoring of physical activity in patients with chronic obstructive pulmonary disease. J Med Internet Res 16, e14 3(4)
14. Hoyle RH, Robinson JC (2004) Mediated and moderated effects in social psychological research: measurement, design, and analysis issues. In: The SAGE handbook of methods in social psychology, pp 213–234
15. Brownley KA, Hurwitz BE, Schneiderman N (2000) Cardiovascular psychophysiology. In: Cacioppo T, Tassinary LG, Berntson GG (eds), Handbook of p, Cambridge University Press, New York
16. Schneider F, Martin J, Hapfelmeier A, Jordan D, Schneider G, Schulz CM (2017) The validity of linear and non-linear heart rate metrics as workload indicators of emergency physicians. PLoS ONE 12(11):e0188635

17. McDuff DJ, Hernandez J, Gontarek S, Picard RW (2016) COGCAM: contact-free measurement of cognitive stress during computer tasks with a digital camera. In: Proceedings of the 2016 CHI conference on human factors computing system. ACM, San Jose, pp 4000–4004
18. Åstrand PO (2003) Textbook of work physiology: physiological bases of exercise, 4th edn. Human Kinetics, US
19. Acharya UR, Kannathal N, Sing OW, Ping LY, Chua T (2004) Heart rate analysis in normal subjects of various age groups. Biomed Eng Online 3(1):1–8
20. Leiter MP, Bakker AB, Maslach C (2014) Burnout at work: a psychological perspective. Psychology Press, London
21. Uusitalo A, Mets T, Martinmäki K, Mauno S, Kinnunen U, Rusko H (2011) Heart rate variability related to effort at work. Appl Ergon 42(6):830–838

# Non-pharmacological Interventions for People with Dementia: Design Recommendations from an Ergonomics Perspective

Gubing Wang(✉), Armagan Albayrak, Johan Molenbroek, and Tischa van der Cammen

Delft University of Technology, 2628 CE Delft, The Netherlands
g.wang-2@tudlft.nl

**Abstract.** Non-pharmacological interventions have been applied to manage Behavioural and Psychological Symptoms of Dementia (BPSD). However, these interventions have not been assessed from an ergonomics perspective. Ergonomics has investigated the age-related capability changes in terms of sensory, cognition and movement aspects. This study aims to review the existing non-pharmacological interventions for BPSD targeting nursing home residents and generate design recommendations based on the domain of ergonomics in ageing. The electronic databases MEDLINE, EMBASE, PsycINFO were searched for studies which applied non-pharmacological interventions for treating BPSD in nursing home residents. A total of 67 studies met the inclusion criteria; from which 16 types of interventions were identified. Within these intervention types, the main capabilities required from the interventions for People with Dementia (PwD) were identified. The interventions were then categorized into sensory-, cognition-, and movement-oriented according to the main capabilities. Design recommendations were then generated for the interventions with knowledge from the domain of ergonomics in ageing.

**Keywords:** Ageing population · Sensory capabilities · Cognition
Movement · Behavioural and Psychological Symptoms of Dementia
Intervention design

## 1 Introduction

Antipsychotic medication is frequently prescribed to relieve Behavioural and Psychological Symptoms of Dementia (BPSD). However, they have low efficacy with serious side effects such as increased mortality and stroke occurrence [1]. Consequently, non-pharmacological interventions have been developed with the attempt to reduce the use of antipsychotic medicine in the management of BPSD. Several systematic reviews have investigated the effectiveness of these interventions and found the level of evidence for all these interventions is generally low [2–7].

This study proposes that improving intervention designs is one approach to increase the level of evidence, and these intervention designs could be potentially improved by applying knowledge from the domain of ergonomics in ageing. This domain of ergonomics has investigated the capability changes of healthy individuals as they age


S. Bagnara et al. (Eds.): IEA 2018, AISC 826, pp. 112–122, 2019.
https://doi.org/10.1007/978-3-319-96065-4_15

[8], and the majority of People with Dementia (PwD) belong to the elderly age group. For instance, in the United States 96% of people with Alzheimer's Dementia are above the age of 65 [9]. This means that changes in capabilities because of ageing is also applicable for the majority of PwD.

The age-related capabilities investigated so far have been classified into sensory, cognition and movement aspects [10]. Some capabilities were found to be reduced (e.g. vision, auditory, working memory) [11–14] while some were reported to be relatively intact in the elderly (e.g. semantic memory) [15]. These findings have been implemented into the adaptation of working place, products and services for the elderly with reported improvement in both their performance and well-being [12]. The domain of ergonomics in ageing could be extended to designing non-pharmacological interventions for PwD. Specifically, the PwD should have the capabilities to participate for the results of the interventions to be reliable.

Previous reviews have not assessed these interventions based on the capabilities of PwD. Besides, these reviews have not distinguished between the interventions targeting PwD from those targeting caregivers [2–7]. As only the capabilities of the PwD are of interest, this study focuses on interventions targeting PwD. Therefore, our study aims to first review literatures on interventions for BPSD targeting nursing home residents and age-related capability changes reported in ergonomics, then generate design recommendations for these interventions by aligning with the capabilities to improve the intervention designs.

## 2 Method

This study was conducted in several steps. Firstly, studies on interventions for BPSD targeting nursing home residents from 1998 to 2018 were systematically reviewed. Only studies written in English and published in peer-reviewed journals were included. The literature search was implemented in three electronic databases: MEDLINE, PsycINFO and EMBASE. The intervention types were then identified from these studies. After that, the capabilities that PwD should have to be able to participate in each intervention type were identified. The interventions were then categorized into sensory-, cognition-, and movement-oriented according to the main capability they required from the PwD. The ergonomics literature on age-related capability changes was reviewed. The findings in age-related capability changes was applied to generate design recommendations for the interventions. The steps are summarized in Fig. 1.

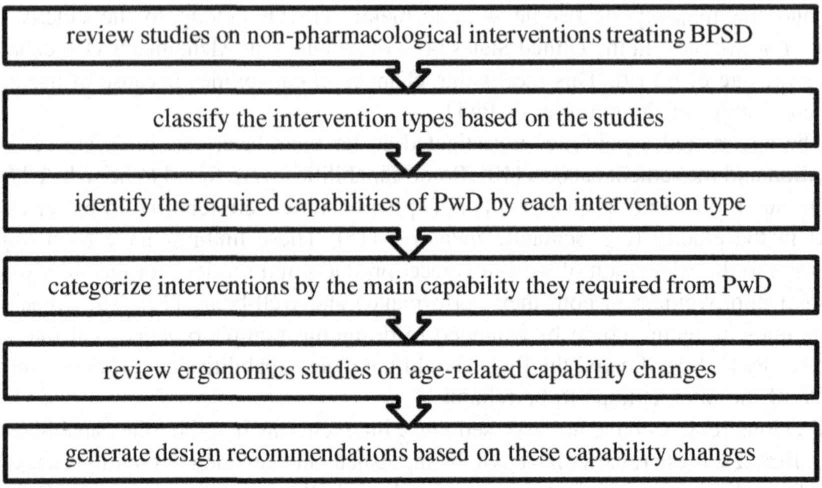


**Fig. 1.** A flow diagram of this study

## 3 Results

After the systematic review, 67 studies met the inclusion criteria. We found 16 intervention types in total. An overview of these studies is shown in Fig. 2. The number of studies conducted on one intervention type is indicated between brackets. The references to these studies are available upon request.

Sensory-oriented interventions have been investigated by more studies than cognition- and movement-oriented interventions. Specifically, 9 intervention types require the PwD to have certain level of sensory capability, 5 intervention types require the PwD to have certain level of cognitive capability and 2 intervention types require the PwD to have certain level of movement capability. The sensory-oriented interventions were investigated by 47 studies, while the cognition- and movement- oriented interventions were examined by 11 and 9 studies respectively. Within sensory-oriented interventions, music therapy has been evaluated by 21 studies, which is the most-investigated intervention. Several interventions have been evaluated by one study only.

A few studies have compared the effectiveness of two intervention types (e.g. music therapy vs simulated presence therapy [16]); however, no study has compared two designs of the same intervention type (e.g. music therapy with generic music or preferred music by PwD).

The age-related capability changes reported in the sensory aspect are auditory, olfactory, visual and somatosensory capabilities. The capability changes described in the cognition aspect are attention, memory, and executive control. The capability changes stated in the movement aspect are balance, muscular strength, movement speed, and locomotion. The changes of these capabilities with age will be described in the Discussion Section.

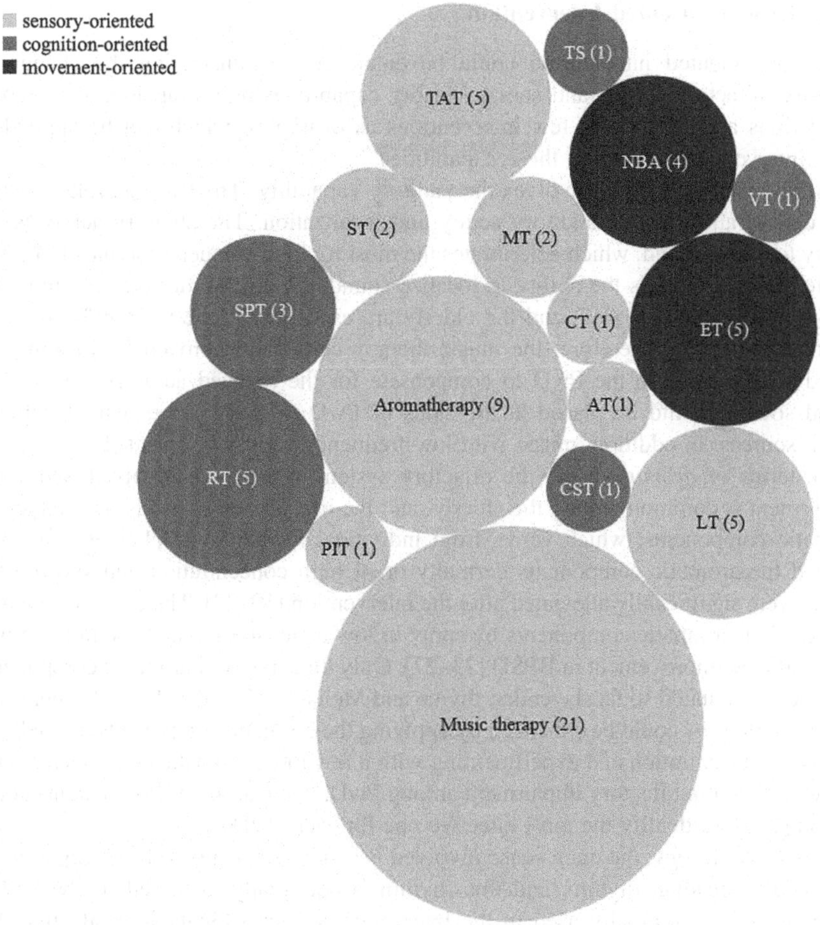


**Fig. 2.** Overview of non-pharmacological interventions for BPSD after categorization. AT = Animal Therapy; CST = Cognitive Stimulation Therapy; CT = Clowning Therapy; ET = Exercise Therapy; LT = Light Therapy; MT = Massage Therapy; NBA = Nature-Based Activities; PIT = Positive Image Therapy; RT = Reminiscence Therapy; SPT = Simulated Presence Therapy; ST = Snoezelen Therapy; TAT = Technology-Assisted Therapy; TS = Time Slips; VT = Validation Therapy.

## 4 Discussion

The outcome of this study is to generate design recommendations based on findings from ergonomics in ageing to potentially improve the interventions for BPSD targeting nursing home residents. In general, sensory-oriented interventions have larger variety and have been relatively more evaluated than cognition- and movement- oriented interventions. The results also revealed that the effect of intervention designs on effectiveness needs to be explored. The design recommendations for intervention designs are discussed below in terms of the intervention categories.

### 4.1 Sensory-Oriented Intervention

A sensory-oriented intervention could be enhanced by considering the change of auditory, olfactory, visual and somatosensory capabilities due to ageing. The recommendations are given for a few interventions as examples, which can be applied to other interventions involving these capabilities.

*Music therapy* mainly involves the auditory capability. The two age-related auditory deteriorations are in auditory acuity and localization. The auditory acuity is the ability to detect sound, which deteriorates the most for high-frequency sounds [14]. The auditory localization is the ability to localize sound, which also declines the most for high-frequency sounds [17], and the elderly are especially prone to front/back localization errors [18]. Therefore, the music therapy could be improved by locating the sound source close to the PwD to compensate for the reduced auditory acuity. The sound source should be placed at the sides of PwD to avoid issues with localizing sound sources. In addition, music with low frequency should be selected.

In terms of *aromatherapy*, the olfactory system is the most involved. Olfactory impairment is common among the elderly, and the impairment depends on the type of aromatic components, which varies from individual to individual [12]. Four studies applied the aromatic components dermally or of high concentration and showed the BPSD were significantly alleviated after the intervention [19–22]. The other five studies delivered the aromatic components by spray in low concentration and found there was no significant improvement in BPSD [23–27]. Only three types of aromatic components have been evaluated so far: lavender, thyme and Melissa officinalis [19–27]. Therefore, the *aromatherapy* could be enhanced by applying the aromatic components dermally or of a high concentration and experimenting with a few more aromatic components. Due to the variation in olfactory impairment among PwD, a few aromatic components could be compared to identify the most effective one for each PwD.

For *light therapy*, the main sense involved is vision. Exposure to bright light could restore the circadian rhythm, and this rhythm is commonly disturbed in PwD [28]. Artificial light is normally used in the interventions with a higher intensity than the standard light [29]. Visual deterioration of the elderly means that more illumination is required for them to see adequately [12]. Therefore, besides circadian rhythm restoration, another possible benefit of this therapy would be that it allows PwD to see their surroundings more clearly. The capability of elderly adults to adapt to darker conditions is reduced [11], and their susceptibility to glare is higher than that of young adults [30]. Thus, the *light therapy* could be improved by reducing the brightness in the room gradually at the end of the intervention to allow PwD time to adapt to the standard light condition. In addition, the emitted light should be diffused to create ambient light and reflecting surfaces (e.g. mirrors) should be avoided to reduce glare in the therapeutic room.

A *Snoezelen therapy*, also called multisensory therapy, gently stimulates the senses of sight, hearing, smell, and touch of the PwD [31]. In terms of sight, the PwD are exposed to lighting conditions with a wider wavelength spectrum than in *light therapy*. The variety of colors used in the therapy aims to engage the visual system of PwD [32]. It has been found that the elderly have more difficulty discriminating and perceiving short wavelength light [33]. Therefore, it is recommended that objects with short-wavelength lighting should be separated from backgrounds with similar lighting conditions even though their colours are different (e.g. blue and violet). The elderly have been found to have difficulty in sensing the weight of objects, which results in more variability in maintaining constant force when grasping an object [13]. This capability change could increase the number of dropping incidents which disturb the therapy. Thus, the objects which are easy-to-hold and drop-proof are advised to be used in the therapy. The recommendations of *music therapy* and *aromatherapy* could also be applied, as *Snoezelen Therapy* also employs the auditory and olfactory systems.

## 4.2 Cognition-Oriented Intervention

A cognition-oriented intervention could be improved by incorporating the changes in attention, memory, and executive control during ageing. The recommendations given in this section are applicable to all cognition-oriented interventions.

In terms of attention, the elderly have reduced capability to inhibit irrelevant information [34], thus, it is of importance to remove distractors during the therapy. A dedicated room is advised to help PwD to concentrate more during the therapy. Some prompts could be designed to redraw the attention of the PwD back to the intervention for the interventions not conducted by therapists. For instance, in *simulated presence therapy*, the PwD is mainly guided by the audio or video recording instead of the therapist. The PwD could get distracted during the intervention which might not be noticed immediately, thus affecting the outcome of the intervention. Even if a therapist notices and reminds the PwD, this reminder would disturb the flow of the therapy. Therefore, embedding some prompts in the audio or video recording would help PwD to stay focused on the therapy without disturbances.

There is a decline in the working memory [13], episodic memory [35] and prospective memory [36] in the elderly except the semantic memory [15]. The *reminiscence therapy* and *simulated presence therapy* have already utilized this capability change and structured conversations with PwD based on their semantic memories. For example, old photos, nostalgic music, and voices of family members were applied to stimulate the semantic memory in these therapies. Due to the decline in the working memory [13], the elderly might forget the beginning of a sentence by the time the sentence is finished for a long sentence. Similarly, as the ability of language comprehension depends on the interaction of working memory and semantic memory, it is also difficult for the elderly to comprehend a sentence in which many words and clauses bisect the subject and verb [13]. Therefore, the instructions and questions to the PwD should be simplified and shortened during the interventions. Due to the decline in the episodic memory [35], the PwD could forget the topic of their talk during the conversation, which could trigger negative emotions in them. Hence, it is recommended to constantly remind PwD about what they have been talking about without interrupting the conversation. A memory aid could be developed based on the technology of real-time transcription, which could remind the PwD about what has been said in the conversation in a written form. Due to the reduction in prospective memory [36], it is advisable to minimize conversations related to future planning during interventions.

Lastly, decline in executive control has been found in the elderly [37]. Executive control encompasses a range of cognitive activities such as planning, sequencing and multitasking. Thus, it is recommended that the activities in these interventions are designed to involve short-term feedbacks, less steps, and one cognitive task at a time.

### 4.3 Movement-Oriented Intervention

A movement-oriented intervention could be enhanced by considering balance, muscular strength, movement speed and locomotion in the elderly. Movement-oriented interventions are more likely to cause physical injuries than sensory- and cognition-oriented interventions. As a result, it is important to consider these capability changes to avoid injuries. The recommendations given in this section are applicable to all movement-oriented interventions.

In terms of balance, postural sway increases with age [38]. Falls due to postural sway is a serious problem for the elderly [39]. This age-related postural sway is greater when the person is standing on a more compliant flooring [40]. Consequently, the interventions are recommended to be carried out on a hard flooring to minimize the risk of fall. Protective clothing should be worn by the PwD in the meantime in case any fall incidents. The muscular strength drops with age [41]. It is essential to be aware of the lowered strength and endurance limits of PwD to design suitable and safe postures and movements for them. Water therapy, such as swimming pool exercises, could be incorporated for PwD whose muscular strength is inadequate for lifting their body weights. In addition, the elderly are slower in movement than young adults [42], hence, the therapists should demonstrate the postures and movements at a rhythm slow enough for PwD. Regarding locomotion, the elderly have shorter steps and increase in the time that both feet are on the ground [43]. This gait change is the strategy of the elderly to maintain balance. This strategy should be incorporated to the design of postures and movements in the interventions.

### 4.4 Selection of Intervention

Selecting the suitable intervention for each PwD is also an essential component of intervention design. According to ergonomics in ageing, the heterogeneity of various capabilities is larger in the elderly in comparison with that in young adults [10]. In terms of managing BPSD in a nursing home, it is unlikely that one intervention could be suitable for all residents; e.g. music therapy might not be suitable for PwD with auditory impairments. Therefore, it is advised to create a capability profile for each PwD, and the suitable intervention for a PwD could be identified by matching the capabilities required in the intervention with the remaining capabilities of the PwD. The changes in capabilities over time for each PwD could also be tracked with the profile so that other suitable interventions could be identified as the disease progresses.

## 5 Conclusion

This study provides an overview of important considerations for interventions treating BPSD from an ergonomics perspective. The interventions for BPSD from 1998 to 2018 were reviewed and categorized into sensory-, cognition- and movement-oriented. Most of capabilities in these three aspects have been found to decline with age apart from semantic memory. As the majority of PwD belong to the elderly age group, the interventions for BPSD could potentially be improved by aligning with the capabilities of the elderly in view of these age-related changes. A summary of design recommendations based on the capabilities involved in the interventions is shown in Table 1. This study establishes a start point of utilizing ergonomics findings in ageing to design non-pharmacological interventions for BPSD targeting nursing home residents. In the future, the design recommendations stated in this paper should be validated in clinical trials.

**Table 1.** Summary of recommendations for non-pharmacological interventions treating BPSD

| Intervention categories | Involved capabilities | Recommendations |
|---|---|---|
| Sensory-oriented | Auditory | Locate sound sources closer and by the sides of PwD; use low-frequency sounds |
| | Olfactory | Apply aromatic components dermally or of high concentration; experiment with various aromatic components and compare their effects |
| | Visual | Reduce brightness gradually at the end of intervention; diffuse emitted light; avoid reflecting surfaces; apply short-wavelength light carefully |
| | Somatosensory | Use objects to be easy-to-hold and drop-proof |
| Cognition-oriented | Attention | Conduct interventions in a dedicated room; embed prompts in the interventions |
| | Memory | Stimulate semantic memory; use short and simple sentences; use memory aid for PwD; avoid questions about future |
| | Executive control | Minimize activities involving planning, sequencing and multitasking |
| Movement-oriented | Balance | Carry out movement on hard flooring with protective clothing |
| | Muscular strength | Use movement with lower intensity |
| | Movement speed | Use movement with slower rhythm |
| | Locomotion | Use movement mimicking the locomotion of the elderly |

## References


1. Gill SS, Bronskill SE, Normand S-LT, Anderson GM, Sykora K, Lam K, Bell CM, Lee PE, Fischer HD, Herrmann N, Gurwitz JH, Rochon PA (2007) Antipsychotic Drug Use and Mortality in Older Adults with Dementia. Ann Intern Med 146:775. https://doi.org/10.7326/0003-4819-146-11-200706050-00006
2. Kales HC, Gitlin LN, Lyketsos CG (2015) Assessment and management of behavioral and psychological symptoms of dementia. BMJ 350:h369. https://doi.org/10.1136/BMJ.H369
3. Seitz DP, Brisbin S, Herrmann N, Rapoport MJ, Wilson K, Gill SS, Rines J, Le Clair K, Conn D (2012) Efficacy and feasibility of nonpharmacological interventions for neuropsychiatric symptoms of dementia in long term care: a systematic review. J Am Med Dir Assoc 13:503. https://doi.org/10.1016/j.jamda.2011.12.059
4. Cohen-Mansfield J (2001) Nonpharmacologic interventions for inappropriate behaviors in dementia: a review, summary, and critique. Am J Geriatr Psychiatry 9:361–381. https://doi.org/10.1097/00019442-200111000-00005
5. Whear R, Coon JT, Bethel A, Abbott R, Stein K, Garside R (2014) What is the impact of using outdoor spaces such as gardens on the physical and mental well-being of those with dementia? a systematic review of quantitative and qualitative evidence. J Am Med Dir Assoc 15:697–705
6. Agency for Healthcare Research and Quality (2016) Nonpharmacologic Interventions for Agitation and Aggression in Dementia. Nonpharmacologic Interv Agit Aggress Dement
7. Konno R, Kang HS, Makimoto K (2012) The best evidence for minimizing resistance-to-care during assisted personal care for older adults with dementia in nursing homes: a systematic review. JBI Database Syst Rev Implement Reports 10:4622–4632. https://doi.org/10.11124/jbisrir-2012-431
8. Boot WR, Nichols TA, Rogers WA, Fisk AD (2012) Design for Aging. In: Salvendy G (ed) Handbook of human factors and ergonomics. Wiley, Hoboken, pp 1442–1471
9. Alzheimer's association (2017) Alzheimer's Disease Facts and Figures
10. Freudenthal A (1999) The design of home appliances for young and old consumers. Delft University of Technology
11. Jackson GR, Owsley C, McGwin G (1999) Aging and dark adaptation. Vis Res 39:3975–3982. https://doi.org/10.1016/S0042-6989(99)00092-9
12. Farage MA, Miller KW, Ajayi F, Hutchins D (2012) Design principles to accommodate older adults. Glob J Health Sci 4:2–25. https://doi.org/10.5539/gjhs.v4n2p2
13. Voelcker-Rehage C, Stronge A, Alberts J (2006) Age-related differences in working memory and force control under dual-task conditions. Aging Neuropsychol Cogn 13:366–384. https://doi.org/10.1080/138255890969339
14. Pichora-Fuller MK, Singh G (2006) Effects of age on auditory and cognitive processing: implications for hearing aid fitting and audiologic rehabilitation. Trends Amplif 10:29–59. https://doi.org/10.1177/108471380601000103
15. Zacks RT, Hasher L, Li KZH (2000) Human memory. In: Salthouse TA, Craik FIM (eds) Handbook of aging and cognition. Lawrence Erlbaum, Mahwah, pp 293–357
16. Garland K, Beer E, Eppingstall B, O'Connor DW (2007) A comparison of two treatments of agitated behavior in nursing home residents with dementia: simulated family presence and preferred music. Am J Geriatr Psychiatry 15:514–521. https://doi.org/10.1097/01.JGP.0000249388.37080.b4
17. Lorenzi C, Gatehouse S, Lever C (1999) Sound localization in noise in hearing-impaired listeners. J Acoust Soc Am 105:3454. https://doi.org/10.1121/1.424672

18. Abel SM, Giguère C, Consoli A, Papsin BC (2000) The effect of aging on horizontal plane sound localization. J Acoust Soc Am 108:743–752
19. Lin PW, Chan W, Ng BF, Lam LC (2007) Efficacy of aromatherapy (Lavandula angustifolia) as an intervention for agitated behaviours in Chinese older persons with dementia: a cross-over randomized trial. Int J Geriatr Psychiatry 22:405–410. https://doi.org/10.1002/gps.1688
20. Akhondzadeh S, Noroozian M, Mohammadi M, Ohadinia S, Jamshidi AH, Khani M (2003) Melissa officinalis extract in the treatment of patients with mild to moderate Alzheimer's disease: a double blind, randomised, placebo controlled trial. J Neurol Neurosurg Psychiatry 74:863–866
21. Ballard CG, O'brien JT, Reichelt K, Perry EK (2002) Aromatherapy as a safe and effective treatment for the management of agitation in severe dementia: the results of a double blind, placebo controlled trial 63(7): 553–8
22. Smallwood J, Brown R, Coulter F, Irvine E, Copland C (2001) Aromatherapy and behaviour disturbances in dementia: a randomized controlled trial. Int J Geriatr Psychiatry 16:1010–1013
23. Yoshiyama K, Arita H, Suzuki J (2015) The effect of aroma hand massage therapy for people with dementia. J Altern Complement Med 21:759–765. https://doi.org/10.1089/acm.2015.0158
24. O'Connor DW, Eppingstall B, Taffe J, van der Ploeg ES (2013) A randomized, controlled cross-over trial of dermally-applied lavender (Lavandula angustifolia) oil as a treatment of agitated behaviour in dementia. BMC Complement Altern Med 13:315. https://doi.org/10.1186/1472-6882-13-315
25. Fu CY, Moyle W (2013) A randomised controlled trial of the use of aromatherapy and hand massage to reduce disruptive behaviour in people with dementia. BMC Complement Altern Med 13:165. https://doi.org/10.1186/1472-6882-13-165
26. Snow LA, Hovanec L, Brandt J (2004) A controlled trial of aromatherapy for agitation in nursing home patients with dementia. J Altern Complement Med 10:431–437. https://doi.org/10.1089/1075553041323696
27. Holmes C, Hopkins V, Hensford C, MacLaughlin V, Wilkinson D, Rosenvinge H (2002) Lavender oil as a treatment for agitated behaviour in severe dementia: a placebo controlled study. Int J Geriatr Psychiatry 17:305–308. https://doi.org/10.1002/gps.593
28. Ancoli-Israel S, Gehrman P, Martin JL, Shochat T, Marler M, Corey-Bloom J, Levi L (2003) Increased light exposure consolidates sleep and strengthens circadian rhythms in severe Alzheimer's disease patients. Behav Sleep Med 1:22–36. https://doi.org/10.1207/S15402010BSM0101_4
29. Lyketsos CG, Veiel LL, Baker A (1999) A randomized, controlled trial of bright light therapy for agitated behaviors in dementia patients residing in long-term care. Int J Geriatr Psychiatry 14: 520–525. http://dx.doi.org/10.1002/%28SICI%291099-1166%28199907%2914:7%3C520::AID-GPS983%3E3.0.CO;2-M
30. van den Berg TJTP, (René) van Rijn LJ, Kaper-Bongers R, Vonhoff DJ, Völker-Dieben HJ, Grabner G, Nischler C, Emesz M, Wilhelm H, Gamer D, Schuster A, Franssen L, Wit GC de, Coppens JE (2009) Disability glare in the aging eye. assessment and impact on driving. J Optom 2: 112–118. https://doi.org/10.3921/joptom.2009.112
31. Minner D, Hoffstetter P, Casey L (2004) Snoezelen activity: The good shepherd nursing home experience. J Nurs Care Qual 19:343–348
32. van Weert JCM, van Dulmen AM, Spreeuwenberg PMM, Bensing JM, Ribbe MW (2005) The effects of the implementation of snoezelen on the quality of working life in psychogeriatric care. Int psychogeriatrics 17:407–427

33. Said FS, Weale RA (1959) The variation with age of the spectral transmissivity of the living human crystalline lens. Gerontologia 3:213–231
34. Stoltzfus ER, Hasher L, Zacks RT, Ulivi MS, Goldstein D (1993) Investigations of inhibition and interference in younger and older adults. J Gerontol 48:P179–P188
35. Tulving E (2002) Episodic memory: from mind to brain. Annu Rev Psychol 53:1–25. https://doi.org/10.1146/annurev.psych.53.100901.135114
36. Park DC, Hertzog C, Kidder DP, Morrell RW, Mayhorn CB (1997) Effect of age on event-based and time-based prospective memory. Psychol Aging 12:314–327
37. Resnick SM, Pham DL, Kraut MA, Zonderman AB, Davatzikos C (2003) Longitudinal magnetic resonance imaging studies of older adults: a shrinking brain. J Neurosci 23:3295–3301
38. Kristinsdottir EK, Fransson PA, Magnusson M (2001) Changes in postural control in healthy elderly subjects are related to vibration sensation, vision and vestibular asymmetry. Acta Otolaryngol 121:700–706
39. Horak FB, Shupert CL, Mirka A (1989) Components of postural dyscontrol in the elderly: a review. Neurobiol Aging 10:727–738
40. Redfern MS, Moore PL, Yarsky CM (1997) The influence of flooring on standing balance among older persons. Hum Factors J Hum Factors Ergon Soc 39:445–455. https://doi.org/10.1518/001872097778827043
41. Ketcham CJ, Stelmach GE (2004) Movement control in the older adult. Technol Adapt Aging
42. Lajoie Y, Teasdale N, Bard C, Fleury M (1996) Attentional demands for walking: age-related changes. Adv Psychol 114:235–256. https://doi.org/10.1016/S0166-4115(96)80011-2
43. Lockhart TE (1997) Ability of elderly people to traverse slippery walking surfaces. In: Proceedings of the human factors and ergonomics society annual meeting. Human factors and ergonomics society, Inc

# How to Help Older Adults Learn Smartphone Applications? A Case Study of Instructional Design for Video Training

Fengli Liu and Jia Zhou(✉)

Department of Industrial Engineering, Chongqing University, Chongqing 400044, People's Republic of China
zhoujia07@gmail.com

**Abstract.** Video training is a useful way for older adults to learn to use smartphone applications, but the instructional design of adapting age-related changes is necessary to improve learning effectiveness. This study investigates the influence of visual cues and tapping methods on older adults' intention to use, ease of learning, satisfaction, and task completion time when learning how to use smartphone applications through instructional videos. Twenty-four older adults learned smartphone applications using two tapping methods (the tapping with/without validation method) on three types of instructional videos with different visual cues (red rectangle, cartoon finger, and real finger). The results indicated that use of a cartoon finger contributed to higher intention to use, higher ease of learning, higher satisfaction, and shorter task completion time compared with use of a red rectangle or a real finger. Moreover, older adults preferred the tapping with validation method rather than that without validation method. These findings will be a useful reference for designers of instructional videos and developers of smartphones.

**Keywords:** Older adults · Video training · Visual cues · Tapping methods Smartphone applications

## 1 Introduction

With the fast development of openly available online courses and video-sharing websites, video training has become widely used in daily life (e.g. Coursera, Udacity, and edX). It is a particularly promising option for older adults learning how to operate smart phone applications (apps), when no one is around to teach them directly. The animation of instructional videos is useful for learning tasks that involve a series of actions or processes, specifically ones that are difficult for users to imagine [1]. However, nowadays, most instructional videos either do not target older adults or do not fully take into account the characteristics of older people. When older adults learn smartphone applications using instructional videos, their learning effects are influenced by age-related changes in perception, cognition, and movement abilities [2]. There are two main problems. First, older adults usually find it difficult to follow instructions and simultaneously watch the changes on the touch screens of smart phones, particularly in instructional videos. Second, older adults are prone to unintended tapping errors and


S. Bagnara et al. (Eds.): IEA 2018, AISC 826, pp. 123–136, 2019.
https://doi.org/10.1007/978-3-319-96065-4_16

may not have enough time to obtain and understand the information before tapping on to the next step because tapping is a very short time-lapse gesture.

To deal with the first problem, adding visual cues in instructional videos is a practical way to grab the attention of older adults [1]. There are three common kinds of visual cues: prompt box, cartoon finger, and real finger. These are typical features used in on-boarding tutorials to provide a good first-time user experience and, thus, might compensate for an age-related decline in visual abilities, such as motion sensibility for example. However, in existing studies there is no clear answer as to which visual cue is more suitable for older adults in video training.

As for the second problem, the tapping methods used on smart phones influences the ability of older adults to learn smartphone applications (not limited to video training) and this applies to other situations. The proportion of older adults using smart phones is lower than that of other age groups because they face more difficulties when using touch screens [3]. In fact, touch interaction on a touchscreen is not so easy. The tapping method required by smartphones needs accurate tapping and well-developed motor abilities and there is a high possibility of unexpected tapping errors. Older adults frequently tend to miss their intended targets because of parallax effects and the large contact area of each finger [4]. In addition, this situation is especially serious for older 'novice' users. Inaccurate tapping makes it more likely that older novice users will carry out some unrelated or wrong operations, which is a source of confusion in the learning process. There are three suggested methods to address the gap between intended and actual tapping locations. First, a calibration mechanism is provided to prevent inaccurate tapping. Second, appropriate visual feedback is used to indicate where the users touch the screen, even when they have missed all of the interactive targets. Third, dissociating selection and validation phases of the tapping gesture to give users enough time to get and understand information about available functions [5].

This study aims to investigate the influence of different visual cues and tapping methods on older adults' performance and preference when they learn smartphone applications through using instructional videos. The results of this study indicate that better choices of visual cues and tapping methods can improve the effectiveness of video training for older adults, which may provide useful guidance for designers of instructional videos and developers of smartphones.

## 2 Literature Review

### 2.1 Visual Cues

Age-related changes in attention ability influences the design of instructional videos for older adults. An older adult's ability to maintain attention for a long period of time decreases with age [6]. Therefore, adding visual cues to attract attention [7] is needed to enhance visibility and discoverability on screen [8].

Visual cues can be used to direct a learner's attention to the thematically important information on visual displays [9]. It is well documented that visual cues are effective in directing a learner's attention [10–13] and learning is enhanced by reducing irrelevant visual search activities [14–20]. For example, Kühl, Scheiter, and Gerjets [21] conducted a study to investigate whether learning from dynamic and two presentation formats for static visualizations can be enhanced by visual cues. The results indicated visual cues would enhance learning with dynamic and static visualizations, so that learners in the cued conditions would outperform learners in the uncued conditions, particularly for pictorial recall and transfer tasks. Similarly, Yung and Paas [22] also investigated the effect of a pedagogical agent with visual cues on students' learning performance, cognitive load, and instructional efficiency. The results regarding learning and learning efficiency indicated that visual cues were strong enough to direct learners' attention.

Although most researchers indicated visual cues are effective in directing learners' attention and learning effects are improved by reducing irrelevant visual search activities, the results of other studies argued the effectiveness of visual cues had some restrictions. For example, the visual cue was not effective when combined with cued narrations [23] or it was only effective for several seconds before the effect disappeared [24]. Besides, most of the existing studies did not involve older adults and did not conduct a comparison between different forms of visual cues. Therefore, the effectiveness of visual cues in older adults' video training performance still remains questionable.

### 2.2 Tapping Methods

Age-related changes in visual and motor abilities influence the selection of tapping methods for older adults. An older adult's ability to deal with details and to accurately move decreases with age [25]. Tapping on a screen to activate targets was understood pretty soon but some users, especially the older adults, had a persistent problem in timing the gesture [26]. Thus, tapping interaction for older adults had a lower accuracy than a drag-and-drop interaction [4]. Furthermore, novice older adults were a little slower to accomplish tactile interaction compared with young adults [27] and experienced older adults [28]. Therefore, adding corrections to avoid and correct tapping errors and allowing more time to get and decide upon information about the taps that are available are necessary for novice older adults to improve tapping accuracy.

A correction mechanism might be a way to improve tapping accuracy [4]. However, practical correction is very difficult because of the handheld nature of smart phones. Users frequently tilt and rotate their devices, depending on their actual situations.

Feedback was used to indicate and prevent errors and help the interaction [29] and this was another way to improve tapping accuracy. There were three kinds of feedback. Firstly, visual feedback would help users to know whether the intended target had been touched [4]. A prominent feature of visual feedback is that it could explicitly indicate the current screen mode [30]. Secondly, auditory feedback may be a valuable non-visual cue for support gestures of interaction [31] and it proved to be useful in improving tapping performance on a touchscreen [30]. Thirdly, tactile feedback was considered distracting for novice older users and its effect was less than that of auditory feedback [32]. In addition, due to these age-related changes, multi-modal feedback might provide more benefits to older adults who are unfamiliar with touchscreen devices.

Apart from providing a correction mechanism or feedback, dissociating the selection and validation phases of the tapping gesture into two separate phases might also be an effective way of improving tapping accuracy. Chêne, Pillot, and Chaumon [5] conducted an experiment to investigate the effect of dissociating selection and validation into two separate phases for the tactile interaction of older adults. The results indicated that the separation of the selection from the validation phase was an appropriate approach. Furthermore, compared with the correction mechanism or feedback, dissociating selection and validation phases has the distinct advantage that the user has enough time to choose between the validation phase and other selection options, after initial selection has been achieved. Therefore, this study explored the effects of dissociating the selection and validation phases on the tapping method for older adults, when learning how to use smartphone applications using video training.

# 3 Methodology

## 3.1 Variables

The two independent variables are visual cues and tapping methods, which are within-subject variables. Visual cues have three levels: red rectangle, cartoon finger, and real finger (as shown in Fig. 1). Tapping methods have two levels: tapping without validation and tapping with validation. The tapping with validation method refers to the situation where the selection and validation phases are dissociated, i.e. separated one from each other in the tapping procedure. Therefore, tapping with validation makes the user choose between the validation phase and other information after first selection phase achievement.

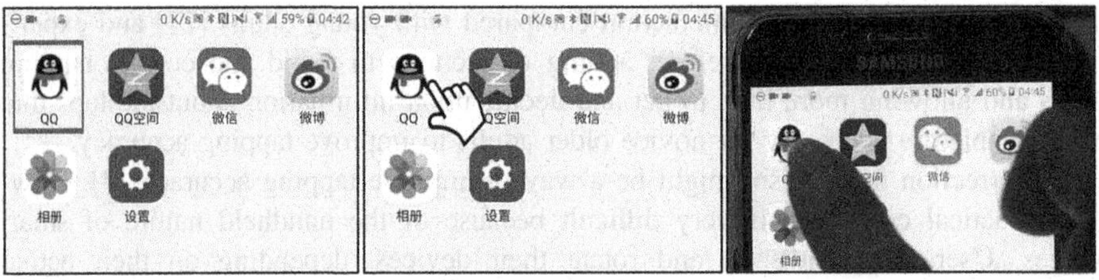


Red rectangle Cartoon finger Real finger

**Fig. 1.** Three types of visual cues (Color figure online)

There are four dependent variables: intention to use, ease of learning, satisfaction, and task completion time. Intention to use is measured using one item from the research of Davis [33]. Ease of learning is measured using two items from the research of Renaud and Van Biljon [34]. Satisfaction is measured using existing satisfaction questionnaires [35]. All items are presented in Appendix. These items were first translated from Chinese, according to the requirements of the study. Then, to ensure the consistency of Chinese and English expressions, a back translation was conducted.

Each item was rated using a seven-point Likert scale: from strongly disagree to strongly agree. Task completion time is measured from the moment when the participants clicked the icon of the prototypes to the moment when they finished all of the steps and this was recorded by screen recording software (DC Recorder).

### 3.2 Participants

A total of 24 older adults participated in the experiment (20 females and 4 males). They were recruited from the Zhandong Road community, in the Shapingba district in Chongqing, China. The age of the participants ranged from 60 to 79 years old (Mean = 66.29, SD = 5.034). A total of 83.3% of older adults has a middle school or high school diploma. All of the participants are smartphone users and 79.2% of them use a smartphone no more than four hours per day. When participants had some questions in using smartphone applications, asking for help from other people around them was a common way for them. Their children often were the primary helpers, but 62.5% of participants did not live with their children. In addition, the participants who lived with their children sometimes also could not timely get their children' help because of various reasons. Therefore, participants are highly interested in the video training that teaches them to use smartphone applications.

### 3.3 Task

There were 12 tasks about smartphone applications and these tasks were designed by considering the usage needs of older adults and their lifestyles. Half of the tasks required five operational steps and the other half required four operational steps (as shown in Table 1). Moreover, visual cues and tapping methods were both within-subject variables, so a 3 $\times$ 2 factorial design was used in this experiment. For each visual cue, participants were asked to learn smartphone apps by using two tapping methods. Therefore, participants needed to learn smartphone apps in six situations in total. The order of the six situations was counterbalanced to decrease the learning effects. In each situation, participants were asked to complete two random tasks that had different operational steps. Each task was only used in one out of the six situations. In general, each participant needed to complete 12 different learning tasks using instructional videos.

### 3.4 Prototype and Video Material

Three prototypes were developed using Axure RP 8.0, including two prototypes with different tapping methods and one training prototype specifically designed for tapping with validation. For each tapping method, three types of instructional video (with three different visual cues) were made. Thus, 72 instructional videos were used in this experiment, in total. To make participants learn smartphone apps effectively, the instructional videos used a step-by-step teaching method. However, according to two pilot tests, the step-by-step instructional videos might be not enough for participants to have an overall grasp of the order of the operational steps because there is a pause

**Table 1.** Experimental tasks about smartphone applications

| | Learning tasks |
|---|---|
| Tasks with five operational steps | Add Lao Wang as a friend in QQ |
| | Delete friend Xiao Wang in QQ |
| | Clear chat history of the family in WeChat |
| | Quit the elderly health care community WeChat group |
| | Delete a photo |
| | Recover a deleted photo |
| Tasks with four operational steps | Collect the Sina Microblog of people's network |
| | Cancel attention to the Sina Microblog of Xiao Ming |
| | Hide the dynamic of Xiao Zhang in Qzone |
| | Cancel attention to the Qzone of Xiao Li |
| | Turn off data roaming |
| | Turn off personal hotspot |

between the steps. Therefore, for each step-by-step instructional video, a corresponding video without the pause was made to show the full operations.

### 3.5 Equipment and Procedures

A smartphone (Samsung Galaxy C5, 5.2 in., 1920 × 1080 pixel resolution, Android 6.0.1), a laptop (MacBook Air, 13 in., 1440 × 900 pixel resolution, macOS High Sierra 10.13.2), and a camera (Sony FDR-AX40) were used in this experiment. The two prototypes were running on the smartphone. Instructional videos of smartphone apps were filmed using the camera. The laptop was then used to play these instructional videos.

Participants signed consent forms and filled in questionnaires with their demographic information and their experience with mobile phones and video learning. The experimenter briefly introduced the experimental system being used and then the formal experiment was conducted.

The formal experimental process was carried out as follows: first of all, the participants randomly selected two learning tasks from the twelve learning tasks available. These two learning tasks were no longer used in subsequent phases of the experiment and comprised one four-operation-step task and one five-operation-step task. Secondly, participants needed to learn the two learning tasks through instructional videos. In this process, participants for each learning task had the choice of watching the video twice. In the first instance, participants watched the step-by-step operation video and then they watched the full operation video after that. Next, the participants completed questionnaires about their intention to use, ease of learning and satisfaction and then completed the two experimental tasks that had just been learned. After that, participants were asked to repeat the above steps to complete the other groups of experiments. The experimental scenario is shown in Fig. 2. At the end of the experiment, the participants were interviewed about their preferences and other subjective views.

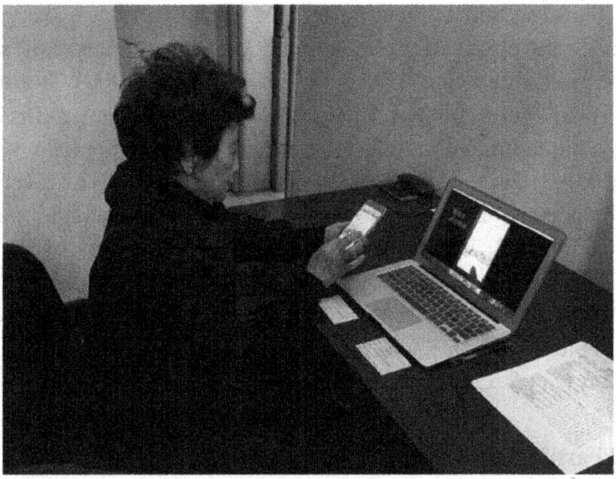

**Fig. 2.** Participants learn to use smartphone applications using instructional videos

# 4 Results

## 4.1 Intention to Use

The influence of visual cues and tapping methods on intention to use was analyzed using a two-way repeated analysis of variance (ANOVA), as shown in Fig. 3. Interaction effects and main effects on intention to use were tested. The results of ANOVA indicated that the interaction effect between visual cues and tapping methods was not significant on intention to use ($F_{(2, 46)} = 1.255$, $p = 0.295$).

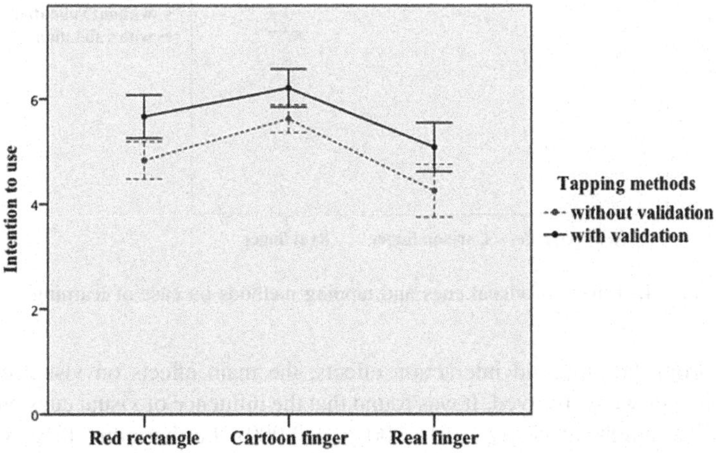


**Fig. 3.** Effects of visual cues and tapping methods on intention to use

Apart from the aforesaid interaction effects, the main effects on visual cues and tapping methods were observed. It was found that the influence of visual cues on intention to use was significant ($F_{(2, 46)} = 11.761$, $p = 0.000$). Furthermore, three levels of visual cues were compared using the LSD test. The results indicated that participants had a higher intention to use for cartoon finger compared with red rectangle ($t = 3.352$, $p = 0.003$) and real finger ($t = 4.664$, $p = 0.000$). However, there was no significant difference on intention to use between red rectangle and real finger ($t = 1.963$, $p = 0.062$). As for tapping methods, the difference of the tapping with validation and the tapping without validation was significant on intention to use ($F_{(1, 23)} = 35.599$, $p = 0.000$). The results of the LSD test indicated that participants for the tapping with validation had a higher level of intention to use than the tapping without validation ($t = 5.952$, $p = 0.000$).

### 4.2 Ease of Learning

The influence of visual cues and tapping methods on ease of learning was analyzed using a two-way repeated ANOVA, as shown in Fig. 4. Interaction effects and main effects on ease of learning were tested. The results of ANOVA indicated that the interaction effect between visual cues and tapping methods was not significant on ease of learning ($F_{(2, 46)} = 0.365$, $p = 0.696$).

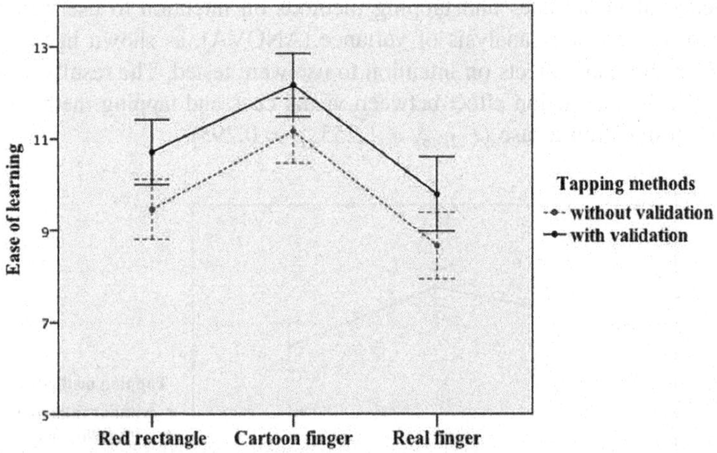


**Fig. 4.** Effects of visual cues and tapping methods on ease of learning

Apart from the aforesaid interaction effects, the main effects on visual cues and tapping methods were observed. It was found that the influence of visual cues on ease of learning was significant ($F_{(2, 46)} = 16.200, p = 0.000$). Furthermore, three levels of visual cues were compared using the LSD test. The results indicated that participants had a higher ease of learning for cartoon finger compared with red rectangle ($t = 4.642$, $p = 0.000$) and real finger ($t = 5.254$, $p = 0.000$). However, there was no significant difference on ease of learning between red rectangle and real finger ($t = 1.761$, $p = 0.091$). As for tapping methods, the difference in the effects of tapping with validation and tapping without validation on ease of learning was significant ($F_{(1, 23)} = 22.326$, $p = 0.000$). The results of the LSD test indicated that participants perceived that the tapping with validation method made it easier to learn mobile apps when compared to the tapping without validation method ($t = 4.727$, $p = 0.000$).

### 4.3 Satisfaction

The influence of visual cues and tapping methods on satisfaction was analyzed using a two-way repeated ANOVA, as shown in Fig. 5. Interaction effects and main effects on satisfaction were tested. The results of ANOVA indicated that the interaction effect between visual cues and tapping methods on satisfaction was not significant ($F_{(2, 46)} = 2.335$, $p = 0.108$).

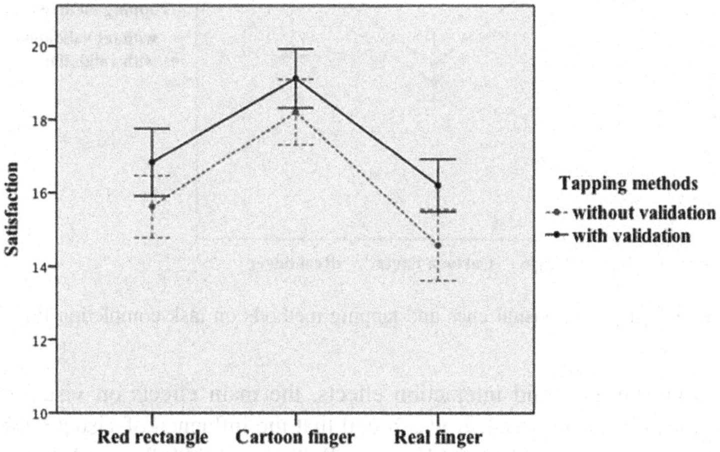


**Fig. 5.** Effects of visual cues and tapping methods on satisfaction

Apart from the aforesaid interaction effects, the main effects on visual cues and tapping methods were observed. It was found that the influence of visual cues on satisfaction was significant ($F_{(2, 46)} = 18.635$, $p = 0.000$). Furthermore, three levels of visual cues were compared using the LSD test. The results indicated that participants had a higher satisfaction for cartoon finger compared with red rectangle ($t = 4.752$, $p = 0.000$) and real finger ($t = 5.915$, $p = 0.000$). However, there was no significant difference on satisfaction between red rectangle and real finger ($t = 1.386$, $p = 0.179$). As for tapping methods, the difference of the tapping with validation and the tapping without validation method on satisfaction was significant ($F_{(1, 23)} = 27.094$, $p = 0.000$). The results of the LSD test indicated that participants for the tapping with validation method had a higher satisfaction than the tapping without validation method ($t = 5.208$, $p = 0.000$).

## 4.4 Task Completion Time

The influence of visual cues and tapping methods on task completion time was analyzed using a two-way repeated ANOVA, as shown in Fig. 6. Interaction effects and main effects on task completion time were tested. Mauchly's test of sphericity indicated that sphericity was violated ($df = 2$, $p = 0.020$) in analyzing interaction effects, so the Huynh-Feldt correction was made. The results of ANOVA indicated that the interaction effect between visual cues and tapping methods on task completion time was not significant ($F_{(1.630, 37.489)} = 0.232$, $p = 0.749$).

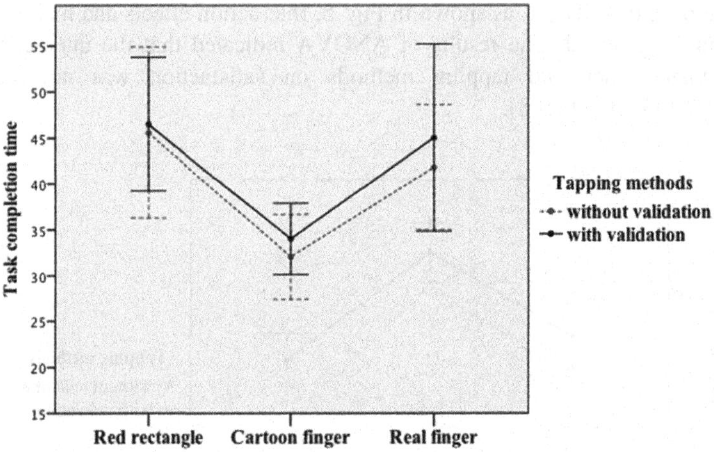


**Fig. 6.** Effects of visual cues and tapping methods on task completion time

Apart from the aforesaid interaction effects, the main effects on visual cues and tapping methods were observed. It was found that the influence of visual cues on task completion time was significant ($F_{(2, 46)} = 9.905$, $p = 0.000$). Furthermore, three levels of visual cues were compared using the LSD test. The results indicated that participants spent less task completion time using cartoon finger compared with red rectangle ($t = 868$, $p = 0.000$) and real finger ($t = 3.784$, $p = 0.001$). However, there was no significant difference on task completion time between red rectangle and real finger ($t = 0.701$, $p = 0.490$). As for tapping methods, the difference of the tapping with validation and the tapping without validation method on task completion time was significant ($F_{(1, 23)} = 1.854$, $p = 0.186$).

# 5 Discussion

Visual cues are very useful for older adults to improve the effects of video learning but it is unclear which visual cue is more suitable for older adults. The results of this experiment indicated that use of a cartoon finger contributed to a higher intention to use, greater ease of learning and satisfaction and less task completion time when compared with using a red rectangle or real finger. One possible reason is that a cartoon finger has better directivity than a red rectangle; it blocks less of the video picture and gives a clearer picture than a real finger. In addition, the results of the interviews indicated that 75% of participants ranked the cartoon finger as the most preferred visual cue among the three visual cues available because they thought cartoon finger was more obvious and easier to understand than the other visual cues. 16.67% of participants ranked red rectangle as the most preferred visual cue among the three visual cues because they thought red was more obvious. However, more participants thought that the red rectangle was more likely to be seen as a background color rather than a visual cue. Only 8.33% of participants ranked the real finger as the most preferred visual cue among the three visual cues because most participants thought that the real finger was too big and blocked too much of the screen and did not give a clear direction. In general, the results of the interview agreed with the results of the experiment.

As for tapping methods, the tapping with validation method contributed to a higher intention to use, ease of learning, and satisfaction than that without validation method. The results agreed with the findings of Chêne, Pillot, and Chaumon [5] that showed that the separation of the selection and validation phase was considered to be most appropriate for older adults. Moreover, the tapping with validation method gives users enough time to choose between the validation phase and other selection options, so it makes older adults more daring to operate in the process of learning to use smartphone applications.

However, the difference between the tapping with validation method and the tapping without validation method on task completion time was not significant. One possible reason for this is that the tapping with validation method has one more validation phase than the tapping without validation method, which may greatly reduce the advantage of tapping with validation over tapping without validation. The number of operations in the tapping with validation method was twice as much as tapping without validation, but tapping with validation only increased task completion time by 5.0% when compared to tapping without validation. Therefore, the tapping with validation method contributed to a slightly greater task completion time than the tapping without validation method but the results of the interview indicated that most participants (91.67%) were still willing to use the tapping with validation method when learning smartphone applications. They thought that the tapping with validation method gave them a deeper impression and reduced the probability of their making a mistake in the learning process. In the tapping without validation method, if participants made a mistake, they could choose to tap the back button, but most of them thought that tapping the back button was more difficult to remember and understand than the validation process. Besides, participants reported that ease of learning was more important for them than task completion time when learning smartphone applications. Of course, the results of the interview also indicated that once those participants had learned to use these smartphone apps, all of them preferred to use the tapping without validation method because they thought that was faster and more convenient.

## 6 Conclusion

This study investigated the effects of visual cues and tapping methods on video learning for older adult, in terms of intention to use, ease of learning, satisfaction and task completion time. Participants used instructional videos that had three different types of visual cues and two different types of tapping methods to learn smartphone apps. There were four main findings:

First, participants had higher intention to use and satisfaction for the instructional video with a cartoon finger compared with those with a red rectangle or a real finger.

Second, participants thought it was easier to learn smartphone apps using the instructional video with a cartoon finger compared with those with a red rectangle or a real finger.

Third, participants had higher satisfaction for the instructional video with a cartoon finger compared with those with a red rectangle or a real finger.

Fourth, the tapping with validation method performed better for the participants than that without validation method, in terms of intention to use, ease of learning and satisfaction.

Two limitations of this study should be noted. First, out of twenty-four participants, only three participants were male, which may distort the results of the experiment. Future studies could provide a better balance in the ratio of male to female subjects. Second, this study recruited participants with similar backgrounds, which may also bias the results. Future studies should extend this experiment to the wider population.

**Acknowledgments.** This work was supported by funding from the National Natural Science Foundation of China (Grants nos. 71661167006) and Chongqing Municipal Natural Science Foundation (cstc2016jcyjA0406).

## Appendix

Intension to use (Davis, 1986):

1. I intend to use this learning way in the future.
   Ease of learning (Renaud & Van Biljon, 2008):
   In this way of learning,
2. It is easy to learn smart phone applications.
3. I want to learn how to use smart phone applications.
   Satisfaction (Lewis, 1995):
4. Overall, I am satisfied with the ease of learning smartphone applications through this way of learning.
5. Overall, I am satisfied with the amount of time it took to learn smart phone applications through this way of learning.
6. Overall, I am satisfied with the support information available when learning smart phone applications.

## References


1. Plaisant C, Shneiderman B (2005) Show me! guidelines for producing recorded demonstrations. In: 2005 IEEE symposium on visual languages and human-centric computing, pp 171–178. IEEE
2. Boot WR, Nichols TA, Rogers WA, Fisk AD (2012) Design for aging. In: Handbook of human factors and ergonomics, Fourth Edn, pp 1442–1471
3. Hwangbo H, Yoon SH, Jin BS, Han YS, Ji YG (2013) A study of pointing performance of elderly users on smartphones. Int J Hum Comput Interact 29:604–618
4. Kobayashi M, Hiyama A, Miura T, Asakawa C, Hirose M, Ifukube T (2011) Elderly user evaluation of mobile touchscreen interactions. In: IFIP conference on human-computer interaction, pp 83–99. Springer, Heidelberg
5. Chêne D, Pillot V, Chaumon MÉB (2016) Tactile interaction for novice user. In: International conference on human aspects of IT for the aged population, pp 412–423
6. Commodari E, Guarnera M (2008) Attention and aging. Aging Clin Exp Res 20:578–584
7. Mayer RE, Moreno R (2003) Nine ways to reduce cognitive load in multimedia learning. Educ Psychol 38:43–52
8. Norman DA, Nielsen J (2010) Gestural interfaces: a step backward in usability. Interactions 17: 46–49
9. Lin L, Atkinson RK, Savenye WC, Nelson BC (2016) Effects of visual cues and self-explanation prompts: empirical evidence in a multimedia environment. Interact Learn Environ 24:799–813
10. de Koning BB, Tabbers HK, Rikers RM, Paas F (2009) Towards a framework for attention cueing in instructional animations: guidelines for research and design. Educ Psychol Rev 21:113–140
11. de Koning BB, Tabbers HK, Rikers RM, Paas F (2010) Attention guidance in learning from a complex animation: seeing is understanding? Learn Instr 20:111–122
12. Ozcelik E, Arslan-Ari I, Cagiltay K (2010) Why does signaling enhance multimedia learning? evidence from eye movements. Comput Hum Behav 26:110–117
13. Ozcelik E, Karakus T, Kursun E, Cagiltay K (2009) An eye-tracking study of how color coding affects multimedia learning. Comput Educ 53:445–453
14. Amadieu F, Mariné C, Laimay C (2011) The attention-guiding effect and cognitive load in the comprehension of animations. Comput Hum Behav 27:36–40
15. Boucheix J-M, Guignard H (2005) What animated illustrations conditions can improve technical document comprehension in young students? format, signaling and control of the presentation. Eur J Psychol Educ 20:369–388
16. de Koning BB, Tabbers HK, Rikers RM, Paas F (2007) Attention cueing as a means to enhance learning from an animation. Appl Cogn Psychol 21:731–746
17. de Koning BB, Tabbers HK, Rikers RM, Paas F (2010) Learning by generating vs. receiving instructional explanations: two approaches to enhance attention cueing in animations. Comput Educ 55:681–691
18. Jamet E, Gavota M, Quaireau C (2008) Attention guiding in multimedia learning. Learn Instr 18:135–145
19. Lin L, Atkinson RK (2011) Using animations and visual cueing to support learning of scientific concepts and processes. Comput Educ 56:650–658
20. Wouters P, Paas F, van Merriënboer JJ (2009) Observational learning from animated models: effects of modality and reflection on transfer. Contemp Educ Psychol 34:1–8
21. Kühl T, Scheiter K, Gerjets P (2012) Enhancing learning from dynamic and static visualizations by means of cueing

22. Yung HI, Paas F (2015) Effects of cueing by a pedagogical agent in an instructional animation: a cognitive load approach. Educ Technol Soc 18:153–160
23. Mautone PD, Mayer RE (2001) Signaling as a cognitive guide in multimedia learning. J Educ Psychol 93:377
24. Lowe R, Boucheix J-M (2011) Cueing complex animations: does direction of attention foster learning processes? Learn Instr 21:650–663
25. Fisk AD, Czaja SJ, Rogers WA, Charness N, Sharit J (2009) Designing for older adults: principles and creative human factors approaches. CRC press, Boca Raton
26. Leonardi C, Albertini A, Pianesi F, Zancanaro M (2010) An exploratory study of a touch-based gestural interface for elderly. In: Proceedings of the 6th nordic conference on human-computer interaction: extending boundaries, pp 845–850. ACM
27. Stößel C, Wandke H, Blessing L (2009) An evaluation of finger-gesture interaction on mobile devices for elderly users. Prospektive Gestaltung von Mensch-Technik-Interaktion 8:470–475
28. Motti LG, Vigouroux N, Gorce P (2015) Improving accessibility of tactile interaction for older users: lowering accuracy requirements to support drag-and-drop interaction. Procedia Comput Sci 67:366–375
29. Gorce P, Nadine V, Motti L (2017) Interaction techniques for older adults using touchscreen devices: a literature review from 2000 to 2013. J d'Interaction Personne-Système 3
30. Harada S, Sato D, Takagi H, Asakawa C (2013) Characteristics of elderly user behavior on mobile multi-touch devices. In: IFIP conference on human-computer interaction, pp 323–341. Springer, Heidelberg
31. Leonard VK, Jacko JA, Pizzimenti JJ (2005) An exploratory investigation of handheld computer interaction for older adults with visual impairments. In: Proceedings of the 7th international ACM SIGACCESS conference on computers and accessibility, pp 12–19. ACM
32. Lee J-H, Poliakoff E, Spence C (2009) The effect of multimodal feedback presented via a touch screen on the performance of older adults. In: International conference on haptic and audio interaction design, pp 128–135. Springer, Heidelberg
33. Davis FD (1989) Perceived usefulness, perceived ease of use, and user acceptance of information technology. MIS Q 13:319–340
34. Renaud K, Van Biljon J (2008) Predicting technology acceptance and adoption by the elderly: a qualitative study. In: Proceedings of the 2008 annual research conference of the South African institute of computer scientists and information technologists on IT research in developing countries: riding the wave of technology, pp 210–219. ACM
35. Lewis JR (1995) IBM computer usability satisfaction questionnaires: psychometric evaluation and instructions for use. Int J Hum Comput Interact 7:57–78

# Design as a Provocation to Support Discussion About Euthanasia: The Plug

Marije De Haas(✉), Gyuchan Thomas Jun, and Sue Hignett

Loughborough University, Loughborough, UK
m.de-haas@lboro.ac.uk

**Abstract.** Dementia affects 47 million people worldwide [1]. It is a collection or consequence of many illnesses with symptoms including deterioration in memory, thinking and behaviour; it is a terminal disease. The fear of dementia may lead people to signing an Advance Euthanasia Directive (AED). AEDs are rarely adhered to because the dementia symptoms conflict with the due care criteria; a person requesting euthanasia must be able to confirm the request at time of death and must be undergoing hopeless suffering. Once dementia has progressed, the euthanasia 'wish' can no longer be confirmed, and assessing suffering in a person with dementia is hard. This creates difficulties for physicians supporting patient wishes. Speculative Design is described as a way to prototype other realities [2]. This paper describes a Speculative Design to explore patient autonomy for end-of-life decisions in dementia. A short video was developed to imagine the AED as an implant that would trigger a swift and painless death, once the conditions described in the AED were reached. Data were collected at the DementiaLab conference in Dortmund, Germany, September 2017. The workshop was attended by 15 participants of varying ages and backgrounds. The results found that the Speculative Design had potential to aid discussion between stakeholders, without each party needing to be a specialist. It sparked debate, but with a caveat about the importance of boundaries for awareness of the wider context and sensitivity to inherent bias.

**Keywords:** Dementia · Euthanasia · Speculative Design · Autonomy

## 1 Introduction

Decision making for a good death in dementia is complex. This paper describes the use of Speculative Design for complex decision-making to explore why and how a person making an autonomous decision to opt for euthanasia can proceed (or not) with this decision when they have been diagnosed with dementia.

Dementia affects 47 million people worldwide with 9.9 million new cases each year [1]. Dementia is a collection or consequence of many illnesses, including Parkinson's disease, vascular dementia and Alzheimer's disease. There is a set of similar symptoms in which there is deterioration in memory, thinking and behaviour; it is a terminal disease.

Euthanasia has many definitions from the Greek origins of 'good death' or 'easy death' [3] to the Nazi euphemism for the deliberate killings of physically, mentally, and


S. Bagnara et al. (Eds.): IEA 2018, AISC 826, pp. 137–152, 2019.
https://doi.org/10.1007/978-3-319-96065-4_17

emotionally handicapped people, leaving the term with extremely negative connotations [4]. The definition used in this paper is "*The act of assisting someone who is terminally ill and whose suffering is unbearable and untreatable, to be in control of the manner of their dying*" [2].

As euthanasia is illegal in most of the world, this paper will use the Dutch guidelines and legal framework [5], which states "*euthanasia is not punishable if the attending physician acts in accordance with the statutory due care criteria. These criteria hold that: there should be a voluntary and well-considered request, the patient's suffering should be unbearable and hopeless, the patient should be informed about their situation, there are no reasonable alternatives, an independent physician should be consulted, and the method should be medically and technically appropriate*" [6]. Euthanasia for people living with dementia is a complex issue because the symptoms clash with the due care criteria for euthanasia and unbearable suffering is difficult to assess in dementia [7–9]. It is hard for a person living with dementia to consent to euthanasia at the point of death because of the decline in their cognitive functioning [10]. In 2017 only three people with advanced dementia received euthanasia versus 166 cases of euthanasia in early stages of the disease, out of a total of 6,585 euthanasia cases in 2017 [11].

The dilemma between the need for consent and the challenges in obtaining it is addressed in this paper by offering a fictional solution (speculative design) as a framework for stimulating and supporting discussion.

## 2 Literature Summary

The systematic literature review used a seven-stage framework based on the PRISMA statement [12] for eligibility, search, identification of relevant papers from title and abstract, selection and retrieval of papers, appraisal and synthesis.

Seven databases were searched (Medline, Science Direct, Web of Science, PsychArticles, Cochrane Library, Scopus and PubMed) using the following search terms and criteria.

- *String search:* euthanasia OR "assisted suicide" OR "physician assisted suicide" AND dementia OR Alzheimer AND planning.
- *Date range:* 1994–2017. This range was selected as 1994 is when the Oregon Death with Dignity Act (ODDA) was passed, it specifies that a physician may prescribe lethal medication that is to be used to hasten death for competent, terminally ill persons who voluntarily request it [13].
- *Language:* Limited to Dutch and English as accessible literature and as the research is based on the Dutch legal framework.
- *Geography:* To include geographies where assisted dying is legal: Netherlands, Belgium and some US States.

As part of the inclusion/exclusion criteria the stance was taken that death is final so research literature about objections to euthanasia based on religious belief were not included. The literature was critically appraised using MMAT [14] and categorised into the following themes: Suffering, Autonomy and Planned Death.

This paper will only discuss the 'Autonomy' theme, which refers to control at the end-of-life whereby the only way to exercise control is by writing an Advance (Euthanasia) Directive.

### 2.1 Autonomy

*"One of the most important ethical principles in medicine is respect for each patient's autonomy. When this principle conflicts with others, it should almost always take precedence"* [15].

Being in control about ones' own end-of-life is a way to experience autonomy. Schroepfer, Noh, and Kavanaugh [16] report that terminally ill people want to be in control over decision making, independence, mental attitude, instrumental activities of daily living and relationships. Creating an advance directive can give a person control over their end-of-life [17]. The option of assisted dying may also give a sense of control. Legalization of assisted dying may have a therapeutic benefit for terminally ill patients, who often report feeling more at peace merely by knowing that they have the option to end their lives when they want to [19, 20].

However, if assisted dying is not an option, some people may choose to die by suicide. The effect of suicide on the people left behind can be much worse than a planned death. Families reported being better prepared for their loved ones death where people have requested assisted dying, and better able to accept it than those whose loved one has died 'naturally' of a terminal illness [21]. Not everyone wants to be in control of their own end-of-life decision; Cicirelli [22] studied end-of-life decisions for older people and found that approximately one third of participants (n = 388) favoured deferring end-of-life decisions to someone else, such as a family member, close friend, or physician. It has also been suggested that patient autonomy at the end of life does not exist as what is in the interests of the patient and what is in the interests of the surrounding family and friends will go hand-in-hand [23].

An advance directive is a tool used in planning for end-of-life. It is a document used to make provisions for health care decisions in the event that, in the future, the person becomes unable to make those decisions. Advance euthanasia directives in dementia are rarely complied with even though patient suffering may be judged to be extreme [11]. The way people adjust to suffering, a 'response shift', is sometimes argued to be the reason that dementia patients contradict earlier preferences, rendering advance directives meaningless. Jongsma et al. [24] argue that a response shift is a change in self-evaluation of quality of life. As dementia patients lack the ability to self-evaluate, this results in complexities in measuring quality of life or even having an opinion on it. They conclude that advance directives cannot simply be ignored, though they also say that directives should not be blindly adhered to either.

Another major obstacle in advance planning and dementia is the personality change that is associated with dementia: *"The core of the argument revolves around the undeniable change in personality, and arguably even identity, between the competent person who executed the directive and the incompetent person who will be affected by it"* [25]. This can place a huge strain on physicians and health care proxies, who have to make life-and-death decisions on behalf of the person who wrote the advance directive. Essentially an advance directive is the formerly competent person asking his/her proxies to ignore their demented self. Several authors question if this is a fair question to ask loved ones [7, 9, 26].

In practice advance directives are rarely adhered to in advanced dementia, which limits their role in advance care planning and end-of-life care of people with advanced dementia. Advance directives for euthanasia may raise false expectations and, in addition, place too much responsibility on elderly care physicians and relatives [8, 11, 27, 28]. However, some literature *did* approach advance directives as a tool for adequate advance care planning in dementia. Burlá et al. [18] suggested that the living will (advance directive) can be presented to the patient in the early days of their diagnosis and Flew [29] proposed a specific advance euthanasia directive that should be adhered to, even in advanced dementia. Others are aware of the problems with advance directives in dementia and propose solutions. For example, Gastmans and de Lepeleire [30] claim that, in an ethical evaluation of euthanasia, the dignity of the human person, relational autonomy, quality of life and care must be observed. They introduced the concept of 'relational autonomy' to give more control to close family/friends and social context of the person with dementia.

Menzel and Steinbock [31] describe identity in reference to Dworkins' 'critical interests'; these can be described as life values and go beyond 'experiential interests' which only exist in the here and the now. The critical interests shape a person and describe the kind of person they are and want to be – these are the interests that should be protected in an advance directive. This causes a dilemma: if the experiential interests of the person with dementia are not violated once dementia takes hold, but conflict with their critical interests. The authors propose a sliding scale solution, where autonomy is weighed against capacity of enjoyment, on a case by case assessment. Advance directives give people control over their lives once they themselves are no longer capable; *"the way they die is an important reflection on the way they lived"* and should be taken into consideration [31].

## 2.2 Interpretation of the Literature: Outlining the Problem Space

It is important that patients and their carers understand that their advance directives will not easily apply in dementia. There may be false hope attached to these directives, that may be reassuring to the person diagnosed with dementia while they are still cognitively sound, but likely to be a source of much distress to their proxies once the disease has progressed, and this document is largely ignored. A conflict arises between a person's 'critical interests' and 'experiential interests'. It can be said that people living with dementia, certainly in the later stages of the disease, live in the moment. Their personhood, based on their critical interests, their life values, ceases to exist to the person living with dementia, but not to their proxies.

The case in favour of adhering to an advance euthanasia directive based on a person's autonomy, does not take into account the implementation of the directive. Is it ethical to ask carers and clinicians to act upon an advance directive when the facts, in that time, do not endorse the earlier directive? The responsibility to adhere to a euthanasia request lies with a general practitioner (GP). In the Netherlands, a person usually has the same GP over a long period of time (near place of residence). The GP should know the person well, and will have discussed the advance directive.

Additionally, a second specialist physician will be involved to carefully assess the request. However, this raises the question whether euthanasia decisions should be made by physicians alone, as there are more dimensions beyond a physical death that could require other specialist capacities [32, 33].

The key points from the literature review are:

- People should be able to make an autonomous decision about their end-of-life; they can exert control by making an advance (euthanasia) directive.
- Advance euthanasia directive are not adhered to in dementia, because the disease presents with personality change and the disease makes suffering impossible to assess.
- The parties having to execute the wish expressed in an advance directive are faced with a difficult moral decision; do they respect the person who has written the directive, or the 'new' person the directive is about?

## 3 Speculative Design as a Method to Explore this Debate

The term Speculative Design was coined by Dunne and Raby [34] as design used to stimulate discussion and debate amongst designers, industry and the public about the social, cultural and ethical implications of existing and emerging technologies. Design Fiction is described by Bleecker [2] as a thoughtful exploration of speculative scenarios; a way to prototype other realities; this practise has also been called Speculative Design, Critical Design, Design Probes and Discursive Design. All these design research practices are similar in that there are no commercial constraints, all use fiction to present a diegetic alternative to existing issues, and prototypes as a method of enquiry [35]; for this research the term Speculative Design will be used. Ways of collecting data from Speculative Designs vary greatly. Speculative Designs are often placed in an exhibition context and left for public debate [35], or used as a tool to aid discussion [36, 37]. Tanenbaum [38] positions design fiction as storytelling *"Situating a new technology within a narrative forces us to grapple with questions of ethics, values, social perspectives, causality, politics, psychology, and emotions"*. These stories are important, as the prototypes created exist only within these stories, and this is precisely what makes them fictional [39]. For this research, Speculative Design is approached as a practical thought experiment. A thought experiment considers a hypothesis for the purpose of thinking through consequences and thought experiments are frequently used in philosophy and physics. The thought experiment can make the offered choice more real to result in a different kind of discussion [40].

There is no specific method on how to construct a successful speculation, but there are a few guidelines: A design speculation is a concept about a possible future. This speculation can be critical about a likely future, or it can be more like a 'what if' scenario for a desirable future [34, 41]. It is suggested that a speculation should sit in-between normal life and fiction. The story should be probable and credible, the viewer should be able to *"suspend their disbelief"* about the proposed prototype [42]. Auger [35] proposes that the speculation should offer a bridge between reality and the fictional element of the concept; in order to get the audience engaged, provocations can be used but they must be dealt with carefully, especially for controversial subjects (such as death), as the provocation can lead to revulsion or shock. He calls this *"managing the uncanny"*, shifting focus between familiarity and the proposed idea are ways to manage the experience of the uncanny. In this research, within the context of euthanasia and dementia, design is used as an anchor point between the different stakeholders; people with dementia, non-professional care-givers and professional care-givers.

## 4 Design Decisions: How the Speculation Was Constructed

This section will explain how the Speculative Design was constructed to illustrate the problem space. The designed prototypes aimed to make the euthanasia in dementia debate more tangible and accessible.

**Speculative Design: The AED-Plug**

This Speculative Design is called the Advance Euthanasia Directive Plug (Fig. 1; https://vimeo.com/231854700; http://aed-plug.com/). The proposed design is an implant that, once the conditions are reached as described in the Advance Euthanasia Directive, would trigger a swift and painless death.

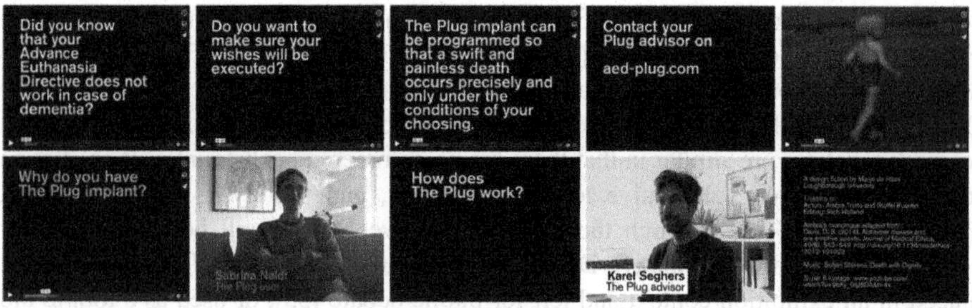


**Fig. 1.** Screenshots from the video about The Plug

This speculation was developed from the literature on Advance Euthanasia Directives for dementia. As discussed earlier, some authors argue that if an autonomous person has made an Advance Euthanasia Directive, then this should be adhered to. This adherence at all cost to an Advance Euthanasia Directive can put a lot of strain on clinicians having to enact this order. The Speculative Design explores an idea to remove this strain with a new kind of advance euthanasia directive, the *"AED-Plug"*, implanted in a fully cognitively competent person. Should this person develop dementia, they can be sure their euthanasia wish would be complied with, without upsetting clinicians, or putting stress on their proxies.

A Design Speculation requires a connection to exist between the audience perception of their world and the fictional element of the concept [35]. In order to make the speculation credible, and be taken seriously, the audience member must be able to believe in the possibility of its existence. This prototype was crafted in such a way that it could already be in existence, using contemporary media and messages (Fig. 1). It was presented as an advertorial; walking a fine line between documentary and commerce.

**The Video (Design Speculation)**

*Introduction*

A black background and a 'neutral' typeface was used (Akzidenz Grotesk). The black background was chosen to communicate death, with a large size type, as the message is generally aimed at an older generation [43].

*Music*

The soundtrack used was entitled "Death with Dignity", by Sufjan Stevens. The music is melancholy yet positive, filled with hope. The reveal of the title in the end credits is important as the title of the track is relevant to the topic.

*Super 8 Footage*

The super 8 footage is introduced, an intentional break in the message, and to conceptually visualise a sense of (memories) lost.

*Actor 1 (Sabrina Naldi)*

- *Casting:* An Italian woman in European setting, well-educated and well-travelled.
- *Setting:* Set in Sabrina's home, artworks on the wall, a telescope in the background, to indicate that Sabrina is a woman of the world.
- *Monologue:* Sabrina is passionate and well informed. She has clearly given the subject matter serious thought and made a rational decision. Some distance is created by the language used, demonstrating that she clearly knows what she is talking about. Her eloquence is convincing with the intention that the viewer can identify with her.
- *Filming:* The camera was positioned at an angle and Sabrina is engaged in conversation with the interviewer. Filmed as a single shot the intention was that complete story behind Sabrina's rationale is revealed to the viewer.

*Actor 2 (Karel Seghers)*

- *Casting:* Karel is a Dutch physician who has taken on a new role as AED-Plug advisor. Karel is pragmatic, working within the framework of the Dutch legal system, doing what he believes is the right thing.
- *Setting:* Karel's office, but not too formal, Karel is wearing informal clothes and his office hints of a family life with children's drawings on the wall.
- *Dialogue:* Karel explains in lay terms how the plug works, referencing familiar medical implants (pacemaker), to create familiarity.
- *Filming:* The camera is positioned at an angle and Karel is engaged in conversation with the interviewer. The film was cut at several points to give the idea that perhaps the way the Plug works is a little more complicated than this edit shows.

*Credits*

The credits reveal that this is a fictional scenario.

# 5 Review of the Speculative Design

The Speculative Design video was used to aid discussion in a workshop at the DementiaLab conference in Dortmund, Germany, September 2017. The workshop was attended by 15 participants of varying ages and backgrounds. All were active within dementia as designers, professional care-givers and/or academics. This workshop was run as a pilot study, to test the validity of the concept and to inform how data collection could best be done on this speculation with a group of carefully selected participants. Future participant criteria are that the participants need to be Dutch (or living in the Netherlands): a cultural acceptance of euthanasia as a common practise, who have either personal of professional experience with Dementia. Results of this research will be discussed during the conference. Ethical approval was given from Loughborough University.

## 5.1 Workshop Structure and Questions

The workshop was structured into three sections with the speculation carefully introduced to avoid the "uncanny" [35]. The first section was an introduction to dementia and dying. It featured subjective images from a caregiver's perspective on the disease (Fig. 2), expressing that this disease can be tough. A summary of the literature about

**Fig. 2.** Subjective image with quotes from Per Lundgren, Dementia nurse for over 37 years (interviewed by author on 18.02.16). Image: https://www.designboom.com/art/anish-kapoor-descension-public-art-fund-brooklyn-bridge-park-new-york-02-17-2017/

euthanasia and dementia was given, focussing on autonomy and advance euthanasia directives.

The second part of the workshop was designed to help the participants become more comfortable talking about death and dying and featured three activities. The first activity was drawn from a podcast on "how to die a good death" [44]. Participants were asked to choose their preferred demise from the four most common ways to die in western society; heart disease, stroke, dementia and (lung) cancer [45]. This was followed with an exercise discussing celebrity deaths (including Princess Diana, Robin Williams) where the participants were asked rate the quality of four different deaths; an accident, a suicide, cancer and old age. Finally, in this section, the participants were asked to create their ideal death scene in Lego.

The third part of the workshop introduced the Speculative Design video (Sect. 4). After watching video, the participants were asked to discuss three questions:

1. What factors, other than patient autonomy, should be considered for end-of-life decisions in dementia?
2. Who should be involved in making these end-of-life decisions?
3. Is Speculative Design a useful tool to further this debate?

Data were collected to critically appraise the Design Speculation and consider methods to further the speculation within the wider context of the euthanasia in dementia debate.

### 5.2 Workshop Results

The responses to the three questions are outlined below.

**What factors, other than patient autonomy, should be considered for end-of-life decisions in dementia?**
Opinions differed greatly for this question and a lot of concerns were raised. For example

- Was dementia really that bad?
- Why would people want to take their own lives?
- Was it not our responsibility to make sure people wouldn't want to die?

Some participants felt dementia wouldn't be that bad, whereas others found it more problematic and felt strongly that there should be a choice: *"Living with dementia doesn't have to be negative – but I, for myself, would like to have a choice. And I, now at this point, don't want to be that person [with dementia], no matter how happy I am going to be at that point, because that is not me. I felt very strongly with this woman in the video, who talked about these two personalities, for me it felt very relatable."*

The change in personality was addressed with arguments that personality change could be part of personal growth, and that a person could not make a decision now about their future self. It was concluded that The Plug would have to be an option, not an obligation. The decision would have to lie with the person diagnosed with dementia, meaning patient autonomy should be the main consideration.

**Who should be involved in making these end-of-life decisions?**
Discussions included extended life, questioning decisions to keep people with dementia alive *"... in hospital people try to keep them alive as much as possible like, you know, feeding them through tubes, that's so invasive, is it really needed? It is also horrible thinking, will she die from starving? I mean, and is ... I don't know if ... for instance ... in those moments, the relatives could have some kind of say."*

Another participant introduced the concept of an Artificial Intelligence to make the decision for you leading to interesting thoughts: *"You could hack the Plug... Kill them... Or keep them alive!"*

An important issue was raised about changing the decision *"the problem with the human heart and the human mind is that it's not constant and we change our minds and I'm afraid that I might change my mind and someone is not going to [help me]"*.

**How useful is Design Fiction as a tool to further this debate?**
The offered speculation triggered debate but and raised new questions. The speculation was successful in the way it clearly communicated the complicated issues extracted from the literature, and helped to move away from the original dilemma: 'the need for consent the difficulty obtaining this consent', to new questions such as 'the ethical implications of giving authority to an artificial devise' to 'how to deal with a response shift' to 'a right to die'. A lot of positive feedback was received about the workshop with the design speculation making people think differently about end-of-life. Some people felt confronted and some even angry, but all were happy to be challenged. Additional questions and comments were received after the workshop as the subject matter was further considered including a question about economics, and the impact that the cost of caring for dementia could have on decisions *"a strange power conflict"*. The designed speculation was successful in improving the accessibility (and discussion) for the complexities surrounding euthanasia in dementia. The workshop participants grasped complicated ethical dilemmas after watching a short video. It must be noted that the speculation, and the results it obtained will inherently be biased.

## 6 Developing the Design Speculation: From 'What if?' to 'then What?'

To give the Design Speculation more depth, the next stage was to follow the 'what if' of the proposed implant (The Plug), to the 'then what?' A website was designed using more of the discussion points from the workshop. In the workshop people wondered what it would look like, how would it actually work and what would happen in case of (technical) failure. A website was developed to give the impression that The Plug was commercially available (http://aed-plug.com/) with a contemporary style, lay terminology for accessibility and common menu items. The website sections include:

1. ***What is The Plug?***
   This section on the website refers to research literature to give an evidence base.
2. ***How does it work?***
   The Plug is a small implant that functions 'like a reverse pace-maker'. Here we show what the implant looks like, and shows a personalised box on how one would

receive this Plug (Fig. 3). A simple illustration (Fig. 4) explains how The Plug would work with information received from sensors and medical data. This section is expanded to show how the information gathered from the sensors can make decisions based on a Boolean string type of query (Fig. 5).

**Fig. 3.** What does the AED-Plug look like?

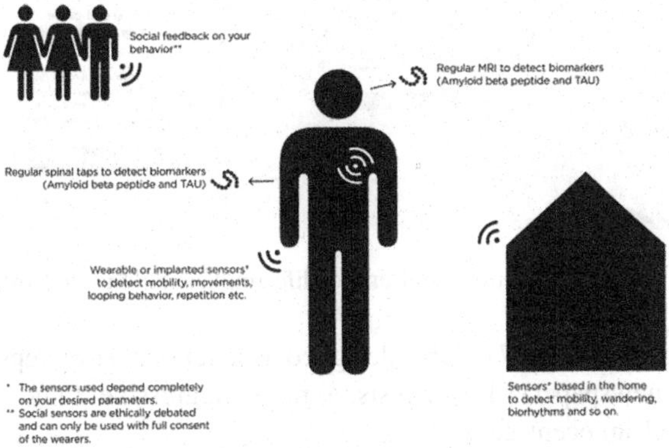


**Fig. 4.** How does the Plug know?

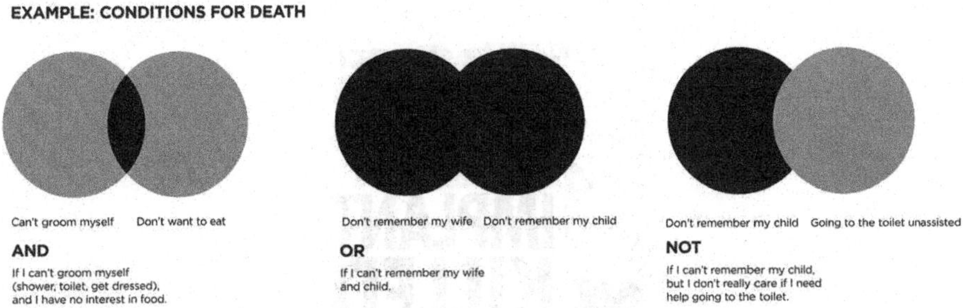


**Fig. 5.** Decision-making process for the Plug.

3. ***Frequently Asked Questions***
These answer questions raised in the workshop, such as "can I change my mind?", "will there be pain?" and "what if it doesn't work?"
4. ***The Plug in the News***
Two very different media stories were created for inclusion on the website to show how The Plug could go wrong (Fig. 6), or could be perceived to go wrong (Fig. 7).
The first story (Fig. 6) is shown as a fictional design hosted by the BBC to add credibility. It reports a case of The Plug not working to illustrate the complexity for specific conditions that may not be met by sensor-based technologies.

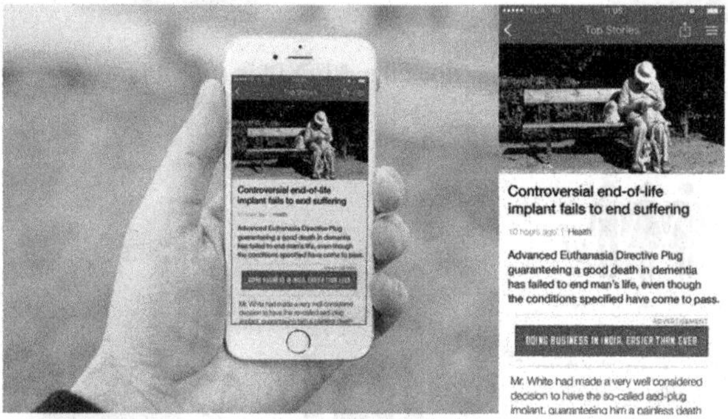


**Fig. 6.** Story 1: 'Controversial end-of-life implant fails to end suffering'

The second story (Fig. 7) is also designed as a fictional story reported in The Sun (UK tabloid newspaper). It suggests a more aggressive Plug design with the potential to kill innocent people.

The full design speculation (video and website) will encourage discussion between participants including people living with dementia, non-professional and professional care-givers.

**Fig. 7.** Story 2: 'Implant killed mother'

## 7 Discussion

In creating a Design Speculation to explore the autonomy theme in the euthanasia for dementia debate, there is a need to acknowledge bias. The focus was to explore the dilemma between making an Advance Euthanasia Directive and then assuming it will be enacted with two contrasting perspectives:

- An Advance Euthanasia Directive is sufficient safeguarding for future end-of-life decisions in dementia.
- An Advance Euthanasia Directive cannot be adhered to in dementia, because the disease presents with personality change and additionally it makes suffering impossible to measure.

This research has attempted to explore the challenges faced in the moral dilemma for the physician having to enact the directive. The Design Speculation has attempted to move the responsibility of that final decision from the physician to the person with dementia. The danger with presenting this Design Speculation in the wider context of euthanasia in dementia is that it presents inherent bias by enabling the possibility of euthanasia in dementia, therefore the framing is important to clarify that the purpose is to explore *autonomy*.

As a critique of the method (Speculative Design) one issue is the credibility (and quality) of the Design Speculation, so the way that it is crafted may have an impact on how it is perceived. To address this potential bias, it is important to document and reflect on the purpose of speculation. Few guidelines exist on how to create a *good* Speculative Design and it is important to "suspend disbelief" [42] – Auger [35] has suggested guidelines on how to achieve this. However, further research is needed to develop ways to critically access the quality of design speculations. This is a challenge as there may be little or no comparative material within the same context so a critical review of literature on the crafting of design fictions may offer some ideas and insights [35, 36, 46]. This paper has tried to describe how the speculation was designed and why the specific design decisions were made, as a knowledge contribution to future guidelines/education on crafting evidence-based design speculations. However, as with any design, quality can only really be judged in relation to the context and purpose (usability, functionality) of the design created.

## 8 Conclusion

The euthanasia in dementia debate is at an impasse. The intention for producing a Speculative Design was to offer a new perspective on this dilemma by designing a platform for reframing questions. Further testing is required to explore the value of this speculative design in reframing these questions, especially with the key stakeholders: people living with dementia, non-professional and professional care-givers.

However, this research has shown that there are benefits in presenting a dilemma, such as euthanasia in dementia, in a different format. The Speculative Design has potential to aid discussion between various stakeholders, without each party needing to be a specialist. The use of a provocative speculation was found to spark debate, but a caveat is the importance of boundaries whereby stakeholders are made aware of the greater context of the problem space so that inherent bias is addressed.

## References


1. WHO (2017). http://www.who.int/mediacentre/factsheets/fs362/en/
2. Bleecker J (2009) Design fiction: a short essay on design, science, fact and fiction. Near Future Laboratory, March, p 49
3. dictionary.com (2017)
4. Euthanasia, Wikipedia (2017). https://en.wikipedia.org/wiki/Euthanasia
5. Learner's dictionary (2018). http://learnersdictionary.com/definition/euthanasia
6. Dutch Euthanasia Act (2002). https://www.rijksoverheid.nl/onderwerpen/levenseinde-en-euthanasie/euthanasie
7. Buiting HM, Gevers JKM, Rietjens JAC, Onwuteaka-Philipsen BD, van der Maas PJ, van der Heide A, van Delden JJM (2008) Dutch criteria of due care for physician-assisted dying in medical practice: a physician perspective. J Med Ethics 34(9):e12. http://doi.org/10.1136/jme.2008.024976
8. Hertogh CMPM (2009) The role of advance euthanasia directives as an aid to communication and shared decision-making in dementia. J Med Ethics 35(2):100–103. https://doi.org/10.1136/jme.2007.024109
9. Rietjens JAC, van Tol DG, Schermer M, van der Heide A (2009) Judgement of suffering in the case of a euthanasia request in The Netherlands. J Med Ethics 35(8):502–507. 6p. http://doi.org/10.1136/jme.2008.028779
10. Emanuel EJ (1999) What is the great benefit of legalizing euthanasia or physician-assisted suicide? Ethics 109(3):629–642. https://doi.org/10.1086/233925
11. Rurup ML, Onwuteaka-Philipsen BD, Van Der Heide A, Van Der Wal G, Van Der Maas PJ (2005) Physicians' experiences with demented patients with advance euthanasia directives in the Netherlands. J Am Geriatr Soc 53(7):1138–1144. https://doi.org/10.1111/j.1532-5415.2005.53354.x
12. NRC Handelsblad (2018). https://www.nrc.nl/nieuws/2018/03/08/het-om-wil-nu-zelf-grenzen-euthanasie-onderzoeken-a1594903
13. www.prisma-statement.org (2018)
14. Fenn DS, Ganzini L (1999) Attitudes of Oregon psychologists toward physician-assisted suicide and the Oregon Death With Dignity Act. Prof Psychol Res Pract 30(3):235–244. https://doi.org/10.1037/0735-7028.30.3.235
15. MMAT (2018). http://mixedmethodsappraisaltoolpublic.pbworks.com/w/file/fetch/84371689/MMAT%202011%20criteria%20and%20tutorial%202011-06-29updated2014.08.21.pdf
16. Billings JA (2011) Double effect: a useful rule that alone cannot justify hastening death. J Med Ethics 37(7):437–440. https://doi.org/10.1136/jme.2010.041160
17. Schroepfer TA, Noh H, Kavanaugh M (2009) The myriad strategies for seeking control in the dying process. Gerontologist 49(6):755–766. https://doi.org/10.1093/geront/gnp060
18. Burlá C, Rego G, Nunes R (2014) Alzheimer, dementia and the living will: a proposal. Med Health Care Philos 17(3):389–395. https://doi.org/10.1007/s11019-014-9559-8

19. Rosenfeld B (2000) Assisted suicide, depression, and the right to die. Psychol Pub Policy Law 6(2):467–488. https://doi.org/10.1037//1076-8971.6.2.467
20. Brock DW (2000) Misconceived sources of opposition to physician-assisted suicide. Psychol Public Policy Law Off Law Rev Univ Ariz Coll Law Univ Miami Sch Law 6(2):305–313. https://doi.org/10.1037/1076-8971.6.2.305
21. Carlson WL, Ong TD (2014) Suicide in later life. Failed treatment or rational choice? Clin Geriatr Med 30(3). http://doi.org/10.1016/j.cger.2014.04.009
22. Cicirelli VG (1997) Relationship of psychosocial and background variables to older adults' end-of-life decisions. Psychol Aging 12(1):72–83. https://doi.org/10.1037/0882-7974.12.1.72
23. Hardwig J (1997) Is there a duty to die? Hastings Cent Rep 27(2):34–42
24. Jongsma KR, Sprangers MAG, van de Vathorst S (2016) The implausibility of response shifts in dementia patients. J Med Ethics. medethics-2015-102889. http://doi.org/10.1136/medethics-2015-102889
25. Davis DS (2014) Alzheimer disease and pre-emptive suicide. J Med Ethics 40(8):543–549. https://doi.org/10.1136/medethics-2012-101022
26. Bernheim JL, Distelmans W, Mullie A, Ashby MA (2014) Questions and answers on the Belgian model of integral end-of-life care: experiment? Prototype? J Bioethical Inquiry 11 (4):507–529. https://doi.org/10.1007/s11673-014-9554-z
27. De Boer ME, Dröes RM, Jonker C, Eefsting JA, Hertogh CMPM (2011) Advance directives for euthanasia in dementia: how do they affect resident care in Dutch nursing homes? Experiences of physicians and relatives. J Am Geriatr Soc 59(6):989–996. https://doi.org/10.1111/j.1532-5415.2011.03414.x
28. Kouwenhoven PSC, Raijmakers NJH, van Delden JJM, Rietjens JAC, van Tol DG, van de Vathorst S, van Thiel GJMW (2015) Opinions about euthanasia and advanced dementia: a qualitative study among Dutch physicians and members of the general public. BMC Med Ethics 16(1):1–6. 6p. http://doi.org/10.1186/1472-6939-16-7
29. Flew A (1999) Advance directives are the solution to Dr Campbell's problem for voluntary euthanasia. J Med Ethics 25(3):245–246
30. Gastmans C, De Lepeleire J (2010) Living to the bitter end? A personalist approach to euthanasia in persons with severe dementia. Bioethics 24(2):78–86. https://doi.org/10.1111/j.1467-8519.2008.00708.x
31. Menzel P, Steinbock B (2013) Advance directives, dementia, and physician-assisted death. J Law Med Ethics 41(2):484–500. https://doi.org/10.1111/jlme.12057
32. Bosshard G, Broeckaert B, Clark D, Materstvedt LJ, Gordijn B, Müller-Busch HC (2008) A role for doctors in assisted dying? An analysis of legal regulations and medical professional positions in six European countries. J Med Ethics 34(1):28–32. https://doi.org/10.1136/jme.2006.018911
33. Daly P (2015) Palliative sedation, foregoing life-sustaining treatment, and aid-in-dying: what is the difference? Theor Med Bioeth 36(3):197–213. https://doi.org/10.1007/s11017-015-9329-5
34. Dunne A, Raby F (2013) Speculative everything: design, fiction, and social dreaming. MIT Press. ISBN 0262019841 9780262019842
35. Auger J (2013) Speculative design: crafting the speculation. Digital Creativity 24(1):11–35. https://doi.org/10.1080/14626268.2013.767276
36. Tsekleves E, Darby A, Whicher A, Swiatek P (2017) Co-designing design fictions : a new approach for debating and priming future healthcare technologies and services. Arch Des Res 30(2): 5–21
37. Malpass M (2017) Critical design in context, history, theory, and practices. Bloomsbury, London. ISBN 978-1-4725-7518-0

38. Tanenbaum J (2014) Design fictional interactions: why HCI should care about stories, pp 22–23. http://doi.org/10.1145/2648414
39. Lindley J, Coulton P (2016) Pushing the limits of design fiction : the case for fictional research papers
40. Stanford Encyclopedia of Philosophy (2014). https://plato.stanford.edu/entries/thought-experiment/
41. Blythe M (2014). Research through design fiction : narrative in real and imaginary abstracts
42. Sterling B (2009) Design fiction. Interactions 16(3):20. https://doi.org/10.1145/1516016.1516021
43. Fonts.com (2017). https://www.fonts.com/content/learning/fyti/situational-typography/designing-for-seniors
44. The Guardian Podcast (2016) Utopia 2016: how to die a good death. https://www.theguardian.com/membership/audio/2016/nov/01/utopia-2016-how-to-die-a-good-death-guardian-live-event-podcast. Minutes 15:17–16:27
45. WHO (2017) The top ten causes of death. http://www.who.int/mediacentre/factsheets/fs310/en/index1.html
46. Schulte BF, Marshall P, Cox AL (2016) Homes for life: a design fiction probe. In: Proceedings of the 9th nordic conference on human-computer interaction, pp 80:1–80:10. http://doi.org/10.1145/2971485.2993925

# Aging Effects of Inner Character Space and Line Space of Japanese Language

Nana Itoh(✉) and Ken Sagawa

National Institute of Advanced Industrial Science and Technology (AIST), Tsukuba 305-8566, Japan
nana-itoh@aist.go.jp

**Abstract.** This study used Japanese sentences to investigate how inner character space, line space, and reader age differences affect readability. Results can suggest good layout of sentences for older adults. Japanese sentence experiments revealed that adequate inner character, line space, and number of characters provide good readability.

**Keywords:** Age · Inner character space · Line space · Readability

## 1 Introduction

Reading is a very common and important activity for older adults to obtain correct information such as instructions for medications, instruction manuals, and insurance agreements. Such text, however, is not always designed to be readable by people at any age or under any viewing conditions. Despite this importance, no general method exists to design sentence layouts in terms of visibility and readability or to compensate for age effects.

Now in ISO TC159, standardization of the "minimum legible font size" is being considered based on experimental data (Fig. 1). In the background of this standardization, age differences of visual acuity have been found. Acuity differs with the visual distance, luminance of the letters to be read, and font complexity [1, 2].

An earlier study of "minimum legible font size" estimation demonstrated a useful method to scale legibility at a supra-threshold level. A "Good" level of legibility, which is regarded as comfortable reading, is obtained with a font that is twice as large as the minimum legible font size, irrespective of the viewing conditions [3].

It remains unclear how sentences should be designed for readability, especially when dealing with different inner character spaces, line spaces, and numbers of characters. It is also questionable how those layout factors are related to the "minimum legible font size."

This study used Japanese characters and sentences to investigate the effects of inner character space, line length, and the number of characters for readability.

This study used Japanese sentences to investigate the influences of inner character space, line spacing (see Fig. 2), number of characters, and reader age on readability. Results are expected to suggest a good layout of sentences for older adults.


S. Bagnara et al. (Eds.): IEA 2018, AISC 826, pp. 153–157, 2019.
https://doi.org/10.1007/978-3-319-96065-4_18

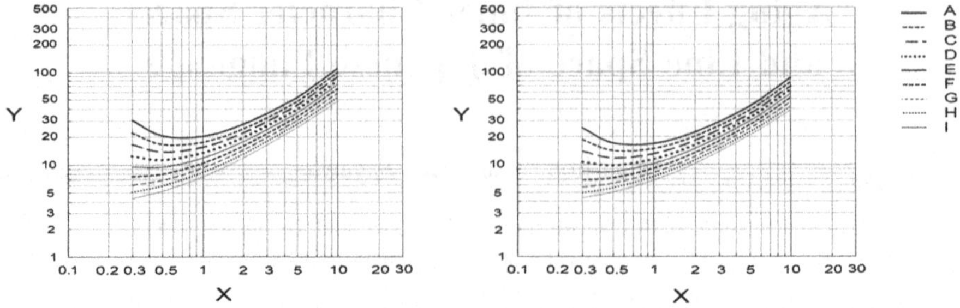


**Fig. 1.** Legible font sizes of respective age groups [1]. A–F signify age groups of 10s–70s: left, serif; right; san-serif font.

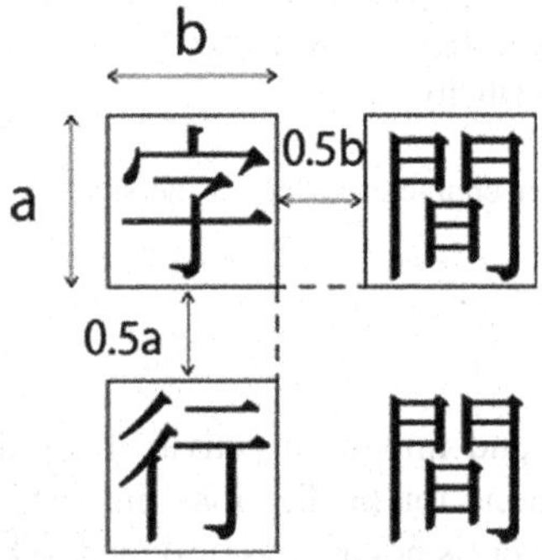


**Fig. 2.** Inner character space, line space: a, Character height, 0.5a, inner character space, half size of character height; b, Character width, 0.5b, Line space, half size of character width.

## 2 Method

### 2.1 Inner Character Space and Line Space

A total of about 192 of Japanese sentences with variable font size (12 point, 18 point), variable line length (10, 20, 40, 60 characters/line), variable inter-character spacing, and variable inner line spacing were presented on a 27-in. monitor. Conditions of inner-line space, line space, and the number of characters in a line are presented in Table 1. All sample sentences were collected from column articles of newspapers for elementary school students to unify the level of understanding. The viewing distance from the monitor to the subject was fixed as 0.5 m.

After reading each sample sentence, subjects were asked to judge the following impression factors: 1. Readability, 2. Fragmentation, 3. Neutrality, 4. Ease of following words and lines, and 5. Fatigue. For each factor, subjects were asked to try to read articles presented on the monitor and judge them using a five-point scale from 1 (very bad) to 5 (very good) with no time restriction. As described in this paper, results related to “Readability” will be discussed.

**Table 1.** Numbers of sentences used in this experiment and conditions of inner character space, line space, and point size of characters. Four sentences means that four numbers of characters (10, 20, 40, 60 characters/line) were presented one time each. Eight sentences means that four numbers of characters were presented twice each.

| | Serif font | Line Space | | | | Serif font | Line Space | | | |
|---|---|---|---|---|---|---|---|---|---|---|
| | 12 point | 0.0 | 0.5 | 1.0 | 2.0 | 18 point | 0.0 | 0.5 | 1.0 | 2.0 |
| Inner Character Space | 0.0 | 4 | 4 | 4 | 4 | 0.0 | 4 | 8 | 4 | 4 |
| | 0.5 | 0 | 0 | 0 | 0 | 0.5 | 0 | 4 | 0 | 0 |
| | 1.0 | 0 | 0 | 0 | 0 | 1.0 | 0 | 4 | 0 | 0 |
| | 2.0 | 0 | 2 | 0 | 0 | 2.0 | 0 | 4 | 0 | 0 |
| | San-serif font | Line Space | | | | San-serif font | Line Space | | | |
| | 12 point | 0.0 | 0.5 | 1.0 | 2.0 | 18 point | 0.0 | 0.5 | 1.0 | 2.0 |
| Inner Character Space | 0.0 | 4 | 8 | 4 | 4 | 0.0 | 8 | 12 | 8 | 8 |
| | 0.5 | 0 | 4 | 0 | 0 | 0.5 | 4 | 8 | 4 | 4 |
| | 1.0 | 0 | 4 | 0 | 0 | 1.0 | 4 | 8 | 4 | 4 |
| | 2.0 | 0 | 4 | 0 | 0 | 2.0 | 4 | 8 | 4 | 4 |

?? > In the table, make sure to use a space before POINT 18 point NOT 18 point etc.

## 3 Results

### 3.1 Inner Character Spaces

Figure 3 presents the change of readability according to the inner character space with 0.5 line spacing. Results show that readability is best when no inner character space exists: when only a small gap of margin exists between characters. As the inner character space gets wider, the subjective evaluation became lower. The same tendency was found for both older and younger subjects.

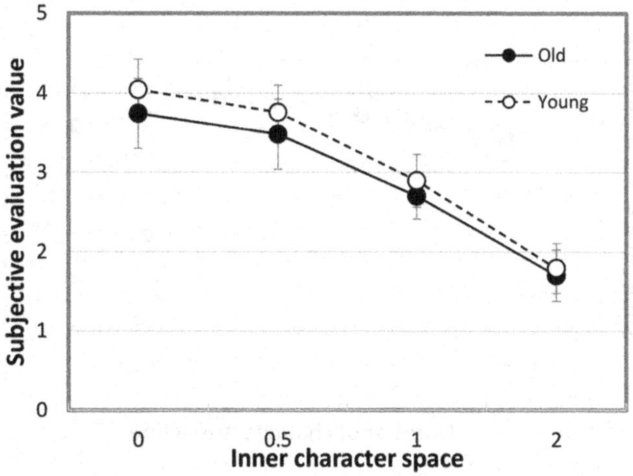


**Fig. 3.** Readability effects of inner character space (line space 0.5).

As portrayed in Fig. 4, readability worsened as the line space decreased. No significant difference was found between line spaces 1 and 2. According to this result, a certain line space works for reading of sentences smoothly, but when the line space became equal to the character height, the readability score reached a peak.

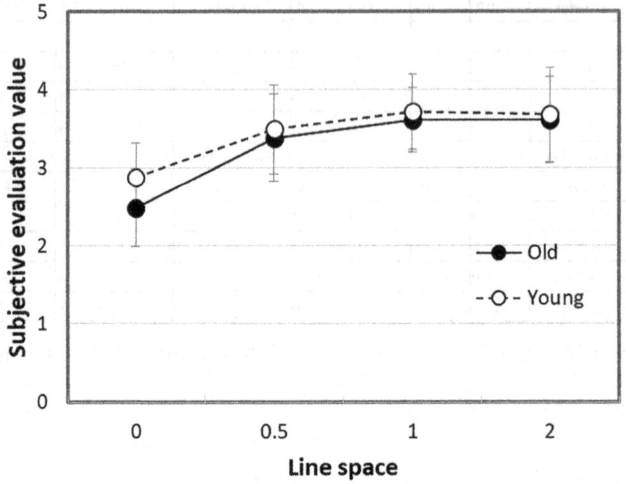


**Fig. 4.** Readability effects of line space (inner character space 0).

Figure 5 shows subjective evaluation according to the numbers of characters in a line. According to the data, conditions of the 40 characters are the highest subjective evaluation score for both older and younger subjects.

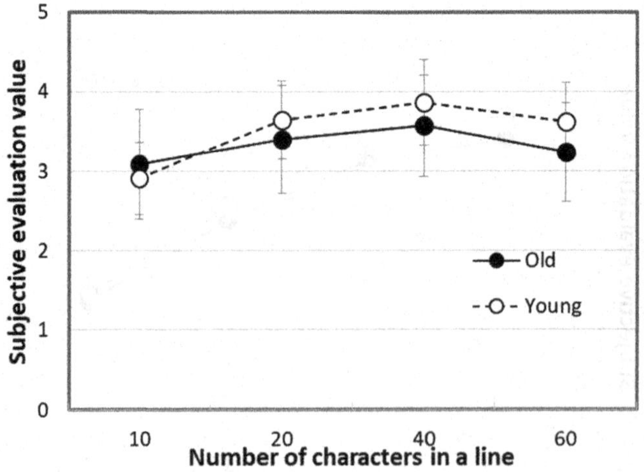


**Fig. 5.** Readability effects of line space (inner character space 0).

## 4 Discussion and Conclusion

According to these results, to attain good readability, it is desirable to have (1) inner character space of 0, (2) line space of 1 or more, and (3) number of characters in a line of 40. However, in this study, only two character sizes were used. It remains unclear how character size affects the appropriate numbers of characters in a line. The appropriate numbers of characters in a line might differ according to the character size if readability is related to the numbers of characters within a certain size of the visual field. In addition, when considering European languages, not only those variable factors, but also inner word spaces should be considered.

Additional investigations are necessary to confirm the design guidelines of layout sentences, being based on these data and together with data taken on the legibility of single characters. This might be a useful guideline for designing documents that include many sentences such as manuals or agreements.

## References

1. Sagawa K (2002) Visual function of older people and visibility of traffic signs. Gerontechnology 1:296–299
2. Sagawa K, Ujike H, Sasaki T (2003) Legibility of japanese characters and sentences as a function of age. In: Proceedings of the IEA 2003, vol 7, pp 496–499
3. Sagawa K, Itoh N (2006) Legible font size of a Japanese single character for older people. In: Proceedings of the IEA 2006. CD-ROM

# Digital Making as an Opportunity for Social Inclusion

Marita Canina(✉) and Carmen Bruno(✉)

Politecnico di Milano, Via Durando 38/a, 20156 Milan, Italy
{marita.canina, carmen.bruno}@polimi.it

**Abstract.** Within the framework of the EU-funded project 'Digital Do-It-Yourself (DiDIY), we have explored the dynamics facilitating the acquisition of different competencies in this practice.

Rooted in design and construction, the digital making activities often emphasize the acquisition of problem-solving, critical thinking, creativity, cross-disciplinarily and collaboration.

Starting from the results of the Digital DIY project has come to light that the developed process can also facilitate inclusivity. The contemporary phenomenon of "digital making" linked to "make" enabled by digital technologies (e.g., Web 2.0, 3D printers) is a meaningful context for social inclusion through interpersonal productive activities.

In this paper, we will first define such competencies as a result of a comparative study of the literature analysis. The analysis of the current scenario of digital DIY as a social innovation phenomenon enabled us to define a model through which it was possible to identify the fundamental dynamics and factors for skilling. The same model can allow the replication and adaptation of such dynamics into a different environment, such as the social inclusion.

We will conclude with the proposal of transferring the skilling dynamics identified in digital DIY to promote social inclusion of elderly.

The DiDIY co-design model, developed during the EU projects, includes the development of tools that facilitate the involvement of people in the design process merging digital making and skills improvement. The social empowerment and the individual creativity and self-improvement skills are crucial elements to avoid social exclusion.

**Keywords:** Social inclusion · Elderly · Digital making

## 1 Introduction

The EU funded project Digital Do It Yourself (DiDIY) [1] explored and validated the importance of the development of key competences such as creativity, collaboration, critical thinking, etc.… for people empowerment and the society growth. From this research emerges that digital making can be a powerful tool to develop such competences and to facilitate social inclusiveness.

DiDIY is a new human-centric, socio-technological phenomenon, enabled and rapidly evolving thanks to the widespread social availability of affordable technological tools.


S. Bagnara et al. (Eds.): IEA 2018, AISC 826, pp. 158–167, 2019.
https://doi.org/10.1007/978-3-319-96065-4_19

The EU research has developed a multidisciplinary vision of this recent phenomenon analysing how it has affected the world of work, education, creativity and intellectual property. Within the project, Politecnico di Milano dealt with the realization of a model that represents the key dynamics and competences (i.e. creativity, critical thinking, collaboration) developed through the practice of collaborative digital making and that could be transferred in various research fields [2].

This model integrates the following three expressions of Digital DIY: a phenomenon of social innovation due to its collaborative and sharing nature; a social practice involving artefacts, skills and motivations of practitioners; a creative and cognitive process for the development of skills [3].

Considering the collected results, it is essential to identify how to adopt these results to facilitate the social inclusion of the elderly, which represents a relevant category of the population. From one side they are at risk of social exclusion and loneliness and from the other they are custodians of a rich cultural and educational heritage.

To this end, the contemporary phenomenon of "digital making" represents an opportunity for inclusion, exploiting the knowledge traditionally held by the elderly in terms of doing, producing, and repairing. Contemporary digital making (that use technology such as 3D printers, Arduino) becomes the interface of the knowledge sharing between the elderly and the makers, considered as the craftsmen of the future, who have lost the traditional manual craft skills.

The literature has indeed putted an emphasis on the risks of this phenomenon in terms of manual skills loss and cultural heritage.

The key point therefore becomes the enhancement of the elderly skills, while stimulating their learning and their active involvement in social collaboration activities to bring their tradition and experience out.

The collaboration and sharing of one's own knowledge is an opportunity of growth and enrichment for the person and the community [4]. By using appropriate tools, it is indeed possible to include the elderly in dynamics of collaboration while enhancing their richness of knowledge. The social inclusion can offer them new opportunities for growth and self-awareness in new areas, by reflecting on and sharing the experiences to make them feasible also to others.

To this end, the paper aims to show a way to support social inclusion of the elderly in the field of digital making by using tools borrowed from Design Thinking and Creative Thinking. These tools, specific for the digital DIY field, are useful to encourage dynamics of skills learning and sharing [3].

Digital DIY represents a point of intersection between:

- the needs of a social category at risk of exclusion;
- the risks of deskilling brought by a contemporary phenomenon;
- a population which risks of losing a substantial part of its cultural baggage.

To guide the reader in the understanding of the research, in Sect. 2 we present the reflection on the skilling opportunity provided by DiDIY and the de-skilling risks. In Sect. 3 we discuss the benefits and limits of using digital making for the social inclusion of elderly people. In Sect. 4 we finally present the process and the tools that enable digital making as a means for facilitating inclusion. Benefit, limits and further steps of the research are discussed in the conclusion.

## 2 Skilling and De-skilling Through Digital DIY

The current trend of self-production (i.e. Do-It-Yourself or DIY) [5] is emblematic of the contemporary attitude to making and its investigation represented an opportunity for a better understanding of the dynamics underpinning the acquisition of the 21st century competences[1].

This set of competences comprises not only relevant knowledge and skills, but also a range of personal qualities and the ability to perform adequately and flexibly in well-known and unknown situations. The acquisition of different forms of knowledge and skills is considered of fundamental importance for people to face the complexity of contemporary age and to thrive as tomorrow's leaders, workers, and citizens in a constantly changing world and never-ending learning process.

From the literature we interpreted three categories of key competences for the next century citizen:

- Cognitive: i.e. skills including cognitive abilities such as critical thinking, creativity and problem solving;
- Intrapersonal: i.e. skills that occur within a person's own mind such as intrinsic motivation, self-development and self-management;
- Interpersonal: i.e. skills used to interact with people such as communicating and collaborating, adaptability and flexibility, system thinking.

The current socio-technical trend of self-production and making facilitated by digital media represents an opportunity for the engagement of a wider audience in the development of the 21st century skills [6, 7]. A number of researchers and educational leaders see in the digital DIY the potential to engage people in personally compelling, creative investigations of the material and social world [8].

Making encourages a deep engagement with content, critical thinking, problem solving and collaboration while sparking curiosity [9].

DiDIY practice is driven by strong motivational aspects that are believed to be crucial for sustaining the practice over time. The practitioner is supposed to persevere (or being strongly motivated) in overcoming the difficulties related to self-organization and the use of spare time on the one hand, and on the other social interactions when collaborating and participating. Moreover, DiDIY is about activities carried out collaboratively (the plural form of "you", also known as "Do It With Others", DIWO, or "Do It Together", DIT) and transdisciplinary. From a shared point of view, one can find almost always some form of collaboration, as even the individual maker builds on previous knowledge produced by others and shares within (online) communities.

[1] Such key competences have been widely defined and work programmes have been activated to promote their application among the educational and work fields both in Europe (i.e. the Lifelong Learning programme edited by the European Commission (2006/962/EC)) [6] and across United Stated (i.e. the Partnership for the 21st century skills, National Research Council) [7].

The collaboration, both with peers – i.e., other DiDIYers – and facilitators – and the sharing of knowledge, experiences, spaces, and projects is a critical enabler to create and keep alive a community which shares the same ethics and system ecology.

As a consequence, it is agreed that making fosters lifelong learning by encouraging learning by doing [9]. However, the spreading of digital fabrication raises arguments on its potentially deskilling effect [10]: "On the one hand, these technologies are said to encourage passive consumers to engage in creative making process in their spare time without having to pick up years of craft learning – reskilling, whilst on the other, they are said to automate making processes previously requiring craft skill – deskilling." [11].

The main argument on the deskilling effect refers to the highly digital nature of the creative process through technologies such as 3D printers, CNC mills and laser cutters. This undermines the ability for the practitioner to experience material qualities (e.g. hardness) and manufacturability (e.g. lathing, melting), and to learn through hand making, thus flattening the three-dimensional knowledge of hand making to the bi-dimensional realm. The ultimate effect is the development of a creative process which is led by a virtual idea disconnected from the material world.

The analysis of the current scenario of digital DIY and the several workshops run by Politecnico di Milano within the framework of the EU funded DiDIY project, allowed us to explore and validate the importance of DiDIY as a powerful tool to empowering people with competences such as creativity, critical thinking, collaboration, … The research enabled us to consider it as a significant context to facilitate social inclusion.

## 3 Social Inclusion and Digital Making: What Are the Benefits and the Barriers?

There is empirical evidence that the risk of becoming socially excluded is widespread among older people, particularly among those who have left the labor market, and that their respective risk is even rising with age [12].

Social exclusion is linked to two main factors, poverty and social isolation, and often represents a silent process of drifting to the margins of the social context.

Aging well is recognized as a privilege [13] and a challenge of the 21st century (EU 2007). In this century the elderly will constitute 1/3 of the Italian (ISTAT 2015) and the worldwide population [14].

In view of the demographic and societal changes, the social inclusion of the elderly is a topic of growing importance as well as the strategies to promote different activities among older people. However, given the increasing life expectancy, it must be taken into account that older people are already a heterogeneous population group and are increasingly becoming so. This means that old age is characterised by a growing diversity in lifestyles, values and specific chances and challenges. Consequently, older people's resources in terms of finances, health and social contacts are also extremely diverse and decisively influence the personal scope for autonomy, active participation and the assumption of responsibilities in old age.

In "Empowering people, driving change, Social Innovation in the European Union" and "Strategic Implementation Plan for the European Innovation partnership on active and healthy ageing", social inclusion is indicated as one of the priority strategy. Social policies aim at involving olds in active life.

The Successful Aging model [15] identifies the elderly people basic needs that allow the improvement of their quality of life with proactive actions (Fig. 1).

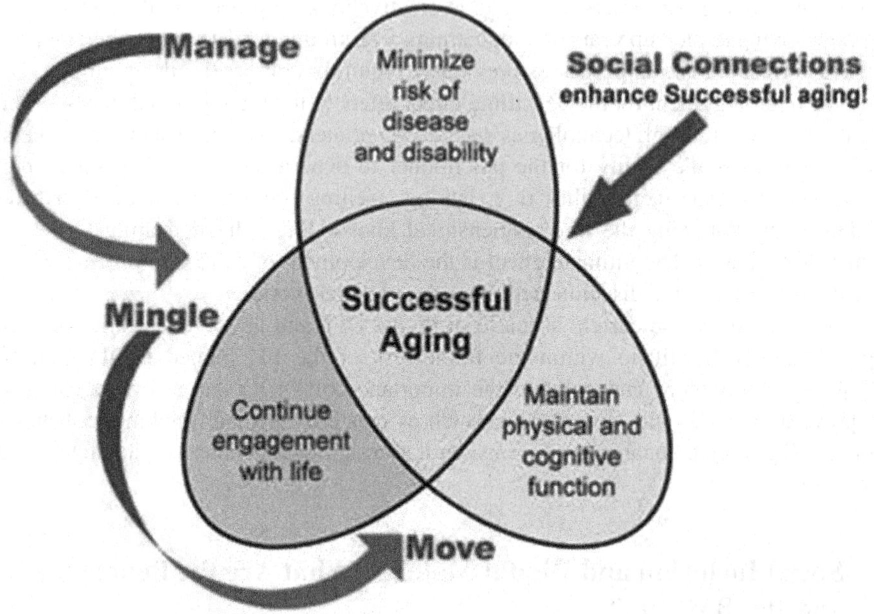


**Fig. 1.** Successful Aging model from Kahana et al. 2003

Specifically, their active commitment through interpersonal relationships and the exchange of good practices can become an opportunity for reflection, comparison and enrichment. This represents the key for their social inclusion.

In searching for measures to promote the social inclusion of the older population, we do believe that the contemporary phenomenon of "digital making" represents a significant context to facilitate social inclusion through interpersonal production activities [16].

Makerspaces (i.e. FabLabs) are digital making places where people develop ideas following the logics of "commons-based peer production" [17] and interact to share skills, generate knowledge and produce goods with digital technologies, for themselves and for the community.

This bottom up innovation, able to connect through making [18] and identified as the third industrial revolution by Rifkin (2011), represents a fertile context for the elderly inclusion.

The typical skills of making owned by the elderly could be applied and enhanced for the development of projects in the makerspaces, keeping them physically and psychologically active. Their participation in digital production activities would support an intergenerational transfer of a wealth of technical and experiential skills that young people are losing, and which represent an added value for the labour market of the next generation, for the social cohesion and the active citizenship (EU 2006, P21C 2008).

The risk of losing manual and intellectual skills in digital making is widely debated in the literature [10, 19], and this match would contribute not only to increase the elderly well-being, but also to improve the abilities of the new generations.

Digital making is here considered a context in which the exchange of knowledge and skills fosters an intergenerational dialogue, the inclusion and active participation of the elderly in the community, the preservation of cultural heritage and potential technological and cultural innovation.

Despite governments across the globe have declared their commitment to building a people-centred, inclusive and development-oriented Information Society, a significant proportion of the global population remains 'digitally excluded'. The majority of those who fall into this category are the elderly, and yet digital technologies offer enormous potential benefits to this sector of the population [20]. In particular, numerous investments for the digitalization of the elderly population have been done (see the EU2020 initiatives), but such users are still far away from digital making places [10].

Age itself is not a barrier to using digital technologies, and although older people tend to face other barriers to access such as cost, skills or disability, research suggests that many simply do not perceive the relevance of these technologies to themselves [20]. Discussions also surround barriers to digital technology use for older adults, the codification of digital technology use within society, and how older adults use digital technology in a facilitative and inclusive way to empower themselves and protect them from the negative effects of the digital divide [21].

The crucial challenge is the identification of methods and tools that enable the active involvement of the elderly in collaborative digital making activities and that allow the expression of tacit skills through a "learning by doing" approach. The use of design tools and methods is therefore strategic because they are able to interpret the people needs and satisfying them by using an approach that places the individual at the centre of the design action (Human-Centred Design). The co-design tools that actively involve those who benefit from the design interventions, encourage the development of ideas and the exchange of skills between the elderly and the digital makers community during a projects realization. There is therefore a swift from a more traditional design for the elderly to a design with the elderly [22].

## 4 Digital DIY Co-design Process as a Tool to Promote Social Inclusion in Old Age

As a main result of the explorative and generative workshops run during the DiDIY project, the Politecnico di Milano research team developed a toolkit and guidelines called "Co-design in the Digital DIY scenario" [23]; a (co)design-driven tools specific for Digital DIY that can help people to put into practice the potentialities of this social innovation phenomenon.

The toolkit includes all the tools and activities based on the co-design process for digital DIY, built with and for non-designer, it comprises the Discovery activity designed to explore the DiDIY phenomenon and the "DiDIY Factors Stimuli" tool highlighting the DiDIY potentialities to consider during the project. This tool is the most significant ones due to its specific content and distinctive flexibility (Fig. 2).

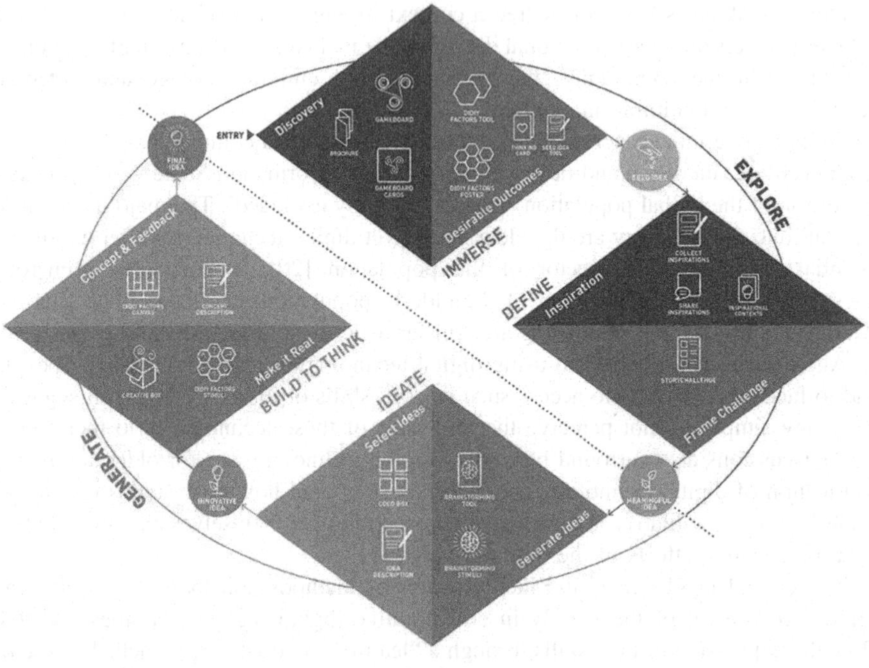


**Fig. 2.** The Digital DIY co-design process from the "Co-design in the Digital DIY scenario" toolkit and guidelines by IDEActivity (Color figure online)

Based on the knowledge gained, the research team designed a specific co-design process that allow the active involvement and collaboration of ageing people with multidisciplinary competencies.

The objective is to promote the social inclusion of the elderly by organizing digital making activities with the local community, that could activate processes of skills sharing facilitated by the co-design tools. This approach focuses on the analysis of people (Human-Centered Design) and the use of a Co-Design approach.

To guarantee this objective it is necessary to work on three parallel levels of action. The first level is based on the creation of meeting, training and collaboration experiences between the elderly and digital makers to allow the exchange of intergenerational skills. Design experiences, exploiting the potential of production activities with digital technologies, could be structured to allow the actors involved to know each other's needs, abilities and languages, and to develop a project of common interest. The second level of action contributes to preserve the skills and cultural heritage of the elderly (i.e. craftworking techniques and sartorial cuts) by transferring them to the younger generations of local makers and communities, especially during collaborative design experiences. The collaborative context of digital making facilitates the fusion of digital techniques and knowledge with the (often) analogical ones of the elderly thus generating potential innovation. The third level must promote the use, development and implementation of co-design tools for social inclusion.

The activities are structured through participatory experiences of training and co-design with the elderly inside the makerspaces.

The first step is the selection of the elderly by giving importance at their manual craft skills such as wood, metal and ceramic processing; food preparation using traditional techniques and tools; tailoring and electronic skills. After this selection it is necessary to interviewing them to better understand their stories and to identify the crucial factors for facilitating the skills transfer dynamics.

The activities begin with the Explore stage of the design process (see the blue area in the figure), in which the elderly people are introduced to the new digital manufacturing technologies within the makerspaces through a series of workshop describing how such CNC machines (e.g. 3D printers and CNC milling machines) and IT (e.g. Arduino/Genuino) works. Beside the digital literacy it is important to organize moment of discussion about their familiarity with these technologies, the spirit of collaboration in digital making, the expected benefits and difficulties of a more active involvement. The process foresees a Discovery activity within the IMMERSE step in which the analysis of significant DiDIY projects helps them to explore the phenomenon, to comprehend the context of DiDIY, and to highlight the potentialities, the benefits and the innovative features of DiDIY.

These workshops are enriched with practical exercises on the digital concept learned that allow the skills transfer by using storytelling technique. This technique is used to bring out the elderly impressions, emotions, experiences and the information related to their interests. This is one of the most delicate phases that allows to collect, share and integrate their life experiences and skills with those of the makers. It is a necessary step for the definition of the subsequent co-design activities. The tools collected in the toolkit are therefore essential to anyone who wants to activate a related social inclusion project. It indeed supports the management of all the design process phases.

After the Explore stage, the process proceeds to the Generate stage (see the orange area in the figure), with collaborative activities between the elderly and the digital makers on projects of common interest, in fields such as carpentry, clothing and sewing, jewellery, agriculture and gardening, cooking. The elderly are divided into subgroups based on the topics of interest and then matched with the digital makers to begin the co-design journey. The project-based activity is a design collaborative experience of digital production which allows groups of elderly and digital makers to express themselves and acquire skills, facilitated by the supervision of the makerspace experts. The aim of the co-design activities is to lead the elderly through a participatory design experience that allows them to express themselves while acquiring new technical skills.

During this stage, people creativity and group collaboration, can be stimulated by using the tools from the IDEActivity toolkit, mentioned above.

Activities format that differ for typology and length, should be considered according with the interests and the skills of the elderly and the makers emerged during the previous activities and in relation with the cultural featured of the territory. The format flexibility guarantees the occurring of the natural creative processes of a makerspace thus avoiding the risk of a modus operandi distortion with consequences on the effectiveness of each makerspace.

## 5 Conclusion

A fundamental future development of the research is the implementation of the model including an analysis of replicability. The analysis of the activities dynamics could allow to verify and implement the theoretical model. These dynamics include how the factors enabling digital DIY (such as technology, motivation and collaboration) have influenced both positively and negatively the inclusion of the elderly in digital making activities considering the specific people profiles (i.e. gender, age, ability), in different step of the design process (i.e. ideas generation, ideas selection, product creation) or in relation to the type of activity carried out (i.e. continuous or occasional). The resulting optimized model could then be used for studies in other socio-geographical contexts or for promoting social inclusion through actions.

Furthermore, a replicability analysis should be performed to identify other possible socio-cultural contexts that could benefit from the implementation of the research experience. It is important to bear in mind that often the solutions cannot be replicated but they can easily be reconstructed and adapted, basing on the context and needs diversity.

## References

1. European Project Digital Do It Yourself (DiDIY). http://www.didiy.eu/
2. Bruno C, Salvia G, Canina M (2016) Digital self-production as a means to improve education. In: 10th international technology, education and development conference 2016, Valencia, pp 2304–2310
3. Salvia G, Bruno C, Canina M (2016) Skilling and learning through digital Do-It-Yourself: the role of (Co)Design. In: DRS2016: Design+ Research+ society-future-focused thinking. Design Research Society, Brighton, pp 2077–2089
4. Atkinson P (2006) Do it yourself: democracy and design. J Des Hist 19(1):10
5. Anderson C (2012) Makers: the new industrial revolution. McClelland & Stewart, Toronto
6. European Commission (2007) Key competences for lifelong learning. European Reference Framework0
7. The Partnership of 21st Century Skills (2008) 21st century skills, education & competitiveness. A resource and policy guide
8. Vossoughi S, Bevan B (2014) Making and tinkering: a review of the literature. In: National Research Council Committee on out of school time STEM, pp 1–55
9. Peppler K, Bender S (2013) Maker movement spreads innovation one project at a time. Phi Delta Kappan 95(3):6

10. Hielscher S, Smith A (2014) Community-based digital fabrication workshops: a review of the research literature (No. 08). SPRU Working Paper Series
11. Ree R (2011) Master thesis: 3D printing: convergences, frictions, fluidity, University of Toronto
12. Naegele G, Schnabel E (2010) Measures for social inclusion of the elderly: the case of volunteering. Working paper. European foundation for the improvement of living and working conditions, Ireland. https://www.eurofound.europa.eu/publications/report/2010/quality-of-life-social-policies/measures-for-social-inclusion-of-the-elderly-the-case-of-volunteering-working-paper. Accessed 28 May 2018
13. World Health Organization (2002) The world health report 2002: reducing risks, promoting healthy life. World Health Organization, Geneva
14. United Nation (2002) Report of the second world assembly on ageing Madrid, New York, 8–12 April 2002. http://www.un.org/en/ga/search/view_doc.asp?symbol=A/CONF.197/9
15. Kahana E, Kahana B, Kercher K (2003) Ageing Int 28:155
16. Murray R, Caulier-Grice J, Mulgan G (2010) The open book of social innovation. Nesta and Young Foundation, London
17. Benkler Y, Nissenbaum H (2006) Commons-based peer production and virtue. J Polit Philos 14(4):394–419
18. Gauntlett D (2011) Making is connecting: the social meaning of creativity, from DIY and knitting to YouTube and Web 2.0. Polity Press, London
19. Soderberg J (2013) Automating amateurs in the 3D printing community. Work Organ Labour Globalisation 7(1):124–139
20. Olphert CW, Damodaran L, May AJ (2010) Towards digital inclusion-engaging older people in the 'digital world'. In: Accessible design in the digital world conference 2005, Dundee, Scotland. https://pdfs.semanticscholar.org/b312/dac0a90bbdfeffc8b4348a5793593fbc58a1.pdf. Accessed 28 May 2018
21. Hill R, Betts L, Gardner S (2015) Older adults' experiences and perceptions of digital technology: (Dis)empowerment, wellbeing, and inclusion. Comput Hum Behav 48:415–423
22. Sanders E, Chan P (2007) Emerging trends in design research. In: IASDR07 - International Association of Societies of Design Research, The Hong Kong Polytechnic University, School of Design, Hong Kong
23. IDEActivity: Codesign in the Digital DIY scenario. Toolkit and guidelines (2017). http://www.ideactivity.polimi.it/toolkits/. Accessed 18 Dec 2017

# Evidence Based Data for the Design of Rotary Control Elements for Fine Motor Adjustment Tasks with Respect to the Elderly User

Peter Schmid(✉), Benedikt Janny, and Thomas Maier

Institute for Engineering Design and Industrial Design (IKTD), Department of Industrial Design Engineering, University of Stuttgart, Pfaffenwaldring 9, 70569 Stuttgart, Germany
peter.schmid@iktd.uni-stuttgart.de

**Abstract.** With increasing age, the loss of sensorimotor skills plays an important role regarding human-machine interaction. With diminishing eyesight, it becomes increasingly difficult to operate an interface. Referring to this, presenting information via the haptic channel in critical situations can relieve the elderly user. Therefore two experimental studies were conducted. In both experimental studies the subjects had to perform a simple adjustment task with a rotary control element in the context of driving. In the first study optimal locking angle and operation torque values were detected. In the second study, elderly users had to perform adjustment tasks using control elements with varying torque curves. The purpose of this study was to identify which coding feature is suitable for marking a preferred value on a scale or for a menu change. Based on the first study, optimal locking angle and operation torque values are presented. The second study shows that elderly users prefer significant coding features to choose a preferred value or to perform a change. The combination of a rotation angle magnification and an increase in torque also leads to a better task performance. For the indication of a menu change elderly users prefer the combination of rotation angle magnification and torque increase.

**Keywords:** Haptics · Adaptive human-machine interface
Rotary control element · Fine motor adjustment tasks · Design for the elderly

## 1 Introduction

An observable trend in human-machine interface design both in the capital and in the consumer goods sector shows the machine complexity of operating functions. With a simultaneous reduction of the physical control elements, the machine complexity continuously increases [1, 2]. As a consequence the usability of human-machine interfaces is reduced and might lead to an increasing number of operating errors.

Most information of technical products is therefore usually transmitted exclusively audiovisually, which leads to an overload of the human perception and information processing capability. More unused channels of perception would be available [3].

By adapting the shape and structure, adaptive rotary control elements offer a way to reduce the perception overload [4, 5]. In combination with haptic feedback, an adaptive


S. Bagnara et al. (Eds.): IEA 2018, AISC 826, pp. 168–177, 2019.
https://doi.org/10.1007/978-3-319-96065-4_20

rotary control element can be adapted to different tasks, so that the visual perception channel can be relieved [3]. A falling saw tooth shape is preferred by users in terms of greatest comfort and precision impression [1, 5, 6]. By varying characteristic parameters such as torque amplitude or locking angle, many different haptic functions can be created and used for a multimodal and relieving information transfer. In this study various torque characteristics are examined with respect to their suitability as a haptic coding feature for the relief of the visual perception channel. A torque function with an amplitude change at a defined point, a rotary angle change and a combination of an amplitude and rotary angle change can be used to detect a point of interest, e.g. the center of a scale or a preferred value or a menu change.

Investigations show that the visual perception channel can be significantly supported by changes in the torque amplitude, the locking angle or by high torque blockers at the end of a selection list [3]. But further studies are necessary to identify a preferred value, a center marker or a menu change especially for special focus groups.

This research study investigates the use of variable controlling torques for information transfer concerning the elderly. Coding features for marking a center mark, a preferred value and a menu change are investigated. In this study the user age was of special interest since the age-related loss of perception as well as the decrease in cognitive performance cause accelerated aging with increasing age [7]. The study contributes to optimizing and increasing the operating safety of human-machine interfaces.

# 2 Methods

## 2.1 Participants

Altogether, 61 subjects participated in the experiment in two time slots. Especially 31 subjects (19 male and 12 female) participated in the first experiment aged from 55 to 82 years (mean = 66.06, SD = 6.98) and 30 volunteers (16 male and 14 female) in the second experiment aged from 55 to 78 years (mean = 63.20 SD = 5.58). The criteria for participation were possession of a driving license and a minimum age of 55 years in order to get a representative age range. For all subjects the total time of driving license possession ranges from 30 to 58 years. In the first study 27% and in the second study 29% deal regularly with a rotary control element in their car.

## 2.2 Apparatus

During the two studies two tasks had to be fulfilled. A primary task and a simultaneously performed second task. The primary task is a simple driving simulator (Fig. 1). The driving simulation used is the Lane Change Test (LCT). The LCT driving scene is projected on a screen 50 cm in front of the participants. In the LCT simulation a participant drives at a constant speed of 60 km/h on a straight, three-lane road on which no other cars are present. Signs on both sides of the road indicate the participant to change lanes. The information appears 40 m ahead of the sign. While driving the LCT software records the deviation from a normative path is recorded. Consequently,

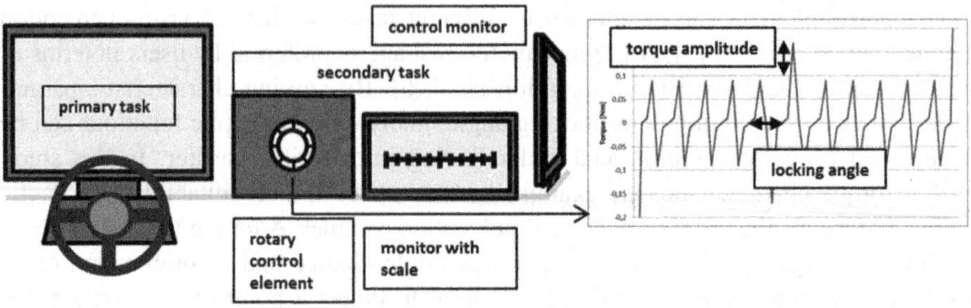


**Fig. 1.** Test bench for primary and secondary task

late perception of signs (or missed signs), slow lane change and poor lane-keeping result in greater deviation. A LCT analysis software is used to compute the mean deviation for the participants. In addition to the primary task a secondary task has to be fulfilled which consists of setting a defined value along a ten-step scale using the rotary control element of the test bench. In the front area of the test bench, a rotary control element is mounted, which represents the human-machine interface. The rotary control element sits on a shaft and can be exchanged by means of a screw connection. A DC motor from Maxon Motor AG generates different torque-angle profiles. The maximum torque of the engine is 0.606 Nm. The drive shaft of the motor is connected to a rotary encoder (ROD 426 from Dr. Johannes Heidenhain GmbH) which measures the angle of rotation with a resolution of 2500 increments per revolution and transmits the measurement signals directly to a dSPACE measurement and control board (DS 1102 DSP Controller Board). The measuring signals are processed in real time via the freely programmable controller board. The control and torque-rotation profiles are programmed with MATLAB Simulink and are controlled by dSpace's ControlDesk software.

# 3 Investigations

## 3.1 Initial Parameters

Table 1 shows the initial parameters in terms of torque functions. The complete investigation is carried out in laboratories of the IKTD. At the beginning of the study, a questionnaire collects demographic data. In addition to questions about fine motor activities, data on driving performance and experience with central control elements in cars are collected. Subsequently, the sensorimotor and cognitive capabilities of the subjects are determined by means of the Wiener Test System using the Trail-Making-Test according to Langensteinbach (TMT-L), a Reaction Test (RT) and the Motor Performance Series (MLS).

The main task of the subjects is to set up a defined value along a 10-step scale (secondary task) while performing the LCT (primary task). After each run, objective and subjective parameters are recorded. Optimal locking angle and operation torque values are detected using the usability measures of task performance, positioning time,

**Table 1.** Torque function for the first investigation concerning initial parameters

| Torque function | Rotation [°] | Torque [Nm] | Torque function | Rotation [°] | Torque [Nm] |
|---|---|---|---|---|---|
| V01 | 19 | 0,072 | V09 | 30 | 0,072 |
| V02 | | 0,090 | V10 | | 0,090 |
| V03 | | 0,113 | V11 | | 0,113 |
| V04 | | 0,141 | V12 | | 0,141 |
| V05 | 24 | 0,072 | V13 | 38 | 0,072 |
| V06 | | 0,090 | V14 | | 0,090 |
| V07 | | 0,113 | V15 | | 0,113 |
| V08 | | 0,141 | V16 | | 0,141 |

subjective satisfaction and precision. Performance is measured in terms of the number of tasks completed and the operating time is measured through the Control Desk software by dSpace. LCT driving performance is measured in term of mean deviation. A rating scale is used to detect the user satisfaction during the task fulfillment. The total duration of the experiment is about 2 h.

### 3.2 Operating Performance and Semantic Preference

The second investigation consists of two experimental parts. The complete examination is carried out in laboratories of the IKTD. The duration of the experiment is about 2 h in total. At the beginning of the study, a questionnaire collects demographic data. In addition to questions about fine motor activities, data on driving performance and experience with central control elements in cars are collected. Subsequently, the sensorimotor and cognitive capabilities of the subjects are determined according to the first investigation (initial parameters).

In the second study, elderly users have to perform adjustment tasks using control elements with varying torque curves. The operating task and the time of execution are announced during the driving simulation via an acoustic signal. The goal is to do both the driving and the adjustment task as quickly as possible and to assess the haptic feedback after each pass. In this study the operating performance consists of a total of 28 repetitions (Table 2). The experiment starts with a training phase, in which the individual steps are explained exactly by the experimenter. As initial torque 0.09 Nm is used for the initial locking angel 30°. To investigate the operating performance, each subject is given 14 symmetrical and asymmetric torque curves (grouped into 4 torque amplitude jumps, 2 detent angle changes and the combination thereof). They had to perceive the center mark as quickly as possible and then should turn two turning points further. In a symmetrical course, the locking angle change before and after the locking point is the same (36° and 47°). In an asymmetrical course of the locking angle after locking point is greater (36° or 47°) compared to the locking angle before locking point (30°).

**Table 2.** Torque functions for the second investigation concerning the torque coding features

| Symmetrical coding feature | | | | Asymmetrical coding feature | | | |
|---|---|---|---|---|---|---|---|
| Torque function | Torque [Nm] | Torque jump | Locking angle [°] | Torque function | Torque [Nm] | Torque jump | Locking angle [°] |
| V01 | 0.09 | 1.2 | 0° | V15 | 0.09 | 1.2 | 0° |
| V02 | 0.09 | 1.5 | 0° | V16 | 0.09 | 1.5 | 0° |
| V03 | 0.09 | 1.875 | 0° | V17 | 0.09 | 1.875 | 0° |
| V04 | 0.09 | 2.343 | 0° | V18 | 0.09 | 2.343 | 0° |
| V05 | 0.09 | 1 | 30°–36° | V19 | 0.09 | 1 | 30°–36° |
| V06 | 0.09 | 1 | 30°–47° | V20 | 0.09 | 1 | 30°–47° |
| V07 | 0.09 | 1.2 | 30°–36° | V21 | 0.09 | 1.2 | 30°–36° |
| V08 | 0.09 | 1.5 | | V22 | 0.09 | 1.5 | |
| V09 | 0.09 | 1.875 | | V23 | 0.09 | 1.875 | |
| V10 | 0.09 | 2.343 | | V24 | 0.09 | 2.343 | |
| V11 | 0.09 | 1.2 | 30°–47° | V25 | 0.09 | 1.2 | 30°–47° |
| V12 | 0.09 | 1.5 | | V26 | 0.09 | 1.5 | |
| V13 | 0.09 | 1.875 | | V27 | 0.09 | 1.875 | |
| V14 | 0.09 | 2.343 | | V28 | 0.09 | 2.343 | |

# 4 Results

## 4.1 Initial Parameters

Based on task performance, it can be seen in Fig. 2 that the tasks are more reliably fulfilled by the test subjects using a locking angle of 30°. On average, the task fulfillment of this angle level goes up to 97%. Concerning the two smaller locking angles (19° and 24°), the task performance fluctuates significantly. Here, the performance varies between 57% and 90%. The best performance is achieved using feature V15 (100%). With regard to positioning time (Fig. 2), it can be seen that with increasing torque amplitude, the positioning time increases. The smallest positioning time is achieved using torque function V01, V05 and V10, 1.9 s each. Torque function V12 leads to the largest positioning time with 2.3 s. In terms of severity, feature V09 performs best. Figure 3 shows the classification of the precision of the features. Features with only one large torque amplitude (V03 and V04) or a larger detent angle (V13 or V14) are evaluated as less precise. Feature V10 is rated most accurate (94%) in this study. The lowest precision rate is achieved using features V03 and V04 with 75%. The lane deviation of the subjects during the operating task is also shown in Fig. 3. Very high lane deviations occur with torque function V01, V04, V09 and V15. The smallest lane deviations occur using torque functions V06 and V10.

## 4.2 Operating Performance/User Performance

The task performance for examining the operating performance can be seen in Figs. 4 and 5. It can be seen that the tasks are better fulfilled with a larger torque amplitude

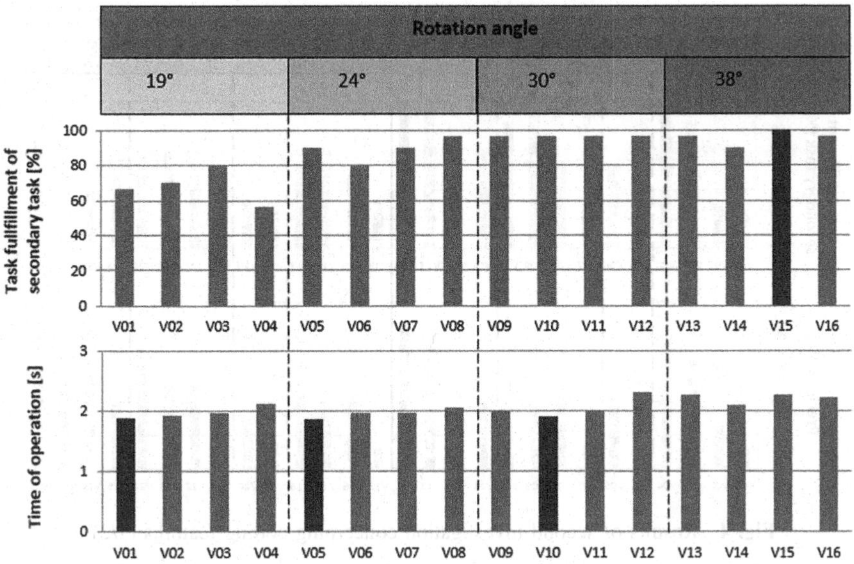


**Fig. 2.** Results of the first investigation concerning the initial parameters (1/2)

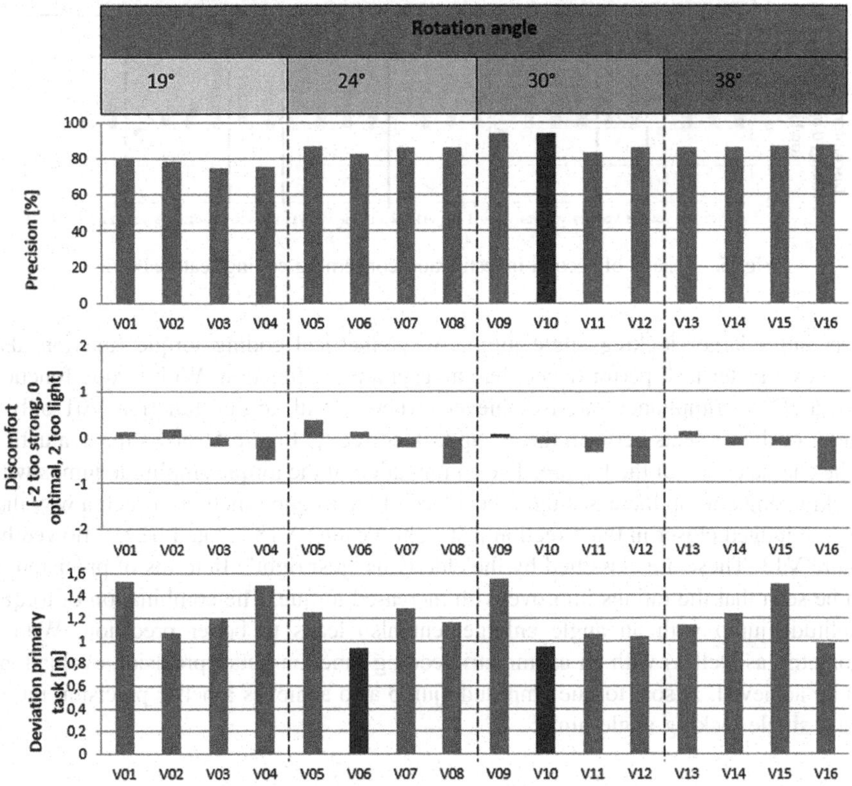


**Fig. 3.** Results of the first investigation concerning the initial parameters (2/2)

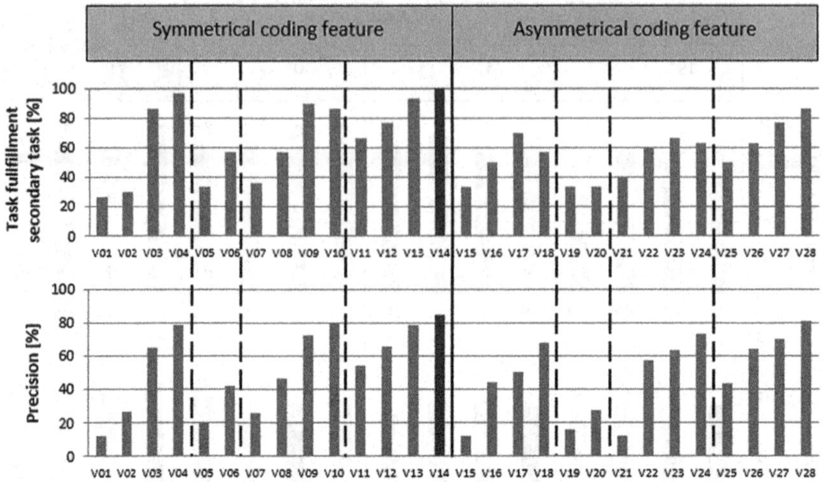


**Fig. 4.** Results of second investigation concerning coding features (1/2)

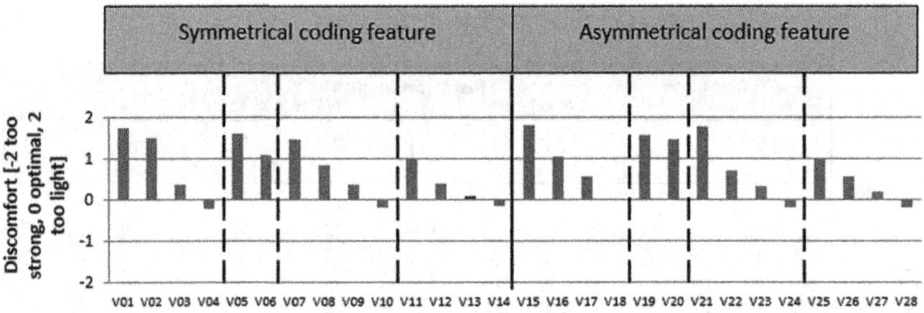


**Fig. 5.** Results of second investigation concerning coding features (2/2)

jump and a larger locking angle jump. A symmetrical coding torque function also achieves a better task performance than an asymmetric function. With torque function V14, a 100% compliance rate is achieved, whereas with torque function V01 only a quarter of the subjects perform the set up task correctly. Figure 5 shows the evaluation of the characteristic of the feature. The combination of the torque amplitude jumps with a locking angle jump increases the acceptance of the torque function in such a way that they are judged closer in the direction just right. Feature V18 is rated best, followed by feature V13. These are classified by the elderly as "just right". In terms of precision, it can be seen that the ratings improve with increased torque. The combination of torque amplitude jump with an angle enlargement also leads to better precision. With a symmetric as well as with an asymmetric coding function better precision evaluations can be achieved. A sole torque amplitude jump also achieves a better precision rating than a single locking angle jump.

# 5 Discussion

## 5.1 Initial Parameters

A significant correlation between the execution of the secondary task and the sensorimotor and cognitive capabilities of the subjects (TMT-L, RT, MLS) cannot be found. In order to identify significant differences between the individual experimental characteristics with regard to the positioning time, the LCT deviation, the precision and the discomfort in investigation 1, a Friedman test is conducted, followed by a Dunn Bonferroni's post-hoc test. The evaluation is evaluated with the statistics program SPSS Statistics 22 (IBM). A summary of significant (+: $p < 0.05$) and highly significant (*: $p < 0.01$) differences of the individual experimental characteristics are shown in Fig. 6. There are no significant differences in precision. Based on Fig. 6 it can be stated that no significant differences are found in the criteria V06, V10 and V11 for the examination criteria. With regard to the perception threshold value, an operation characteristic with a torque function of 0.09 Nm and a locking angle of 30° for the second test series is selected as the initial parameter for the coding possibilities on the basis of this test series in the case of the elderly. With a short positioning time (t = 1.9 s), this feature achieves a performance of 93.5%. The discomfort is classified by the older subjects as just right and the task performance is described precise (precision: 94%). Because of its high efficiency and effectivity according to a high user satisfactory this test parameter is chosen as the initial parameter for the second study.

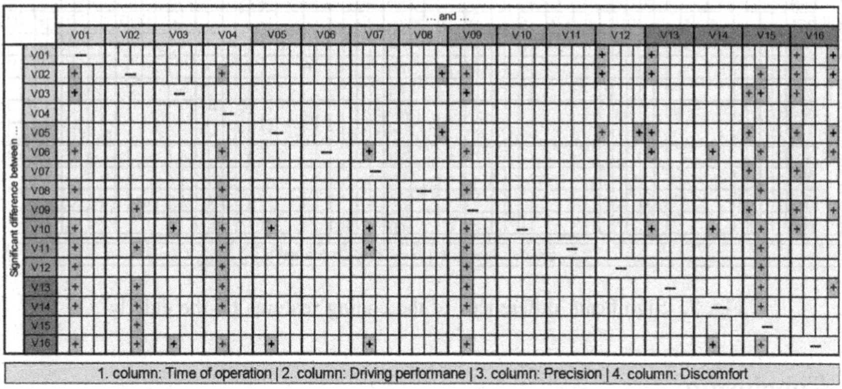


**Fig. 6.** Significant differences between the initial parameters concerning the elderly

## 5.2 Operation Performance/User Performance Concerning the Second Study

The data is analysed by friedman test. Post hoc tests are performed using bonferroni correction. A significant correlation between the execution of the secondary task and the sensorimotor and cognitive capabilities of the subjects cannot be found. The differentiation test is conducted in this part of the evaluation with regard to the discomfort and the precision. The significant (+: $p < 0.05$) and highly significant (*: $p < 0.01$) differences of the individual experimental characteristics are shown in Fig. 7. Significant differences are found in task fulfillment which is completed more successfully with increasing torque. It can also be deduced that the task performance increases with a combination of torque amplitude jump and locking angle jump. Based on the test results an increase in the locking angle or the torque amplitude jump improves the haptic feedback for the elderly and thus leads to a better job performance. A symmetrical course of the feature also contributes to a better task performance. With regard to the task fulfillment the best fulfillment rate is achieved with course V14. The characteristic expression is used for the comfort evaluation of the setting task. Considering the evaluation of the feature expression, the torque amplitude jumps V13 and V27 are classified as "just right, whereas the features with a small torque amplitude jump are judged rather than too easy. There are significant differences between low and high torque amplitude jumps (Fig. 7). In terms of comfort, older subjects rated V18 best. The precision of the setting task is rated better with increasing torque amplitude as well as by combining a larger locking angle with a torque amplitude increase. For older subjects, an angular magnification leads to a better precision assessment. The subjects rated feature V14 (85%) best.

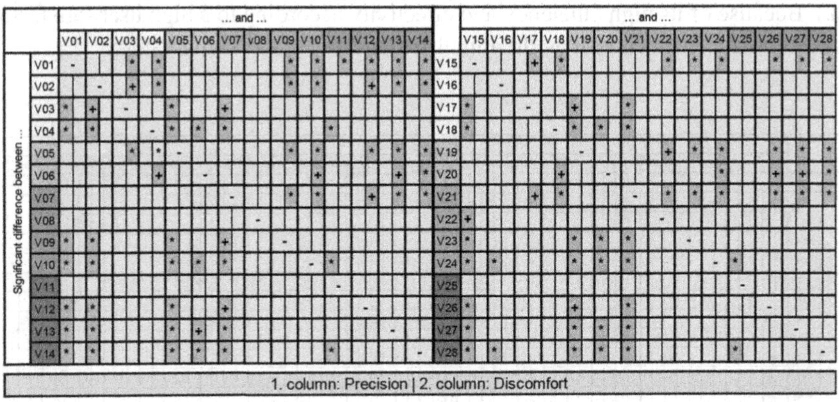


**Fig. 7.** Significant differences of the coding features by the elderly

## 6 Conclusion

This study has shown that the coding of rotary actuators is suitable for assisting the user in the execution of a secondary task. With regard to the haptic perception, torque amplitude of 0.09 Nm at a locking angle of 30° is best fulfilled by this study by the elderly. It can be shown that elderly users prefer significant coding features to choose a preferred value or to perform a change. Low level features, such as a sole rotation angle increase or a small torque increase were hardly perceived by the subjects. The combination of a rotation angle magnification and an increase in torque also leads to a better task performance. For the indication of a menu change elderly users prefer the combination of rotation angle magnification and torque increase. For the assignment of the feature V14 for a middle or a preferred value or the feature V28 for a change, however, the implementation of a further evaluation study is necessary.

**Acknowledgements.** This work is founded by the German Research Foundation (DFG) within the scope of the research project “Personalization options for hand-operated interfaces for the transgenerational user”, grant MA 4210/6-1.

## References

1. Hampel T (2011) Untersuchungen und Gestalthinweise für adaptive multifunktionale Stellteile mit aktiver haptischer Rückmeldung. Universität Stuttgart, Stuttgart. Institut für Konstruktionstechnik und Technisches Design, Forschungs- und Lehrgebiet Technisches Design, Dissertation
2. Winterholler J, Schulz C, Maier T (2017) Multimodal information transfer by means of adaptive controlling torques during primary and secondary task. In: Soares M, Rebelo F (eds) Ergonomics in design. Methods and techniques. CRC Press Taylor & Francis Group, Boca Raton, pp 3–20
3. Petrov A (2012) Usability-Optimierung durch adaptive Bediensysteme. Dissertation, Universität Stuttgart, Institute for Engineering Design and Industrial Design - Dept. Industrial Design Engineering
4. Winterholler J, Böhle J, Chmara K, Maier T (2014) Information coding by means of adaptive controlling torques. In: Yamamoto S (ed) Human interface and the management of information. information and knowledge design and evaluation, HCII 2014, Part I. LNCS, vol 8521. Springer, Cham, Heidelberg, pp 271–280
5. Anguelov N (2009) Haptische und akustische Kenngrößen zur Objektivierung und Optimierung der Wertanmutung. Dissertation, Technische Universität Dresden
6. Winterholler J, Böhle J, Maier T (2013) Informationskodierung mittels variabler Stellmomente im nutzerzentrieren Design. In: Spath D, Bertsche B, Binz H (eds) Stuttgarter Symposium für Produktentwicklung. Stuttgart
7. Fisk A, Rogers W, Charness N (2009) Designing for older adults. Principles and creative human factors approaches (Human factors & aging series), vol 2. CRC Press/Taylor & Francis, Boca Raton

# Evaluation of Comfortable Using Jerk Method During Transfer Caring

Ken Ikuhisa[1(✉)], Xiaodan Lu[2], Tomoko Ota[2], Hiroyuki Hamada[1], Noriyuki Kida[1], and Akihiko Goto[3]

[1] Kyoto Institute of Technology, Matsugasaki, Sakyo-ku, Kyoto, Japan
m7651002@edu.kit.ac.jp
[2] Chuo Business Group Co., Ltd., 1-6-6 Funakoshicho, Chuo-ku, Osaka, Japan
[3] Osaka Sangyo University, 3-1-1 Nakagaito, Daito-shi, Osaka 574-8530, Japan

**Abstract.** This paper focuses on the comfort of the care-receiver. In this study expert and non-expert caregiver in transfer caring with slide board are investigated. The device for 3-D motion analysis used in the experiment is MAC3D SYSTEM, an optical motion capture system manufactured by Motion Analysis Corporation. Coordinate data of each time of the each marker was collected from the data of motion capture. Data from reflection makers in-stalled on the head of the care-receiver was collected and recorded in this experiments. The transfer operation was divided into four processes by the difference of operation. The parameter Jerk in the Z direction over time in Process 3 (actual transport process) was evaluated. In physics, Jerk is the rate of change of acceleration and if Jerk=0, the direction of the acceleration would be changed. Too many changes of acceleration direction represent unsteady in the process of transferring the patient. Therefore, the stability of care-receiver during transfer caring was evaluated by the number of Jerk=0 in this study.

**Keywords:** Expert caregiver · 3-D motion analysis · Jerk=0

## 1 Introduction

The trend of aging in Asia is rising from 10% in 2009 to 30% in 2050. In an aging society, it poses many problems both to caregivers and to care-receivers. Therefore, reducing the burden on caregivers and comfort of care- receivers is important. Transfer caring had already been studied. [1–3] Caregiver usually has transferred the care-receiver without the help of any tool. Since transferring is a frequently performed in the caring center, it may seriously damage the back of the caregiver. However, by using the slide board, which can share a large portion of the weight of the care-receiver, the pressure on the caregiver is thus reduced and the transferring is carried out in a much more efficient manner. Expert movements is smooth [4] and smoothness of movement has been demonstrated in several studies [5–7] that use Jerk method. In this paper was focused on the comfort of the care-receiver in transfer caring with slide board, to evaluate between expert and non-expert caregivers by Jerk.


S. Bagnara et al. (Eds.): IEA 2018, AISC 826, pp. 178–188, 2019.
https://doi.org/10.1007/978-3-319-96065-4_21

## 2 Method

### 2.1 Subjects

As shown in Table 1, it was defined as an expert or non-expert person in care by a slide board. There were 8 trials (4 trials of expert caregiver and 4 trials of non-expert caregiver). The care-receiver, played by a college student, is presumed to be partially paralyzed. All experiments concerned were conducted under the consent of the participants. In the experiment, the caregiver is asked to simulate the process of moving the patient from the wheelchair to the bed. Focusing on the reaction of patients, this experiment mainly collects data from reflection makers installed on the head of the care-receiver.

**Table 1.** Subject conditions

| | Sex | Age | Care experience | Slide board use |
|---|---|---|---|---|
| Expert | Male | 35 years old | 8 and a half years | 5 year |
| Non-expert | Male | 28 years old | 6 years | None |

### 2.2 Measuring Equipment

The device used in the experiment is MAC3D SYSTEM, an optical motion capture system manufactured by Motion Analysis Corporation. When capturing data, the working frequency of the device is set at 120 Hz. In the experiment was installed 21 infrared reflection makers on the skilled caregiver, 27 on the unskilled caregiver and 7 on the patient. In establishing the coordinate system, it was defined the long side of the bed as the X axis, the short side as the Y axis and the vertical direction as the Z axis.

## 3 The Four Processes of the Experiment

In order to better analyze the transfer caring, it was divided into four processes. Process 1 is the first preparation phase before actually transferring, called the posture adjusting phase before transferring. Figure 1(a) shows a photograph of the operation. When caregiver touch care-receiver's body it was counted as the beginning. When the caregiver touched the slide board, it was counted as the end of the project 1 and the start of the process 2. Process 2 is the stage called the slide board inserting phase. Figure 1(b) shows a photograph of the operation. During process 2 caregiver first inserted one end of the slide board into the gap between care receiver's hip and wheelchair, and then put the other end to the bed, forming a skateboard bridge. When the caregiver completed the inserting of slide board, his hand off the slide board, which is the end of the process 2, and the beginning of the process 3. Process 3 is the process of moving a care-receiver from

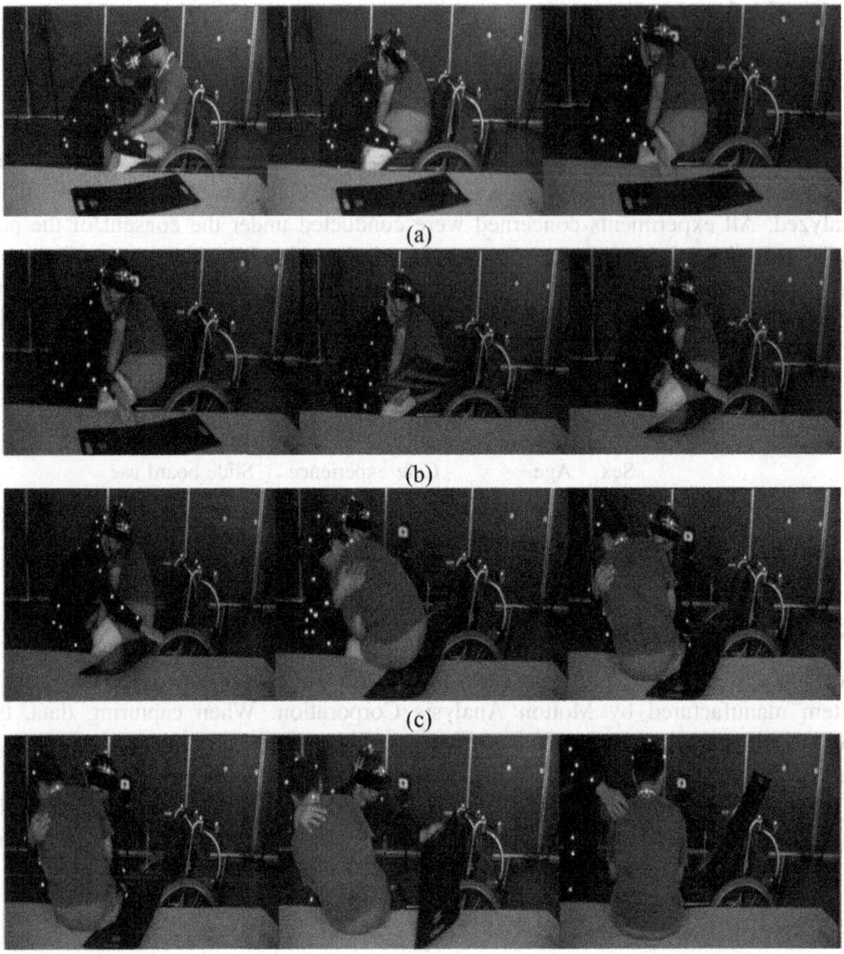

(a) (b) (c) (d)

**Fig. 1.** The Four Processes of the Experiment ((a) Process 1: the posture adjusting phase before transferring, (b) Process 2: the slide board inserting phase, (c) Process 3: the actual transferring phase, (d) Process 4: the posture adjusting phase after actual transferring)

wheelchair to bed with a slide board, which called the actual transferring phase. Figure 1 (c) shows a photograph of the operation. When the hip of care receiver set down to the bed, the process 3 ended and began process 4. Process 4 is the posture adjusting phase after actual transferring. In this phase the actual transfer process had been completed, this process is the final phase of the entire transferring process. Figure 1(d) shows a photograph of the operation. In this process, caregiver help care-receiver to adjust his posture on bed, so that he feel comfortable and safe. When the posture adjustment work is completed, care-receiver sitting on bed, caregiver stand on the bed side, hand off the care-receiver, process 4 end, and the entire transferring process is completed.

### 3.1 Result of Process Analysis

The results of process analysis are shown Fig. 2. In process 1, there is no significant difference in time between expert and non-expert caregiver. In Process 2, process 3, process 4, expert caregiver has significance, and care time is short. Four trials of the track diagram of care-receiver head on the horizontal level by expert caregiver is shown Fig. 3. Four trials of the track diagram of care-receiver head on the horizontal level by non-expert caregiver is shown Fig. 4. Four trials of the track diagram of care-receiver's head in the Z direction by expert caregiver that changes with time was shown Fig. 5. Four trials of the track diagram of care-receiver's head in the Z direction by non-expert caregiver that changes with time was shown Fig. 6.

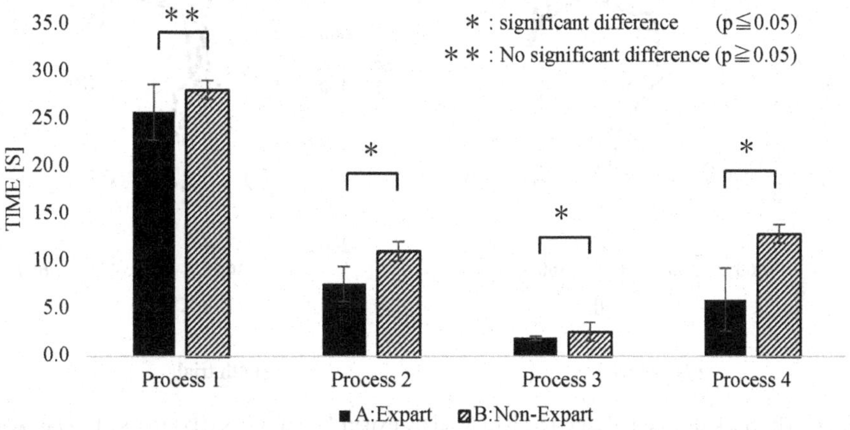


**Fig. 2.** Time taken for each of the four processes of transfer caring (Process 1: the posture adjusting phase, Process 2: the slide board inserting phase, Process 3: the actual transferring phase, Process 4: the posture adjusting phase)

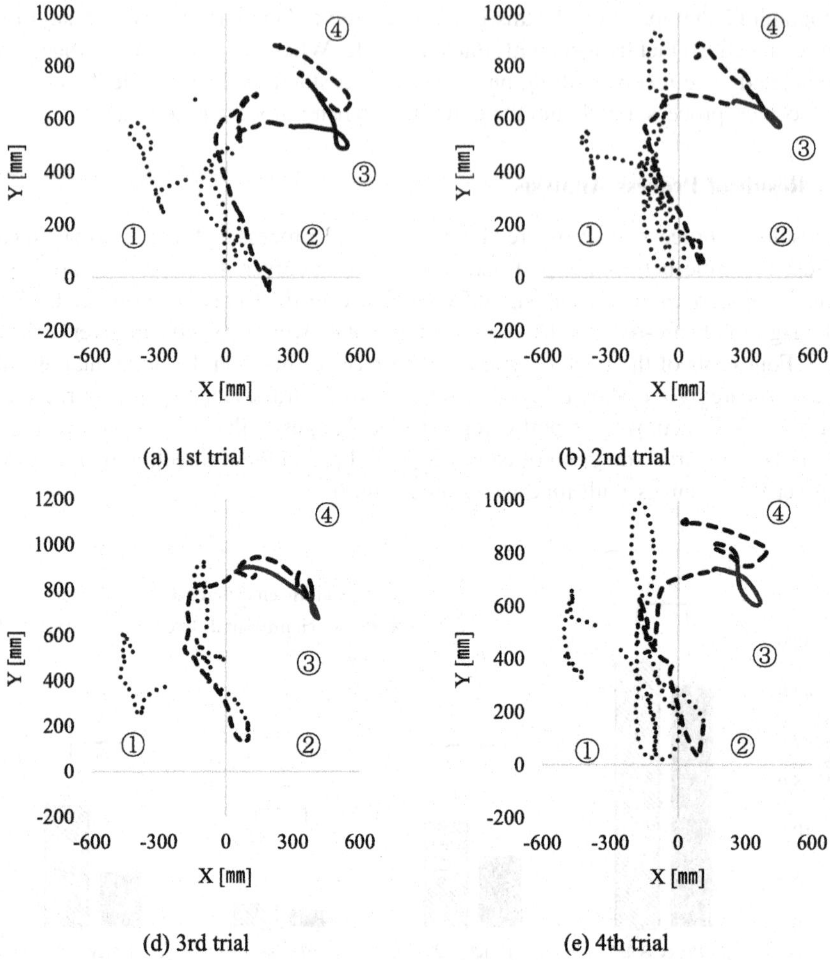


**Fig. 3.** The track diagram of care-receiver head by expert In four trials (① process 1, ② process 2, ③ process 3, ④ process 4)

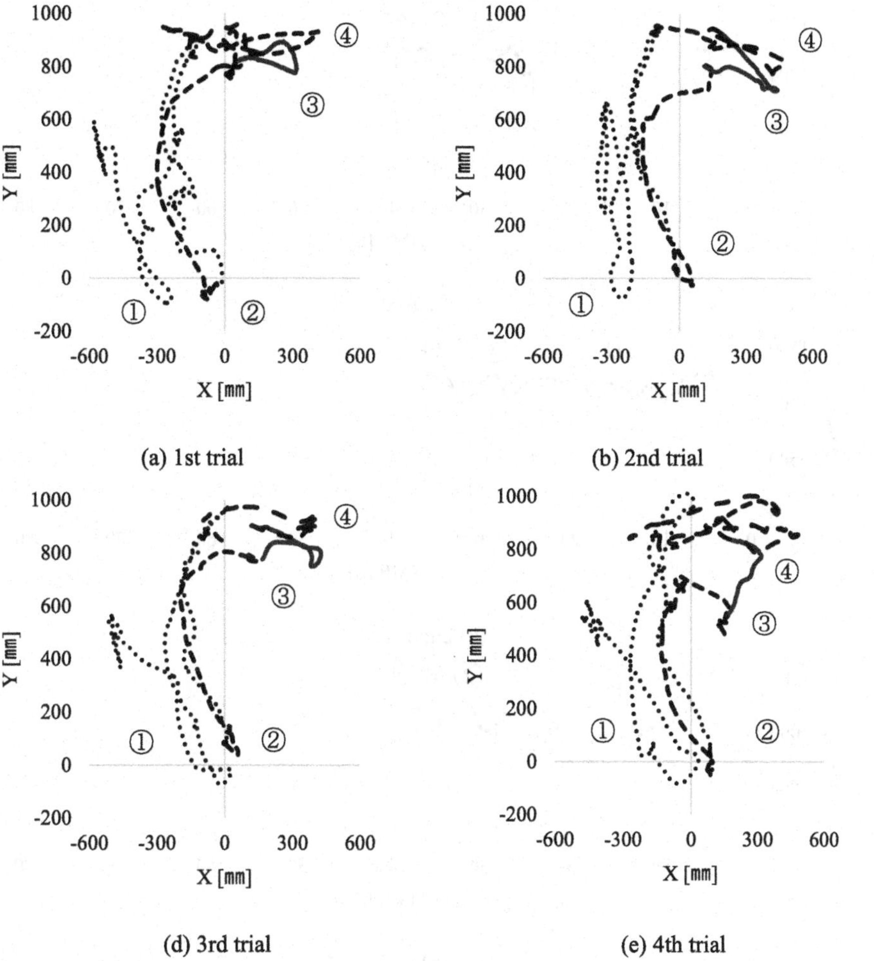


**Fig. 4.** The track diagram of care-receiver head by non-expert in four trials (① process 1, ② process 2, ③ process 3, ④ process 4)

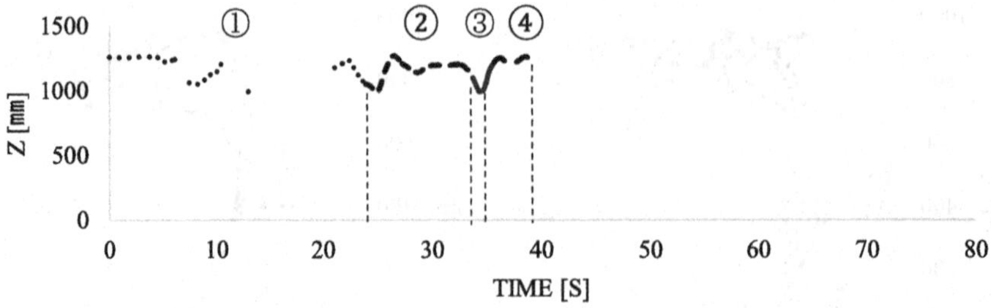


(a) 1st trial

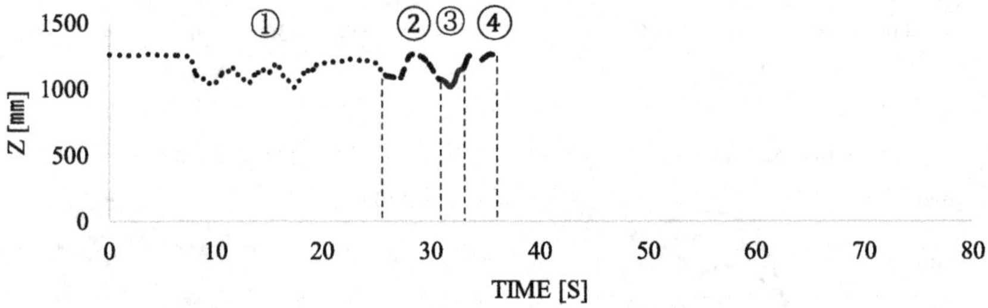


(b) 2nd trial

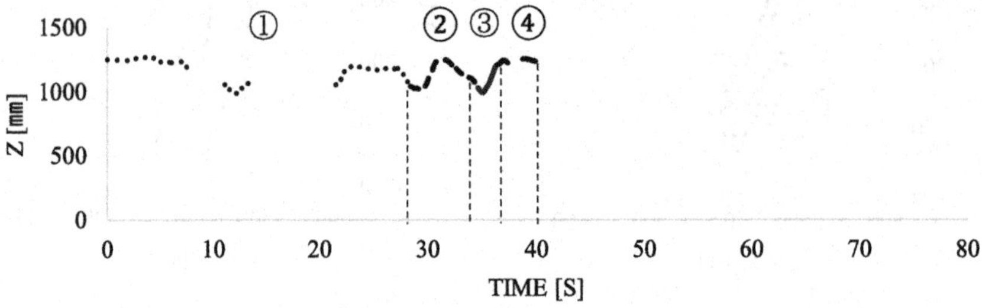


(c) 3rd trial

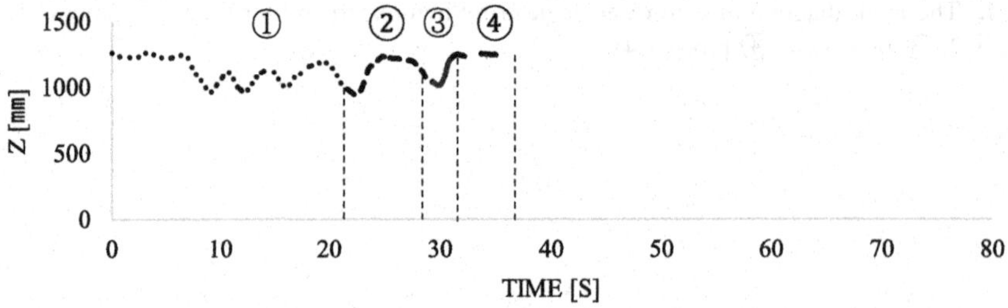


(d) 4th trial

**Fig. 5.** 4th trial the track of care-receiver's head in the Z direction by expert in four trials (① process 1, ② process 2, ③ process 3, ④ process 4)

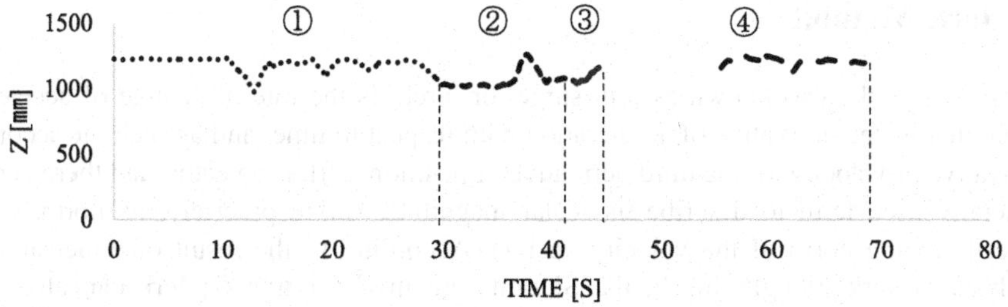


(a) 1st trial

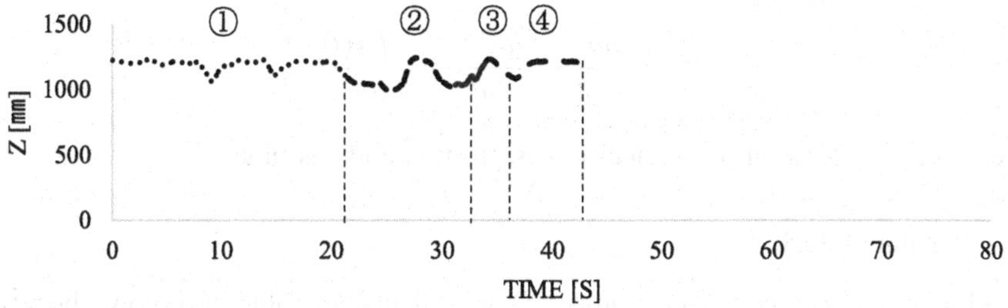


(b) 2nd trial

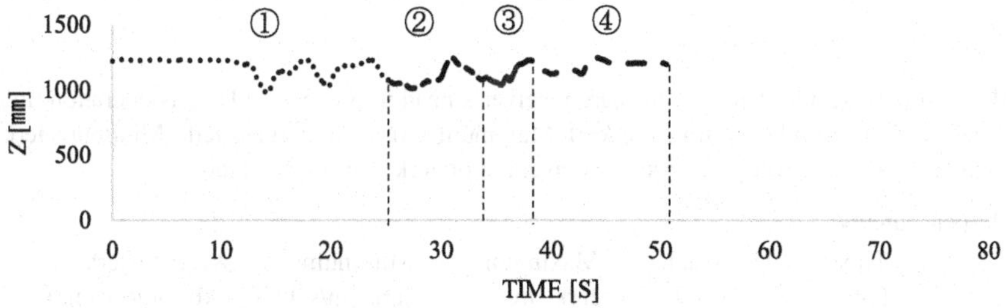


(c) 3rd trial

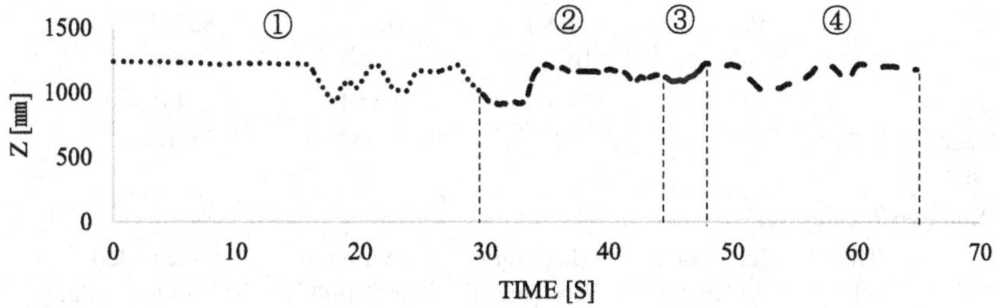


(d) 4th trial

**Fig. 6.** 4th trial the track of care-receiver's head in the Z direction by non-expert in four trials (① process 1, ② process 2, ③ process 3, ④ process 4)

## 4 Jerk Method

In physics, Jerk, also known as jolt, surge, or lurch, is the rate of change of acceleration; that is, the derivative of acceleration with respect to time, and as such the second derivative of velocity, or the third derivative of position. Jerk is a vector, and there is no generally used term to describe its scalar magnitude (more precisely, its norm, e.g. "speed" as the norm of the velocity vector). According to the result of dimensional analysis of Jerk, [length/time$^3$], the SI units are m/s$^3$ (or m·s$^{-3}$); Jerk can also be expressed in standard gravity per second (g/s). Jerk can be formulated in any of the following equivalent ways:

$$\mathrm{J(t)} = \frac{da(t)}{dt} = \frac{d^2v(t)}{dt^2} = \frac{d^3r(t)}{dt^3} \tag{1}$$

Where, $a$ is acceleration, $v$ is velocity, $r$ is position and $t$ is time.

### 4.1 Results of Jerk

Jerk of the care-receiver's head in process 3 was calculated. Table 2(a) shows the result of Jerk for 4 trials by expert caregiver. Table 2(b) shows the result of Jerk for 4 trials by non-expert caregiver. Comparison of the average for 4 trials between expert and non-

**Table 2.** 4 trials result of Jerk of the care-receiver's head in process 3 (Time is operation times, Jerk reach 0 is the number of times Jerk=0, Maximum Jerk is the largest Jerk, Minimum Jerk is the smallest Jerk, and Average Jerk is the average of Jerk's absolute value)

(a) Expert caregiver

| | Time [s] | Jerk reach 0 [Times] | Maximum Jerk [m/s$^3$] | Minimum Jerk [m/s$^3$] | Average Jerk (Absolute value) [m/s$^3$] |
|---|---|---|---|---|---|
| Trial 1 | 1.9 | 182 | 3180 | −3430 | 74592 |
| Trial 2 | 1.9 | 180 | 4994 | −6118 | 85763 |
| Trial 3 | 1.8 | 160 | 10312 | −8393 | 132614 |
| Trial 4 | 1.7 | 143 | 3263 | −3391 | 73451 |
| Average (4 tials) | 1.8 | 166 | 5437 | −5333 | 91605 |

(b) Non-expert caregiver

| | Time [s] | Jerk reach 0 [Times] | Maximum Jerk [m/s$^3$] | Minimum Jerk [m/s$^3$] | Average Jerk (Absolute value) [m/s$^3$] |
|---|---|---|---|---|---|
| Trial 1 | 2.1 | 186 | 9014 | −9669 | 135890 |
| Trial 2 | 2.8 | 233 | 16559 | −9219 | 84233 |
| Trial 3 | 2.6 | 230 | 14728 | −12757 | 84920 |
| Trial 4 | 2.6 | 230 | 4691 | −5663 | 55735 |
| Average (4 tials) | 2.5 | 220 | 11248 | −9327 | 90194 |

expert. The number of Jerk reaching 0 (Jerk=0) was less for expert caregiver than that of non-expert caregiver. Expert caregiver have a lower maximum Jerk value than that of non-expert caregiver. Expert caregiver have a higher minimum Jerk value than that of non-expert caregiver.

## 5 Discussion

When the value of Jerk=0, the acceleration would be reached its minimum or maximum value. In other words, when Jerk is zero, the direction of the acceleration would be changed. Too many changes of acceleration direction represent unsteady in the process of transferring the patient. If the acceleration changes too frequently, it could be inferred that the transfer process is not steady enough. Therefore, the stability of care-receiver during transfer caring was evaluated by the number of Jerk=0 in this study. From the results, Jerk reaching zero of expert caregiver was less than non-expert caregiver. Therefore, it was thought that expert caregiver was comfortably transferring care as compared with non-expert caregiver.

## 6 Conclusion

In this experiment, motion capture technology and 3-D analysis technology are employed to analyze the trajectory, height and Jerk of each set of data. This study clarified the difference between expert and non-expert caregiver in transfer caring using slide board. In the process analysis, the transfer nursing care was divided into four processes. Based on the results, we clarified the difference in speed of operation between expert and non-expert caregiver. Furthermore, it was analyzed by Jerk in process 3. As a result the number of times that expert caregiver had Jerk reaching zero was less than non-expert caregiver.

## References


1. Ito M, Yoshikawa T, Goto A, Ota T, Hamada H (2015) Effect of care gesture on transfer care behavior in elderly nursing home in Japan. In: Duffy V (ed) Digital human modeling. Applications in health, safety, ergonomics and risk management: ergonomics and health, DHM 2015. LNCS, vol 9185. Springer, Cham, pp 127–132
2. Ito M, Takai Y, Goto A, Kuwahara N (2014) Research on senior response to transfer assistance between wheelchair and bed. In: Duffy VG (ed.) Digital human modeling. Applications in health, safety, ergonomics and risk management, DHM 2014. LNCS, vol 8529. Springer, Cham, pp 558–566
3. Koshino Y, Ohno Y, Hashimoto M, Yoshida M (2007) Evaluation parameters for care-giving motions. J Phys Ther Sci 19:299–306
4. Nelson WL (1983) Physical principles for economies of skilled movements. Biol Cybern 46 (2):135–147

5. Flash T, Hogan N (1985) The coordination of arm movements: an experimentally confirmed mathematical model. J Neurosci 5:1688–1703
6. Hogan N (1988) Planning and execution of multijoint movements. Can J Physiol Pharmacol 66:508–517
7. Schneider K, Zernicke RF (1989) Jerk-cost modulations during the practice of rapid arm movements. Biol Cybern 60:221–230

# Towards an Age-Differentiated Assessment of Physical Work Strain

Matthias Wolf(✉) and Christian Ramsauer

Graz University of Technology, Rechbauerstrasse 12, 8010 Graz, Austria
Matthias.wolf@tugraz.at

**Abstract.** Companies in different industries are all facing the same challenge: demographic change. This leads to a higher share of elderly workers in most workplaces. When designing workstations or evaluating risks at workplaces, it is necessary to take into account age-related changes in physical and sensory skills and in cognitive or mental capabilities. Although there is a lot of data available in literature concerning age-related changes in human performance prerequisites, only a few methods consider the factor age. For example, the REFA-method uses a factor considering age when calculation maximum forces. However most ergonomic standards and occupational risk assessment methods used in industry do not consider this data and the specific needs of an ageing workforce. The aim of this paper is to introduce a work evaluation framework which was developed based on literature and will be compared to practical findings from a field study conducted with a manufacturing company and a food retailer in Austria. Therefore, a three step approach was carried out: First the changes which humans experience when they get older were elaborated based on an extensive literature analysis and age-critical factors, that should be covered in ergonomic risk assessment methods, were derived. Second existing risk assessment methods that cover physical occupational risks were collated and examined, considering their coverage of age-relevant factors. Finally, all findings were combined to an age-differentiated workplace risk screening method. As a result, we developed a method that should allow to reveal existing problem-workplaces for elderly workers and point out the need for age-related adaption at them. Furthermore, it reveals existing age-appropriate workplaces, where workers should be able to work their whole working life.

**Keywords:** Industrial engineering · Demographic change
Age-based risk assessment

## 1 Introduction

The industry is a central driver for economic growth, societal wealth and improved standard of living in Austria. In the past, technology, particularly automation, has been the key driver for increasing manufacturing's productivity. However due to globalization and the ongoing demographic change, as well as changing market demands, the manufacturing industry is now facing new challenges. an increase in shorter product life cycles, in product variety as well as in aggregation of work contents and in productivity requirements, let many companies see themselves compelled to continuously restructure


S. Bagnara et al. (Eds.): IEA 2018, AISC 826, pp. 189–205, 2019.
https://doi.org/10.1007/978-3-319-96065-4_22

their production processes in order to remain competitive [1]. Instead of only focusing on automation and mass production, it now becomes more and more necessary to focus on the human as the central part of the production system [2, 3].

Due to changes of the working environment along with the current demographic challenge in Europe, many companies face the problem of coping those changes with a continuously aging workforce. In addition, increasing (skilled) workforce shortage in many occupations[1] will also force different industries, to apply more elderly workers. To successfully cope with these new challenges future work and production processes also have to be designed, specifically for age-related changes of the performance conditions of the employees [4]. Not having developed sustainable concepts to an aging workforce, companies risk losing their productivity and innovation [5].

Despite of ongoing automation and increasing technological possibilities to reduce physically demanding work, still most frequent reasons for sick leave days are musculoskeletal disorders (MSD) resulting from physical overloading [6]. The identification and assessment of physical stress thus takes on a high priority, because it enables an estimation of MSD risk potential at workplaces. Knowing and assessing the physical strains at work moreover is the basis for deriving measures to reduce stress and thus an important contribution to decrease negative consequences, especially in the face of an on average ageing population. Still, objective instruments and methods presenting a tangible and holistic framework for a systematic age-differentiated work evaluation are missing [7].

## 2 Demographic Change and Changing Workforce Abilities

Due to a complex combination of social and technological improvements like better prosperity, wealthy diet, better medical and social care, and human adapted work conditions the anticipated average (working) life increases [8–10]. Many physical, sensory and cognitive skills and abilities change during the aging process which can lead to a mismatch in the workforce's workability and the demands at their work places. Important tendencies of general human abilities' age-related changes are summarized in Fig. 1. Particularly decreasing physical capabilities (e.g. muscular strength or fine motoric skills) decreasing sensory abilities (e.g. sense of hearing or vision) changing cognitive and mental functions (e.g. mental flexibility, speed of information processing, declining function of short term memory) and declining amount of information which can be processed at the same time could negatively influence the industrial workability. Further changes connected in literature to elderly workers are a smaller willingness to learn new things or adapt to changes or to work with new operating means and technologies. On the other hand, caused by increasing life and job experience and changing problem solving strategies, many skills like leadership skills, self-organization skills, the ability to judge in new or complex situations, problem solving skills, social and teamwork-skills, etc. increase with age. (cp. [12–25] and Fig. 1). The arrows next to the different changes the changing

[1] Cp. Austrian skilled workforce radar: http://fk-monitoring.at/.

characteristics indicate the direction of change. Upwards arrows indicate an increase or improvement of the specific characteristic, red arrows indicate a decrease or worsening and yellow arrows define characteristics that stay on the same level with increasing age.

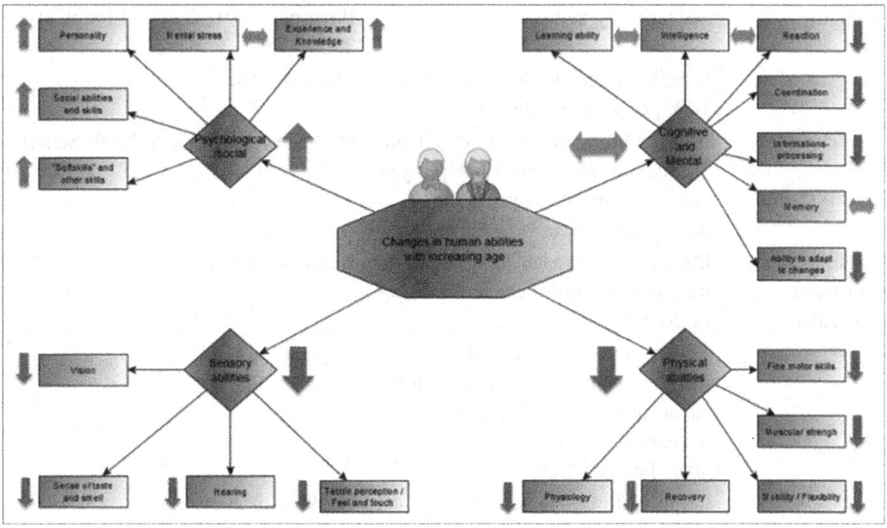


**Fig. 1.** General trends in development of human abilities and capabilities when ageing [26]

Studies evaluating the effect of age on worker abilities in their natural environment typically, encounter the so called "healthy worker effect". Healthy workers tend to continue working, whereas those with poorer health have a tendency to retire early or transfer to a less physically demanding job. The isolated influence of aging upon physical work capacity among the workforce is therefore often compromised due to the tendency of healthier and stronger workers remaining in physically demanding jobs longer [27–29].

Even though the changes within human physical abilities that are encountered with aging, are influenced by genetics and individual lifestyle, the work environment as well as several compensation mechanisms caused by increasing work experience (cp.e.g. [30]), it can be stated that most physical abilities tend to decrease when getting older. For example, Baines et al. [31] developed an age related human performance model for discrete event simulation (DES), where they use an age-related decrease in performance, adding up to a total performance loss of 35% till the age of 65 years in terms of cycle times in assembly [31]. Whereas other authors found evidence for increasing total productivity with higher age [32]. However, Table 1 summarizes main findings from literature review concerning quantifiable age relevant decreasing changes in human capabilities relevant for manufacturing and manual work.

**Table 1.** Age related changes of physical human abilities. (Adapted by [33])

| Decline in abilities | Example | Source |
|---|---|---|
| Peak force (Muscular Strength) | It peaks (100%) between the ages of 25 and 30. By age 40, 95%; by age 50, 85%; by age 65, 75% | Vitasalo et al. [50]; Aoyagi and Shephard [34]; De Zwart et al. [38]; Ilmarinen 2001; Gall and Parkhouse 2004 |
| | It peaks at age 20, then decreases by 10% per decade until 60 | Mazzeo 2000 |
| | It peaks between the age of 20 and 30, then decreases by 1.5% per year age of 50–60. Afterwards the decline rate raises to 3% per year | Hollmann and Strüder 2009; Scherf [47] |
| Dynamic actions with force (Muscular Power) | Decline 10% greater than decline in muscular strength between the ages of 20–80 | Kenny et al. 2008 |
| | The leg forces are needed for dynamic work (e.g. lifting from the squat, pushing, carrying). It decreases by about 5% per decade after the age of 20 | Lindle et al. 1997 |
| | Gripping forces peak around the age of 35 and thereafter decrease by about 5% per year. At age 65 75–80% of the maximum value remains | Hank et al. 2009 |
| Endurance (Aerobic Capacity) | 10% loss per decade. In machine-paced tasks, the standard rate demands 80% of the sustainable aerobic capacity of a 40-year-old worker | Shephard 1999 |
| | It peaks in the 20 s, and declines by 1% per year thereafter | De Zwart et al. 1995 |
| | At age 65 it is 70% of that at age 25 | Kowalski-Trakofler et al. 2005 |
| | It decreases by about 5–10% per decade after the age of 25 years. After the age of 40, by about 10–15% per decade | Heath et al. 1981 |
| | At age 40 cardiac output is 95% of that at age 30 | Goldich 1995 |
| Reaction and movement Times (Responses) | Aged 66-or-over are 30% slower than aged 18–30 | Tolin and Simon 1968 |
| Awkward postures (Flexibility) | It decreases 20% to 30% from age 30 to 70 | Johns and Wright [44]; Chapman et al. 1992 |
| | Flexibility decreases after the age of 25. By age 35 90%, by age 50 80%, by age 70, 70%. | Brown and Miller 1998 |

As can be seen in Table 1 most physical human abilities have a significant decline over age. On that basis it is clear that also occupational strains differ individually. Basic concept that explains age-differentiated or individualized work evaluation is the stress strain concept [51]. According to the stress-strain-concept, work tasks needing abilities with decreasing performance factors, as described above, can lead to a higher strain of the elderly worker under the same stress level [52]. The effect of negative work-system influences on age-related changing human abilities needs to be pointed out when referring to an age-differentiated design of work-systems. Increased strains in the course of life can lead, besides a decrease of the performance and the physical well-being [53], to a higher fatigue of the elderly employee. Furthermore, it can provoke signs of wear and damages manifesting in the form of inactive periods caused by illness or even lead to health impairments like MSD [54].

## 3 Age Differentiated Workplace Evaluation

By combining the changes in physical factors with possible hazards of work, age-critical factors can be derived. according to the stress strain concept these factors are that areas which have to be evaluated in a different way when designing work for elderly people. Figure 2 summarizes age critical factors identified from literature. Possible occupational hazards are clustered into eleven categories which have to be considered when assessing workplaces according to [55]. The list was supplemented with age critical factors according to literature [56, 57]. Further factors which are influenced by age relevant changes are marked in red.

| | | | | | | | | |
|---|---|---|---|---|---|---|---|---|
| 1 | Mechanical hazards | 1.1 Unprotected moving machine parts | 1.2 Parts with dangerous surfaces | 1.3 Mooving transport and work equipment | 1.4 Uncontrolled moving parts | 1.5 Falling, stumbling slipping, kinking | 1.6 Crashing / Falling | |
| 2 | Electronic hazards | 2.1 Electric shock | 2.2 Electrical arc | 2.3 Electrostatic charges | | | | |
| 3 | hazardous substances | 3.1 Skin contact with hazardous substances | 3.2 Inhalation of hazardous substances | 3.3 Ingestion of substances | 3.4 Physico-chemical hazard | | | |
| 4 | Biological working materials | 4.1 Infection hazard by pathogenic microorganisms | 4.2 Sensitizing and toxic effects of microorganisms | | | Factors changing with age | | |
| 5 | Fire and explosion hazard | 5.1 Combustible solids, gases, Liquids, | 5.2 Explosive atmosphere | 5.3 Explosives | | | | |
| 6 | Thermal hazards | 6.1 Hot media / surfaces | 6.2 Cold media / surfaces | | | | | |
| 7 | Special physical factors | 7.1 Noise | 7.2 Ultrasound, Infrasound | 7.3 Whole-body vibration | 7.4 Hand-arm vibrations | 7.5 Optical radiation (UV, IR, laser) | 7.6 Ionizing radiation (e.g., X-rays) | 7.7 Electro-magnetic fields |
| 8 | working environment | 8.1 Climate (heat, cold) | 8.2 Light | 8.3 Suffocation or drowning | | | | |
| 9 | Physical work load | 9.1 Heavy dynamic work, combination of static + dynamic work | 9.2 Unilateral dynamic work, body movement | 9.3 Holding work + Postural work (forced postures) | 9.4 Load handling | 9.5 Repeating uniform movements (finger-hand-arm, foot-leg) | 9.6 Working while sitting, standing, or walking, squatting, kneeing or lying | |
| 10 | Mental and psychological factors | 10.1 Inadequately designed work task | 10.2 Inadequately designed work organization | 10.3 Inadequately designed social conditions | 10.4 Inadequately designed work-places and environments | 10.5 Dequalification, outdated knowledge, learning cessation | | |
| 11 | Other hazards | 11.1 By animals | 11.2 Through plants + herbal products | 11.3 By people | 11.6 Human - machine cooperation | | | |

**Fig. 2.** Age-critical aspects of working hazards based on [55, 56] (Color figure online)

As can be seen in Fig. 2 all physical as well as all mental and psychological factors are considered age-dependent. Further all factors concerning sensory skills like seeing or hearing are judged as age-dependent as well. In this paper only physical hazards are considered further. For physical ergonomic risk assessment there are fife traditional domains to be considered in literature [23, 58], namely, the (1) energy spent on a work day or per minute (active metabolic rate), (2) body postures and body movements at the work place including awkward postures, (3) action forces or thrust forces, (4) load handling and (5) repetitive activities.

The **working energy sales** is a value which can be used to evaluate activities not covered by other assessment methods. For example, the energy which is spend on maintaining unnatural body postures can be estimated via the energy sales. Also energy spend on movements as walking or on climbing a ladder can be taken into account by considering the working energy expends. Equations for calculation working energy sales for different activates were developed by Spitzer and Hettinger (1964) [59], Garg et al. (1978), and Genaidy und Asfour (1987) [60]. In occupations were heavy dynamic work can occur it is necessary by law to document the energy expenditure in Austria. Therefor a tool called "Schwerarbeitsrechner" [61] was developed. The metabolic rate of a person is linked to its heart beat frequency and therefor to its aerobic capacity or oxygen consumption. The maximum possible oxygen consumption again is a factor highly significant depended on age as shown in Table 1 and in [62].

Unnatural **body postures** or forces working postures are known to impose problem to elderly workers. Especially working overheadad height or working in kneeling position can be very strain full in higher age. The evaluation of such positions is possible via checklists or the OWAS method. Screening approaches for assessing body postures dependent of the time executed are included in German methods like the "New production Worksheet (NPW)" or the "European assembly worksheet (EAWS)", rated by a point system, as well as in the "assembly worksheet light (AWS), where directly the risk level is derived.

To evaluate the risk deriving from **load handling** the "key indicator method (KIM)" for manual material handling and the KIM for pushing and pulling are tools very often in Germany and Austria. The risk deriving from load manipulation is calculated based on key indicators like execution time, weight of the load, execution conditions and body posture [63]. Further, when dealing with many different loads the "multiple Lasten tool (MLT)" [64] is a useful supplement which was developed to evaluate tasks in logistics were several different loads have to be transported. Also the MLT provides an algorithm to calculate a combined risk value for pulling and lifting activities [64]. ON a more detailed level the international standards ISO 11228 or EN 1005 as well NIOSH tool can be used to calculate ergonomic risks from manual load handling. Load limits can again be a function of worker's age while the maximum load suggested also decline with age [66, 67]. Also it is stated in the KIM that loads which lead to an evaluation result in between the values of 25–50 (yellow area of the scale) can lead to an overburden of less resilient workers, which are defined to be persons younger than 20 years and older than 40 years as well as low skilled workers [65].

**Repetitive actions** are one of the main cause for problems within arms, hand and fingers. Typical assessment methods for these domain are again a KIM which calculate the risk based on movements per minute, execution conditions, frequency of task changes, time of execution, and ergonomic factors [68]. Age based evaluation of repetitive action should take into account the increasing time needed for reactions and the execution of movements of an elderly worker compared to a younger one [69, 70].

Figure 3 summarizes the different sources of ergonomic risks at the workplace, ergonomic standards and methods used to evaluate this risks, limits from common assessment methods and age based changes that can increase risks at the particular activity.

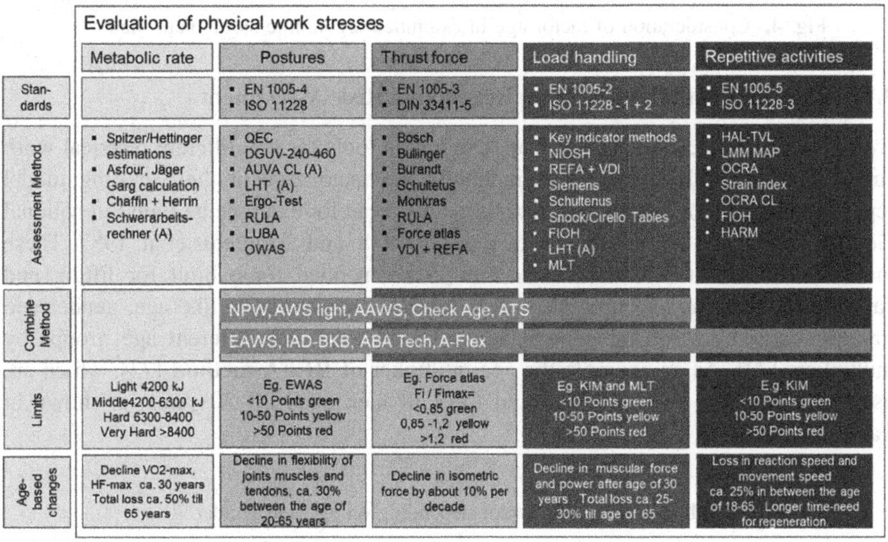
Evaluation of physical work stresses
Metabolic rate
Postures
Thrust force
Load handling
Repetitive activities
Stan-dards
Assessment Method
Combine Method
Limits
Age-based changes
EN 1005-4
ISO 11228
EN 1005-3
DIN 33411-5
EN 1005-2
ISO 11228 - 1 + 2
EN 1005-5
ISO 11228-3
Spitzer/Hettinger estimations
Asfour, Jäger Garg calculation
Chaffin + Herrin
Schwerarbeits-rechner (A)
QEC
DGUV-240-460
AUVA CL (A)
LHT (A)
Ergo-Test
RULA
LUBA
OWAS
Bosch
Bullinger
Burandt
Schultetus
Monkras
RULA
Force atlas
VDI + REFA
Key indicator methods
NIOSH
REFA + VDI
Siemens
Schultenus
Snook/Cirello Tables
FIOH
LHT (A)
MLT
HAL-TVL
LMM MAP
OCRA
Strain index
OCRA CL
FIOH
HARM
NPW, AWS light, AAWS, Check Age, ATS
EAWS, IAD-BKB, ABA Tech, A-Flex
Light 4200 kJ
Middle4200-6300 kJ
Hard 6300-8400
Very Hard >8400
Eg. EWAS
<10 Points green
10-50 Points yellow
>50 Points red
Eg. Force atlas
Fi / Fimax=
<0,85 green
0,85 -1,2 yellow
>1,2 red
Eg. KIM and MLT
<10 Points green
10-50 Points yellow
>50 Points red
E.g. KIM
<10 Points green
10-50 Points yellow
>50 Points red
Decline VO2-max, HF-max ca. 30 years Total loss ca. 50% till 65 years
Decline in flexibility of joints muscles and tendons, ca. 30% between the age of 20-65 years
Decline in isometric force by about 10% per decade
Decline in muscular force and power after age of 30 years. Total loss ca. 25-30% till age of 65
Loss in reaction speed and movement speed ca. 25% in between the age of 18-65. Longer time-need for regeneration

**Fig. 3.** Different areas of evaluation, the limit values allowed and age related changes in the different areas (based on [71])

Knowing the factors which should be assessed differently when talking about elderly workforce, a literature review at existing assessment methods and the consideration of age-relevant aspects within them was carried out. In total 42 ergonomic workplace assessment methods from different countries and physical strain areas were assed concerning the coverage of factor gender (1), age (2), worker limitations e.g. back pain (3) different physical workability (4) and perceived work strain (5).

The result in Fig. 4 shows that only a few tools considered these individual factors. While 15 assessment methods diver in the grading between male and female workers only 5 consider age relevant factors or the perceived stresses at workplace, 3 consider different work abilities or skill levels and only 2 consider limitations existing at the workforce. The rating partly considered was given if the assessment methods inform that the factor can be an important influence, but no further information or consideration is suggested. Methods which do include age relevant grading are described in the subsequent section.

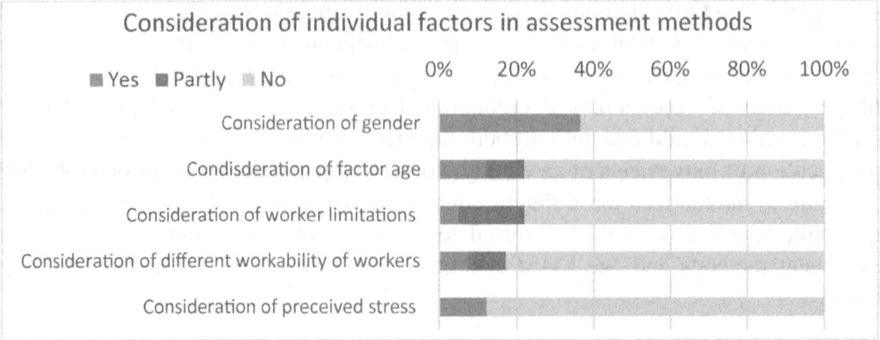


**Fig. 4.** Consideration of factor age in examined ergonomic assessment methods

### 3.1 Age Differentiated Tools for Ergonomic Risk Assessment

Even though there exist many of work evaluation tools for the different physical work stresses the are only a few tools including the factor age in their grading model (cp. Fig. 3 and [71]). Age-differentiating methods can for example be found in national German approaches based on work of Burandt 1978, and Schultetus et al. 1987. These methods are used to calculate maximum recommended force limit for lifting and carrying activities and focuses also on individual characteristics like age, gender and training status by lowering general advised load limits for different age groups by multipliers [72]. As an example the age-multiplier of **REFA-method** [73], which are used to reduce the average maximum force of men aged 20–30 years, is shown in Table 2.

**Table 2.** Weighting factor depending on age and gender 70

| Age | Male | Female |
|---|---|---|
| 15–18 | 0,70 | 0,50 |
| 19–35 | 1 | 0,6 |
| 36–45 | 0,95 | 0,55 |
| 46–55 | 0,85 | 0,5 |
| >55 | 0,8 | 0,4 |

Also **international standards** like the **ISO 11228** and **EN 1005** use such factors when calculating maximum allowed forces which reduce maximum weight force limits for different percentiles of people with different age [72]. But other than the classical methods CEN and ISO standards use a more statistical approach and focus on "intended user" or "general working" populations and their distributions in force capabilities. All approaches calculate a maximum recommended force limit. The ratio of the force to be exerted and the maximum recommended force limit serves a basis for assigning a risk zone according to the 3 zone rating system as described in EN 614-1:2006 [72].

Newer methods also considering age-differentiated evaluation of work places are the "**Age-differentiated task analysis and screening method (ATS)**" [7] which uses the so-called "Chemnitz age model" [11] in a six-step approach to identify age-critical activities during production planning. The aim of the approach is the collection and subsequent reduction of age-related stress. For this purpose, the work requirements imposed by an assembly process are compared with empirical age profiles of production-relevant abilities, and the age-criticality of the examined assembly process is evaluated.

The "**assembly specific force atlas**" [74] which is manly designed for assessing thrust forces (hole body, hand arm) in different body postures in the automotive industry. The risk of the action observed is calculated by comparing the force actual needed to the maximum recommended force. The recommended force is dependent on maximum isometric forces which were measured and tabulated for certain groups, direction of force excretion, frequency of force exertion, body position while applying the force, as well as gender and age of the worker [74]. To consider age, the authors, suggest correction factors considering two different age groups, namely the young workers aged 26-35 years and an older group of workers aged 46–60 years [75].

The so called "**Check-Age**" [71] a German workplace screening approach which has its roots in the "**assembly worksheet light (AWS-light)**", also an evaluation tool for quick screening risks in assembly industry [76], as well as in two key indicator methods [65, 68]. The authors added an age-differentiated assessment by increasing the risk levels at physical assessments (load handling and execution conditions) by one category if elderly workers are working at the workplace. Further assessment parts for containing yes/no questions for several age-specific factors like necessity of maximum forces or necessity of high movement speed were added. In addition, risk ratings for seeing, hearing, sensing and information processing were added where the age-specific rating also increases the risk level by one category.

"**A-Flex**" [77] is a method developed in Germany especially for the assessment of work places in steel industry considering elderly workers. The method composes of three pillars the age structure analysis, the skill need analysis and the age differentiated risk assessment. For risk assessment the different KIM are used as well as a developed checklist rating strains from body postures and movements, load handling, and environment conditions in a high middle low schema dependent on the time executed or based on limit values derived from international standards and legal regulations in Germany. Besides of physical factors also psychological and mental factors are covered in the assessment [77] (Table 3).

Summing up there are some ergonomic risk assessment method already considering the factor age in a way. Never the less most methods which do an age differentiated assessment only make a difference between young and old and do not take into account individual differences within persons or within age classes, as shown in EN 1005 or the REFA method. Therefor an approach is suggested which considers the dependency of people's average skill level at a specific age (similar to Keil [7]) by applying age dependent multipliers to the risk values (similar to REFA and KIM for pushing and Pulling for females) while taking risk values from well-known and validated methods (KIM, MLT, AWS, EWAS).

**Table 3.** Age-differentiated work evaluation methods and age critical factors included

| Evaluation method | Domain | Age factors covered | Method of consideration |
|---|---|---|---|
| REFA | Load handling | Load handling | Factor for reducing the maximum isometric force |
| ISO 11228<br>EN 1005 | Load handling | Load handling | Correction factor dependent on age groups for reference weight |
| A-Flex | Body postures+ movements, load handling, repetitive actions, work environment | Load handling, body postures and movement, sensory tasks | Three categories (high, middle, low) for assessment. Manual handling is assessed with KIM. Other aspects are limited by time of execution |
| Check-age | Load handling, body postures and movement, work environment | Load handling, body postures and movement, sensory tasks, information | Generally increasing risk level by one category. Yes/no questions where yes answers lead to a need for detailed assessment |
| Force atlas | Thrust forces | Static force | Measured correction factor group 45+ compared to group 25–45 |

## 4 The Approach Developed

For an age differentiated evaluation of risks at work places we decided to model the factor age with mathematical functions where possible. Similar to the calculation of risk factors in the traditional German approaches a decline in abilities caused by age is used as a multiplier lowering the recommended value. For the evaluation of an age-differentiated risk factor at workplace level several methods were implemented in an easy to use excel file. For applying the different declines of abilities in age-differentiated risk calculations mathematical functions for these changes in mean values by age were derived according to the values mentioned in Table 1.

**Load Handling.** For calculating of the decline of force the values gathered in table one for muscular strength were transferred in a mathematical formula using mean values for 5-year age groups starting the decline at age of 30 years. The factor calculated by this formula is the on average remaining force at a certain age. This factor is applied as a divisor on the result of the KIM for Lifting carrying and pushing/pulling to calculate the age-differentiated risk value. Further instead of using the general value it is also possible to manually apply a specified value for the actual remaining muscular strength in comparison to an average 30-year-old person, which could be determined for example from a company doctor. Therefor an individual risk value can be deriver for material handling depending on either workers age or his remaining muscular power. Figure 5 shows the decline in muscular strength and the formula used to calculate it in the model.

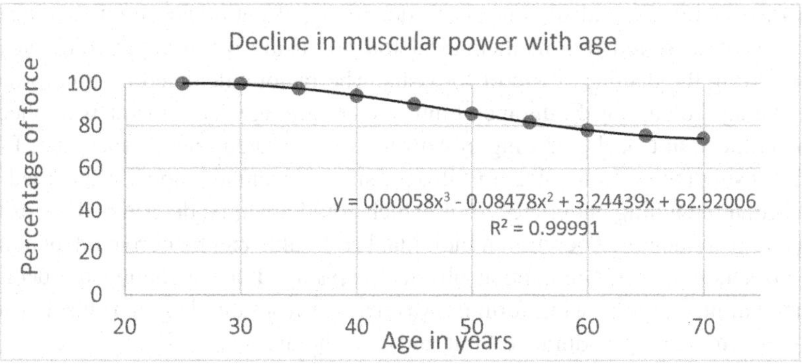


**Fig. 5.** Decline in muscular power with age

Applying the multiplier as mentioned on the result of the KIM increases the risk value by an age-dependent amount. Similar to the ATS method described earlier this can be used to derive age limits or special limits for workers with special impairments of physical abilities. The results for risk calculation with the KIM or the MLT is shown in Fig. 5. For calculating the general risk level, the MLT was used. The risk value resulting for manual material handling is composed of some pushing activities and some lifting and carrying activities. The resulting risk value added up to 37 (42,5) points. Figure 6 shows the risk values for different aged based on the approach suggested. It can be seen that by applying the calculated factor the KMI result raises to 45 (52) points for a 55-year-old and to 49 (56) for 65-year-old men.

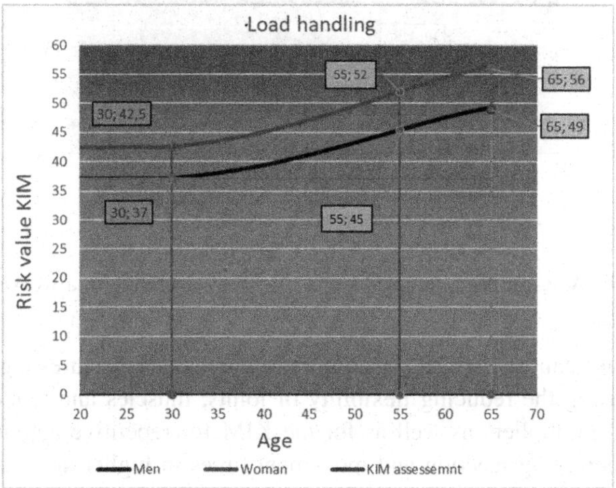


**Fig. 6.** Age differentiated assessment of load handling

**Endurance Limit** For calculating of the decline in of endurance limit the data from Table 1 was transferred in a mathematical formula using mean values for 5-year age groups starting the decline at age of 25 years. The factor calculated by this formula is the on average remaining Endurance limit at a certain age. This factor is applied as a divisor on the result of the grading system for work related energy spending. Further instead of using the general value it is also possible to manually apply a specified value for the actual remaining endurance limit, which could again be determined for example from a company doctor. Therefor an individual risk value can be derived depending on either workers age or his remaining physical capacity. For calculating the decline in endurance limit. The following formula was derived from the data in Table 1 where Y refers to the remaining endurance limit and x to the age in years.

$$Y = -0,0072 * x^2 - 0,598 * x + 119,62 \tag{1}$$

Figure 7 shows that at a workplace with an energy consumption of 4180 kJ per workday a 30-year-old person works at 75% of its endurance limit while a 55 year old person already works at 105% of its capacity and therefor over the limit.

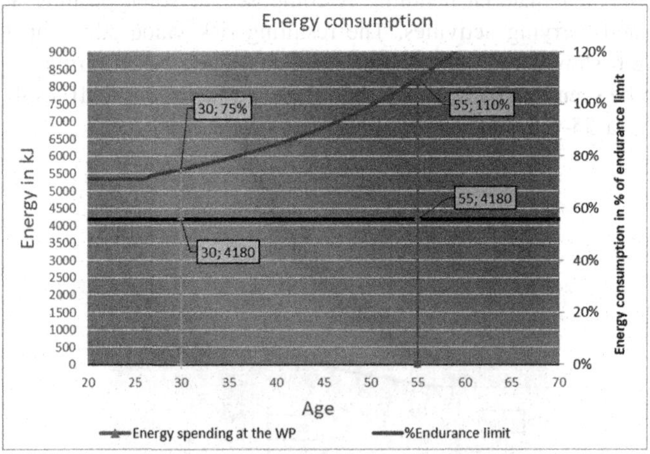


**Fig. 7.** Age-differentiated evaluation of energy spending at a workplace

Similar values can be derived for point ratings for body postures e.g. from EWAS worksheet by using the reducing flexibility of joints, muscles and tendons for evaluating age based multipliers as well as for the KIM for repetitive activities when calculation with increasing reaction and movement times in higher age.

## 5 Summary and Outlook

Based on literature on age related changes of human abilities an approach to adapt ergonomic risk assessment methods to the changing needs in times of demographic change was suggested. The derived method is based on very common assessment tools used in many industrial companies in Austria and Germany. The tool for assessing age differentiated risks is currently tested within the two company partners. The results gathered with the developed approach will be compared with the available company data. This data gathered beforehand in interviews will be excessed and examined to get insights concerning the workforce's age-structure, the extent of sick-leave and the reasons which caused employees to quit work as well as the highest strains at their workplaces. Comparing results of the observations and evaluations with sick leaves and problems of elderly workers in the company will be to validate the approach developed.

## References


1. Kinkel S, Friedewald M, Hüsing B, Lay G, Lindner R (2008) Arbeiten in der Zukunft. Strukturen und Trends der Industriearbeit Berlin
2. Bauernhansl T et al (2014) Industrie 4.0 in Produktion, Automatisierung und Logistik. Anwendung, Technologien, Migration. 1. Aufl. Springer, Wiesbaden
3. Botthof A (2015) Zukunft der Arbeit im Kontext von Autonomik und Industrie 4.0. In: Botthof Au, Hartmann EA, Zukunft der Arbeit in Industrie 4.0. Springer Vieweg. Berlin, S. 4–6
4. Spanner-Ulmer B, Keil M (2009) Konsequenzen des demographischen Wandels für zukünftige Produktions und Technologieabläufe am Beispiel der Automobilindustrie (Consequences of demographic change for future production and technological automotive industry). Zeitschrift Industrie Management 2(2009):17–20
5. Brandenburg U, Domschke JP (2007) Die Zukunft sieht alt aus, Herausforderungen des demografischen Wandels für das Personalmanagement (The future looks old, challenges of demographic change for human resources management), Wiesbaden
6. Marschall J, Hildebrandt S, Sydow H, Nolting H-D (2017) Gesundheitsreport 2017: Analyse der Arbeitsunfähigkeitsdaten. Update: Schlafstörungen (1. Auflage). Beiträge zur Gesundheitsökonomie und Versorgungsforschung, vol 16
7. Keil M, Spanner-Ulmer B (2010) Conception of a task analysis and screening-method for identifying age-critical fields of activity on the basis of the Chemnitz Age Model. In: Karwowski K, Salvendy G, (eds) Advances in human factors, ergonomics and safety in manufacturing and service industries, 3rd international conference on applied human factors and ergonomics, Boca Raton, USA, pp 393–400
8. Riley J (2001) Rising life expectancy: a global history. Cambridge University Press, Cambridge
9. WHO: World Health Statistics 2016: Monitoring health for the SDGs. http://www.who.int/gho/publications/world_health_statistics/2016/EN_WHS2016_AnnexB.pdf?ua=1. Accessed 01 May 2018
10. WHO: Austria: WHO statistical profile online. http://www.who.int/gho/countries/autpdf?ua=1. Accessed 01 May 2018

11. Keil M, Spanner-Ulmer B (2009) Chemnitz age model – an interdisciplinary critical performance factors. In: Research basic approach to characterize age congress 17th world ergonomics – IEA 2009, 9th–14th August, Beijing, China
12. Biermann H, Weißmantel H Regelkatalog SENSI-Geräte: bedienerfreundlich und barrierefrei durch das richtige
13. Design (1997) Inst. für Elektromechanische Konstruktionen, Darmstadt. http://www.emk.tu-darmstadt.de/~weissmantel/sensi/sensi.html. Accessed 20 July 2015
14. Blumberger W et al (2004) "Arbeit – Alter – Anerkennung". IBE-Research Report, vol 2, Linz
15. Ilmarinen J, Tempel J (2002) Working ability 2010 – What can we do so that you stay healthy?, Hamburg
16. Ilmarinen J (2006) Towards a longer work life! Ageing and the quality of work life in the European Union, Finnish institute of occupational health, ministry of social affairs and health, and Juhani Ilmarinen
17. Kleindienst M et al (June 2016) Demographic change and its implications for ergonomic standardization. Book of proceedings of ergonomics 2016, Zadar
18. Kruse A et al (1987) Gerontologie-eine interdisziplinäre Wissenschaft. Bayrischer Monatsspiegel Verlagsgesellschaft mbH, Munich
19. Prasch M (2010) Integration leistungsgewandelter Mitarbeiter in die variantenreiche Serienmontage. Herbert Utz Publishing Company GmbH, Munich
20. Rensing L, Rippe V (2012) Altern-zelluläre und molekulare Grundlagen, körperliche Veränderungen und Erkrankungen. Therapieansätze. Springer, Berlin Heidelberg
21. Richter G et al (2012) Demografischer Wandel im Arbeitsleben. Bundesanstalt für Arbeitsschutz und Arbeitsmedizin (eds) Dortmund. www.baua.de/de/Publikationen/Fachbeiträge/Artikel30.html. Accessed 28 July 2015
22. Riedel S et al (2012) Einflüsse altersabhängiger Veränderungen von Bedienpersonen auf die sichere Nutzung von Handmaschinen. In: Bundesanstalt für Arbeitsschutz und Arbeitsmedizin (eds) Research project F 2118, Dortmund Berlin Dresden. https://www.baua.de/de/Publikationen/Fachbeitraege/F2118.pdf. Accessed 20 July 2015
23. Schlick C et al (2010) Arbeitswissenschaft. Springer, Heidelberg
24. Weger G (2006) Alter(n)svielfalt im Betrieb. In Alter(n)svielfalt im Betrieb-Strategien und Maßnahmen für eine nachhaltige Unternehmenspolitik in kleinen und mittleren Unternehmen, Arbeitsbereich Gender and Diversity in Organizations. Wirtschaftsuniversität Wien (eds) 1st edn, pp 22–29
25. Winkler R (2005) Ältere Menschen als Ressource für Wirtschaft und Gesellschaft von morgen. In: Clemens W, Höpflinger F, Winkler R (eds) Arbeit in späteren Lebensphasen-Sackgassen, Perspektiven, Visionen. Haupt Publishing Company, Bern pp 84–104
26. Wolf M, Kleindienst M, Ramsauer C, Zierler C, Winter E (2018) Current and future industrial challenges: demographic change and measures for elderly workers in industry 4.0. In: Annals of the Faculty of Engineering Hunedoara - international journal of engineering, 16, pp 67–76
27. Ilmarinen J, Rantanen J (1999) Promotion of work ability during ageing. Am J Ind Med 36:21–23
28. Gall B, Parkhouse W (2004) Changes in physical capacity as a function of age in heavy manual work. Ergonomics 47:671–687
29. Aittomaki A, Lahelma E, Roos E, Leino-Arjas P, Martikainen P (2005) Gender differences in the association of age with physical workload and functioning. Occup Environ Med 62:95–100
30. Giniger S, Despenszieri A, Eisenberg J (1983) Age, experience, and performance on speed and skill jobs in an applied setting. J Appl Psychol 68:469–475

31. Baines et al (2004) Humans: the missing link in manufacturing simulation?
32. Shephard RJ (2000) Aging and productivity: some physiological issues. Int J Indus Ergon 25:535–545
33. Boenzi F et al (2015) Modeling workforce aging in job rotation problems. IFAC-PapersOnLine 48–3:604–609
34. Aoyagi Y, Shephard RJ (1992) Aging and muscle function. Sports Med 14:376–396
35. Baines T, Mason S, Siebers PO, Ladbrook J (2004) Humans: the missing link in manufacturing simulation? Simul Model Pract Theory 12(7–8):515–526
36. Baines TS, Asch R, Hadfield L, Mason JP, Fletcher S, Kay JM (2005) Towards
37. Chapman EA, de Vries HA, Swezey R (1992) Joint stiffness: effects of exercise on young and old men. J Geront 27:218–221
38. De Zwart B, Frings-Dresen M, Van Dijk F (1995) Physical workload and the aging worker: a review of the literature. Int Arch Occup Environ Health 68:1–12
39. Gall B, Parkhouse W (2004) Changes in physical capacity as a function of age in heavy manual work. Ergonomics 47(6):671
40. Goldich RL (1995) Military retirement and personnel management: should active duty military careers be lengthened? In: Congressional research service, Washington
41. Hank K, Jürges H, Schupp J, Wagner GG (2009) Isometrische Greifkraft und sozialgerontologische Forschung: Ergebnisse und Analysepotentiale des SHARE und SOEP [Isometric grip strength and social gerontological research: results and analytic potentials of SHARE and SOEP]. Z Gerontol Geriatr 42(2):117–126. https://doi.org/10.1007/s00391-008-0537-8
42. Heath G, Hagberg J, Ehsani A, Holloszy JO (1981) A physiological comparison of young and older endurance athletes. J Appl Physiol 51:634–640
43. Hollmann W, Strüder HK (2009) Sportmedizin: Grundlagen für körperliche Aktivität, Training und Präventivmedizin (5, völlig neu bearbeitete und erw. Aufl). Stuttgart. Schattauer, New York
44. Johns RJ, Wright V (1962) Relative importance of various tissues in joint stiffness. J Appl Physiol 17:824–828
45. Kenny GP, Yardley JE, Martineau L, Jay O (2008) Physical work capacity in older adults: implications for the aging worker. Am J Ind Med 51(8):610–625
46. Kowalski-Trakofler KM, Steiner LJ, Schwerha DJ (2005) Safety considerations for the aging workforce. Saf Sci 43:779–793
47. Scherf C (2014) Entwicklung, Herstellung und Evaluation des Modularen Alterssimulationsanzugs eXtra (MAX) (Dissertation). Technische Universität Chemnitz, Chemnitz
48. Shephard RJ (1999) Age and physical work capacity. Exp Aging Res 25(4):331–343
49. Tolin P, Simon JR (1968) Effect of task complexity and stimulus duration on perceptual-motor performance of two disparate age groups. Ergonomics 11:283–290
50. Vitasalo J, Era P, Leskinen A, Heikkenen E (1985) Muscular strength profiles and anthropometry in random samples of men aged 31–35, 51–55, and 71–75. Ergonomics 11 (3):283–290
51. Rohmert W (1984) Belastungs-Beanspruchungs-Konzept (stress-strain concept). Zeitschrift für Arbeitswissenschaft 4(1984):193–200
52. [18] Keil M, Hensel R, Spanner-Ulmer B (2010) Process model elements adjusted to abilities
53. [19] Schmidtke H (1993) Teil 3. Belastung und Beanspruchung. (stress and strain). In: Schmidtke H, Ergonomie (Ergonomics), 3. Aufl., pp 110–116, München
54. Schlick C, Bruder R, Luczak H (2010) Arbeitswissenschaft (Human Factors and Ergonomics), 3. Aufl, Heidelberg
55. BGMH 2016: BGMH Information 102:Beurteilung von Gefährdungen und Belastungen (2016). https://www.bghm.de/fileadmin/user_upload/Arbeitsschuetzer/Gesetze_Vorschriften/Informationen/BGHM-I_102.pdf. Assessed 30 May 2018

56. Szymanski, H.: Die alterssensible Gefährdungsbeurteilung- Basis für eine zeitgemäße Arbeitsgestaltung. REFA-Nachrichten 6/2006
57. Kleindienst M, Wolf M, Ramsauer C, Winter E, Zierler C (2016) Demographic Change and its implications for ergonomic standardization. In: Book of proceedings of the 6th international ergonomics conference: ERGONOMICS 2016 - focus on synergy. Croatian ergonomics society, pp 179–188
58. Börner K, Bullinger-Hoffmann AC (2017) Alter(n)sgerechte Arbeitsplatzgestaltung – Prävention von Anfang an. Betriebliche Prävention 6:240–245
59. Spitzer H, Heftinger T (1964) Tafeln für den Kalorienumsatz bei kör- perlicher Arbeit. Sonderheft der REFA- Nachrichten, REFA-Verband für Arbeits- studien (Hrsg), Beuth-Vertrieb GmbH, Berlin
60. Bongwald O, Luttmann A, Laurig W (1995) Leitfaden für die Beurteilung von Hebe- und Tragetätigkeiten. Hauptverband der gewerblichen Berufsgenossenschaften (HVBG). Sankt Augustin
61. WKO. https://www.wko.at/service/arbeitsrecht-sozialrecht/Schwerarbeitsrechner.html. Accessed 30 May 2018
62. Zülch G Makroergonomische Probleme im Lager- und Transportbereich in: Griefahn B. et al Deutsche Gesellschaft für Arbeitsmedizin und Umweltmedizin e.V. 50. Wissenschaftliche Jahrestagung Aachen 2010, pp 400–408
63. Brandstädt, F (2014) The Key Indicator Methods (KIM) - risk assessment of physical workload on screening level. In: XX. Weltkongress für Sicherheit und Gesundheit bei der Arbeit 2014: Globales Forum Prävention: Unsere Vision: Prävention nachhaltig gestalten, 24–27 August 2014, Frankfurt, Programm, F02.14 DGUV, Berlin
64. Schaub K et al (2010) Das Multiple-Lasten-Tool: integrierte Bewertung unterschiedlicher Arten manueller Lastenhandhabung in. In: Mensch- und prozessorientierte Arbeitsgestaltung im Fahrzeugbau, Herbstkonferenz 2010 der Gesellschaft für Arbeitswissenschaft, 23–24 September 2010, Wolfsburg. GfA Press, Dortmund
65. BAUA 2001: Leitmerkmalmethode zur Beurteilung von Heben, Halten, Tragen. https://www.baua.de/DE/Themen/Arbeitsgestaltung-im-Betrieb/Physische-Belastung/Leitmerkmalmethode/pdf/LMM-Heben-Halten-Tragen.pdf?__blob=publicationFile. Accessed 30 May 2018
66. Köck PG, Pedersen E (1969) Er- mittlung von Grenzwerten für Männer und Frauen bei Hebe- und Tragearbeiten in Industrie und Gewerbe. Forschungs- auftrag der Arbeitsgemeinschaft zum Studium von Arbeitsbelastungen, durch- geführt vom Arbeitswissenschaftlichen Institut der Technischen HochschuleWien
67. Hecktor K, Schaub K, Jäger M (2014) Biomechanische Gefährdungsbeurteilung bei Montagearbeitsplätzen. Zeitschrift für. Arbeitwissenschaft, 68:7–17
68. Klussmann A et al (2017) Risk assessment of manual handling operations at work with the key indicator method (KIM-MHO). Determination of criterion validity regarding the prevalence of musculoskeletal symptoms and clinical conditions within a cross-sectional study. In: BMC musculoskeletal disorders, vol 18, Ausgabe 184. https://doi.org/10.1186/s12891-017-1542-0
69. Hodgkins J (1962) Influence of age on the speed of reaction and movement in females. J Gerontol 173:385–389
70. Scholz H (1964) Wechselbezihunge zwischen alter und Leistung. In: Arbeits- und betriebskundliche Reiche1. Bund Verlag, Köln, pp 9–34
71. Börner K, Löffler T, Bullinger-Hoffmann C (2017) CheckAge – Screening-Verfahren für die Bewertung alter(n)sgerechter Arbeitsplätze. Reihe aw&I Report, Heft 2, Verlag aw&l - Wissenschaft und Praxis, Chemnitz

72. Schaub et al (2015) Development and testing of a screening approach for the evaluation of forceful operations in industry. In: Proceedings of 19th triennial congress of the IEA, Melbourne
73. REFA Chemicals Expert Committee REFA (1987)
74. Schaub K et al (2015) The assembly specific force atlas. Hum Factors Ergon Manufact Serv Indus 25(3):329–339
75. Wakula J et al (2009) Der montagespezifische Kraftatlas (BGIA-Report 3/2009). Hrsg.: Deutsche Gesetzliche Unfallversicherung (DGUV), Berlin
76. Kugler M et al (2010) KoBRA – Kooperationsprogramm zu normativen Management von Belastungen bei körperlicher Arbeit. Ergonomie in der Industrie – aber wie?. Darmstadt. http://www.kobra-projekt.de/download/aws-light. Assessed 30 May 2018
77. Szymanski H, Lange A Den demografischen Wandel in der Eisen- und Stahlindustrie gestalten -- eine Handlungshilfe zur alter(n)sgerechten Arbeitsgestaltung. www.ergo-stahl.de. Assessed 30 May 2018

# Social Medial as Facilitator of Self-value Realization for Elderly

Dongjuan Xiao(✉) and Miaosen Gong

School of Design, Jiangnan University of Technology, No. 1800, Lihu Road, Binhu District, Wuxi 214122, China
dj.xiao2016@jiangnan.edu.cn

**Abstract.** Aging is a problem all over the world. It is a process involving many changes in physiology, psychology and society. "Active aging" has gradually become the world's policy framework for coping with aging. On the one hand, it pays attention to the physical and mental health of the elderly. On the other hand, it fully explores the value of the elderly to the society. Intelligent technology development appears to be a lot of obstacles for the elderly, the elderly people feel helpless in the face of new technology, but if we can effectively use the advantages of intelligent technology, such as mobile technology to help the elderly properly adapt to the society, let them can enjoy the convenience of high-tech life. With the help of social media, elderly can enjoy better service and more ways can be created for them to realize their self-value and play their own values for the benefit of society. In this article, the author will show some cases of how to apply design thinking to product and service design in active aging, such as innovative service for the elderly through mobile Internet technology. Through deep user research and interviews with elderly and youngers, we found that some of the elderly have specialty in doing something, such as teaching English, knitting, playing instruments, photographing and so on. And on the other hand, some young people want to learn new skills from the elderly, so we can build a platform and service system that young people and elderly can communicate with each other. In this paper will describe one case studies of how to build a platform for elderly to decrease loneliness and empower elderly people and provide them with a sense of social involve and greater control and self-efficacy.

**Keywords:** Social media · Self-efficacy · Active aging

## 1 Introduction

At present, China has entered the rapid development period of population aging. According to the data published by the state's Aging office, by the end of 2017, the number of over 60 years old people will account for 17.3% of the total population. It is estimated that the number of elderly population in China is expected to reach a peak of 487 million by 2050. The needs of the elderly have been raised to a level of great concern. As the age increases, elderly people are suffering physiological decline and limited mobility, meanwhile, their response time becomes sluggish. This restriction of mobility often leads to less participation in social activities, which often increases


S. Bagnara et al. (Eds.): IEA 2018, AISC 826, pp. 206–212, 2019.
https://doi.org/10.1007/978-3-319-96065-4_23

loneliness and reduces satisfaction with life. As we all known that, for the elderly, their health and well-being depend largely on their emotional and social relationships with their family and friends.

To a certain extent, high technology separates older people from society. Older people are afraid of technology and are reluctant to learn new technologies and accept new things. But on the other hand, technology can play an important role in helping the elderly create new social network and maintain these social relationships. It can improve their quality of life by reducing their sense of isolation. Use of social media tools can help us to redefine the role of elderly in today's global society for educating the youth more effectively [1, 2].

By determining the user environments that are suitable for the elderly, including web page accessibility, interface design and real social life transformation, this article proposes the factors for a social media website, the factors for the elderly to use social media platforms, a social media platform design that can be easily used by the elderly and design factors suitable for the elderly.

The objective of this paper is to propose a framework for a social knowledge management and build a platform for elderly to decrease loneliness and empower elderly people and provide them with a sense of social involve and greater control and self-efficacy.

## 2 Literature Review

### 2.1 Social Media

With the rapid development of technology, the term "Web 2" and "user-generated content" appeared along with the evolution of social media in 2004.

Social media refers to Internet based applications based on ideological and technological of Web 2, which is the developer and end-users can utilize the World Wide Web together by continuously modified by all users in a participatory. There is no agreed definition for social media, but there are four commonalities of current social media services is identified by researchers, such as Web 2.0 Internet-based, user-generated content, specific user profiles, and connecting a profile with those of other individuals or groups [3].

Social media can be understood as internet-based applications that create links among users and user-generated content in online environments. The most activity performed among the user is to access and log into the social network to interact with other people via blogs, web forums, social bookmarking sites, photo and video sharing communities, content communities, social networking sites, and virtual games with the goal to consume, co-create, share, and modify content generated by the same users. Nowadays, the most popular form of Social Media is the Online Social Network (OSN), such as Facebook, WhatsApp, LinkedIn, Skype, Google+, Instagram, Twitter and Snapchat [5, 6]. While in China the most popular OSN is WeChat with almost one billion users, which is second only to the number of Facebook users.

WeChat is a Chinese multi-purpose messaging, social media and mobile payment app developed by Tencent. It had over 889 million monthly active users in 2016, 90% of whom were Chinese [4]. One of the Webchat's core aim is to encourage people to feel more connected by sharing details of their personal lives as a means of socializing with others.

WeChat supports different instant messaging methods, including text message, voice message, walkie talkie and stickers. Users can send previously saved or live pictures and videos, name cards of other users, coupons, lucky money packages, or current GPS locations with friends either individually or in a group chat.

"Moments" is WeChat's brand name for its social feed of friend's updates. Moments allows users to post images, post text, post comments, share music, share articles and post "likes."

WeChat users can register as an official account, which enables them to push feeds to subscribers, interact with subscribers and provide them with services [4].

### 2.2 Social Media and Elderly People

Technology can, however, also be seen as providing opportunity for building new patterns of family and social bonding for elderly. Prevention and relieve of social isolation and loneliness are the key factors to improve the quality of life of the elderly. The wide availability of mobile communications networks, smart phones and tablets facilitate the elderly to commutation and share information with others through the use of text, voice and image media. In dealing with loneliness issues, social media has the potential to decrease loneliness and increase perceived social support, sense of belonging and feelings of connectedness. The elderly population have recently shown a special enthusiasm for the adoption of new network tools that allow them to share, with a growing network of contacts, links, photos, videos, news and status updates, and consequently improve their skills and opportunities for communication, information searching, knowledge sharing, and relationship building [6, 7].

WeChat application is the most popular social media among urban elders in China. According to official data published by TenCent Company in 2017, there are 50 million monthly active users above 50 years old on WeChat. The use of WeChat text and voice is more than 80%, the utilization rate of "Moments" reached 77%, and the utilization rate of group chat reached 61.7%. Through the 'Moments', they can understand the dynamic life of their children and relatives and friends in order to meet the strong emotional needs of their children and relatives. They can express their feelings and wishes at anytime and anywhere to meet the needs of their attention and concern [4].

Active and healthy aging requires maintaining social activities and healthy relationships with others.

Good social connection can prevent aging with anxiety, depression or mental problem. Wellbeing is an important issue for elderly person and it has accepted that social interaction and social activities or positive social life styles can increase the overall wellbeing of elderly. The elderly felt positive attitude toward life and self-acceptance as their social involvement increased [8, 9].

In addition, elders have more life experience, as they have had a long life, they have much knowledge, and if they don't transfer it to others it will disappear or die with them. They can guide the younger generation in many aspects of life. Therefore, the use of these highly precious elderly wisdom is indeed very important for all of us and for future generations. In general, elderly can be a good source of advice for younger people, particularly in the areas of relationships and in cases of uncertainty. Unfortunately, as a society, we do not readily afford cross generational opportunities for such social interactions and mutual growth. Eliminating existing structural age barriers (for example, providing opportunities for education for all ages, working and leisure opportunities) and increasing communication between generations may lead us to a healthier ageing society. Therefore, if we are concerned about the development of wisdom, we also need to work together to provide creative opportunities, promote its development, and at the same time expand social benefits [9]. The use of social media may be a meaningful move in this direction.

So, it is necessary to explore more opportunities for elderly to enjoy a happy and meaningful life through social media platform which can combine with offline community, which can decrease loneliness and empower elderly people and provide them with a sense of connectedness and greater control and self-efficacy.

## 3 Research and Methodology

Both qualitative and quantitative research method were adopted in this study. Although social media is becoming more and more popular with the elderly, we still find that the elderly and young people still have their own circle of friends, and they have few opportunities to communicate with each other. Social media make communication between people more convenient, but intergenerational communication is still a serious problem. How to use social media to promote intergenerational communication is the problem we need to study.

Therefore, to test the hypotheses we first distributed online questionnaires to young people (18–35 years old) about the related problems of communication with the elderly, and eventually got 48 valid questionnaires.

Because the senior people are not good at online questionnaire, so we did a deep interview with 3 elderly people in order to study their lifestyle, to know that what we could offer to them to improve their lives and as to whether they are willing to participate actively to share their life experience and wisdom with the others?

## 4 Findings and Discussion

From the questionnaire, it shows that young people are willing to communicate with the elderly, and they are willing to chat with the elderly and teach the elderly better use of intelligent products. 75% of the participants thought they could learn something from the elderly.

**Table 1.** Demographic of participants.

| Demographic | Mrs. Zhang | Mr. Wang | Mrs. Zhou |
|---|---|---|---|
| Gender | Female | Male | Female |
| Age | 65 | 72 | 81 |
| Healthy condition | Diabetes | Healthy | Physical inconvenient |
| Hobby | Knitte sweater | Swimming Playing Taiji | Reading and talking in English |
| Status | Live with her adult kids | Live with his wife | Live with her husband |

There are 10 older people take part in our interview. We had several open-ended questions for the older participants. Such as "Would you like to communication with young people?", "Are you willing to help them if they have some questions to ask you?", "Would you like ask for help from young people?" "Do you familiar with social media?" We chose the typical 3 participants basic demographic in Table 1.

"It would be great if we have the opportunities to communicate with the young people, but the opportunity is too rare. We seldom talk with the young eve our own children. I like swimming and flying kites, and playing Taiji, I hope there are more opportunities to do some exercise together with young people together. I am good at using WeChat, but most of my online friends are almost the same age with me" 72 years old Mr. Wang said.

"I like knit with wool yarn, I can weave different style of sweaters, but young people think that is very old fashioning. I hope I can have opportunities to teach young people do the knitting" 65-year-old Mrs. Zhang told us.

"I was an English teacher before I retired. Now I still like talking English, especially with young people, but I do not have that chance, I often use WeChat to look through international news". Said Mrs. Zhou, an 81 years lady.

From the investigation, we find that only the social media itself cannot promote the solving of intergeneration problem, the opportunity for young people and elderly to interaction and communication with other still very limit. So, we were planning to design a social platform which can combine online and offline community activity together.

According to the above three participants, we set up a personal image who is having specialties, such as English speaking, knitted and playing Taiji. These skills or specialties should be seen in an online platform. One of the WeChat function is called "Subscription" allowing push content and notification updates for subscribed followers displayed in the subscription area. This subscription platform can show different stories and activity pictures. The community can use this platform to announce different activities which both interested in young people and elderly, such as learn English, learn Taiji or other interesting things.

We analysis the motivation both of young people, elderly, and community organizer (non-profit institution), and profit company. The motivation matrix is showed as Fig. 1.

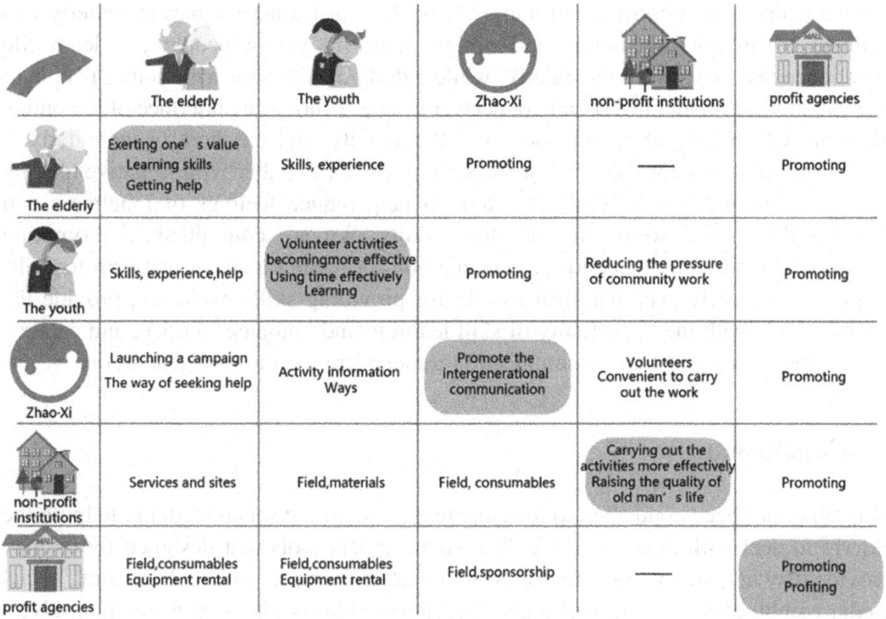
The elderly
The youth
Zhao·Xi
non-profit institutions
profit agencies
The elderly
Exerting one' s value
Learning skills
Getting help
Skills, experience
Promoting
—
Promoting
The youth
Skills, experience,help
Volunteer activities becomingmore effective
Using time effectively
Learning
Promoting
Reducing the pressure of community work
Promoting
Zhao·Xi
Launching a campaign
The way of seeking help
Activity information Ways
Promote the intergenerational communication
Volunteers
Convenient to carry out the work
Promoting
non-profit institutions
Services and sites
Field,materials
Field, consumables
Carrying out the activities more effectively
Raising the quality of old man' s life
Promoting
profit agencies
Field,consumables
Equipment rental
Field,consumables
Equipment rental
Field,sponsorship
—
Promoting
Profiting

**Fig. 1.** Motivation matrix

**Fig. 2.** Subscription interface

We called this subscription a name "Zhao Xi", in Chinese it means "elderly and younger". Through this online and offline platform, young people can learn life experience and skills from the elderly, while elderly can interact with young people to show their wisdom and feel social connected. Figure 2 shows the interface of the online platform, which can public information of the activity, and the work of the elderly.

Information and communication technology (ICT), together with initiatives led by local communities can be used as a tool to help reduce feelings of loneliness and increase the mental well-being of older adults. We are committed to promoting intergenerational communication and interaction, providing a social platform for self-value of the elderly. The platform mainly for providing skills exchange, provide the young people with the opportunity of skill learning and volunteer service, and create a more harmonious and free communication atmosphere for the two generation.

## 5 Conclusion

This study has been conducted to investigate the role of the social media in helping the elderly to deal with their daily life. The social media tools that designed by us may provide may help in lending a reason for our elderly members to live. In general, social media combined with community activity allows older adults to express themselves, participate in discussions and stay in contact with society. Participating in social networks can empower elderly people and provide them with a sense of connectedness and greater control and self-efficacy.

# Spatial Optimization of Bedroom Area for Effective Elderly Patient Handling

Priyanka Rawal(✉)

IDC, School of Design, Indian Institute of Technology Bombay,
Mumbai 400076, India
priyanka24rawal@gmail.com

**Abstract.** Increasing population age has become a fact for most of the countries world over. The number of people aged 65 and above is estimated to increase by 65% in the next 25 years, with doubling of the number of people aged greater than 85 years. Ageing restricts many physical and cognitive abilities of the elderly. This demands greater attention towards healthcare of this section of the population.

This study attempts to come up with ways to make best use of the space around the elderly patients especially in the bedroom area and various elements in it for better handling, more so if the persons are disabled or dependent for ambulation on others (caregivers and nurses). The course of this study started with investigation of points of bodily pain of caregivers and nurses while handling the elderly patients in constrained or non-customized spaces. This was accomplished using a Standard Nordic Questionnaire, validated by Kuroinka, to measure the prevalence of Musculoskeletal Disorder (MSD) in caregivers at homes and nurses in hospitals. Further, assistance of a professional physiotherapist was sought to understand standard procedures of handling patients who are dependent on others for ambulation. Hierarchical Task Analysis (HTA) and REBA were carried out next, which used videography and photography of these procedures as a tool to evaluate postural stress during the procedures. Furthermore, Anthropometric Data Analysis was performed to decide the optimized space for the patient handling and transfer task. For all the calculations, Indian Anthropometric data was considered.

Local hospitals of Jabalpur, India served as settings for this study. According to that, space requirement for the individual activities have been identified and derived for future design of bedrooms free space.

**Keywords:** Patient handling · Patient transfer
Space optimization and bedroom space

## 1 Introduction

In a UN backed study on the wellbeing of the elderly in a rapidly ageing world, out of 91 countries, India stands a cheerless 73rd. Sweden was said to be the best place to grow old and Afghanistan, the worst. This list is based on data from the World Health Organization (WHO) and other agencies on the older people's income, health, education, employment and their environments [1].


S. Bagnara et al. (Eds.): IEA 2018, AISC 826, pp. 213–222, 2019.
https://doi.org/10.1007/978-3-319-96065-4_24

According to the United Nation Population Fund, India is facing an elderly population time bomb, as its number of old people will triple by 2050. India's present population is 1.2 billion, which is expected to increase above 1.6 billion by 2050 [2]. By then it will become the world's largest nation, and would overtake China's ageing population. According to the report, two-third population over 60 suffered a chronic ailment in 2011, which is expected to increase by more than 200 million by 2050, due to which the current hospital and welfare services are not sufficient for the growing number of aged people [2].

There are 728 old age homes in India and detailed information of 547 homes is available. Out of these, 325 homes are free of cost while 95 old age homes are on pay & stay basis, 116 homes have both free as well as pay & stay facilities and 11 homes have no information. A total of 278 old age homes all over the country are available for the sick and 101 homes are exclusively for women. Kerala has 124 old age homes, which is highest in any state [3]. A financially independent senior citizen now prefers to stay in retirement resort instead of old age homes, which are overcrowded and unsafe [3].

All these population trends and imminent burdening of the healthcare system call for a greater focus on the care for elderly. This study attempts to take a step forward in this direction, by taking up especially the case of elderly patients who are dependent on others for transfer and ambulation. Ensuring these patients better designed spaces around them will ensure least physical and mental distress during tasks like patient handling through stretchers and wheelchairs etc.

### 1.1 Patient Transfer: Ergonomic Approach in Transfer System

Injury during patient handling, either to the patient or to the caregiver, is not new. In a landmark report titled Survey of Manual Handling in 1968 Repatriation Institutions, Ferguson reported that the strain injuries due to manual handling of patient is responsible for lost time accidents in the repatriation department [4]. He decided to collaborate various professions to eliminate the strain hazard, to ensure safety and from design and engineering stages. He specifically called for system design and equipment design on the ergonomic basis for human operators, which is standardized and reviewed continuously [4].

The degree of dependency and functional capacities of patients is key factor for determining the handling procedures and other essential considerations for developing a functional layout and design of rooms and buildings. It is important to consider possible changes over a time in functional capacity of patients/residents and future occupants of a room and building [5]. Even though there are many number of categories based on dependencies and assistance, the space requirements and design demands for the elderly care facility rooms are quite similar [4].

Ergonomics based injury prevention programs are successful in reducing work related injuries and cost in healthcare industries. Studies show that ergonomic approaches in patient transfer system have reduced staff injury from those involved in patient transfer system from 20% to 80%, and reduce the lost time due to injury and also reduce workers compensation costs. In last 35 years, intensive training of patient transfer has decreased back injuries of staff, who are directly involved in the transfer task [6]. When health care systems apply innovative approaches to reduce occupational injury risk reduction, it benefits the caregivers, patients and other staffs, which helps to gain a higher quality of work life for healthcare, which eventually result into higher staff quality and improve quality of care for patient and reduce turnover. All these benefits can be achieved through a well-designed ergonomic management program. Many other non-health care industries are taking benefits of the ergonomic approach in their fields [6].

### 1.2 Spaces and Environment

The patient handling tasks which include lifting, turning, pulling and positioning of patients can lead to fatigue, muscle strain and cause injury like back pain. Lifting a patient is far more difficult compared to lifting the box that has handles to hold [7]. Unlike a box, a patient's body is not evenly distributed and the mass is asymmetric and bulky. Handling of patient can be unpredictable, depending upon the degree of dependency of patient and many more factors like muscle spasms, loss of balance etc.

With this the home and hospital environment adds to the complexity of patient handling process [8]. Safe access to a patient in a clutter or unmanaged space becomes difficult, especially bedside in a small room and also small bathroom. For nursing staff and caregiver it becomes difficult to position themselves properly when assisting dependent patients in small spaces like washrooms. Crowded room of patient leads to awkward posture of caregiver while gaining access to patient in bed. The environment and patient become unpredictable while handling different tasks [7, 9].

A transfer must be safe and comfortable, and should neither cause pain nor discomfort to the patients or to the nurses. The nurses should choose a recommended work technique that not only is in line with the patient's preferences and needs, but also prevents the nurses from putting excessive load on the musculoskeletal system that might lead to MSD [7]. Several methods for performing a patient transfer task have been described. According to one recommended method the nurses have to consider their own capabilities, the patient's abilities, needs and preferences, as well as the possibilities and limitations of the environment, and then choose an optimal transfer technique accordingly [8, 10].

## 2 Methods

For the purpose of data collection, three methods namely questionnaire, interview and direct observation were used. The Standard Nordic Questionnaire was used to determine the need and problems of caregivers and patient at home and nurses at hospitals. Interview was conducted for getting deeper insights into the problems, which were not covered, in the questionnaire. Tasks performed in hospitals and homes by nurses, physiotherapists and caregivers were observed directly, to get more reliable data in the user's environment. Wherever allowed, videos and photos were clicked for HTA, which provided input data for postural analysis and anthropometric analysis.

# 3 Result and Discussion

Caregivers between the age group of 35 years to 55 years, which included 16 females, and 4 males, participated in the study. From Standard Nordic Questionnaire, it was found that 65% of the participants were facing the lower back pain due to patient handling and transfer tasks. Neck pain came a closer second at 55%, with shoulder pain at 45%, knees pain at 40% and hands following at 30%. During Interview, the Physiotherapist explained the task while performing it in a local hospital in Jabalpur, India. This information was used in HTA flow chart for deeper evaluation. REBA score of tasks performed by caregiver at home was 10 and for the same task REBA score for physiotherapist was 07. In both cases, changes were required but in case of caregiver, change was found necessary sooner. Based on tasks performed by physiotherapist, space and bed dimensions were calculated by anthropometry.

## 3.1 Hierarchical Task Analysis

Hierarchical Task Analysis (HTA) was used as a tool to get deeper understanding of the patient handling and transfer task in hospitals. Physiotherapist demonstrated patient handling and transfer methods and explained biomechanics of the body. First task in patient handling is to give 'clear instructions of the procedure to be carried out to the patient' and second is to 'follow bio-mechanic standards'. These should be carried out during every procedure and sub-procedure under HTA. The other sub-procedure could be performed in any order as per requirement. Photo grabs from the videos were used to support the study. Explanation for each relevant patient handling procedure is given below.

### 3.1.1 Roll the Patient on Bed

Rolling the patient was one of the sub-procedure of patient handling and transfer process. This process is repeated many times while handling patient properly. The Physiotherapist demonstrated the procedure of rolling the patient (Fig. 1). First, he firmly lifted the left leg and placed it on the right leg to roll the patient on his left side (Figs. 1(1) and 1(2)). Then he folded the patient's hands for safe rolling and to avoid any discomfort to the patient (Fig. 1(3)) and then placed the hands under the scapula (Fig. 1(4)) and next, the patient was rolled towards the physiotherapist i.e. on his left side (Figs. 1(5) and 1(6)) and finally patient was brought back again to his initial position (Fig. 1(7)). This process was used in many more sub-procedures like making occupied bed, side lying etc.

Other similar sub-procedures are - Comfortable lying on bed, lying on side/side lying, making bed, transferring the patient from bed to wheelchair, transferring the patient from wheelchair to bed, transferring the patient from wheelchair to chair, transferring the patient from chair to wheelchair and repositioning of patient in chair/wheelchair.

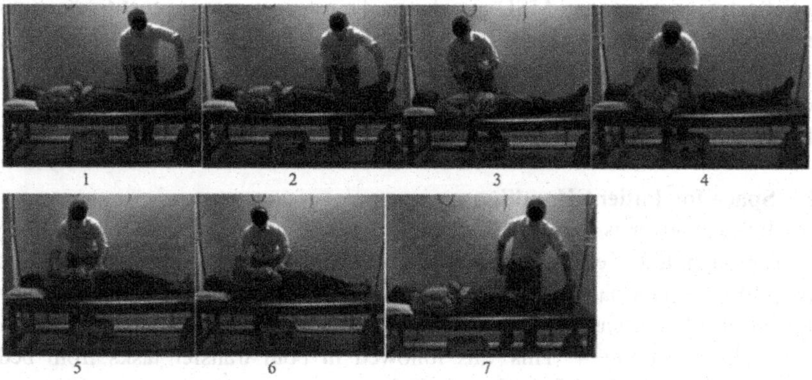


**Fig. 1.** Sequence of physiologist rolling a patient

### 3.2 Posture Analysis - REBA

Photos from the HTA were considered for the postural analysis of the most physically demanding procedure. Task performed by physiotherapist and untrained caregiver was considered to calculate the REBA score for both. Lifting the patient in both cases was considered to calculate REBA score (Fig. 2).

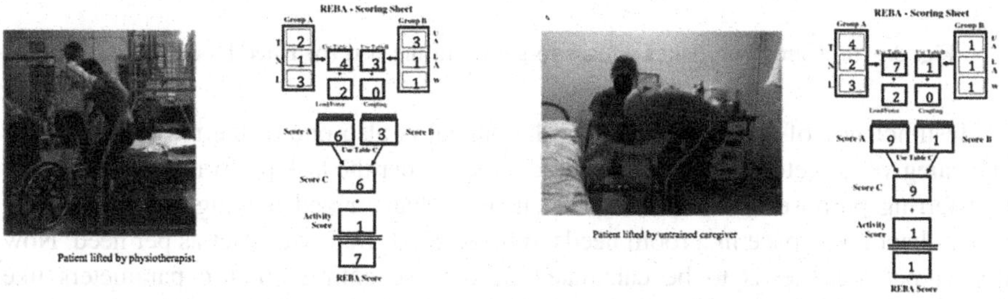

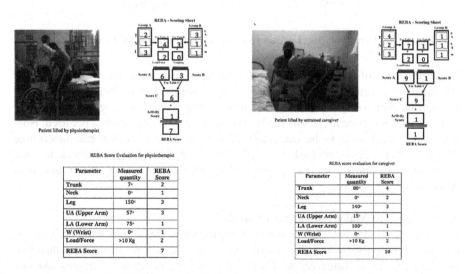

REBA Score Evaluation for physiotherapist

| Parameter | Measured quantity | REBA Score |
|---|---|---|
| Trunk | 7° | 2 |
| Neck | 0° | 1 |
| Leg | 150° | 3 |
| UA (Upper Arm) | 57° | 3 |
| LA (Lower Arm) | 75° | 1 |
| W (Wrist) | 0° | 1 |
| Load/Force | >10 Kg | 2 |
| REBA Score | | 7 |

REBA score evaluation for caregiver

| Parameter | Measured quantity | REBA Score |
|---|---|---|
| Trunk | 80° | 4 |
| Neck | 0° | 2 |
| Leg | 140° | 3 |
| UA (Upper Arm) | 15° | 1 |
| LA (Lower Arm) | 100° | 1 |
| W (Wrist) | 0° | 1 |
| Load/Force | >10 Kg | 2 |
| REBA Score | | 10 |

**Fig. 2.** REBA score for Physiotherapist and untrained caregiver

### 3.3 Anthropometry

In this section, space required for patient handling and transfer task was calculated. From HTA and observation, it was found that all the patient handling tasks if are performed in a particular manner by following ergonomics and weight handling rules can be effective in injury and MSD avoidance. Transfer process is repetitive or similar, when transferring resident/patient to and forth from the bed or wheelchair or static comfortable chair. Patient handling task is a complex process; optimizing space for all percentile people is a challenging job. Values have been decided by accounting all points and constraints.

### 3.3.1 Space for Patient Handling

In Fig. 3(a), caregiver is making the patient to get up in bed, as demonstrated, where patient is brought to the edge of the bed and is facilitated by caregiver to get up in bed. In Fig. 3(b), caregiver has blocked the knee of patient using his own his knee to avoid movement of leg and has supported the scapula of the patient as demonstrated in all transfer tasks as in Fig. 4. This was followed in both transfer tasks from bed to wheelchair/chair and wheelchair/chair to bed.

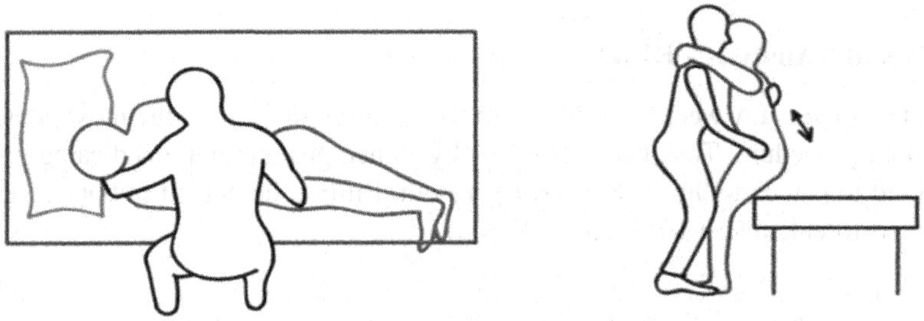

**Fig. 3.** (a) Caregiver makes patient to get up in the bed (b) Knee blocking activity

Insufficiency of space for handling the patient is also one of the reasons for nurse and caregiver to get MSD. Space required varies as per the task performed. It can be for transferring purpose to and forth wheelchair or chair or bed making etc. So for safe patient handling, space in a room needs to be decided. Bed size is set as per need. Now free carpet area needs to be calculated from these anthropometric parameters like Maximum body depth, relaxed, Forward step length, Foot length, Buttock to knee length, normal sitting, Buttock to popliteal length, normal sitting, Waist and Span Akimbo.

### 3.3.2 Space to Make Patient Sit up in Bed

Caregiver needs to stand near the bed for this procedure. Here, 'maximum body depth, relaxed' of 95th percentile male is considered as no one requires more space than him i.e. 409 mm. But the caregiver needs to bend or move forward or sometimes take one leg behind for better hold on floor and for that "forward step length" is considered again of 95th percentile for the same reason to provide sufficient space for all i.e. 799 mm.

Space for maximum body depth, relaxed and forward step length was considered for 95th percentile and maximum space out of these parameters is 799 mm, which was sufficient for both the postures of caregiver's body.

Optimized value for space to make patient up in bed: 799 mm.

### 3.3.3 Space for Transferring the Patient

In Fig. 4, to and forth transfer between wheelchair and bed as demonstrated by physiotherapist was used to calculate the space for this task. In the Fig. 4(1) caregiver has supported the scapula of the patient and blocked the movement of the knee by placing his knee in front of the patient's knee, to facilitate the patient to get up or sit as per procedure at hand. Similarly in Fig. 4(3), caregiver has supported the scapula and blocked the movement of knee and facilitated patient to sit and get up as required. In Fig. 4(2), caregiver and patient moving in small steps to reach wheelchair or bed as per the procedure requirement.

In Fig. 5, caregiver is positioning the patient in wheelchair as explained by physiotherapist by placing legs towards patient. In Fig. 6, caregiver is holding the patient by supporting on scapula of the patient and knee blocking.

**Fig. 4.** Transferring patient between wheelchair and bed

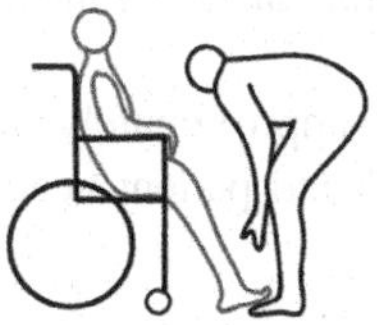

**Fig. 5.** Repositioning patient in wheelchair

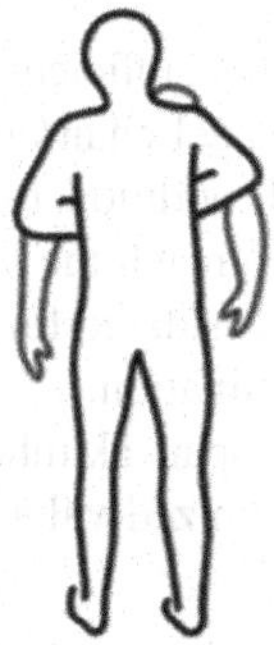

**Fig. 6.** Standing position while transfer

In Fig. 5, caregiver is positioning the patient in wheelchair as explained by physiotherapist by placing legs towards patient. In Fig. 6, caregiver is holding the patient by supporting on scapula of the patient and knee blocking.

Space for transferring the patient from bed to wheelchair or chair was same as both tasks required same steps to perform the task. To and forth transfer between wheelchair to bed or bed to wheel-chair again required same space near wheel chair and bed, as same task is repeated but in reverse manner.

95th percentile male maximum body depth, relaxed = 409 mm.

2 × 95th percentile male maximum body depth, relaxed = 2 × 409 = 818 mm.

But after that, caregiver needed space to make patient stand out of bed by blocking patient's knee with his knee. As direct data is not available for this position so again "Forward step length" and maximum body depth, relaxed of 95th percentile considered as space of "Forward step length" plus space for patient to stand i.e. "Maximum body depth, relaxed".

95th percentile male Forward step length = 799 mm.

95th percentile male maximum body depth, relaxed = 409 mm.

Space to make patient stand out of bed = 95th percentile male (Forward step length + maximum body depth, relaxed) = (799 + 409) mm = 1208 mm.

Next step was to make patient to reach the bed or wheel chair in small steps by holding the scapula and for that "Span akimbo" of 95th percentile person was considered, as caregiver and patient were required to provide a space of Span akimbo of 95th percentile person.

95th percentile male Span akimbo = 959 mm.

As soon as patient reached the chair/wheelchair or bed, caregiver and patient again needed to perform the same task as performed before i.e. knee blocking and make person sit, but in reverse order. If transfer of patient performed to and forth to nearby chair or wheelchair then same space can be utilized for both tasks i.e. to make patient up and make patient to sit with some allowance to rotate the patient to reach to right location, so allowance of hand and foot 1208 mm + 192 mm = 1400 mm.

Space for caregiver and patient near bed = 95th percentile male (Forward step length + maximum body depth, relaxed) + Allowance = ((799 + 409) + 192) mm = **1400 mm.**

So 1400 mm space was sufficient for transfer if wheel-chair/chair was aligned to the bed. And if bed and wheel-chair/chair were far from each other then space required would be twice and with addition to that space of span akimbo of the person was required during the path to reach the place where patient was needed to be transferred. So 95th Span akimbo is considered for the width of the passage followed by patient and caregiver to reach the destination.

95th percentile male Span akimbo = 959 mm, approximately 960 mm and with allowance 1000 mm. Optimized value for caregiver and patient near bed: 1400 mm x 1000 mm.

### 3.3.4 Space for Wheelchair

Space for patient handling and transfer was calculated but with that wheelchair also took space in the room. Figure 7 shows dimensions of an existing wheelchair in local hospital of Jabalpur, India. Total width of the wheelchair was 460 mm and length was 1170 mm and height 990 mm i.e. 1170 mm × 460 mm × 990 mm.

Above discussions are for the space required for transferring the patient but space is also required by wheel chair to align to the bed, to take the patient out of the room and to bring the patient inside room. And for all these procedures, space for movement and rotation of wheelchair is required. Usually people buy wheelchair or get it from hospital or in some cases from market, so modification or customization is not easy compared to customizing the bed as they can order bed and customize it as per their requirement.

So wheel chair size is kept as it is, so that space as per existing wheelchair can be calculated for easy transfer and patient handling.

***The Turning Space for Wheelchair***

The turning space for wheelchair should be either circular area or T-shaped passage. A circular space should have 1500 mm diameter with the allowance of knee and toe clearance and for the caregiver if patient cannot drive the wheelchair on his own (Fig. 8).

**Fig. 7.** Existing Dimensions of the Wheelchair

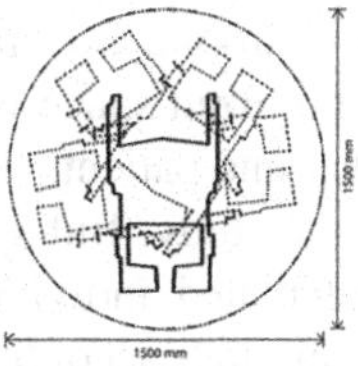


**Fig. 8.** Turning space for wheelchair

Length of wheelchair + 95th percentile Male foot length = (1170 + 274) mm = 1444 mm.

1444 mm and allowance for dynamicity for caregiver movement, 1500 mm space is considered the appropriate space for the turning of the patient and department of public works also set 1500 mm diameter circular space for wheelchair turning for architectural guideline for design of accessible buildings (facilities for disabled persons).

***Minimum Clear Floor Space for Wheelchair***

Minimum clear floor space for one wheelchair movement in one direction then width of passage should be width of wheelchair plus allowance for elbow for that "Maximum body breadth (relaxed)" can be considered of 95th percentile, so 619 mm with allowance on both sides for elbow.

95th male Maximum body breadth (relaxed) + 2 (allowance for elbow) = 619 mm + 100 mm = 719 mm.

For safe driving of the wheelchair for new patient who has just started using wheelchair, little more space is required as wheelchair is not running on rail. So 20 mm on both sides as extra space for new learner of wheelchair = 719 mm + 40 mm = 759 mm approximately 760 mm. So 760 mm minimum clear space for passage for movement in the room and to get access to products in the room.

Wheelchair required more clearance than normal person walking through the door, so door dimension needed to be fixed for wheelchair. Minimum clear floor space can be considered but if person needs to take a turn immediately after going out of the room then more clearance required for wheelchair. So 800 mm was decided as width of the door with extra allowance of 20 mm on both sides for easy turning.

## 4 Conclusion

This study was based on the underpinning assumption that caregivers and patients can benefit if sufficient space for patient handling and transfer would be provided. Conclusion of the literature review confirmed that space plays an important role in performing these tasks. Even though the caregivers and nurses are well aware of the recommended procedures, shortage of adequate space can lead to the discomfort to the patient as well as nurses. Proper patient handling practices do contribute positively to the patient and caregiver safety and avoids MSD. Findings of the study confirmed that caregivers face problems after some duration of continued performance of these tasks. Area norms have standards for the plinth area of the building, but there are no standard dimensions for the bedroom. So on the basis of recommended patient handling and transfer practices, optimized space was proposed for the task in bedrooms to maintain the safety of caregiver and patient. However, not all possible patients and environments were considered in the study. But space required for some important patient handling procedures were calculated individually and can be implemented in any bedroom, which proved to be enough according to the results.

## References


1. India no country for old men: UN report - NDTV. http://www.ndtv.com/article/world/india-no-country-for-old-men-un-report-426545. Accessed 17 Feb 2018
2. India facing elderly population time bomb. http://www.telegraph.co.uk/news/worldnews/asia/india/9690781/India-facing-elderly-population-time-bomb.html. Accessed 10 Feb 2018
3. NGO works for Senior Citizens helping Old Age in India with Best NPO employment & top volunteers at jobs. http://dadadadi.org/old-age-homes-in-India.html. Accessed 19 Feb 2018
4. Begg F et al (December 1999) Designing Workplaces for safer handling of patients/residents. Victorian Workcover Authority
5. Evidence-Based Practices for Safe Patient Handling and Movement. http://www.nursingworld.org/MainMenuCategories/ANAMarketplace/ANAPeriodicals/OJIN/TableofContents/Volume92004/No3Sept04/EvidenceBasedPractices.aspx. Accessed 9 Mar 2018
6. Nelson A, Fragala G et al (October 2001) Patient care ergonomics resource guide: safe patient handling and movement. Developed by the Patient Safety Center of Inquiry (Tampa, FL), Veterans Health Administration and Department of Defense
7. Kneafsey R (February 2012) An exploration of the contribution of nurses and care. University of Birmingham Research Archive
8. Johnsson C (2005) The Patient transfer task, method for assessing work technique. Department of nursing Karolinska institute, Stockholm, Sweden. ISBN 91-7140-262-4
9. A guide to designing workplaces for safer handling of people for health, aged care, rehabilitation and disability facilities. Workspace victoria, September 2007
10. de Castro B (2006) Guidelines for a safe practice environment. Washington State Nurses Association, ESHB 1672

# A Sustainable Working Life for All Ages – The swAge-Model

Kerstin Nilsson(✉)

Division of Occupational and Environmental Medicine, Faculty of Medicine, Lund University, Lund, Sweden
kerstin.nilsson@med.lu.se

## 1 Introduction

The demographic shift is fast becoming a global challenge and an important public health concern. The proportion of people over 65 years is currently 10% around the world, but it is expected to rise to 22% by the year 2050 [1]. Populations ageing rapidly is widely seen as one of the most powerful transformative forces affecting society over the next four decades [1–4]. The ageing population is probably one of the most significant threats to global wealth because of the potential strains it imposes on the robustness of social welfare, public health, and economic prosperity. The challenges and opportunities posed by this demographic shift are therefore of special interest to sustainable societies and require the implementation of policies to help people stay healthy and active while ageing, but also to work longer providing the welfare systems.

It is obvious that the main factor that distinguishes older workers from other employees is their age. However, age could be defined in different manner. A good psychological work environment and improved control over their lives also help people to feel younger [5], and older people who work in a healthy psychological environment often demonstrate better mental and physical health than their retired counterparts of the same age [6]. Also, an individual's work life affects their aging and their ability to continue working in old age. Urgent action is needed at different levels which impact a sustainable work life for all ages, in combination with healthy ageing, aimed to increase the possibility to work in an extended working life.

The objectives of this paper is to describe the framework of a theoretical model of sustainable working life for all ages - the swAge-model. The model describe how different age conceptions associated to determinant factors to work and of importance to a sustainable extended working life. The intention of the swAge-model is to be a tool in the work of understanding how to make the work life more sustainable and healthier for all ages.


S. Bagnara et al. (Eds.): IEA 2018, AISC 826, pp. 223–230, 2019.
https://doi.org/10.1007/978-3-319-96065-4_25

## 2 Method and Material

The swAge-model is developed by the grounded theory method from more than 30 studies in the research groups of healthy work places, older workers and aging during 15 years. The method and material in these studies were qualitative studies of deep interviews, focus group interviews and observations with older workers (55–74 years of age), younger workers, managers on different levels, employer, HR-professionals, unions and occupational health care; quantitative studies with surveys to employees in different age groups and managers on different levels and in different sectors, register studies of the total population in Sweden; intervention studies on different work places and in different sectors.

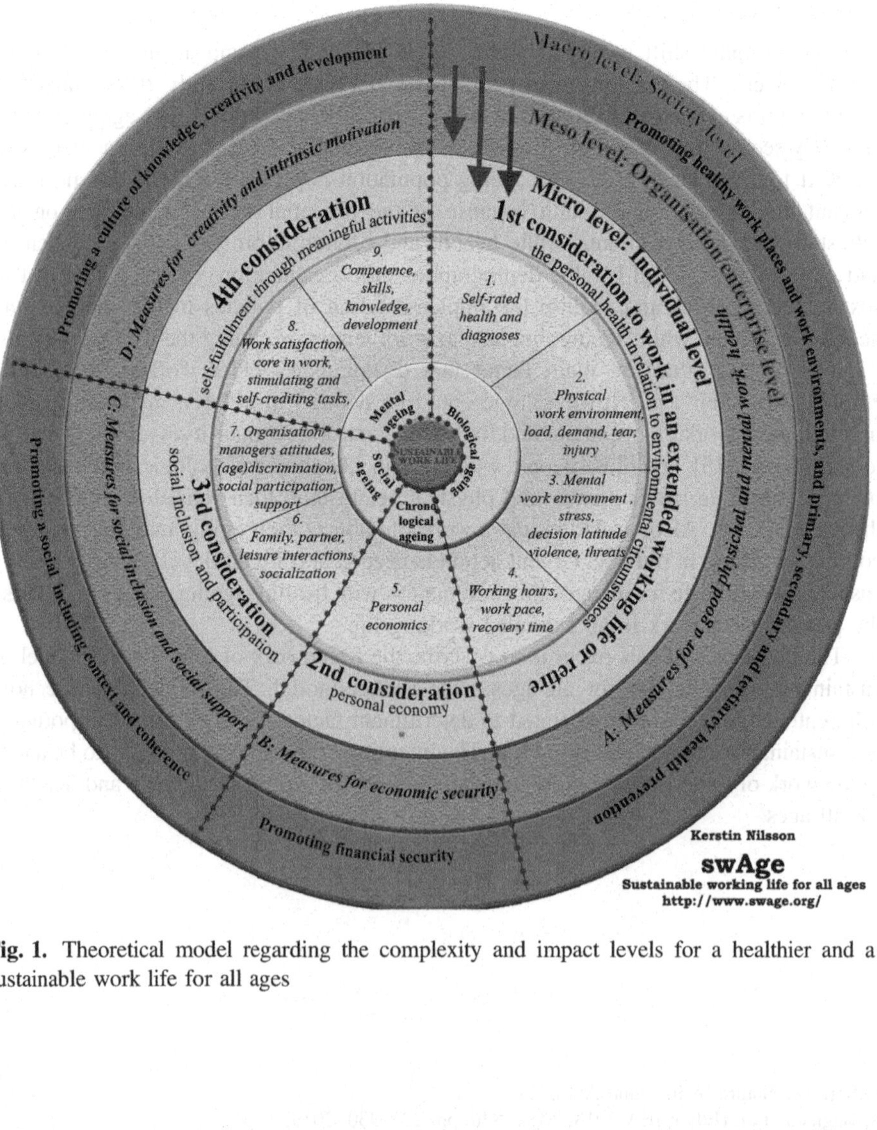


**Fig. 1.** Theoretical model regarding the complexity and impact levels for a healthier and a sustainable work life for all ages

## 3 The swAge-Model

The swAge-model describe three influence levels of importance to a sustainable extend working life: micro level, meso level and macro level.

### 3.1 Micro Level

Four different conceptualizations of ageing are related to work-life participation: employees' *biological ageing, chronological ageing, social ageing,* and *mental ageing,* which affect work-life participation to an older age [7–10]. These four ageing concepts are therefore in the inner circle in the swAge-model (Fig. 1).

There are many factors which determine if individuals can and want to participate in a working life [7–25]. These factors could be sorted into nine areas:

1. Employee self-rated health and diagnoses
2. Employee physical work environment and injury prevention
3. Employee mental work environment, stress, effort/reward balance, violence and threats
4. Employee working hours, the pace at work, and recovery time
5. Employee personal finances
6. Employee family situation, partner, leisure interactions, and socialization
7. Employee experience of organisational and managerial attitudes, employee social participation in the work group, discrimination, and social support
8. Employee-stimulating and self-crediting tasks, core in work, and work satisfaction
9. Employee-competence, skills, knowledge, and opportunities for development

These nine areas of determinants to an sustainable working life are related to the four different conceptualizations of ageing in the swAge-model [7–10] (Fig. 1).

Before deciding whether or not to continue working in an extended-work life or to retire, individuals take four factors into account [7–12]. These four considerations are:

1. The employee considers if their own self-rated health and diagnoses (in relation to the physical work environment, the mental work environment and the working hours, the pace at work and the recovery time) is sufficient as employed, or if their health would be better as a retiree.
2. The employee considers if their personal financial situation would be better by being employed or by being retired.
3. The employee considers if the options for contextual participation and social inclusion with others are best served by being employed or retired.
4. The employee considers if the opportunity for meaningful and self-crediting activities is best served as employed or as retired.

The four considerations extended-work life or to retire are linked to the nine areas of determinants for a sustainable working life and are also linked to the four conceptualizations of ageing in the swAge-model (Fig. 1).

### 3.2 Meso Level

An individual's ability to extend their work life is influenced by organisations and enterprises and measures and attitudes towards the ageing work force [7–35]. To enable a sustainable work life for all ages, organisations and enterprises have to be aware of the nine areas of importance for work-life participation, their association to the four different conceptualizations of ageing, and the four factors affecting older workers who are considering retirement. To generate sustainable work life's for older workers, organisations and enterprises needs to implement measures and strategies related to those factors within their level of the system (Fig. 1). The four specific areas and common ground measures for organisations and enterprises are:

- *Measures to provide a good physical and mental work environment for employees by thinking through:*
  - an organisational culture that promotes the use of ergonomic aids
  - rotation, variation, and a change of duties to reduce physical and mental demands
  - work schedules
  - the pace of work
  - reducing work stress
  - effort/reward balance
  - reducing violence and threats in the work environment
  - physical activity for maintenance of the body, and to retain and increase mental and physical good health
  - importance of a good diet for good (working) health
  - occupational health services and support to prevent work-environment issues and to increase good health in the workplace
- *Measures to provide social and personal financial security by thinking through:*
  - salary and economic benefits
  - work environmental security and risk assessments to lessen the incidence of health problems and the reduced economic benefits owing to sick leave, disability pensions, and unemployment
  - measures to enable competence development and continued employability so that the elderly are able to provide for themselves by working until a higher age
- *Measures for social inclusion and social support in the work situation by thinking through:*
  - employee social inclusion in the work group
  - decrease risk assessment for negative attitudes and age discrimination
  - social support to increase the older employee's self-esteem in the organisation
  - to utilize older employees' mentoring and calming effects on the group as an important productivity investment
  - work schedules that are tailored to individual needs, participation in family activities, leisure and hobbies
- *Measures designed to encourage creativity and intrinsic work motivation by thinking through:*
  - access to competence development regardless of age

- organizational culture to include older workers in new projects and to take advantage of older employees' experiences and knowledge
- increasing motivation and employability
- the rotation and change of tasks to improve motivation and job satisfaction.

### 3.3 Macro Level

Societies, today, focus mainly on chronological ageing in a desire to increase the retirement age with regard to statutory pension systems, i.e., individuals working beyond 60–65 years-of-age, and do not attend as much intention to biological, mental and societal ageing in relation to extended working life [7–38]. Societies also need to consider economic-related regulations which are important in discussing extended work life. Measures that affect good biological ageing, social ageing, and mental ageing is also needed. Measures introduced by society that only focus on individuals extending their work life will be more effective if also implemented at the meso level – where organisations and enterprises are involved. Measures in society are also needed to provide incentives and inducements for organisations and enterprises wishing to keep their employees until an older age and to provide a sustainable work life for the increasing numbers of older workers in modern society. If a society wants to encourage a longer sustainable work life, it must implement methods, strategies, suitable legislation, regulations, economic incentives as well as opportunities that meet and conform to all four measuring fields at the individual and organisational/enterprise levels in the theoretical model (Fig. 1).

## 4 Summary

There are many complex challenges associated with a sustainable work life to an older age. The suggestion is that this theoretical model could be an applied tool for societies, employers, managers, occupational health care/HR personnel, and individuals charged with developing strategies for healthy ageing in an inclusive and sustainable extended work life. The nine important areas for work-life participation, four conceptualisations of ageing, and the four considerations for work-life participation need to be weighed. Measures and initiatives need to take into account the complex inter-relationships between the society level and the individual level with the intent of making work life more sustainable and healthier to an older age. Measures are also needed from the society level to the organisational/enterprise level as well as from the organisational/enterprise level down to the individual level. However, the swAge-model and measuring assumptions are based and developed upon research in Swedish circumstances, but also by reviewing studies from other countries. The proposal is to use and test this theoretical model and the developed measurable areas in intervention projects at the macro, meso and micro levels in different countries. The hypothesis is that this theoretical model could be used in most countries to highlight the urgent need for action at different levels which impact healthy ageing and a sustainable work life for all ages. This theoretical model, the swAge-model, is also a contribution to the critical debate concerning our responsibility to make work life sustainable for all age groups.

**Conflict of Interest:** No conflict of interest.

## References


1. World Economic Forum (2016) Global Agenda Council on Ageing. https://www.weforum.org/communities/global-agenda-council-on-ageing/
2. Rechel B, Grundy E, Robine J-M et al (2013) Ageing in the European union. Lancet 181:1312–1322
3. UNFPA & HelpAge International (2012) Ageing in the twenty-first century: a celebration and a challenge. United Nations Population Fund (UNFPA), New York & HelpAge International, London
4. OECD (2016) Health at a glance: Europe 2016. State of health in the EU Cycle. OECD Publishing, Paris. https://doi.org/10.1787/9789264265592-en
5. Kunze F, Raes AML, Bruch H (2015) It matter how old you feel: antecedents and consequences of age in organizations. J Appl Psychol 100(5):1511–1526
6. Beehr TA, Glazer S, Nielson NL, Farmer SJ (2000) Work and nonwork predictors of employees' retirement age. J Vocat Behav 57:206–225
7. Nilsson K (2016) Conceptualization of ageing in relation to factors of importance for extending working life – a review. Scandinavian J Publ Health 44:490–505
8. Nilsson K (2017) Bästföredatum på arbetskraften? – olika åldersbegrepps betydelse för äldre i arbetslivet [Best before date on the workforce? - Different age concepts important for older people in the workplace]. In: Krekula C, Johansson B (eds) Ålder, makt och organisering – teori och empiri. Studentlitteratur, Lund (in Swedish)
9. Nilsson K (2013) To work or not to work in an extended working life? Factors in working and retirement decision. Lund University, Lund, 4
10. Nilsson K (2017) The ability and desire to extend working life. In: Vingård E (ed) Healthy workplaces for men and women in all ages. Swedish Work Environment Authority, Stockholm, pp 30–49
11. Nilsson K, Rignell-Hydbom A, Rylander L (2011) Factors influencing the decision to extend working life or to retire. Scandinavian J Work Environ Health 37(6):473–480
12. Nilsson K (2012) Why work beyond 65? discourse on the decision to continue working or retire early. Nord J Working Life Stud 2(3):7–28
13. Nilsson K (2005) Vem kan och vill arbeta till 65 år eller längre? En studie av anställda inom hälso- och sjukvården. [Who can and want to work until 65 years or beyond? A study with employed in health and medical care]. Arbete och hälsa, 14 (in Swedish)
14. Nilsson K (2006) Äldre medarbetares attityder till ett långt arbetsliv. Skillnader mellan olika yrkesgrupper inom hälso- och sjukvården. [Older workers attitude to an extended working life. Differences between occupations in health and medical care.] Arbetsliv i omvandling, 10. (in Swedish)
15. Nilsson K (2003) Förlängt arbetsliv – En litteraturstudie av faktorer med betydelse för förlängt arbetsliv som alternativ till tidig pensionsavgång. [Extended working life – A literature review on factors important to an extended working life as an alternative to early retirement]. Swedish National Institute of Working life, Malmö (in Swedish)
16. Nilsson K (2005) Pension eller arbetsliv? [Pension or working life?]. Swedish National Institute in Working life, Malmö (in Swedish)

17. Caffaro F, Lundqvist P, Micheletti Cremasco M, Nilsson K, Pinzke S, Cavallo E (2018) Being a farmer at old age: an ergonomic analysis of work-related risks in a group of Swedish farmers aged 65 and over. J Agromedicine 23:78–91
18. Hovbrandt C, Håkansson C, Karlsson G, Albin M, Nilsson K (2017) Prerequisites and driving forces behind an extended working life among older workers. Scandinavian J Occup Therapy 1–13. https://doi.org/10.1080/11038128.2017.1409800
19. Nilsson K (2017) The influence of work environmental and motivation factors on seniors' attitudes to an extended working life or to retire. A cross sectional study with employees 55–74 years of age. Open J Soc Sci 5:30–41
20. Gyllensten K, Wentz K, Håkansson C, Nilsson K (2018) Older assistant nurses' motivation for a full or extended working life. Ageing and Society
21. Nilsson K (2017) Hållbart arbetsliv inom hälso- och sjukvården – studie om hur 11 902 medarbetare upplever sin arbetssituation och möjlighet att arbeta i ett förlängt arbetsliv [Sustainable working life in health care – survey with 11 902 employees regarding their experience of their work situation and possibility to work in an extended working life]. Arbets- och miljömedicin, Lunds universitet: Rapport nr 13/2017
22. Nilsson K, Pinzke S, Lundqvist P (2010) Occupational injuries to senior farmers in Sweden. J Agric Saf Health 16(1):19–29
23. Nilsson K (2015) Kön och ålderspension – en tvärsnittsstudie om skillnader mellan män och kvinnor i att kunna och vilja arbeta till 65 år eller längre. [Gender and age pension - a cross-sectional study about the differences between men and women to be able and willing to work to 65 years or longer] Occupational and Environmental medicine, Lund University: Report 4/2015 (in Swedish)
24. Nilsson K, Rignell-Hydbom A, Rylander L (2016) How is self-rated health and diagnosed disease associate with early or deferred retirement: a cross sectional study with employees aged 55–64. BMC Publ Health 16:886. https://doi.org/10.1186/s12889-016-3438-6
25. Nilsson K (2003) Arbetstillfredsställelse hos äldre läkare och sjuksköterskor. [Work satisfaction among older physician and nurses]. Swedish National Institute of Working life, Malmö (in Swedish)
26. Bengtsson E, Nilsson K (2004) Äldre medarbetare. [Older worker]. Swedish National Institute of Working life, Malmö (in Swedish)
27. Nilsson K (2018) Managers' attitudes to their older employees - a cross-sectional study. WORK J Prev Assess Rehabil 59(1):49–58
28. Nilsson E, Nilsson K (2017) The transfer of knowledge between younger and older employees in the health and medical care: an intervention study. Open J Soc Sci 5:71–96
29. Nilsson K (2017) Active and healthy ageing at work. - A qualitative study with employees 55–63 years and their managers. Open J Soc Sci 5:13–29
30. Hörnstedt K, Nilsson K, Albin M, Håkansson C (2017) Managers' perceptions of older workers and an extended working life in Sweden. Int J Gerontol Geriatr Res 1:14–20
31. Nilsson K (2007) Chefers attityder till äldre medarbetare inom kommunen. [Municipal managers attitude to their older workers]. Swedish National Institute of Working life, Stockholm, 4
32. Nilsson E, Nilsson K (2017) Time for caring? elderly care employees' occupational activities in the cross draft between their work priorities, "must-do's" and meaningfulness. Int J Care Coord 20(1–2):8–16
33. Nilsson K (2011) Attitudes of managers and older employees to each other and the effects on the decision to extended working life. In: Ennals R, Salomon RH (eds) Older workers in a sustainable society. Labor, education and society, pp 147–156. Peter Lang Verlag, Frankfurt
34. Pinzke S, Nilsson K, Lundqvist P (2012) Tractor accidents in Swedish traffic. Work J Prev Assess Rehabil 41:5317–5323

35. Nilsson K (2017) Interventions to reduce injuries among older workers: a review of evaluated intervention projects. WORK J Prev Assess Rehabil 55(2):471–480
36. Nilsson K, Östergren P-O, Kadefors R, Albin M (2016) Has the participation of older employees in the workforce increased? Study of the total Swedish population regarding exit working life. Scandinavian J Publ Health 44:506–516
37. Kadefors R, Nilsson K, Rylander L, Östergren P-O, Albin M (2017) Occupation, gender and work-life exits: a Swedish population study. Ageing Soc 1–18
38. Fridriksson, JF, Tómasson K, Midtsundstad T, Sivesind Mehlum I, Hilsen AI, Nilsson K, Albin M, Poulsen OM (2017) Working environment and work retention. Nordiska ministerrådet TemaNord, Copenhagen, nr. 559:1–121

# Gender and Work

# Occupational Well-Being of Policewomen in India and China: Scope of Ergonomic Design Interventions

Shilpi Bora(✉), Debkumar Chakrabarti, and Abhirup Chatterjee

Ergonomics Laboratory, Department of Design, IIT Guwahati,
Guwahati 781039, Assam, India
{shilpi.bora,dc,drachatterjee}@iitg.ernet.in

**Abstract.** Impact of environmental stress on occupation well-being of women engaged in police service requires thorough understanding, which is deemed obvious to inspire them to join police organization The objective of this study was to assess environmental stress at the workplace and consequences on occupational well-being of policewomen in two types of regions – Guwahati, (a Metro city of Assam, also its capital city) and Hangzhou, (China). The survey was conducted on 43 policewomen from All Women Police Station (AWPS) of Guwahati and 31 policewomen from Hangzhou city of China. The questionnaire was designed using standard procedures (including analysis of reliability and validity by Cronbach's alpha) to analyses job stress and occupational well-being. This piece of research attempted to elucidate the real-time scenario and the need of corrective ergonomic design interventions, which when implemented, might reduce the workplace stress, thus making Indian police organizations more women-friendly.

**Keywords:** Indian policewomen · Chinese policewomen
Ergonomic interventions · Job satisfaction · On-job stress · Basic amenities

## 1 Introduction

Government is inducting more and more women into police as a part of its empowerment obliged; a survey conducted among those serving has found that they still tackle with lack of basic amenities like toilets, want of privacy, accommodation and lack of staff. These personnel have to go thirsty for long hours while on duty as there are scarcely any toilets around. A familiar problem faced by women coming to the woman police station in the Guwahati city Assam as compared to China is the lack of basic amenities including toilets, a place to sit, availability of drinking water, etc. For improving the working condition of women personnel and augment the representation of women in the police force to tackle the specific requirements of women in the fast changing society. The exploration and quantification of occupational stress at the workplace, and its consequence on occupational well-being, could also lead to changes in extant policies and norms. This might also be contributed to by ergonomic design intervention/design modification of existing infrastructure of police stations. Differences in security situations between regions could also influence environmental stress at the workplace/workstation for policewomen.


S. Bagnara et al. (Eds.): IEA 2018, AISC 826, pp. 233–244, 2019.
https://doi.org/10.1007/978-3-319-96065-4_26

With the intention of a comparative assessment of occupational well-being perspectives of policewomen in India (like, in Assam) with some other the BRIC country (deemed to be at a similar stage of newly advanced economic development), China was selected to serve the purpose. For this, Hangzhou, PRC was selected for collecting data, depending on availability of policewomen, scope of administering the questionnaire and conducting the personal interviews. The economic growth of China (five times the Gross Domestic Product of India) could also be the reason for greater recruitment and better job benefits of women in police services, including pay and welfares, along with other occupations. Greater number of women at workplace (in China, as compared to that in India) might be the likely factor for fair work distribution among the workforce, leading to reduction in overall work stress. The objective of this study was to assess environmental stress at the workplace and consequences on occupational well-being of policewomen in two types of regions – Guwahati, (a Metro city of Assam, also its capital city) and Hangzhou, (China). The study also attempted to address some ergonomic design interventions to improve the situation at all women police stations in Guwahati.

## 2 Methodology

A questionnaire-based survey was designed to assess stress of policewomen at workplace, factors contributing to the stress and whether they underwrote the affecting of their occupational well-being. The survey was conducted on 43 policewomen from All Women Police Station (AWPS) of Guwahati and 31 policewomen from Hangzhou city of China (WPS). The questionnaire was designed using standard procedures (including analysis of reliability and validity by Cronbach's alpha) to analyses job stress and occupational well-being. The questionnaire consisted of two main parts. The first part of the questionnaire had three sub-parts in turn viz. 1A, 1B, and 1C considering participants' response for subjective assessment of workplace and on-job amenities respectively. The part 2 of the questionnaire incorporated the satisfaction index of the policewomen.

## 3 Results and Discussion

Analysis of the observations revealed that, there was varied opinions among policewomen with respect to views on requirement of basic amenities such as toilets for women and other womanhood related aspects for both locations of study. Policewomen in Hangzhou opined that there was less stress at workplace, while policewomen in AWPS Guwahati proposed that these are important for their occupational improvement. Further, due to the security concern in Hangzhou, policewomen urged for more safety at the police station and implement some basic amenities in the workplace. With regard to job satisfaction and promotional avenues, the policewomen in both places had mixed response. A need of modern equipment and up-to-date facilities was also perceived. In terms of occupational stress, resources and night shift policy, the response was more or less neutral. In addition, policewomen also think that presence of child/day care center would not increase their satisfaction level. This spoke of that, issues related to work environment might be more important, as they already had sufficient infrastructure in place.

Greater number of women at workplace (in China, as compared to that in India) might be the likely factor for fair work distribution among the workforce, leading to reduction in overall work stress. Respondents were neutral when enquired about on-job stress. The uniqueness of policewomen in China was moderately less disparity among the responses, as compared to India. However, despite higher staff and availability of equipment's, there were agreement on lack of specific washrooms, childcare units as well as modular van for policewomen in China. There was also agreement on unfair work environment. Though, during interview, they introduced the modular van (Fig. 1) allotted which was self-sufficient in inside. As observed from comparison of occupational stress between WPS Hangzhou, China and AWPS Guwahati, India (after intervention, see Table 1, there was relatively much less insufficiency of resources, staff and modern equipment with policewomen in China. The interviews revealed that, there were around 400 policewomen in AWPS for metropolitan areas of China, much higher compared to 60 in AWPS Guwahati. However, after intervention in AWPS Guwahati, the assertiveness of the responses improved, though with some degree of dissatisfaction with respect to lack of resources. Views regarding discriminating/uncomfortable work environment at AWPS Guwahati improved after intervention with somewhat better perception of the overall occupational well-being than that of AWPS earlier. It indicated that this ergonomics based design intervention was perceived to be satisfactory by the policewomen in India. Respondents were, however, neutral when enquired about the job stress.

(a) (b)

**Fig. 1.** Identity of police vehicles (a) Mobile police station of China (b) Police Car in Assam, India.

Table 1 showed that, for most of questions for job burn out, responses of AWPS Guwahati after intervention were quite similar to those of WPS Hangzhou. Policewomen in China apparently disagreed with aspects like burn-out due to work. This could likely be because of greater number of policewomen in WPS Hangzhou than AWPS Guwahati. Further, with regard to feeling exhausted of routine work, the policewomen in China had somewhat scattered response (more disagreed than agreed) as compared to strong agreement with that by policewomen of AWPS Guwahati.

**Table 1.** Comparative observations on exposure to occupational and environmental stress and perceived well-being of policewomen across the AWPS Pan Bazar Guwahati after intervention (AI) and WPS Hangzhou (China)

| Q Sl No | AWPS Guwahati (AI) | | | | | WPS Hangzhou (China) | | | | |
|---|---|---|---|---|---|---|---|---|---|---|
| | (1) | (2) | (3) | (4) | (5) | (1) | (2) | (3) | (4) | (5) |
| Q1 | 43 | | | | | 27 | 4 | | | |
| Q2 | 43 | | | | | | | | 27 | 4 |
| Q3 | 43 | | | | | | | | 28 | 3 |
| Q4 | | **25** | | **18** | | | | 29 | 2 | |
| Q5 | 43 | | | | | | | 30 | 1 | |
| Q6 | 18 | | 16 | | 9 | | | 27 | 4 | |
| Q7 | 26 | 17 | | | | | 2 | 29 | | |
| Q8 | 21 | 14 | | 8 | | | | 31 | | |
| Q9 | | 16 | | 11 | 16 | | | 31 | | |
| Q10 | 20 | 23 | | | | | | 31 | | |
| Q11 | 43 | | | | | 2 | 29 | | | |
| Q12 | 43 | | | | | | | | 30 | 1 |
| Q13 | 8 | 12 | 18 | 5 | | | | 31 | | |
| Q14 | | | | 43 | | | | 31 | | |
| Q15 | | 24 | | 19 | | 1 | 30 | | | |
| Q16 | | **6** | | **37** | | | 29 | 2 | | |
| Q17 | 43 | | | | | | | | 31 | |
| Q18 | 27 | 10 | | 6 | | | 28 | 3 | | |
| Q19 | | | | | 43 | | 28 | 3 | | |
| Q20 | 43 | | | | | | 28 | 3 | | |
| Q21 | 43 | | | | | | 30 | 1 | | |
| Q22 | 43 | | | | | | 25 | 6 | | |
| Q23 | | | | | 43 | | 31 | | | |
| Q24 | | | | | 43 | | 31 | | | |
| Q25 | | | | | 43 | | 31 | | | |
| Q26 | | | | | 43 | | 31 | | | |
| Q27 | | | | | 43 | | | | 29 | 2 |

**(1)** Represents Strongly Agree, **(2)** Agree, **(3)** Neutral, **(4)** Disagree and **(5)** Strongly Disagree

Tabulated representation of the questionnaire-based survey on exposure to occupational and environmental stress and perceived well-being. Q 1–27 represents the questions from the questionnaire (Part 1A, explained in the methodology section). (1) Law enforcement is generally regarded as a masculine profile, therefore we who are inducted in this job, felt that convenience is equally important for us. (2) Administrative over shifting is common. (3) Staff shortages cause stress. (4) Lack of resources cause stress. (5) In equal sharing of work responsibilities cause stress. (6) Shift work causes stress for special cases like pregnancy, expecting mother, lactating mother, menstruation period. (7) Traumatic events affects psychophysical health. (8) Social life outside the job is impacted by duty regimen (9) Occupation-related health issues in special cases like pregnancy, expecting mother, lactating mother, menstruation period. (10) Not finding time to stay in good physical condition. (11) Feelings like you are always on the job and other responsibilities are compromised. (12) Working beyond working hours brings boredom. (13) Noisy work area. (14) Frequent interruptions brings disturbance in the work place. (15) Inadequate or poor quality equipment/maintenance. (16) Unfair work environment in this job. (17) Lack of a modern system/apparatus on duty. (18) Occupational health issues (e.g. back pain, neck pain, and joint pain). (19) A good infrastructure brings satisfactions while doing work. (20) Lack of resources in professional/promotional. (21) Working alone at night is risky and I don't feel good. (22) Prolong standing affects physical health. (23) Lack of separate modular convenience/prompt service utilities in every police station. (24) Basic amenities like isolated/separate restrooms and child care units are still a major requirement for women police personnel. (25) Lack of residential accommodation which is seen as one of the major impediments faced by women in joining police force. (26) While I am involved in outdoor activities such as patrolling, security duty on several occasions, touring in and outside the district where mobile convenience facility is a compulsory requirement. (27) Crèches/day care centre in the police station for working mother will help them to take care of their children.

**Table 2.** Comparative observations on on-job burnout of policewomen across AWPS Pan Bazar Guwahati after intervention and WPS Hangzhou (China)

| Q Sl No | AWPS Guwahati (AI) | | | | | WPS Hangzhou (China) | | | | |
|---|---|---|---|---|---|---|---|---|---|---|
| | **(1)** | **(2)** | **(3)** | **(4)** | **(5)** | **(1)** | **(2)** | **(3)** | **(4)** | **(5)** |
| Q1 | | **3** | **40** | | | | | 31 | | |
| Q2 | | | | 26 | 17 | 7 | 24 | | | |
| Q3 | | | | 26 | 17 | | | | 28 | 3 |
| Q4 | 20 | 23 | | | | 4 | 26 | 1 | | |
| Q5 | 20 | 9 | 4 | 5 | 5 | 3 | 25 | 3 | | |
| Q6 | 43 | | | | | | 29 | 1 | 1 | |

**(1)** Represents Strongly Agree, **(2)** Agree, **(3)** Neutral, **(4)** Disagree and **(5)** Strongly Disagree
Tabulated representation of the questionnaire-based survey on job burnout. Q 1–6 represents the statements from the questionnaire (Part 1B, explained in the methodology section). (1) My work is emotionally exhaustive. (2) I feel burnt out because of my work. (3) My work frustrates me. (4) I feel burn out at the end of the working day. (5) I feel exhausted in the morning only by the thought of another similar day at work. (6) I feel quite energetic while passing time with family, friends and relations.

Table 2 reflected that, for questions related to job promotion as well as rules and procedures, the responses were grossly similar for WPS Hangzhou and AWPS Guwahati (after intervention). Policewomen in China appeared to agree strongly with regards to be paid fairly and service benefits as compared to other organizations. This was likely because of higher economic growth in China and also higher income of women in general as compared to that in India. Supplementary to that, in response to the question related to supervisor's competency, the policewomen in Hangzhou was neutral as compared to mixed responses of policewomen in AWPS Guwahati.

Comparison of the outcomes of the survey (Tables 3 and 4) for satisfaction index conducted on policewomen in AWPS Guwahati (after intervention) and WPS Hangzhou represented comprehensive improvement in response to aspects related to public attitude towards women, lack of separate utilities, separate modular van and better work environment among policewomen in AWPS Guwahati. With these enrichments after intervention, the policewomen from Guwahati appeared to have some more positive perceptions than those from Hangzhou. However, for queries related to accommodation and child care center, the response was more or less similar for personnel of both locations. Design interventions at AWPS Guwahati indeed augmented the satisfaction index among policewomen. This indicated that further design interventions could be even more useful in uplifting work environment and thereby productivity of policewomen in India at large.

**Table 3.** Comparative observations on job satisfaction of policewomen across AWPS Pan Bazar Guwahati after intervention (AI) and WPS Hangzhou (China)

| Q Sl No | AWPS Guwahati (AI) | | | | | WPS Hangzhou (China) | | | | |
|---|---|---|---|---|---|---|---|---|---|---|
| | **(1)** | **(2)** | **(3)** | **(4)** | **(5)** | **(1)** | **(2)** | **(3)** | **(4)** | **(5)** |
| Q1 | 2 | | 3 | 15 | 23 | 1 | 26 | 3 | 1 | |
| Q2 | 4 | 17 | 12 | 10 | | | | 31 | | |
| Q3 | | 37 | 1 | 5 | | | | 31 | | |
| Q4 | | 10 | | 11 | 22 | 3 | 28 | | | |
| Q5 | | 8 | 5 | 13 | 11 | | | 4 | 27 | |
| Q6 | | | | 36 | 7 | | | | 29 | 2 |

**(1)** Represents Strongly Agree, **(2)** Agree, **(3)** Neutral, **(4)** Disagree and **(5)** Strongly Disagree
Tabulated representation of the questionnaire-based survey on job satisfaction. Q 1–6 represents the statements from the questionnaire (Part 1C, explained in the methodology section). (1) I feel I am being paid a fair amount for the work I do. (2) My supervisor is quite competent in doing his/her job. (3) When I do a good job, I receive the recognition for it that I should receive. (4) The benefits we receive are as good as most other organizations offer. (5) Many of our rules and procedures make doing a good job simple. (6) Those who do well on the job stand a fair chance of being promoted.

**Table 4.** Comparative observations on exposure to satisfaction index of policewomen across AWPS Pan Bazar Guwahati after intervention (AI) and WPS Hangzhou (China)

| Q Sl No | AWPS Guwahati (AI) | | | | WPS Hangzhou (China) | | | |
|---|---|---|---|---|---|---|---|---|
| | Yes | | No | | Yes | | No | |
| | BI | AI | BI | AI | BI | AI | BI | AI |
| Q1 | 33 | 10 | 7 | 36 | 31 | | | |
| Q2 | 43 | | | 43 | 31 | | | |
| Q3 | 9 | 34 | 34 | 9 | 31 | | | |
| Q4 | | | 43 | 43 | 31 | | | |
| Q5 | | | 43 | 43 | 31 | | | |
| Q6 | 43 | | | 43 | 31 | | | |
| Q7 | 43 | | | 43 | 31 | | | |
| Q8 | 43 | | | 43 | 31 | | | |

Tabulated representation of the questionnaire-based survey on satisfaction index. Q No 1–8 represents the questions from the questionnaire (Part 2, explained in the methodology section). (1) Public attitude towards women police is awkward. (2) Lack of separate utility facilities in police stations. (3) Problems related to training. (4) Govt accommodation for womanhood related issues. (5) Difficulties faced in upbringing of children – day care center is essential. (6) Need to have a better working environment in terms of infrastructure. (7) Provision of separate toilet facility at all offices/outpost. (8) A modular mobile convenience facility while outdoor duty an immediate need.

However, for queries related to accommodation and child care center, the response was more or less similar for personnel of both locations. Design interventions at AWPS Guwahati indeed augmented the satisfaction index among policewomen. This indicated that further design interventions could be even more useful in uplifting work environment and thereby productivity of policewomen in India at large.

(a) (b)

**Fig. 2.** Picture representing women police in (a) Hangzhou, China, and Assam, India (b)

The mobile police station seen in Hangzhou (PRC) was advantageous for maintaining law and order in heavily populated areas in the city. The vehicle was comprehensively equipped with cameras at the top and windows having bullet proof glass (as identified from discussion with the police). The van was greatly useful for patrolling even during night. However, this van did not have provisions for separate washrooms/restrooms for women, but the common washroom was clean and hygienic enough, and anodyne for use of women also. As compared to that in Hangzhou (Fig. 1), police car in Assam (Fig. 2(b)) was relatively simpler, containing mainly red/blue beckon, siren and wireless communication system. Unlike, mobile utility van in Hangzhou, police car in India lacks sophisticated instrumentation (cameras, LED displays, etc.). Policewomen in Hangzhou (Fig. 2(a)) appeared to be equipped with better uniforms than those policewomen in India (Fig. 2(b)), where design of uniform for policemen and policewomen were almost the same. Policewomen in Hangzhou had all basic equipment required for patrolling at different places. On the other hand, policewomen in India lacked of basic equipment required for patrolling. Similarly, vehicle for patrolling was also found to be more sophisticated in China with presence of red beacon, shelf to hold necessary equipment and siren (Fig. 3(a)). For policewomen in India (Fig. 3(b)), the vehicles were not usually provided. Even if they were provided, they were relatively simple without any special provision for holding equipment.

(a) (b)

**Fig. 3.** Women Police with their two wheelers in Hangzhou Women Police with their two wheelers in China (a) (b) Motor bike and Scooty in India (source: http://indiagirlsonbike.blogspot.com/2016/11/india-lady-police-bike-ride.html#) Scooty (Image courtesy: BCCL; obtained from article (dated April, 2013). Weblink: http://www.idiva.com/news-work-life/maharashtra-has-most-female-cops/20797)

## 4 Ergonomic Design Intervention at All Women Police Station (AWPS) Assam, India

A vivid, interactive interview with policewomen of AWPS (Pan Bazar, Guwahati) and the questionnaires they responded to, revealed a gross mixed trend of opinions regarding the current scenario in terms of exposure to environmental and occupational stress and perceived well-being, conveniences and job satisfaction. The survey reports, collected before proposing the ergonomic design interventions, revealed a gross dissatisfaction identifying some real-time inadequacies pertaining to womanness issues like basic amenities, privacy, proper sitting area, convenience facilities etc. Ergonomic interventions embrace the concern of entire work place, work methods as well as the work organization, etc. Moreover, ergonomic intervention would probably facilitate the individuals for prevention of various health hazards which may cause due to improper working environment. The present study proposed implementation of some useful ergonomic interventions approached for the AWPS (All Women Police Station) to manoeuvre the physical susceptibilities relevant to physical hazards. The intervention was designed after survey of the present situation through the questionnaire in detail following personnel interview with working individuals (highlighting various occupational and environmental issues in their workplace). During the investigation, lack of basic amenities, poor design and age-old furniture, along with improper sitting posture was found to be the prime reason behind the reported issue. Therefore, a design incorporating suitable ergonomic interventions was proposed for the existing AWPS, which they implemented accordingly (Fig. 4). In addition to it, individuals were made aware of proper sitting posture in order to work comfortably for prolonged time.

Analysis of the questionnaire revealed that, the police stations as workstation had a significant influence on employee's efficiency. With a virtuous work environment, women police could accomplish stress-free workplace, which could in turn enhance the overall efficiency. The workplace before and after implementing ergonomic design interventions was shown in the Fig. 4. Analysis of the situation in AWPS before the intervention elucidated that, the space was too clumsy, lacking proper sitting area with age-old furniture; while after implementation of the proposed intervention, fully modernized furniture at office, sitting area and counselling desk were allotted in the workplace for better working conditions of the women police personnel.

Workplace of the AWPS suffered from low maintenance, resulting in the negative consequences, and thereby reducing the job satisfaction. Subsidence of these risk factors was the major goal to render utmost precautionary approaches in the work environment. Several respondents stated that policing is not an easy occupation for females because of insufficiency of resources, separate arrangement and amenities and communal gravities (Haider 2015). Figure 4 showed the overview of furniture that were required for women with knee pain (elderly), prolonged sitting and pregnant women. This furniture could be utilized for AWPS for reducing various stress among women.

In many of the developing countries, it is evident that the office workstation design is at the initial stage as long as a police station is concerned. The Parliamentary Committee of India on 'Empowerment of Women' also documented the working conditions of women in the police force (in its 2013–14 and 2014–15 reports) referring to this lack of facilities for the women. The Committee articulated that, these issues can only be tackled through persistent efforts and constant follow up by the government along with time bound action plans. Occupational stress occurred due to lack of amenities and resources available in the AWPS.

One of the foremost apprehension stated by the policewomen was the necessity to improve privacy in the workplace (Bora et al. 2016; Bora et al. 2017; Times of India 2014). The survey found the almost all the policewomen suffered from occupational health consequences like back pain, neck pain, joint pain mostly due unavailability of resources in the workstation, most importantly, proper furniture.

Figure 4 shows the comparison of the interior of AWPS before and after implementation of certain design interventions. These interventions were similar to the one proposed further. It could be observed that there was improved and upgraded basic amenities like toilets, modern furniture in both office and visitors' room. In addition, as was observed visually, the cleanliness and hygiene also improved considerably. These interventions were designed based on the difficulties perceived while surveying the police personnel. Though, some interventions were implemented, amenities related to child care, reception area, restrooms, and canteen were not modified, which while interacting with policewomen at AWPS, was found to be because of lack of space. In order to quantify and assess their satisfaction after implementation, questionnaire was administered again after intervention.

(a) Women Police Station – before study

(b) Women Police Station– now

(c) Rest room was common for all, and inside the prisoners' cell

(d) Modernized rest rooms for police, built outside the prisoners' cell

(e) Office furniture's – before study

(f) Office furniture's – now

(g) Office and sitting area – before study

(h) Office and sitting area – now

**Fig. 4.** Ergonomic interventions proposed for workplace improvements of Women Police Station, Pan bazar, Guwahati and some of their implementations towards facilitating the workplace environment and basic amenities. The Figure depict the conditions prior to ((a), (c), (e), (g)) and after ((b), (d), (f), (h)) ergonomic interventions.

### 4.1 Statistical Interpretation of Improvements Due to Ergonomic Design Interventions AWPS

The responses of policewomen to satisfaction index (through Part 2 of the questionnaire) across the various police stations were subjected to statistical analyses using Spearman's ranked correlation to explore the reliance (if any) of physical implementation of the ergonomic design intervention on the corresponding subjective perception, vs that of 2D sketches of intervention epitomized theoretically. Table 5 represented the correlational observations (r) along with their significance (P) for different locations and conditions.

In case of AWPS, there was a significant direct correlation after intervention (AWPS/AI, r = 0.12, P < 0.05) with r2 being 0.0146; while before interventions (AWPS/BI), r showed to be 0.01 (NS), registering no correlation at all. The correlation with the changes after intervention for AWPS was significant, perhaps because of the physical implementation of ergonomics design interventions in AWPS.

**Table 5.** Representation of correlation properties (r) for AWPS showing the significance of satisfaction index of policewomen across the various police stations through the Questionnaire. (Part 2)

| Condition | r | P | Significance | $CI_{95(L)}$ | $CI_{95(U)}$ |
|---|---|---|---|---|---|
| AWPS/BI | 0.01 | 0.915 | NS | –0.110 | 0.123 |
| AWPS/AI | 0.12 | 0.04 | * | 0.004 | 0.234 |

AWPS: All Women Police Station, BI: Before Intervention, AI: after intervention; r: Spearman rank correlation; $CI_{95(L)}$ and $CI_{95(U)}$: Lower and Upper limits of 95% Confidence Interval respectively; NS: Not Significant, * = P < 0.05.

## 5 Conclusion

This study analyses occupational well-being and job stress at police stations in both Indi and China. Further, it also evaluates certain design interventions in all women and common police stations. From the comparison of Policewomen at AWPS Pan Bazar, Guwahati (after intervention) and WPS Hangzhou (China), it was found that the satisfaction level is higher in China compare to AWPS, Guwahati because of the higher economic growth in China and possibility higher expenditure towards maintenance of police station. Further, the job stress is lower in China due to significantly higher number of women staff as compared to that of India. In terms of facilities too, conditions were significantly better in China. However, despite this, China also lacks modular van facilities for women. Regarding evaluation of intervention, there was significant improvement in basic amenities. Based on statistical correlations, it was found that changes after intervention for AWPS was significant, perhaps because of the physical implementation of ergonomics design interventions in AWPS.

## References

Bora S, Chatterjee A, Rani P, Chakrabarti D (2016) On-the-job stress: interventions to improve the occupational well-being of policewomen in Assam, India. J Int Women's Stud 18(1):260–272

Haider A (2015) Needs and Challenges for Women Police: Study of Islamabad Capital Territory (ICT) Police

Parliament of Lok Sabha Publications, 2013–2014. Committee of Women empowerment: First report on "Working Condition of Women Police". Lok Sabha, Delhi Secretariat

Parliament of Lok Sabha Publications, 2014–2015. Committee of Women empowerment: Second report on "Working Condition of Women Police". Lok Sabha, Delhi Secretariat

Bora S, Chatterjee A, Karmakar S, Chakrabarti D (January 2017) Implementation of ergonomic design interventions to improve workplace amenities for Assam policewomen. In: International conference on research into design, pp 219–229. Springer, Singapore

# Why Do We Often Forget Gender During Ergonomic Interventions?

Karen Messing(✉), Mélanie Lefrançois, and Johanne Saint-Charles

CINBIOSE, Université du Québec à Montréal, Montréal, QC H3C3P8, Canada
{messing.karen, saint-charles.johanne}@uqam.ca, lefrancoismelanie@yahoo.ca

**Abstract.** The Université du Québec à Montréal has signed agreements with community groups providing access to help with research and training initiatives. In this context, a union-university partnership performed research aimed at improving occupational health and gender equality. Over its 17-year lifetime, researchers responded to twenty union requests for action-oriented ergonomics research on jobs marked by a gender division of tasks and health and safety risks. However, during several interventions, gender was "forgotten" and ergonomists concentrated on general health and safety issues, at least initially. Reasons for this blindness arise from workplace constraints but also from the very nature of ergonomics. Integrating observations of gender and other dimensions of social relations into ergonomic analysis and intervention is necessary, but not simple, and fraught with obstacles.

**Keywords:** Gender · Ergonomic intervention · Social relations

## 1 A Union-University Partnership for Occupational Health and Gender Equality

Ergonomists have suggested considering gender at each stage of work analysis, from the initial request for intervention to devising solutions [1–4]. It is hoped that such consideration will lead to the improvement of workers' health while protecting women's access to professional equality. Researchers at the CINBIOSE research center, based at the Université du Québec à Montréal (UQAM), have been able to profit from a context favourable to collaboration with women workers to improve health and equality. During the late 1970s, three Quebec labour unions established a formal agreement with UQAM, which agreed to provide training and research in response to union requests [5; preface]. The agreement is administered by the university outreach service (Service aux collectivés) where a full-time employee is assigned to reconciling the needs and constraints of researchers and unions. A joint university-community steering committee makes sure that the projects meet both community needs and university standards. The union agreement, most recently renewed in 2017, has given rise to hundreds of research and training projects and enriched the research of professors and students [6].


S. Bagnara et al. (Eds.): IEA 2018, AISC 826, pp. 245–250, 2019.
https://doi.org/10.1007/978-3-319-96065-4_27

From 1993 to 2010, a union-university partnership performed research specifically aimed at improving occupational health and gender equality. Called "L'invisible qui fait mal"a (IQFM), the team was composed of researchers in ergonomics and legal sciences as well as union representatives from women's affairs or occupational health and safety [5, 7]. The research activities were supported by government funds solicited through the usual academic channels and led to publications in books and learned journals as well as transformations of local working conditions and changes in public policy. The IQFM partnership came to the end of its formal life in 2009 and has been succeeded by a new series of collaborations involving early-career researchers [8].

## 2 Union Requests for Studies of Women's Jobs

Over its lifetime, ergonomics researchers from IQFM responded to twenty union requests for action-oriented research on jobs marked by a gender division of tasks and health and safety risks. The studies were usually undertaken in partnership with both local unions and the national union confederations, sometimes with the collaboration of management. Following our standard approach to ergonomic analysis [9], the studies had a data collection sequence that began by meetings with key actors, followed by preliminary observations and interviews, individual validation exchanges, systematic observations and collective validation sessions with groups of workers. The studies included observations ranging from 25 to 100 h, spanning months or years of field presence. Preliminary observations and interviews with key stakeholders were aimed at understanding and situating the initial request and identifying possible determinants of problems to investigate using systematic observations. Whether related to physical, cognitive, emotional or interpersonal strategies, systematic observations of work activity were based on a detailed observation grid, measurement tools (Kronos, Actogram) or other tailor-made tools developed as needed.

## 3 Four Types of Interventions in Women's Jobs

Studies from the earliest time of the partnership focused on recognition of "invisible" activities and risks in jobs traditionally assigned to women. Work activity analysis, with its careful documentation of observable constraints, readily lends itself to this type of demonstration, and many preceding and subsequent studies succeeded in gaining recognition for hidden skills and risks among office workers [10], crab processors [11] and print shop workers [4]. Later in the partnership, some interventions sought to improve the entry and retention of women in jobs traditionally done by men. These usually resulted in suggestions for changes in equipment, job design and human resource management. However, retention of women in the jobs during the studies was generally very poor and the loss of women may even have been exacerbated by the studies, since they generally resulted in the women becoming more conscious of the obstacles they faced.

Somewhat later, the unions asked us to gather evidence on newly emerging forms of work. The focus was on increased precarity in women's work rather than gender as such, although gender stereotypes could have contributed to rendering the effects of precarity on work less visible. The results of these observations underlined how job precarity interacted with work activity. For instance, on-call and contract teachers and nurses were observed to have reduced operational leeway because of lack of information due to intermittent contact with their coworkers: on-call nurses struggled to find equipment and build relationships with colleagues and patients, while contract teachers raced from school to school without being able to develop pedagogical approaches. Precarity could also weaken social support, for example through competition for scarce "regular" jobs, through too-frequent team shuffling, or through lack of respect from pupils or clients [12]. For those whose precarity stemmed from minority status (immigrant workers in a clothing factory), observations revealed that both women and men had language-related difficulties in protecting their health, with some pronounced effects on women [13, 14].

The last set of studies initiated during the partnership concerned work-family articulation, due to government policy discussions going on at the time. Again, the emphasis was not specifically on gender but on the work-family interface. The observations showed that work-family articulation strategies were embedded in work activity, due to family responsibilities that manifested during work hours, to work schedules and responsibilities that invaded family time, or to work execution strategies aimed at preserving energy for home responsibilities. The observations revealed the complexity and the invisibility of these informal strategies in workplaces with minimal or no formal policy or negotiated clauses favouring work-family balance. Collectively, again, these studies were used by the unions to orient and advance union and government policy and collective bargaining in the area of work-family articulation.

## 4 The Problem

During the partnership, in several instances, we were surprised to note that gender was "forgotten" during IQFM ergonomic interventions. For example, three of us observed cleaners for several shifts, reporting on how some surfaces were hard to dust and other surfaces were too cluttered to mop, but completely losing the focus on the division of tasks and work activity by gender. Another time, an ergonomist, member of a steering committee of an IQFM study on health care aides, publicly rejected the idea of a gendered division of tasks and risks in this population, six months after commenting on versions of and approving a study extensively documenting the division [15]; he had forgotten the results he had endorsed. In the studies of work-family articulation and precariousness, we have caught ourselves repeatedly reminding each other to think about gender, only to have the subject wander back to the specifics of the work situation under study.

## 5 Five Reasons for Unconscious Gender Blindness

Reasons for this blindness may be fivefold: (1) Ergonomists, particularly those in the work activity analysis tradition [9], are trained to analyse work, not workers, and to observe the work of a variety of workers so that individual differences "disappear." (2) Ergonomic intervention is directed toward identification of risks and immediate transformation of work activity; gender is not usually perceived as a risk or a transformable element. (3) Gender usually denotes relational phenomena rather than characteristics of individuals [16]. Although statistical analysis of large samples can reveal differences in - for example - women's and men's size, shape, flexibility, endurance, family responsibilities and pay grade, such differences do not always leap to the ergonomists' eyes when observing the work activity of individual workers. Moreover, within these data sets, women and men do not fall into totally distinct categories; individual women can be stronger, taller, have fewer family obligations, and be more senior than their male colleagues [17]. The interactions between minds, bodies and specific, multidimensional work stations and multifaceted task demands cannot be simply predicted by gender or sex; they are mediated, for example, by worksite and task design. (4) Workplace participants, including many ergonomists, have expressed fear that gender analysis will lead to gender stereotyping. By and large, women in non-traditional jobs may maintain that they work and are treated exactly like their male counterparts, despite any observations suggesting the opposite. Rather than concentrating on differences, enlarging all workers' operational leeway by means of tool and equipment adaptations is often safer for women workers than initiating discussions of sex and gender during these interventions. (5) Gender and other power relations are often taboo subjects in the workplace [18, 19]; it is unlikely that employers will accept changes in these relations as an outcome of an ergonomic intervention [20].

## 6 Going Forward

We conclude that observations of women and men at work can provide compelling evidence that can alter perceptions of gender, potentially leading to transformations inside the workplace, in workplace policies and in the wider political and social sphere. However, interventions run the risk of ignoring gender when it is relevant, of conflating gender with other aspects of work assignments, and of reinforcing gender stereotypes. Integrating gender during work analysis and transformation requires considerable delicacy if both health and equality at work are to be served. Frequent, frank discussions among ergonomists and with partners can be helpful.

**Notes**

a. Literally, "The invisible that hurts".

## References


1. Messing K (1999) La pertinence de tenir compte du sexe des “opérateurs” dans les études ergonomiques: Bilan de recherches. Perspectives interdisciplinaires sur le travail et la santé (PISTES) 1(1). http://journals.openedition.org/pistes/3840
2. Vézina N, Chatigny C, Calvet B (2016) L’intervention ergonomique: que fait-on des caractéristiques personnelles comme le sexe et le genre? Perspectives interdisciplinaires sur le travail et la santé (PISTES) 18(2). http://journals.openedition.org/pistes/4847
3. Laberge M, Caroly S (2016) Prendre en compte le sexe et le genre dans le choix des situations à analyser: un enjeu pour l’intervention ergonomique. Perspectives interdisciplinaires sur le travail et la santé (PISTES) 18(2). http://journals.openedition.org/pistes/4902
4. Chappert F, Théry L (2016) Égalité entre les femmes et les hommes et santé au travail: Comment le genre transforme-t-il l’intervention sur les conditions de travail? Perspectives interdisciplinaires sur le travail et la santé (PISTES) 18(2). http://journals.openedition.org/pistes/4882
5. Messing K (2014) Pain and prejudice: what science can learn about work from the people who do it. BTL Books, Toronto
6. Blanc M, Fontaine C, Kurtzman L, Lizée M, Vanier C, van Schendel V (2011) L’UQAM dans la Cité : la contribution du Service aux collectivités. Université du Québec à Montréal, Montréal, Canada. https://sac.uqam.ca/upload/files/UQAM_dans_la_cite.pdf. Consulted 17 December 2017
7. Messing K, Lippel K (2013) L’invisible qui fait mal: Un partenariat pour le droit à la santé des travailleuses. Travail Genre Sociétés 29:31–48
8. Riel J, Saint-Charles J, Messing K (2017) Women’s occupational health: resisting when we can. New Solutions J Environ Occup Health Policy 27(3):279–283
9. St-Vincent M, Vézina N, Bellemare M, Denis D, Ledoux É, Imbeau D (2014) Ergonomic intervention. Institut de Recherche Robert-Sauvé en Santé et en Sécurité du Travail (IRSST), Montréal
10. Teiger C, Bernier C (1992) Ergonomic analysis of work activity of data entry clerks in the computerized service sector can reveal unrecognized skills. Women Health 18(3):67–77
11. Major M-E, Vézina N (2017) The organization of working time: developing an understanding and action plan to promote workers’ health in a seasonal work context. New Solutions J Environ Occup Health Policy 27(3):403–423
12. Seifert AM, Messing K, Chatigny C, Riel J (2007) Precarious employment conditions affect work content in education and social work: results of work analyses. Int J Law Psychiatry 30 (4–5):299–310
13. Premji S, Lippel K, Messing K (2008) “We work by the second!” Piecework remuneration and occupational health and safety from an ethnicity- and gender-sensitive perspective Perspectives interdisciplinaires sur le travail et la santé 10(1). http://journals.openedition.org/pistes/2193
14. Premji S, Duguay P, Messing K, Lippel K (2010) Are immigrants, ethnic and linguistic minorities over-represented in jobs with a high level of compensated risk? results from a Montréal, Canada study using census and workers’compensation data. Am J Ind Med 53 (9):875–885
15. Messing K (2017) A feminist intervention that hurt women: biological differences, ergonomics and occupational health. New Solutions J Occup Environ Health Policy 27 (3):304–318

16. Philipps S (2011) Including gender in public health research. Publ Health Rep 126, supp. 3:16–21
17. Messing K, Lefrançois M, Tissot F (2016) Genre et statistiques: Est-ce que l'analyse de grappes peut nous aider à comprendre la place du genre dans la recherche de solutions pour l'articulation travail-famille? Perspectives interdisciplinaires sur le travail et la santé 18(2). http://journals.openedition.org/pistes/4854
18. Lacomblez M, Ollagnier E, Teiger C (2016) Les ergonomes peuvent-ils rester borgnes? À propos de la relation intervention-formation-genre. Perspectives interdisciplinaires sur le travail et la santé 18(2). http://journals.openedition.org/pistes/4829
19. Riel J, Major M-E (2017) The challenges of mobilizing workers on gender issues. lessons from two studies about the occupational health of teachers in Québec. New Solutions J Environ Occup Health Policy 27(3):284–303
20. Chappert F, Messing K, Peltier E, Riel J (2014) Conditions de travail et parcours dans l'entreprise: vers une transformation qui intègre l'ergonomie et le genre? Revue multidisciplinaire sur l'emploi, le syndicalisme et le travail (REMEST) 9(2). http://www.remest.ca/pages/documents/4-Chappert_REMEST_Vol9No2_2014-VERSIONKaren-1.pdf

# Facets of the Precariousness of Women's Work: Outsourcing and Informal Activity

Viviane Herculani Cardillo, Flávia Lima Traldi, and Sandra F. Bezerra Gemma(✉)

School of Applied Sciences, University of Campinas, Limeira, São Paulo, Brazil
sandra.gemma@gmail.com

**Abstract.** Due to the exhaustion of the pattern of accumulation, the transformations in contemporary capitalism demonstrate the strengthening of neoliberalism, changes in the organization of work and individuals, especially in relation to women. It is known that flexibilization within the scope of work assumes multiple formulations, either by the new forms of hiring, use of time, remuneration of work, solution of conflicts, or by the way of thinking the organization of time and space of work. With regard to the gender cut in outsourcing and informality, the few studies that have been carried out show a difference between women and men, with women being subjected to a more insecure condition, with higher turnover, lower wages and a major noncompliance with labor rights. In view of the context presented, the objective was to analyze the precariousness of women's work, in the outsourcing and informal activity instances observed in the work of outsourced school food handlers in the city of Campinas-SP, and in the scope of semi-jewels manufacture in the Municipality of Limeira-SP. These findings are the result of two researches carried out in 2017 by researchers from the FCA/UNICAMP Interdisciplinary Master's Program in Human and Applied Social Sciences. In both surveys it was possible to observe the harsh daily reality of several women who work through outsourced companies (in the case of school food handlers) or in their homes (in the manufacture of semi-jewels). These workers face situations like low pay, intensified work, bullying issues and precarious work environments.

**Keywords:** Women's work · Outsourcing · Informal work · Semi-jewels School food handler

## 1 Introduction

Due to the exhaustion of the pattern of accumulation, the transformations occurring in contemporary capitalism demonstrate the strengthening of neoliberalism, changes in the organization of work and individuals, especially in relation to women (Krein and Castro 2015).

The crisis of the 1970s brought with it a new economic and political order, resulting in increasing economic liberalization, internationalization of the production of goods and services, redefinition of the role of the State and productive restructuring, generating tension in the sense of establishing a social regulation of labor more flexible and with a lower level of social protection (Krein and Castro 2015).


S. Bagnara et al. (Eds.): IEA 2018, AISC 826, pp. 251–258, 2019.
https://doi.org/10.1007/978-3-319-96065-4_28

From these macro and microeconomic transformations, the development of world capitalism began to take hold with the toyotist model of work organization, benefiting new configurations of organizational management and stimuli to the technological advance. Throughout its worldwide expansion, the toyotist model spread through the big companies, choosing flexible and lean companies as standard, excellence in productive processes, higher quality and smaller contingent of labor force (Antunes 2009).

Flexibilization of work, among the many transformations that the world of work has undergone, is the one that persists in reinventing itself, either through new forms of hiring, use of time, remuneration of work, resolution of conflicts, or the way of thinking of the organization of time and space of work (Krein and Castro 2015).

One of the forms of flexibilization of the work is through outsourcing, being used by the companies as a management strategy, with the objective of making their economic activities more cheap and flexible (Krein and Castro 2015). In Brazil, in the last two decades, it has been observed an epidemic of outsourcing that "contaminated" from the industry to the public service, not limited to the activities-average but also present in the end-activities. Through this model of work organization, we have working and salary conditions that define first-class workers (composed of directly hired workers) and second category (composed of outsourced workers). The differences between the first and second category workers stand out for worse wages; shorter training time offered by contracting companies; longer working hours; higher turnover in employment; more intense and dangerous work, being more vulnerable to accidents and illness due to work (Antunes and Druck 2013).

If outsourcing in the private sector justifies itself as a management strategy, with the objective of making its economic activities more cheap and flexible, in the public sector the logic is similar, representing an efficient way of dismantling the social content of the state and its privatization, being the main tool found by neoliberalism to attack the pillars of a social and democratic State (Druck 2017).

Regarding the issue of women's work and the outsourcing of the public service, there is a preponderance that the service segment, because it presents characteristics such as low remuneration, repetitive work, little qualification requirement, where women are precisely inserted, are the main targets of outsourcing (Nogueira 2017).

Another form of work flexibility is centered on informality, which brings discussions aimed at understanding the phenomenon of non-inclusion of the less favored population in the productive process. Since the 1980s, this debate has taken on new forms as a reflection of the transformations of capitalism and the growth of unemployment. It is known that the informal labor expansion movement had a decisive impact on the service sector, traditionally more susceptible in certain branches to the proliferation of this type of activity (Costa 2010).

The new urban informality that expands contributes to an even greater heterogeneity of the labor market, marked by the precarious conditions of work and life, denial of the most elementary principles of citizenship, perpetual reproduction of poverty and social inequalities (Dombrowski et al. 2000).

In relation to the gender cut linked to the work precariousness, the few studies that have been done show a difference between women and men, with women being subjected to a more insecure condition, with higher turnover, lower wages and a great noncompliance with labor rights (Hirata 2005). On average, two-thirds of the work-force in the confectionery, food, financial, health and education sectors are women, reinforcing the labor market segmentation thesis (Krein and Castro 2015).

The sexual division of labor is the form of social division of labor resulting from social relations of sex; this form is adapted historically and to each society. It has as its characteristics the priority destination of men to the productive sphere and of women to the reproductive sphere and, simultaneously, the apprehension by men of the functions of strong social added value (political, religious, military, among others) (Kergoat 2009).

It is known that the relations established between men and women are not merely a reflection of economic relations, but they are translated into representations and symbols with which men and women face their daily lives (Souza-Lobo 2011, p. 173). The role of women in family care, the emergence of motherhood, the supposed "fragile sex" and so many other pressures point to the weakening of female labor, precarious and diminished by dominant instances.

In view of the context presented, the objective was to analyze the precariousness of female work, in the instances of outsourcing and informal activity, observed in the work of outsourced school food handlers of the Municipality of Campinas-SP, and in the scope of the manufacture of semi-jewels in the Municipality of Limeira- SP, Brazil. These findings are the result of two researches carried out in 2017 by members of the FCA/UNICAMP Interdisciplinary Master's Program in Human and Applied Social Sciences.

These researches carried out by Ergonomics, Psychology and Nutrition authors were developed using an adaptation of the Ergonomic Work Analysis (EWA) method. The main objective of the first study was to analyze the role of school food handlers as actors implementing the National School Feeding Program (PNAE). The other study prepared in the city of Limeira-SP, was based on the purpose of analyzing the experiences of pleasure and suffering of the workers who perform activities with semi-jewels.

## 2 Methodology

The researches carried out with the school food handlers and the women who work in the production of semi-jewels are qualitative, using as an adaptation method of the Ergonomic Work Analysis (EWA). The approach taken here is that of the Francophone current of ergonomics, which aims to make explicit the real work through a set of foundations, methods and practices articulated in a way pertinent to the singularities and specificities of work situations.

The application of the EWA method is a construction that, starting from an initial demand, takes singular forms with the development of each step (Guérin et al. 2001). The objective of this approach is to study the real work activity, taking into account the participation of the subjects and the overall work situation, that includes the variability of both technology and production and of workers (Abrahão et al. 2009).

The method consists of the following steps: (1) Analysis of initial demand, (2) Collection of information about the company, (3) Survey of population characteristics, (4) Choice of situation for analysis (5) Analysis of the task, (6) Global and open observations of the activity, (7) Elaboration of pre-diagnosis, (8) Systematic observations, (9) Validation, (10) Diagnosis, (11) Recommendations and transformations (Abrahão et al. 2009).

In the case of the investigations under discussion in this article, the steps of the EWA method, as described above, were not fully implemented. Only the steps that were necessary were used, in the construction of a path that took as close to the objectives, both general and specific, to which the research was proposed.

The adaptation of the EWA method was therefore carried out through the following steps: (1) Analysis of the initial demand, (2) Collection of information about the company, (3) Survey of the characteristics of the population, (4) Choice of situation for analysis, (5) Task analysis, (6) Global and open observations of the activity, (7) Elaboration of pre-diagnosis.

It is important to clarify that although there was a dialogue with workers, leaders and owners, seeking to understand their problems in the scope of work, there was no initial demand in these surveys, that is, there was no demand that first came from the actors involved in the work analyzed. There existed here, an instruction of demand based on an academic interest.

Below, in Figs. 1 and 2 the scheme of adaptation of the EWA method carried out in the surveys.

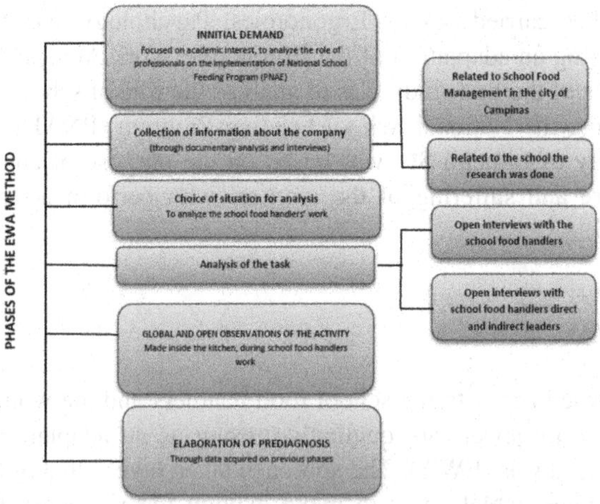


**Fig. 1.** Adaptation of the EWA method, carried out in the research with the outsourced school food handlers of the Municipality of Campinas/SP Source: Own authorship

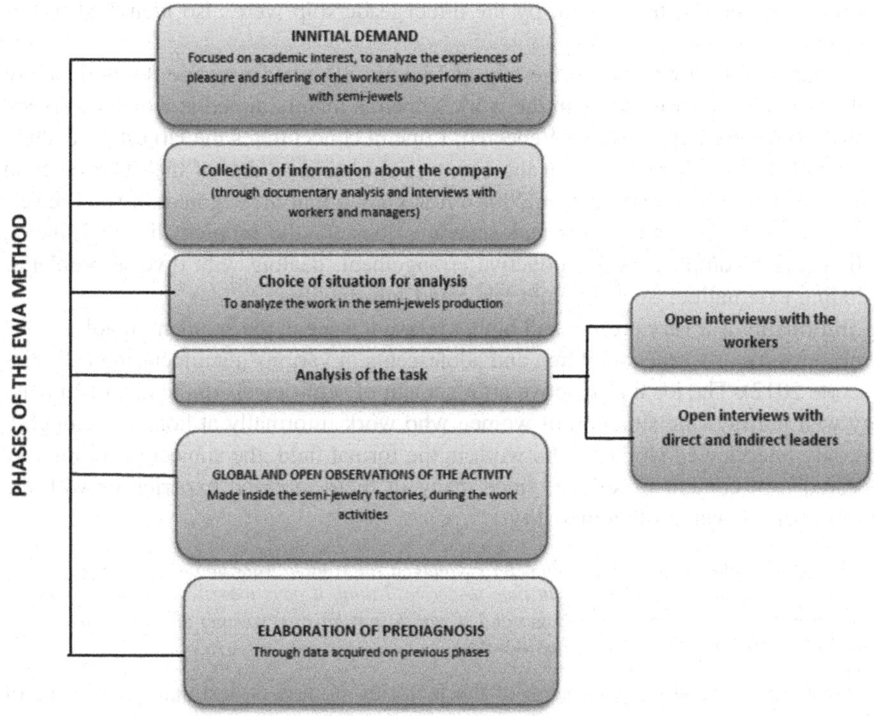


**Fig. 2.** Adaptation of the EWA method carried out in the research with the workers of the semi-jewel factories of the Municipality of Limeira/SP Source: Own authorship

## 3 Results and Discussion

Among the aspects observed in the two studies cited, it stands out as a reflection of the precariousness of outsourced and informal work to low pay, intensified work, precarious environments and issues of moral harassment.

The research carried out with outsourced school food handlers from the Municipality of Campinas/SP, located in three different schools of the State network, showed the precariousness of the work. Regarding the profile of the professionals participating in the research, all were female, aged 25 to 65 years, married and with children. As for schooling, only one professional has completed higher education. As for professional experience, six of the nine participants started working life during adolescence. Most reported previous experience with domestic work, whether as a cook, domestic or cleaning, some of these experiences already being in outsourced companies.

As for the workplace, the presence of precarious environments, in relation to the kitchen infrastructure, lack of equipment and utensils and an extensive bureaucracy to be attended, which greatly limited the margin of maneuver of the workers, were noted. These situations presented as a hindrance in the accomplishment of the activities of these professionals. The intensified work, the low remuneration, as well as the double working day and the harassment by the direct leadership were also identified in this research.

In the survey carried out in the city of Limeira-SP, we can notice the presence of another kind of precariousness in the work, which is mainly aimed at women workers.

Known as the Capital of the Veneered, Limeira concentrates the largest production and export of costume jewelry in the country, operating as one of the main poles in Latin America (Vilela and Ferreira 2008). It is known that the segment of semi-jewels brings gains in economic terms and generates jobs for the municipality of Limeira, while being a complex local productive arrangement, dealing with diverse problems, including externalities such as child labor and informality.

Informality is a very present and highly relevant issue in the segment, involving the domestic work of women, children and adolescents in various manufacturing activities (Lacorte 2012). The local productive arrangement of semi-jewels in the city of Limeira very well portrays the situation of women who work informally at home. Although it was only interviewed workers who work in the formal field, the subject of informality appeared as a content of speech, since many of them reported experiences with the work of semi-jewels in this modality.

> *"At home you make your own schedule, for example, if you want to wake up late, you work until later. At the firm you have the right time to arrive. I think it pays more to work for a firm, because if you have a service you get and if you do not have it, you get the same way (an employee's report from the shipping industry)."*

Much of the productive process of the factories studied is destined to this type of informality. In this segment, many stages of production, such as assembly, nailing and welding of accessories, are not carried out in factories, but are intended for informal household work.

Beatriz Bragotto Dermond street, in Limeira-SP, popularly called Rua da Alegria (Happiness street), is known for the large number of people who carry out activities in their homes. Located in the Jardim Boa Vista neighborhood, this street is the destination of many factories that outsource the workforce and production. Companies already know the houses that best perform the work required and at the time imposed. However, although the factories send the models they wish drawn on paper, it often happens that the work is not performed as expected, it does not come out well. In cases where the factories are dissatisfied with the final product performed in the houses, bags containing the unwanted parts are returned and unpaid until the required quality is reached.

Due to the routine of accelerated production, we also noticed the existence of an informality in the formality. That is, the possibility of a factory worker carrying out activities to be performed at home after work, and thus, to advance the formal production demand. Because of the volume of requests under time constraints, it is understood that such reality becomes necessary to reach the established deadlines. Faced with this is the workload of 8 h a day and at the end of the itinerary, thousands of pieces are taken with the worker home, to be delivered ready the next day. Making anticipations of formal factory work beyond entailing a non-rest, results in a "job competition" with the demands of family life.

In addition, it is perceived that the precarization of this type of work reflects in the gain of the workers. The remuneration of the activity is extremely low, being able to reach a production of approximately 4,000 ready-made parts per day and remuneration of R $ 3.00 to R $ 5.00 per thousand assembled (Vilela and Ferreira 2008). As a result of the products manufactured and/or the peculiarity of the production management, the work with semi-jewels is markedly repetitive, which demands visual acuity, patience, dexterity and attention, since it is necessary to have precision of movements when dealing with thousands of pieces of different sizes. This resumes the concept of sexual division of labor (Hirata and Kergoat 2007), which assigns different activities to men and women.

## 4 Conclusion

As can be seen, there is a precariousness of work brought about by the new forms of accumulation of capital, so that the flexibilization of labor, expressed through outsourcing and informal activity, has profound implications for workers and especially for women.

In the researches that portray the work of outsourced school food handlers from the Municipality of Campinas-SP and the manufacture of semi-jewels in the Municipality of Limeira-SP, it is possible to observe the theory contemplating the daily reality of several women who work through outsourced companies or in their homes.

In the case of school food handlers, it is noticed above all that the work environment, as well as the lack of structure and equipment for the accomplishment of the same inserts itself as a difficult one in the performance of the activities. Harassment at work coming from the direct bosses also characterized another element that reaffirms the presence of precarious work in this segment.

In the fabrication of semi-jewels, the informality present in the household work of women and girls stands out, which perform their activities in places improvised for that. There are also low wages and double working hours, in which women are divided between domestic and work activities.

Therefore, in the cases highlighted, common situations are identified in different fields of work, which corroborate and reiterate the precariousness resulting from outsourcing and informal work.

**Acknowledgment.** FAPESP - Project 2014/25829-0; UNICAMP/FAEPEX for the masters' scholarship.

## References

Abrahão J, Sznelwar L, Silvino A, Sarmet M, Pinho D (2009) Introdução à ergonomia: da prática à teoria. Blucher, São Paulo

Antunes R (2009) Os sentidos do trabalho. Boitempo, São Paulo

Antunes R, Druck G (2013) A terceirização como regra? Revista tribunal superior do trabalho 79 (4):214–231 Brasília

Costa MS (2010) Trabalho informal: um problema estrutural básico no entendimento das desigualdades na sociedade brasileira. Caderno CRH, Salvador, 23(58):171–190

Druck G (2017) Terceirização no serviço público: múltiplas formas de precarização do trabalho. In: Navarro, v. l.; Lourenço EAS (orgs) O avesso do trabalho IV - Terceirização precarização e adoecimento no mundo do trabalho, 1ed. Outras Expressões, São Paulo

Dombrowski O, Jacobsen K, Martins, R (2000) Introdução. In: Jakobsen K, (organizador) Mapa do trabalho informal: perfil socioeconômico dos trabalhadores informais na cidade de São Paulo. Fundação Perseu Abramo, São Paulo

Guérin F, Laville A, Daniellou F, Duraffourg J, Kerguelen A (2001) Compreender o trabalho para transformá-lo: a prática da ergonomia. Edgard Blucher, São Paulo

Hirata H (2005) Globalização, trabalho e gênero. Rev Pol Públ 9(1):111–128

Hirata H, Kergoat D (2007) Novas configurações da divisão sexual do trabalho. cadernos de pesquisa, 37(132):595–609

Kergoat D (2009) Divisão sexual do trabalho e relações sociais de sexo. In: Hirata H, Laborie F et al Dicionário crítico do feminismo. Edunesp, São Paulo

Krein D, Castro B (2015) As formas flexíveis de contratação e a divisão sexual do trabalho. análise, n. 6. Disponível em: http://library.fes.de/pdffiles/bueros/brasilien/12084.pdf

Lacorte LEA (2012) Construção de políticas públicas em rede intersetorial para a erradicação do trabalho infantil em Limeira-SP. 2012, f. 172. Dissertação de mestrado universidade de São Paulo. Faculdade de saúde pública, São Paulo

Nogueira CMA (2017) terceirização das mulheres no setor público: algumas notas introdutórias. In: Navarro VL, Lourenço EAS (orgs) O avesso do trabalho IV - Terceirização precarização e adoecimento no mundo do trabalho, 1ed. Outras Expressões, São Paulo

Souza-Lobo E (2011) A classe operária tem dois sexos: trabalho, dominação e resistência. 2º edição. Editora Fundação Perseu Abramo, São Paulo

Vilela R, Ferreira M (2008) Nem tudo brilha na produção de jóias de Limeira - SP. Revista Produção 18(1):183–194 São Paulo

# Revealing the Hidden Processes Behind Discrimination Against Part-Time Teachers in France: A Lever for Improving Their Situation

D. Cau-Bareille[1(✉)], C. Teiger[2], and S. Volkoff[3]

[1] Université Lyon 2, ECP, CREAPT, Lyon, France
dominique.cau-bareille@univ-lyon2.fr
[2] GRESHTO CRTD CNAM, Paris, France
moufcat@gmail.com
[3] CREAPT CNAM, Paris, France
serge.volkoff.visiteur@lecnam.net

**Abstract.** This research on the working conditions for teachers working part-time in secondary schools in France (often women) aims to reveal certain discriminatory practices by the education authorities, which often go unnoticed. This study adopts an original approach by combining an ergonomic approach to work and the psychological model of the system of activities. This combination of methods – interviews (20) - timetable analysis (210) – questionnaires on teachers' experiences at work (106) produces a systemic approach to the activity, situated in organisational and institutional contexts, and a diachronic perspective on health and professional activity. Our findings show that the forms of commitment to work change according to changes in teachers' health and their private lives. Paradoxically, instead of improving, the working conditions of part-time teachers often deteriorate, and their reputation within the institution is damaged. They are subject to pressure and, more or less hidden, discrimination from their line management, as well as their colleagues. These initial findings call for further investigation of the need to gender the diachronic approach to health and the work activity. Better understanding the conflicts between different spheres of life and how they impact health and careers will reveal how the decisions made regarding the temporal organisation of the activity influence the careers of both women and men. In particular, it will draw out the mechanisms behind the early-onset fatigue experienced by women, who are most affected by the need to find an improved work-life balance.

**Keywords:** Teachers · Part-time · Gender

## 1 Introduction

In several European Union countries part-time working is the norm for women: 80% of women in the Netherlands, over half in the United Kingdom, 40–50% in Germany, in Sweden and in Belgium [1]. This is not the case in France where 24% of the workforce work part time. A part-time job is considered to be "abnormal" [2] and full-time work


S. Bagnara et al. (Eds.): IEA 2018, AISC 826, pp. 259–268, 2019.
https://doi.org/10.1007/978-3-319-96065-4_29

remains the reference; this part-time work is therefore stigmatised and devalued. It is generally considered to concern women more than men, and in this case is socially tolerated [3].

In the teaching profession there are two forms of elective part-time work: *"legally-entitled"* part-time work[1] and *"authorised"* part-time work,[2] requests for which can be refused based on the needs in terms of workload. In both cases, teachers can choose to temporarily reduce their working hours with the option to return to full-time work when they wish. Just over one secondary school teacher in ten works part time, and there are more women than men working part time (15%/5%) [4], which is very different from other European countries [5]. According to Ernst [6], the same proportion of teachers working part-time work 50% of a full-time schedule as those working 80% (22%/25%). This phenomenon peaks for teachers aged 34 years, mostly women with young children (*"legally-entitled"* part-time working). Part-time work is seen as a tool for finding a better work-life balance during a phase of life where this is difficult to achieve. There is a second peak at the end of teachers' careers (*"authorised"* part-time working): in this case it is used as a strategy for remaining in good health until retirement, and for continuing to enjoy working without tiring to the point of exhaustion [7].

However, we will demonstrate that although part-time working meets a genuine need and allows these teachers to continue doing their job whilst managing personal or health issues, these teachers are nonetheless subject to more or less hidden processes of discrimination. Within the education system, teachers' line managers often associate the choice to work part time with a form of disengagement from their work. These situations can generate severe tensions within schools.

## 2 The Initial Request: Career Inequalities and Workplace Harassment Affecting Women More Than Men

In 2011, the leaders of a trade union women's group, who have observed career inequalities, a pay gap between men and women, and issues of harassment against women in some schools, asked us to conduct research into the day-to-day work of teachers, by examining the specific impact of social gender relations on their working conditions and experience at work.

The results of that study [8] revealed specific issues for part-time teachers, the subject of this paper. Paradoxically, although one might suppose that part-time working would improve their quality of life at work and work-life balance, these teachers actually report a deterioration in their working conditions, and of significant pressure exerted by head teachers and/or their teaching colleagues creating conflictual situations. We attempted to understand the challenges surrounding this part-time work within schools, and the difficulties experienced by teachers, in order to shine a light on these

[1] "Legally-entitled part-time work": to raise a child under 3 years old, to care for a spouse, to care for a dependent child or parent, or for preventive medical reasons.

[2] "Authorised part-time work": for reasons other than those cited above (to prepare an exam, participate in professional training or to exercise another professional activity in parallel, for example).

issues, which are very rarely discussed in the teaching or trade union literature. To this end, different work analysis tools and data collection methods were used as part of a multidisciplinary approach.

## 3 An Analytical Approach at the Intersection Between the Ergonomic Work Analysis Model and the System of Activities Model

This research as a whole forms part of an ergonomic approach aiming to develop both a systemic approach to teachers' work through the restitution of the organisational and institutional contexts, and a diachronic approach to health [9] and activity which captures the changes in forms of commitment to work in relation to the changes in people's health and their professional and private lives. Following in the footsteps of Buchmann et al. [10], we wanted to develop analysis methods which go beyond the temporal framework of real-time observation, without dropping this altogether, but by giving it a new role within a broader model for understanding the relationships between health and work. This led us to conduct our study over a long period of time which corresponds to teachers' career paths. This makes it possible to access the temporal dynamics of the changes that operate in work over the time-scale of a career.

We also used the psycho-sociological model of activity system analysis [11] which adopts a synchronic approach to forms of subjectivity, which we believe is well-adapted to tackling gender relations at work. In this model, each individual searches for strategies to best combine the three spheres of their lives - professional, private and social - which result in different pathways emerging due to adjustments in behaviours in one area of life made due to the needs in the other areas at a given moment in time. In this context, understanding the different forms of commitment in the professional environment requires examining the connections the subject establishes between the different behaviours in this sphere of life, and those developed in other areas of life. This model identifies the links of inter-significance between behaviours in the process of plural socialisation [12] and analyses the resulting decisions made, which may differ according to whether the subject is male or female. It extends the scope of the analysis to determinants outside of work, which can be characterised in terms of the resources, but also the constraints, influencing the relationship with work, and the experience of work and of its constraints, which can lead to conflict.

This research was built over three phases and used several investigation methods, following the example of previous ergonomics research studies [13]. Our analytical point of entry was not the activity as it takes place on a day-to-day basis, but the relationship with the organisational and temporal constraints of work from a diachronic perspective. We therefore conducted our study over three phases, prioritising the interview phases (individual and group) which were followed by a broader questionnaire phase.

Phase 1 involved conducting individual exploratory interviews (60) with teaching staff in different schools in order to clarify the problems raised. The interviews lasted approximately 1.5 h and were conducted at the teacher's school or home. The teachers interviewed were invited to go back over their teaching career, the different schools they have taught in, the reasons for any changes of school, and the reason for deciding to work part time. The discussion then turned to their current working conditions, the impact of their part-time working on their activity, the timetables, their relationships with their colleagues and their line management, the preparation time for their lessons and their work-life balance. Out of the 60 teachers interviewed, 16 were working part time.

Phase 2 involved group interviews by subject lasting approximately 3 h (five focus groups composed of around 10 people, which met twice), in order to discuss the themes identified during the individual interviews.

In phase 3 we worked on the scale of the whole school, a secondary school[3] in a large city, at which we conducted three types of investigations: interviews with a range of stakeholders (line managers, teachers including six working part time), a detailed analysis of the timetables of all teaching staff at the school (N = 210) which appeared to be the focus of much of the tension expressed during the interview phases, and a survey of all the teaching staff conducted via questionnaire. The themes tackled in the questionnaires were: the teacher's career path prior to joining their current school, the reasons for any career breaks or requests for part-time working, teaching activities outside of the school, responsibilities within the school, expectations regarding their timetable and overtime, the atmosphere in the school, relationships with their colleagues and line management, relationships with pupils, health issues, the impact of their professional activity on their personal life. Out of the 210 questionnaires handed out, 106 were returned: 69 from women and 37 from men. The answers underwent statistical analysis (frequency tables and cross tabulation, as well as multiple correspondence analysis) and were analysed according to gender, age, the teacher's family structure and the proportion of working hours. This paper will only present the standout results from these analyses in relation to the qualitative findings from the study.

Regarding the questionnaire survey, 13 of the questionnaires returned came from part-time teachers. In the interviews over the three phases, we met with 24 people working part time: 18 women and 6 men.

## 4 Results

### 4.1 The Reasons for Reducing Working Hours

In the questionnaire survey, out of the 13 respondents working part time, 8 stated they were *"too tired to work full time"* (in addition to which two mentioned *"health problems")*; 6 wanted to *"have more time for their families"* and 4 "more time for themselves".[4] These issues correspond to those raised in the interviews, out of which two main reasons for reducing working hours stand out in particular: "*to take care of young children*" and "*health problems/fatigue*", with the two reasons sometimes given by the same person.

[3] A "lycée" in France corresponds to years 11, 12 and 13 of UK secondary schools.

[4] The respondents were able to give more than one answer.

The people who work part time to *"take care of their children"* are mainly women. This corresponds to part-time working to which they are legally entitled, requests for which the education authorities cannot refuse. The decision to work part time may result from parental choice regarding their children's upbringing, or from extreme fatigue linked to the difficulties of juggling work and raising young children. These results corroborate the findings in the literature on part-time working [6–14]. Two men also put forward this argument, but also systematically gave an additional reason: for one it was to "*exercise another professional activity*" (39 years old), for the other it was to "*better carry out their teaching work"* (40 years old). The "*taking care of children*" aspect is therefore not so important in the men's choices as in the women's choices; a relatively unsurprising result [14].

Other teachers request "authorised" part-time working: these requests generally concern teachers at the end of their career or people with health issues, and less frequently people who want to take exams or exercise a professional activity outside of the school and/or outside of the teaching profession. Two men at the end of their careers also opted for part-time working following a "professional trauma": one had to confront a pupil who spat in his face (60 years old) and the other was hit by a parent during a lesson (53 years old). Working part time was a way of distancing themselves from their work. These situations are not uncommon in the teaching profession and affect both men and women; they necessarily have an impact on the teacher's decision to continue working full time or not.

## 4.2 Working Part Time Does not Indicate a Lack of Commitment

In the typological analysis of the answers to the questionnaires, conducted using multiple correspondence analysis, the part-time teachers appeared alongside profiles which we labelled "distant", with variables reflecting the rejection (for personal reasons or fear of taking on excessive responsibility) of coordination roles. However, in the interviews we realised that part-time work is not a sign of withdrawal or professional disengagement.

The vast majority of part-time teachers interviewed (90%) stated that they used the time freed up from teaching to prepare their lessons, for cross-cutting projects, or to conduct research. "*Since I have been working part time, I spend much more time each week - when I'm not preparing lessons - conducting research and doing lots of other things; so as far as I'm concerned my commitment is intact. I don't feel like I'm doing anything differently in my job*" (F, 49 years old). For some people, working part time allows them to do their job better, with less time pressure and a better work-life balance. Part-time working is rarely associated with an actual reduction in workload: "*Part-time working is an illusion*", as one teacher describes it (F, 41 years old). The downsides of this illusion of part-time working, already highlighted by Vogel [1], are rarely raised and debated. The decision to reduce working hours is not taken with all the information in hand and often leads to disappointment. Whilst part-time teachers do not feel that they *"lack professional commitment"* [3], they often feel that their head teacher or colleagues consider them as becoming disengaged from their work and the school's preoccupations, as the institutional requirements and constraints have become increasingly strained over the successive reforms to the education system.

## 4.3 Part-Time Working Poses a Problem in the Current Context Given the Reduced Resources Allocated to Schools

The successive reforms affecting secondary education in the last few years have contributed to a multi-form densification of the activity for all stakeholders in the school, teachers, managers, and administrative staff. This is linked to two main factors: the first concerns the continuing reduction of human and financial resources allocated to schools, which requires teachers to work longer hours; the second is the multiplication of assessments, without having thought about or budgeted for the human resources required to implement them. The schools therefore count on the goodwill and availability of teachers to do this. The line management is therefore faced with the difficulty task of getting teachers to accept this additional work, on top of their already heavy workload. Furthermore, the situation for part-time teachers complicates matters, as their status prevents them from doing overtime which therefore consistently falls to their colleagues. This creates tensions and is often the root cause of pressures placed on teachers, which they consider to be unbearable.

## 4.4 Multiple Consequences

Deciding to reduce their working hours exposes these teachers to a number of different consequences.

It is often difficult to get their head teacher to accept their part-time working. As they are considered to have more free time than their full-time colleagues, they are often solicited to do the additional tasks their status allows: acting as form teacher, invigilating exams, accompanying groups on school trips… all of which reduce the time they have for themselves. "*When you ask to work part time, it's because you want to spend more time with your family. There are a lot of head teachers who don't understand that. (…) So it's not very well perceived by the head teacher who thinks, "Oh yes, you can never ask them to do anything!"*" (F, 36 years old). Many part-time teachers endeavour to avoid these tasks. Out of the questionnaire respondents working part time, 3 out of 13 are form teachers for a class, whereas half of the full-time staff take on this responsibility; none of the part-time staff expressed a desire to take on more responsibility whereas one in four full-time teachers did.

Various forms of "reprisals" resulting from this situation were mentioned by 90% of the part-time teachers questioned, who complained that they had *"rubbish"* timetables since they went part time. They feel they are considered to be the moveable factor that head teachers use to fill the holes in pupils' timetables, but without taking into account the teachers' wishes. Indeed, going part time does not necessarily free up any days or half-days in their week: *"I don't think you get such a good timetable working part time as working full time. In the first draft of the timetable I was given in September I was in every day, I had two hours in the morning and two hours in the afternoon."* (F, 54 years old). We were able to confirm this in the analysis of the teachers' timetables for the secondary school we studied in detail. This form of timetabling does not necessarily lead to the best work-life balance. "*I was constantly thinking: Careful, you need to get ready, you mustn't be late! In half an hour you'll need to leave again.*" (F, 41 years old).

These teachers end up in a paradoxical situation: the splintering of their timetables makes it difficult to juggle between their personal and professional lives, and raises time management issues, whereas the reason for going part time in the first place was to improve their work-life balance. This explains the tears and distress when the timetables are given out at the start of the year: *"When we get our timetables and people realise that their hours are divided up in such a way that the impact of the part-time working is not felt, then people actually do burst into tears on the in-service training days leading up to the new school year, I saw this happen again this year."* (F, 56 years old). This was confirmed by the head teacher of the study school. This problem is compounded for parents with young children as in addition to the reduction in income related to the reduced hours, if they have a divided timetable they may still have to pay for childcare every day of the week.

The same is true for those who wish to take on additional responsibilities, such as coordination roles: the teachers questioned have the impression these roles are given primarily to full-time teachers, people we "*can count on*" due to their commitment to the school and their willingness to take on additional tasks. The criterion of time availability seems to take priority over skills and competence.

Out of the part-time teachers met with, 90% explained that following their decision to go "part time", their line management took away the "*good classes*" they had, such as those preparing exams (*baccalaureat*) or the post-bac classes preparing for higher education qualifications, classes they don't necessarily get back when the return to full-time working.

However, there is frequently pressure from full-time colleagues regarding the allocation of the most difficult classes. Part-time teachers often find themselves forced to take on the most difficult classes, on the pretext that they have more time to recover after lessons. "*When negotiating who gets which classes, we hear things like: you're part time, you've got less hours, it'll be easier for you if you take on a few more year 11 classes*" (F, 54 years old). There are also times at which certain colleagues question the legitimacy of part-time teachers to have their say on the subject taught: "*I had a colleague who once said at a meeting, that given the time I spent in the school I wasn't really concerned by the problem in hand!*" (F, 60 years old). This reveals the fragility of the teaching staff who have to take on increasingly heavy workloads. Part-time working is not just an individual issue, it is also a collective and organisational issue.

This all contributes to a feeling that "part-time staff" are considered differently from "full-time staff": "*One of the consequences of part-time working, is that you go down in your line management's estimation.*" (F, 32 years old). For three part-time staff this took the form of a lower administrative rating awarded by the head teacher after they went part time. This rating affects the final rating obtained by the teacher which determines their career progression and salary. "*I think that the fact that I had children, and went part time, put the brakes on my career, particularly at the start. I didn't get very good appraisals from the head teachers I worked for when I was part time, because I didn't get involved in the way they wanted me to.*" (F, 43 years old). The head teacher of the school studied in this case also proceeds in this manner. This differential treatment is the cause of a feeling of injustice and resentment between teachers and sometimes leads to people giving up their part-time work. These situations are rarely discussed by teaching staff or trade unions, which leave the teachers affected alone with their problems.

Finally, if we analyse the impact of gender on the experience of part-time working we observe that the timetabling issues and class allocation issues affect both men and women. However, whilst part-time working for women raising young children conforms to a social stereotype considered to be legitimate, the situation for men is rather different. Often their request meets with incomprehension and results in unpleasant comments from their colleagues and line management. In the same way, going part-time due to fatigue, health issues or exhaustion are less well-accepted when they concern men.

## 5 Discussion

In contrast to what one might expect, going part-time in secondary education does not necessarily result in an improvement in working conditions and a better work-life balance. It often marginalises teachers within an education system designed around full-time working and in which there are changing expectations in terms of the time teachers spend in school. Within the close framework of the work organisation, there are inequalities in terms of treatment, which can have an impact on the teaching activity itself (classes, levels), on the temporal organisation of the activity (timetables) and on mental health. Understanding these factors requires observing the work up close, analysing the changes operating within the work organisations, meeting with teachers, and encouraging them to express their opinions on their timetables which is a taboo subject between colleagues and which they negotiate with their line management. Without a forum within the school at which they can express the difficulties they face, certain tense or abusive situations may go unnoticed by the local trade union representatives who cannot then provide their support to the teachers concerned.

It therefore seems important to improve the dialogue around labour relations within schools not only with regards to working hours, but also around the densification of work revealed by this part-time working, the criteria for quality work, and teacher assessments. This requires raising trade union and head teachers' awareness of the issues of part-time working. As Angeloff suggests [3], it is important to put into place more personalised support for part-time employees, precisely because they are marginalised by their circumstances. Fundamentally, this supposes considering part-time working as a sub-category, in relation of the majority of full-time staff, and not as a separate category in its own right. Following our intervention, the head teacher of the school where we conducted our research changed how he went about preparing the timetables. He now first integrates the part-time staff and for the first year ever there have been no timetabling issues at the start of the school year. Revealing dysfunctions can sustainably transform working conditions.

From a theoretical and methodological perspective, our research has provided arguments supporting the cross-referencing of analysis models and methods in order to fully grasp the complexity of work situations and better integrate external determinants, in particular family factors, of a person's work, experience of work and health. Whilst ergonomics prefers to use instant analysis tools on a very fine scale in order to understand the work activity, there are problems it cannot tackle and which require the use of other approaches. The system activity analysis model, adopts a processual, diachronic perspective in order to better understand the impact life events have on the work activity and the subsequent impact of changes in the person's professional life on managing their work-life balance. This better accounts for the influence of social roles on the professional activity, the negotiations around working conditions, and more broadly on career paths, and issues of fatigue and health relating to balancing the person's professional and private lives. From this point of view, our analysis invites a gendered, diachronic approach to health in work and the work activity as proposed by Volkoff and Molinié [10] in order to better account for the conflicts between different spheres of life and their impact on people's health and careers. This both better accounts for the mechanisms by which women experience early-onset fatigue due to the work-life balance issues which still affect them disproportionately [15, 16], and better accounts for the impact decisions made regarding the temporal organisation of the activity have on people's careers, thus shedding light on the mechanisms of selection and self-sabotage of career opportunities which contribute to the observed end-of-career pay gap.

## References

1. Vogel L (2012) Les pièges du temps partiel. Dossier spécial 10/28 HesaMag#05
2. Maruani M, Michon F (1998) Les normes de la dérégulation: questions sur le travail à temps partiel. Economies et Sociétés, série AB 20(3/98):127–166
3. Angeloff T (2009) Genre, organisation du travail et temps partiel. In: Chappert F (ed) Genre et conditions de travail. Mixité, organisation du travail, santé et gestion des âges, Etudes et documents, ANACT, pp 66–84
4. Hilary S, Louvet A (2014) Enseignants de collège et lycée publics en 2013: panorama d'un métier exercé par 380 000 personnes. France, portrait social, Edition 2014, 38 p
5. Siniscalco MT (2002) Un profil statistique de la profession d'enseignant. Organisation des Nations Unies pour l'Education, la Science et la Culture/BIT, 82 p
6. Ernst E (2013) En 2010, 5,5 millions de salariés travaillent dans la fonction publique. Insee Première 1442, 4 p
7. Cau-Bareille D (2009) Vécu du travail et santé des enseignants en fin de carrière: une approche ergonomique. Centre d'Études de l'Emploi, Rapport de recherche 56, 65 p
8. Cau-Bareille D, Jarty J (2014) Trajectoires/Itinéraires et rapports de genre dans l'enseignement du second degré. Rapport de recherches SNES, 165 p. https://www.snes.edu/IMG/pdf/rapport_final_snes_groupe_femmes.pdf
9. Volkoff S, Molinié AF (2011) L'écheveau des liens santé travail, et le fil de l'âge. In: Degenne A, Marry C, Moulin S (eds) Les catégories sociales et leurs frontières. Presses de l'Université de Laval, Québec, pp 323–344
10. Buchmann W, Mardon C, Volkoff S, Archambault C (2018) Peut-on élaborer une approche ergonomique du «temps long»? Perspectives interdisciplinaires sur le travail et la santé, 20-1, 2018, 28 p. Journals.openedition.org/pistes/5565
11. Curie J (2002) Parcours professionnels et interdépendances des domaines de vie. Educ Permanente 150:23–32
12. Almudever B, Le Blanc A, Hajjar V (2013) Construction du sens du travail et processus de personnalisation: l'étude du transfert d'acquis d'expériences et des dynamiques de projet. In: Baubion-Broye A, Dupuy R, Prêteur Y (eds) Penser la socialisation en psychologie, Erès, Toulouse, pp 171–185

13. Teiger C, David H (2003) L'interdisciplinarité ergonomie et sociologie, une histoire inachevée. Travail et emploi 94:11–30
14. Jarty J (2009) Les usages de la flexibilité temporelle chez les enseignantes du secondaire. Temporalités, 9, 18 p. http://temporalites.revues.org/index1057.html
15. Messing K, Östlin P (2006) Gender equality, work and health: a review of the evidence. World Health Organization, Geneva, p 46
16. Cau-Bareille D (2014) Estratégias de trabalho e dificuldades dos professores em fim de carreira: elementos para uma abordagem sob o prisma do género. Laboreal 10(1):59–78

# Interventions for Improving Working Environment in Home Care Work in Sweden – Preliminary Findings from the First Year: A Gender Perspective

Britt Östlund[1(✉)], Charlotte Holgersson[1], Rydenfält Christofer[2], Inger Arvidsson[3], Gerd Johansson[2], and Roger Persson[4]

[1] KTH, Royal Institute of Technology, Stockholm, Sweden
brittost@kth.se, charlotte.holgersson@itm.kth.se
[2] Department of Design Sciences, Lund University, Lund, Sweden
{christofer.rydenfalt,gerd.johansson}@design.lth.se
[3] Occupational Environmental Medicine, Lund University, Lund, Sweden
inger.arvidsson@med.lu.se
[4] Department of Psychology, Lund University, Lund, Sweden
roger.persson@psy.lu.se

**Abstract.** Home care services are an important part of the Swedish social welfare system. Considering the size of the sector and the increasing future needs for home care, due to growing elderly populations, it is worrisome that problems in terms of injuries, sick leave and staff turnover appear common in this occupation. A problem for managing the situation is, however, that the current knowledgebase is fragmented and not much developed being characterized by a homogenous workforce in terms of gender, dominated by women. A common perception is that this work is low skilled and something that can be done part time in parallel with household work. To improve the working environment in home care work, we initiated a project in which one of the goals was to create a better overview of published results of interventions and examples from community practice. Accordingly, we undertook a systematic review of the scientific literature with the purpose to find and map practical examples of interventions. The preliminary result suggests that there is a considerable lack of knowledge and often lacking, or poor, analyses of the consequences of the unequal gender balance in home care work. Interventions could be grouped into four types of interventions: scheduling, education and training, organizational change and digitization. Interestingly, it seems as if single problems at the workplace level to a larger extent are covered in the scientific literature while problems on the system level are more seldom addressed.

**Keywords:** Home care · Gender · Interventions


S. Bagnara et al. (Eds.): IEA 2018, AISC 826, pp. 269–277, 2019.
https://doi.org/10.1007/978-3-319-96065-4_30

# 1 Introduction

Discussions on how to improve the working environment in the home care sector in Sweden is typically contextualized by debates around challenges such as growing older populations; the restructuring of care from nursing homes to the homes of older and disabled citizens; and the lack of employed and educated home care professionals. Since local governments and regional county councils are responsible for home care services, the Swedish Association of Local Authorities and Regions has present strategies to meet these challenges [17]. A number of governmental committees have also been launched to review these challenges. The conclusions are that welfare services need to be streamlined and made more efficient in order to sustain the supply of skills, and that new technology, full time employments and prolonged work life will decrease recruitment problems and growing care needs [12, 17].

The current discourse is strongly attached to the technological developments in society. In particular, there is a strong emphasis on new technology as a catalyst for the developments in home care and what solutions can lower the costs [12] in relation with patient safety and care damages [18]. Following the European discourse the concept of "welfare technology" was swiftly adopted in the Nordic countries and similarly thought of as a way to make the home care sector more efficient [8, 14]. However, the existing documents on the subject matter, even those published by the Association of Local Authorities and Regions, discuss the challenges with welfare technology almost exclusively in relation to hospital care. Thus, very little effort is spent addressing problems found in local home care. In addition available reports on highly aggregated levels such as OECD, and the Oxford study which presents analyses of what kind of jobs can be automized as a consequence of digitization, does not easily translate to the concrete level, that is, improving our understanding of what happens "down on the floor" [17]. In fact, home care is seldom spoken of in its own right. When home care is addressed it is typically mentioned just as part of the organization or as part of the information system that can improve efficiency [12].

Yet, problems exist and are well-known in this sector. For example, the labor union "Kommunal", that organize the majority of home care workers in Sweden, have raised issues related to lack of resources, lack of management support and low status of the profession [10]. Research results confirm that these problems are linked to the large number of injuries, sick leaves, staff turnover, dissatisfied employees [2, 4, 5, 22]. In 2015, the Swedish Research Council for Health, Working Life and Welfare was assigned by the government to prepare a call for research in female dominated sectors. The background was that Sweden has a gender segregated labor market, where for example, only three of the 30 largest professions in Sweden have an even gender distribution [19] and women are on sick leave to larger extent than men. The highest degree of sick leaves is reported among employees in service and care. This is mainly due to the differences in work environment between professions and therefore also between women and men.

In a recent review of research regarding work environment in women dominated occupations [9] it was concluded that many of the women dominated industries are found in care services, education and health and medical care. In 2014, the most female-dominated occupation was assistant nurses, home help services, home based personal care and nursing homes with 93 percent of women and 7 percent of men [19]. The review found that these occupations are in many ways characterized by physical and mental strain that is higher than in other parts of the labor market [9]. Women are to higher extent on sick leave compared to men, mainly due to physical and mental problems. The studies in the review highlight a connection between these problems and factors in the work environment, such as psycho-social and physical strain. Women and men appear to be equally prone to suffer from work injuries. When exposed to the same physical or mental strain, the same increase in risk of injuries due to strain and mental problems appears for women and men. However, women and men are often not exposed in the same way since they work in different industries and occupations; they have different tasks, and are differently exposed due to the design of the workplace and work equipment. Negative stress is particularly present in women dominated occupations compared to both the male dominated occupations and the gender-balanced occupations. However, the physical demands are also high, and are getting closer to those in the very male dominated occupations (ibid.).

In addition, work conditions often differ between different tasks. A study involving the inspection of municipal operations found that there were considerable differences between operations that were associated with women and those associated with men [20]. A clear difference was found between home care services where mainly women worked and technical services were mainly men worked. Home care services had for example more employees per manager, worse communication with decision-makers, more part-time employed, inferior quality cars and less support functions (ibid.).

Research also indicates that the increased demands on effectiveness and results within health- and medical care following privatization and/or new public management (NPM) strategies in Sweden, have involved negative changes in working conditions for women. The monitoring of work and work rate intensity has increased. Moreover, the focus on results has led to increasing demands on emotional labor. In fact, research shows that emotional work and emotional demanding work are factors considered as psycho-social risks for women working in the health care sector [22].

## 1.1 Our Project and the Aim of the Paper

In November 2016 our project "To create a better work environment in the home care sector - participatory organizational change in practice" received funding from the Swedish Research Council for Health, Working Life and Welfare. The project plan includes making a scientific review of interventions in home care; mapping interventions and practical examples from Sweden's 290 local municipalities; and making subsequent interventions on a selected group of local communities in the south of Sweden. The project includes competence in gender studies, ergonomics, physiology therapy, psychology and sociology; and entails several external partners and/or actors such as the labor union Kommunal; four local communities and one private company delivering home help services.

In this paper we present and discuss two questions from our synthesis so far of research and practice. Specifically, we focus on two questions:

(1) How much importance has been attributed to gender when explaining the work conditions in the home care sector?
(2) Is increased gender awareness a part of the solution to the problems in the home care sector?

# 2 Methods

## 2.1 A Scientific Review

The scientific review undertaken in the first year of the project was based on previous studies indicating that there are needs for comprehensive improvements in the home care work environment. The review aimed at mapping out existing interventions in home care in order to improve the work environment and working conditions, including two research questions: (1) What organizational interventions have been implemented? (2) What effects have these had on the work environment? The review were based on the study design of PRISMA 2000 structured search methodology taking into consideration that paving the way through published articles is an iterative process and at the same time systematic [23]. The selection process extends from articles that were published in English between the years 2008–2017 about caring up to interventions in home-care-work-like organizations with employed home care workers. Starting with 2715 articles in the first screening, the final selection included 18 articles. These articles were read carefully and the analysis of them showed that they covered four main topics: (a) trying to solve lack of education and training, (b) trying out new ways of scheduling working hours, (c) trying out new ways of organizing home care or (d) introducing new digital applications. Gender or the fact that home care is gender segregated was discussed in one paper only. This single paper reported on the evaluation of a political reform to make home care workers work full-time [6]. Two other papers suggests that a participatory approach can be an important factor in making interventions successful. Self-reports turned out to be one of what was the most effective measure aside with active collaboration crossing professional boundaries [11, 13].

## 2.2 A Survey to Local Communities

The responsibilities for home care work in Sweden lays within the local government (kommunen). In Sweden there are 290 municipalities ran by local governments. In the first year a short survey was sent out to all Swedish municipalities distributed to their home care units. The survey covered two general questions concerning change and equality, respectively:

(a) Have you made any changes that affect the home care work/work environment in a positive way during the last three years? If you have, what was made and why? What was the result? Where you satisfied with the result? In what way were the home care workers involved? Is there any evaluations we can take part of?

(b) Have these changes affected the equality in the organization?

For this paper the second question is the most important, considering gender and equality aspects. The response rate was 179 units in which 259 projects aiming at changes was reported. The result is now the basis for interviews in selected municipalities. It should be noted that these 179 units does not represent one local municipality each, there can be several units within one municipality.

## 2.3 A Survey to Home Care Workers

In addition to the first survey a second survey was directed to the members of the labor union "Kommunal" that organizes the majority of home care workers in Sweden. The questions was this time distributed as an add-on to Kommunals ordinary membership mail that, in theory, reaches circa 80 000 members. The response rate was extremely low as only 65 on-line responses were returned.

The survey was very straight forward as it only covered (1) whether any positive change had occurred at the workplace and the type of change, (2) what it consisted of and what it was that made it positive, (3) whether the employee had been involved in the design of the change.

These 65 responses included no gender aspects or comments on gender equality. Some comments were made in relation to employee participation in relation to conducting changes at work. Yet, the variability in responses was high. Some responders experienced high degree of participation while other did not experience any involvement in changes affecting the working environment.

# 3 Results

The results from the literature review showed that the many articles had an emphasis on patient safety and hospital care as opposed to home care. In addition, the scientific literature seem to address single problems more frequently than multi-dimensional (i.e. complex) problems and problems on the system level.

Collectively, the results from the two data collections showed that improved equality was reported on some occasions; including the fact that more men and men borne outside Sweden tend to get employed in the home care sector which is expected to make the sector less female dominant; that access to the manager is leveled out between different groups of employees; that employees have more power compared to before; that there are plans to evaluate equality; that women has got access to clothes and cars to the same extent as men; and some report that they have an equal working place. However, the majority of the answers see no effect on neither equality nor report initiatives that focus on gender.

Of the 259 interventions that were listed by the 179 responding units in the first data collection ten overall themes were identified to be subject to change: working life education, ergonomics, digital support system, care for care receivers, premises, organization of work, scheduling, collaboration with home health care, security and useful vehicles.

## 4 Discussion

To sum up, gender aspects does not seem to be paid much attention in neither of the review and the two surveys. There is a considerable lack of knowledge and often poor analyses of the consequences of the unequal gender balance in home care work. The scientific review that turned out to cover four different topics did not problematize gender, neither were needs for gender awareness pointed out as an explanation for poor working conditions for example low wages, high number of injuries, sick leave and staff turnover. The surveys to the home care workers and to the municipalities do not add anything to the picture with the exception of a few answers from local home care units. With this background it is obvious that the awareness of gender and its consequences for equality is low both in local municipalities and among home care workers. Gender does not seem to be an important aspect to explain the working life conditions in the home care sector. Home care in general and gender in particular is invisible. That being said, do we have reasons to assume that gender aspects are related to the working life problems in the home care sector?

Starting answering that question, it should be noted that the home care sector is one of the most gender segregated and most women dominated sector in Sweden and a sector that employs a large number of people (Statistics Sweden 2016). In addition, Sweden is an interesting country to study when it comes to home care work since this is one of the core professional in the development of general welfare and that statistics exist that make it possible to follow the development of occupation, sick-leave and proportion of men and women. The gender imbalance in this sector and the problems pointed out seem to reinforce attitudes that home care work is unskilled and simple, and as something women can do part time in parallel with the work at home in their own household [1, 22]. Other publications provide evidence for that gender imbalance has consequences for the degree of influence, leadership, division of work and representation of male and female [16, 21]. Australian studies show that such devaluing reproduced by a familial logic of care sustain home care as a low pay sector [15] and it is likely that this situation contributes to low status and affect working life in home care negatively. The tendency that male immigrants are entering this sector which was pointed out in the survey to local home care units is a new trend that should be investigated. Will this change the gender balance and if so, in what way? Or will this maintain the attitudes of this being a low status job?

A potential hidden gender aspect that should also be discussed is the lack of response from Kommunal members. Does that mirror an uninterest of the home care workers or is it a sign of that it is not worthwhile answering the survey? Is it a sign of that one does not trust to be able to influence? Swedish studies show tendencies that health was affected by gender inequality but that the home careers being interviewed did not see gendered division as a gender equality issue [7].

Another important aspect is that published articles on care work prove that there is an emphasis on patient safety and hospital care. A great number of articles excluded from the scientific review were about nursing but not home care. One explanation why nursing dominates can be that it is framed by legislations and rules regulating product developments and standards in hospitals which is lacking in the home care sector.

In addition home care work is taken place in the homes of people, more or less in non-controlled environment, similar to home care workers own household. For the home care workers, they become part of an unregulated and to a great extent under-researched context. The home becomes both a working environment and a home which turn the home into a space difficult to regulate. These aspects add to the invisibility of the home care sector.

A third result of the analysis of the review and the surveys was that single problems seems, to a large extent, be covered in the scientific literature while problems on the system level are more seldom addressed. The MTO- perspective (Man-Technology-Organization) for example or the HTO-perspective (Health-Technology-Organization) is not used for understanding the home care working environment problems [4, 13].

## 5 Conclusions and Future Work

In our synthesis of research and practice we have observed that the home care sector´s invisibility in research, public debate and policy making seem to be sustained by conservative attitudes to women's work and societal positions. These attitudes appear to be propagated among the home care workers themselves with consequences on their working life conditions and wages. However, the lacking research on specific problems in the home care sector also raises questions as to what extent also researchers share conservative values in their choice of questions; or to what extent researchers follow "the path of least resistance". It is plausible that the invisibility of the home care sector in research actually reflect the many practical and logistic problems that are associated with doing research in this sector. For example, the research participants are geographically dispersed and the possibilities to effectively intervene in someone's home may be quite difficult compared with a hospital ward. In addition, there are also questions pertaining to ethics and legal issues.

In any event, continuing understanding these mechanisms and research challenges are of utmost importance to change the working life conditions for the better, upgrade home care as a profession and make it visible as one of the core sectors to provide welfare.

## References

1. Andersson K (2012) Paradoxes of gender in elderly care: the case of men as care workers in Sweden. NORA Nord J Feminist Gend Res 20(3):166–181
2. Arvidsson I, Gremark J, Granqvist L, Enquist H, Andersson J, Renglin J (2017) Arbetsmiljö och hälsa hos personal inom hemvård och särskilt boende [Working environment and health for employees in home care and elderly living]. Report. Arbets- och miljömedicin Syd, no 9
3. Czuba LR, Sommerich CM, Lavender SA (2012) Ergonomic and safety risk factors in home health care: exploration and assessment of alternative interventions. Work 42:341–353
4. Dellve L, Williamsson A, Strömgren M, Holden RJ, Eriksson A (2015) Lean implementation at different levels in Swedish hospitals: the importance for working conditions and stress. Int J Hum Factors Ergon 3(3–4):235–253

5. Denton M, Zeytinoglu IU, Davies S (2002) Job stress and job dissatisfaction of home care workers in the context of healthcare restructuring. Int J Health Serv 32:327–357
6. Ede L, Rantakeisu U (2015) Managing organized insecurity: the consequences for care workers of deregulated working conditions in elderly care. Nord J Work Life Stud 5(2):55
7. Elwer S, Alex L, Hammarström A (2012) Gender (in)equality among employees in elder care: implications for health. Int J Equity Health 11:1–1
8. European Commission (2016) Blueprint digital transformation of health and care for the ageing society. Report. http://bit.ly/2j4gxCg. Accessed 03 May 2018
9. FORTE (Swedish Research Council for Health, Working Life and Welfare) (2016) Förstudie inför utlysning av forskningsmedel rörande arbetsmiljön i kvinnodominerade sektorer [Prestudy before call for funding of research concerning working environment in women dominated sectors]. Report dnr. 2015-01433 https://forte.se/app/uploads/2016/03/rapport-arbetsmiljo-kvinnodominerade-sektorer.pdf. Accessed 26 Apr 2018
10. Kommunal (2018) Personal som stannar. En rapport om arbetsmiljön i äldreomsorgen [Employers who stay. A report on the working environment in elderly care]
11. Håland E, Rösstad T, Osmundsen TC (2015) Care pathways as boundary objects between primary and secondary care: experiences from Norwegian home care servives. Health 19 (6):635–651
12. LEV (Long-term Demand for Welfare Services) (2010–2017) Från vård till hälsa – om värdet av digitalisering [From health care to health –on the value of digitization] Program report. Institute for Future Studies, Stockholm
13. Fredrik N (2018) Bortom IT. Den komplexa vårdens digitalisering. Om hälsa i en digital tid [Beyond IT. The complex digitization of health care. About health in times of digitization]. LEV rapport 4. Institute for Future Studies, Stockholm
14. Nordic Welfare Center (2012). http://www.nordicwelfare.org/PageFiles/9530/e-helsealliasen23102012.pdf. Accessed 05 Mar 2018
15. Palmer E, Eveline J (2012) Sustaining low pay in aged care work. Gend Work Organ 19 (3):254–275
16. Regnö K (2013) Det osynliggjorda ledarskapet. Kvinnliga chefer i majoritet [The invisible leadership. Female managers in majority]. Dissertation. KTH, Royal Institute of Technology, Stockholm
17. SKL, Swedish Association of Local Authorities and Regions (2018) Sveriges viktigaste jobb finns i välfärden [Sweden's most important jobs are in welfare]. Rekryteringsrapport
18. Socialstyrelsen, The National Board of Health and Welfare (2017) Tillståndet och utvecklingen inom hälso- och sjukvård [State of the art and developments concerning health care]. Report
19. Statistics Sweden (2016) Stora skillnader i lön mellan högst och lägst betalda yrkena [Big differences between highest and lowest salaries]. https://www.scb.se/hitta-statistik/sverige-i-siffror/utbildning-jobb-och-pengar/mest-och-minst-betalda-yrkena/. Accessed 02 May 2018
20. Swedish Work Environment Authority (2014) Projektrapport – Inspektioner av kvinno- och mansdominerad kommunal verksamhet, hemtjänst och teknisk förvaltning [Project report – Inspection of female and male dominated municipal occupations, home care and technology]. Rapport 3. https://www.av.se/globalassets/filer/publikationer/rapporter/projektrapport-inspektioner-kvinno-mansdominerad-hemtjanst-teknisk-kunskapssammanstallning-rap-2014-03.pdf?hl=kommunal%20verksamhet. Accessed 26 Apr 2018
21. Wahl A (2013) Gendering management. In: Sandberg Å (ed.) Nordic lights. Work, management and welfare in Scandinavia. SNS förlag, Stockholm

22. Vänje A (2015) Sick Leave-A signal of unequal work organizations? Gender perspectives on work environment and work organizations in the health care sector: a knowledge review. Nord J Work Life Stud 5(4):85
23. Moher D, Liberati A, Tetzlaff J, Altman DG (2009) The PRISMA Group. Preferred reporting items for systematic reviews and meta-analyses: the PRISMA statement. PLoS Med 6(7): e1000097

# Analysis of Posture Adopted by Female Kolhapuri Chappal (Footwear) Manufacturing Workers India

Urmi Salve(✉) and Ganesh Jadhav

Department of Design, Indian Institute of Technology Guwahati, North Guwahati, Amingaon 781039, Assam, India
meeturmi@gmail.com

**Abstract.** Footwear industry is an important sector of leather industries in India, distributed in various locations. It is divided into organized and unorganized (mainly craft based). Kolhapuri Chappal (footwear) is prominent among all craft based footwear manufacturing. This craft which has significant presence in local and national economy (export and domestic market), is a cottage based industry and mainly family entrepreneurship. The entire manufacturing process is divided into various steps and steps are divided among male and female workers of the house. Female workers are mostly involved in less forceful activities. But while doing, female workers occupy apparently various non-optimal postures which may lead to development of musculoskeletal and other occupational disorders. To understand the above a study was taken up for identifying the working postural load and its effect. The current working conditions and the frequencies of MSD symptoms of 51 female workers were evaluated. Musculoskeletal health data were collected through standard questionnaire and RULA was used for postural load analysis. SPSS version 20.0 was used for statistical data analysis. Age, height and weight of the subjects were 33.6 (±13.17) years, 152.3 (±3.2) cm and 51 (±4.3) kg respectively. The results of postural load analysis revealed that there is need for effective implementation of intervention program and some changes are required immediately with respect to their current workstations. Further it has been identified that there is a high prevalence of musculoskeletal pain among female workers at low back followed by neck, knee and upper-back. Further in understanding the relation with the postural load and types of musculoskeletal pain, it was found that there are association with RULA grand score and prevalence of low-back pain. The above result is comparable with other available literature of similar types of industry. As this occupation is very specific in nature it requires further investigation in detail.

**Keywords:** Women · Musculoskeletal disorders · Footwear industry

## 1 Introduction

The footwear manufacturing is an important section of leather industry in India. India ranks second among the footwear manufacturing countries next to China (Szubert et al. 2001). The footwear manufacturing industry occupies a place of prominence in the


S. Bagnara et al. (Eds.): IEA 2018, AISC 826, pp. 278–286, 2019.
https://doi.org/10.1007/978-3-319-96065-4_31

Indian economy and there is huge potential for export and employment (Gangopadhyay et al. 2011). Footwear industry in India is divided into two categories. One is organized and other is unorganized. Unorganized footwear sector is mainly craft based footwear; such as Mojari, Jaipuri, juttis, cane & bamboo made footwear, Zari etc. Kolhapuri footwear (Chappal) is one of the important handicrafts of Maharashtra, India (Salve et al. 2017). It is handmade footwear which is being locally manufactured by local habitat of Kolhapur region using various traditional methods.

Kolhapuri footwear is produced in nearby towns and villages of Kolhapur area situated at Maharashtra, India. Every household contributes in the process of this footwear production. The entire family produces approximately 35 to 45 pairs of footwear per week. This craft has an ancient history and slowly in modern days it started taking shape of commercial product during 19$^{th}$ century. It is being used as a status symbol by various local politician and elite class people. Along with domestic market, it has a significant existence in international export category.

As mentioned earlier, this craft is mainly a cottage industry and it includes a small group of working population where no formal contracts or regulation is being followed (Loewenson 2002). It has been further observed that these kind of informal sectors are neglected by part of economies. In order to improve quality of worker's life and to improve productivity major attention should be given to such kind of family enterprise business.

Kolhapuri footwear manufacturing techniques and processes are very traditional and conventional in nature. It involves various manual activities such as bottom making, skiving, punching, polishing, pattern cutting, attachment of heels, ears and upper bottom making, stitching, finishing and final assembling. But these activities are distributed among male and female members of the family according to the heaviness of the job type. Less strenuous jobs such as veni making, finishing, assembling etc. are being done by female members of the family. Although job categories are different, the workstations are almost similar for all jobs and for all genders. Required tools and necessary materials are being available from local markets. Materials such as adhesive glue, colour of the product are made from soil of local river and locally available flower respectively. To add on aesthetics of the product, some time they use some tree seeds in between two layers of soles. Lime stone is being used to smoothen the leather quality. In addition, traditionally they use sisal leaf (Cactus) for stitching.

Various workstation used by the female workers are presented in Fig. 1.

**Fig. 1.** Various workstations

While carrying out manufacturing activities all female workers occupy a seated working position on the floor. They use any wooden block which is readily available as their workstation. These workstations don't have any standardization. Used tools are also very primitive and traditional in nature. Few of them are presented in Fig. 2.

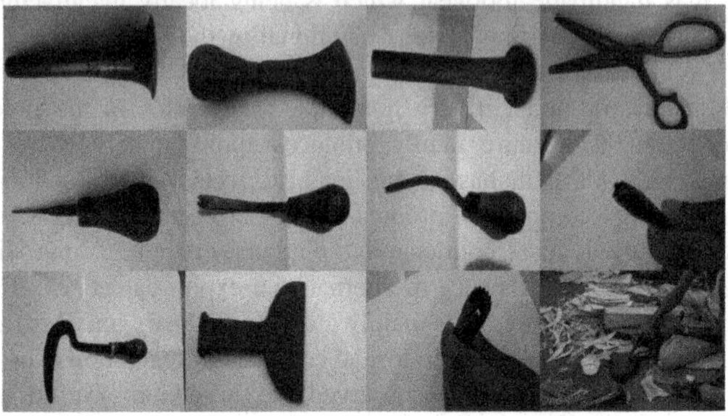

**Fig. 2.** Tools used during manufacturing

Working hours of these workers are not fixed and not regulated by any legislation. Further, these are family enterprise and family income is depended on the number of footwear produced per day. Therefore, workers work at a stretch for long hours, minimum 3–4 h. Working with awkward posture for long hours may lead to various work-related musculoskeletal disorders (WRMSD). There are various studies which reports the prevalence of various WRMSD among cottage industry workers in India. But the nature of the job of Kolhapuri footwear manufacturing is very different from other reported cottage industrial activities and worker's nature (specifically female). Although there studies available which clearly mentioned that long working hours in bending posture creates WRMSD, but there is no such study available which can explain causal job specific (Kolhapuri footwear manufacturing) MSD and its impact on the workers occupational health.

In view of the above a study was formulated with an objective to identify postural load among the female workers and its relation with prevalence of work-related musculoskeletal disorders.

## 2 Methodology

**Study location:** The whole study emphasized on Kolhapuri footwear manufacturing. Workers from fifteen co-operative societies participated in this study.

**Sample size:** Total 51 female workers participated in this study. Among them 20 were involved in stitching, 5 in punching and 26 various other parts.

**Study tool:** Two major tools were used for data collection. Nordic Musculoskeletal Questionnaire (NMQ) was used to understand the prevalence rate of various MSD, its' nature, possible causal factor, frequency and intensity (Kuorinka et al. 1987). Using body map discomfort chart (Corlett and Bishop 1976) various body map were rated using six points scale (where 1 = no pain, 2 = very low pain, 3 = low pain, 4 = moderate pain, 5 = high pain, 6 = very high pain). Rapid Upper Limb Assessment (RULA) was used to understand overall and specific body-parts' postural load (McAtamney and Corlett 1993).

**Informed consent:** After describing study objective and protocol a signed informed consent was obtained from each participant.

**Statistics:** Statistical analysis of the data was performed with SPSS software version 20.0.

# 3 Result

Demographic data of the 51 female Kolhapuri footwear manufacturing workers are presented in Table 1. The workers age were ranging from 18 to 66 years (mean = 42.61 years; SD = ±14.21 years). The average height and weight of the workers were 165.35 ± 4.85 cm and 58.45 ± 9.75 kg respectively.

**Table 1.** Mean and Standard Deviation (SD) for demographic parameters age, height and weight

| | Age (years) | Height (cm) | Weight (kg) |
|---|---|---|---|
| Mean | 42.61 | 165.35 | 58.45 |
| SD | ±14.21 | ±4.85 | ±9.75 |

Socioeconomic data of the subjects was represented in Figs. 3 and 4. Figure 3 represents the marital status of the subjects. Among 51 subjects, 46 (90%) were married and 5 were unmarried (10%).

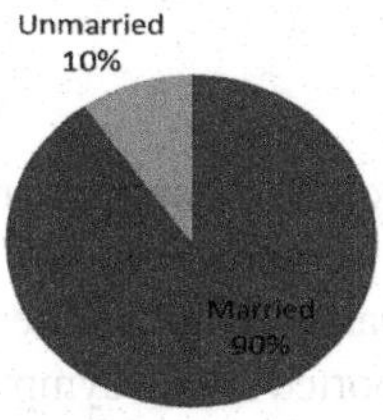


**Fig. 3.** Marital status of the subjects

Figure 4 represents the state of education among female workers. Majority (63%) of the participants were illiterate. 27% received only primary education and only 10% got secondary education.

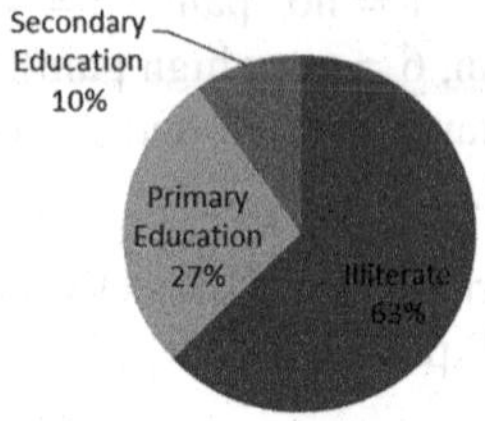


**Fig. 4.** Educational status of the participants

Table 2 explains job details of the participants of the study. It reveals that participants average work experience was 14.7(±8.2) years. Their daily working hours were reported as 7.7(±1.6) hours excluding domestic activities. 34.2% participants reported about continuous working without break (>10%) for approximately 2 h. Whereas, 31.6% participants were reported, 1–2 h continue work at a stretch.

**Table 2.** Job details of the subjects

| Job description | |
|---|---|
| Job experience (years) | |
| Mean (SD) | 14.7 (±8.2) |
| Range | 1–34 |
| Daily working hours (hours) | |
| Mean (SD) | 7.7 (±1.6) |
| Duration of continuous work without break (>10 min) | |
| <1 h | 34.2% |
| 1–2 h | 31.6% |
| >2 h | 34.2% |

Pont prevalence of musculoskeletal symptoms in different body regions of female footwear manufacturing workers is shown in the Table 3. These symptoms were defined as pain in different body parts and difficulties in activities for those body parts. There are 80% of the workers reported MSD symptoms during last 12 months. Most commonly affected body regions of the workers were Neck (88.5%), shoulder (74.2%), wrist and hand (88.5%), upper back (71.4%). Disruption of normal activities due to discomfort was found among 61.2% of the respondents.

**Table 3.** Musculoskeletal Disorder prevalence and its' effect on disruption of normal activities

| Body Region | Prevalence of symptoms n (%) | Disruption of normal activities due to symptoms n (%) |
|---|---|---|
| Neck | 31 (88.5) | 24 (68.5) |
| Shoulders | 26 (74.2) | 18 (47.3) |
| Elbows | 11 (31.4) | 2 (5.7) |
| Wrists/Hands | 31 (88.5) | 21 (60) |
| Low back | 23 (65.7) | 17 (48.5) |
| Upper back | 25 (71.4) | 14 (40) |
| Hips | 9 (25.7) | 1 (2.8) |
| Knees | 17 (48.5) | 28 (80) |
| feet | 7 (20.4) | 4 (11) |

Intensity of MSD problems were rated using 'six point' scale as mentioned in methodology section. Result of the same in percentage is presented in Fig. 5. It reveals that

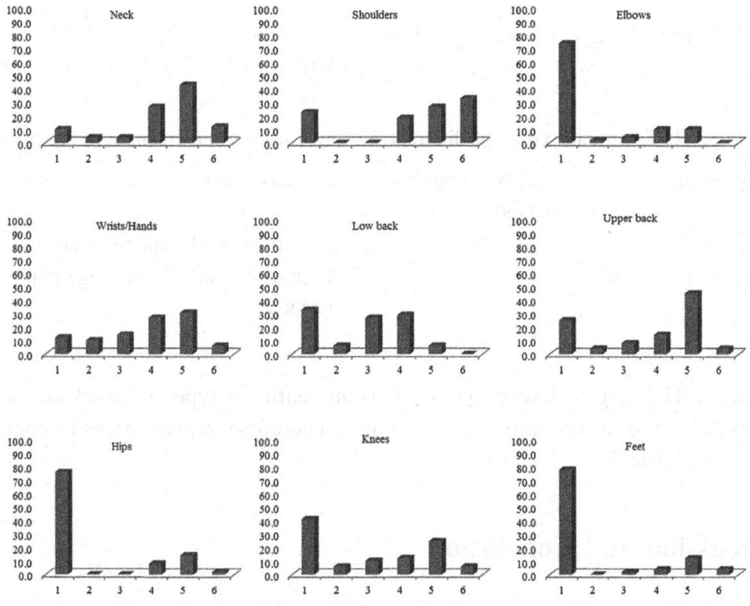


**Fig. 5.** Intensity level of musculoskeletal discomfort

Table 4 represents the posture evaluation report. It revealed that most of the workstations were not conducive to continue the job for longer hours.

**Table 4.** Distribution of RULA scoring (n = 49)

| RULA score | Upper arms n (%) | Lower Arms n (%) | Wrists n (%) | Score A | Neck n (%) | Trunk n (%) | Legs n (%) | Score B | Grand Score n (%) |
|---|---|---|---|---|---|---|---|---|---|
| 1 | – | 4 (11.4) | – | – | – | – | 20 (57.1) | – | – |
| 2 | 7 (20) | 16 (45.7) | – | – | 4 (11.4) | 6 (17.1) | 10 (28.5) | – | – |
| 3 | 16 (45.7) | 15 (42.8) | 14 (40) | 4 (11.4) | 14 (40) | 17 (48.5) | 4 (11.4) | 1 (2.8) | – |
| 4 | 11 (31.4) | – | 21 (60) | 11 (31.4) | 15 (42.8) | 11 (31.4) | 1 (2.8) | 4 (11.4) | 4 (11.4) |
| 5 | 1 (2.8) | – | – | 8 (22.8) | 1 (2.8) | – | – | – | 2 (5.71) |
| 6 | – | – | – | 8 (22.8) | 1 (2.8) | – | – | 12 (34.2) | 4 (11.4) |
| 7 | – | – | – | 4 (11.4) | – | – | – | 10 (28.5) | 24 (68.5) |
| 8 | – | – | – | – | – | – | – | 8 (22.8) | |
| Mean (SD) | 3.1 (0.8) | 2.4 (0.7) | 3.6 (0.5) | 5.9 (1.3) | 3.4 (0.9) | 3.2 (0.7) | 1.6 (0.8) | 6.0 (2.0) | 6.4 (1.0) |

**Table 5.** RULA grand scores of different postures of footwear manufacturing process

| Working posture | RULA grand score | Activity level |
|---|---|---|
| Stitching | 7 | Investigate and implement change |
| Preparation of various Parts | 4.4 | Further investigation change may be needed |

Average RULA grand score while working with all types of workstations need either investigation immediately or near future. Therefore, a need arises to analyze the workstation (Table 5).

## 4 Discussion and Conclusion

The purpose of present study was to evaluate the working postures of female Kolhapuri footwear manufacturing workers with respect to their working conditions. Most of the common symptoms were reported in shoulder, neck, upper back, lower back and knees. Average RULA grand score was 6.4 which indicates, there is a need for ergonomic intervention. Current study results further revealed that work related pain also affects their normal activities. The number of variables such as daily working hours, duration of continuous work without rest, job experience was found to be associated with musculoskeletal symptoms in some body regions.

The result of the current study further indicates that wrist pain, low back pain, shoulder pain, and neck pain were highly prevalent among the female workers involved in Kolhapuri footwear manufacturing activities. In a recent study conducted among Iranian hand sewn shoe workers (Dianat and Salimi 2014) it was found that the prolonged daily working hours contributed to greater number of reports of upper back symptoms. This result is very similar to the current study results. Wang et al. (2007) in their study observed that female sewing machine operators work for long duration which may cause various MSD.

From the present study it is apparently found that those workers who have less experience find more difficulty while performing their task. From field observation and discussion with workers it is reported that the stitching activity was monotonous (subjective response) and repetitive in nature. Although, detailed analysis is not being reported in present paper, it is subject to further scope of this project.

Present study tried to further analyze is there any specific association between nature of job and types of prevalence of MSD. It revealed that workers who were associated with stitching operations suffer from symptoms such as neck pain, eye strain, wrist/hand pain, knee pain etc.

Further from the daily working hours analysis it was clear that workers had frequent time of long duration of work without breaks. Long duration of sedentary work without break has been shown to be associated with neck problems among sewing machine operators (Wang et al. 2007), which is also being reported in present study. In another study (Aghali et al. 2012), it was found that prevalence of disorders of cervical area, shoulders with hands, vertebral column, back, knees, thigh with feet were higher in sewing operators due to poor work posture. In present study it was also reported that working posture is not at all optimal in nature which may be one of the cause of development of MSD symptom in above mentioned body parts.

In some other study where working population was bag manufacturing, reported that training and awareness programme is necessary to reduce the prevalence rate of MSD among the workers (Giri et al. 2012). In another study (Tiwari 2005), suggested that regular work rest cycles have positive influences on eyestrain in footwear manufacturing units.

From the present study it is very clear that there is an immediate requirement to provide ergonomic intervention for improvement of productivity and workers occupational health. As suggested by O'Neill (2000) "due to ergonomic constraints (limited infrastructure support to ergonomic activities and interventions, lack of appropriate anthropometric data) in developing countries the need for simple low cost practical solutions to improve working conditions should be implemented" there is a requirement to study further and provide the low cost solution which will help to maintain the identity of the craft and also help to improve productivity and occupational health.

**Acknowledgement.** The authors express their truthful appreciation to all those footwear manufacturing workers who rendered enormous assistance during the completion of this study. Also they express their sincere gratitude to all the students who helps the research team to collect field data. Further they express their gratitude to Mr. Hemant Shete for his contribution during this whole project.

## References


Aghali M, Asilian H, Parinaz P (2012) Evaluation of musculoskeletal disorders in sewing machine operators of a shoe manufacturing factory in Iran. J Pak Med 62:20–25

Dianat I, Salimi A (2014) Working conditions of Iranian hand-sewn shoe workers and associations with musculoskeletal symptoms. Ergonomics 4:602–611

Corlett EN, Bishop RP (1976) A technique for assessing postural discomfort. Ergonomics 19 (2):175–182

Gangopadhyay S, Ara T, Dev S, Ghoshal G, Das T (2011) An occupational health study of the footwear manufacturing workers of Kolkata, India. Ethno Med 5(1):11–15

Giri PA, Meshram PV, Kasbe AM (2012) Socio-demographic determinants and morbidity profile of people engaged in bag making occupation in an urban slum of Mumbai, India. Natl. J Community Med 3:601–606

Kuorinka I, Jonsson B, Kilbom A, Vinterberg H, Sorensen FB, Andersson G, Jorgensen K (1987) Standardised nordic questionnaires for the analysis of musculoskeletal symptoms. Appl Ergon 18:233–237

Loewenson R (2002) Occupational hazards in the informal sector: a global perspective. In: Health effects of new labour market, pp 329–342

McAtamney L, Corlett E (1993) RULA: a survey method for the investigation of work related upper limb disorders. Appl Ergon 24(2):91–99

O'Neill DH (2000) Ergonomics in industrially developing countries: does its application differ from that in industrially advanced countries? Appl Ergon 31:631–640

Salve UR, Jadhav GS, Shete HK (2017) Design solution of shoe sole (base of the footwear) preparation in traditional hand sewn footwear manufacturing: a case study on Kolhapuri Chappal. In: Advances in ergonomics in design, vol 588. Springer, Cham, pp 995–1013

Szubert Z, Wilczynska U, Sobala W (2001) Health risk among workers employed in rubber footwear plant. Med Pr 52(6):409–416

Tiwari R (2005) Child labour in footwear industry: possible occupational health hazards. Indian J Occup Environ Med 9:7–9

Wang P, Rampel D, Harrison R, Chan J, Ritz B (2007) Work-organizational and personal factors associated with upper body musculoskeletal disorders among sewing machine operators. Occup Environ Med 64:806–813

# Self-management Process After a Work Accident: A Gender Analysis

Liliana Cunha[1(✉)], Cláudia Pereira[2], Marta Santos[1], and Marianne Lacomblez[1]

[1] Centro de Psicologia da Universidade do Porto, FPCEUP, Universidade do Porto, Porto, Portugal
lcunha@fpce.up.pt

[2] Faculdade de Psicologia e de Ciências da Educação da Universidade do Porto, Porto, Portugal

**Abstract.** The return to the professional activity after a work accident corresponds to a complex process, whose path often reveals inequality factors, especially if we look at it through the gender lens.

Under the scope of the project "Return-to-work after an accident: to overcome obstacles" carried out in Portugal together with the National Association of Disabled Workers Injured On-the-job, with 371 participants, the circumstances and the consequences of the work accidents were analyzed.

Although the gender dimension was not considered at the origin of the investigation, the study's methodological layout took into account the representativeness of men and women for the purposes of sample constitution and findings' analysis – findings that show a clear inequality especially in the moments prior to the accident and in the reinsertion process.

Consequently, there seems to be a tendency towards the precariousness in the job and working conditions after the accident: for men, apart from the salary, the unemployment allowance emerges as the main source of income, while in the case of women the sick pay prevails as income source. From the perspective of the working conditions of those who return to the same company after the accident, most participants do not get a readjusted workstation or work schedule and there is a sense that the job is on the line, reported particularly by women.

**Keywords:** Work accident · Paths · Gender

## 1 Background

There is still a gender segregation in the labor market leading to different work situations for men and women [1]. Well, if men and women are not equally represented in the various activity sectors, do not perform the same activities, do not have the same paths, what is common and what is different in the journey after experiencing a work accident?

Indeed, a work accident is a turning point in any professional path, at any age, for any worker. Analyzing the work accidents based on this assumption means that they are considered breaking moments in the continuity of a given path as it was meant to


S. Bagnara et al. (Eds.): IEA 2018, AISC 826, pp. 287–293, 2019.
https://doi.org/10.1007/978-3-319-96065-4_32

be; those moments can therefore be taken as significant turning points that exist throughout life [2], or as moments that intersect the various life contexts [3].

Having as reference the study we hereby present on this matter, developed in Portugal, we call for a reflection that takes into account the specificities of the post-accident paths. The aim is to refocus the discussion about the absence of scientific data to enable a better surveillance regarding work accidents, and, in particular, regarding the inequalities where the gender dimension is the determinant factor.

## 2 Methodology

### 2.1 Goals

One of the purposes of the study Return-to-work after an accident: to overcome obstacles [4], launched by the National Association of Disabled Workers Injured On-the-job (ANDST), and funded by the National Institute for Rehabilitation, P. I., was the screening of the impacts the accident causes in the professional, personal and family aspects of the injured person. In this text, we will privilege an analysis of the post-accident paths and of the impacts such an event had on the job and working conditions of the participant workers.

### 2.2 Participants

The participants were selected from a database provided by the ANDST, comprising members and former members, seeking to follow principles of randomness and both gender and region representativeness, according to the national distribution of work accidents available at the time (year of 2010), though ensuring the allocation of cases to compensate the sub-representation of women.

The study's sample included 371 participants, 287 men and 84 women. 71.3% of the participants were aged 45–54 years, and 30.2% were aged 35–44 years.

A gender analysis reveals a higher percentage of women aged 45–54 years when the accident happened (39.4%), compared to 31.9% of men. Regarding seniority, 78.1% of the sample has been with the company for 1–10 years when the accident happened, followed by 61.8% for 11–20 years. There is a higher percentage of women with a higher level of seniority, over 21 years (27%, compared to 20.4% for men).

### 2.3 Procedure

The methodology was two-fold and combined a qualitative and a quantitative approach. It included documents collection and analysis (legislation on post-accident path, national statistics on work accidents); 10 individual interviews to explore and reconstruct the post-accident path, including the creation of path maps [5]; and the preparation of a questionnaire sent and filled in by all the participants, members and former ANDST members.

The individual interviews (audio-recorded; an average of one hour and thirty minutes per interview), focused mainly on the attempt to understand the journey and the paths after a work accident. This reflexive exercise turned out to play a major role on the adjustment of the structure of the survey made by the use of a questionnaire.

The questionnaire is divided into 3 main parts: sociodemographic characterization; the accidents' circumstances and consequences (including, to name but a few, the questions of the Nottingham Health Profile[1]); take stock of the impact caused by the accident on the job and on the personal/family/social life. The preparation of the questionnaire did not include a gender perspective that would shed light on the exploration of how specific the post-accident paths are for men and women.

The questionnaires were sent and returned by mail and were filled in autonomously by the participants. The SPSS software (version 24) was used to analyze the data. The statistical analysis that was carried out revealed how the job and working conditions after the accident are different for men and women. It also underlined the importance of considering this gender dimension in the construction of research tools.

## 3 Findings

### 3.1 Activity Sectors and Types of Accidents

Men and women tend to perform their activity in different companies and production sectors. Consequently, the type of accident and the type of injury also reveal such singularities.

**Table 1.** Activity sector, gender-related accident and injury

| | Men | Women |
|---|---|---|
| Activity sector | • Construction industry (21%)<br>• Transportation and Storage (24.7%)<br>• Manufacturing Industry (17.2%) | • Health and Social Support Activities (21.2%)<br>• Manufacturing Industry (15.4%)<br>• Domestic Work (15.4%) |
| Type of accident | The accidents are mainly related to the loss of machine control, tools handling, and means of transport (46.7%) | Women report accidents related to slips, falls (38%) and body movements subjected to physical constraints (25.3%) |
| Type of injury | Injuries in the upper limbs, hands, torso, back, belly, and lower limbs | Higher incidence of injuries in the upper and lower limbs |

According to the information on Table 1, Manufacturing Industries is the only sector common to both men and women who participate on this study. A more detailed analysis reveals that it does not refer to the exact same type of industries: while for men the metallurgical sector is identified more often, women have a higher presence in the textile and clothing industry. Working in different sectors, with dissimilar tasks, conditions, and work tools as resources at use lead to accidents and injuries that are also unlike. While men suffer accidents that are particularly related to the contact with machines/work tools, causing injuries "distributed" for several body parts, women's

[1] *Perfil de Saúde Nottingham* [Nottingham Health Profile] – Portuguese version, Centro de Estudos e Investigação em Saúde, 1997.

accidents result from situations such as falls or body movements subject to physical constraints, so the incidence of their injuries is higher on the upper and lower limbs.

## 3.2 Job and Working Conditions After the Accident

As far as the job and working paths are concerned, it was noticeable a global tendency after the accident for their deterioration (check Fig. 1).

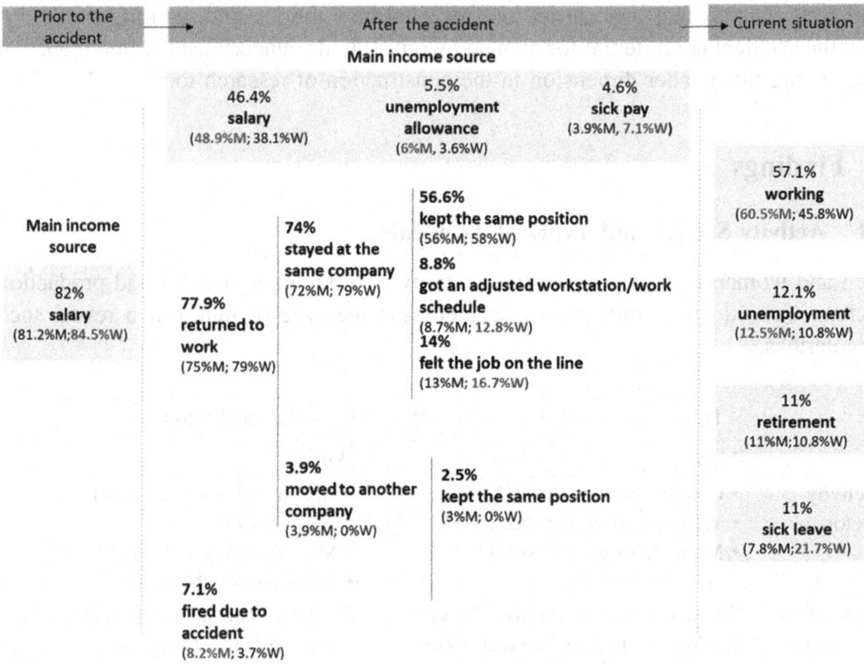


**Fig. 1.** Job and working conditions prior to and after the accident, plus current situation, per gender.

The following aspects emerge from a thorough analysis to the data on Fig. 1:

1. Prior to the accident the salary was the main income source for most participants (no significant differences between men and women – 81.2% and 84.5%, respectively), however after the accident there are other sources of main income depending on the gender: unemployment allowance in the case of men (6%), and sick pay particularly for women (7.1%);
2. Concerning the current job situation, the majority of the participants that are currently working are men (60.5%); there is a higher percentage of women on sick leave (21.7%);
3. For the majority of the participants there is no adjustment in the workstation or the work schedule (8.7% for men and 12.8% for women), and there is also a feeling that the job may be on the line, reported in particular by women (16.7% compared to 13% registered for men);

This deterioration in the job/working conditions goes beyond the workstation/work schedule, as it encompasses, on the one hand, the impossibility to accomplish a job well done and, on the other hand, the absence of recognition/help from coworkers and/or supervisors. In the context of the interviews, certain verbalizations illustrate this situation, for instance, "I was never 100% accepted by some of my colleagues. They never looked straight at me, as if saying, «he just does what he wants whenever he wants»" (former construction industry builder's laborer; 49 years old, with an "absolute incapacity to perform any given job"); "I started feeling left aside … as if saying, «go away, you're no longer useful to me»" (former hairdresser; 59 years old, currently retired).

### 3.3 Health State Perception After the Accident

In addition to these accidents and injuries, the results from the Nottingham Health Profile show that most accidents' after-effects are not visible nor diagnosed. Moreover, they occur in the form of pain (61.9% in the case of women and 50.9% in the case of men) in items such as "I feel unbearable pain" and "I feel pain during the night"; and also in the form of difficulties in physical mobility (54.8% for women and 39% for men), illustrated by the answers to the items "It is hard for me to go up and down stairs and steps" and "I find it hard to bend down". These manifestations are slightly more expressive in the case of women (Fig. 2).

The health problems that leave invisible sequels behind, such as pain, reported by women more than by men, bring new demands to the after-accident path. These symptoms not only last in time, but they are incapacitating; they are not tangible nor easily diagnosed; their acknowledgment is frequently compromised, and it is then up to the individual to manage such symptoms. For that reason, given the type of accidents and sequels felt by the women in particular, it is understandable that, after the accident, the women take sick leave more than men do (check Fig. 1: 21.7% compared to 7.8% for men).

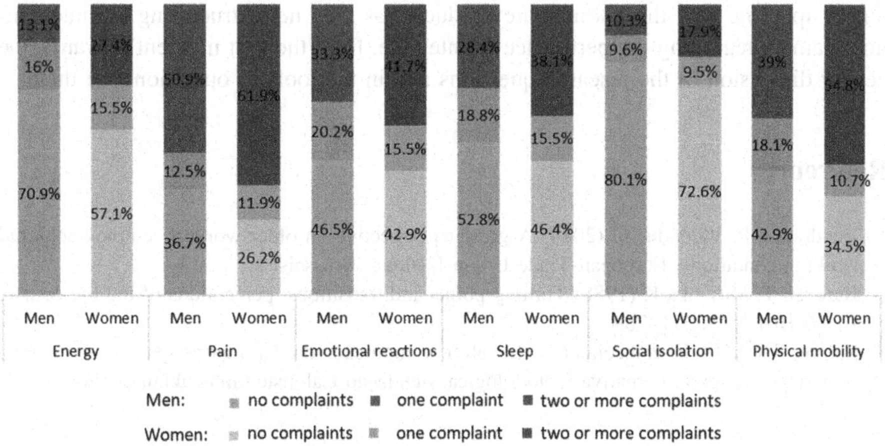


**Fig. 2.** Analysis to the no. of complaints reported by men and women according to the dimensions of the Nottingham Health Profile

## 4 Conclusions

From the participant's point of view, returning to work after the accident proves the accident occurred due to the working conditions, the post-accident path, however, is the beginning of a completely new process of constraints and difficulties that these workers have to face. Quite often, that process requires the individual to manage the post-accident path in a relatively lonely way.

Hence, despite the overall findings show painful and degrading situations for men and women alike, women do seem to be more frequently in a more vulnerable situation, which is explained by their job and working conditions, by the types of accidents and injuries they suffer, but also due to the impact these accidents have on health. Long-lasting pain is a manifestation of disability, the pain prevails in the working context and outside work, causing a silent incapacity.

Considering these particular workers, they were not always alone; they sought and got help from ANDST. In fact, this study was carried out with members of this Association alone. We acknowledge it is a limitation of the study itself, but it also reveals other gender inequalities: there are more male than female members, and the actors from this institution have less experience in the analysis of actual accident situations and of the support to claim the female victim's rights. The differences highlighted in this study are significant, so it reinforces and brings to the center of the discussion the importance of introducing the gender dimension in a broader analysis to work accidents, and of issuing regular official data from the liable institutions.

Admitting gender as a research option made it possible to understand, in a more complete and incisive way, the complexity of the circumstances and consequences of the work accidents. It also gave a different visibility to certain issues that deserve broader dissemination, so to identify intervention avenues for the liable institutions on this field (e.g. Courts, Insurance Companies, Working Conditions Authority), in and out the working contexts, aiming at improving the working conditions for all the workers, either injured or not, and, to a larger extent, enhancing the conditions to express citizenship at work and beyond. Ahead of this public action, it is important to reflect upon the way the findings are produced, as they help structuring the intervention, namely regarding the pertinence of integrate, from the first moment onwards, the gender dimension in the research questions and in the tools to operationalize them.

## References

1. Vendramin P, Valenduc G (2014) A gender perspective on older workers' employment and working conditions. European Trade Union Institute, Brussels
2. Hareven T, Masaoka K (1988) Turning points and transitions: perceptions of the life course. J Fam Hist 13(3):271–289
3. Ramos S (2010) Envelhecimento, trabalho e cognição: do laboratório para o terreno na construção de uma alternativa metodológica. Fundação Calouste Gulbenkian, Lisboa

4. Cunha L, Santos M, Pereira C (2017) Return to employment after a work-related accident: how he/she defines himself/herself in relation to others. ETUI Seminar: His and Hers occupational hazards, health, justice and prevention actors. ITUH, Brussels
5. Pereira C, Rodrigues V (2018, accepted) A reconstituição de experiências após um acidente de trabalho: mapas de percurso com instrumento de análise. Laboreal, 14(1)

# Challenges in Implementing Inclusive Participatory OHS Management Approaches Sensitive to Gender and Diversity Issues

Valérie Lederer(✉) and Jessica Riel

Département de relations industrielles, Université du Québec en Outaouais, C.P. 1250, succursale Hull, Gatineau, QC J8X 3X7, Canada
valerie.lederer@uqo.ca

**Abstract.** Purpose: Although the effectiveness of participatory approaches in occupational health and safety (OHS) is, in theory, widely recognized today [1, 2], their implementation is sometimes guided by power relations or by the interests of the most vocal stakeholders to the detriment of others. Several challenges remain to implement it fully in the workplace and to achieve the active, inclusive and representative participation of all stakeholders in terms of gender and, more broadly, diversity (ex. ethnicity, language, age, (dis)abilities) [3, 4]. The research question of the present study was: What are the favorable and unfavorable conditions for implementing inclusive participatory OHS management approaches that integrate gender and diversity issues?

Methods: We conducted an iterative scoping review of the literature following Arksey and O'Malley's guidelines [5]. We searched several databases including Pubmed, PsycInfo, Web of Science and Google Scholar for discussion or research on diversity management in participatory approaches in ergonomics and in OHS management. Papers were screened for inclusion based on their title and abstract at first and based on full text, as a second step. Information were extracted on composition/functioning of the work group, values/beliefs/attitudes towards gender and diversity, social climate, financial/human/operational resources. We performed a qualitative thematic analysis and synthetized the evidence and the questions/hypotheses/practical considerations raised in the discussion of the topic. From this analysis, we derived a conceptual framework.

Results: This research demonstrated the potential for participatory approaches sensitive to gender and diversity to overcome some of the limits of the classical OHS management approaches. We synthetized and discussed the principal barriers and facilitators to implementing inclusive participatory approaches in OHS and proposed a conceptual framework for the implementation of inclusive participatory approaches. Three dimensions were identified to a successful implementation of inclusive participatory approaches sensitive to gender/diversity: (a) establishment of a working group representative of the diversity within the organization, (b) integration of issues related to gender/diversity into the discussions, (c) establishment of a shared equitable decision-making process (ex. commitment of all stakeholders, exchange of information, skills acquisition in order to balance power relations). Various barriers and facilitators were identified.


S. Bagnara et al. (Eds.): IEA 2018, AISC 826, pp. 294–295, 2019.
https://doi.org/10.1007/978-3-319-96065-4_33

Conclusions: Understanding and addressing the principal barriers and facilitators to a greater inclusion and diversity in participatory approaches could ultimately lead to improved participatory interventions to everyone's benefit.

**Keywords:** Participatory approach · Gender · Implementation

## References

1. Van Eerd D et al (2010) Process and implementation of participatory ergonomic interventions: a systematic review. Ergonomics 53(10):1153–1166
2. Wilkinson A (2010) The Oxford handbook of participation in organizations. Oxford University Press, Oxford
3. Franche R-L et al (2005) Workplace-based return-to-work interventions: a systematic review of the quantitative literature. J Occup Rehabil 15(4):607–631
4. Johnstone S, Ackers P (2015) Finding a voice at work?: new perspectives on employment relations. Oxford University Press, Oxford
5. Arksey H, O'Malley L (2005) Scoping studies: towards a methodological framework. Int J Soc Res Methodol 8(1):19–32

# When Being a Woman Represents a Major Risk of Commuting Accidents?

Silvana Salerno[1(✉)] and Claudia Giliberti[2]

[1] ENEA - National Agency for New Technologies, Energy and Sustainable Economic Development, Rome, Italy
silvana.salerno@enea.it

[2] INAIL-DIT - National Institute for Insurance Against Accidents at Work, Rome, Italy
c.giliberti@inail.it

**Abstract.** Since 2010, National Institute for Insurance against Accidents at Work (INAIL) recognized commuting accidents as work injury. Commuting is the leading cause of death among Italian and immigrant women. Commuting disabilities have been studied from INAIL records within the last five years (2012–2016). Results show how women present more disabilities in commuting than men (53.2% vs 46.8% men, $p < 0,001$). Men show more disabilities due to *commuting with vehicle* (39.8% vs 36.8% women) while *commuting without vehicle* is more frequent among women (16.4% vs 7% men, $p < 0,05$) with more permanent disabilities and severe ones (20.4% vs 9% men, $p < 0.001$; severe 16.5% vs 7.2% men, $p < 0,001$). Skeleton-motor apparatus is the most affected among women (87.7% vs 80.7% men, $p < 0.001$) in all work injury disabilities. Lower limbs are the women body site mainly injured (40.4% women vs 38.6% men, $p < 0.01$) after upper limbs.

More disabilities due to commuting, particularly without vehicle, can be explained with women more home-work-home walking together with double burden, low socio-economic status, work and environmental conditions.

Ergonomics should deal towards commuting gender-oriented prevention, in terms of home-work-home journey concept (sidewalks, lighting, etc), work organization (family/work balance, flexible working hours, limited night shift work, standing and sitting posture, dual tasks, etc.), specific women's health prevention (osteoporosis, strengthening lower limbs, wearing suitable shoes for walking and working, etc.).

**Keywords:** Commuting · Gender · Disabilities

## 1 Introduction

European Transport Safety Council showed how in 2016 road accidents victims due to home-work-home journey or work traveling were more than 10.000 [1]. Nevertheless only fourteen European countries recognize commuting as a work accident [2]. This is the reason why real and up to date European commuting accidents statistics are far to be complete. European Union promulgated three Directives on this issue: 1. On the *Introduction of measures to encourage improvements in the safety and health of*


S. Bagnara et al. (Eds.): IEA 2018, AISC 826, pp. 296–307, 2019.
https://doi.org/10.1007/978-3-319-96065-4_34

*workers at work* (Directive 1989/391/EEC); 2. *Concerning certain aspect of the organization of working time* (Directive 2003/88/EC); *3. On the framework for the deployment of Intelligent Transport Systems in the field of road transport and for interfaces with other modes of transport* (Directive 2010/40/EU) providing precautionary indications to be implemented in the workplace in terms of: reduction of working hours and increase of flexibility, respecting rest, breaks, annual holidays, promotion of good practices for pedestrians and cyclists. Each country has developed specific prevention strategies such as: the creation of canteens, the limitation of long breaks that favor the return home, the incentive to use public transport, company shuttles, car pooling and car sharing etc.

In Italy, since 2010, *commuting* as accident at work is recognized by the National Institute for Insurance against Accidents at Work (INAIL). INAIL defines "commuting accident" as the accident during the home-work-home journey or the journey from one workplace to another, or the journey for the consumption of meals when canteen is not available. All INAIL work accident statistics consider the involvement of a vehicle as follows: "*commuting accidents*" with or without the involvement of a vehicle and "*during work accidents*" with or without the involvement of a vehicle.

In Italy, commuting accidents represent the leading cause of death at work for Italian and immigrant women, as shown in a previous study where the main causes of deaths were identified [3]. A Finnish study showed also how commuting accident frequency was 1.4 times higher among women than men and European Occupational Safety and Health (OSHA-EU) confirmed commuting as a major risk for women [4].

The phenomenon of commuting is therefore relevant for the health of women and prevention-oriented ergonomics is crucial for the improvement of their working conditions.

Desired outcome of this study is to analyze gender differences in commuting disability records from INAIL Statistical Database in order to improve gender oriented ergonomics.

## 2 Materials and Methods

INAIL Statistical Database on disabilities in all employment settings (Industry and services, Agriculture, State) by type of work accident (*during work with and without vehicle*, *commuting accident with and without* vehicle) was analyzed by gender and country of birth (Italy or not) and by grade of severity (2009–2015 updated to October 31st, 2016) [5].

Disabilities due to physical or psychological impairment are divided into: *temporary disability (more than three days)* and *permanent disability (life long) (1%–100%)*. A compensation is paid, proportional to the severity of the damage suffered, in terms of days of work lost and work capacity. Permanent versus temporary disabilities have been analyzed in detail in the last five years (2012–2016 updated to October 31st, 2017).

Impairments due to permanent disability have been divided into five increasing ranks of severity: 1. *Poor disabilities* with modest or little change in working life (1%–15%); 2. *Medium/low disabilities* with relevant change in working life (16%–25%); 3. *Medium/high disabilities* with heavy working life impairment (26%–50%) 5. *Severe disabilities* with few or no ability at work (51%–100%) [6]. Medium/high/severe disabilities (26%–100%) have been selected and compared with the less severe ones (<26%) by gender and country of birth (Italy or not).

The most frequent type and body site disability have been studied using INAIL Disability Data Base (available since 2010) that provides work accidents (not specific data for commuting) and occupational disease records [7]. The main differences by gender and country of birth (Italy or not) were analyzed through the $2 \times 2$ table and statistical significance by chi-square test applied ($p < 0.05$).

# 3 Results

## 3.1 Disabilities Due to All Injuries at Work (2009–2015) for Gender and Country of Birth

Table 1 shows all work accident disabilities (temporary and permanent) among Italian and immigrant women. "*During work" accidents without vehicles* represent the first cause of women disability (Italian women 66% vs immigrant women 65%, $p < 0.05$), *commuting accidents (with and without vehicle)* follows at the second rank (29% both Italian and immigrant women).

Italian women have more *commuting without vehicle* disabilities (12% Italian vs 8% immigrant women, $p < 0.001$); immigrant women more *commuting with vehicle* (21% immigrant women vs 17% Italian women, $p < 0.001$). Disabilities *"during work with vehicle*" have the lowest rank among women (Italian women 5% vs 6% immigrant women, $p < 0.05$). In Table 2, the same picture is shown for Italian and immigrant men. At the first rank "*during work without vehicle"* disabilities, significantly higher among immigrant men (80% immigrant men vs 77% of Italian men, $p < 0.001$). Commuting injuries follow at the second rank (16% Italian men vs 14% immigrant men, $p < 0.05$). More disabilities due to *commuting with vehicle* are shown among men (14% Italian men vs 12% immigrant men, $p < 0.05$), and no difference in commuting without vehicle (2%). "*During work with vehicle"* disabilities are more frequent among Italian men (7% Italian men vs 6% immigrant men, $p < 0.001$).

## 3.2 Disabilities Due to Commuting Injuries (2012–2016)

Table 3 shows the latest available disability data due to commuting accidents per gender (women N. 172.003, men N. 151.409, total N. 323.412). Overall commuting with vehicle presents more disabilities 76.6% than without vehicle 23.4%. Women have more disabilities due to commuting (53.2% vs 46.8%, $p < 0.0001$) and higher disabilities due to commuting without vehicle (16.4 women vs 7% men, $p < 0.05$).

Table 4 shows the differences by country of birth and gender for the same disabilities. Italian women (32.2% Italian women vs 33.1% Italian men, $p < 0.05$) and immigrant women (4.6% immigrant women vs 6.7% immigrant men, $p < 0.001$) have lower *commuting with vehicle* disabilities than men. On the other hand, disability due to *commuting without vehicle* is more common among Italian women (14.7% Italian women vs 6.1% Italian men, $p < 0.001$) and immigrant women (1.6% immigrant women vs 0.9% immigrant men, $p < 0.001$).

Table 5 shows commuting accidents disability as temporary and permanent and, among permanent, the severe ones. Temporary disabilities are 77.4% of all events (N. 250.476). Women have more temporary disabilities (80.4% women vs 74% men, $p < 0.001$) than permanent (19.6% women vs 26% men, $p < 0.001$) and among permanent few severe disabilities (1.7% women vs 4.2% men, $p < 0.001$).

Gender and country of birth differences are shown in Table 6, resulting with the same trend among Italians for temporary disabilities (80.3% women vs 73% men, $p < 0.001$) and immigrants (81.2% women vs 79.6% men, $p < 0.001$) showing more temporary disabilities among women. Permanent disabilities among Italians (19.7% women vs 27% men $p < 0.001$) and immigrants (18.8% women vs 20.4% men, <p 0.001) show again less permanent disabilities among women. Severe disabilities, among the permanent ones, reinforce the trend (1.6% Italian women vs 4.2% Italian men, $p < 0.001$; 2.3% immigrant women vs 4.3% immigrant men, $p < 0.001$). Immigrant women show more severe disabilities than Italian women (2.3% women immigrants vs 1.6% Italian women, $p < 0.01$). Italian men if compared with immigrants show more permanent disabilities (27% Italian men vs 20.4% immigrant men, $p < 0.001$) and a similar picture of severe disabilities (4.2% Italian men vs 4.3% immigrant men).

Table 7 shows all permanent disabilities per gender and type of commuting accidents with or without vehicle (N. 72.936, 22.6% of all events). *Commuting with vehicle* represents the major cause of all permanent disabilities (70.6% vs 29.4%) particularly among men (44.9% vs 25.7% women, $p < 0.001$). All women show more permanent disabilities due to *commuting without vehicle* (20.4% women vs 9% men, $p < 0.001$).

Same results per gender and country of birth are shown in Table 8. Permanent disabilities due to *commuting without vehicle* are more frequent both among Italian women (18.6% women vs 8.1% men, $p < 0.001$) and immigrant women (1.8% women vs 0.9% men, $p < 0.001$). Italian men and women show more disabilities due to *commuting with vehicle* (38.9% Italian men vs 5.9% immigrant men, $p < 0.001$; 22.3% Italian women and 3.4% immigrant women, $p < 0.001$).

Severe disabilities (26%–100%) represent only 3% of all permanent disabilities (N. 2216 of all events N. 72936) mostly due to *commuting with vehicle* (2.7% with vehicle vs 0.3% without vehicle). Severe disabilities are less frequent among women (0.8%) than men (2.2%). All severe commuting disabilities with and without vehicle (N. 2216) are shown in Table 9. *Commuting with vehicle* is the most frequent severe disability (90.5%) compared with *commuting without vehicle* (9.5%). *Commuting with vehicle* is the most frequent cause of severe disabilities among men (with vehicle 92.8% men vs 83.5% women, $p < 0.001$) while women suffer more for severe disabilities when *commuting without vehicle* (16.5% women vs 7.2% men, $p < 0.001$). Country of birth differences on severe disabilities are shown in Table 10. Italian men show more severe disabilities when *commuting with vehicle* than Italian women (61.1% men vs 18.3% women, $p < 0.001$), the same result among immigrant men towards immigrant women (7.8% immigrant men vs 3.1% immigrant women). Italian women are more affected by severe disabilities when *commuting without a vehicle* (3.5% women vs 3.4% Italian men, $p < 0.001$). This is not the same among women immigrant women that have less severe disabilities when *commuting without a vehicle* (0.7% women immigrant vs 1.9% men immigrant, $p < 0.01$).

Geographical distribution of disabilities and age of workers have been also studied for the last five years (2012–2016). Commuting disabilities (with and without vehicle) are more frequent in Italian North-West regions (both for women and men without differences. Age differences have been found per gender and type of commuting. *Commuting with vehicle* is more frequent in the interval age 35–39 and 40–44 without gender differences. *Commuting without vehicle* is more frequent in older age women (50–59) than men (40–49). Age interval for women in commuting without vehicle is increasing in the last years (55–59) while men are mainly in 40–44 age interval. Specific gender data on immigrant age were not available.

### 3.3 Disabilities Database: Type of Pathologies

Gender differences in all disabilities compensation due to work injuries or occupational diseases (N. 610138, women N. 87.348 (14.3%) and men N. 522.790 (85.7%)) are shown in Table 11. Of all disability compensation, work injuries represent 57.6% and occupational diseases 42.4%. Gender differences show how women received more work injuries compensation (61.1% vs 57.1% men, $p < 0.001$) than occupational diseases (38.9% women vs 42.9% men, $p < 0.001$). Severe disability is again more frequent among men 4.1% (21566) than women 2.4% (n. 2090) ($p < 0.001$) (table not shown). Inequalities due to disabilities compensation should be verified in a specific study.

Type of pathology (motor, psycho-sensorial, cardiorespiratory and others) per gender due to work injuries is reported in Table 12. Skeleton-motor pathologies are more frequent among women (87.7% women vs 80.7% men, $p < 0.001$) as other diverse pathologies (1.4% women vs 1% men, $p < 0.001$). Psycho-sensorial disabilities are more frequent among men (17.2% men vs 10.1% women) as cardiorespiratory (1.1% men vs 0.8% women). Of all women pathologies due to work injuries (N. 53354), lower limbs (N. 18894, 35.4%) and upper limbs (n. 18.300, 34.3%) are in the highest rank. Men (N. 298251) present mainly upper limbs pathologies (n. 104164, 34.9%) followed by lower limbs (92956, 31.2%) (table not shown).

In Table 13 gender differences in body site pathologies due to work injuries are also shown. Skeleton-motor site disabilities distribution show again how women are mainly affected by lower limbs pathologies (40.4% women vs 38.6% men $p < 0.01$) and upper limbs among men (43.2% men vs 39.1% women $p < 0.001$).

Psycho-sensorial site disabilities distribution show more psychological and neurological disabilities (61.2% women vs 42% men) among women and more sensorial disabilities (58% men vs 38.8% women) among men. Undefined diverse pathologies are more frequent among women (64.3% vs 47.9%, $p < 0.001$).

**Table 1.** Women - work injury disabilities per cause and country of birth (INAIL Database 2009–2015)

| Work disabilities 2009–2015 N. 64.219 | Italy women | | Not Italy women | | Total | |
|---|---|---|---|---|---|---|
| | N. | % | N. | % | N. | % |
| Commuting | *16913* | *29* | *2074* | *29* | *18987* | *30* |
| With vehicle | 9918 | 17 | 1481 | 21** | 11399 | 18 |
| Without vehicle | 6995 | 12** | 593 | 8 | 7588 | 12 |
| During work | 37999 | 66* | 4542 | 65 | 42541 | 66 |
| During work with vehicle | 2309 | 5 | 382 | 6* | 2691 | 4 |
| *Total* | *57221* | *100* | *6998* | *100* | *64219* | *100* |

Country of birth differences among women *p < 0.05; **p < 0.001

**Table 2.** Men - work injury disabilities per cause and country of birth (INAIL Database 2009–2015)

| Work disabilities 2009–2015 N. 193.667 | Italy men | | Not Italy men | | Total | |
|---|---|---|---|---|---|---|
| | N. | % | N. | % | N. | % |
| Commuting | *27396* | *16** | *3170* | *14* | *30563* | *16* |
| With vehicle | 23359 | 14* | 2746 | 12 | 26105 | 14 |
| Without vehicle | 4034 | 2 | 424 | 2 | 4458 | 2 |
| During work | 131597 | 77 | 18898 | 80** | 150495 | 77 |
| During work with vehicle | 11309 | 7** | 1300 | 6 | 12609 | 7 |
| *Total* | *170299* | *100* | *23368* | *100* | *193667* | *100* |

Country of birth differences among men *p < 0.05; **p < 0.001

**Table 3.** Gender differences in *commuting disabilities* (INAIL Data Base 2012–2016)

| 2012–2016 N. 323.412 | Women N. 172.003 | | Men N. 151.409 | | Total N. 323.412 | |
|---|---|---|---|---|---|---|
| Commuting | N. | % | N. | % | N. | % |
| With vehicle | 119072 | 36.8 | 128680 | 39.8* | 247752 | 76.6 |
| Without vehicle | 52931 | 16.4* | 22729 | 7.0 | 75660 | 23.4 |
| *Total* | *172003* | *53.2*** | *151409* | *46.8* | *323412* | *100* |

Gender differences *p < 0.05; **p < 0.001

**Table 4.** Country of birth differences in *commuting disabilities* (INAIL Data Base 2012–2016)

| 2012–2016 N. 323.412 | Women | | | | Men | | | |
|---|---|---|---|---|---|---|---|---|
| Commuting | Italy N. | % | Not Italy N. | % | Italy N. | % | Not Italy N. | % |
| With vehicle | 104256 | 32.2 | 14816 | 4.6 | 107068 | 33.1* | 21612 | 6.7* |
| Without vehicle | 47656 | 14.7* | 5275 | 1.6* | 19862 | 6.1 | 2867 | 0.9 |
| *Total* | *151912* | *46.9** | *20091* | *6.2* | *126930* | *39.2* | *24479* | *7.6** |

Gender differences within the same country of birth *p < 0.05; p < 0.001

**Table 5.** Gender differences in commuting per *temporary and permanent disabilities* (INAIL Data Base 2012–2016)

| 2012–2016 N. 323.412 | Women N. 172.003 | | Men N. 151.409 | | Total N. 323.412 | |
|---|---|---|---|---|---|---|
| Commuting | N. | % | N. | % | N. | % |
| Temporary disabilities | 138377 | 80.4* | 112099 | 74 | 250476 | 77.4 |
| Permanent disabilities of whom: | 33626 | 19.6 | 39310 | 26* | 72936 | 22.6 |
| *Severe disabilities* | *571* | *1.7* | *1645* | *4.2** | *2216* | *3.0* |
| *Total* | *172003* | *100* | *151409* | *100* | *323412* | *100* |

Gender differences *p < 0.001

**Table 6.** Country of birth differences in commuting per *temporary and permanent disabilities* (INAIL Data Base 2012–2016)

| 2012–2016 N. 323.412 | Women | | | | Men | | | |
|---|---|---|---|---|---|---|---|---|
| Commuting | Italy N. | % | Not Italy N. | % | Italy N. | % | Not Italy N. | % |
| Temporary disabilities | 122059 | 80.3* | 16318 | 81.2* | 92616 | 73 | 19483 | 79.6 |
| Permanent disabilities of whom: | 29853 | 19.7 | 3773 | 18.8 | 34314 | 27* | 4996 | 20.4* |
| Severe disabilities | 485 | 1.6 | 86 | 2.3 | 1429 | 4.2* | 216 | 4.3* |
| *Total* | *151912* | *100* | *20091* | *100* | *126930* | *100* | *24479* | *100* |

Gender differences within the same country of birth *p < 0.001

**Table 7.** Gender differences in *permanent commuting disabilities* (INAIL Data Base 2012–2016)

| 2012–2016 N. 72.936 | Women | | Men | | Total | |
|---|---|---|---|---|---|---|
| Commuting | N. | % | N. | % | N. | % |
| With vehicle | 18725 | 25.7 | 32754 | 44.9* | 51479 | 70.6 |
| Without vehicle | 14901 | 20.4* | 6556 | 9.0 | 21457 | 29.4 |
| *Total* | *33626* | *46.1* | *39310* | *53.9* | *72936* | *100* |

Gender differences *p < 0.001

**Table 8.** Country of birth differences in *permanent commuting disabilities* (INAIL Data Base 2012–2016)

| 2012–2016 N. 72.936 | Women | | | | Men | | | |
|---|---|---|---|---|---|---|---|---|
| Commuting | Italy N. | % | Not Italy N. | % | Italy N. | % | Not Italy N. | % |
| With vehicle | 16250 | 22.3* | 2475 | 3.4 | 28420 | 38.9* | 4334 | 5.9 |
| Without vehicle | 13603 | 18.6* | 1298 | 1.8* | 5894 | 8.1 | 662 | 0.9 |
| *Total* | *29853* | *40.9* | *3773* | *5.2* | *34314* | *47* | *4996* | *6.8* |

Gender differences within the same country of birth * p < 0.001

**Table 9.** Gender differences in *severe commuting disabilities* (grade of severity 26%–100%) (INAIL Data Base 2012–2016)

| 2012–2016 N. 2216 | Women | | Men | | Total | |
|---|---|---|---|---|---|---|
| Commuting | N. | % | N. | % | N. | % |
| With vehicle | 477 | 83.5 | 1528 | 92.8* | 2005 | 90.5 |
| Without vehicle | 94 | 16.5* | 117 | 7.2 | 211 | 9.5 |
| *Total* | *571* | *100* | *1645* | *100* | *2216* | *100* |

Gender differences *p < 0.001

**Table 10.** Country of birth and gender differences in *severe commuting disabilities* (grade of severity 26–100%) (INAIL Data Base 2012–2016)

| 2012–2016 N. 2216 | Women | | | | Men | | | |
|---|---|---|---|---|---|---|---|---|
| Commuting | Italy N. | % | Not Italy N. | % | Italy N. | % | Not Italy N. | % |
| With vehicle | 407 | 18.3 | 70 | 3.1 | 1354 | 61.1* | 174 | 7.8 |
| Without vehicle | 78 | 3.5* | 16 | 0.7 | 75 | 3.4 | 42 | 1.9 |
| *Total* | *485* | *21.9* | *86* | *3.8* | *1429* | *64.5* | *216* | *9.8* |

Gender differences within the same country of birth *p < 0.001

**Table 11.** Gender differences in disabilities compensation due to *work injuries and occupational diseases* (INAIL Disabilities Data base, 2010)

| | Women N. | % | Men N | % | Total N. | % |
|---|---|---|---|---|---|---|
| Work injuries | 53354 | 61.1* | 298251 | 57.1 | 351605 | 57.6 |
| Occupational diseases | 33994 | 38.9 | 224539 | 42.9* | 258533 | 42.4 |
| *Total* | *87348* | *100* | *522790* | *100* | *610138* | *100* |

Gender differences *p < 0.001

**Table 12.** Gender differences in pathologies compensation due to *work injuries* (INAIL Disabilities Data Base, 2010)

| | Women N. | % | Men N. | % | Total N. | % |
|---|---|---|---|---|---|---|
| Skeleton-motor | 46774 | 87.7* | 240688 | 80.7 | 287462 | 100 |
| Psycho-sensorial | 5409 | 10.1 | 51196 | 17.2 | 56605 | 100 |
| Cardio-respiratory | 418 | 0.8 | 3314 | 1.1 | 3732 | 100 |
| Others pathologies | 753 | 1.4* | 3053 | 1 | 3806 | 100 |
| *All* | *53354* | *100* | *298251* | *100* | *351605* | *100* |

Gender differences *p < 0.001

**Table 13.** Gender differences in body site pathologies due to work injuries (INAIL Disabilities Data Base, 2010)

| | | Women N. | % | Men N. | % | Total | % |
|---|---|---|---|---|---|---|---|
| Skeleton-motor | Head and neck | 1633 | 3.5 | 8092 | 3.3 | 9725 | 3.4 |
| | Upper limbs | 18300 | 39.1 | 104164 | 43.2* | 122464 | 42.6 |
| | Trunk and others | 7947 | 17 | 35476 | 14.7 | 43423 | 15.1 |
| | Lower limbs | 18894 | 40.4* | 92956 | 38.6 | 11850 | 38.9 |
| | *All* | *46774* | *100* | *240688* | *100* | *287462* | *100* |
| Psycho-sensorial | Psychological and neurological | 3308 | 61.2 | 21501 | 42 | 24809 | 43.8 |
| | Sensorial | 2101 | 38.8 | 29695 | 58 | 31796 | 56.2 |
| | *All* | *5409* | *100* | *51196* | *100* | *56605* | *100* |
| Cardio-respiratory | Cardio-circulatory | 122 | 29.2 | 1076 | 32.5 | 1198 | 32.1 |
| | Respiratory | 296 | 70.8 | 2238 | 67.5 | 2534 | 67.9 |
| | *All* | *418* | *100* | *3314* | *100* | *3732* | *100* |
| Other pathologies | *All* | 753 | 64.3* | 3053 | 47.9 | 3806 | 100 |
| | *Total* | *53354* | *15.2* | *298251* | *84.8* | *351605* | *100* |

Gender differences *p < 0.01–0.001

## 4 Discussion

This study shows how commuting accidents represent the second cause of disability among Italian and immigrant women after *"during work"* accidents. Previous studies have already identified commuting accident as the first cause of death at work among women. "*During work*" represents the first cause of disability and death among Italian and immigrant men [3, 8]. In this figure immigrants have the highest risk conditions as a recent Italian epidemiological study confirmed [9].

Gender and country of birth differences found in this study have not been observed in other researches. This is probably due to poor gender and state of immigration statistics at national and international level [10].

Women have less permanent affections and more temporary disabilities than men. Inequalities due to disabilities compensation should be verified in a specific study.

These disabilities are recorded and compensated when the worker needs more than 3 days to recover. Less than 3 days work accidents disabilities are not recorded and, for this reason, underestimated although they also represent an impairment towards women's health. The results show also how is osteoarticular apparatus, particularly lower limbs, the main target of women work-related injuries. During the last years, in fact, women chronic osteoarticular diseases had shown an increase with an important reduction in their healthy-life expectancy [11].

Italian and immigrant women show more disabilities due to "commuting accidents without vehicles" than men. Working women more often than men walk to or from work without using any vehicle. This exposure may lead to a higher risk [4] of slips and falls that can cause lower limb trauma and/or fracture leading to temporary or permanent disabilities [12]. Moreover, women lower limb can be weakened by osteoporosis that increases with ageing, together with varicose veins, hallux valgus and other deformations of foot [13]. Unsuitable footwear and heeled shoes, required in some typically female jobs [14], can result in a higher risk of disability when running or walking on uneven paths, with adverse (rain, lightning, hailstorm) or extreme weather events more frequent for climate change [15]. A recent Spanish study confirms that leg injuries are also prevalent in Spanish women and are mainly due to walking and/or running with consequent falls at the same or different level [2]. Furthermore, standing posture whole working time, a typical condition in women's job, increase the risk for various lower limb diseases [16].

Poor economic status, together with poor public transportation and urban mobility increase the need for longer walking paths home-work-home particularly among women [17, 18]. Longer paths with an increased risk of injury may also be due to women employment in more than one job a day. Poor and scarcely enlightened walk ways represent an additional risk for women employed in night and shift works [10, 19]. Walking path home - work - home can also increase the risk of sexual harassment particularly in night shifts. In 2014 INAIL compensated for the first time a sexual assault as "commuting accident without vehicle". This was the case of a woman cleaner who was harassed during her journey back from work [20].

Multitasking due to women work-life balance can increase the risk of commuting accidents when walking to or from work. Using a mobile phone while walking is a typical condition of a dangerous dual reconciliation task [21, 22].

Women disabilities in "commuting without vehicle" are more frequent in older age. This can be due to the retirement reform (2011) that increased the employment among older Italian women. Ageing at work is another condition of risk of commuting because of poor body balance when walking.

Single woman disability due to *commuting without vehicle* should be searched in a specific individual INAIL records analysis.

Difficulties in family and social life and the resulting psychophysical stress associated with home management, children, the elderly and the possible separation from the spouse and/or condition of singles can negatively affect the attention both in driving the vehicle and in the home-work journeys on foot, increasing the risk of injury [23, 24].

## 5 Conclusion

Being a woman represents a major risk of disability or death in commuting with or without vehicle. Women permanent disabilities are more frequent in commuting without vehicle.

A gender commuting ergonomics in terms of work organization (family/work balance, flexible working hours, limited night shift work, standing and sitting posture, dual tasks, etc.) and women's health (prevention of osteoporosis, strengthening lower limbs, wearing and affording suitable shoes for walking and working, etc.) should be considered in the welfare system in order to ameliorate healthy life expectancy and reducing social expenditure.

## References

1. European transport safety council. https://etsc.eu/etsc-announces-2017-work-related-road-safety-award-winners/. Accessed 28 May 2018
2. Camino Lopez MA, Gonzalez Alcantara OJ, Fontaneda I (2017) Gender differences in commuting injuries in Spain and their impact on injury prevention. Biomed Res Intl 2017:3834827
3. Giliberti C, Salerno S (2016) Gender differences and commuting accidents in Italy: INAIL Data Base analysis on fatalities (2009–2013). Med Lav 107(6):462–472
4. European Agency for Safety and Health at Work (2013) Schneider E. New risks and trends in the safety and health of women at work. https://osha.europa.eu/en/tools-and-publications/publications/reports/new-risks-and-trends-in-the-safety-and-health-of-women-at-work. Accessed 28 May 2018
5. http://bancadaticsa.inail.it/bancadaticsa/login.asp. Accessed 28 May 2018
6. Rossi P (2013) Raggruppamento delle menomazioni ai fini di un classamento statistico. Rivista Italiana di Medicina Legale no 3
7. http://apponline.inail.it/DisabiliApp/Login.do;JSESSIONID_APO=4_7yt2P2nNXFwXHbHabFK7JauoN-JJx5Gl2ffnXBkjJ7gUpJnbZB!-1878024728!1574823583. Accessed 28 May 2018

8. Salerno S, Brusco A, Bucciarelli A, Giliberti C (2017) La prevention des morts pour accidents trajet des femmes en Italie: un defi pour l'ergonomie. 52ème Congrès de la Société d'Ergonomie de Langue Française SELF 2017 Toulouse, 20–22 Septembre 2017, pp 373–376
9. Bena A, Giraudo M (2017) The health of foreign workers: an Italian and international priority. Recenti Prog Med 108(7):303–306
10. Schneider E. EU-OSHA (2015) Gender at work and varying forms of exposure. http://www.etui.org/Events/Women-s-health-and-work-Sharing-knowledge-and-experiences-to-enhance-women-s-working-conditions-and-gender-equality. Accessed 28 May 2018
11. Gennaro V1, Ghirga G, Corradi L (2012) In Italy, healthy life expectancy drop dramatically: from 2004 to 2008 there was a 10 years drop among newborn girls. Ital J Pediatr 38:19
12. Inail D, Mochi S (2008) Lavoro e infortuni: le differenze di genere. https://www.inail.it/cs/internet/docs/ucm_071044.pdf. Accessed 28 May 2018
13. Garcia MG, Graf M, Läubli T (2017) Lower limb pain among workers: a cross-sectional analysis of the fifth European Working Conditions Survey. Int Arch Occup Environ Health 90(7):575–585
14. Blanchette MG, Brault JR, Powers CM (2011) The influence of heel height on utilized coefficient of friction during walking. Gait Posture 34(1):107–110
15. Lundgren K1, Kuklane K, Gao C, Holmér I (2013) Effects of heat stress on working populations when facing climate change. Ind Health 51(1):3–15
16. Hours M, Fort E, Charbotel B, Chiron M (2001) Jobs at risk fo work-related road crashes: an analysis of the causalities from the Rhone Road Trauma Registry (France). Saf Sci 49:1270–1276
17. Morency P, Strauss J, Pépin F, Tessier F, Grondines J (2018) Traveling by bus instead of car on urban major roads: safety benefits for vehicle occupants, pedestrians, and cyclists. J Urban Health 95:196–207
18. Stevenson M, Thompson J, de Sá TH, Ewing R, Mohan D, McClure R, Roberts I, Tiwari G, Giles-Corti B, Sun X, Wallace M, Woodcock J (2016) Land use, transport, and population health: estimating the health benefits of compact cities. Lancet 388(10062):2925–2935
19. Hill JD, Boyle LN (2007) Driver stress as influenced by driving maneuvers and roadway conditions. Transp Res Part F Traffic Psychol Behav 10:177–186
20. Salerno S, Dimitri L, Livigni L, Magrini A, Talamanca IF (2015) Mental health in hospital. Analysis of conditions of risk by department, age and gender, for the creation of best practices for the health of nurses. G Ital Med Lav Ergon 37(1):46–55
21. Giliberti C, Figà Talamanca I, Salerno S (2014) Ergonomic design for young users of mobile phones. In: Rebelo F, Soares M (eds) Advances in ergonomics in design, usability & special populations part II. Applied human factors and ergonomics, AHFE conference 2014
22. Lee JH, Lee MH (2018) The effects of smartphone multitasking on gait and dynamic balance. J Phys Ther Sci 30(2):293–296
23. Costa G, Pickup L, Di MV (1988) Commuting-a further stress factor for working people: evidence from the European Community. I. A review. Int Arch Occup Environ Health 60 (5):371–376
24. Santamariña-Rubio E, Pérez K, Olabarria M, Novoa AM (2014) Gender differences in road traffic injury rate using time travelled as a measure of exposure. Accid Anal Prev 65:1–7

# Telecommuting in Academia – Associations with Staff's Health and Well-Being

Marina Heiden[1(✉)], Linda Richardsson[1], Birgitta Wiitavaara[1], and Eva Boman[2]

[1] Centre for Musculoskeletal Research, Department of Occupational and Public Health Sciences, Faculty of Health and Occupational Studies, University of Gävle, 801 76 Gävle, Sweden
marina.heiden@hig.se

[2] Department of Social Work and Psychology, Faculty of Health and Occupational Studies, University of Gävle, 801 76 Gävle, Sweden

**Abstract.** The ability to telecommute has changed working life for staff at universities and colleges. Although the opportunity to work away from the office at any time gives workers more freedom to manage their work, it also imposes higher demands on workers to set limits to their work. The aim of this ongoing study is to determine if there is an optimal amount of telecommuting for male and female academics with respect to perceived health, work stress, recovery, work-life balance, and work motivation. A web-based survey is currently being conducted among lecturers and professors at Swedish universities and colleges. Results so far show that perceived fatigue and stress associated with indistinct organization and conflicts are higher among academics that telecommute to a larger extent. The results also show that female academics are more fatigued and stressed at work than male academics, but this does not seem to be related to the extent of telecommuting performed.

**Keywords:** Telework · Lecturer · Work stress

## 1 Introduction

The ability to telecommute has changed working life for staff at universities and colleges. Although the opportunity to work away from the office at any time gives workers more freedom to manage their work, it also imposes higher demands on workers to set limits to their work. To date, little is known about how telecommuting affects teaching and research staff, and how it should be practiced to maintain healthy, productive and motivated personnel. In an ongoing study, we aim to determine if there is an optimal amount of telecommuting for male and female academics with respect to perceived health, stress, recovery, work-life balance, and work motivation.


S. Bagnara et al. (Eds.): IEA 2018, AISC 826, pp. 308–312, 2019.
https://doi.org/10.1007/978-3-319-96065-4_35

**Table 1.** Background characteristics of the participants.

| | | Proportion | Mean | SD |
|---|---|---|---|---|
| Age (years) | | | 49.0 | 9.8 |
| Gender | Male | 38.5 | | |
| | Female | 61.5 | | |
| Marital status | Living alone | 21.7 | | |
| | Living with partner | 78.3 | | |
| Children at home | | 52.8 | | |
| Profession | Adjunct | 45.6 | | |
| | Lecturer | 45.0 | | |
| | Professor | 9.4 | | |
| Type of employment | Temporary | 12.9 | | |
| | Permanent | 87.1 | | |
| Extent of employment (%) | | | 94.3 | 17.4 |
| Work content | Teaching (%) | | 62.2 | 31.4 |
| | Research (%) | | 22.9 | 25.4 |
| | Management (%) | | 7.8 | 17.4 |
| | Other (%) | | 10.5 | 14.1 |
| Commuting time (min) | | | 67.5 | 91.3 |
| Extent of telecommuting | Less than once/month | 16.2 | | |
| | Several times/month | 41.7 | | |
| | Several times/week | 42.1 | | |

SD = standard deviation.

## 2 Methods

A web-based survey is conducted among lecturers and professors at Swedish universities and colleges. The questionnaires included in the survey are the General Health Questionnaire [1] for assessing health, Work Stress Questionnaire [2] for assessing work-related stress, validated items for assessing recovery [3], parts of Copenhagen Psychosocial Questionnaire [4] for assessing work-life balance, and the Basic Need Satisfaction Scale [5] for assessing work motivation. In addition to these questionnaires, background questions are included about age, gender, family situation, type and extent of employment, work content, commuting time and the extent of telecommuting performed. Background characteristics of the participants included so far (n = 309) are shown in Table 1. Data have been analyzed with multivariate (MANOVA) and univariate (ANOVA) analyses of variance to determine whether ratings of health, stress, recovery, work-life balance, and work motivation differ depending on the extent of telecommuting performed. The level of significance was set to $p < 0.05$.

# 3 Results

In crude models with the single independent variable *telecommuting* (less than once per month/several times per month/several times per week), the MANOVA was significant (Table 2). ANOVAs showed that this difference was attributed to differences in fatigue and stress associated with indistinct organization and conflicts (Table 3). When *gender* was added to the MANOVA model, it too was significant, but there was no interaction effect between *gender* and *telecommuting* (Table 2). In ANOVAs, the results for *telecommuting* remained similar, and *gender* differences were identified in fatigue as well as stress associated with indistinct organization and conflicts, individual demands and commitment, and influence at work (Table 3). After adjusting for age, type of employment, commuting time and amount of teaching time, the MANOVA results did not change, and marginally different results were found in the ANOVAs (Tables 2 and 3). Among the covariates, age, type of employment, and amount of teaching time were significant.

**Table 2.** Results from multivariate analyses of variance.

| | Tcom | Gender | Tcom*Gender | Age | EmployType | ComTime | TeachTime |
|---|---|---|---|---|---|---|---|
| Wilks' λ | 0.889 | | | | | | |
| F | 1.6 | | | | | | |
| P | **0.035** | | | | | | |
| Wilks' λ | 0.872 | 0.891 | 0.912 | | | | |
| F | 1.9 | 3.2 | 1.3 | | | | |
| P | **0.009** | **<0.001** | 0.192 | | | | |
| Wilks' λ | 0.854 | 0.915 | 0.907 | 0.917 | 0.926 | 0.966 | 0.924 |
| F | 2.2 | 2.4 | 1.3 | 2.4 | 2.1 | 0.9 | 2.2 |
| P | **0.002** | **0.006** | 0.139 | **0.021** | **0.021** | 0.517 | **0.018** |

Significant differences are shown in bold. Tcom = telecommuting (less than once per month/several times per month/several times per week); Tcom*Gender = interaction between telecommuting and gender; EmployType = type of employment (temporary/permanent position); ComTime = commuting time (minutes); TeachTime (% teaching).

Significant differences are shown in bold. Tcom = telecommuting (less than once per month/several times per month/several times per week); Tcom*Gender = interaction between telecommuting and gender; EmployType = type of employment (temporary/permanent position); ComTime = commuting time (minutes); TeachTime (% teaching); Health = index from General Health Questionnaire; Autonomy, Competence, Relatedness = indices from Basic Need Satisfaction Scale; Org/Conflict, Dem/Com, Influence, W-L interfer = indices of stress associated with indistinct organization and conflicts, individual demands and commitment, influence at work and work to leisure time interference from Work Stress Questionnaire; Rest, Fatigue = indices of recovery; W-L balance = index of work-life balance from Copenhagen Psychosocial Questionnaire.

**Table 3.** Results from univariate analyses of variance (p-values).

| | Tcom | Gender | Tcom*Gender | Age | EmployType | ComTime | TeachTime |
|---|---|---|---|---|---|---|---|
| Health | 0.803 | | | | | | |
| Autonomy | 0.900 | | | | | | |
| Competence | 0.417 | | | | | | |
| Relatedness | 0.341 | | | | | | |
| Org/Conflict | **0.019** | | | | | | |
| Dem/Com | 0.415 | | | | | | |
| Influence | 0.440 | | | | | | |
| W-L interfer | 0.478 | | | | | | |
| Rest | 0.176 | | | | | | |
| Fatigue | **0.005** | | | | | | |
| W-L balance | 0.285 | | | | | | |
| Health | 0.713 | 0.632 | 0.151 | | | | |
| Autonomy | 0.979 | 0.232 | 0.215 | | | | |
| Competence | 0.519 | 0.621 | **0.022** | | | | |
| Relatedness | 0.267 | 0.160 | 0.499 | | | | |
| Org/Conflict | **0.027** | **0.002** | 0.800 | | | | |
| Dem/Com | 0.401 | **<0.001** | 0.827 | | | | |
| Influence | 0.541 | **0.030** | 0.632 | | | | |
| W-L interfer | 0.508 | 0.389 | 0.987 | | | | |
| Rest | 0.175 | 0.102 | 0.796 | | | | |
| Fatigue | **0.004** | **0.006** | 0.699 | | | | |
| W-L balance | 0.368 | 0.293 | 0.525 | | | | |
| Health | 0.441 | 0.879 | 0.118 | **0.017** | 0.616 | 0.081 | 0.797 |
| Autonomy | 0.768 | 0.408 | 0.211 | 0.054 | 0.824 | **0.047** | 0.652 |
| Competence | 0.269 | 0.598 | **0.013** | 0.158 | **0.019** | 0.439 | 0.157 |
| Relatedness | 0.578 | 0.197 | 0.561 | 0.411 | 0.080 | 0.690 | 0.518 |
| Org/Conflict | **0.023** | **0.010** | 0.892 | 0.083 | **0.028** | 0.421 | 0.073 |
| Dem/Com | 0.435 | **0.004** | 0.752 | **<0.001** | **0.040** | 0.199 | 0.068 |
| Influence | 0.728 | 0.066 | 0.643 | 0.452 | 0.495 | 0.155 | 0,300 |
| W-L interfer | 0.631 | 0.462 | 0.948 | 0.236 | 0.339 | 0.155 | **0.034** |
| Rest | 0.424 | 0.287 | 0.784 | **0.005** | 0.300 | **0.033** | 0.205 |
| Fatigue | **0.007** | **0.032** | 0.663 | **0.004** | 0.454 | 0.808 | 0.748 |
| W-L balance | 0.328 | 0.479 | 0.503 | 0.060 | 0.062 | 0.253 | **0.020** |

## 4 Conclusion

The preliminary findings suggest that perceived fatigue and stress associated with indistinct organization and conflicts increase as the extent of telecommuting becomes larger, but their cause and effect relationship cannot be deduced.

## References

1. Hardy GE, Shapiro DA, Haynes CE, Rick JE (1999) Validation of the General Health Questionnaire-12 using a sample of employees from England's health care services. Psychol Assess 11(2):159–165
2. Holmgren K, Hensing G, Dahlin-Ivanoff S (2009) Development of a questionnaire assessing work-related stress in women–identifying individuals who risk being put on sick leave. Disabil Rehabil 31(4):284–292
3. Lindfors P (2002) Psychophysiological aspects of stress, health and wellbeing in teleworking women and men. Dissertation, Stockholm University, Sweden
4. Pejtersen JH, Kristensen TS, Borg V, Bjorner JB (2010) The second version of the Copenhagen Psychosocial Questionnaire. Scand J Pub Health 38(3 suppl):8–24
5. Deci EL, Ryan RM (2000) The 'what' and 'why' of goal pursuits: human needs and the self-determination of behavior. Psychol Inq 11:319–338

# Meghalaya Tourism: A Study on Women's Attitudes and Perceptions Towards the Cultural Exposure and Interaction in the Context of Meghalaya Tourism

Wanrisa Bok Kharkongor(✉), Abhirup Chatterjee, and Debkumar Chakrabarti

Department of Design, Indian Institute of Technology, Guwahati, India
wanrisa89@gmail.com, abhirup2k4@gmail.com, dc@iitg.ernet.in

**Abstract.** In a matriarchal society like Meghalaya where women are given high value of importance and independence it is seen that their involvement and exposure to work in various spheres is increasing over the years. The matriarchal way of thinking and upbringing has made the women in Meghalaya to inherit a confident, amiable and open-minded persona, which makes them more likely to interact effortlessly with people. There is very less research done on women in a matriarchal society in the field of tourism. There is need to understand and interpret their behavioural patterns and extent of willingness to share their culture with the tourists. Data Collection on 120 indigenous men and women of Meghalaya was done to find out their views on the context of Meghalaya Tourism and the extent of how culture is shared as well as how the level of interaction between the natives and tourists is achieved through tourism was done.

The Results showed that the women natives of Meghalaya have a positive approach to interact and share their culture with the tourists but have no platform to do so. This shows that there is a need to design an interactive platform where culture which is the essence of a place can be shared to the tourists and provide them with a fulfilling and excellent authentic experience that they can carry when they leave the place.

**Keywords:** Meghalaya tourism · Women behavioural patterns
Cultural exposure · Interaction

## 1 Introduction

Tourism is one of the world's largest and fastest growing industries. In many countries it acts as an engine for development through foreign exchange earnings and the creation of direct and indirect employment. Tourism contributes 5% of the world's GDP and 7% of jobs worldwide. It accounts for 6% of the world's exports and 30% of the world's exports in services. Research shows the different ways in which tourism can contribute to economic growth, poverty reduction and community development. However, less attention has been paid to the unequal ways in which the benefits of


S. Bagnara et al. (Eds.): IEA 2018, AISC 826, pp. 313–326, 2019.
https://doi.org/10.1007/978-3-319-96065-4_36

tourism are distributed between men and women, particularly in the developing world. [1] Studies have shown the role of women in their workplace but fewer studies have shown the role of women in the tourism sector. In the tourism industry, the percentage of women who work in the industry is high, but their function is dominated by unskilled, low-paid jobs [2] Tourism jobs for women are frequently low-skilled; part-time; seasonal; involve long, irregular and unsocial hours; and are characterised by subcontracting and flexible/casual working [2–5].

Tourism also enables women to interact with people from outside their own community, increasing their knowledge of gender roles in other contexts and potentially leading them to reflect on their own situation [2] Tourism involves processes which are constructed out of complex and varied social realities and relations that are often hierarchical and usually unequal. Gender relations are one element of this complex. Whether we examine divisions of labour, the social construction of landscape (both natural and human influenced), how societies construct the cultural 'other', or the realities of the experiences of tourist and host it is possible to examine issues of relationships, differences and inequalities resulting from tourism-related processes in terms of gender relations. This allows us to concentrate on women's and men's differential experiences, constructions and consumption of tourism [6].

Past literature highlights how creating a strong relationship between tourism and culture can help destinations become more attractive as well as more competitive. Culture is increasingly an important element of the tourism product, which also creates distinctiveness in a crowded global marketplace. At the same time, tourism provides an important means of enhancing culture and creating income, which can support and strengthen cultural heritage, cultural production and creativity. Creating a strong relationship between tourism and culture can therefore help destinations to become more attractive and competitive as locations to live, visit, work and invest in [7–10].

Recently, many researchers to look at individuals' social gains and impacts from tourism development, evaluating their participants' quality of life [11] or looking at how much their participants support tourism development [12] have used social exchange theory. However, there has been limited literature addressing a deeper understanding of social gains by assessing how these may be gendered [9].

The interaction of women with tourists is minimal. [13] Socio-structural and cultural factors can influence how women perceive tourism. Rural Women in rural Tourism are a more recent research interest in the Tourism literature than that of the research on women and tourism in general. [14] This paper tries to explore women's experiences of tourism and their perception towards tourism in the rural areas of Meghalaya. It also tries to compare the psychosocial factors between women and men who are exposed to the tourists in terms of their interaction as well as in sharing their culture.

# 2 Methodology

## 2.1 Survey Design

In the current study, qualitative research methods were employed to attain an understanding of the behavior and attitudes of the women of Meghalaya who are exposed to the tourists. This study was done on the famous tourist spots of Meghalaya.

A structured and standardized questionnaire was designed to fulfill the objectives stated above. The questionnaire was divided into sections which will be described later. Reference was taken from a few papers that focus mainly on the natives' behavior and attitude towards tourism and on the tourist as a whole. Data regarding the tourists' perception and attitude of the place and the people; as well as the impact of tourism on the socio-economic status of the state from the natives' perspective were recorded in the questionnaire. Information regarding the demographic profile of respondents was recorded which includes their name, age, education, tribe, annual income and residence. On a 5-point Likert scale (where 1 represents strongly disagree, 2 disagree, 3 neither agree nor disagree, 4 agree and 5 strongly agree), the natives rated their responses to the overall tourism service which includes their outlook, experience, perception, impacts and influence of tourism on the state. Apart from the questionnaires a face-to-face interview consisting of a set of open-ended questions was conducted with the female officials working in the tourism department of the state. Responses on whether a need for interaction and cultural exposure is entailed in the tourism sector were recorded.

The questionnaire tries to extract the information from the respondents for determining and analyzing the physiological factors that affect the women's behavior and attitudes towards tourism and their readability to interact with the tourists. The same is being tested for reliability using Cronbach's alpha. After validation of the reliability of the questionnaire, they were scored in terms of the number of responses; and the generated data then underwent independent sample T test. For every case, statistical significance was set at $P \leq 0.05$.

A sample size of n = 120 Natives of Meghalaya, with 60 males and 60 females including all the three tribes Khasi, Jaintia and Garo were selected through non-probability sampling.

The data was collected from the famous tourist spots, which includes Cherrapunjee, Mawlynnong and the capital, Shillong city. The collected data were subjected to appropriate statistical analyses using Statistical Package for Social Sciences (SPSS for windows) and GraphPad InStat 3 to explore any significant change in opinion across time or location.

## 2.2 Questionnaire Design

The questionnaires designed for the natives of Meghalaya was divided into four sections:

excluding the section which includes questions based on the basic information to get the demographic data [15–17].

The sections are as follows:

Section I: Natives' attitudes and perception on Meghalaya Tourism
Section II: Impacts of Tourism
Section III: Tourists' interaction with the natives
Section IV: General Information

The period to complete the data collection is three months. The data was collected from all the three tribes of Meghalaya which includes Khasi Jaintia and Garo. This was done in East Khasi Hills (Mawlynnong. Cherrapunjee, Shillong), West Khasi Hills, East Jaintia Hills (Nartiang, Sohshrieh) and on the Garos residing in the capital city Shillong. Data was also collected from the government officials who work in the Tourism sector and also from the natives who are in touch with the tourists on a daily basis. A soft copy of the questionnaire was also sent through mail to the natives. Hotels and resorts where data was collected include the following (Fig. 1 and Table 1):

**Table 1.** Hotels and resorts participating in the survey

| Sl. no | A. Cherrapunjee | B. Laban | C. Police Bazaar |
|---|---|---|---|
| 1. | Jiva Resort | Pine Brook | Pine Borough |
| 2. | Me-Me ai Resort | Bonnie Resort | Hotel Heritage |
| 3. | Cordial Resort | Nalangri Resort | Silk Route |
| 4. | Coniferous Resort | | Pinewood Hotel |
| 5. | Roots (Orchid Hotel) | | Pegasus Hotel |
| 6. | | | Hilltop Hotel |

A descriptive and comparative analysis was conducted on the data by statistical means of analysing the data obtained from the respondents. Therefore, the results that are outlined and discussed include a summary of the demographic profiles of the female age groups and the percentage distribution of the tribes, as well as plotting and discussing the means of the responses and comparing them with the age group and gender. The independent sample T-Test done on both the male and female subjects is depicted in the results of the study.

**Fig. 1.** From top: Tourism Department of Meghalaya, Resorts and restaurants where data was collected

# 3 Results and Discussion

The reliability of the questionnaires regarding the women's perception and attitude towards tourism and tourists; as well as the impact of tourism was examined by subjecting them through Cronbach's Alpha ('$\alpha$') for their reliability and internal consistency of the scales. An $\alpha > 0.89$ is considered to have an excellent reliability and validity, while an $\alpha > 0.79$ is very good and $\alpha > 0.69$ is considered average (not acceptable enough).

The questionnaire given to the natives showed good reliability with the Cronbach's Alpha of 0.856, ($\alpha$ based on standardized items = 0.864 for 22 questions). The high Cronbach Alpha values for the questionnaires show a good internal consistency of the questionnaire and shows how well and reliable the questions are to test or measure what it should.

Content analysis was conducted to show the differences between male and female responses to tourism. The data collected also showed behavioural attributes of women belonging to different tribes as well as different age groups. It is expected to see that women of Meghalaya are as much open towards tourism and the tourists as much as male are.

## 3.1 Summary of the Demographic Profile of the Respondents

In the demographic composition of the sample surveyed, questions relating to the tribe and age of the respondents were asked to find out how each tribe differed in their behaviour towards tourism and also how the responses varied with the age range.

The survey showed a high percentage of respondents in the age group of 18 to 25. The respondents lack the basic qualification in the field of Tourism but are exposed to tourists visiting the place. A few are owners of hotels, guesthouses and bed and breakfasts others work in the government sector of Tourism (Fig. 2).

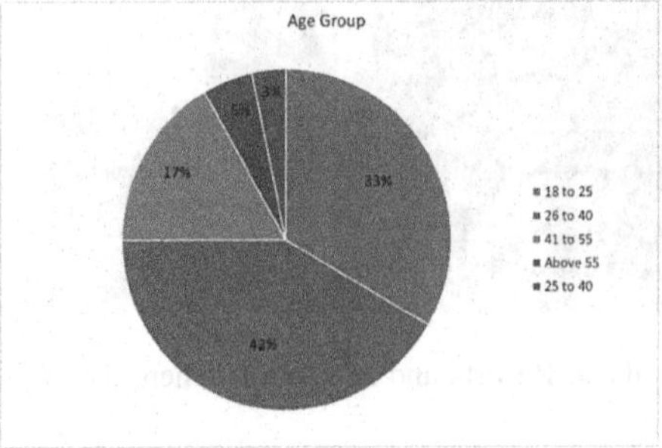

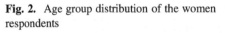

Fig. 2. Age group distribution of the women respondents

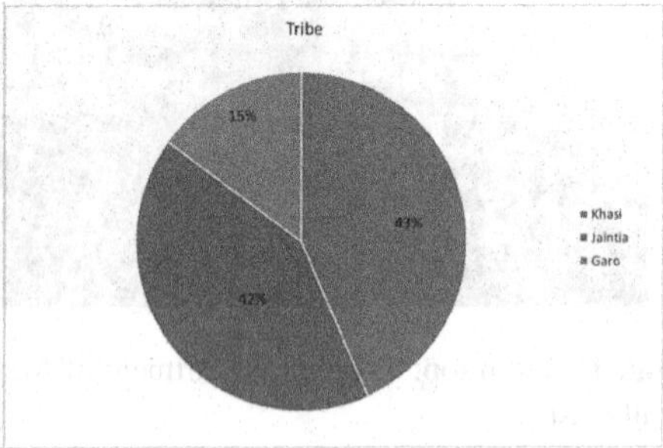


Fig. 3. Tribal distribution

**Means and Percentage Distribution of the Response Based on Gender**

The subjects were asked to rate the overall Tourism Sector in general in a likert scale of 1 to 5. "1" being Very Poor to "5" Very Good. In this first section, they were asked to rate, the following in the Likert scale:

a. Interaction with the natives of Meghalaya
b. Support from the local community
c. Contributions towards the society
d. Facilitations from the Government/NGOs.

The comparative study between male and female as shown in Fig. 3 showed that the means are very close for both the genders. In terms of interaction with the tourists, female are more interactive and open in communicating with the tourists compared to the males. Both Genders agree that the local community support tourism in the state. This shows that tourism has a potential to grow in hilly areas as well. An interview was conducted with the headman of Mawlynnong Village, which is known to be the cleanest village in Asia. He quoted "*Tourism has brought an immense contribution to the village in terms of infrastructure and the overall development of the place*" he also quoted that *"There is a big scope of job opportunities because of Tourism, locals are able to open shops and restaurants"*. Not only has tourism exposed the place to visitors but it has also contributed to the society as whole (Fig. 4).

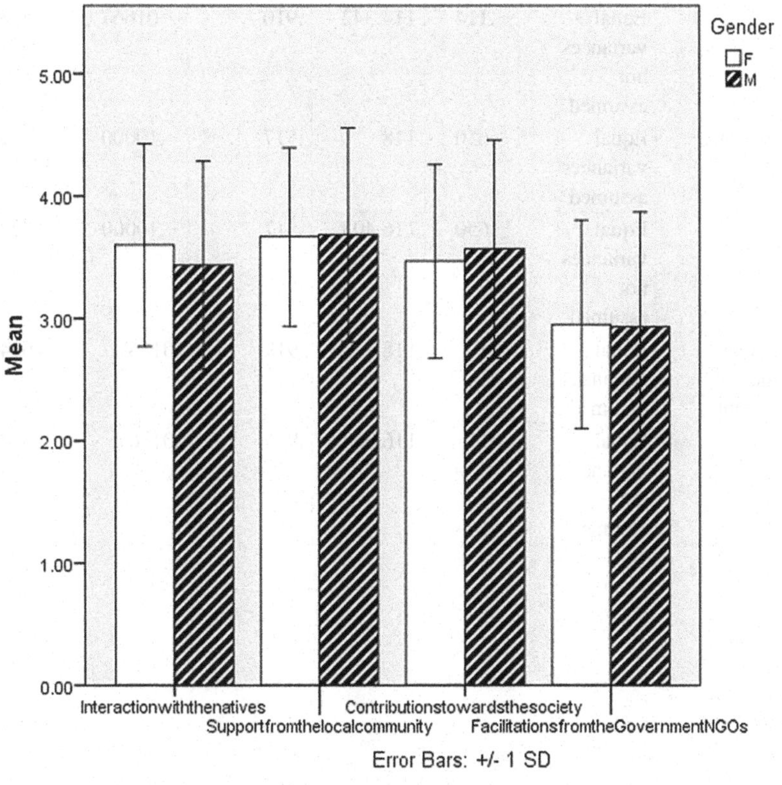


**Fig. 4.** Mean plot of Gender

**Independent Samples T-Test on Male and Female Samples**

The independent sample T-test done on both the samples showed the significance to be less that the significant value p = 0.05 hence accepting the Null Hypothesis that both men and women of Meghalaya attitudes and perception are the same towards the tourism industry (Tables 2 and 3).

**Table 2.** Independent Samples T-Test on male and female samples on overall perception of Tourism

| | | T-test for equality of means | | | | |
|---|---|---|---|---|---|---|
| | | t | df | Sig. (2-tailed) | Mean difference | Std. error difference |
| Interaction with the natives | Equal variances assumed | 1.088 | 118 | .279 | .16667 | .15324 |
| | Equal variances not assumed | 1.088 | 117.907 | .279 | .16667 | .15324 |
| Support from the local community | Equal variances assumed | −.114 | 118 | .910 | −.01667 | .14683 |
| | Equal variances not assumed | −.114 | 114.342 | .910 | −.01667 | .14683 |
| Contributions towards the society | Equal variances assumed | −.650 | 118 | .517 | −.10000 | .15374 |
| | Equal variances not assumed | −.650 | 116.403 | .517 | −.10000 | .15374 |
| Facilitations from the Government NGOs | Equal variances assumed | .102 | 118 | .919 | .01667 | .16346 |
| | Equal variances not assumed | .102 | 116.968 | .919 | .01667 | .16346 |

**Table 3.** Independent Samples T-Test on male and female samples on cultural exposure and interaction

| | | T-test for equality of means | | | | |
|---|---|---|---|---|---|---|
| | | t | df | Sig. (2-tailed) | Mean difference | Std. error difference |
| Are you interested in cultural exchange/sharing your culture with others | Equal variances assumed | −1.750 | 118 | .083 | −.26667 | .15241 |
| | Equal variances not assumed | −1.750 | 114.233 | .083 | −.26667 | .15241 |
| What level of interaction have you had with the tourists so far | Equal variances assumed | −1.723 | 118 | .088 | −.31667 | .18380 |
| | Equal variances not assumed | −1.723 | 117.437 | .088 | −.31667 | .18380 |
| If there are provisions for entertainment oriented activities at any tourist spot which can serve as a platform for interaction (cultural exchange) shall you be interested to participate | Equal variances assumed | −.356 | 118 | .723 | −.06667 | .18733 |
| | Equal variances not assumed | −.356 | 115.644 | .723 | −.06667 | .18733 |

## Means and Percentage Distribution of the Response Based on Age Group of Female Respondents

Tourism can bring a change in the community and landscape/environment of place; however, there is also a need to understand that displaying ethnicity is a concern. The term 'cultural tourism' is frequently used in this sense. Some of this takes the form of visits to societies that have not been affected by industrialization and western commercial values, something that may represent a 'vanishing life-style'. Walle (1998) uses the term 'cultural tourism' to refer solely to that, which relates to the culture of 'ethnic groups and hinterland peoples' living usually in small-scale societies, relatively untouched by western values. [18]. Meghalaya is situated in the North-Eastern part of India and because of the hilly terrain and lack of transportation facilities, tourists influx is least expected. Nevertheless, the natives who are exposed to the tourists have a hope of sharing their culture. The data showed that 83% of the female subjects are interested to share their culture. The mean of the responses of all the age groups was also plotted showing the decrease of interest to share the culture with age.

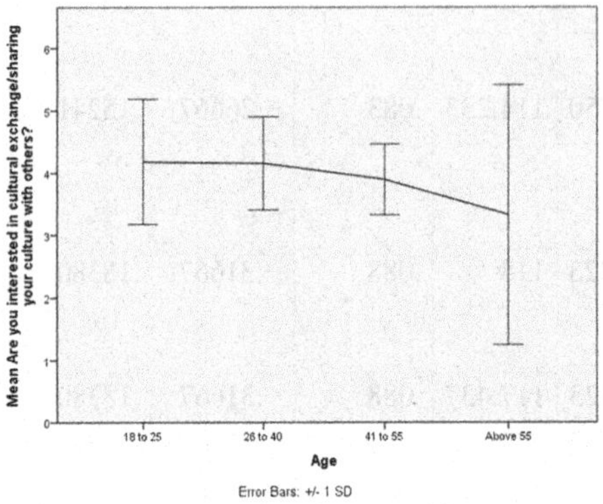


**Fig. 5.** Mean of cultural exchange based on age

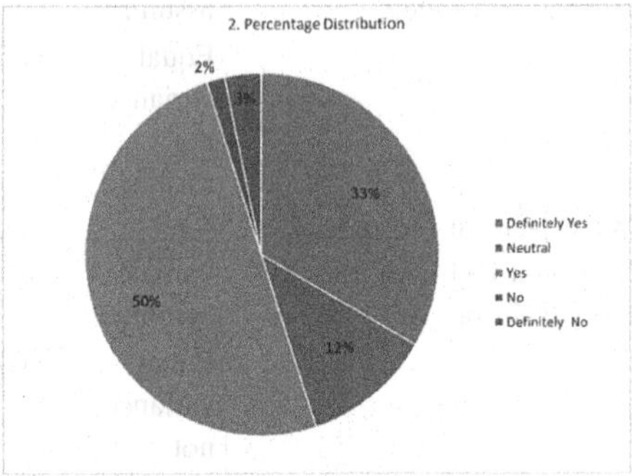


**Fig. 6.** Percentage distribution female response for cultural exchange

Tourism can have both a positive and negative impact on the community. Literature supports the belief that the tourism industry provides a range of income-generating opportunities for women.

The debate on impact of tourism on women generally is limited to seeing women as victims, in terms of either sex work or advertising. Nevertheless, few attempts have been made to examine the particular experience of women as hosts, entrepreneurs, craftspeople or even as observers of the tourist scene [19] At every tourist spot there is always an opportunity for the local women to open food stores, cafes, hotels and resorts. Especially for women earning on a daily wage. It is seen that there is a distinctive stockpile of crops produce, which are harvested extensively for profit or subsistence. Handicrafts and Handloom also finds a place at the tourist spots, as every tourist loves to carry a token of remembrance of the place. In Fig. 5 when asked about the impacts of tourism in the local community it is seen that in terms of job opportunities the tourism industry has brought a significant impact. As stated earlier as tourism in Meghalaya is centred mainly in the wettest places on earth Mawsynram and Cherrapunjee, the cleanest village of Asia Mawlynnong and areas around it like Dawki so the environment is less affected in other places because of tourism (Figs. 6 and 7).

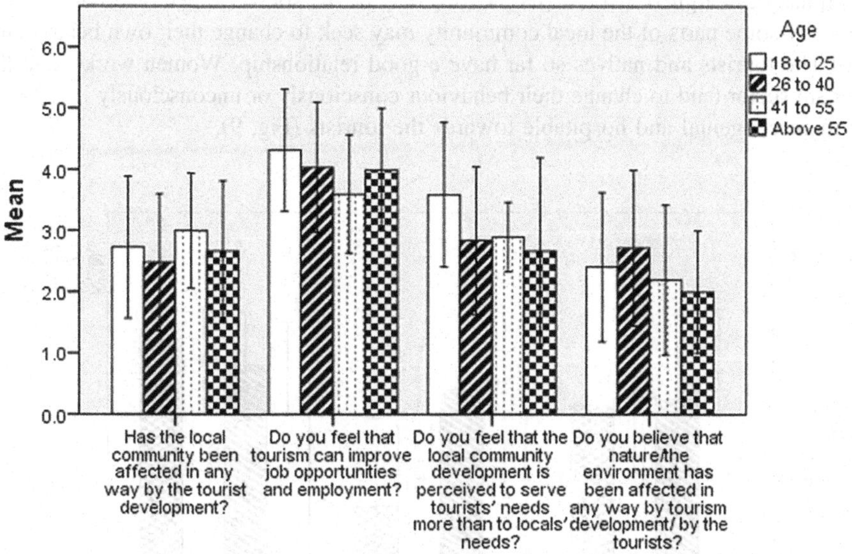


**Fig. 7.** Mean of responses on the impacts of tourism on the local community

As these women are in contact with the tourists, it is important to understand how they feel about their personal safety as well as their overall relationship with the tourists. The means of the responses regarding their sense of safety show a lower score on the Likert scale. There is lack of security measures provided for women workers which in turn reflects on their overall experience with the tourists (Fig. 8).

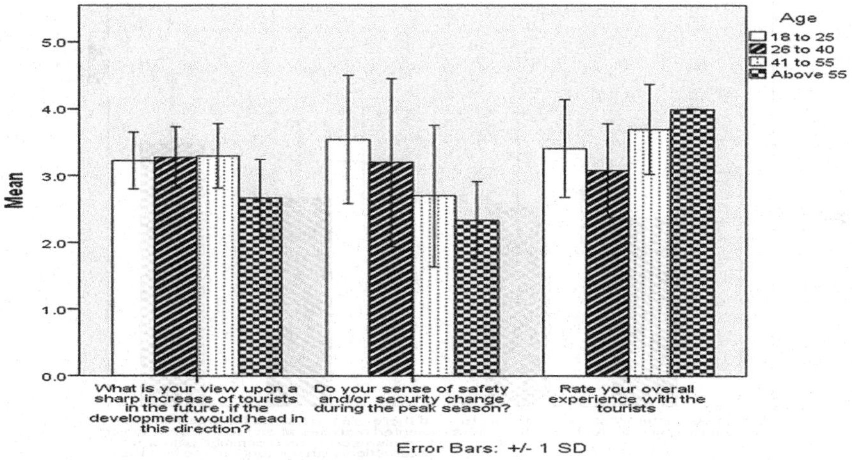


**Fig. 8.** Mean of responses on safety, overall experience with the tourists

It has been argued that tourists' actions may be the source of demonstration effects whereby some parts of the local community may seek to change their own behaviours. [20] The tourists and natives so far have a good relationship. Women working in the tourism sector tend to change their behaviour consciously or unconsciously in order to be more congenial and hospitable towards the tourists (Fig. 9).

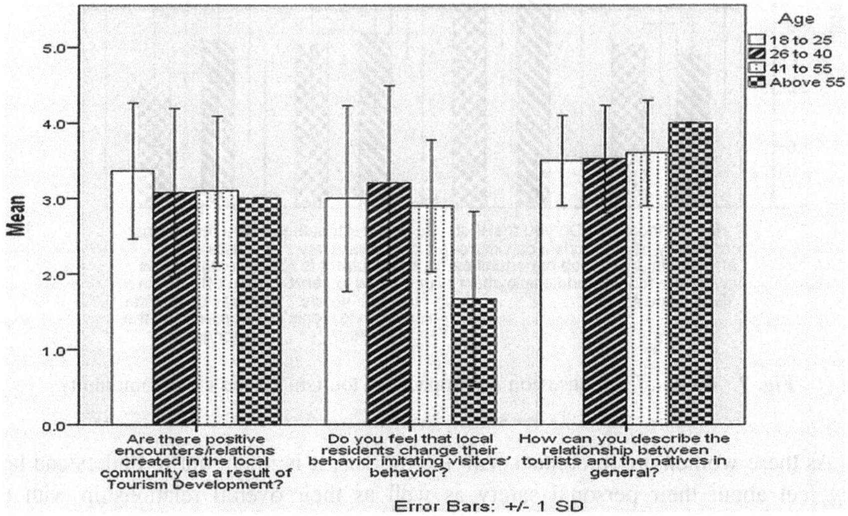


**Fig. 9.** Mean of responses on relationship between tourists and natives

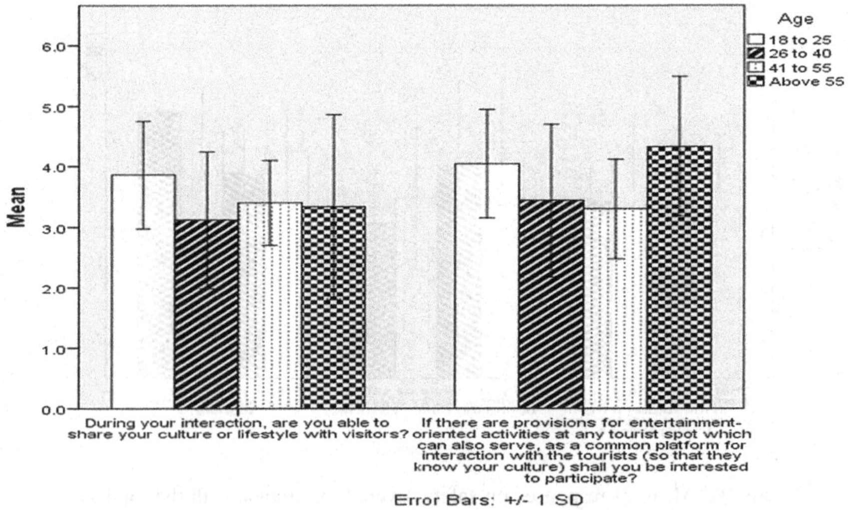


**Fig. 10.** Mean of responses on cultural exchange and interaction

During the interaction between the tourists and the women natives the extent of cultural exposure was recorded and analyzed, the means of the responses are shown below. The bar diagram shows that sharing of culture during the interaction is moderate. The second bar shows a relatively positive response when the natives were asked if they would be interested to participate in a platform where they can interact and share their culture. It can be inferred that the reason behind the lack of cultural exposure is not the lack of interest from the natives but it can be because of the lack of a platform (Fig. 10).

## 4 Conclusion

As stated earlier the fact that Meghalaya being a matriarchal society where women are socially privileged and take charge and control of the property in a family, reflects on the responses given by the female natives when interacting with the tourists. The independent sample T-Test on both the male and female samples showed negligible differences between the two. Thus accepting the Null Hypothesis. Women in the tourism industry are as very much unconstrained, confident and forthright when communicating and interacting with the tourists as much as men are. This provides a comfortable and hospitable tourist environment a lot more conducive. There is also a remarkable potential of job opportunities for women in Tourism for all age ranges as noted from the high scoring responses. This can be one way of encouraging new and more unexplored places in the state to be supportive and open towards tourism.

Given below are a few features from the responses that can be taken into consideration:

1. Lack of security and safety measures for women workers
2. Skill Training and development for women working in the tourism sector
3. Lack of model or a conceptual framework in understanding how culture can be displayed in a limited period of time
4. Lack of platform for interaction and cultural exchange to outsiders
5. A need to bring an awareness of the tangible and non-tangible developments in the community as a result of tourism
6. Communication problem of women workers, results in the loss of business as quoted by one of the subjects *"A translator should be arranged beforehand so that the interaction becomes easy to understand, hence also creating employment"*

Considering all these factors there is a need and more scope to present a collaborative system to bring about an improved satisfaction for both the natives as well as the tourists. There is scope to perform an in depth study on the psychosocial factors and constraints of women workers towards sharing their culture and preserving it.

## References


1. Global Report on Women in Tourism 2010 Preliminary Findings (2011) World Tourism Organization (UNWTO) and the United Nations Entity for Gender Equality and the Employment of Women (UN Women), pp 2–8
2. Cave P, Kilic S (2010) The role of women in tourism employment with special reference to Antalya, Turkey. J Hosp Market Manage 19:1–7
3. Dumbrăveanu D, Light D, Young C, Chapman A (2016) Exploring women's employment in tourism under state-socialism: experiences of tourism work in socialist Romania. Tour Stud 16:4–20
4. Chant S (1997) Gender and tourism employment in Mexico and the Philippines. In: Sinclair T (ed) Gender, work and tourism. Routledge, London, pp 120–179
5. Scott J (1997) Chances and choices: women and tourism in Northern Cyprus. In: Sinclair T (ed) Gender, work and tourism. Routledge, London, pp 60–90
6. Kinnaird V (1996) Issues of women in tourism development debate: an understanding derek hall. Tour Manage 17(2):6–14
7. Alexandros A (2003) The convergence process in heritage tourism. Ann Tour Res 30(4): 795–812
8. Arzeni S (2009) Foreword in the impact of culture on tourism. OECD, Paris, pp 1–17
9. Kladou S (2013) Cultural destinations and the role of gender in sustainable tourism development: focusing on handicraft entrepreneurs. In: Critical tourism conference V, Sarajevo, pp 3–9
10. Palmer C (1999) Tourism and the symbols of identity. Tour Manage 20:313–321
11. Tovar C, Lockwood M (2008) Social impacts of tourism: an Australian regional case study. Int J Tour Res 10(4):365–378
12. Huh C, Vogt CA (2008) Changes in residents' attitudes toward tourism over time: a cohort analytical approach. J Travel Res 46(4):446–455
13. Suhitra C (1998) Issues of women in tourism development debate: an understanding. Equations, pp 35–39
14. Ahuja B, Cooper M (2004) Women's role in Indian rural tourism: towards a social-infrastructure model for rural development, pp 1–8
15. Szell AB (2012) Attitudes and perceptions of local residents and tourists toward the protected area of Retezat National Park, Romania, Master's theses. Western Michigan University, Kalamazoo, Michigan, pp 10–118
16. Skipper TL (2009) Understanding tourist-host interactions and their influence on quality tourism experiences, Master theses. Wilfrid Laurier University, pp 22–143
17. Marzuki A (2012) Local residents' perceptions towards economic impacts of tourism development in Phuket, vol 60, no 2, pp 199–212
18. Hughes HL (2002) Culture and tourism: a framework for further analysis. Manag Leisure 7:164–175
19. Anisah D, Vyasha H, Frinwei NA (2016) Women in tourism: experiences and challenges faced by owners of small accommodation establishments. Afr J Hosp Tour Leisure 5:2–15
20. Huimin G, Chris R (2008) Place attachment, identity and community impacts of tourism—the case of a Beijing hutong. Tour Manage 29:637–647

# A Bio-cooperative Robotic System to Ensure Ergonomic Postures During Upper Limb Rehabilitation in Occupational Contexts

F. Scotto di Luzio[1(✉)], F. Cordella[1], C. Lauretti[1], D. Simonetti[1], S. Sterzi[2], F. Draicchio[3], and L. Zollo[1]

[1] Research Unit of Biomedical Robotics and Biomicrosystems, Department of Engineering, Università Campus Bio-Medico di Roma, Via Àlvaro del Portillo 21, 00128 Rome, Italy
f.scottodiluzio@unicampus.it

[2] Unit of Physical and Rehabilitation Medicine, Department of Medicine, Università Campus Bio-Medico di Roma, Via Àlvaro del Portillo 21, 00128 Rome, Italy

[3] Department of Occupational and Environmental Medicine, Epidemiology and Hygiene, INAIL, Via Fontana Candida 1, Monte Porzio Catone, Rome, Italy

**Abstract.** Inappropriate work conditions represent the main cause for upper limb musculoskeletal disorders in many working professions. In this context, robotics and novel technologies might represent a new frontier of devices able to treat musculoskeletal disorders. This paper aims at proposing and preliminary testing a bio-cooperative robotic platform for upper limb rehabilitation composed of a redundant anthropomorphic manipulator, an active arm gravity support and a multimodal interface. With the proposed platform it is possible to extract performance and muscular fatigue indicators and accordingly adapt the level of assistance, provided by the anthropomorphic robot arm, and of arm support. Furthermore, it was verified if the use of the proposed platform allowed subjects to execute highly controlled movements while maintaining an ergonomic posture able to limit the trunk compensatory movements during reaching. A preliminary study on 8 healthy subjects was carried out and the Rapid Upper Limb Assessment test was adopted to assess the subject's upperlimb posture during the rehabilitation task. The obtained results are encouraging for extending the study for rehabilitation in occupational contexts of patients with upper limb musculoskeletal pathologies.

**Keywords:** Robotic system · Ergonomic postures · Occupational contexts
Adaptive control · Human-robot interaction

## 1 Introduction

Inappropriate work conditions represent the main cause for upper limb musculoskeletal disorders in many working professions [1]. The intensity of the gestures together with the often improper movements and postures can lead to long-lasting injuries, especially


S. Bagnara et al. (Eds.): IEA 2018, AISC 826, pp. 327–336, 2019.
https://doi.org/10.1007/978-3-319-96065-4_37

in the shoulder/elbow district [2]. In this context, robotics and novel technologies might represent a new frontier of devices able to treat musculoskeletal disorders (MSDs) to accelerate recovery of work-injured people and quick reinsertion in their workplace [3, 4]. MSDs affect workers in different occupations that can be exposed to risk factors like heavy loads and incorrect body postures. Robots could allow to perform task-specific and highly-precise movements as well as to monitor limb kinematics and human-robot interaction. A few robotic systems have been proposed in the literature to rehabilitate patients with musculoskeletal diseases. In [5], an exoskeleton-based arm rehabilitation robotic device, i.e. the Armeo Spring, is adopted to rehabilitate 17 geriatric patients with proximal humeral fractures. Subjects have been asked to perform simulated activities of daily living with the aid of a virtual environment. In [6], a six degrees of freedom (DoFs) robotic arm has been used to rehabilitate arm mobility after forearm and elbow fractures. 3 patients participated in the study: they interacted with a virtual environment scenario by means of the end-effector of the robot arm and received multimodal feedback during the execution of motion tasks. The preliminary results obtained in these works demonstrated an improvement in the range of motion and in the forces applied by the impaired limb. Nevertheless, the significance of the results is limited due to the low number of treated patients.

Bio-cooperative systems could represent an important solution for the future of musculoskeletal disorders rehabilitation. They represent the new generation of robotic platforms that promote a bidirectional interaction between the robot and the patient based on multimodal interfaces [7]. These platforms use multisensory monitoring systems to objectively assess patient performance during therapy and correspondingly dynamically adapt robot behavior.

An important aspect to be considered during rehabilitation is related to the gravity effect due to the weight of the upper limb. Supporting the weight of the subject's impaired arm appears to be a key point in rehabilitation limiting the unhealthy effects of abnormal synergies thus permitting a greater range of motion. In it has been demonstrated that a gravity compensation strategy based on sling suspension led to an improvement of arm function of stroke patients after 9 weeks of training. Therefore, the sole application of gravity compensation might be a valuable strategy to foster functional improvements.

Moreover, the use of monitoring systems could allow supervising and, possibly, ensuring a correct and ergonomic posture during the execution of the tasks. In fact, to monitor during rehabilitation the natural compensation performed by the patients of movement deficits such as muscle weakness, abnormal postural adjustment and loss of inter-joint coordination is of paramount importance. When these deficits are encountered during robot-aided therapy, which is a situation in which the patients are not always monitored by the therapist, the patients naturally compensate by recruiting unaffected muscles and joints [8, 9]. Such a compensation can compromise patient recovery and introduce additional problems [10].

Several methods have been adopted in the literature to detect compensation movements during robot-aided rehabilitation. They comprise marker-based and marker-less vision-based devices [11], wearable sensors, such as magneto-inertial units (IMUs) [12] and electromagnetic sensors [13].

Therefore, it is evident the need for a complete platform that can (i) safely provide assistance-as-needed during rehabilitation; (ii) support the weight of the subject's impaired arm on the basis of the patient's state; (iii) avoid negative effect on subject posture. Objective of this paper is to propose and preliminary test on healthy subjects a bio-cooperative robotic platform for rehabilitation of patients with upper limb musculoskeletal pathologies able to satisfy all these requirements. Such a system consists of a redundant anthropomorphic manipulator, a purposely developed active arm gravity support, a multimodal interface and a virtual reality. The robot-aided rehabilitation aims at restoring the full joints Range of Motion (RoM) and muscle strength with an ergonomic posture. Personalization of the treatment is a prerogative of the proposed system; in fact, physiological and sensory-motor measurements are included into the control loop to create a rehabilitative scenario that is tailored to the specific subjects' needs.

This paper is organized as follows: Sects. 2.1 and 2.2 provide a brief overview of the proposed bio-cooperative platform, in Sect. 2.3 and in Sect. 2.4 the module to acquire sEMG, evaluate muscular fatigue and the posture analysis approach are described, respectively. Experimental setup and protocol are presented in Sect. 3. Section 4 is devoted to the experimental results and discussion. Finally, conclusions are drawn in Sect. 5.

# 2 Materials and Methods

## 2.1 The Bio-cooperative Platform

The proposed bio-cooperative platform, presented in Fig. 1, is composed of three main blocks:

- Robot Module (RM): it is composed of an anthropomorphic robot arm (i.e. the Kuka Light Weight Robot 4+) attached to the human subject wrist. The sensors embedded in the robotic arm are used to monitor patient movements in terms of position/force. These information are used to provide assistance-as-needed, as described in the following, by supporting subject wrist in reaching desired positions.
- Adaptive Arm-Weight Support Module (AWSM): it is made of a motorized arm weight support that guarantees the correct level of limb support with an adaptive control based on elbow 3D position and level of muscular fatigue. This information are provided by M-IMUs positioned on the impaired arm and sEMG electrodes positioned as explained in Sects. 2.3 and 2.4.
- Virtual reality: it reproduces the task to be performed and gives the user a continuous feedback on him/her motion performance (in terms of error between the subject hand avatar position and the target).

The proposed architecture allows dynamically supporting subjects in order to provide the correct amount of assistance with an assistance-as-needed approach. Furthermore, the proposed platform is designed to guarantee the correct ergonomic posture during the execution of the task, as demonstrated in Sect. 4.

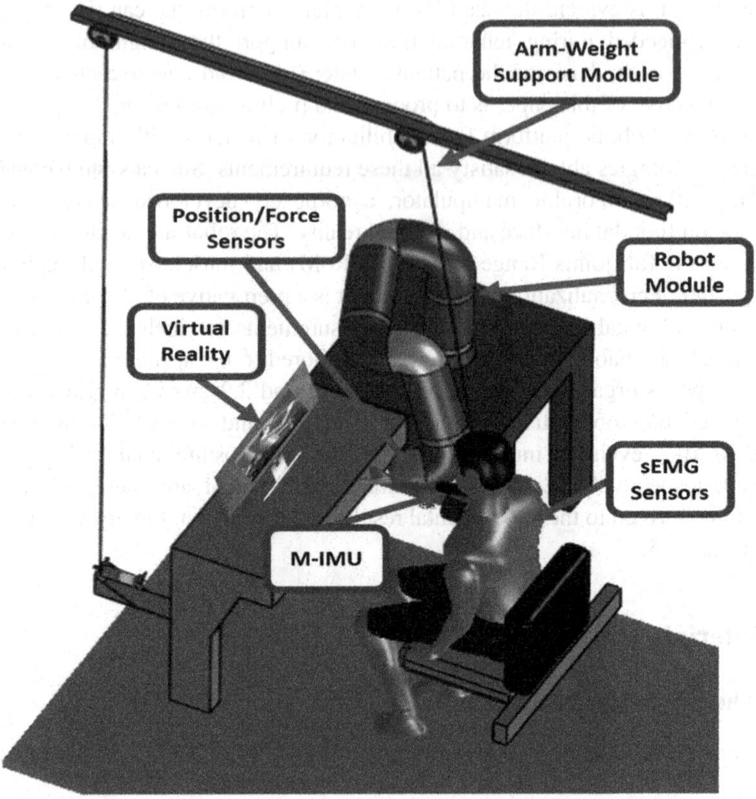


**Fig. 1.** An overall view of the proposed bio-cooperative platform.

## 2.2 RM and AWSM Control

The RM is controlled with an impedance control based on a threshold algorithm with three levels of stiffness $(K_r)$ and task durations $(t)$, according to the following equations

$$C_{k_r} = \frac{1}{2}\alpha + \frac{1}{8}q_{corr1,4} + \frac{1}{8}q_{corr2,4} + \frac{1}{8}UMF + \frac{1}{8}UPF \tag{1}$$

$$C_t = \frac{1}{2}MAPR + \frac{1}{8}q_{corr1,4} + \frac{1}{8}q_{corr2,4} + \frac{1}{8}UMF + \frac{1}{8}UPF \tag{2}$$

where $\alpha$ is the aiming angle, *MAPR* is the mean arrest period ratio, $q_{corri,j}$ is inter-joint coordination and *UMF* and *UPF* are useful mean force and useful peak force, respectively (see [14] for more details).

The AWSM is based on a torque control with gravity compensation. The input of this module are (i) the subject 3D elbow position, reconstructed by means of the Augmented Inverse Kinematics approach proposed in [15], which is used to compute the desired crankshaft position ($q_d$); (ii) the level of necessary support ($L_s$) chosen on the basis of muscular fatigue indicators computed on sEMG signals, as described in Sect. 2.3. The correct level of torque to be provided by the AWS to substain subject limb is expressed by

$$\tau(q) = K_p(q_d - q) + K_d \frac{d}{dt}(q_d - q) + L_s \tau_{max} \cos(q_d - q) \tag{3}$$

where $q$ is the current crankshaft position obtained from encoder, $\tau_{max}$ is the maximum level of torque evaluated at the beginning of the task, $K_p$ and $K_d$ are the control proportional and derivative gains, respectively.

### 2.3 Fatigue Assessment

The proposed architecture integrates a module to acquire sEMG and estimate muscular fatigue. sEMG signals are recorded at 1 kHz from 7 upper limb muscles: biceps brachii (BB), the lateral triceps (LT), the anterior deltoid (AD), the lateral deltoid (LA), the posterior deltoid (PD), the pectoralis major (PM), the upper trapezius (UT). Electrodes for each muscle are placed according to SENIAM guidelines [16]. sEMG data are filtered using a sixth order bandpass Butterworth filter to remove noise with cutoff frequencies (30,450) Hz. The filtered sEMG signal is normalized with respect to the Maximum Voluntary Contraction (MVC). sEMG signals of the contraction phases are used to compute Dimitrov's Spectral Fatigue Index (DI) defined as

$$DI_i = \frac{\int_{f_1}^{f_2} f^{-1} * PS_i(f) * df}{\int_{f_1}^{f_2} f^5 * PS_i(f) * df} \qquad i = 1, \ldots, 8 \tag{4}$$

where $PS_i(f)$ is the $i$ - signal power spectrum and $f_1$ and $f_2$ are the lowest and the highest frequency of the bandwidth.

In order to establish the correct level of support to be provided to the subjects, a fatigue indicator $C_m$ is introduced. It considers the DI estimated for each muscle, normalized with respect to their maximum and then weighted as follows

$$C_m = \frac{1}{14}\left(DI_{BB} + DI_{LT} + \frac{1}{2}DI_{AD} + \frac{3}{2}DI_{LA} + DI_{PD} + DI_{PM} + DI_{UT}\right) \tag{5}$$

The level of support $L_s$ is determined on the basis of the $C_m$ value as follows

$$L_s = \begin{cases} 0 \text{ } if \text{ } C_m < 0.20 \\ 1 \text{ } if \text{ } 0.20 \leq C_m < 0.40 \\ 2 \text{ } if \text{ } 0.40 \leq C_m < 0.60 \\ 3 \text{ } if \text{ } 0.60 \leq C_m < 0.80 \\ 4 \text{ } if \text{ } 0.80 \leq C_m < 1. \end{cases} \tag{6}$$

### 2.4 Posture Analysis

A set of criteria for upper limb posture during working called Rapid Upper Limb Assessment (RULA) [17] is adopted in order to evaluate subject posture. This approach allows performing a rapid assessment of angles without the need of special equipment. The RULA score has been computed by using trunk, shoulder and neck angles measured on the subjects during task execution with the aid of the bio-cooperative platform. The approach proposed in [18, 19] has been adapted for evaluating trunk, shoulder and neck behaviour.

Four M-IMUs have been used to detect user's motion. They are placed on the chair where the subject is seated and on the user's trunk, head and impaired arm in order to evaluate the shoulder flexion/extension (F/E), abduction/adduction (A/A) and intra/extra rotation (I/E) angles, the trunk F/E, lateral F/E and left/right rotation and the head roll, pitch and yaw. The sensor placed on the chair acts as reference for the sensors positioned on the subject. The M-IMUs orientation, expressed in terms of rotation matrix, is computed between the sensor-fixed coordinate system and a earth-fixed reference coordinate system G. For instance, the head angles are computed as follows. Let us to define $IMU_1$ and $IMU_2$ as the fixed coordinate systems of the sensors placed on the subject chair and head, respectively. The rotation matrix between these two coordinate system is

$$R_{IMU_1}^{IMU_2} = R_{IMU_1}^{G} R_{G}^{IMU_2} \quad (7)$$

The set of Euler angles (i.e. the head roll, pitch and yaw angles) are extracted by solving the inverse problem.

The elbow F/E has been computed by applying the Augmented Inverse Kinematics approach proposed in [15] and the wrist angle is fixed during the whole trial, due to the flange adopted for connecting the subject wrist with the robot arm end-effector.

## 3 Experimental Setup and Protocol

The proposed bio-cooperative platform, shown in Fig. 1, is composed of the anthropomorphic robotic arm, the actuated arm-weight support and a multimodal interface. The robotic arm is the Kuka Light Weight Robot 4+, which is characterized by 7 Degrees of Freedom (DoFs) and embeds position and torque sensors at joints. The arm-weight module actuation system is based on an EC-max 40 brushless Maxon Motor. Finally, an ergonomic brace for arm support enables to set correct fitting depending on patients requirements. The multimodal interface is composed of sEMG sensors, to monitor subject muscular activity and extract information about muscular fatigue, and M-IMUs (i.e. XSens MTw), which are integrated in the system with the twofold aim of reconstructing the upper-limb kinematics to control the AWS and of monitoring the behavior of subject trunk, neck and shoulder during therapy. The M-IMU and robot sensors data are acquired at 100 Hz. A virtual reality is included in the platform to help subjects to know the action to be performed and to monitor his/her performance during task execution.

Eight healthy subjects have been involved in this study. Each subject performed the task sitting on a chair, in front of a virtual reality, with his/her right arm positioned in an ergonomic backing for the arm and with the wrist connected to the end-effector of the robot. Each subject performed 16 point-to-point reaching movements in the 3D space. In order to understand if the proposed platform influences the subjects posture, subjects performed the assigned task in two different conditions: (1) with full assistance provided by the platform (i.e. 100% of arm weight support and high gain for the robotic arm); (2) without assistance provided by the platform (i.e. 0% of arm weight support and low gain for the robotic arm). Subject's upper limb kinematics and their posture are reconstructed by means of the M-IMU sensors.

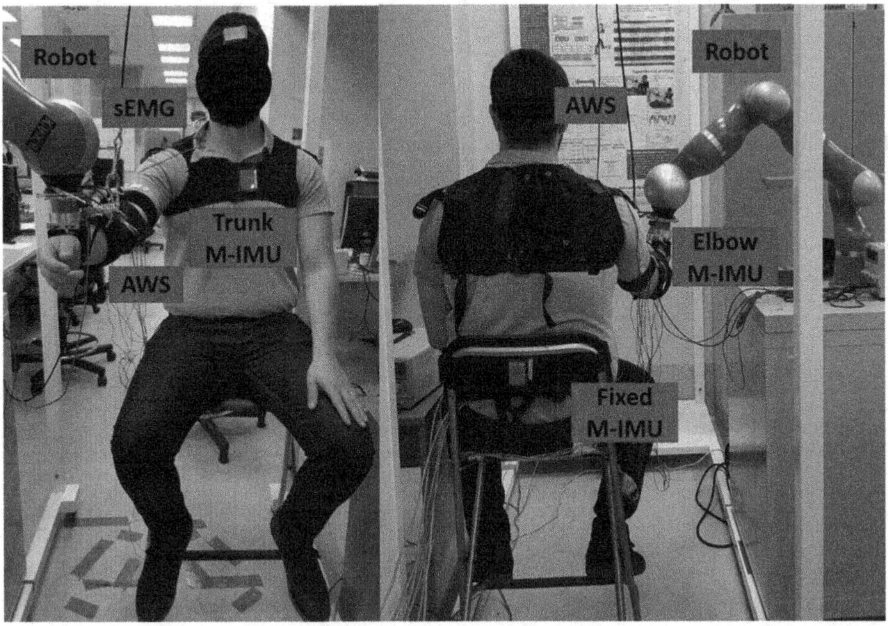


**Fig. 2.** A frontal and posterior view of the experimental setup.

# 4 Results and Discussions

The results of one representative subject are reported, but similar results have been obtained for the other subjects. The robot end-effector position during task execution with full assistance is shown in Fig. 2. The 3D Cartesian coordinates of the elbow in the two conditions are shown in Fig. 3. As evident the difference between the two working conditions is small (i.e. the maximum error is along the X-axis and is less than 0.05 m). The same behavior was obtained for the other joints. It means that the level of assistance provided by the platform does not influence the subject's posture. During the two conditions, the muscular fatigue was evaluated. In particular, the mean value and standard deviation, computed on the 8 subjects, of the muscular fatigue indicator $C_m$ and of the corresponding level of AWS $L_s$ are $C_m = 0.63 \pm 0.15$ and $L_s = 2$ or 3, with full assistance, and $C_m = 0.22 \pm 0.13$ and $L_s = 0$ or 1, without assistance (Fig. 4).

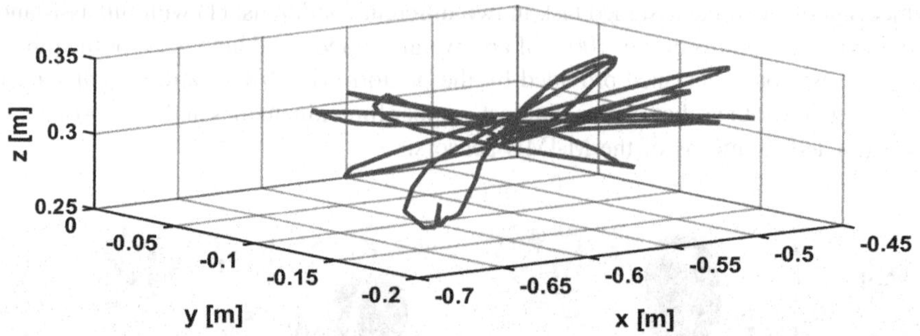


**Fig. 3.** Robot end-effector position with full assistance for a representative subject during point-to-point reaching movements.

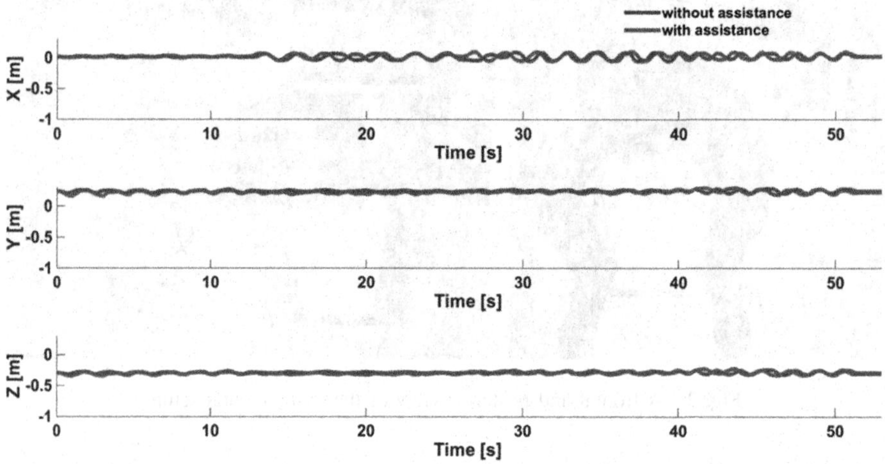


**Fig. 4.** Elbow position with and without assistance for a representative subject point-to-point reaching movements.

The subject's trunk, head, shoulder, elbow and wrist angles were evaluated during the trials. In order to assess if the platform negatively influences the posture, the RULA score was computed on these values. For all the subjects, the shoulder F/E angle is comprised between 0° and 20° the shoulder I/E is comprised between −15° and 15°, the elbow F/E is comprised between 5° and 35°, the wrist is fixed with a null angle, the neck pitch is comprised between 0° and 10°, the neck roll and yaw are less than 5°, the trunk F/E is comprised between 0° and 20° the trunk lateral F/E and left/right rotation are less than 5°. Legs and feet are supported and the load to be held by the subjects is null. Therefore, the RULA score is equal to 2, which correspond to an acceptable posture.

## 5 Conclusion

In this work a bio-cooperative robot to guarantee ergonomic postures during upper limb rehabilitation in occupational context is presented. The proposed system allows to tune the level of assistance on the basis of patient performance and his/her muscular fatigue in order to build a patient-tailored rehabilitation. The experimental results demonstrated that the proposed platform does not negatively influence the subject posture. Future activities will include the validation of the proposed system with patients affected by shoulder disorders. In addition, the integration in the control architecture of the information about posture will be envisaged to modify the level of assistance also on the basis of the compensatory movements.

**Acknowledgement.** This work was supported partly by the Italian Institute for Labour Accidents (INAIL) with the RehabRobo@work (CUP: C82F17000040001), PCR 1/2 (CUP: E57B16000160005) and PPR AS 1/3 (CUP: E57B16000160005) projects and partly by the European Project H2020/AIDE: Adaptive Multimodal Interfaces to Assist Disabled People in Daily Activities (CUP: J42I15000030006).

## References

1. Hämäläinen P, Saarela KL, Takala J (2009) Global trend according to estimated number of occupational accidents and fatal work-related diseases at region and country level. J Saf Res 40(2):125–139
2. Dembe AE, Erickson JB, Delbos R (2004) Predictors of work-related injuries and illnesses: national survey findings. J Occup Environ Hyg 1(8):542–550
3. Hakim RM, Tunis BG, Ross MD (2017) Rehabilitation robotics for the upper extremity: review with new directions for orthopaedic disorders. Disabil Rehabil Assist Technol 12 (8):765–771
4. Wang Q, Markopoulos P, Yu B, Chen W, Timmermans A (2017) Interactivewearable systems for upper body rehabilitation: a systematic review. J Neuroeng Rehabil 14(1):20
5. Schwickert L, Klenk J, Stahler A, Lindemann U (2011) Robotic-assisted rehabilitation of proximal humerus fractures in virtual environments. Zeitschrift für Gerotologie und Geriatrie 44:387–392. https://doi.org/10.1007/s00391-011-0258-2
6. Padilla-Castaneda MA, Sotgiu E, Frisoli A, Bergamasco M, Orsini P, Martiradonna A, Olivieri S, Mazzinghi G, Laddaga C (2013) A virtual reality system for robotic-assisted orthopedic rehabilitation of forearm and elbow fractures. In: IEEE/RSJ international conference on intelligent robots and systems
7. Simonetti D, Zollo L, Papaleo E, Carpino G, Guglielmelli E (2016) Multi-modal adaptive interfaces for 3D robot-mediated upper limb neurorehabilitation: an overview of bio-cooperative systems. Robot Auton Syst 85:62–72
8. Cirstea MC, Levin MF (2000) Compensatory strategies for reaching in stroke. Brain 123 (5):940–953

9. Harvey RL (2014) Stroke recovery and rehabilitation. Demos Medical Publishing, New York
10. Alankus G, Kelleher C (2012) Reducing compensatory motions in video games for stroke rehabilitation. In: Proceedings of the ACM annual conference on human factors computing system (CHI), pp 2049–2058
11. Zhi YX, Lukasik M, Li MH, Dolatabadi E, Wang RH, Taati B (2017) Automatic detection of compensation during robotic stroke rehabilitation therapy. Rehabilitation devices and systems
12. Wong MY, Wong MS (2009) Measurement of postural change in trunk movements using three sensor modules. IEEE Trans Instrum Meas 58(8):2737–2742
13. Maduri A, Wilson SE (2009) Lumbar position sense with extreme lumbar angle. J Electromyogr Kinesiol 19(4):607–613
14. Papaleo E, Zollo L, Spedaliere L, Guglielmelli E (2013) Patient-tailored adaptive robotic system for upper-limb rehabilitation. In: 2013 IEEE international conference on robotics and automation (ICRA), pp 3860–3865. IEEE
15. Papaleo E, Zollo L, Garcia-Aracil N, Badesa F, Morales R, Mazzoleni S et al (2015) Upper-limb kinematic reconstruction during stroke robot-aided therapy. In: Medical & Biological Engineering & Computing, pp 815–828. Springer
16. Hermens HJ et al (1999) European recommendations for surface electromyography. Roessingh Res Dev 8(2):13–54
17. McAtamney L, Corlett EN (1993) RULA: a survey method for the investigation of work-related upper limb disorders. Appl Ergon 24(2):91–99
18. Lauretti C, Davalli A, Sacchetti R, Guglielmelli E, Zollo L (2016) Fusion of M-IMU and EMG signals for the control of trans-humeral prostheses. In: IEEERAS/EMBS international conference on biomedical robotics and biomechatronics
19. Lauretti C et al (2017) Comparative performance analysis of M-IMU/EMG and voice user interfaces for assistive robots. In: 2017 international conference on rehabilitation robotics (ICORR). IEEE

# Interrelationship Between Dietary Intake, Bone Mineral Density and Incidence of the Development of Musculoskeletal Disorders in College Students

Amanpreet Kaur(✉) and Ajita Dsingh

Department of Sport Sciences, Punjabi University, Patiala, Punjab, India
aman.kaur221@gmail.com

**Abstract.** For most college going students, participation in sport and exercise provides many health benefits. Unfortunately some individuals do not ingest sufficient energy to meet the demands of physical exertion and all physiological functions due to irregular eating habits. This leads to a state of low energy availability (LEA) which can negatively affect bone health, cardiovascular function, and reproduction, gastrointestinal and mental health. The three most researched health consequences of LEA are the clinical end points eating disorders and disordered eating, musculoskeletal disorders, and stress fractures (due to loss of bone health) which occur on a spectrum from health to disease. [Purpose] Identifying female students along the spectrum can result in early detection which in turn can prevent these conditions from progressing and reaching the clinical end points. [Subjects and Methods] 100 participants (18–25 years of age) were recruited according to the selection criteria and ethical clearance. Outcome measures were Eating Attitudes Test- 26 (EAT-26), questionnaire regarding the participant's exercise patterns, bone mineral density and Nordic questionnaire for musculoskeletal disorders. [Results] Pearson correlation test was applied on the variables to get positive linear correlation between physical activity and bone mineral density and negative correlation between scores from EAT-26 and bone mineral density. Results from Nordic questionnaire showed that 63% of the subjects had experienced musculoskeletal disorders, most of which was regarding low back. In the present study 58% of females reported having an injury in the past year. [Conclusion] Disordered eating certainly affects the overall well-being of a female by affecting the bone health. Dietary counselling and physical activity can help improve the health status in the college going females. Controlling their attitude towards health through proper awareness can also improve their quality of life.

**Keywords:** Disordered eating · Musculoskeletal disorders
Bone mineral density


S. Bagnara et al. (Eds.): IEA 2018, AISC 826, pp. 337–343, 2019.
https://doi.org/10.1007/978-3-319-96065-4_38

# 1 Introduction

Participation in sport and exercise provides always proves to have many health benefits. Physical activity (PA), sport and exercise are considered as part of a healthy lifestyle worldwide. It is known that Physical Activity promotes bone health, improves blood lipid profiles and helps maintain muscle mass (Pate et al. 1995). However, there are certain situations in which exercise can have a negative effect on health, specifically bone health, lipid profiles and hormones; ironically some of the very beneficial health outcomes that exercise promote (Mountjoy et al. 2014). These effects can only be seen when an energy deficit exists is severe enough that the body has insufficient energy to meet the needs of physical exertion and normal physiological functioning. Unfortunately some individuals do not get required energy to meet the needs of training and all physiological functions. This results in a state of low energy availability (LEA) which can affect normal functioning of all body systems such as musculoskeletal system in terms of bone health, cardiovascular system, reproduction, gastrointestinal and nervous system. The health consequences of LEA include disordered eating, menstrual irregularity, and stress fractures which occur on a spectrum from health to disease. Identifying young females along the spectrum allows for early detection which may prevent these conditions from progressing and reaching the clinical end points (Nattiv et al. 2007). This will allow young females of all demographics to enjoy exercise, whilst maintaining good health and maximize their sporting performance. At present, the extent to which Indian young females are at risk of LEA is unknown. This information is required to help appropriately direct resources into early detection of individuals at risk of LEA, which is crucial in protecting their current and long term health. Taking bone health into consideration, according to global estimates, about 200 million women suffer from osteoporosis worldwide (Cooper et al. 1999). Low BMD leads to limitations in mobility along with high medical cost. Low bone mineral density (BMD) manifesting as fragile bones mainly comprise of osteoporosis and osteopenia (NIH 2001).

OSTEOPOROSIS: Osteoporosis is a disease characterized by low bone mass and micro architectural deterioration of bone tissue leading to enhanced skeletal fragility and increased risk of fracture (Bouillon et al. 1991). An expert panel convened by World Health Organization (WHO) has established the following diagnostic criteria (Kanis et al. 1994).

a. Normal: bone mineral density (BMD) which is no more than 1 standard deviation (SD) below the mean of young adults.
b. Osteopenia: BMD between 1 and 2.5 SD below the mean of young adults.
c. Osteoporosis: BMD more than 2.5 SD below the mean of young adults plus one or more fragility fractures.

Prevalence of low BMD varies according to age, sex, ethnicity and type of skeletal bone (NIH 2001). To date, the gold standard technique for diagnosing low BMD is the dual energy x-ray absorptiometry (DXA) which has high predictive validity (sensitivity and specificity) (Moayyeri et al. 2012).

### 1.1 Need of the Study

The need of this study is to identify female students along the spectrum (Disordered eating, musculoskeletal disorders, and bone mineral density) which can result in early detection which in turn can prevent these conditions from progressing and reaching the clinical end points.

### 1.2 Aim and Objectives of the Study

Aims and objectives:

1. To determine the prevalence of Disordered eating, musculoskeletal disorders, Bone mineral density among college going females.
2. To determine the correlation between Disordered eating, and bone mineral density among college going females.

### 1.3 Hypothesis

Young females may develop these three components namely disordered eating, musculoskeletal disorders, and bone mineral density which is inter-related somehow.

## 2 Methodology

### 2.1 Study Design

This study used a non-random sample, cross-sectional research design that consisted of correlational and descriptive statistical analysis that determined the prevalence of the disordered eating, musculoskeletal disorders, and bone mineral density in young females. This study examined if there is a relationship between these three attributes.

### 2.2 Study Duration

6 months.

### 2.3 Population Sample

Total 100 female college going subjects from nursing colleges.

### 2.4 Inclusion Criteria

18–25 years of age.

### 2.5 Exclusion Criteria

History of any neurological or metabolic disorders, subjects who are not cooperative to procedure, pregnancy, long term oral steroid use, history of thyroid disorder, pituitary tumour, cardiovascular disease.

### 2.6 Procedure

**Measurement of Data**

After the approval of proposal and registration, informed consent was obtained from each subject and all procedures of study was explained to them.

**Protocol**

Sample (n = 100)

Eating Attitudes Test- 26 (EAT-26).

Nordic Questionnaire for musculoskeletal disorders.

Bone mineral densitometer.

Data collection

## 3 Results

### 3.1 Participant Characteristics

Young females' self-reported age, height, weight, and BMI (mean ± SD) were 20.7 ± 1.3 years, 165.2 ± 7.0 cm, 59.4 ± 8.9 kg, and 21.8 ± 2.9 kg/m2, respectively. Age of menarche was 12.4 ± 1.2 years (Table 1).

**Table 1.** Respondent anthropometric and demographic characteristic.

| Characteristic total (n = 100) | |
|---|---|
| Weight (kg)[a,b] | 59.4 (8.9) |
| Height (m)[a,b] | 1.65 (0.07) |
| Body mass index ($kg.m^2$)[b,c] | 21.8 (2.9) |
| Age (y)[a] | 20.7 (1.3) |
| Exercise per week (h)[a,b] | 8.5 (5.0) |
| Body mass index ($kg.m^2$)[c,d] | Underweight (<18.5)[c,d] 12 (7.8) |
| | Healthy range (18.5–24.9)[c,d] 67 (43.1) |
| | Overweight (25.0–29.9)[c,d] 19 (18.6) |
| | Obese ($\geq$ 30.0)[c,d] 2 (0.1) |

[a]Self-reported
[b]Mean (Standard deviation)
[c]Calculated from self-reported height and weight
[d]Number (percentage)
Note: Percentages may not add up to 100% due to rounding.

### 3.2 Prevalence Estimates of the Components

Of the 100 young females in the sample, 35 (35%) met the criteria for disordered eating, and 26 (26%) had low BMD for their age (WHO criteria).

Results from Nordic questionnaire showed that 63% of the subjects had experienced musculoskeletal disorders, most of which was regarding low back (72%).

Pearson correlation test was applied on the variables to get positive linear correlation ($r^2$ = +0.972, p = 0.05) between physical activity and bone mineral density and negative linear correlation ($r^2$ = −0.72, p = 0.05) between scores from EAT-26 and bone mineral density (Fig. 1).

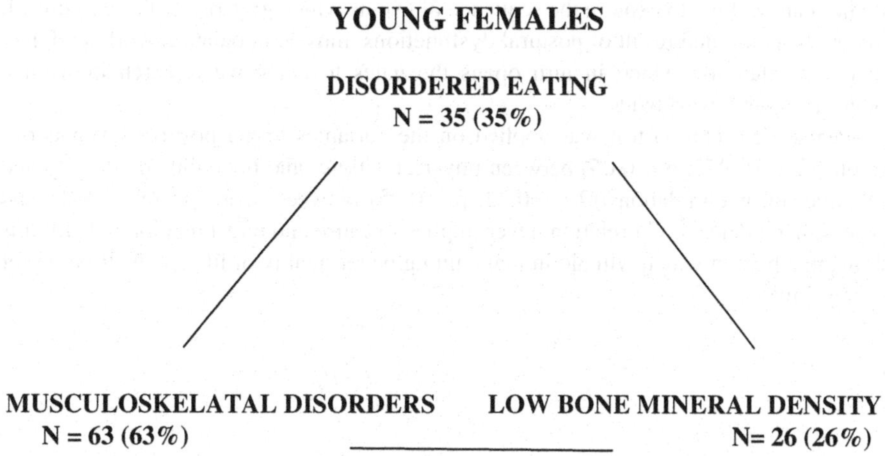


**Fig. 1.** Prevalence of disordered eating, musculoskeletal disorders, and bone mineral density in young females

## 4 Discussion

In general, all girls and women are encouraged to participate in physical activities because benefits hold more water than risks. However, disordered eating, musculoskeletal disorders, and low bone mineral density in young females make physically active and inactive females prone to many health risks. A real potential is there to prevent disordered eating, musculoskeletal disorders, and low bone mineral density in young females by formulation of early awareness or educating program which should be focused on teaching the young females about risks of disordered eating and low energy availability.

The current study had many imitations. First of all, a cross-sectional design used to allow data to be collected from both athletes and non-athletes. The limitation of using a cross-sectional design is that it does not allow data to be collected over a period of time to determine if changes in any kind of pressure or depression among young females affect the risk of developing low energy availability or musculoskeletal disorders. Apart from that, due to the cross sectional design we have no way of determining the sequence of events that lead to our finding However, the current cross sectional study has made a pavement to recognize a number of associations that may exist. Therefore combined with findings from previous research the results of this study can prove to be useful in generating hypotheses and topics to investigate in future research.

## 5 Conclusion

A significant number of young females 35 (35%) met the criteria for disordered eating, and 26 (26%) had low BMD for their age (WHO criteria). Given the high prevalence, proper awareness has the potential to prevent these components in young females, therefore, improving health and averting long-term complications.

Results from Nordic questionnaire showed that 63% of the subjects had experienced musculoskeletal disorders, most of which was regarding low back (72%). Further studies can be done to explore the sequence of events which give rise to these figures. It can be the combined result of postural dysfunctions, muscle imbalance, work load, low energy availability, which in turn opens the gates to extensive research in women health programs worldwide.

Pearson correlation test was applied on the variables to get positive linear correlation ($r2 = +0.972$, $p = 0.05$) between physical activity and bone mineral density and negative linear correlation ($r2 = -0.72$, $p = 0.05$) between scores from EAT-26 and bone mineral density. Co-relation never implies to cause and effect relation but still it is showing a pattern which will aid in maintaining better quality of life and well-being for young females.

## References

Bouillon P, Burckhardt P, Christiansen C et al (1991) Consensus development conference: prophylaxis and treatment of osteoporosis. Am J Med 90:107–110

Cooper C, Campion G, Melton LJ 3rd (1999) Hip fractures in the elderly: a world-wide projection. Osteoporos Int 22:285–289

Kanis JA, Melton J, Christiansen C, Johnston CC, Kjaltaev N (1994) The diagnosis of osteoporosis. J Bone Miner Res 9:1137–1141

Moayyeri A, Adams JE, Adler RA et al (2012) Quantitative ultrasound of the heel and fracture risk assessment: an updated meta-analysis. Osteoporos Int 23:143–153

Mountjoy M, Sundgot-Borgen J, Burke L, Carter S, Constantini N, Lebrun C, Ljungqvist A (2014) The IOC consensus statement: beyond the Female Athlete Triad-Relative Energy Deficiency in Sport (RED-S). Br J Sports Med 48(7):491–497

Nattiv A, Loucks AB, Manore MM, Sanborn CF, Sundgot- Borgen J, Warren MP (2007) American college of sports medicine position stand. The female athlete triad. Med Sci Sports Exerc 39:1867–1882

NIH (2001) Consensus development panel on osteoporosis, prevention, diagnosis and therapy. JAMA 285:785–795

Pate RR, Pratt M, Blair SN, Haskell WL, Macera CA, Bouchard C, King AC (1995) Physical activity and public health: a recommendation from the Centers for Disease Control and Prevention and the American College of Sports Medicine. JAMA 273(5):402–407

World Health Organization (1994) Assessment of fracture risk and its application to screening for postmenopausal osteoporosis, Geneva, Switzerland

# Teaching with Gender Perspective Oriented to the Training of Women Students of Interactive Design and Interactive Technologies

Mercado Colin Lucila(✉) and Rodea Chávez Alejandro

Universidad Autónoma Metropolitana, Unidad Cuajimalpa, Mexico City, Mexico
{lmercado, arodea}@correo.cua.uam.mx

**Abstract.** There's needed of studies for teaching with gender perspective oriented to train women students of Interactive Technologies and Interactive Design, whom will have the task of developing easy-to-use, useful and pertinent systems, considering transition towards a more social design, characterized by giving people voice, centering results of design processes in users; thinking about action's sustainable consequences, or focusing on services, rather than products. At present, society is characterized by marked economic, social and gender differences, visible in the phenomenon known as digital divide, in which same opportunities for access and use of digital technologies do not exist for all individuals. Economic and geographical access obstacles, as well as difficulties of using new technologies are widening the digital divide between people; differentiating between those who may or may not acquire a series of products or services, between the countries that develop technology and the countries that adopt it, as well as between men and women who use technologies; that is to say between "what some do and others cannot do". Women have been lagged respect to men in the digital domain. These differences have been built from multiple social factors, giving cultural, social and gender legitimacy. Despite equal access to Information Technologies, skills valuation between genders is distinctive, remaining masculine the imaginary representations about computer expert. Visualizing scenarios for a gender education, which social self-concept and self-image of women develop positively, promoting benefiting experiences to students that develop products designed by women, by making efforts to prevent digital divide by gender.

**Keywords:** Gender divide · Interactive design and technology women students
Teaching scenarios

## 1 Towards a Sustainable Society

Currently, it is recognized that one of the great problems of humanity is the global environmental imbalance, which has been unleashed during the process to satisfy the individual needs of materials, exploiting in an excessive way the biological resources. "In other words, what is at stake now is nothing less than the capacity of our planet to


S. Bagnara et al. (Eds.): IEA 2018, AISC 826, pp. 344–353, 2019.
https://doi.org/10.1007/978-3-319-96065-4_39

sustain life in the years to come, which confronts us not only with the dilemma of our survival as well as specie, but also poses the fundamental question of whether it will be materially possible one day to well living" [1].

Although in various sectors such as industry and education the practice of Sustainability has mostly focused its efforts to work towards economic and environmental factors, society has struggled little by little to promote a transition in which the equitable use of the natural resources by the societies respected, being more and more frequent that one reads to the Sustainability like a discipline that adopts thoughts of social, cultural, political, generational and environmental equity.

Currently, four fundamental pillars of sustainability are recognized: environmental, economic, cultural and social. Economic dimension refers to the need to plan the economy according to material and immaterial needs. Environmental dimension focuses on the need to preserve and favor natural resources and ecosystems, efficiently managing their productivity without breaking the link between societies and their culture with the exploitation of resources; Cultural dimension strives for the respect and consideration of the different ways of life, customs, beliefs and values in force at a given moment, manifested in the relationship of the culture of the groups and their environment, shaping the identity of social groups and nations, contributing to the eradication of poverty, to inclusive, equitable and human-centered development. Social dimension refers to the diversity of social relationships that are established in a sustainable community.

Being closely linked, interconnected and interdependent, the four pillars of sustainability must be addressed simultaneously to achieve the welfare of societies.

The concern to move towards sustainable societies has led member countries of the United Nations (UN) to raise the Sustainable Development Goals, which are a set of shared proposals that aim to be a guide to solve some of the challenges of humanity about people, the planet, prosperity, peace and associations.

There are multiple problems identified as priorities linked to poverty, education, inequality, access to water, and many others that afflict societies and the planet.

In the period from 2000 to 2015, efforts focused on eight Millennium Development Goals (MDGs): Eradicate extreme poverty and hunger, achieve universal primary education, promote gender equality and empower women, reduce infant mortality, improve maternal health, combat HIV/AIDS, malaria and other diseases, ensure environmental sustainability and foster a global partnership for development. The MDGs constitute one of the most significant efforts in contemporary history to help those most in need [2].

### 1.1 MDGs y SDGs Mexico

Regarding the 51 indicators in which Mexico committed efforts, at the conclusion of the MDG work, total compliance was reported in 37 of them. One of these goals met with particular relevance to this context is to ensure that primary education by the year 2014 reaches 95.9% of children completed a full cycle of primary education. Eliminate inequalities between the sexes in primary and secondary education at all levels of education by the year 2015. With regard to promoting gender equality and women's autonomy, Mexico has achieved more equitable levels, among which it stands out the increase in female enrollment in higher education, from 75 women for every 100 men in 1990, to 93.3 women for every 100 men in 2014, and for the period 2016–2017 the ratio women and men in higher education was 0.982:1 [3].

In the new UN agenda (2016–2030) there are 17 Sustainable Development Goals (SDGs, also known as Global Objectives) that aim to continue the work done with the MDGs and in some areas, according to what was reported by the UN there were notable achievements, however, inequalities persist. The SDGs are a universal call for the adoption of measures for all people to enjoy peace and prosperity [4].

Consistent with the complexity that characterizes sustainability, the SDGs are interrelated, so the key to one's success will often involve the intervention and development of the other, as they address multidimensional problems. The SDGs show the main global problems such as social exclusion, education, inequitable exploitation of natural resources, hunger, overpopulation, inequality between people and groups, violence, among others. All of them are multi-factorial problems.

Among these SDGs are particularly relevant to this research, both to ensure an inclusive and equitable quality education and to promote lifelong learning opportunities for all, as well as to achieve gender equality and the empowerment of all women and girls. However, "the countries of Latin America show alarming social and economic indicators. Far from diminishing, marginality, unemployment, poverty and social violence tend to increase and deepen. Huge proportions of the population (ranging from 20 to 50% according to different countries and indicators) live in conditions of exclusion, marked by a set of deficits: housing, food, education, access to goods and services. Overcoming these social problems is probably the greatest political and economic challenge of local governments. It is, at the same time, the largest social debt in the region" [5].

For Mexico, the technology-development-dependency relationship is one of the strongest criticisms due to the enormous dependence on technology generated in other parts of the world. This problem, seen in a broad way, has among other detonators situations of economic, social, political and educational order. "The reasons can be grouped into three large groups. First, the lack of investment of entrepreneurs in research and technological development (R&D) due to various factors: lack of vision; poor linkage with academic sectors; it is not considered profitable to invest in R&D; lack of financial support; it is cheaper to import technology instead of developing it. Second, low support for R&D by government institutions for various reasons: budget constraints; lack of vision or ignorance of the rulers; corruption; political motives and so on. Finally, the limited presence of human capital focused on R&D caused by the lack of an adequate education and by the flight of talents abroad" [6]. Medina also establishes as one of the reasons for the country's technological dependence the lack of linkage with the academic sector and the scarcity of human resources, due to the lack of adequate education that focuses on R&D. Both factors involve Universities as trainers of human capital.

## 2 Technologic Education in Mexico

In the National Survey of Science and Technology [7] it is reported that 18.3% of the sample surveyed "does not know" (or cannot) mention a word related to technology. Basically, what were reported "instead of" were devices (cell phones, tablets, phones, and computers), all items that also are familiar and desired by young people. This is explained through another of the test items that inquire about the means by which Mexicans acquire knowledge about technology, indicating that 69.7% do it through television, which places it as the main broadcaster, followed by Internet with 33.7%. These statistics allow seeing that television is training Mexican children and young people, where basic school, secondary school and high school are failing.

This argument is consistent with what was reported by the Organization for Economic Cooperation and Development (OECD), where only 17% of young people between 25 and 64 years old in Mexico had completed higher education in 2016. While it is contrasting that in 2015, 32% of the new students entering higher education chose areas related to science, technology, engineering and mathematics, that is, 5% points more than the OECD average. This places Mexico among the first six countries of the OECD with respect to students who choose careers in the field of sciences. Another encouraging news is that there are now more young people enrolled in technical training programs while completing their baccalaureate studies; situation that the OECD attributes to the educational policy of the Mexican government to promote technological education [8].

### 2.1 Gender Inequality

The 2008 Statistics on Gender Inequality and Violence against Women affirm that both in Mexico and around the world, women are treated unequally by the State and society as a whole, on the basis of historical discrimination, also affirming that at the national level there is no equality of treatment or opportunities between men and women.

However, Mexico´s trend seems to be closing the gap and for 2016–2017, the ratio men and women with higher level studies was practically the same, 1:0.982.

**Table 1.** Percentage distribution of the undergraduate population by sex, according to the 10 professions with the most enrollment, 2012–2013 school year. Source: Own development based on data from Zubieta et al. (2015)

| Priority | Percentage men | Profession | Percentage women | Profession |
|---|---|---|---|---|
| 1 | 7.6 | Law | 7.6 | Law |
| 2 | 4.3 | Psychology | 5.6 | Psychology |
| 3 | 3.5 | Administration | 3.8 | Administration |
| 4 | 3.0 | Business Admin. | 3.2 | Nursing |
| 5 | 2.9 | Industrial Eng. | 2.5 | Business Admin. |
| 6 | 2.3 | Comp. Systems Eng. | 2.4 | Pedagogy |
| 7 | 2.0 | Architecture | 2.1 | Preschool Education |
| 8 | 1.9 | Civil Engineering | 2.0 | Primary School Ed. |
| 9 | 1.6 | Mechatronics Eng. | 1.9 | Industrial Eng. |
| 10 | 1.4 | Mechanics Eng. | 1.8 | Business Mgmt. Eng. |

Nonetheless, gender differences exist in the selection of professions. Although the 5 most requested professions (law, psychology, administration, industrial engineering and business administration) [9] appear in the options of both genders, there are important differences between both genders regarding their other options (Table 1).

Basically, the data show that both men (18.4%) and women (20.2%) choose to study in their first four disciplinary options such as Law, Psychology and Administration. On the other hand, men choose from among their top ten options to pursue higher education, five careers related to technology, but being their 5th and up to the 10th option, those only represents 10.1%. The case of women is even more extreme, since women choose technology-related degrees until their 9th and 10th options, which represents only 3.7% (Table 2).

**Table 2.** Percentage distribution of the undergraduate population in its first 10 career options, differentiated by Social Sciences/Administration vs. Exact Sciences. Source: Own development based on data from statistics of higher education. School Year 2012–2013. ANUIES

| | Percentage relative to 100% of male undergraduate students | Percentage relative to 100% of female undergraduate students | Percentage relative to both gender |
|---|---|---|---|
| Soc. Sc./ Admin. | 20.4% | 29.2% | 77.82% |
| Exact Sciences | 10.1% | 3.7% | 22.18% |

The demand of the university student population of both genders shows that, among the first ten study options, on average 77.82% enter Social and Administrative Sciences, while 22.18% enter Exact Sciences.

The National Survey of Science and Technology reports that with respect to the interest of Mexicans in technology, 38.9% of respondents answered "something," but adding the answer "a lot" both add up to just over half of the sample (51.4%). And that it is considered that generating Mexican technology is "very important" (84.6%) and "regular" (9.2%). By age, young people aged 15 to 24 responded very important (60%), while from 25 to 34 years old, 63.1%, where the group aged 65 and over 26.9%. From the data presented we could infer that young people between 15 and 24 years are in the process of entering or staying in a higher education program, while the group of 65 and elder are many who have formed recent generations; this being the case, there is a perceived lack of interest of elder people that could be transmitted to younger generations.

It stands out, however, the fact that 74.3% of the interviewees do not know any technology developed by Mexicans.

With respect to disciplines such as Design, the National Autonomous University of Mexico (UNAM) reports that for the 2013–2014 school year, of the total number of aspiring students, only one in ten were accepted, of whom 34% were men and 66% women [10]. On the other hand, the Autonomous Metropolitan University (UAM) reports that for 2016 school year, 26 men (34.2%) and 50 women (65.8%) signed up for the Design degree; while to the degree in Information Technologies and Systems, there were 41 men (61.2%) and 20 women (38.8%); and in Computer Engineering, were 50 men (75.5%) and 18 women (26.5%).

With this panorama, it is observed that the entrance of women to disciplines related to the use and generation of technologies for interactive design is low, compared to that of men, approximately at a ratio of 3:7, except for the Bachelor of Design, where the reason is favorable to women in a 1:3 ratio.

Design professionals have much to offer for the development of interactive systems (including those related to sensors, actuators, virtual and augmented reality, code generation, programming, etc., but also objective evaluation technologies such as biofeedbacking, eyetracking, neurofeedbacking, or subjective ones, as user evaluations, from affective, anthropometric, cultural or functional). However, this is a field of action little explored in many of the educational institutions in Mexico. Such a situation depending on various factors such as the schools or institutions where it is taught, the theories it handles, or the sub disciplines themselves that it concentrates on. So far there is no data to help define the percentage of women who opt for optional subjects related to technology in their professional training, seeking to specialize their training in the field of interactive systems design.

The country has technological lag, and although the figures say that little by little a greater number of students select technological degrees, it is clear that this is not happening in the same way for men and women.

Based on the data reported, we know that women have the same opportunity to access higher education than men, but could it be a matter of gender or personality that discourages or marginalizes women to enter the University in these areas of knowledge.

But what factors would explain the apparent "disinterest" or lack of opportunities for women to access degrees and jobs in which knowing, using and developing technology is paramount. This situation is multidimensional; it can be explained from the perspective of technological backwardness and its negative impact on the low inclusion of women in the labor field, the limited development and technological investment or gender roles prevailing in the family, educational and social environment.

## 2.2 Gender Roles and Technology Profile

"The category of gender is a cultural construction that establishes the characteristics that are socially assigned to men and women according to their sex" [11].

An important factor for the construction of the personality of men and women is that of self-concept, "which can be defined as the image that each subject has of their person, reflecting their experiences and the ways in which these experiences are interpreted. There is a cognitive component of self-concept: self-image. Since the self-concept is largely a cognitive structure that contains images of who we are, what we want to be and what we manifest and want to manifest to others" [12].

Self-concept and stereotypes are important factors in the perception of what men and women know and what they can do and, therefore, influence the aspirations that individuals have in different areas of life.

Stereotypes are learned, which may imply that the social mechanisms that disseminate them can be modified in social, family and school environments through formal, non-formal and informal education, which transmit values of equality, access to the possibilities of communication, information, knowledge and social recognition of achievements, regardless of gender.

School environments are not free of gender stereotypes, in fact, several studies agree that: The group dynamics in class and the interventions of teachers influence the way women learn computer skills; Teachers pay more attention, time and encouragement to men in the use of the computer; There are disputes between girls and boys about access to machines, due to the selection of games, programs or browsing themes, with males generally winning; Women get more nervous when they have to handle computers in the presence of other people, especially if they are male [13]. This is related to the pressures they receive from the boys, who often disqualify, ridicule and try to dominate the team, demonstrating greater competition or even physical strength; although girls can use the computer equally, in terms of time and skill, imaginary representations of the computer expert remain masculine.

Thus, a set of social representations has been constructed around the difference between the sexes, which have legitimized the power of the male over the female sex. Such representations crystallize in language and precisely because of the symbolic character of language; it becomes a privileged means of analysis for the study of social representations of gender [14].

Gender roles are dynamic ideas that change over time, being multi-variable structures are impacted by various factors, cultural, social, economic, educational and political changes.

## 3 Sampling Frame

The survey period was from February 2018 to April 2018 in the facilities and groups of students of the Universidad Autónoma Metropolitana-Cuajimalpa. Using a quantitative discourse analysis as a strategy, a survey was applied in which students were asked to write five words that they associate with the "woman-technology" relationship with the intention of observing if there are prevailing patterns in personality references (self-concept) of female students that help identify the areas that are most relevant in the academic life of students of interactive systems. And therefore to propose an investigation that helps to define the effect that the perception on his personality has on the construction of his self-concept and the effect of this on his school performance.

For this research, the target population was composed of a total of 95 Higher Education students whose Teaching Learning Units were related to the generation of technological products from the Bachelor's Degrees in Information Technologies and Systems (36), Design (25 students) and Computer Engineering (27 students).

### 3.1 Categories of Analysis

The self-concept test and form (AF5), created by García and Musitu [15], was used as a reference for the definition of the categories. Called Autoconcepto Forma 5 or AF5, is a multidimensional model based on the theoretical model of Shavelson, Hubner and Stanton. With this questionnaire a five-dimensional self-concept evaluation is carried out: academic/professional, social, emotional, family and physical.

The numerous mentions of positive terms in the field of Academics, shows the relevance of school performance in their role as students, which speaks of expectations and comparison frameworks between students. On the other hand, the low mention of positive emotional aspects (seen globally) shows the emotional distance in the reading that students have on the total conformation of their self-concept, where personal values and beliefs are not very visible. The social, as a form of recognition of the individual, is subordinated to academic performance (Fig. 1).

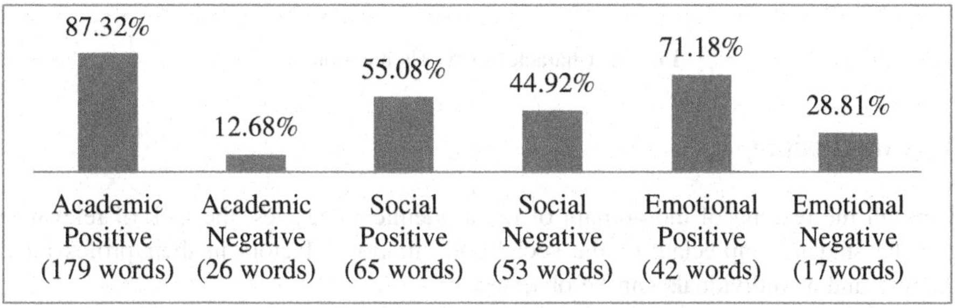


**Fig. 1.** Positive and negative percentage of self concept categories.

The mention of positive academic terms shows a cultural change, which presents the social recognition of the qualities they possess and the achievements made by women as part of the student community (Fig. 2).

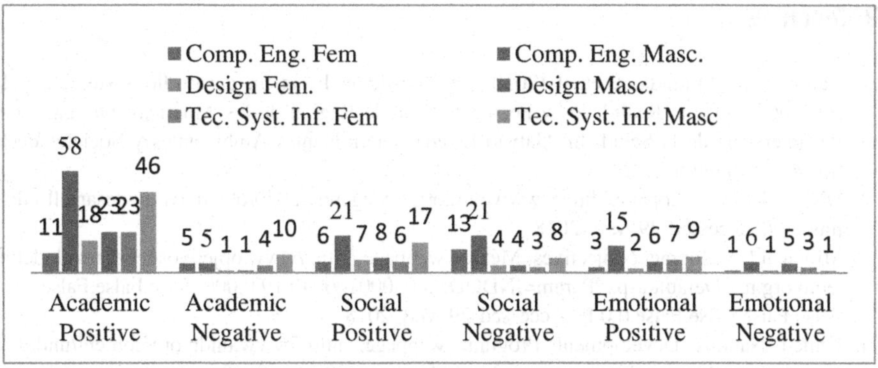


**Fig. 2.** Positive and negative self concept words per gender and career.

A striking factor is the high number of responses that "do not apply" (71 words out of a total of 495). This can be explained as the terms mentioned in this item are linked to devices (phones, computers, apps, etc.) or to professional activities in the field of technology (developers, programmers, etc.), family terms that market messages through television they have propagated by associating them with technology. It also highlights that, although the positive responses are by far the dominant ones, their reason for the negative words (3: 1 approximately) shows opportunities for intervention (Fig. 3).

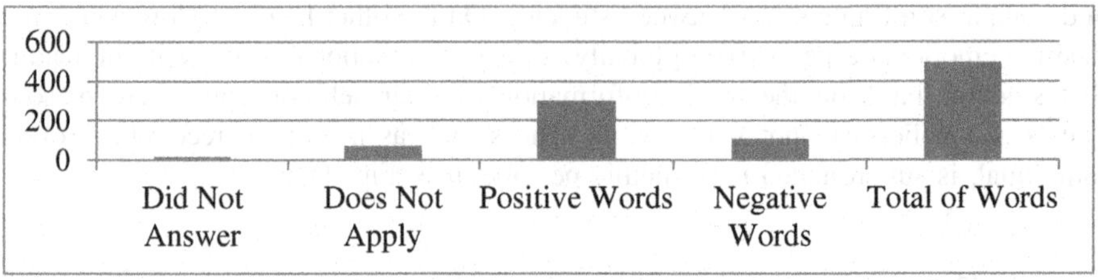


**Fig. 3.** Character of words mentioned.

## 4 Conclusions

Through the reading of the domain of the academic categories, the lack of relevance that the students appreciate of the social and emotional factors in their professional training and as individuals can be observed.

It is necessary to look for other mechanisms for linking students with technology beyond television media, to increase the culture of the technological field.

It is important to link technology with social sustainability to trigger the relevance of giving voice and design focused people, that is, identify what students need to exponentiate their skills, knowledge and attitudes, achieving an integral development of themselves.

## References

1. Carvalho de Miranda A et al (2010) La Transición hacia un desarrollo sostenible y la soberanía humana. Realidades y Perspectivas de la Región de las Américas. Organización Panamericana de la Salud. In: Galvao L (ed) Determinantes Ambientales y Sociales de la Salud, Washington D.C
2. ONU México webpage. http://www.onu.org.mx/agenda-2030/objetivos-de-desarrollo-del-milenio/. Accessed 29 May 2018
3. Millenial Development objectives, México webpage. http://www.objetivosdedesarrollodelmilenio.org.mx/Default.aspx?Param=INDODM003000100040,19,0,000,False,False,False,False,False,False,False,0,0,E. Accessed 29 May 2018
4. United Nations Development Program webpage. http://www.undp.org/content/undp/es/home/sustainable-development-goals.html. Accessed 29 May 2018
5. Voices in Phoenix webpage. http://www.vocesenelfenix.com/content/sistemas-tecnol%C3%B3gicos-para-el-desarrollo-inclusivo-sustentable. Accessed 29 May 2018

6. Medina Ramírez S (2004) La Dependencia Tecnológica en México. Revista Economía Informa UNAM 330:73–81
7. Franco J (2015) Ciencia y Tecnología: Una mirada Ciudadana. Encuesta Nacional de Educación Colección Los Mexicanos vistos por sí mismos, Los grandes temas. Nacionales. UNAM, México
8. Weforum Webpage. https://www.weforum.org/es/agenda/2017/09/el-preocupante-nivel-educativo-en-mexico. Accessed 29 May 2018
9. Zubieta García, J et al (2015) Datos procedentes ANUIES, "estadísticas de educación superior. Ciclo escolar 2012–2013" en: Educación. Las paradojas de un sistema excluyente. Encuesta Nacional de Educación, UNAM
10. UNAM Educative Offer Webpage. http://oferta.unam.mx/carreras/45/diseno-y-comunicacion-visual. Accessed 29 May 2018
11. Lamas M La (1996) construcción cultural de la diferencia Sexual. Porrúa and PUEG, México
12. González MC, Tourón J (1992) Autoconcepto y Rendimiento Escolar. EUNSA, Pamplona
13. Bonder G (2007) Las nuevas tecnologías de información y las mujeres: Reflexiones Necesarias. Potencialidades y Límites para la Alfabetización Informática. CEPAL, Naciones Unidas. Santiago de Chile
14. Flores J (2015) Comunidades, instituciones, visión de la existencia, identidad e ideología. In: Galeana P, Vargas Becerra P (eds) Géneros Asimétrico. Representaciones y Percepciones del Imaginario colectivo. Encuesta Nacional de Género. Colección Los Mexicanos vistos por sí mismos, Los grandes temas nacionales. UNAM, México
15. García F, Musitu G (2014) Manual AF-5 Autoconcepto Forma 5, 4th edn. TEA Ediciones, Madrid

# Gendered Indicators in OHS: A Number to Convince and Transform Public Policies

Florence Chappert(✉)

ANACT, 192 Avenue Thiers, 69006 Lyon, France
f.chappert@anact.fr

Up until 2008, the Anact-Aract network[1] used to be asked by companies to look at the gender mix of occupations, in a context marked by a shortage of labour. The question asked by metallurgy, building and motorway operation companies was: "What working conditions do we need to provide so that women can integrate male occupations?"

In 2009, with the economic crisis having drastically reduced this type of request, Anact's joint Board backed by the Women's Rights and Equality Department (*Service du droit des femmes et pour l'égalité*), decided to include the "gender-based approach" in its methods for the improvement of working conditions. The hypothesis which was made at that time was that prevention of some occupational health problems could be improved if there was understanding and taking into account of different working situations between men and women.

When, in 2009, we looked for gender-differentiated occupational health and safety data and research results on the issue of "gender and working conditions", we discovered the gaps in this area. We therefore had the same view as Karen Messing in Quebec, which formed the title of her 1998 book: *One-Eyed Science: Occupational Health and Women Workers*[2].

## 1 Some Statistics Which Influence Public Policies: More and More Working Accidents for Women for 16 Years

Since 2012 Anact has published a gender-differentiated statistical analysis using data on recognised occupational accidents and diseases supplied by the French health insurance fund for employees, the *Caisse nationale d'assurance maladie des travailleurs salariés* (CNAMTS).

The last statistic photograph published in 2017[3] reveals that although women are affected by half as many accidents as men in 2015, the drop over 15 years in the number of working accidents (reduction of 16% between 2000 and 2015) masks an asymmetric evolution between the sexes, namely a decrease of 30% for men in all

[1] The French Agence nationale pour l'amélioration des conditions de travail (National Agency for Improved Working Conditions) is a joint public institution, answering to the Ministry of Labour, which has the task of promoting innovative approaches to improve working conditions. It manages the network of Aracts or joint regional associations, which are spread throughout the country.

[2] Messing (1998).

[3] Chappert and Therry (2017).


S. Bagnara et al. (Eds.): IEA 2018, AISC 826, pp. 354–362, 2019.
https://doi.org/10.1007/978-3-319-96065-4_40

sectors of activity, compared to an increase of 33% for women, particulary in those sectors where women predominate.

We believe that, over these 15 years, women entered the labour market in France in growth sectors (like medical, healthcare and social sectors, but also bank/insurance or trade/retail), but in jobs that exposed them to inadequately assessed and recognised risk factors and in a context where prevention policies do not seem to be effective enough for the activities carried out by women.

Furthermore, again during these 15 years, the number of occupational diseases reported and recognised (with 80% being musculoskeletal problems) increased for women (+155%) at nearly twice the rate for men (+80%). In 2015, slightly more occupational diseases were recognised for men.

Work carried out in companies and studies into absenteeism (illness and accident, excluding paternity and maternity leave) show that, in France, women have 30% more absences than men according to the Department for Research, Studies and Statistics (*Direction de l'animation de la recherche, des études et des statistiques*)[4] of the Ministery of Labour. Work carried out in companies by the Anact-Aract network indicates that the role of working conditions is fundamental, even if part of this difference of four days absence between women and men per year is due to sick leave taken prior to maternity leave. As part of our work in companies, we have shown that there is no correlation between the level of absence and the number of children; only the "separated, divorced, widowed" situation is associated with more absences for women and men also.

## 2 Another Framework «Whereas, Everything Is not Equal»: A Key Understanding of Health Inequalities Between Women and Men

It must be noted that even the epidemiological analysis based on an "all else being equal" approach were not necessarily relevant for understanding the differences, in terms of occupational health, between women and men. We note that, in companies, women and men are in "all else being unequal" work situations, as they are not in the same occupations, do not work under the same conditions, do not follow the same career paths and do not enjoy the same work-life balance conditions.

Anact-Aract's network's interventions in companies have progressively allowed us to build a framework for analysis of occupational health inequalities between women and men with four areas of analysis: work organization and gender mix, working conditions, career paths and working time.

This model involves four areas of analysis which can explain different effects of work on health for women and men (cf Fig. 1).

[4] "Les absences au travail des salariés pour raisons de santé: un rôle important des conditions de travail", Dares Analyse, No 9, February 2013. See www.fonction-publique.gouv.fr/files/files/publications/hors_collections/Absences-raisons-de-sante-2013.pdf.

1. *Work organisation and gender mix*: women and men do not pursue the same occupations and do not have the same jobs;
2. *Work*: women and men are exposed to different risk and difficulty factors, which to a degree become apparent through differing health effects, especially in jobs where women predominate;
3. *Career paths*: women and men do not have the same career paths;
4. *Organisation Time*: women and men are not subject to the same working time constraints and do not pursue the same activities "outside of work", especially at home.

All these elements help to explain the different effects of work on the health of women and men (see Fig. 1).

A diagnosis a small company[5]

The involvement of the Lower Normandy Aract in a printing works formed the starting point of the model for understanding health inequalities between women and men. Despite investing in its machinery, the company was trying to understand why the female workers were suffering more musculoskeletal problems and were therefore more absent than the men.

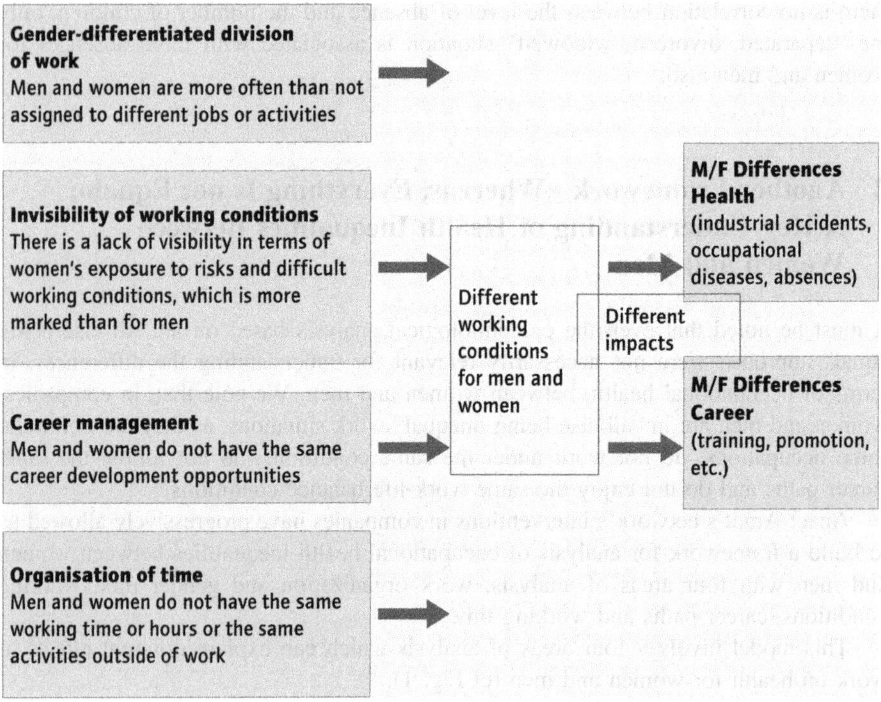


**Fig. 1.** Anact-Aract Gender Model: 4 areas of analysis

[5] For more information: Chappert et al. (2014).

The working conditions adviser, who was also an ergonomist, pointed out that, even though they worked in the same place, the female and male workers were not doing the same thing. In the same workshop, the women had access to four different types of job, and the men to nine. He also showed that the activities to which women were assigned, particularly finishing support where they were over-represented, were particularly demanding: repeatedly carrying small loads totalling up to 11 tonnes per day; fast-paced repetitive movements that involved grasping very "weighty" tomes.

All the stakeholders in the company were astonished that the women's jobs had become tougher than the men's, which, following automation, now mainly involved monitoring the machinery from their operator positions.

However, the diagnosis was further refined by taking a demographic approach to the staff data, which showed differing career development for women and men who had started in the same finishing support job. After three years, the men would move on from this beginner's job in the company to other opportunities within or outside the company. Some were also dismissed because they could not stand the pace or were not performing well in this woman's job, whereas the women remained in the same job until retirement or dismissal due to no longer being fit for work. The book cutting work, which allowed for quicker career development, was reserved for men because it was extremely difficult.

Lastly, the adviser showed that the system of rest breaks and their financial compensation for the two sexes were unfavourable to women: the women had to stop for unpaid rest breaks, whereas the men had negotiated a bonus when the automated machinery was installed that could not be stopped! When the results were passed on to the committee on hygiene, safety and working conditions, the women cried when they discovered these injustices.

Fortunately, the company did not stop at this diagnosis and took action, which is rare because many companies go no further than the findings, leaving a feeling of powerlessness. Firstly, the printing works realised that it could not manage the difficult working conditions simply through the lever of recruitment. It worked upstream with the book suppliers to limit the strain on the upper limbs and the carrying of loads, and redesigned the workstations. It also looked again at the issue of women's career paths: by recognising their skills in finishing support and their responsibility for the smooth operation of the production line, it allowed women to move on to operating assistant or machine operator jobs.

Following this experiment in this Normandy printing works, the Anact-Aract network has helped a lot of other companies where advisers have put on their "gender glasses". These actions have allowed a model for analysing occupational health inequalities between men and women to be gradually developed. This model has since been used to also analyse career and pay gaps, with the goal of gender equality in the workplace.

An experiment in a big company

As part of long-term action, Anact has sought to understand in a big logistics company why female workers are absent 30% more than male workers (50% more for those who were recently recruited), taking into account that sick leave before maternity leave explains only a very small part of this gap. Even though in this company, there is the same number of male and female workers who do the same jobs and are paid almost the same.

The executive management hesitate a lot to deal with this subject of gender absenteeism gap. Their first reaction was to tell us: "You will open Pandora's Box"

And when we asked them «What do you mean?». They gave us two answers: «First of all pace, and secondly schedule». In fact, the pace of work was designed originally for fit young man and the schedule was designed to work very early in the morning, typical of part time work, which doesn't work for parents who have children at home».

As to the question «Why women are more absent than men?» the ready made answer we heard was that women are absent because of their advancing age or from trade unions particularly because of management style. But with our gender diagnosis we revealed 20 hypotheses and secondly we checked 16 of them over the 4 areas of analysis.

First of all, the equipment, such as sorting racks, or the modes of transport, such as bicycles or cars, are unsuited to women's sizes (designed for the average size of the whole population, or for men of an average size) and are therefore a source of musculoskeletal problems and more frequent unfitness for work among women. Gender has also opened possibilities of organizational reasons. We have shown for example the negative effects of seniority rule on the attribution for sectors or time off because women were recruited after men in this company. The physical demands of work are exactly the same for men and women, or actually perhaps more demanding for women given that, due to the seniority rule, female workers are over-represented in sectors where there are more parcels to be carried. Given their physical makeup (less muscle strength etc.) and their lower level of physical fitness, female workers have to make a much greater physical effort than male workers.

Therefore, when women started work in this company, the work organisation did not change enough to take into account the different work and life situations for women and men. And this unsuitability lead to extra absenteeism for women. We are far from conclusions made in previous studies carried out in this company which showed age and gender as the principal causes of absenteeism. This doesn't make sense because we can't change the data intrinsically linked to people.

## 3 Anact Network Gender Contribution for New Acts and Policies

The statistical data obtained and the experiments conducted in a number of companies have lent weight to the Anact-Aract network and encouraged legislative developments. We can say today that Anact network studies have led to legal evolution.

Our contribution to public and parliamentary debate has allowed the subject of gender occupational health at work to emerge. Following our first publication about gender data of occupational health accidents, we have been interviewed as Anact' experts several times by National Assembly, and Senate since 2013.

And during the discussions about the Equality law in 2014, many Minister's of Parliament and Senators who listened to us intervened to file amendments, so that the gender question be included in Risk Evaluation. It is not the Ministry of Labour but the Ministry of Women's Rights which insisted. It is really the elected members who listened to us and who were impressed by our results who lead to two improvements of the equality law in the final text in 2014.

Firstly, the Act on real equality between men and women adopted in France on August 4th 2014 makes two changes to the regulations. Firstly, in terms of gender equality in the workplace, companies with more than 50 employees are now required to produce "gender-based occupational health and safety indicators similar to career or pay indicators".

Secondly, in terms of risk prevention, it is now stated that «Risk Evaluation has to take into account the different impacts of risk exposure according to sex». It would be wrong to read this article too quickly, with an essentialist vision, as this could lead to the conclusion that women should be excluded from certain tasks, positions and occupations due to their specific characteristics, which would be discriminatory. Up to now, in France, the only legislative provisions that can now lay down restrictions with regard to women's health and safety are those intended to protect pregnancy and maternity. This new provision on taking into account the gender-differentiated impact of exposure to risks during risk assessments stems from the different conditions of exposure to risks for women and men. This is because, even in one company, men and women do not have the same occupations, career paths and activities outside of work, and even the same work does not necessarily have the same impact on the health for women and men. The new legislative provision in France therefore reinforces and supplements the existing legal framework with a view to improving the single risk assessment document, which is mandatory in companies, and to adapting preventive actions and making them more effective.

On the other hand, the Modernisation Act of the Health System in 2016 requires that the Social Security (*Caisse nationale d'assurance maladie des travailleurs salariés* CNAMTS) and occupational health doctors now have to carry out gender data analysis every year in their report.

But it needs to be understood that we never proposed a modification of the law. We never thought that it could go so far. The result is that our approach has contributed to improving equality policies but also health and prevention policies. We were surprised to discover in the media the improvement in the Health Law at the end of 2015, specifically, the measures for our Social Security to produce and publish gender data in it's management annual report and also for the occupational health doctors to do the same for their annual report.

This is not directly linked to Anact intervention but it is part of the directive given by the Ministry of Women's Rights so that an integrated equality approach be carried out in all areas of public action.

## 4 Progress but a Lot of Vigilance Needed

The aim of those new provisions is to adapt systems of work and prevention policies, by taking into account the different conditions of exposure for women and men. Our experience in companies has, however, shown that it is still risky in France to reveal gender-differentiated occupational health data that is unfavourable to women because this can reinforce the prejudices and stereotypes that are deeply rooted in this country ("women are more fragile and cannot withstand pressure as well") and lead to women being discriminated against by being dismissed or no longer recruited.

However, for those working for equality, introducing the health issue runs the risk of women being disadvantaged and discriminated against in the labour market. Taking the gender issue into account can be an opportunity to improve ergonomics and ease difficult working conditions or improve work organisation and prevention. The questions raised by taking gender into account in occupational health and safety are complex and even taboo in certain respects, in an apparent context of neutrality and egalitarianism in France.

A lot of vigilance points should be taken into account. Firstly it is still very delicate to use gender occupational health data especially in France. To reveal vulnerability for women in the workplace can be disturbing. There is a strong risk of essentialism to label women by their «nature» (and not men of course), and to use stereotypes such as «women are not adapted to working environment for example to pressure of work» or «women are fragile and their place is with the children». We have a real risk for women's employment with employers who may say consciously or not: «They are too small, our machines are not adapted to their body shape, so we will not recruit them». Or, «women are more absent than men, so we would rather recruit men». Indirect discrimination risks are very high, especially in high a productivity work context, that meaning that everyone would not be able to follow the work place rhythm in a sustainable manner.

In order to avoid that situation, gender diagnoses have to be intensified. For example, for absenteeism, we should go against stereotypes and do a more precise analysis which would allow us to explain that very often, if women are more absent and more injured than men, it is because they are not in the same jobs, in the same career paths, that they stay stuck in the same positions and that we did not take into account their body shape to adapt working tools. That is subversive and uncomfortable. But it has to be done, if not, there is a real risk for women's employment especially for the least qualified.

Secondly, one of the problems is that we hear a double talk which often stops us from improving in that field. On one hand, we are told: «What you are saying is obvious. When we think about occupational health, we cannot reason in term of equality. It is normal that work has different effects on people. It depends on the requirement of the job and on individual physical, biological and morphological characteristics». But as far as the executives are concerned (both trade union and companies) they say: «It is not normal to ask those questions». Work is the same for everyone, rules are the same for everyone. Seniority is a rule which was created to ensure equality between employees. Fathers also look after their children, men also suffered from carrying weights. Politicians and economists are torn between 2 completely opposed ideas, and this misunderstanding prevents us from changing anything.

In the big logistics company already mentioned, we asked occupational health doctors to provide us charts about the bodyshape of employees. We observe that the size of women and men is distributed like a «Gauss curve», like the general French population. Women are on average 1,62 m tall and men 1,75 m. If equipment, tools, and vehicles are calculated on the average size of women and men, we sometimes have only 30% of the employees for whom the equipment is the correct size.

In fact, it is complicated because in a general context of resistance to take into account the concept of equal opportunity: how can we take into account the size difference? Do we propose the same treatment for everyone or a different treatment? Or else can we suggest new standards, better for everyone, which is the solution we recommend. For example to limit weight carrying to 15 kg for everyone and not 25 kg for women and 55 kg for men as is the case at present in the labour law.

But globally, we have noticed for the moment that even if gender occupational health indicators are in the law, it is not normal to look at them as we look at the salary indicators gap. Occupational health stakeholders are also far from including a gender approach in their analysis and recommendations.

The third point is that at every Anact intervention, we have to insist and demonstrate that we are doing with a gender approach what we call «putting on gender glasses» helping to improve working conditions for everyone - women and men. Because of this when we get to the end of an intervention, we never act only for women.

In the case of a logistic's company, we worked to improve schedules for everyone: morning shift for parents, carrying weights, shelf sizes. We are really obliged to explain that at the start. In France, we are very afraid of positive actions and corrective measures. So the idea that a gender diagnosis will help us to improve the situation for everyone, is in itself very important.

Though, the idea is really to say that a gender approach will help us to improve working conditions for everyone. It is very important but it is not so easy to carry out. For example, the day when a working group suggested to use seniority rule alternatively for men and for women (like in France for municipal elections), trade union did not agree because we were questioning the «sacred» and (individual) principal of: «My professional live has been difficult, so it is my right». Trade unions did not want to look at things globaly taking into account discriminatory effects of such a measure on women but also on young people. However, in that case as other one, gender approach was at the beginning, but after «women' causes» which may disappear after. It is paradoxal. Gender is a question at the start, but in the end, changes happened not because of equality reason but for other reasons, for example commercial issues: «The schedule is being changed because of the customer's needs to be served in the afternoon and not in the morning. Though, for me, gender approach is a good tool or method, but after women' causes are not completely defended. I would say we loose gender approach after. But including the gender question can allow us to see things we would not see otherwise.

## 5 Conclusion

However, the evolution of occupational health and safety policies legally acceptable may galvanise the use of "these gender glasses" so that progress is made in health prevention and promotion for everyone, regardless of gender. These new measures should allow for the adaptation of work organisation and prevention policies for a better health at work for women and men. This encourages companies to put in place for example risk assessment taking into account gender or including prevention of sexism and sexual harassment at workplace which is a new issue for both equality and occupational heath.

## References

Messing K (1998) One-eyed science: occupational health and women workers, Temple University Press

Chappert F, Therry P (2017) Photographie statistique des accidents de travail, trajet et maladies professionnelles selon le sexe entre 2001 et 2015 sur. www.anact.fr

Chappert F, Messing K, Peltier E, Riel J (2014) Conditions de travail et parcours dans l'entreprise: vers une transformation qui intègre l'ergonomie et le genre? Revue multidisciplinaire sur l'emploi, le syndicalisme et le travail, 9(2) (www.remest.ca); and the e-learning on this project at www.cestp.aract.fr/-E-learning

# A Brand-New Risk in Japan?—Risks of Industrial Accidents in the Age of Diversity Management and Their Countermeasures

Hongson Shin(✉)

Tokiwa University, 1-430, Miwa, Mito, Ibaraki, Japan
shin@tokiwa.ac.jp

**Abstract.** In order to address the declining labor force problem in Japan, certain activities called "Diversity Management" are now being promoted by the Japanese government. In these activities, many enterprises have adopted Diversity Management for innovation and improvement in productivity by utilizing various human resources including women, people from overseas, the elderly, and people with disabilities. One of the main themes in diversity management, especially in the growth period of enterprise, has been the promotion of female employees and temporary workers including those who are of younger age and elderly.

Some unexpected accidents have been occasionally reported, which have been caused by cultural differences or differences of values. Under these circumstances, notices such as a checklist of safety points and information of incident reports have been provided for enterprises promoting Diversity Management, especially in the case of younger temporary workers and female temporary workers who have burdens of house-holdings.

Also, an internet questionnaire was performed on parents' anxiety in variable risks including nuclear radiation risk, after the accident of Fukushima Daiichi Nuclear Power Plant, 2011 (n = 1500, performed in 2018). Some trends of career choice were revealed in the influence to children's future: the anxiety of parents who have children of elementary school were highest among other groups. These trends may affect their career promotion and decline job-motivation or the feeling of job-satisfaction. Some countermeasures in the viewpoint of life-career balance should be prepared in the case of over-anxiety.

**Keywords:** Diversity management · Brand-new risk · Temporary workers
Career choice

## 1 Introduction

### 1.1 Diversity Management in Japan

In Japan, some severe effects have gradually begun to appear because of the declining labor forces: a reduction in business hours and decrease in the number of stores owing to lack of employees. The declining labor forces have occurred due to falling birthrates, an aging population, and a low level of women's participation. In order to address these problems, certain activities called "Diversity Management" are now being promoted by


S. Bagnara et al. (Eds.): IEA 2018, AISC 826, pp. 363–369, 2019.
https://doi.org/10.1007/978-3-319-96065-4_41

the Japanese government, especially by the Ministry of Economy, Trade and Industry (METI). In these activities, many enterprises have adopted Diversity Management for innovation and improvement in productivity by utilizing various human resources including women, people from overseas, the elderly, and people with disabilities. And one of main themes in diversity management has been promotions of female employees, especially in the growth of recruitment. More attentions should be paid on the most serious problems concerning employments of women's and younger age; so many women and younger workers have been employed as temporary workers. In Japan, temporary workers usually do not receive the benefits in the Equal Employment Opportunity Law. It will be important things to enable as many workers of young age as possible to improve their skills by strengthening efforts that will enable workers to move from temporary to regular working positions, more easily.

## 1.2 Brand-New Risks? From Diversity Management in Japan

Some unexpected accidents have been occasionally reported from several years in Japan, which have been caused by cultural differences or differences of values. Because managers or safety and health controllers never considered those accidents earlier, they were confused and now they need appropriate advice or countermeasures to adjust to new working conditions. Under these circumstances, notices such as a checklist of safety points and information of incident reports may be needed for enterprises promoting Diversity Management, especially in cases of temporary workers in younger age, females who have burdens of house-holdings.

On the other hand, there have been severe problems of employee's mental health in Japan. From the statistics of mental problems, 59.6% of employee have felt "severe strain" in their jobs and in working life (MHLW: Ministry of Health, Labor and Welfare, Japan, 2016). **[The cause of job strain]:** the most prevalent cause of the job strain for employees is "quality & quantity of job" and "human relations". As for human relations problem, its tendency is shown most strongly in female employee. **[Temporary workers]:** "the anxiety about employment" was the most prevalent cause of the job strain among temporary workers. Considering the fact of 58% female workers are temporarily employed, while temporary workers in male worker is 22%, some supporting systems are needed to ease their strain and anxiety. **[Age]:** most frequent and mean age of temporary female workers is around 20–40 years old, some wider viewpoint of family and career choice may be needed.

**[Anxiety of parents and career choice]:** after the eastern Great Earthquake 2011 and accidents of the Fukushima Daiichi Nuclear Power Plant, types of value may become more diversified and the priority between family and job also have been shifted to family in younger age and female workers. These tendencies may affect their career choice in the near future and they may miss the proper opportunity in the career development. Under these circumstances, more information about anxiety of workers in the viewpoint of children's future should be collected to address these problems confirming their detailed demand. An internet-questionnaire was performed on parents' anxiety in variable risks, after the accident of Fukushima Daiichi Nuclear Power Plant, 2011.

# 2 Variable Risks, Safety Points from Information of Incident and Accident Reports in Japan

## 2.1 The Notice of Variable Risks in the of Diversity Management Diversity Management in Japan

From variable occupational statistics such as the cause of strain, the occupational accident (MHLW) and statistics of incident report (JISHA, Japan Industrial Safety & Health Association) over 30 years, several tendencies were revealed. At first, from the statistics of occupational accidents in Japan, 2016, total number of fatality was 928 and 117,910 in fatality and injuries. Decrease in the fatality and injury for 30 years in Japan has received hard-earned praise. On the other hand, some trends in "the Tertiary industries" such as restaurant, mass retailers, hospitality industries, were suggested from the line-up and ranking of accidents and incidents by industries.

**Risks from "the Tertiary industries":** The number of fatality and injury in Tertiary industry keeps increasing annually, which accounts for 46% of total numbers in 2016, while those in construction (12.8%, 2016) and manufacturing industries (22.4%, 2016) keep declining. Other up-to-date tendencies were revealed as for demographic factors: age, gender, foreign workers (workers from overseas). These factors have close and mutual association with the specific industry: the tertiary industry. The highest frequency cause of injuries in Japanese industry was "slip" (27,152), next was "fall" (20,094). These were now gaining prominence as the most popular causes in the Tertiary industries. There were various problems that have to be solved in the Tertiary industry: however, due to high ratio of small-sized or individual business in the industry, there were few opportunities or no communities for information sharing. These problems have lead to difficulties in disseminating safety education and countermeasures.

**Workers from overseas:** other risks of foreign workers are now revealed from recent statistics. The number of accidents keep increasing due to lack of communications or common sense, difference of lifestyles. Also, accidental risks in young part-time jobs workers such as university & high-school students and elderly workers are highest from thousands ratio. In the viewpoint of Diversity management, some countermeasures and proper educations will be needed in the near future.

**Job strain and mental health in temporary workers:** as some tendency about job strain in workers above, severe risks of workers' mental health should be noted: the most prevalent cause of the job strain for female employees is "human relations". Since the ratio of non-regular employment to regular employment is higher in women than in men, non-regular female employees have job strain from both **"employment instability"** and **"human relations"**. Because the manufacturing industry and the tertial industry have higher ratio of female non-regular employees, these industries hold complex risks for job strain. Those who are in charge of occupational safety and health will be required to take further countermeasures against these risks.

Table 1 summarize those data and tendencies as the notice of variable risks in the viewpoint of Diversity Management Diversity management in Japan (see Table 1).

**Table 1.** The notice of variable risks in the viewpoint of Diversity Management (correction from SHIN, 2018)

<table>
<tr><th></th><th colspan="2">variable risks to be noted</th><th>representive industries having variable risks of each group</th></tr>
<tr><td rowspan="2">Age</td><td>elderly workers</td><td>・intended unsafe act / tendency of risk-underestimation<br>・errors due to the decline of physical function</td><td rowspan="3">【Tertiary industry(restaurant business・mass retailer・hospitarity industries )】<br>・high ratio of accidental injury : temporary workers, younger age, foreign workers<br>・The increase of injury risk : part-time job and operation by one person (especially university students and high-school students)<br>・most frequent accidental causes: "Slip" "burn" "injury by cutting"<br>・backyard accidents<br>・due to high ratio of small-sized or individual business in the tertiary industry, there are few opportunities or no communities for information sharing → difficulties in disseminating safety education.</td></tr>
<tr><td>young workers</td><td>・lack of experiences, misjudgement due to immaturity, lack of awareness<br>・highest risk of injury by accident (thousands rate)</td></tr>
<tr><td>temporary workers</td><td colspan="2">・anxiety about employment (strain)<br>・tendency of low validity in personarily test<br>・high ratio of female, foreign, young or elderly workers (possibility of complex risks)<br>・less opportunity for safety educationand improvement in mental health<br>・tend to be less motivated</td></tr>
<tr><td>Female workers</td><td colspan="2">・Cause of occupational strain: human relations in the workplace<br>・possible incomprehension of wide spectrum of values and diverse way of working<br>・Lack of communication and exchange, especailly in women-scarce workplaces<br>・Internationally low evaluation in women's success</td><td rowspan="2">【medical care, nursing, care】<br>・Problem of languages and communications<br>・Difference in common sense<br><br>【manufacturing industry】<br>Accidental risk by the difference in culture and lifestyle of foreign workers<br><br>etc.</td></tr>
<tr><td>Foreign Workers</td><td colspan="2">・problems of communications: language, mis-understandings of orders, accidental risks<br>・accidental risks due to differences in culturs and lifestyles<br>・difficulties in sharing common sense<br>・difficulties in common education and research implementation</td></tr>
</table>

## 2.2 Parents' Decision Making About the Job to Improve Work-Life Balance, Considering Their Children's Future

**Purpose:** this survey was performed to research factors that affect parents risk perception and anxiety about children's future and to confirm children's most affectable developmental period for parents' decisions.

**Methods:** an internet questionnaire was performed through an investigation firm. 3000 members who registered at the investigation firm were requested by e-mail to answer the online questionnaire at the specific URL, which was assigned by the firm. The firm closed the URL when 1500 access were confirmed. Participants received reward points (500 Japanese yen) from the firm, after finishing the questionnaire. This study was approved by the ethics committee of Tokiwa University.

**Participants:** 1500 participants (Male: 739, Female: 761) (see Table 2).

**Number of Child:** None 36.9%, 1 child 16.8%, 2 children 32.9%, 3 children 11.2% (see Fig. 1)

**Table 2.** Numbers of participants (by Age, gender, %)

| Age | 20–29 | | 30–39 | | 40–49 | | 50–59 | | Over 60 (60–79) | | Total | Total % |
|---|---|---|---|---|---|---|---|---|---|---|---|---|
| | Years | Old | Years | Old | Years | Old | Years | Old | Years | Old | | |
| n/% | 195 | 13.0 | 252 | 16.8 | 289 | 19.3 | 246 | 16.4 | 518 | 34.5 | 1500 | 100 |
| Male | 100 | 13.5 | 127 | 17.2 | 145 | 19.6 | 122 | 16.5 | 245 | 33.2 | 739 | 100 |
| Female | 95 | 12.5 | 125 | 16.4 | 144 | 18.9 | 124 | 16.3 | 273 | 35.9 | 761 | 100 |

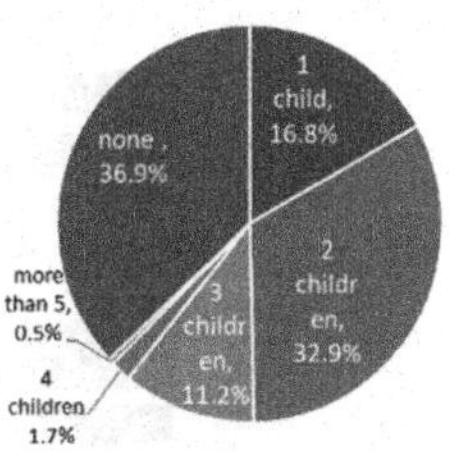


**Fig. 1.** Number of child (n = 1500)

**Period:** 5days: 9th Mar. 2017–13th Mar. 2017

**Questionnaire:** The Questionnaire consisted of 3 parts. [Part 1]: a request of the cooperation to the survey. [Part 2]: items of questionnaire were as follows; anxiety of risk in the Great Eastern Japan, items about preparation for next natural disasters, personality, attitude for nuclear plants, anxiety of children's future, decision making considering children's future. Items were selected from the work of SHIN & Masada, 2000 and some items were selected in order to survey feelings and anxiety of parents those who were caring their children. In this study, 4 items (Q4-3-1, Q4-3-2, Q4-3-4, Q4-3-5) from total questionnaires were reported. Participants were requested to select answer each question from 5 options: "strongly disagree", "disagree a little", "neither agree nor disagree", "agree a little", "strongly agree". Detail questions were as follows,

**Q4-3-1.** I make my own judgement and decision on various problems in everyday life.
**Q.4-3-3.** I feel stress when I make some decision and judgment
**Q4-3-4.** I feel anxiety about children's future
**Q.4-3-5.** My decision may affect my children's future.

[Part 3]: demographic and psychographic items; age, gender, numbers of children, etc.

## 3 Results and Discussions

As a result, highest ratio of "strongly agree" and "agree a little" were observed in the period of elementary school in "**Q.4-3-3.** I feel stress when I make some decision and judgment", "**Q4-3-4.** I feel anxiety about children's future", "**Q.4-3-5.** My decision may affect my children's future". As for "**Q4-3-1.** I make my own judgement and decision on various problems in everyday life", the period of senior high school was highest but the period of elementary school was also higher than others. On the other hand, lowest ratio was observed in the period of university (see Fig. 2).

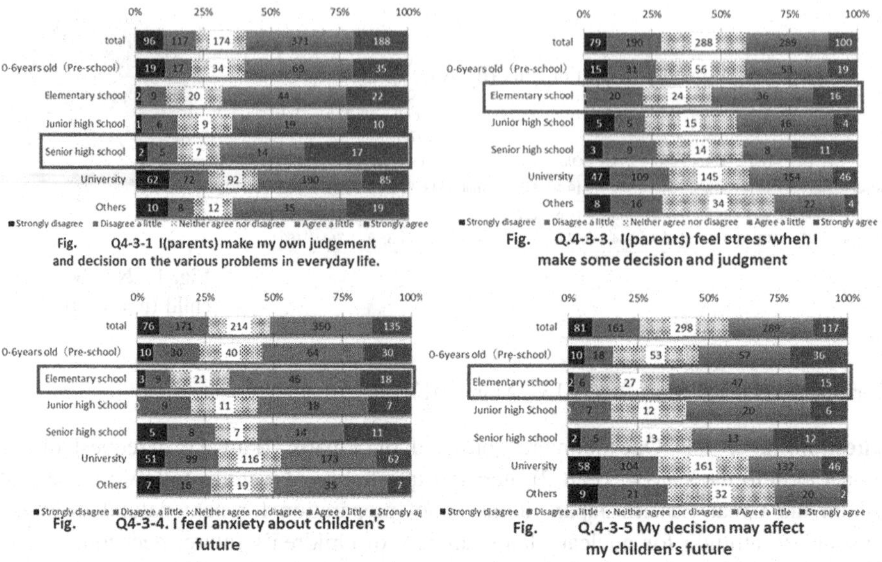


**Fig. 2.** Summary of each question by children's school period.

Highest ratio of "strongly agree" and "agree a little" were observed the period in elementary school. On the other hand, lowest ratio was observed in the period of university. From these results, parents those who have children at the elementary school may resign their job opportunity for children's future.

From these results, children's most influential developmental period on parents' decisions, was the period of elementary school. Also, these results showed that "the factor of children's development" may affect parents' risk perception.

From the viewpoint of diversity management, more attention should be paid on the factor of children's development concerning long-term employment and diversified promotion for who have children. Otherwise, parents who have children at the elementary school feel highest anxiety and strain for their own decisions and judgement and may resign their job opportunity for children's future. These decisions may affect job motivation and will lead to the reduction of human resources.

## 4 Conclusions

In this study, information on accidental risks and current trends of industrial accidents were collected and reported from various resources related to Diversity Management. Results of an internet-questionnaire on the strain and anxiety of workers who are parents have suggested the effectiveness of considering "the factor of children's development" in workers' career choice. These factors may affect their career promotion and decline job-motivation or the feeling of job-satisfaction. Some countermeasures in the viewpoint of life-career balance should be prepared in the case of over-anxiety.

**Acknowledgement.** This work was supported by JSPS KAKENHI Numbers 16K01879 (Grant-in-Aid for Scientific Research (C)).

The former part of this work, especially about statistics and notices about Diversity Management, were also reported on the article, reference no. 2.

A part of this study was presented at the 49th annual conference the Japanese Ergonomics Society, 2018.

## References

1. Shin H, Masada W (2000, in Japanese) A study on risk perception - attitude of residents on the outskirts of nuclear power plants, Japan. Jpn J Ergon 36(4):215–221
2. Shin H (2018, in Japanese) Risks of industrial accidents and their countermeasures in the age of diversity management. Rikkyo Psychol Res 60:15–28
3. Shin H (2014) A survey on the anxiety of preschool children's parents about nuclear power radiation risk and their decision makings, focusing on temporal changes of actions and free descriptions. In: Proceedings of ICAP 2014: international congress of applied psychology, Paris, France
4. MHLW (Ministry of Health and Labor Welfare, Japan). Statistics of industrial accident and job strain. http://www.mhlw.go.jp/toukei/list/dl/h28-46-50_gaikyo.pdf

# Theoretical Bases of New Work Simulator-Based Aptitude Assessment in Vocational Guidance of Students with Disability or Special Educational Needs

Erika Jókai(✉)

Budapest University of Technology and Economics, Budapest 1111, Hungary
jokaie@erg.bme.hu

**Abstract.** In Hungary there is a governmental endeavour to increase the participation of disadvantaged and disabled workers in the free labour market. Nowadays making trades, especially lacking professions, more attractive is in general of high priority. The percentage of unemployed and of those who left their original professions is rather high, but this percentage is even higher among the disabled persons. Since the scope of disabled persons' retained abilities is narrower, and therefore their chances in the free labour market are rather smaller, in their case it is especially important to identify jobs really fitting to their intact abilities. Based on experience, three model components – "main groups of requirements", "ability-demand gap model" and "tentative causal model of input, intermediate and output variables" – have been accepted as theoretical basis for this research. In this paper the first empirical results of our ongoing research concerning vocational guidance of handicapped students and youngsters will also be presented.

**Keywords:** Work simulator · Aptitude assessment · Vocational guidance

## 1 Introduction

Vocational rehabilitation in general, and rehabilitating the younger age group of 15–30 years with disability or special educational needs in particular, is a key issues nowadays in Hungary. The Hungarian state, on the one hand, provides incentives to increase the employment of persons with reduced ability to work, while on the other hand, it penalizes employers if they do not employ a pre-defined number of handicapped persons relative to the total headcount of their employees.

In the period of January–May 2018 we had been participating in a research project aimed to help handicapped students and youth by rethinking the present Hungarian vocational guidance system and its tools. Generally, for the Hungarian young people who are in the process of career choice or career change the National Orientation Portal information provides information on the professions and the requirements. Teachers also support the elementary and secondary school/high school pupils and students by career counseling and career guidance programs (e.g. factory visits). In schools, young people have to indicate which profession or school they wish to learn and accordingly


S. Bagnara et al. (Eds.): IEA 2018, AISC 826, pp. 370–379, 2019.
https://doi.org/10.1007/978-3-319-96065-4_42

they have to attend a medical aptitude test. This aptitude test gives information on the general state of health, physical, mental and sensory capacity of these young people. With the result, the choosen school or training center decides on (or reject of) affiliation. When a young person cannot decide which profession to choose, or he/she is not aware of his/her abilities, he/she can ask for help from a career counselor, who can offer profession alternatives based on the results of the personality and motivation tests, school grades and medical aptitude status and working capacity reports as well.

In the world of labour the working capacity examination usually consists of a personal interview and applying the Assessment Center method, including some practical task simulation, which consists of those tasks that often occur in the given particular work environment.

The assessment methods mentioned above often may contain many subjective assessment criteria, and therefore it is common that young people cannot find and choose professions or careers that are appropriate to their real abilities.

In this pilot research we have integrated a work simulator into the process of personal mentoring as a new evaluation tool for vocational guidance. It is assumed that this service can provide a more effective career orientation service, because if young people choose a career that is appropriate to their interests and real abilities, it is expected that they have fewer failures and fewer will be the school leavers or career changer too.

## 2 Theoretical Models

Our approach is based on a combination of three theoretical models of different elaboration. The first model component describes the main groups of requirements to be employed in a job (Fig. 1). The second model component is the well-known metaphor of the ability-demand gap (ADG), the main message of which is that handicaps can be compensated (Fig. 2). The third model component (Fig. 3) is a tentative causal model of independent variables, intermediate variables, constant parameters (including abilities measured by work simulator) and dependent variables (measures of successful job performance).

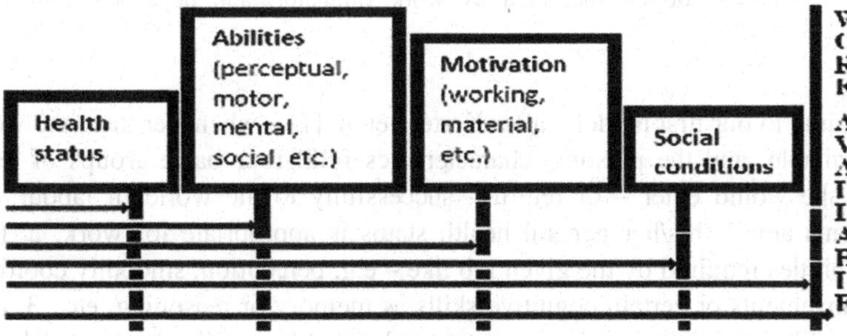


**Fig. 1.** Process of returning to work: the four main groups of requirements, after Kertész et al. [7]

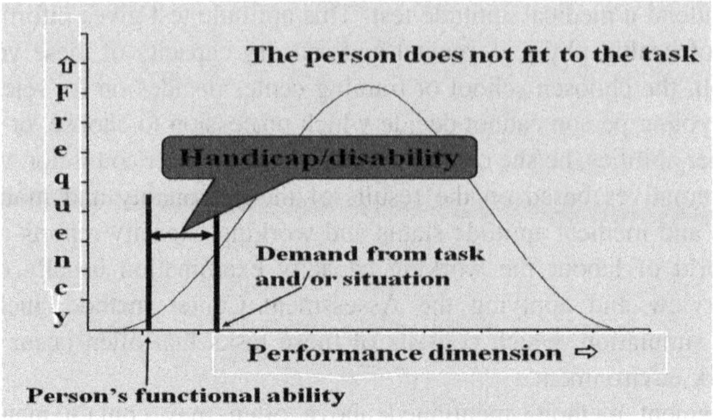


**Fig. 2.** Model of the ability-demand gap (ADG), after Izsó [4]

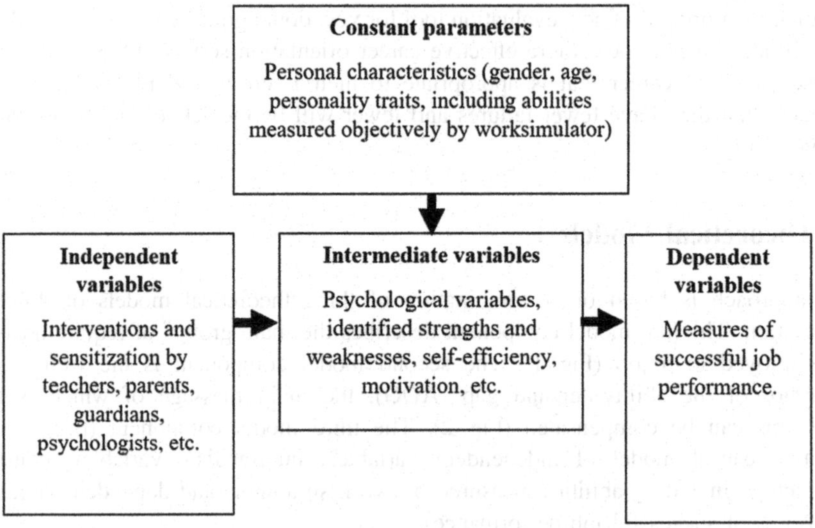


**Fig. 3.** Tentative causal model of independent (input) variables, intermediate variables, constant parameters (including abilities measured by work simulator) and dependent (ouput/target) variables

According to our first model – after Kertész et al. [7] – whenever there are working places available, and the person's characteristics fulfil four basic groups of requirements he/she would enter – or return – successfully to the world of labour. These requirements are: 1. his/her general health status is appropriate for work, 2. having suitable abilities required by the given job like – e.g. perception, smoothly coordinated muscle movements or certain cognitive skills as memory or reasoning, etc., 3. having positive work motivation, 4. supportive and reliable family background, living conditions.

The ability-demand gap model (ADG) is a very positive philosophy and ergonomic design approach. By the perception of this model the relationships between certain given impairment and the resulting handicap (or disability) can be summarised as follows:

- The handicap (or disability) is operationally defined as the difference (gap) between the demand from work task (and/or situation) and the person's actual functional ability, if the latter is the smaller.
- If the demand from task (and/or situation) is smaller than the person's actual functional ability, there is no handicap.
- The gap can be bridged over – among other things – by carefully chosen assistive technologies (AT) and/or appropriate ergonomic design or redesign of working conditions/tools/process.
- All this means that the handicap (or disability) is not only the person's attribute, but of the whole Human–Machine system consisting of the person and his/her AT.
- If we can bridge the gap the handicap disappears: by using AT to raise the person's abilities or abolish barriers by reducing task requirements.

In the Fig. 2, Ability < Demand, therefore handicap exists. If, however, by applying appropriate AT and/or purposeful ergonomic design the relation can get the form Ability > Demand and the handicap disappears.

The third model component below takes the form of a causal flow-chart that hopefully is capable of proving the existing causal relationships.

Since the personality traits of young people cannot yet be considered as stable (constant) – because these traits reach the final state only in the early adulthood –, and the work related abilities measured by the work simulator are dependent on the person's actual physical and emotional state, it is recommended to repeat these measures between the age of 13–25 in every 2 years.

## 3 The Work Simulators in General and the ErgoScope Work Simulator

Before presenting the ErgoScope work simulator specifically, a broader view about work simulators in general will be provided in two steps.

Firstly, just to save room, without writing details about the fundamentals, only some basic source materials are listed here about work simulations and simulators in general. The interested reader can find the basic ideas, concepts and terms concerning work simulators in the following recent books: Vega [13], Thornton and Rupp [10], Terpak [9], Fetzer and Tuzinski [2].

Secondly, the theoretical debates are shortly mentioned here about the concepts of fidelity and validity of a work simulator. The widely accepted results of these debates are that the validity is a more precise and accurate scientific term with already existing mathematical statistical background, and the validity can be used for practical purposes much better than the concepts of fidelity. Some basic publications about this topic are: Kennedy [5], Kennedy and Bhambhani [6], Ting [11], Ting et al. [12], Rustenburg et al. [8], Dahlstrom et al. [1]. The publications listed above are mainly about the validity of the two most widely used general-purpose work simulators: the ERGOS and the Baltimore Therapeutic Equipment Work Simulator.

Now, after Izsó et al. [3], some basic terms will be introduced, later on some researches already done with the ErgoScope work simulator will shortly be presented.

In its broadest sense, simulation is imitation, a kind of abstractions of reality. A work simulator (WS) is a tool for simulating any task, environment, condition, equipment or process related to work. High fidelity WSs utilize very realistic materials, equipment and conditions to represent the work task(s) that the candidate must perform, while low fidelity WSs use materials, equipment and conditions that are less similar to what is used on the real job. Important to emphasize, however, as mentioned above, that validity is a more important concept, than fidelity. While the fidelity is mainly a characteristic of the work simulator as a whole in itself, the validity is very much depends on the details of the equipment, process, situation and task to be simulated. If a WS is used to assess, maintain, or improve the functional capabilities of individuals with disabilities, it can also be considered AT.

The ErgoScope is a new, sophisticated and complex Hungarian made high fidelity WS that has the necessary broad spectrum of evaluation test batteries (or from another aspect, has the necessary broad spectrum of skill developing training facilities).

The ErgoScope consists of three measuring panels, as three independent workstations:

1. Panel 0. for measuring static and dynamic force (in standing position): lifting hutch/box of growing weight to desk height (90 cm) during which measuring pulling/pushing force, etc. (32 measurable parameters)
2. Panel 1. for measuring holding/grasping force, touching/tactile functions, keyboard, knob and "pencil" use etc. (in sitting position, 112 measurable parameters)
3. Panel 2. for measuring efficiency of use of buttons, switches, endurance/loadability, monotony susceptibility, etc. (in standing/walking position, 63 measurable parameters)

That broad variety of task situations and the vast repertory of corresponding measurable performance parameters by the help of which the ErgoScope WS is especially appropriate for skill assessment, skill development and also vocational aptitude tests of physically disabled persons.

## 4 Empirical Researches Performed so Far by the Use of the ErgoScope Worksimulator

As a first part of a planned larger scale research project, during the years 2012–2014 the full scope examination of 173 "healthy" and 59 "disabled" persons was carried out in the laboratory of the Hungarian National Office for Rehabilitation and Social Affairs (NORSA) by the help of the ErgoScope WS. In this sample there were 83 "healthy" and 12 "disabled" young persons (1–30 years). In January–May 2018 we have also measured another 50 disabled and disadvantaged young persons (13–30 years). In the following part of this paper some basic experiences of data analysis will be introduced inspired by our tentative causal model shown in Fig. 3.

The full scale examination with ErgoScope WS consists of as many as 36 different task situations each with 2 to 19 corresponding performance parameters, thus adding up altogether to 203 different measured performance parameters.

## 4.1 The Methodology Used

The data processing of the sample of 50 people is essentially only for the purpose of gaining experiences before launching the planned next step of this research on a much larger scale (involving about 200 new subjects still within 2–3 months). These actual results from this such a small sample can not be applied directly in practice, but may – and actually did – provide us with important experiences utilized in the further steps of this research.

Data processing was performed using IBM SPSS Statistics 23, basically in two respects. First we conducted comparative analyzes within the sample of 50 persons to determine which parameters (measured on the worksimulator) are significally different between the disabilite categories.

Secondly, we compared the parameters for each sub-sample to the values for the group of 173 healthy person. This approach is essential, because the basis of the rehabilitation philosophy is that primarily we focus on the intact capabilities of the people, which can be achieved by referencing the reference values for a healthy population. As the aim is to prepare people for the open labor market, the reference values also have to be characterized by the parameters of the typical intact workers.

Making of sub-samples within the sample of 50 persons, the number of 5 elements was accepted (10% of the total sample size). This is a reasonably accepted lower limit that still makes sense to carry out any mathematical-statistical data processing. Accordingly, the following sub-samples were formed:

1. ICD Q9090, Down syndrome, moderate mental disability: 11 persons
2. ICD F70, Mild mental retardation: 9 persons
3. ICD F90, Attention-deficit hyperactivity disorder: 9 persons
4. ICD H44-H58, Visually impaired: 5 persons

In order to avoid to lessen the number of elements of sub-samples under 5, only those parameters could be analyzed, where preliminary analyzes show no gender differences and it does not have to divide the samples further by gender.

The data analysis consisted of the following steps for each parameter group:

- the Kruskal-Wallis test to analyse if there were significally different parameters between the four sub-sample,
- if there were, than the Mann-Whitney test for each category-pair to identify which parameters are significally different and what is their direction,
- since during multiple pairwise comparisons the danger of alpha-inflation increases, the Bonferroni-correction was applied to cope with this problem (the stricter $p = 0{,}05/6 = 0{,}0083$ criterium was used instead of the original $p = 0{,}05$).

### 4.2 Results and Conclusions

The description and evaluation of all 203 different measured performance parameters would go far beyond the limits of this paper, therefore only the following task situations – and the related groups of measured performance parameters – were selected as examples: touch and handling rotating knob. These parameters reflect different aspects of fine manual motor performance of the studied persons (Fig. 4).

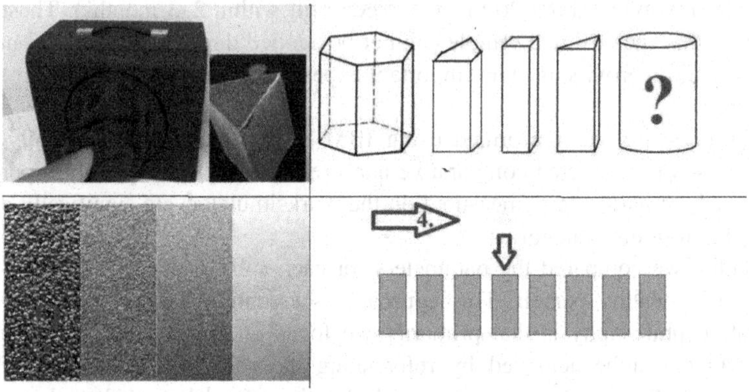


**Fig. 4.** Task 1: examples from the screen of the work simulator

Task 1: Touch

In this task the persons have to select an unseen object in a closed box (by touch only) based on the information on the screen. The WS records the number of correct and incorrect sensations.

The average performances of the whole 50 person "*disabled*" sample can be seen in the Table 1 above, while the corresponding numbers for the four selected disability sub-categories (for correct sensations by right hands only), within this are the following:

Down syndrome: 3.50 (by right hand),
Slight intellectual disabilities: 7.10 (by right hand),
Attention-deficit: 12.50 (by right hand),
Visually impaired: 16.25 (by right hand).

**Table 1.** Touch – number of correct and incorrect sensations

| Sensation | | "Healthy" women and men | | "Disabled" women and men | |
|---|---|---|---|---|---|
| Touch | | Right hand (n = 172) | Left hand (n = 173) | (Right hand) (n = 50) | (Left hand) (n = 50) |
| Correct (Nr) | Mean | 19,730 | 18,710 | 10,680 | 9,640 |
| Incorrect (Nr) | Mean | 3,290 | 3,960 | 6,800 | 6,95 |

By the Mann-Whitney test there was a significant difference between the performance of the youngsters with "Down syndrome" and with "Visually impaired". The difference between the results of the two categories of disability can easily be interpreted: while for young people with Down syndrome the task that needed for knowledge of simple geometrical concepts was unclear, for the young people with visual impairment, the task was just as difficult to accomplish as any other young person with a good vision and understanding (Fig. 5).

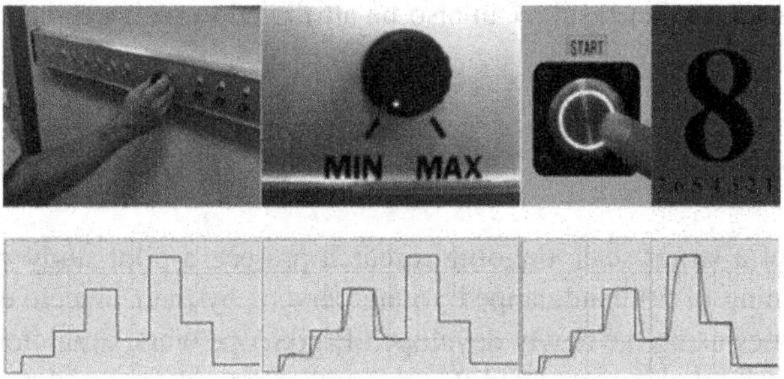


**Fig. 5.** Task 2: handling rotating knob by dominant hand at chest height: "Follow the squared pattern as accurately as possible"

Task 2: Handling rotating knob

In this situation the person has to follow the squared pattern as accurately as possible. He/she has to handle the rotating knob by his/her dominant hand – first at chest height, than above the head. The equipment records squared error, absolute error and maximal difference parameters. The greater is this value, the weaker the performance is.

In this task, good manners and persevering attention are needed. The average performances of the whole 50 person "*disabled*" sample can be seen in the Table 2 above, while the corresponding numbers for the four disability sub-categories (for tasks performed at chest height only) within this are the following:

Down syndrome: 204.95 (for chest height),
Slight intellectual disabilities: 156.32 (for chest height),
Attention-deficit: 28.34 (for chest height),
Visually impaired: 59.01 (for chest height).

**Table 2.** Handling rotating knob, by dominant hand

| Sensation | | "Healthy" women and men | | "Disabled" women and men | |
|---|---|---|---|---|---|
| Position | | Chest height (n = 173) | Above head (n = 173) | Chest height (n = 50) | Above head (n = 50) |
| Absolute error | Mean | 25,669 | 23,713 | 117,070 | 95,830 |

By the Mann-Whitney test there was a significant difference between the performance of the youngsters with "Down syndrome" and with "Attention-deficit". It can be stated, therefore, that the mentally handicapped and the visually impaired young people have worked with much more absolute error (much more inaccurately) than the "healthy" group. Surprisingly, this task was not a big challenge for the young people with attention-deficit. This can be explained by the fact that the task is relatively interesting and short, and the feedback is immediate, so that young people can correct their mistakes immediately, which can generate the feeling of a quick success. Youngsters with Down syndrome, however, could not even interprete how to control the knob, and their weaker results can also be attributed to their eye-hand coordination problems.

## 5 Summary

As a first part of a larger scale ongoing research project, a pilot study was performed aiming at screening of 50 handicapped young persons by their objectively measurable vocational abilities using the newly developed ErgoScope work simulator. In this paper some of the first results are presented that the author (together with the envisaged later results) would like to utilize in the practise based on the three theoretical model components described in the section "Theoretical models".

Based on the first model component titled "main groups of requirements" (Fig. 1), it is clear that perceptual, motor, mental etc. abilities form only a part the requirements needed for successfully putting candidates into man-machine systems belonging to a particular job, Therefore the ergonomic experts have to cooperate closely with other experts (physicians, psychologists, social specialists, etc.).

Based on the second model component titled "Ability-Demand Gap Model" (Fig. 2), it is also clear, that bridging the ability-demand gap (the so called "handicap") could be done either by increasing abilities via applying appropriate AT and/or purposeful ergonomic design, or by decreasing demands via setting lower norms (which can be feasible in working palaces where mainly or exclusively handicapped people work).

Based on the third model component titled "tentative causal model of input, intermediate and output variables" (Fig. 3), there are possibilities that by the help of appropriate mathematical statistical methods causal relationships be identified thus deepening our theoretical and practical understanding of the operation and interactions of real-life man-machines systems containing young handicapped persons. The empirical part of this paper is only a modest first step into this direction.

## References

1. Dahlstrom N, Dekker S, van Winsen R, Nyce J (2009) Fidelity and validity of simulator training. Theor Issues Ergon Sci 10(4):305–314
2. Fetzer M, Tuzinski K (eds) (2013) Simulations for personnel selection, pp 1–268. Springer Science+Business Media, New York. ISBN 978-1-4614-7680-1

3. Izsó L (2015) The significance of cognitive infocommunications in developing assistive technologies for people with non-standard cognitive characteristics (invited plenary talk), 6th IEEE international conference on cognitive infocommunications (CogInfoCom 2015), 19–21 October 2015, Győr, Hungary. http://coginfocom.hu/conference/CogInfoCom15/plenarytalks.html
4. Izsó L, Székely I, Dános L (2015) Possibilities of the ergoscope high fidelity work simulator in skill assessment, skill development and vocational aptitude tests of physically disabled persons ("Best Paper Award" winner conference paper), 13th international conference of the association for the advancement of assistive technology in Europe, 9–12 September, Budapest, Hungary. As book chapter In: Sik-Lányi, C., Hoogerwerf, E.J., Miesenberger, K., Cudd, P. (Editors) Assistive Technology. IOS Press, ISSN 0926-9630 (print), ISSN 1879-8365 (online), pp. 825 – 831
5. Kennedy L (1988) A reliability and validity study of the Baltimore therapeutic equipment work simulator at light, medium and heavy work intensities, p 178. University of Alberta
6. Kennedy LE, Bhambhani YN (1991) The Baltimore therapeutic equipment work simulator: reliability and validity at three work intensities. Arch Phys Yed Rehabil 72:511–516
7. Kertész A, Séllei B, Izsó L (2017) Key factors of disabled people's working motivation: an empirical study based on a Hungarian sample. Periodica Polytechnica, Soc Manage Sci 25 (2):108–116
8. Rustenburg G, Kuijer PPFM, Frings-Dresen MHW (2004) The concurrent validity of the ERGOSTM work simulator and the ergo-kit with respect to maximum lifting capacity. J Occup Rehabil 14(2):107–118
9. Terpak MA (2008) Assessment center strategy and tactics. PennWell Corporartion. ISBN-13 978–1-59370-142-0
10. Thornton GC, Rupp DE (2006) Assessment centers in human resource management: strategies for prediction, diagnosis, and development. Lawrence Erlbaum Associates, Publishers. ISBN 0-8058-5124-0
11. Ting W (1999) The validity of the Baltimore therapeutic equipment work simulator in the measurement of lifting endurance, pp 1–86. Ph.D. thesis, University of Alberta, Department of Occupational Therapy
12. Ting W, Wessel J, Brintnell S, Maikala Y, Bhambhani Y (2001) Validity of the Baltimore therapeutic equipment work simulator in the measurement of lifting endurance in healthy men. Am J Occup Ther 55(2):184–190
13. Vega NG (2002) Factors affecting simulator-training effectiveness. Univ Jyväskylä Stud Educ Psychol Soc Res 162. ISSN 0075-4625, ISBN 951-39-1370-8

# Ergonomic Risk Factors in Women Workers Involved in Handicraft Industry of Patiala District

Ajita Dsingh(✉) and Jeewanjot Kaur

Department of Sport Sciences, Punjabi University, Patiala, Punjab, India
ajitadsinghl@gmail.com

## 1 Introduction

There are many small scale industries in Punjab which help the lower middle class workers to get their employment. Among all, handicraft industry is the mostly widely established small scale industry. It is well established in the Patiala District of Punjab, India and is organized by both government and private sectors. Most of the workers involved in the industry are women. The working hours in both the sectors are long and in most of the cases the workers are employed on the basis of contractual form and get their honorarium on the basis of amount of work done. Thus every worker wants to earn more and try to finish as much as she can. These women workers work very hard at the industries as well as at their home too. Moreover due to poverty and workload stress both at home and their work place they suffer from drudgery also. They have to depend upon a lot of repetitive movements of their joints and have to follow awkward body postures in order to finish the task. It is observed that in these types of industries the workstations which are used are not ergonomically designed. Thus due to prolonged awkward postures and repetitive joint movements for long hours these women may develop certain musculoskeletal disorders.

### 1.1 Aim and Objectives of the Study

Aims and objectives:

1. To identify incorrect posture of female workers belonging to different categories of handicraft industry.
2. To identify the various musculoskeletal disorders developed in this industry.

### 1.2 Hypothesis

1. Repetitive movement during weaving may result in particular type of musculoskeletal disorder.
2. Traditional weaving posture cause back and neck problems.


S. Bagnara et al. (Eds.): IEA 2018, AISC 826, pp. 380–385, 2019.
https://doi.org/10.1007/978-3-319-96065-4_43

# 2 Methodology

## 2.1 Study Design

Experimental.

## 2.2 Study Duration

6 months.

## 2.3 Population Sample

90 workers were chosen who were engaged in various activities of handicraft industry like dari weaving, phulkari weaving, and naale weavings comprising 30 workers in each category.

Dari weaving involves fixed work station where workers have to sit in squat position and bend shoulders for prolonged hours with more exertion on lower back, shoulder and leg musculature.

Phulkari weaving need no fixed workstation as a group of women usually do embroidary while sitting on their individual work which can be carried anywhere with any posture which in turn gives them freedom to opt for awkward postures due to lack of ergonomic knowledge.

Naale weaving have workstation which the worker has to hold with feet resulting in overuse of ankle musculature and repetitive hand movements causing fatigue of hand musculature.

## 2.4 Inclusion Criteria

Workers must work for 3–5 h daily and workers must have working experience of last 5 years.

## 2.5 Exclusion Criteria

1. Pregnant women.
2. Women with CHD.
3. Women with physical disability like polio.

## 2.6 Instrumentation

1. Weighing machine was used to record weight of the subject.
2. Anthropometric rod was used to record stature of each subject.
3. Body mass index of each subject was calculated using the following formula:
4. BMI ($Kg/m^2$) = weight (kg)/Height ($m^2$)

5. Heart rate monitor. It was used to record heart rate during rest, activity and during recovery. On the basis of heart rate physiological cost of work and energy expenditure was calculated.
6. Plum line: To analyze the posture of each subject.

## 3 Results

### 3.1 Participant Characteristics

Table 1 shows the mean and SD values of Physical characteristics of the handicraft workers belonging to different categories. The physical characteristics include age (yrs), height (cms), weight (Kgs) and Body mass index. It is clear from the table that all subjects were almost of same age, height and weight. The mean values of all the categories shows that they were in the normal range of body height and weight according to their age and sex and hence possess normal values of Body mass index.

**Table 1.** Mean and SD values of physical characteristics of the handicraft workers belonging to different categories

| Categories | Age (yrs) | Height (cms) | Weight (Kgs) | BMI |
|---|---|---|---|---|
| Dari weavers (n = 30) | 34.4 ± 7.6 | 153.4 ± 3.8 | 57.7 ± 12.5 | 24.6 ± 5.7 |
| Naale Weavers (n = 30) | 35.7 ± 6.3 | 155.3 ± 5.8 | 61.3 ± 10.7 | 25.3 ± 3.8 |
| Phulkari weavers (n = 30) | 33.5 ± 7.4 | 151.7 ± 5 | 57.4 ± 12.1 | 24.9 ± 5. |

### 3.2 Results of Nordic Questionnaire

Table 2 shows the prevalence and severity of musculoskeletal symptoms in the last 12-months among the study participants. The overall prevalence of musculoskeletal complaints, particularly in the neck (46.6%%), lower back (67.78%), knees (26.67%) and shoulders (25.5%) was relatively high. The prevalence of neck symptoms was significantly highest among Phulkari weavers (60%) followed by Naale weavers (43.3%) and then Dari weavers (36.7%). Low back symptoms were reported highest (80%) by Naale weavers followed by Dari weavers (63.3%) and then Phulkari weavers (60%). The prevalence of work prevention by the subjects who had/trouble in past 12-months was also observed among the workers (Table 2b). It also shows the similar results. Many respondents reported disruption of normal activities due to musculoskeletal symptoms during the last 7-days and this percentage was 38.8% for neck pain, 37.78% for shoulder pain, and 58.89% for lower back pain (Table 3).

The results of the present study are in agreement with several previous studies which have reported the prevalence of musculoskeletal symptoms in different type of handicraft workers. It has also been reported by various studies that gender is also significant factor for neck and shoulder complaints, as females experienced such complaints more frequently than males (Wang et al. 2007) (Table 4).

**Table 2.** Prevalence of musculoskeletal symptoms during the last 12-months among the all subjects

| Body parts | Total subjects | | Dari (n = 30) | | Naale (n = 30) | | Phulkari (n = 30) | |
|---|---|---|---|---|---|---|---|---|
| | N | % | N | % | N | % | N | % |
| Neck | 42 | 46.6 | 11 | 36.7 | 13 | 43.3 | 18 | 60 |
| *Shoulders* | | | | | | | | |
| Right | 3 | 14.4 | 6 | 20 | 5 | 16.7 | 2 | 6.7 |
| Left | 5 | 5.55 | 1 | 3.3 | 1 | 3.3 | 3 | 10 |
| Both | 23 | 25.5 | 14 | 46.7 | 3 | 10 | 6 | 20 |
| *Elbows* | | | | | | | | |
| Right | 2 | 2.22 | 0 | | 1 | 3.3 | 1 | 3.3 |
| Left | | | | | | | | |
| Both | 14 | 15.5 | 2 | 6.7 | 10 | 33.3 | 2 | 6.7 |
| *Wrists/hands* | | | | | | | | |
| Right | 1 | 1.11 | 0 | | 0 | | 1 | 3.3 |
| Left | | | 0 | | 0 | | 0 | |
| Both | 10 | 11.1 | 0 | | 10 | 33.3 | 0 | |
| Upper back | 1 | 1.11 | 0 | | 0 | | 1 | 3.3 |
| Low back | 61 | 67.7 | 19 | 63.3 | 24 | 80 | 18 | 60 |
| Hips/thighs | 1 | 1.11 | 0 | | 0 | | 1 | 3.3 |
| Knees | 24 | 26.6 | 9 | 30 | 9 | 30 | 6 | 20 |
| Ankles/feet | 10 | 11.1 | 1 | 3.3 | 5 | 16.7 | 4 | 13.3 |

**Table 3.** Prevalence of work prevention by the subjects who have/had trouble in past 12 months among the subjects

| Body parts | All (n = 150) | | Dari (n = 30) | | Naalee (n = 30) | | Phulkari (n = 30) | |
|---|---|---|---|---|---|---|---|---|
| | N | % | N | % | N | % | N | % |
| Neck | 37 | 41.11 | 11 | 36.7 | 11 | 36.7 | 15 | 50 |
| Shoulders | 35 | 38.89 | 18 | 60 | 7 | 23.3 | 10 | 33.3 |
| Elbows | 15 | 16.6 | 2 | 6.7 | 11 | 36.7 | 2 | 6.7 |
| Wrists/hands | 11 | 12.2 | 0 | | 10 | 33.3 | 1 | 3.3 |
| Upper back | 1 | 1.1 | 0 | | 0 | 0 | 1 | 3.3 |
| Low back | 55 | 61.1 | 19 | 63.3 | 23 | 76.7 | 13 | 43.3 |
| Hips/thighs | 0 | 0 | 0 | 0 | 0 | | 0 | 0 |
| Knees | 24 | 26.67 | 9 | 30 | 9 | 30 | 6 | 20 |
| Ankles/feet | 9 | 1 | 1 | 3.3 | 5 | 16.7 | 3 | 10 |

**Table 4.** Prevalence of musculoskeletal symptoms during the last 7-days among the subjects.

| Body parts | All (n = 150) | | Dari (n = 30) | | Naalee (n = 30) | | Phulkari (n = 30) | |
|---|---|---|---|---|---|---|---|---|
| Neck | 35 | 38.8 | 11 | 36.7 | 11 | 36.7 | 13 | 43.3 |
| Shoulders | 34 | 37.78 | 18 | 60 | 7 | 23.3 | 9 | 30 |
| Elbows | 15 | 16.6 | 2 | 6.7 | 11 | 36.7 | 2 | 6.7 |
| Wrists/hands | 11 | 12.2 | 0 | | 10 | 33.3 | 1 | 3.3 |
| Upper back | 1 | 1.11 | 0 | | 0 | | 1 | 3.3 |
| Low back | 53 | 58.89 | 19 | 63.3 | 23 | 76.7 | 11 | 37.7 |
| Hips/thighs | 0 | | 0 | | 0 | | 0 | |
| Knees | 23 | 25.56 | 9 | 30 | 9 | 30 | 5 | 16.7 |
| Ankles/feet | 9 | 1.0 | 1 | 3.3 | 5 | 16.7 | 3 | 10 |

## 4 Discussion

This study presents evidence confirming that the work of weavers is strenuous. Consequently, the weavers suffer from musculoskeletal disorders arising from of a number of reasons — the most relevant being the adoption of a constrained sitting posture for prolong time periods. Even in our modern times, the traditional cultural activity of handloom weaving is invaluable, and it is essential that weavers are sufficiently cared for and valued as artisans and employees. Moreover, this profession provides the livelihood of a large section of the working population in Patiala district.

## 5 Conclusion

Women involved in handicraft industries work from morning to evening, usually they adapt same posture throughout work for long period of time without change in posture. Therefore, they reported muscular pain and reduction in capacity of work. Hence, periodic training program should be organized to emphasize on education of the worker regarding musculoskeletal disorder and the importance of rest and pause during their working hours. Moreover, their workstation should be reorganized in order to maintain their proper posture.

## 6 Relevance and Scope

Further research is required to build up effective preventative or ergonomic strategies that may be applied to the handicraft industry to decrease the incidence of occupational diseases. Therefore, interventions toward designing ergonomic weaving workstations and weaving hand tools should also be regarded as a main concern for improving the situation for workers in the handicraft industry. Therefore, there is an immediate need for government cooperation to provide a safer environment and proper ergonomics for weavers in the handicraft industry of Patiala. This study is important not only from the occupational health and financial point of view of the weavers, but also for the sustenance of the aesthetic and cultural value of the handloom weaving profession.

## References


Arphorn S, Limmongkok Y (2008) The modified sculptors workstation in pottery handicraft for reducing muscular fatigue and discomfort. In: The first East Asian ergonomics federation symposium, pp 12–14, Japan

Choobineh A, Hosseini M, Lahmi M, Khani R, Shahnavaz H (2007) Musculoskeletal problems in Iranian hand-woven carpet industry: guidelines for workstation design. Appl Ergon 38:617–624

Gautam K, Bahl N (2010) Occupational health and safety in the informal sector: evidence from the craft sector, pp 1–2. All India Artisans and Craftworkers Welfare Association, New Delhi

Motamedzade M, Choobineh A, Amin M, Arghami S (2007) Ergonomic design of carpet weaving hand tools. Int J Ind Ergon 37:581–587

Motamedzade M, Afshari D, Soltanian A (2014) The impact of ergonomically designed workstations on shoulder EMG activity during carpet weaving. Health Promot Perspect 4 (2):144–150

Mukhopadhyay P, Srivastava S (2010) Ergonomics risk factors in some craft sectors of Jaipur. Off J Hum Factors Ergon Soc Aust. 4–14

Nurmianto E (2008) Ergonomic intervention in handicraft producing operation. In: 9th Asia pasific industrial engineering & management system, Bali, Indonesia, pp 1008–1011

Purnawati S (2008) Occupational health and safety-ergonomics

Roy AK, Medhi JK (2010) Environment, occupational health and safety in the craft sector in India, base line study of selected craft clusters. All India Artisans and Craft Workers Welfare Association, India

Sharma TP, Borthakur SK (2010) Traditional handloom and handicrafts of Sikkim. Indian J Tradit Knowl 9:375–377

Tiwari RR, Pathak MC, Zodpey SP (2003) Low back pain among textile workers. Indian J Occup Environ Med 7(1):27–29

Wang P, Rampel D, Harrison R, Chan J, Ritz B (2007) Work-organizational and personal factors associated with upper body musculoskeletal disorders among sewing machine operators. Occup Environ Med 64:806–813

# Anthropometry

# "La Fabbrica si Misura": An Anthropometric Study of Workers at FCA Italian Plants

Stefania Spada[1], Raffaele Castellone[2(✉)], and Maria Pia Cavatorta[2]

[1] Fiat Chrysler Automobiles - EMEA Region – Manufacturing Planning & Control – Direct Manpower Analysis & Ergonomics, 10135 Turin, Italy
stefania.spada@fcagroup.com

[2] Politecnico di Torino - Department of Mechanical and Aerospace Engineering, 10129 Turin, Italy
{raffaele.castellone,maria.cavatorta}@polito.it

**Abstract.** Nowadays, the use of updated and reliable anthropometric database is of great importance. Anthropometric data play a key role in workstation design, safe use of machineries and protective equipment supplying. The variability of anthropometric data between populations of different ethnic groups, or even different countries or geographical areas of origin, cannot be neglected. The ISO 7250 technical standard reports a collection of statistical data of anthropometric measurements for different nations. As far as the Italian adult population is concerned, the anthropometric survey, referred in the standard, was carried out nearly 30 years ago. The interest and value of a more recent anthropometric database, that could be representative of the Italian working population, was the driver for the project "La Fabbrica si Misura" (The Factory Measures itself).

The present work describes the methodology used to acquire the data on selected FCA plants in different Italian regions. Data collection was aimed in particular at the body measurements that are crucial for the creation of virtual manikins to be used in Digital Human Models, and are not always available in the standard database. A minimum of 3000 volunteer subjects per gender, aged between 18 and 65, participated in the initiative. The paper presents a preliminary statistical analysis on the data for the male sample. An initial comparison with the reference database of the ISO international technical standard is also reported.

**Keywords:** Anthropometry · Human-centered design · Proactive ergonomics

## 1 Introduction

Anthropometry is the study of human body measurements. Typical body measurements include height, weight, body segment lengths, breadths and circumferences. Anthropometry examines how these measurements vary among humans, with reference to several factors like age group, gender, nationality and ethnicity. Differences are generally described in terms of overall body size and bodily proportions, and are known to be greater among ethnic groups. Nonetheless also within the same ethnic group,


S. Bagnara et al. (Eds.): IEA 2018, AISC 826, pp. 389–397, 2019.
https://doi.org/10.1007/978-3-319-96065-4_44

average body sizes and bodily proportions may differ, as it was shown in [1] by comparing the anthropometric data of people from four different regions of East Asia. In addition, various population groups such as civilians, military personnel, students and workers are known to carry differences within the same country [2].

For the purpose of ergonomics design in the industrial field, anthropometric data have fundamental importance, because information about the average human build and its variability is of help in the proper design of workstations, tools and equipment. Although the perfect interaction between the user and a product or a workstation is not always possible, the integration of anthropometric data into ergonomic design ensures an improvement in the use of the product or in the working conditions [2], which helps to increase work performance and productivity, while ensuring the comfort and safety of the user. On the contrary, incorrect considerations or omission of anthropometric data in the design phase can cause psychological problems related to work, physical fatigue, pain or injury, with consequent disorders or musculoskeletal diseases [3, 4].

The current international reference for anthropometric data is ISO 7250-Part 1 and Part 2 [5, 6] that reports a collection of statistical analyzes for anthropometric measurements carried out in different countries. As for the Italian population, the reference in the technical standard is the anthropometric survey: "L'Italia si Misura" [7], carried out on Italian beaches, in the cities of Ancona and Naples, in 1990–1991.

The interest and value of a more recent anthropometric database, that could be representative of the Italian working population, was the driver for a project carried out in cooperation between Fiat Chrysler Automobiles (FCA), Politecnico di Torino and Istituto Nazionale Assicurazione Infortuni sul Lavoro (INAIL). The project, "La Fabbrica si Misura" (FsM, The Factory Measures itself), aims at developing an anthropometric database of the Italian working population in order to support proactive ergonomics in workplace design. In particular, data collections was aimed at the body measurements that are crucial for the creation of virtual manikins to be used in Digital Human Models for early evaluation of the manufacturing workplaces. For this reason, the choice of the anthropometric measures include body dimensions that are not present in the current reference standard. Data collection started in October 2016 and was set to conclude by March 2018.

The present paper describes the methodology used to acquire the anthropometric database on selected FCA plants in different Italian regions and provides a preliminary statistical analysis on the male sample. An initial comparison with the reference database of the international technical standard ISO 7250-2 [6] is also presented.

## 2 Materials and Methods

First phase of the project is data definition and collection. This phase was designed in order to maximize the accuracy and reliability of the anthropometric database, compatibly to time, cost and other limitations.

The preliminary study of the anthropometric survey concerned:

- Sampling
- Anthropometric measurements
- Instrumentation
- The operational procedure for the measurement
- Error checking

Whenever possible, the guidelines given by the international technical norm ISO 15535:2012 [8] were followed. The guidelines define the requirements for creating an anthropometric database.

### 2.1 Sampling

The FsM project aims at creating an anthropometric database representative of the Italian working population. Subjects of the study were workers of FCA Italian plants. In accordance with the international technical standard, the subjects who formed the sample were selected at random among the workers convened in the medical room for periodic medical examinations. Each worker was informed in full detail about aim and nature of the project and was free to decide whether to participate in the measuring campaign, in which case he/she was asked to sign an informed consent.

In consideration of the population size, cost and time required for the investigation, a sample size of 3000 subjects per gender was decided for and the sample size was verified for accuracy. The standard ISO 15535:2012 [8] provides a method to define the number of randomly sampled subjects (N) that is necessary to ensure that a database 5th and 95th percentile estimate the true population 5th and 95th percentiles with a confidence of 95% and with the desired relative accuracy:

$$N = \left(\frac{1,96 \times CV}{a}\right)^2 \times 1,534^2 \quad (1)$$

where:

- *1,96* is the critical value (z value) of a standard normal distribution for a 95% confidence interval
- *CV* is the coefficient of variation, that is the ratio between the standard deviation and the mean of the population for the examined anthropometric measurement
- *a* is the required relative accuracy

For CV estimates, the mean and standard deviation of the Italian population for the body dimension in question were initially taken from the technical standards ISO 7250-2 [6] and were later verified on the actual measurements. The minimum size of the sample was calculated with reference to the body dimension having the largest CV. Results showed that the sample size of 3000 subjects per gender could ensure the desired confidence and accuracy even for the extreme percentiles (1$^{st}$ and 99$^{th}$ percentiles) for all height measurements.

The measuring campaign involved 12 different Italian FCA production plants, located in various Italian regions: Piemonte, Emilia-Romagna, Abruzzo, Lazio, Molise, Campania, Basilicata. The distribution of workers to be measured at each plant was defined in proportion with the working population of the plant. The reason for this choice was twofold. First, it allowed maximizing the speed of data collection, since medical examinations are more frequent in the most populous plants. Secondly, it allowed collecting data that could be representative of various Italian regions and would permit to evaluate the potential influence of the geographic area on the collected data.

**Table 1.** Anthropometric measurements in the FsM project

| Anthropometric measurements | Reference code (ISO 7250) | Instrument | Posture |
|---|---|---|---|
| Body mass | 4.1.1 | Weighing scale | Erect |
| Stature | 4.1.2 | Anthropometer | Erect |
| Eye height | 4.1.3 | Anthropometer | Erect |
| Cervical height | – | Anthropometer | Erect |
| Shoulder height | 4.1.4 | Anthropometer | Erect |
| Elbow height | 4.1.5 | Anthropometer | Erect |
| Trochanteric height | – | Anthropometer | Erect |
| Knuckle height | – | Anthropometer | Erect |
| Tibial height | 4.1.8 | Anthropometer | Erect |
| Body depth | 4.1.10 | Anthropometer | Erect |
| Shoulder biacromial breadth | 4.2.8 | Sliding caliper | Sitting |
| Elbow-to-elbow breadth | 4.2.10 | Sliding caliper | Sitting |
| Hip breadth | 4.2.11 | Sliding caliper | Sitting |

## 2.2 Anthropometric Measurements

Table 1 lists the anthropometric measurements, the relative reference codes in the technical standard ISO 7250-1 [5] (when present), the measuring instrument, and the posture of the subject.

Unlike more complete anthropometric databases (e.g. international technical standards ISO 7250-2 [6]), the project FsM focused on a limited number of anthropometric measures. In the compound dynamics of an industrial context like the production line, the additional time in the medical room required for the anthropometric measurements is quite complex to be dealt with and obviously not free from costs.

In particular, the choice of the anthropometric measures focused on body measurements that are crucial for the creation of virtual manikins to be used in Digital Human Models for early evaluation of the manufacturing workplaces. Indeed, most of the anthropometric points detected in this survey refer to the articular joints that serve to build the kinematic model at the basis of virtual manikins, and of postural angles assessments [9].

The ISO 7250-1 [5] provides the details for a proper measurement. Whenever possible, measurements were taken in accordance with the technical standards. Alternatively, reference was made to the descriptions defined in another important international anthropometric database, the ANSUR [10], that concerns the American military population.

In accordance to [8], a biographic questionnaire was administered to subjects to collect basic demographic descriptors of the subject:

- Year of birth
- Gender
- Region of origin
- Region of origin of the parents
- Safety shoes size (EUR size)

Date of examination and the name of the plant were also reported. To protect the privacy of the subject, all the data were rendered anonymous by associating them to an identification code (ID) that can no longer be traced to the person.

### 2.3 Measurement Instrument

ISO 7250-1 [5] recommends the use of an anthropometer and a sliding caliper for the collection of the selected anthropometric measurements. Each medical room was equipped with the Harpenden anthropometer. This anthropometer is a mountable instrument, which comprises:

- a measuring beam on which a numeric counter slides for the measurement reading;
- 4 beams of different length to vary the measuring height of the instrument and to reach a maximum measurable height of 2072 mm;
- 2 straight branches for the location of anthropometric points.

The Harpenden anthropometer was chosen to obtain a degree of accuracy to the nearest millimeter, not possible with conventional anthropometers, but also for the possibility of assembly and transport. An additional straight branch can be mounted on the measuring beam so that it can be used as a sliding caliper. Body mass was measured through a weighting scale.

### 2.4 The Operational Procedure for the Measurement

At the start of the campaign, doctors and staff involved in data collection were trained in order to eliminate possible doubts about the correct use of the instrumentation and to provide specific references on the precise search of anthropometric landmarks. In addition, the measuring conditions reported in the technical standard [5], concerning subject's clothing and the support surfaces, were specified. In particular, it was recommended that during measurement the subject shall wear minimal clothing and shall be bareheaded and barefoot. Also, it was recommended to use flat, horizontal and non-compressible standing and sitting surfaces.

In order to optimize the procedure of the anthropometric survey, the sequence of the measurements to be carried out was defined step by step and a recording sheet was created to guide the staff in data collection and recording. Details on how to use the anthropometer were also given for each measurement. At first, the staff shall acquire and record heights from the floor, starting from the greatest (stature) down to the smallest (tibial height). Later, staff shall acquire body width and depth by using the anthropometer as a sliding caliper, and the body mass through the weighting scale. In a final step, the medical staff shall insert the measurement for each subject through a worksheet on a specific computer station.

### 2.5 Error Checking

The data collected in the database were analyzed with weekly cadence in order to highlight potential errors during the measurements or data entry. In fact, the measurement itself and the digital worksheet recording operation may be subject to errors.

To this purpose, an automated data control system was implemented, based on body sizes and bodily proportions for the Italian population [6], and thus presumable correlation between the different anthropometric measurements. If one or more anthropometric measurements were signaled as "potentially incorrect", the data ID was indicated to the medical room of the plant. The medical staff would verify possible mistakes in data entry. After the check, the anthropometric measurement was confirmed, corrected or eliminated.

In addition to the periodic error checking during the time of data collection, a further final data check was performed. As recommended in [8], measurement data over $\pm 3\sigma$ from the mean were reviewed for accuracy and eventually deleted from the database.

## 3 Preliminary Statistical Analysis and Results

After eliminating potentially erroneous measurements from the database, the sample was divided by gender and statistically analyzed. This section provides a summary of the data collected and of the statistical analysis results for the male sample.

Data for the male sample were collected from October 2016 to September 2017, and include 3174 male workers aged between 18 and 65. The average age and standard deviation are $43.1 \pm 10.0$ years. Table 2 shows the age distribution of the sample subdivided in five age groups (18–25, 26–35, 36–45, 46–55, and 56–65 years). Although the ISO 15535 [8] suggests a different division of age groups, the chosen subdivision reflects the one adopted in FCA.

Although subjects were randomly selected, Table 2 clearly shows that the distribution of the current working population is not equal among the age groups. Specifically, the age range 36–55 years includes 69% of the whole sample, as it is by far the age group of the majority of workers.

**Table 2.** Male sample age distribution

| Age groups (years) | Male | |
|---|---|---|
| | N | % |
| 18–25 | 226 | 7,1% |
| 26–35 | 465 | 14,7% |
| 36–45 | 1054 | 33,2% |
| 46–55 | 1137 | 35,8% |
| 5–65 | 292 | 9,2% |

Table 3 shows the statistical values of mean and standard deviation of the collected anthropometric measurements. Furthermore, the table reports the values of the anthropometric percentiles used in workstations design (1st, 5th, 95th, 99th) calculated from the actual distribution of individual subjects in the sample.

**Table 3.** Statistical analysis of the male sample (values of all anthropometric measurements are expressed in mm, except for body mass in kg)

| Anthropometric measurements | Mean | SD | P1 | P5 | P95 | P99 |
|---|---|---|---|---|---|---|
| Body mass | 81 | 13 | 56 | 63 | 104 | 120 |
| Stature | 1726 | 71 | 1552 | 1609 | 1840 | 1895 |
| Eye height | 1617 | 70 | 1448 | 1500 | 1727 | 1785 |
| Cervical height | 1496 | 68 | 1333 | 1381 | 1604 | 1652 |
| Shoulder height | 1431 | 67 | 1274 | 1321 | 1540 | 1595 |
| Elbow height | 1064 | 60 | 933 | 970 | 1164 | 1213 |
| Trochanteric height | 905 | 61 | 772 | 810 | 1004 | 1050 |
| Knuckle height | 747 | 46 | 640 | 672 | 826 | 855 |
| Tibial height | 451 | 54 | 332 | 365 | 548 | 589 |
| Body depth | 259 | 45 | 155 | 186 | 336 | 367 |
| Shoulder biacromial breadth | 392 | 44 | 291 | 320 | 462 | 493 |
| Elbow-to-elbow breadth | 474 | 51 | 347 | 388 | 559 | 573 |
| Hip breadth | 350 | 42 | 252 | 280 | 416 | 444 |

It is worthwhile considering a comparison between the data of the acquired sample (FsM) and the reference database from the international technical standard ISO 7250-2 [6]. Table 4 reports the value of mean and standard deviation SD for the anthropometric measurements that are present in both database. The difference Δ between the two mean values, $\Delta_{mean}$, as well as for the P95, $\Delta_{P95}$, is also reported. The difference is calculated as Δ = (FsM - ISO 7250), so that positive Δ values highlight greater measures for the new measurement campaign and vice versa. For the P95, both the FsM and ISO 7250 values refer to the actual distribution of individual subjects in the sample.

**Table 4.** Comparison between mean and standard deviation SD between FsM and ISO 7250-2 (Italian) male samples. Δ values are computed as FsM - ISO 7250

| Anthropometric measurements | FsM | | ISO 7250 (Italian) | | Δ | |
|---|---|---|---|---|---|---|
| | Mean | SD | Mean | SD | $\Delta_{mean}$ | $\Delta_{P95}$ |
| Body mass | 81 | 13 | 76 | 10 | 5 | 11 |
| Stature | 1726 | 71 | 1716 | 69 | 10 | 6 |
| Shoulder height | 1431 | 67 | 1410 | 56 | 21 | 32 |
| Elbow height | 1064 | 60 | 1084 | 51 | −20 | −6 |
| Tibial height | 451 | 54 | 457 | 31 | −6 | 37 |
| Shoulder biacromial breadth | 392 | 44 | 392 | 27 | 0 | 27 |
| Elbow-to-elbow breadth | 474 | 51 | 498 | 45 | −24 | −12 |
| Hip breadth | 350 | 42 | 350 | 28 | 0 | 19 |

As it appears from Table 4, the difference between the mean values is greater for some anthropometric measures than for others. In particular, for Shoulder height, Elbow height and Elbow-to-elbow breadth the difference between the two means exceeds 20 mm. On the other hand, the estimated means for Shoulder biacromial and Hip breadth are the same.

In all cases, the standard deviation is greater in the new measurement campaign; this means a greater variability of the anthropometric measurements and potential greater differences in the estimation of the extreme percentiles.

## 4 Conclusions

The paper presents the initial results of the project “La Fabbrica si Misura” carried out in cooperation between FCA, INAIL and Politecnico di Torino. One of the project aims is to develop an anthropometric database of the Italian working population, in order to support proactive ergonomics in workplace design. Knowledge of anthropometric data is also vital to help in the design of protective or auxiliary devices such as exoskeletons.

A minimum of 3000 subjects per gender, aged between 18 and 65, participated in the initiative. Measurements took place in 12 FCA plants located in various Italian regions between October 2016 and September 2017. Data collection aimed in particular at the body measurements that are crucial for the creation of virtual manikins, such as articular joint locations, and are not always available in standard database.

A preliminary statistical analysis run on the male sample highlights a greater variability in the anthropometric measurement as compared to the data that are present in the standard. Also, mean values exhibit differences that are not always negligible. The large size of the sample allows dividing data into sub-samples of sufficient size to make statistical estimates. Thus, further analyses will be carried out to investigate the potential influence of other parameters such as age and geographical area of the plant.

## References


1. Lin YC, Wang MJJ, Wang EM (2004) The comparisons of anthropometric characteristics among four peoples in East Asia. Appl Ergon 35(2):173–178
2. Pheasant S, Haslegrave CM (2016) Bodyspace: anthropometry, ergonomics and the design of work. CRC Press
3. Bernard BP, Putz-Anderson V (1997) Musculoskeletal disorders and workplace factors; a critical review of epidemiologic evidence for work-related musculoskeletal disorders of the neck, upper extremity, and low back
4. National Research Council: Musculoskeletal disorders and the workplace: low back and upper extremities. National Academies Press (2001)
5. International Standard ISO 7250-1:2008: Basic human body measurements for technological design – Part 1: body measurement definitions and landmarks
6. International Standard ISO/TR 7250-2:2009: Basic human body measurements for technological design – Part 2: statistical summaries of body measurements from individual ISO populations
7. Masali M (a cura di) (2013) L'Italia si Misura. Vent'anni di ricerca (1990–2010). Vademecum antropometrico per il Design e l'Ergonomia, vol II, Collana A misura d'Uomo-Antropometria ed Ergonomia Aracne Editrice S.r.l., Roma, ISBN 978-88-548-5715-5
8. International Standard ISO 15535:2012: General requirements for establishing anthropometric databases
9. Castellone R, Spada S, Sessa F, Cavatorta MP (2017) A simple multibody 2d-model for early postural checks in workplace design. Int J Appl Eng Res 12(23):13451–13461
10. Gordon CC, Churchill T, Clauser CE, Bradtmiller B, McConville JT, Tebbetts I, Walker RA (1989) Anthropometric survey of US Army personnel: summary statistics, interim report for 1988. Anthropology Research Project Inc Yellow Springs OH (1989)

# Anthropometric Data for Biomechanical Hand Model

Kyung-Sun Lee[1(✉)], Myung-Chul Jung[2], Seung-Min Mo[1], and Seung Nam Min[3]

[1] Suncheon Jeil College, Suncheon 57997, Republic of Korea
kyungsunlee81@gmail.com, smmo@suncheon.ac.kr
[2] Ajou University, Suwon 16499, Republic of Korea
mcjung@ajou.ac.kr
[3] Shinsung University, Dangjin 31801, Republic of Korea
msnijnl2@hanmail.ne

**Abstract.** The aim of this study was to investigate the anthropometric data for the segment masses, center of mass (COMs) of the segments of inertia, and radii of gyration are required for the development of the biomechanical hand model. The segment masses were calculated on the basis of the segment volume using a density of 1.1 g/cm$^3$. The segment volume was estimated from the measured length between the participants' distal and proximal joints (segment length) and the diameters of their knuckles. The COMs for the proximal and middle segments and the distal segment were determined by approximating the phalanx by the frustum of a cone and a cylindrical homogeneous rigid body, respectively. The diameters of the knuckles were measured for each participant. We assume that they have a uniform density. The moments of inertia of the proximal and middle segments were determined by approximating the phalanx as the frustum of a conical homogenous rigid body. The diameters of the knuckles were measured for each participant. The moments of inertia of the distal segments were determined by approximating the phalanx as a cylindrical rigid body. The radii of gyration, *Kx*, *Ky*, and *Kz*, of the segment about the x axis, y axis, and z axis are defined as Pytel and Kiusalaas. This information will be provide useful data for development of biomechanical hand model.

## 1 Introduction

Biomechanics is the interdisciplinary study of the mechanical movement and force of the musculoskeletal system. Especially, hand biomechanics models are used various fields such as ergonomics, industrial safety, product design, rehabilitation, robot design, and digital manufacturing simulation. The anthropometric data for the segment masses, moments of inertia, and radii of gyration are required for the development of the biomechanical hand model. Thus, the aim of this study was to investigate the anthropometric data for the segment masses, center of mass (COMs) of the segments of inertia, and radii of gyration are required for the development of the biomechanical hand model.


S. Bagnara et al. (Eds.): IEA 2018, AISC 826, pp. 398–404, 2019.
https://doi.org/10.1007/978-3-319-96065-4_45

## 2 Methods

### 2.1 Segment Masses

The segment masses were calculated on the basis of the segment volume using a density of 1.1 g/cm$^3$ [1]. The segment volume was estimated from the measured length between the participants' distal and proximal joints (segment length) and the diameters of their knuckles. The volumes of the proximal and middle segments were each approximated as that of the frustum of a cone, and the volume of the distal segment was approximated as that of a cylinder. The volumes of the proximal, middle, and distal segments can be determined from the following simple equation [2] (Eqs. 1 and 2) (Fig. 1).

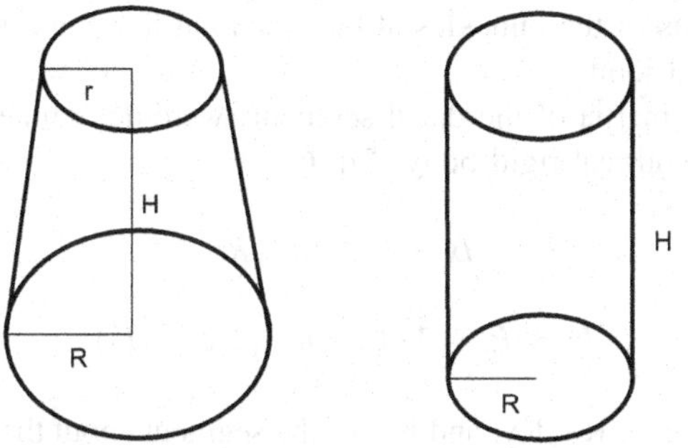


$$V = \pi H/3\left(R^2 + Rr + r^2\right) \quad (1)$$

$$V = \pi HR^2 \quad (2)$$

where V is the volume of the segments, H is the height (segment length), R is the radius of the lower base (the knuckles at the proximal joint), and r is the radius of the upper base (the knuckles at the distal joint). Thus, the segment mass was calculated using segment mass = density (1.1 g/cm$^3$) × segment volume [3].

### 2.2 Center of Masses (COM) of Segments

The COMs for the proximal and middle segments and the distal segment were determined by approximating the phalanx by the frustum of a cone and a cylindrical homogeneous rigid body, respectively. The diameters of the knuckles were measured for each participant. We assume that they have a uniform density [4]. The COMs of the proximal and middle segments (Eq. 3) and distal segment (Eq. 4) can be determined from the following simple equations [5].

$$COM\,(Z) = H/4\left[\left(R^2 + 2Rr + 3r^2\right)/\left(R^2 + Rr + r^2\right)\right] \quad (3)$$

$$COM\,(Z) = H/2 \quad (4)$$

where H is the height (segment length), R is radius of the knuckles at the proximal joint, and r is the radius of the knuckles at the distal joint.

### 2.3 Moments of Inertia and Radius of Gyration

The moments of inertia of the proximal and middle segments were determined by approximating the phalanx as the frustum of a conical homogenous rigid body [6] (Eq. 5). The diameters of the knuckles were measured for each participant.

$$I_z = 3m/10\left[\left(R^5 - r^5\right)/\left(R^3 - r^3\right)\right] \tag{5}$$

where $I_z$ is the principle moment of inertia of a segment, m is the segment mass of each finger, R is the radius of the knuckles at the proximal joint, and r is the radius of the knuckles at the distal joint.

The moments of inertia of the distal segments were determined by approximating the phalanx as a cylindrical rigid body (Eq. 6).

$$Ix = 1/2 \cdot m \cdot R^2$$

$$Iy = Iz = 1/12 \cdot m \cdot \left(3R^2 + H^2\right) \tag{6}$$

The radii of gyration, Kx, Ky, and Kz, of the segment about the x axis, y axis, and z axis are defined as [7] (Eq. 7).

$$Kx = \sqrt{(Ix/m)},\ Ky = \sqrt{(Iy/m)},\ Kz = \sqrt{(Iz/m)} \tag{7}$$

where Ix, Iy, and Iz are the moments of inertia about the x axis, y axis, and z axis, respectively, and m is the segment mass.

### 2.4 Participants

Ten male graduate and undergraduate students participated in the experiment. The averages (standard deviations) of their age, height, and weight were 23.3 (SD 5.2) years, 173.0 (SD 5.0) cm, and 64.2 (SD 23.0) kg, respectively. Table 1 presents their basic anthropometric information. The diameters of the metacarpophalangeal (MCP) joints were calculated using Garrett's formulation [8] (Eq. 8), where a and b are the measured joint breadth and depth. The diameter of the carpometacarpal (CMC) joint at the thumb is hypothesized to be the same as that of the MCP joint of the other fingers.

$$D = 2\pi\sqrt{\left(\left(a^2 + b^2\right)/2\right)} \tag{8}$$

**Table 1.** Basic anthropometric data of participants.

| Participant | Age (year) | Height (cm) | Weight (kg) | Hand length (cm) | Hand circumference (cm) | Hand breadth (cm) |
|---|---|---|---|---|---|---|
| 1 | 20 | 167.0 | 56.0 | 17.4 | 17.7 | 9.6 |
| 2 | 20 | 170.0 | 68.0 | 18.2 | 18.8 | 9.6 |
| 3 | 26 | 167.0 | 65.0 | 15.4 | 19.5 | 9.3 |
| 4 | 21 | 175.0 | 61.0 | 18.5 | 15.3 | 7.8 |
| 5 | 20 | 171.0 | 62.3 | 17.4 | 19.1 | 10.0 |
| 6 | 32 | 180.0 | 85.0 | 19.7 | 19.9 | 8.3 |
| 7 | 20 | 178.0 | 100.0 | 18.8 | 19.5 | 10.2 |
| 8 | 33 | 180.0 | 100.0 | 19.9 | 20.6 | 10.1 |
| 9 | 21 | 170.0 | 65.0 | 18.0 | 17.6 | 8.8 |
| 10 | 20 | 175.0 | 70.1 | 20.0 | 19.7 | 10.4 |

# 3 Results

## 3.1 Segment Masses

The segment masses were calculated on the basis of the segment volumes. The averages (standard deviations) of the distal, middle, and proximal phalangeal segment volumes of the thumb were 8.27 (SD 1.38) $cm^3$, 11.20 (SD 2.93) $cm^3$, and 23.97 (SD 6.03) $cm^3$, respectively. The averages (standard deviations) of the distal, middle, and proximal phalangeal segment masses of the thumb were 9.10 (SD 1.52) g, 12.32 (SD 3.23) g, and 26.37 (SD 6.03) g, respectively.

The averages (standard deviations) of the distal, middle, and proximal phalangeal segment volumes of the index finger were 4.18 (SD 0.65) $cm^3$, 6.43 (SD 1.02) $cm^3$, and 20.71 (SD 2.00) $cm^3$, respectively. The averages (standard deviations) of the distal, middle, and proximal phalangeal segment masses of the index finger were 4.60 (SD 0.71) g, 7.07 (SD 1.13) g, and 20.71 (SD 2.00) g, respectively.

The averages (standard deviations) of the distal, middle, and proximal phalangeal segment volumes of the middle finger were 5.05 (SD 0.74) $cm^3$, 6.58 (SD 0.94) $cm^3$, and 23.62 (SD 2.10) $cm^3$, respectively. The averages (standard deviations) of the distal, middle, and proximal phalangeal segment masses of the middle finger were 5.55 (SD 0.82) g, 7.24 (SD 1.03) g, and 25.98 (SD 2.31) g, respectively.

The averages (standard deviations) of the distal, middle, and proximal phalangeal segment volumes of the ring finger were 4.06 (SD 0.60) cm3, 5.69 (SD 0.78) $cm^3$, and 20.98 (SD 2.8) $cm^3$, respectively. The averages (standard deviations) of the distal, middle, and proximal phalangeal segment masses of the ring finger were 4.47 (SD 0.66) g, 6.26 (SD 0.86) g, and 22.98 (SD 2.81) g, respectively.

The averages (standard deviations) of the distal, middle, and proximal phalangeal segment volumes of the little finger were 2.41 (SD 0.57) $cm^3$, 3.83 (SD 0.81) $cm^3$, and 15.71 (SD 1.71) $cm^3$, respectively. The averages (standard deviations) of the distal, middle, and proximal phalangeal segment masses of the little finger were 2.65 (SD 0.63) g, 4.21 (SD 0.89) g, and 17.28 (SD 1.88) g, respectively. Table 2 lists the mean segment volumes and masses of the thumb and finger.

**Table 2.** Mean value of segment volumes and masses.

| Finger | Distal phalange | | Middle phalange | | Proximal phalange | |
|---|---|---|---|---|---|---|
| | Segment volume ($cm^3$) | Segment mass (g) | Segment volume ($cm^3$) | Segment mass (g) | Segment volume ($cm^3$) | Segment mass (g) |
| Thumb | 8.28 | 9.10 | 11.20 | 12.32 | 23.97 | 26.37 |
| Index | 4.18 | 4.60 | 6.43 | 7.08 | 20.71 | 22.78 |
| Middle | 5.05 | 5.55 | 6.58 | 7.04 | 26.62 | 25.98 |
| Ring | 4.07 | 4.47 | 5.69 | 6.26 | 20.89 | 22.98 |
| Little | 2.41 | 2.65 | 3.83 | 4.21 | 15.71 | 17.28 |

### 3.2 Center of Masses (COM) of Segments

Detailed information on the mean value in centers of mass (COMs) of the segments of the thumb and fingers for participants is presented in Table 3. The averages (standard deviations) of the COMs of the distal, middle, and proximal phalanges of the thumb were 1.37 (SD 0.21) cm, 1.49 (SD 0.11) cm, and 2.14 (SD 0.38) cm, respectively. The averages (standard deviations) of the COMs of the distal, middle, and proximal phalanges of the index finger were 1.11 (SD 0.10) cm, 1.20 (SD 0.12) cm, and 2.03 (SD 0.14) cm, respectively. The averages (standard deviations) of the COMs of the distal, middle, and proximal phalanges of the middle finger were 1.22 (SD 0.13) cm, 1.25 (SD 0.09) cm, and 2.29 (SD 0.16) cm, respectively. The averages (standard deviations) of the COMs of the distal, middle, and proximal phalanges of the ring finger were 1.18 (SD 0.10) cm, 1.24 (SD 0.08) cm, and 2.08 (SD 0.18) cm, respectively. The averages (standard deviations) of the COMs of the distal, middle, and proximal phalanges of the little finger were 0.84 (SD 0.13) cm, 1.09 (SD 0.11) cm, and 1.64 (SD 0.09) cm, respectively.

**Table 3.** Mean value of center of masses (COM).

| Finger | Distal phalange | Middle phalange | Proximal phalange |
|---|---|---|---|
| Thumb | 1.37 | 1.50 | 2.14 |
| Index | 1.11 | 1.20 | 2.03 |
| Middle | 1.22 | 1.25 | 2.29 |
| Ring | 1.18 | 1.24 | 2.07 |
| Little | 0.94 | 1.09 | 1.64 |

### 3.3 Moments of Inertia and Radii of Gyration

Table 4 presents detailed information on the mean value in moments of inertia of the segments of the thumb and fingers for each participant. The averages (standard deviations) of the moments of inertia of the distal, middle, and proximal phalanges of the thumb were 8.26 (SD 3.13) g/cm$^3$, 7.38 (SD 3.60) g/cm$^3$, and 10.59 (SD 4.17) g/cm$^3$, respectively. The averages (standard deviations) of the moments of inertia of the distal, middle, and proximal phalanges of the index finger were 2.62 (SD 0.68) g/cm$^3$, 2.70 (SD 0.79) g/cm$^3$, and 16.98 (SD 2.44) g/cm$^3$, respectively. The averages (standard deviations) of the moments of inertia of the distal, middle, and proximal phalanges of the middle finger were 3.77 (SD 1.09) g/cm$^3$, 2.93 (SD 0.74) g/cm$^3$, and 19.58 (SD 2.53) g/cm$^3$, respectively. The averages (standard deviations) of the moments of inertia of the distal, middle, and proximal phalanges of the ring finger were 2.74 (SD 0.71) g/cm3, 2.18 (SD 0.54) g/cm$^3$, and 16.81 (SD 2.95) g/cm$^3$, respectively. The averages (standard deviations) of the moments of inertia of the distal, middle, and proximal phalanges of the little finger were 0.97 (SD 0.43) g/cm$^3$, 1.15 (SD 0.44) g/cm$^3$, and 11.92 (SD 2.05) g/cm$^3$, respectively.

**Table 4.** Mean value of moments of inertia.

| Finger | Distal phalange | Middle phalange | Proximal phalange |
|---|---|---|---|
| Thumb | 8.26 | 7.38 | 10.59 |
| Index | 2.62 | 2.70 | 16.98 |
| Middle | 3.77 | 2.93 | 19.58 |
| Ring | 2.74 | 2.18 | 16.83 |
| Little | 0.97 | 1.15 | 11.92 |

The averages (standard deviations) of the radii of gyration of the distal, middle, and proximal phalanges of the thumb were 0.93 (SD 0.11) g/cm$^3$, 0.75 (SD 0.08) g/cm$^3$, and 0.63 (SD 0.07) g/cm$^3$, respectively. The averages (standard deviations) of the radii of gyration of the distal, middle, and proximal phalanges of the index finger were 0.75 (SD 0.05), 0.61 (SD 0.04), and 0.86 (SD 0.02), respectively. The averages (standard deviations) of the radii of gyration of the distal, middle, and proximal phalanges of the middle finger were 0.81 (SD 0.07) g/cm$^3$, 0.63 (SD 0.74) g/cm$^3$, and 0.87 (SD 0.02) g/cm$^3$, respectively. The averages (standard deviations) of the radii of gyration of the distal, middle, and proximal phalanges of the ring finger were 0.78 (SD 0.05), 0.59 (SD 0.03), and 0.85 (SD 0.03), respectively. The averages (standard deviations) of the radii of gyration of the distal, middle, and proximal phalanges of the little finger were 0.59 (SD 0.06), 0.52 (SD 0.04), and 0.83 (SD 0.03) (Table 5).

**Table 5.** Mean value of radii of gyration.

| Finger | Distal phalange | Middle phalange | Proximal phalange |
|---|---|---|---|
| Thumb | 0.94 | 0.75 | 0.63 |
| Index | 0.75 | 0.61 | 0.86 |
| Middle | 0.81 | 0.63 | 0.87 |
| Ring | 0.78 | 0.59 | 0.85 |
| Little | 0.59 | 0.52 | 0.83 |

# 4 Discussion

The high incidence and economic cost associated with cumulative trauma disorders of the hand has generated as significant amount of research on the biomechanics of the hand. Thus, the ergonomics and occupational safety fields commonly used biomechanical models of the hand in conducting job analysis, task analysis, hand tools design, hand devices design, and so on. The anthropometric data for the segment masses, centers of mass (COMs) of the segments, moments of inertia, and radii of gyration are required for the development of the biomechanical hand model. Thus, the objective of the this study was to provide the method for collection of the anthropometric data. These anthropometic data were calculated some assumption and simple physic and mathematics equations.

Future research will involve examining various age, gender, and hand size using suggested methods, and these methods and collected data will be identified using MRI and X-ray methods.

**Acknowledgments.** This study was supported by Basic Science Research Program through the National Research Foundation of Korea (NRF) funded by the Ministry of Education (No. NRF-2016R1D1A1B03934542).

This study was supported by a grant from the National Research Foundation of Korea (NRF) (NRF-2015R1C1A1A01055231), which is funded by the Korean government (MEST).

# References

1. Esteki A, Mansour M (1997) A dynamic model of the hand with application in functional neuromuscular stimulation. Ann Biomed Eng 25(3):440–451
2. Stewart J (2008) Calculus: early transcendentals, 5th edn. Thomson Learning, Belmont
3. Dempster WT (1955) Space requirements of the seated operator. WADC-TR-55-159, Aerospace Medical Research Laboratories, Wright-Patterson Air Force Base, OH
4. Kurillo G, Bajd T, Kamnik R (2003) Static analysis of nippers pinch. Int Neuromodul Soc 6 (3):166–175
5. Halliday D, Resnick R, Walker J (2011) Fundamentals of physics, 9th edn. Wiley, New York
6. Myers JA (1962) Handbook of equations for mass and area properties of various geometrical shapes. U.S. Naval Ordnance Test Station
7. Pytel A, Kiusalaas J (1999) Engineering mechanics: dynamics, 2nd edn. Brook/Cole Publishing Company, The Stables
8. Garrett JW (1970) Anthropometry of the hands of male air force flight personnel. Aerospace Medical Research Laboratory, Aerospace Medical Division

# The Definition and Generation of Body Measurements (ISO 8559 Series of Standards)

Youngsuk Lee(✉)

Chonnam National University, Gwangju, Korea
jaabo1994@naver.com

**Abstract.** Clothing sizes have been constantly changing in proportions and dimensions since "ready-to-wear" clothing with standardized sizes appeared in the mid-19th century and, today, sizes across brands and countries continue to vary based on different assumed body shape profiles.

In order to define that size, manufacturers need to know how to measure which body part to ensure the proportions are correct. The ISO 8559 series of standards helps them define "how and where to measure the body".

The ISO 8559 series of standards provides guidelines for clothing manufacturers to develop size and shape profiles based on different populations in order to create all kinds of clothing and mannequins. They have recently been updated to reflect changes in the dynamic clothing sector and eliminate trade barriers by harmonizing size marking and the terms of reference worldwide.

The ISO 8559 series of standards is aimed at increasing customer satisfaction and reducing returns of items as a result of a poor fit. "They will also help to reduce barriers to international trade by providing a universal set of size markings and terms of reference. This will help to simplify information on garment labels for shoppers as, currently, body dimensions are listed on garment sizing labels that do not relate to the body measurements."

This paper is an analysis of measuring body applications in the ISO 8559 series of standards created and developed by ISO/TC133.

**Keywords:** Anthropometry · Clothing · ISO 8559-1 · Measurements

## 1 Introduction

ISO (the International Organization for Standardization) is a worldwide federation of national standards bodies (ISO member bodies). The work of preparing International Standards is normally carried out through ISO technical committees. Each member body interested in a subject for which a technical committee has been established has the right to be represented on that committee. International organizations, governmental and non-governmental, in liaison with ISO, also take part in the work [1]. ISO 8559 series of standards were prepared by Technical Committee ISO/TC 133, Sizing systems and designations for clothes.


S. Bagnara et al. (Eds.): IEA 2018, AISC 826, pp. 405–422, 2019.
https://doi.org/10.1007/978-3-319-96065-4_46

It now comprises three parts, under the general title Size designation of clothes —:

ISO 8559-1 Part 1: Anthropometric definitions for body measurement
ISO 8559-2 Part 2: Primary and secondary dimension indicators
ISO 8559-3 Part 3: Methodology for the creation of body measurement tables and intervals

In this paper, we specify and summarize the concepts and scopes of the ISO 8559 series prepared by ISO/TC 133.

## 2 Scope of ISO 8559 Series of Standards

### 2.1 8559-1: Anthropometric Definitions for Body Measurement

This document is the first of a three-part International Standard (ISO 8559-1, ISO 8559-2 and ISO 8559-3). It comprises the definition and generation of anthropometric measurements that can be used for the creation of size and shape profiles and their application in the field of clothing [2]. It forms a foundation for ISO 8559-2, Primary and secondary dimension indicators, and ISO 8559-3, Methodology for the elaboration of the body measurement tables and intervals.

Reference is made to measurements and procedures in ISO 7250-1.

### 2.2 8559-2: Primary and Secondary Dimension Indicators

This document specifies primary and secondary dimensions for specified types of garments to be used in combination with ISO 8559-1 (anthropometric definitions for body measurement). The primary aim of 8559-2 document is to establish a size designation system that can be used by manufacturers and retailers to indicate to consumers (in a simple, direct and meaningful manner) the body dimensions of the person that the garment is intended to fit. Provided that the size of the person's body (as indicated by the specified dimensions) has been determined in accordance with ISO 8559-1, this designation system will facilitate the choice of garments that fit. This information can be indicated by labelling, etc.

The size designation system is based on body measurements, not garment measurements. The choice of garment measurements is normally determined by the designer and the manufacturers who make appropriate allowances to accommodate the type and position of wear, style, cut and fashion elements of the garment.

These body measurements defined as Primary and secondary dimensions provide the basis for the designation of clothing sizes, independently from the product itself, for garment types as diverse as jackets, skirts, underwear, corsetry (intimate apparel) and headwear. Body measurements are listed separately for men's, women's, boys' and girls' clothing. For infants, body height is used as the primary dimension (see ISO 8559-1) except for headwear, socks, stockings and gloves [3].

Heights and girths are measured in accordance with ISO 8559-1.The primary dimension will constitute the basic size designation system. A number of the secondary dimensions, as given in standard may be used in such a designation system.

### 2.3 8559-3: Methodology for the Creation of Body Measurement Tables and Intervals

This International Standard describes the principles of the establishment of the tables for body measurements, defines the categories of tables (related to intervals), lists the population groups (infants, girls, boys, children women, men) and sub-groups to be used for compiling developing standard ready-to-wear garments. The body measurement tables and intervals are mainly used by the clothing sector to make the development of good fitting products easier and more accurate. The described methodology is mainly based on the application of statistical analysis, using body dimension data. The statistical level has deliberately been kept to a low level in order for the content to be made readily comprehensible to the widest possible readership. This methodology is applicable to various sets of body dimensions. It may be useful to determine intervals for the size designation as described in ISO 8559-2. Values in the tables in this international standard are examples. Garment dimensions are not contained in this document [4].

## 3 Measurements of ISO 8559-1

ISO 8559-1 defined 26 measurements of the landmark points and levels, 5 measurements of lines and planes, 18 measurements of vertical, 7 measurements of breadth, widths and depth, 28 measurements of girth, 23 measurements of distances measured following the surface of the body, 7 measurements of hand and foot, 8 calculated measurements, body mass and shoulder slope.

### 3.1 The Definitions for the Landmark Points and Levels

The measurements included the title, definition how to measure and the comparison with the ISO 7250 definition for the identical measurements.

The examples of measurement are shown from Sects. 3.1.1 to 3.1.8.

#### 3.1.1 Shoulder Point

Most lateral point of the lateral edge of the spine (acromial process) of the scapula, projected vertically to the surface of the skin.

Note 1 to entry: see Fig. 1.

Note 2 to entry: Identical to acromion in ISO 7250-1.

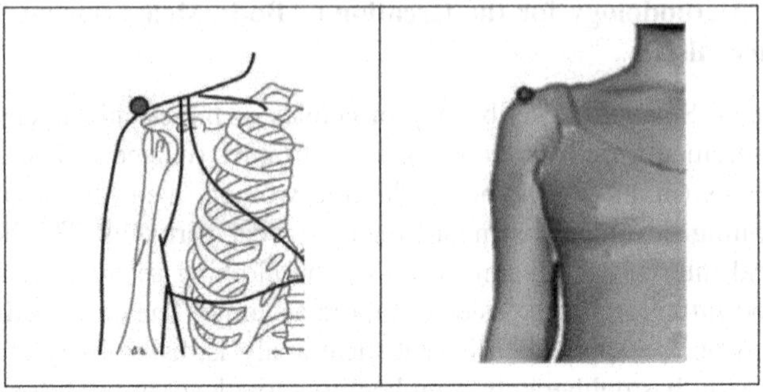

**Fig. 1.** Shoulder point

### 3.1.2 Centre Point of Brow Ridge

Most anterior point of the forehead between the brow ridges in the mid-sagittal plane.

Note 1 to entry: see Fig. 2.

Note 2 to entry: Identical to Glabella in ISO 7250-1.

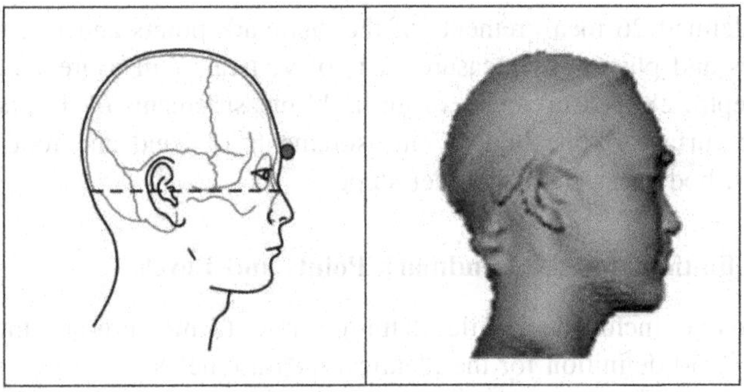

**Fig. 2.** Centre point of brow ridge

### 3.1.3 Tragion

Point of the notch just above the tragus (the small cartilaginous flap in front of the ear hole).

Note 1 to entry: see Fig. 3.

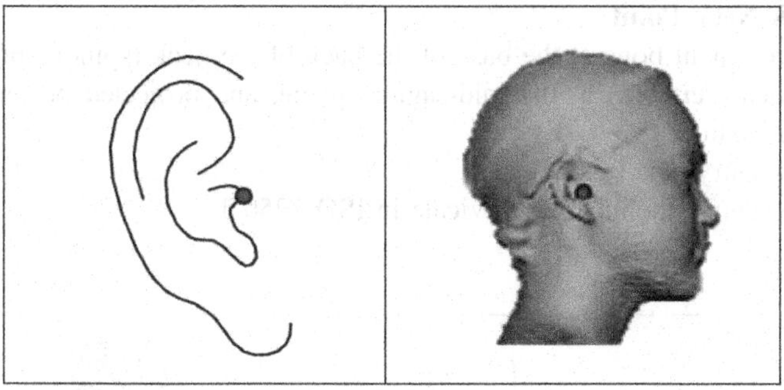

**Fig. 3.** Tragion

### 3.1.4 Orbitale

Lowest point of the lower border of the orbital margin (lower edge of the eye socket).

Note 1 to entry: see Fig. 4.

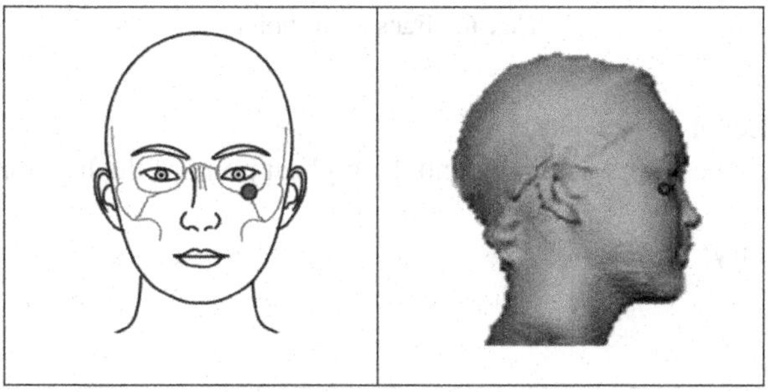

**Fig. 4.** Orbitale

### 3.1.5 Lowest Point of Chin

lowest point of the tip of the chin in the midsagittal plane, projected anteriorly when the head is held in the Frankfurt plane.

Note 1 to entry: see Fig. 5.

Note 2 to entry: Identical to menton in ISO 7250-1.

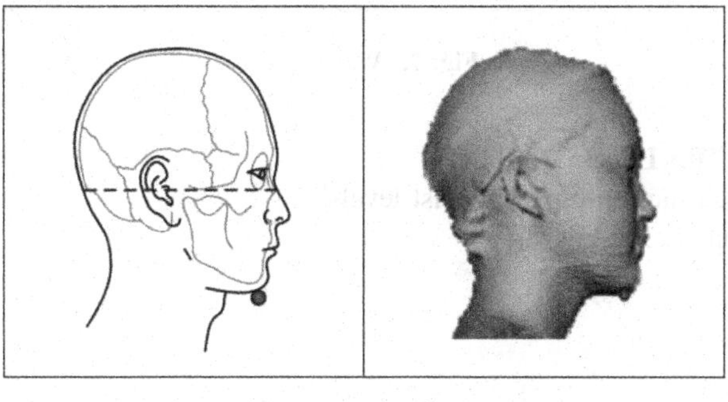

**Fig. 5.** Lowest point of chin

### 3.1.6 Back Neck Point

Tip of the prominent bone at the base of the back of the neck (spinous process of the seventh cervical vertebra) in the mid-sagittal plane, and projected posteriorly to the surface of the skin.

Note 1 to entry: see Fig. 6.

Note 2 to entry: Identical to cervicale in ISO 7250-1.

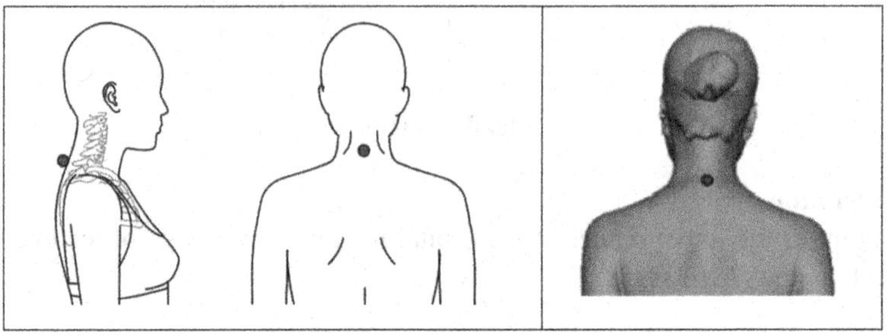

**Fig. 6.** Back neck point

### 3.1.7 Waist Level

Midway between the lowest rib point and the highest point of the hip bone at the side of the body.

Note 1 to entry: see Fig. 7.

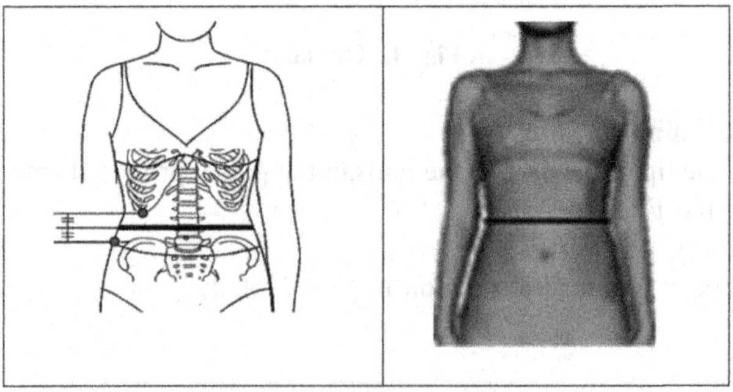

**Fig. 7.** Waist level

### 3.1.8 Upper Hip Level

Midway between the top-hip and waist levels.

Note 1 to entry: see Fig. 8.

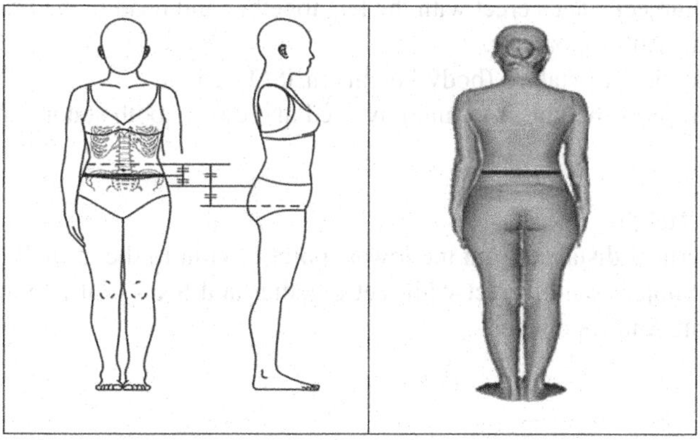

**Fig. 8.** Upper hip level

## 3.2 The Definitions for the Vertical Measurements

The measurements included the title, definition how to measure and the comparison with the ISO 7250 definition for the identical measurements.

The part of the examples are shown from Sects. 3.2.1 to 3.2.4.

### 3.2.1 Stature

Definition: Vertical distance from the highest point of the head in the median line to the ground (see Fig. 9).

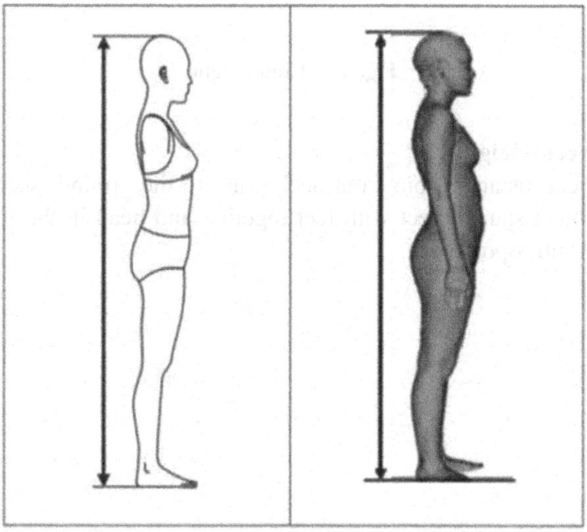

**Fig. 9.** Stature

Position: Subject stands erect with the feet together and head in the Frankfurt plane.
Equipment: Anthropometer.
Note 1 Identical to stature (body height) in ISO 7250-1.
Note 2 The term 'height' is commonly used and covers both stature and recumbent length.

### 3.2.2 Chin Height

Definition: Vertical distance from the lowest point of chin to the ground (see Fig. 10).
Position: Subject stands erect with feet together and head in the Frankfurt plane.
Equipment: Anthropometer.

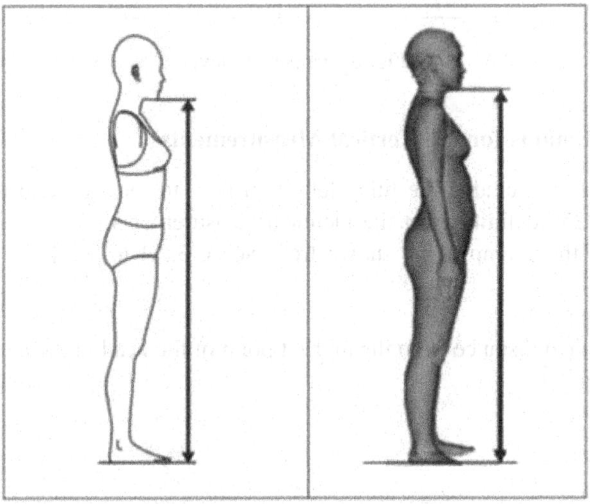

**Fig. 10.** Chin height

### 3.2.3 Front Neck Height

Definition: Vertical distance from front neck point to the ground (see Fig. 11).
Position: Subject stands erect with feet together and head in the Frankfurt plane.
Equipment: Anthropometer.

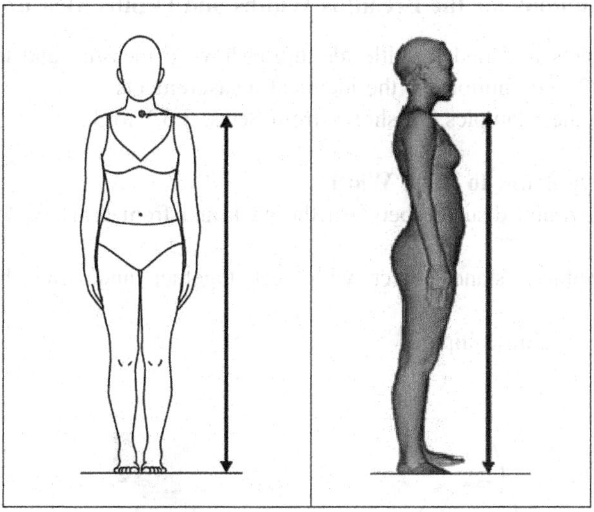

**Fig. 11.** Front neck height

### 3.2.4 Back Neck Height

Definition: Vertical distance from back neck point to the ground (see Fig. 12).

Position: Subject stands erect with feet together and head in the Frankfurt plane.

Equipment: Anthropometer.

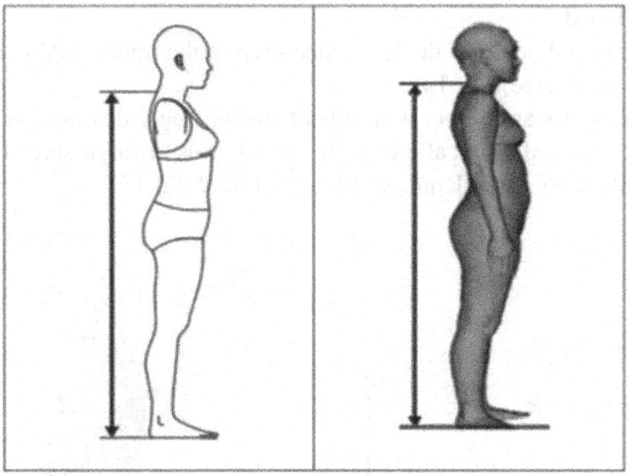

**Fig. 12.** Back neck height

### 3.3 The Definitions for the Breadths Widths and Depths Measurements

The measurements included the title, definition how to measure and the comparison with the ISO 7250 definition for the identical measurements.

The part of the examples are shown from Sects. 3.3.1 to 3.3.3.

#### 3.3.1 Armscye Front to Back Width

Definition: Horizontal distance between the back and front armscye fold points (see Fig. 13).

Position: Subject stands erect with feet together and arms hanging freely downwards.

Equipment: Sliding calliper.

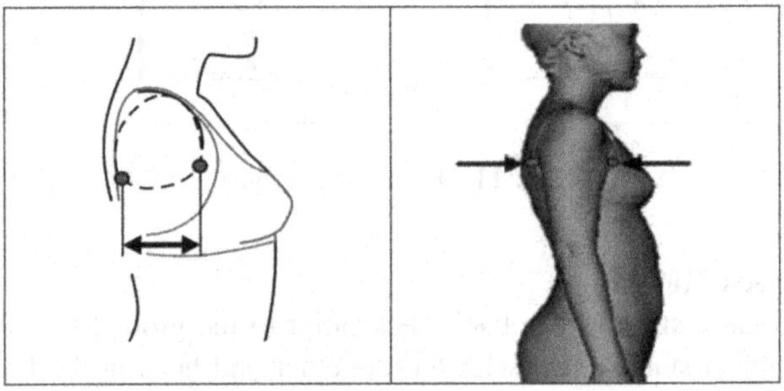

**Fig. 13.** Armscye front to back width

#### 3.3.2 Chest Depth

Definition: Horizontal depth of the torso measured in the midsagittal plane at the level of centre chest point (see Fig. 14).

Position: Subject stands erect with arms hanging freely downwards.

Equipment: Large sliding calliper with curved arms or large spreading calliper.

NOTE Identical to chest depth, standing in ISO 7250-1.

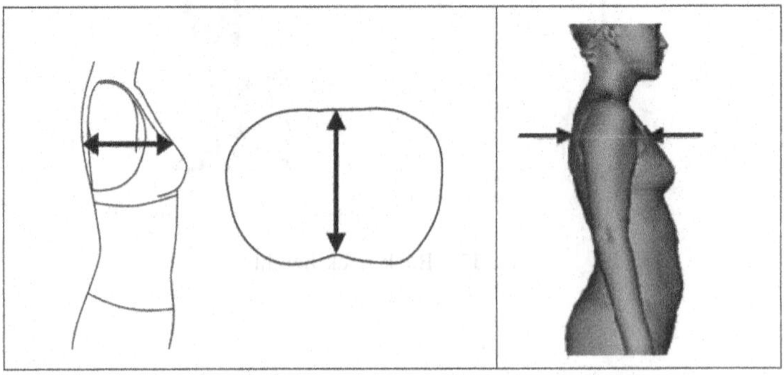

**Fig. 14.** Chest depth

### 3.3.3 Bust Depth

Definition: Maximum horizontal depth of the thorax at the level of the bust point (see Fig. 15).

Position: Subject stands erect with arms hanging freely downwards.

Equipment: Large sliding calliper.

NOTE Identical to thorax depth at the nipple in ISO 7250-1.

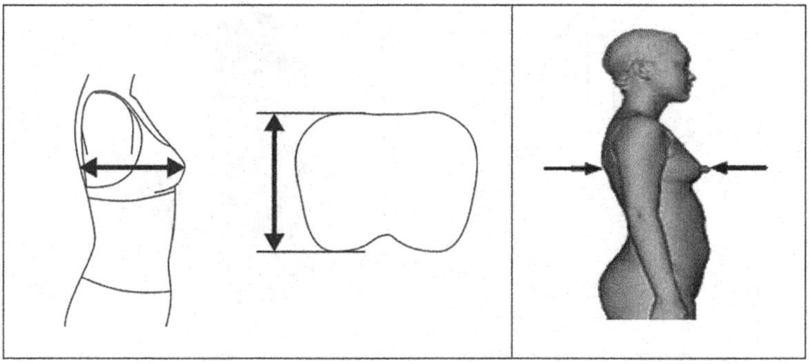

**Fig. 15.** Bust depth

## 3.4 The Definitions for the Girth Measurements

The measurements included the title, definition how to measure and the comparison with the ISO 7250 definition for the identical measurements.

The part of the examples are shown from Sects. 3.4.1 to 3.4.5.

### 3.4.1 Elbow Girth, Arm Bent

Definition: Girth of the elbow at the elbow point (see Fig. 16).

Position: Subject stands erect with the elbow bent at a 90° angle and the hand and fingers extended.

Equipment: Tape measure.

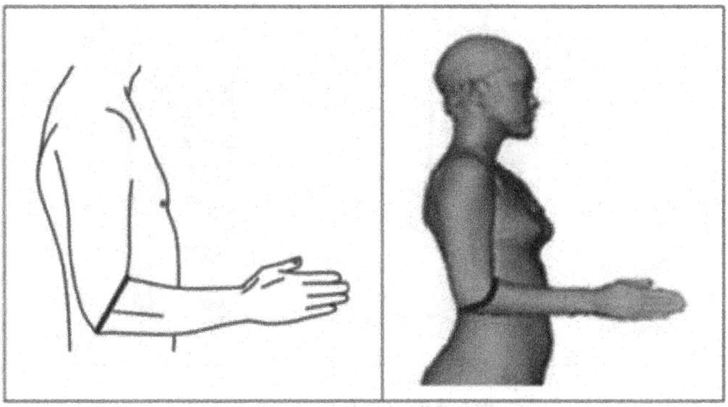

**Fig. 16.** Elbow girth arm bent

### 3.4.2 Wrist Girth

Definition: Girth of the wrist at the level of wrist point (see Fig. 17).

Position: Subject holds forearm horizontal, hand outstretched, fingers extended, palm facing down.

Equipment: Tape measure.

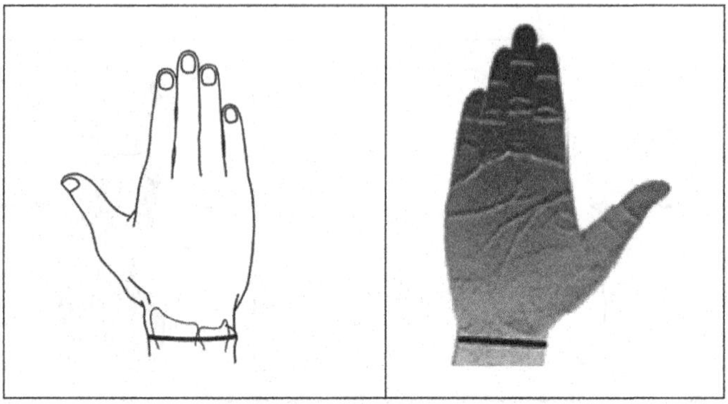

**Fig. 17.** Wrist girth

### 3.4.3 Mid - Thigh Girth

Definition: Horizontal girth of the thigh measured midway between the inside leg level and the centre point of knee cap (see Fig. 18).

Position: Subject stands erect with legs shoulder width apart.

Equipment: Tape measure.

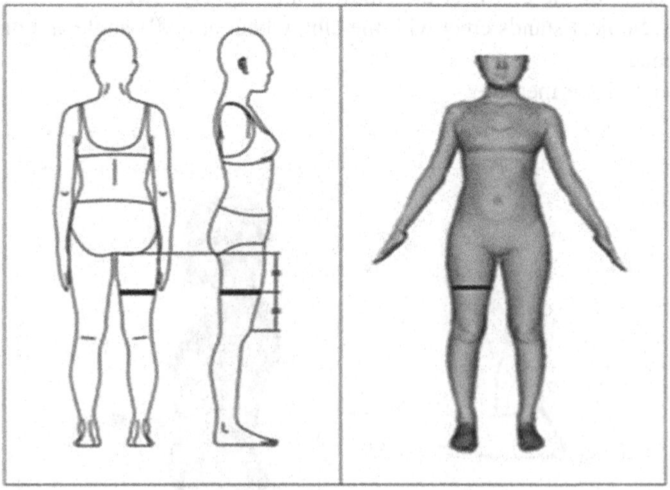

**Fig. 18.** Mid thigh girth

### 3.4.4 Knee Girth

Definition: Horizontal girth of the knee at the level of the centre point of knee-cap (see Fig. 19).

Position: Subject stands erect with legs shoulder width apart.

Equipment: Tape measure.

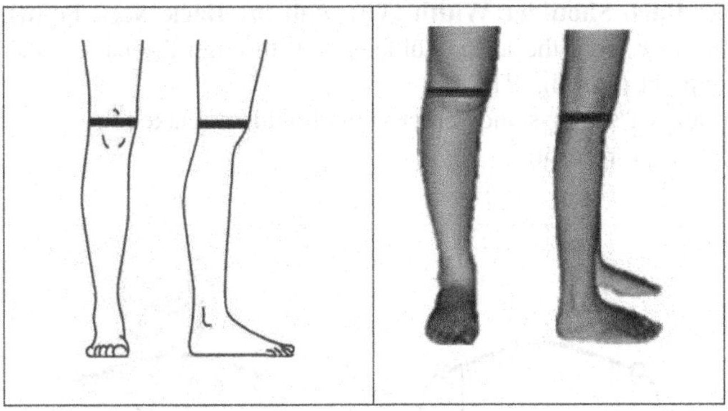

**Fig. 19.** Knee girth

### 3.4.5 Lower Knee Girth

Definition: Horizontal girth of the lower leg just below the patella (see Fig. 20).

Position: Subject stands erect with legs shoulder width apart.

Equipment: Tape measure.

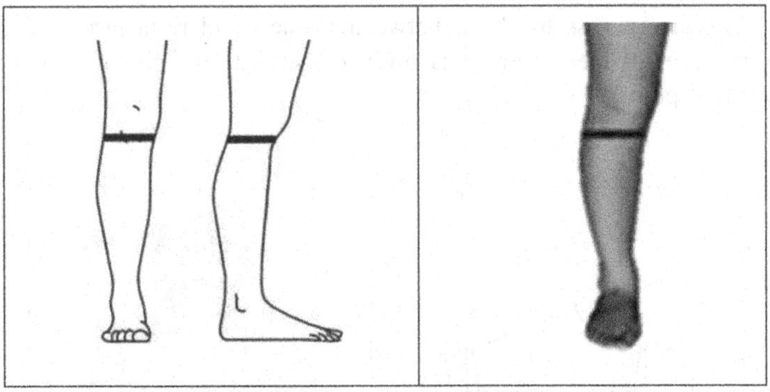

**Fig. 20.** Lower knee girth

### 3.5 The Definitions for the Distances Measured Following the Surface of the Body Measurements

The measurements included the title, definition how to measure and the comparison with the ISO 7250 definition for the identical measurements.

The part of the examples are shown from Sects. 3.5.1 to 3.5.2.

#### 3.5.1 Across Back Shoulder Width (Through the Back Neck Point)

Definition: Distance from the left shoulder point, through the back neck point to the right shoulder point (see Fig. 21).

Position: Subject sits or stands erect with shoulders relaxed.

Equipment: Tape measure.

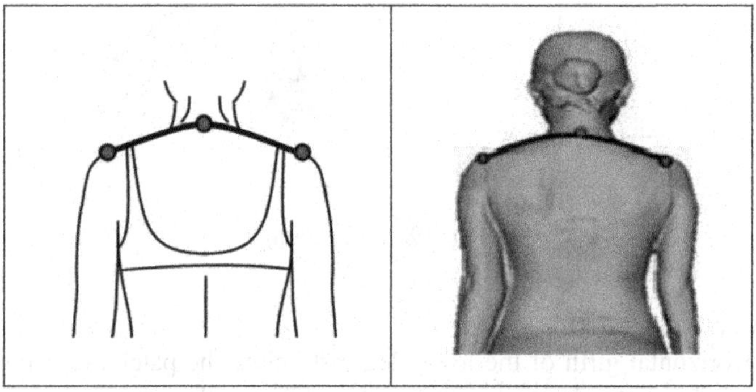

**Fig. 21.** Across back shoulder width (through the back neck point)

#### 3.5.2 Across Back Width

Definition: Distance across the back between the left and right arm scye lines (see Fig. 22). The level of measurement is midway between the shoulder point and the armpit back fold point.

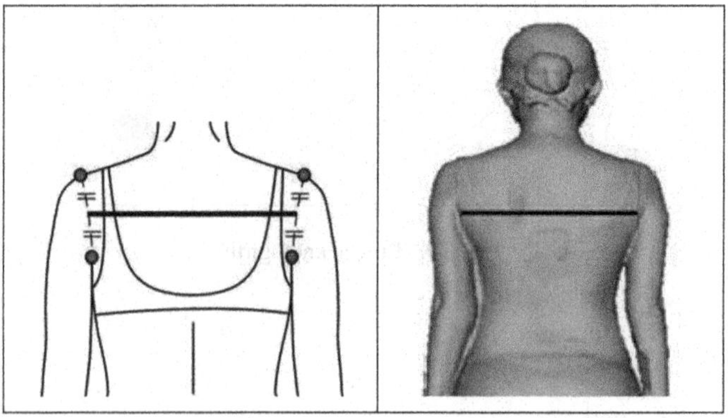

**Fig. 22.** Across back width

Position: Subject sits or stands erect with shoulders relaxed.
Equipment: Tape measure.

### 3.6 The Definitions for the Hand and Foot Measurements

The measurements included the title, definition how to measure and the comparison with the ISO 7250 definition for the identical measurements.

The part of the examples are shown from Sects. 3.6.1 to 3.6.3.

#### 3.6.1 Foot Length

Definition: Distance from rear of the heel to the tip of the longest (first or second) toe, measured parallel to the longitudinal axis of the foot (see Fig. 23).

Position: Subject stands erect with legs shoulder width apart, and weight equally distributed on both feet.

Equipment: Large sliding calliper.

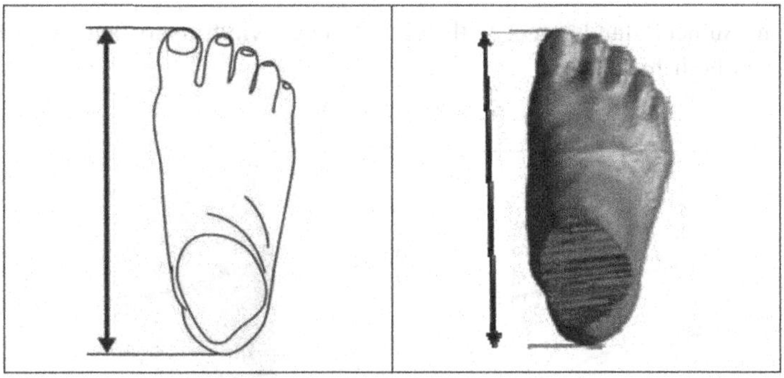

**Fig. 23.** Foot length

#### 3.6.2 Foot Width

Definition: Maximum width of foot across the ball of the foot, measured orthogonal to the longitudinal axis of the foot (see Fig. 24).

Position: Subject stands erect with legs shoulder width apart, and weight equally distributed on both feet.

Equipment: Large sliding calliper.

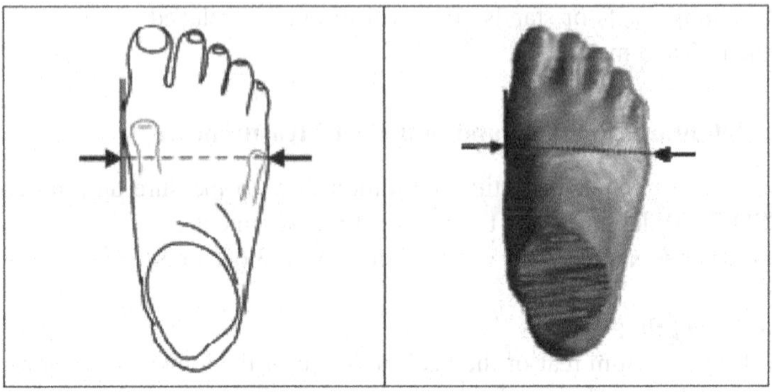

**Fig. 24.** Foot width

### 3.6.3 Foot Girth

Definition: Maximum girth of the foot measured around the ball of the foot (see Fig. 25).

Position: Subject stands erect with legs shoulder width apart, and weight equally distributed on both feet.

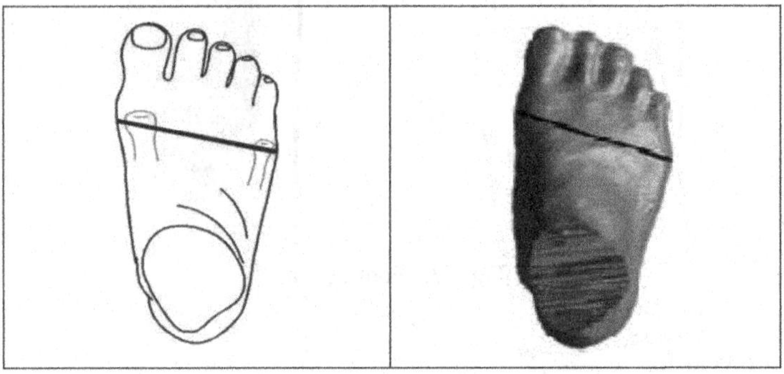

**Fig. 25.** Foot girth

## 4 Discussion

ISO 8559-1 document [2] provides a description of anthropometric measurements that can be used as a basis for the creation of physical and digital anthropometric databases. The list of measurements specified in this document is intended to serve as a guide for practitioners in the field of clothing who are required to apply their knowledge to select population market segments and to create size and shape profiles for the development of all garment types and their equivalent fit mannequins. The list provides a guide for how to take anthropometric measurements, as well as give information to clothing product development teams and fit mannequin manufacturers on the principles of measurement and their underlying anatomical and anthropometrical bases. ISO 8559-1 includes three

Annexes, Annex A describes the use of the pictogram (standardized and modified) based on the selection of most usual body dimensions used for clothing size designation. Annex B describes the standing and seated postures with arm positions related to the measuring conditions. This document is intended to be used in conjunction with national, regional or international regulations or agreements to ensure harmony in defining population groups and to allow comparison of anthropometric data sets.

For the 8559-2 [3], the relevant indication of a men's jacket is designated by the chest girth, indicated on a pictogram. Waist girth and/or body height and/or shoulder width can be given as further optional measurements.

Hand size is a code based on the hand girth in centimetres (ISO 8559-1).

This International Standard describes the principles of the establishment of the tables for body measurements, defines the categories of tables (related to intervals), lists the population groups (infants, girls, boys, children, women, men) and sub-groups to be used for compiling developing standard ready-to-wear garments. The body measurement tables and intervals are mainly used by the clothing sector to make the development of good fitting products easier and more accurate.

The described methodology is mainly based on the application of statistical analysis, using body dimension data. The statistical level has deliberately been kept to a low level in order for the content to be made readily comprehensible to the widest possible readership.

ISO/TC 133 developed IS 8559-3 [4]. In order to size mass-produced clothes, the body size of the intended wearer has to be defined and identified with the nearest size on a table of sizes. In this garment-related system, the body size is defined by scales of the appropriate primary dimensions. A good degree of standardization is achieved by the establishment of open-ended size scales with (fixed or not) intervals in at least the primary control dimension for each garment type. Where body shape is characterized by two primary girth dimensions, the first is placed on fixed scale, while the second (the dependent variable) is not. The processing of body measurement data as described in this International Standard results in the grouping of body sizes appropriate to the studied population concerned. Examples of garment size tables are readily compiled from this information. The frequency distribution of body sizes is a useful means of determining which body sizes are applicable to the bulk of the population. Consequently systems may need to be adjusted, particularly in the case of waist girth for women's wear for which body shape is defined by dimensions other than the waist girth. Distribution of body dimensions may change due to secular changes. However, it may not be necessary to update a size table if the products can accommodate the population. This methodology is applicable to various sets of body dimensions. It may be useful to determine intervals for the size designation as described in ISO 8559-2. Values in the tables in this international standard are examples. Garment dimensions are not contained in this document.

The ISO 8559 series of standards provides guidelines for clothing manufacturers to develop size and shape profiles based on different populations in order to create all kinds of clothing and mannequins. They have recently been updated to reflect changes in the dynamic clothing sector and eliminate trade barriers by harmonizing size marking and the terms of reference worldwide [1].

## References

1. ISO Homepage. http://www.iso.org. Accessed 21 Apr 2017
2. ISO: ISO 8559-1 (2017) Part 1: Anthropometric definitions for body measurement, ISO
3. ISO: ISO 8559-2 (2017) Part 2: Primary and secondary dimension indicators, ISO
4. ISO: ISO 8559-3 (2018) Part 3: Methodology for the elaboration of the body measurement tables and intervals, ISO

# Anthropometric Dimension of Agricultural Workers in North Eastern Thailand

Ekarat Sombatsawat[1], Mark Gregory Robson[2(✉)], and Wattasit Siriwong[1(✉)]

[1] College of Public Health Sciences, Chulalongkorn University, Bangkok 10330, Thailand
ekarat.sombatsawat@gmail.com, wattasit.s@chula.ac.th
[2] School of Environmental and Biological Sciences, Rutgers University, New Brunswick, USA
robson@sebs.rutgers.edu

**Abstract.** Anthropometry is the science of human body measurement which is carried out for designing tools and equipment, not only to achieve performance and productivity but also providing safety and comfort. This study presents anthropometric dimensions of agricultural workers in North-eastern Thailand that carried out in 139 agricultural workers with 15 anthropometric dimensions of standing posture by using the commercial anthropometer set. Statistical analysis were analyzed by mean (±SD). The 5$^{th}$, 50$^{th}$ and 95$^{th}$ percentile values were calculated accordingly. However, the t-tests were used to compare different mean between male and female anthropometric dimensions. The agricultural workers were comprised of 71 males (aged 18–59 years, mean (±SD)) at 44.11 years (±10.02) and 68 females (aged 18–59 years, mean (±SD)) at 45.87 years (±10.83). In addition, there are significant different between male and female agricultural workers in all of anthropometric dimensions ($p < 0.01$) except body weight. Similarly, the comparison mean of anthropometric dimensions in Thai, Taiwanese, Chinese and Singaporean that found Thai male and female are relatively smaller than other countries. Therefore, the finding of this study concluded that anthropometric data were evidently difference between male and female within the region and difference countries which are absolutely necessary for designing of ergonomically machines/tools/equipment of the target users.

**Keywords:** Anthropometry · Standing posture · Agricultural · Workers Thailand

## 1 Introduction

Anthropometry is the science of measurement and the art application to establish the physical geometry, mass properties, and strength capabilities of human body (Del Prado-Lu 2007). Anthropometric body dimensions play significant roles in human-machine interaction, clothing design, ergonomics and architecture because the man-machine interface decides the ultimate performance of the equipment, to achieve better performance and efficiency along with higher comfort and safety to the operators it is imperative to design tools, equipment and workplaces keeping in view of the


S. Bagnara et al. (Eds.): IEA 2018, AISC 826, pp. 423–433, 2019.
https://doi.org/10.1007/978-3-319-96065-4_47

anthropometric data of the agricultural workers (Agrawal et al. 2010). Thais agricultural workers extensively used of manually operated equipment at the beginning with initial land preparation to post-harvest operations. The cultivation of rice as the great majority chief occupation of Thais, that attention needs to be given to their capabilities and limitations during design and operation of various farm equipment (Yadav and Pund 2010). Moreover, it does not contain any anthropometric data of Thai agricultural workers are not considered in the design of agricultural equipment and yet most of the equipment been used are imported from Eastern Asian countries. However, very few study and no updater data of comprehensive anthropometric measurement among agricultural workers in North-eastern Thailand, which is seen as a significant contribution to perform a specific agricultural operation. The data will help to decides on the ultimate performance designing of the agricultural machinery and equipment to increase efficiency safety and comfort. The study aim to presents anthropometric dimensions of agricultural workers in North-eastern Thailand.

## 2 Methods

### 2.1 Study Design and Study Participants

Anthropometric surveys were carried out 139 agricultural workers, that propulsive selected in North-eastern Thailand. In the same way, the participants were randomly selected from Nakhon Ratchasima province in North-eastern Thailand included 71 male and 68 female, aged 18–59 years, more than a year agricultural work experience. During the anthropometric surveys, it was observed that male are main workforces of agriculture in the region. Likewise, their participation of male to female ratio in various agricultural operations, is nearly equal (Agrawal et al. 2010).

### 2.2 Body Dimension

The study was conducted to collected anthropometric dimension of agricultural workers by Anthropometer based on the measurements parameters and techniques were adopted from Pheasant (2003) and ISO7250-1 (2010). Nevertheless, the eleven anthropometric dimensions of standing posture were consisted of (1) stature (body height, (2) eye height, (3) shoulder height, (4) elbow height, (5) iliac spine height, (6) knuckle height, (7) fingertip height, (8) vertical grip reach, (9) span, (10) elbow span, (11) hand length, (12) hand breadth, (13) foot length, (14) foot breadth, in cm and (15) body weight (kg) that were considered adopt for farm equipment design on agricultural workers chosen, presented in Fig. 1.

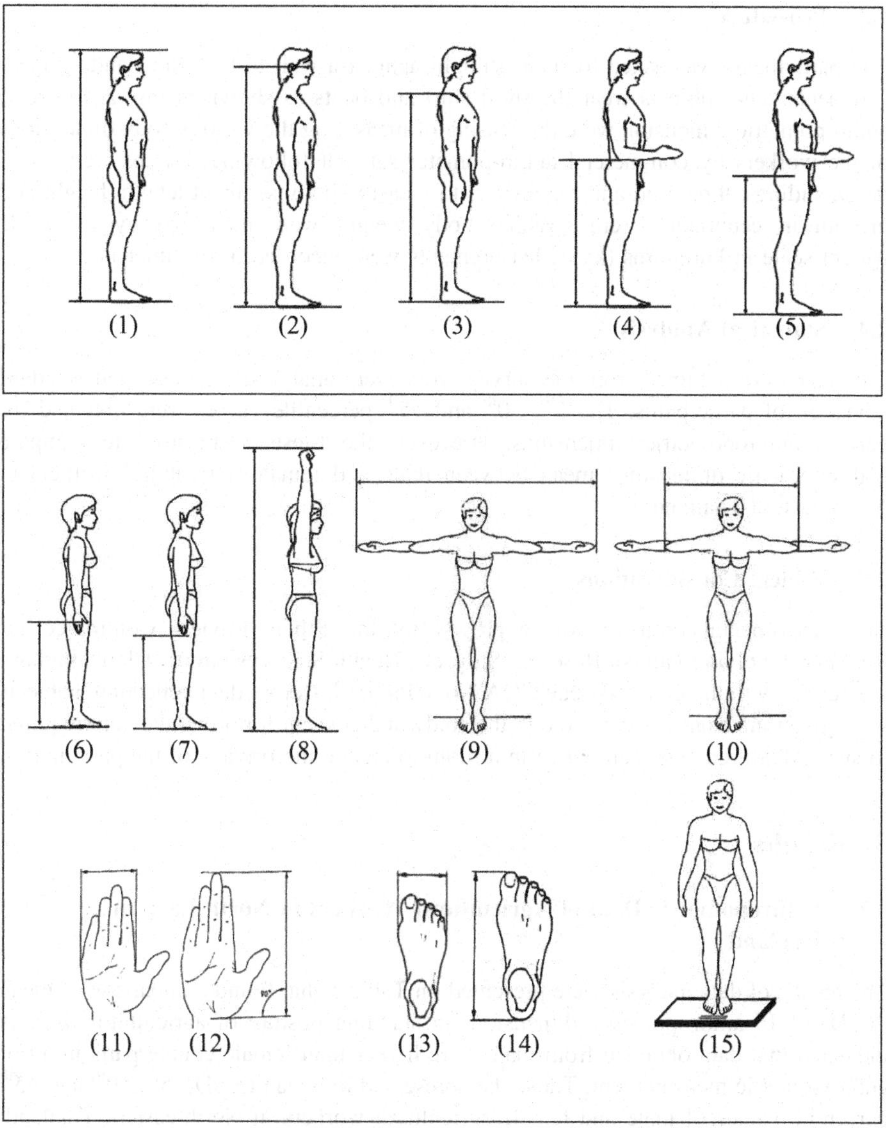


**Fig. 1.** The measurements in standing posture

### 2.3 Procedure

The participants was asked to stand straight against a wall to take the standing measurements. The subjects wear the short shirt and pants as shown in Fig. 2. In order, anthropometric dimension were done using of measuring the various body dimensions of the workers by commercial anthropometer set that following devices were used: tapes, sliding caliper and anthropometers to measure body segment length, height and breadth in centimeters (cm). While, body weight were measured by electronic weight scale in kilograms (kg). Measurements were recorded in centimeters.

### 2.4 Statistical Analysis

The data were obtained from the survey work were analyzed for mean and standard deviation of participants. The 5$^{th}$, 50$^{th}$ and 95$^{th}$ percentile values were obtained for various anthropometric dimensions. However, the t-tests were used to compare anthropometric dimensions mean between male and female agricultural workers in North-eastern Thailand.

### 2.5 Ethical Considerations

The experimental protocol was approved by the Ethics Review Committee for Research Involving Human Research Subjects, Health Sciences Group, Chulalongkorn University with the certified code COA No. 045/2017. Before data collection, subjects were given information, their role in the study and consent form that they are required to sign. Measurements were taken in a room to ensure the privacy of the participants.

## 3 Results

### 3.1 Anthropometric Data of Agricultural Workers in North-Eastern Thailand

The results of data analysis were presented on Table 1 that found comparison of mean (±SD) of 15 anthropometric dimensions in standing posture of agricultural workers indicated that data obtained from male were higher than female counterparts in all of anthropometric measurement. Those data presented by mean (±SD), 5$^{th}$, 50$^{th}$ and 95$^{th}$ percentile values of male and female agricultural workers in North-eastern Thailand. There are slightly increase of variations in the mean and percentile values of 5$^{th}$, 50$^{th}$ and 95$^{th}$ both male and female agricultural workers. Moreover, there are significant different between male and female agricultural workers in stature, eye height, shoulder height, elbow height, iliac spine height, knuckle height, fingertip height, vertical grip reach, span, elbow span, hand breadth, hand length, foot length and foot breadth ($p < 0.01$) except body weight.

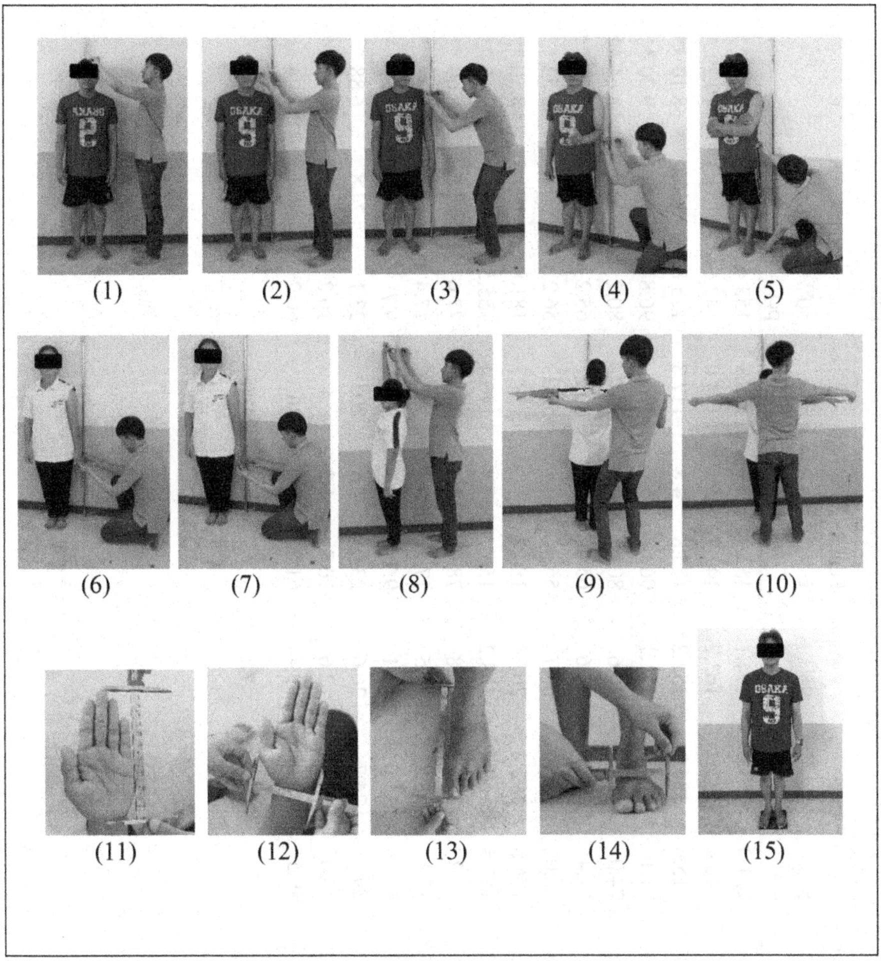


**Fig. 2.** Anthropometric measurements of participant

**Table 1.** Anthropometric data of male and female agricultural workers in North-Eastern Thailand

| Anthropometric measurements (cm) | Male (n = 71) | | | | | Female (n = 68) | | | | | t |
|---|---|---|---|---|---|---|---|---|---|---|---|
| | $\bar{x}$ | SD | $5^{th}$ pct | $50^{th}$ pct | $95^{th}$ pct | $\bar{x}$ | SD | $5^{th}$ pct | $50^{th}$ pct | $95^{th}$ pct | |
| 1. Stature | 163.7 | 6.8 | 151.3 | 163.9 | 174.1 | 153.7 | 5.9 | 145.1 | 153.4 | 164.3 | 9.72 ** |
| 2. Eye height | 153.6 | 6.5 | 140.6 | 153.2 | 163.2 | 142.7 | 5.5 | 134.0 | 142.6 | 153.3 | 10.29 ** |
| 3. Shoulder height | 135.9 | 10.3 | 115.6 | 136.2 | 152.2 | 125.8 | 4.8 | 117.8 | 125.8 | 136.0 | 8.10 ** |
| 4. Elbow height | 103.4 | 5.5 | 91.1 | 104.1 | 111.1 | 96.6 | 5.0 | 89.8 | 96.8 | 103.8 | 8.20 ** |
| 5. Iliac spine height | 82.7 | 4.8 | 74.0 | 83.0 | 90.6 | 80.4 | 4.5 | 73.7 | 80.2 | 86.4 | 3.64 ** |
| 6. Knuckle height | 71.2 | 7.0 | 61.4 | 71.0 | 91.6 | 65.4 | 3.2 | 60.2 | 65.2 | 70.6 | 6.53 ** |
| 7. Fingertip height | 60.0 | 5.8 | 52.0 | 59.7 | 69.7 | 56.0 | 3.1 | 50.3 | 56.2 | 60.8 | 5.52 ** |
| 8. Vertical grip reach | 191.8 | 12.7 | 176.5 | 192.4 | 207.0 | 180.1 | 11.4 | 157.7 | 181.8 | 194.6 | 8.50 ** |
| 9. Span | 170.0 | 9.2 | 157.5 | 169.6 | 188.2 | 156.7 | 7.4 | 146.4 | 155.5 | 168.0 | 9.16 ** |
| 10. Elbow span | 83.6 | 9.5 | 77.2 | 84.7 | 93.8 | 78.0 | 4.4 | 71.0 | 77.6 | 84.5 | 8.69 ** |
| 11. Hand length | 18.6 | 1.3 | 17.1 | 18.8 | 20.5 | 17.4 | 0.8 | 15.8 | 17.4 | 18.9 | 8.78 ** |
| 12. Hand breadth | 10.9 | 0.7 | 9.9 | 11.0 | 12.1 | 9.7 | 0.7 | 8.8 | 9.7 | 10.6 | 12.19 ** |
| 13. Foot length | 24.2 | 2.7 | 19.7 | 24.7 | 26.6 | 23.0 | 1.1 | 20.9 | 23.1 | 24.8 | 6.88 ** |
| 14. Foot breadth | 11.4 | 2.2 | 10.0 | 11.2 | 12.6 | 10.2 | 1.2 | 8.4 | 10.2 | 11.8 | 6.07 ** |
| 15. Body weight (kg) | 64.7 | 11.7 | 49.5 | 63.0 | 89.4 | 61.8 | 9.9 | 46.1 | 60.2 | 80.1 | 1.94 ns |

pct: percentile
ns: not significant, t-test: significant at $p < 0.01$**

### 3.2 Comparison of Anthropometric Data in North-Eastern Thailand with Other Different Asian Countries

The four comparison of anthropometric data of agricultural workers in Thailand from 71 male and 68 female with aged 18–59 years, and other countries as follow Taiwanese from 1,322 male and 799 female with aged 18–65 years (Wang et al. 1999), Chinese from 11,164 male and 11,150 female with aged 18–60 years (China Standards GB/T 10000-1988 1988) and Singaporean from 145 male and 132 female with aged 18–45 years (Chuan et al. 2010). The comparison of male found that Thai is smaller than Taiwanese, Chinese and Singaporean in stature, eye height, shoulder height, elbow height, iliac spine height, knuckle height, fingertip height, vertical grip reach, foot length, span and elbow span. However, the anthropometric data of hand length, hand breadth and foot breadth are similar to other countries. Conversely, body weight of Thai's male is smaller than Taiwanese and Singaporean but higher than Chinese

**Table 2.** Comparison of anthropometric data in North-Eastern Thailand with other different Asian countries

| Anthropometric measurements (cm) | Thai[a] | | Taiwanese[b] | | Chinese[c] | | Singaporean[d] | |
|---|---|---|---|---|---|---|---|---|
| | M | F | M | F | M | F | M | F |
| 1. Stature | 163.7 | 154.0 | 169.9 | 157.3 | 167.8 | 157.0 | 174.0 | 162.0 |
| 2. Eye height | 152.5 | 142.7 | 157.9 | 145.7 | 156.8 | 145.4 | 163.0 | 150.0 |
| 3. Shoulder height | 135.8 | 125.9 | 139.1 | 128.5 | 136.7 | 127.1 | 144.0 | 134.0 |
| 4. Elbow height | 102.1 | 94.3 | 108.8 | 100.7 | 105.4 | 98.7 | 110.0 | 103.0 |
| 5. Iliac spine height | 82.8 | 80.3 | N/A | N/A | N/A | N/A | 97.0 | 89.0 |
| 6. Knuckle height | 70.5 | 65.6 | N/A | N/A | N/A | N/A | 75.0 | 71.0 |
| 7. Fingertip height | 59.9 | 56.1 | 65.7 | 62.0 | N/A | N/A | 65.0 | 62.0 |
| 8. Vertical grip reach | 191.9 | 179.8 | 212.0 | 194.0 | N/A | N/A | 208.0 | 188.0 |
| 9. Span | 83.4 | 77.1 | N/A | N/A | N/A | N/A | 91.0 | 84.0 |
| 10. Elbow span | 169.8 | 157.1 | N/A | N/A | N/A | N/A | 175.0 | 157.0 |
| 11. Hand length | 18.6 | 17.2 | 19.2 | 17.4 | 18.3 | 17.1 | 19.0 | 17.0 |
| 12. Hand breadth | 10.9 | 9.7 | N/A | N/A | N/A | N/A | 9.0 | 7.0 |
| 13. Foot length | 24.2 | 22.8 | N/A | N/A | N/A | N/A | 26.0 | 23.0 |
| 14. Foot breadth | 11.4 | 10.3 | N/A | N/A | N/A | N/A | 10.0 | 9.0 |
| 15. Body weight (kg) | 64.7 | 61.8 | 67.5 | 53.8 | 59.0 | 52.0 | 68.0 | 55.0 |

*M = male and F = female

**a = this study, b = Wang et al. (1999), c = China Standards GB/T 10000-1988 (China Standards 1988) and d = Chuan et al. (2010)

populations. While, comparison of female found that Thai is smaller than Taiwanese, Chinese and Singaporean in stature, eye height, shoulder height, elbow height, iliac spine height, knuckle height, fingertip height, vertical grip reach, span and elbow span. Thai have similarly data of anthropometric in hand length, foot breadth, foot length but higher than other countries in hand breadth and body weight (Table 2).

### 3.3 Average Stature and Body Weight Variations in Four Different Asian Countries

Gender effects on stature in Thai's male and female population is shorter than Taiwanese, Chinese and Singaporean that can be seen in Fig. 3(a). Meanwhile, body weight of Thai's male is similar to Taiwanese and Singaporean, except Chinese. In other hand, weight of Thai's female is higher than other countries, presented in Fig. 3(b).

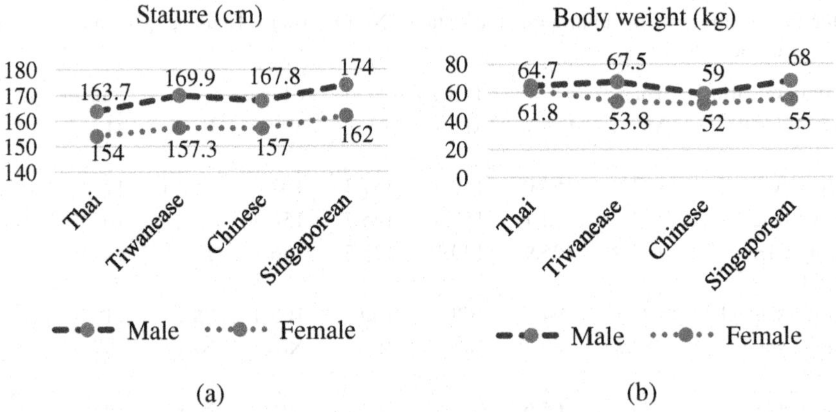


**Fig. 3.** Average stature and body weight variations in four different Asian countries

## 4 Discussion

One hundred thirty-nine agricultural workers presented the comparison of 15 anthropometric dimensions of standing posture between male and female in North-eastern Thailand which stature of male is greater than the female by 9.7 cm and the body weight of male a little bit greater than female by 2.9 kg. However, there are normally different in physical geography of gender between male and female. Similarly with Agrawal et al. (2010) studies on anthropometric considerations for farm tools and machinery design for tribal workers found that the lower body dimensions may lead to having uncomfortable postures adopted while working with implements and machinery leading to low work output. In contrast, other studies found that female is significant smaller than male do not vary much, thus tools and equipment designed based on data collected can effectively be utilized by both male and female agricultural workers within the region (Oduma and Oluka 2017; Kothari 2013). However, further study of body dimension should be include another dimension such as acromial height, elbow height, knee height, waist back length, scapula to waist back length, bi-acromial, are the parameter where anthropometric data can be very useful to designing of ergonomically agricultural equipment, machine, clothing, working postures, especially layout of a tractor driver's workplace (Gite and Yadav 1989). The comparison of male found that Thai is smaller than Taiwanese, Chinese and Singaporean in all of body dimension, except hand length, hand breadth and foot breadth are similar to other countries. The previous studies that found hand length and hand breadth in Taiwanese, Chinese, Japanese, Korean, Iranian and Thai were similarly (Sadeghi et al. 2015 and Klamklay et al. 2008). Furthermore, only body weight in Thai's female is obviously higher than Taiwanese, Chinese and Singaporean that may cause from the different life style of dietary. In particular, higher socioeconomic status implies higher income and is associated with better education, resulting in better nutrition, better health care, and better medical and social services that leads to an increase in overall stature (Chuan et al. 2010; Iseri and Arslan 2009). Different in the demographic proportion of farm-workers in recent decades have shifted the roles of males and females in farm labour; as a result, the designs of the work system and tools must be altered or refined accordingly (Syuaib 2015) to prevent ergonomically health effects agricultural workers.

In Thailand, most of them has been using traditional of hand tools and manually operated equipment which are extensively used and shared between male and female workers that starting from paddy preparation, diggings, cultivating, spraying and harvesting processes. It may be seen that the operator has to apply much force on equipment for better maneuverability and proper operation. Anthropometric database thus will help to incorporating suitable modifications improved in manufacturer related agricultural workers. Even nowadays, the situation has changed from manually tools and equipment used to being manufactured in an agricultural in rural and cities, anthropometric data is far away from the user, and human factors principles application and data are not well understood (Gite and Yadav 1989). Especially, developing countries such as Thailand were imported equipment from manufacturers in other countries such as China, Japan or United States, Germany, should determine whether the design parameters that should be based on anthropometric dimensions are compatible with the anthropometric dimensions of their own population (Nabeel et al. 2008). Thus, the anthropometric data is one of the most important and widely used measure of human body which varies primarily with gender and ethnicity (Shrestha et al. 2009) and it should be concern in agricultural equipment selection of agricultural workers not only within region of Thailand but also the different countries (Dewangan et al. 2005). In addition, this study presents the data of anthropometric in standing posture among agricultural workers for North-eastern region of Thailand which might be a starting point to develop an anthropometric database and apply it in practice to adjustment of design and modify to agricultural equipment. According to the proper matching of machine requirements with the human capabilities is basically necessary for optimum performance of any man–machine system and suit them to the physical characteristics for guarantee safety of agricultural workers (Premkumar et al. 2016). Therefore, anthropometric data of agricultural workers will be used in the improvement of local working conditions to minimize ergonomic problems and related health illnesses in agricultural workers.

## 5 Conclusions

The 15 anthropometric data of Thai agricultural workers is significant different between male and female in various body dimension. In the same way, Thai agricultural workers have relatively smaller dimensions than Taiwanese, Chinese and Singaporean. However, lifestyle nutrition, geographical origin, social and environment, and ethnic composition of populations are some general factors influencing the distribution of anthropometric characteristics. Moreover, a regular updating of anthropometric data is required. Therefore, the survey reveals that the anthropometric dimensions of standing posture is highly recommended that future study should be carried out by concerned authorities in different ethnic group of the Thailand which data will be useful for designing or modifying agricultural machines/tools/equipment suitable for safety and comfort of agricultural workers.

**Funding.** This research was supported by the 100th Anniversary Chulalongkorn University Fund for Doctoral Scholarship. The partial support fund from the Grant for International Research Integration: Chula Research Scholar (GCURS_59_06_79_01) Ratchadaphiseksomphot Endowment Fund is also gratefully acknowledged. This research was also supported in part of by NIEHS sponsored Center Grant P30ES005022 and the New Jersey Agricultural Experiment Station at Rutgers.

## References

Del Prado-Lu, JL (2007) Anthropometric measurement of Filipino manufacturing workers. Int J Ind Ergon 37(6):497–503. https://doi.org/10.1016/j.ergon.2007.02.004

Agrawal KN, Singh KP, Satapathy KK (2010) Anthropometric consideration for farm tools/machinery design for ribal workers of North Eastern India. Agric Eng Int CIGR J 12 (1):143–150

Yadav R, Pund S (2010) Development and ergonomic evaluation of manual weeder. Agric Eng Int CIGR E-J 9:1–9. https://doi.org/10.1.1.517.7772&rep=rep1&type=pdf

Pheasant S (2003) Body space anthropometry, ergonomics and the design of work: second edition. Taylor & Francis Ltd, 11 New Fetter Lane, London EC4P 4EE. Taylor & Francis is an imprint of the Taylor & Francis Group, This edition published in the Taylor & Francis e-Library

ISO 7250-1 (2010) Basic human body measurements for technological design—part 1: body measurement definitions and landmarks, First edition. Thai Industrial Standards Institute. This edition published in the Thai Industrial Standards Institute Library

Wang EMY, Wang MJ, Yeh WY, Shih YC, Lin YC (1999) Development of anthropometric work environment for Taiwanese workers. Int J Ind Ergon 23(1-2):3–8. https://doi.org/10.1016/s0169-8141(97)00095-4

China Standards GB/T 10000-1988 (1988, in simplified Chinese) Human dimensions of Chinese adults. Administration of Technology Supervision, People's Republic of China

Chuan TK, Hartono M, Kumar N (2010) Anthropometry of the Singaporean and Indonesian populations. Int J Ind Ergon 40(6):757–766. https://doi.org/10.1016/j.ergon.2010.05.001

Oduma O, Oluka SI (2017) Comparative analysis of anthropometric dimensions of male and female agricultural workers in South-Eastern Nigeria. Nijotech. 36(1):261–266. https://doi.org/10.4314/njt.v36i1.31

Kothari CR (2013) Research methodology: methods and techniques. Second Revised Edition, pp 185–344. New Age International Publishers, New Delhi

Git LP, Yadav BG (1989) Anthropometric survey for agricultural machinery design: an Indian case study. Appl Ergon 20(3):191–196 https://doi.org/10.1016/0003-6870(89)90076-8

Sadeghi F, Mazloumi A, Kazemi Z (2015) An anthropometric data bank for the Iranian working population with ethnic diversity. Appl Ergon 48:95–103. https://doi.org/10.1016/j.apergo.2014.10.009

Klamklay J, Sungkhapong A, Yodpijit N, Patterson PE (2008) Anthropometry of the southern Thai population. Int J Ind Ergon 38(1):111–118. https://doi.org/10.1016/j.ergon.2007.09.001

Iseri A, Arslan N (2009) Estimated anthropometric measurements of Turkish adults and effects of age and geographical regions. Int J Ind Ergon 39(5):860–865. https://doi.org/10.1016/j.ergon.2009.02.007

Syuaib MF (2015) Anthropometric study of farm workers on Java Island, Indonesia, and its implications for the design of farm tools and equipment. Appl Ergon 51:222–235. https://doi.org/10.1016/j.apergo.2015.05.007

Mandahawi N, Obeidat M, Altarazi SA, Imrhan S (2008) Body anthropometry on a sample of Jordanian males. In: Fowler J, Mason S (eds) Proceedings of the 2008 industrial engineering research conference, pp 1–6, HighBeam Research. IEEE (2008)

Shrestha Om, Bhattacharya S, Jha N, Dhungel S, Jha CB, Shrestha S, Shrestha U (2009) Cranio facial anthropometric measurements among Rai and Limbu community of Sunsari District, Nepal. Nepal Med Coll J 11(3):183–185

Dewangan KN, Prasanna Kumar GV, Suja PL, Choudhury MD (2005) Anthropometric dimensions of farm youth of the North Eastern region of India. Int J Ind Ergon 35(11):979–989. https://doi.org/10.1016/j.ergon.2005.04.003

Premkumari RY, Shirwal S, Veeragoud M, Maski D (2016) Study on anthropometric dimension of women agricultural workers of Hyderabad Karnataka region. IJASR 6(3):359–364

# Anthropometric Survey of Chinese Adult Population

Chayi Zhao(✉), Linghua Ran, Taijie Liu, and Aixian Li

SAMR Key Laboratory of Human Factors and Ergonomics,
China National Institute of Standardization, 4 Zhichun Road, Haidian District,
Beijing 100191, People's Republic of China
{zhaochy, ranlh, liutj}@cnis.gov.cn, liax@cni.gov.cn

**Abstract.** Anthropometric data is the basis of human-centred design and is also widely used in the industrial design of clothing, furniture, sportswear, consumer product, machinery, personal protective equipment, vehicles, architectures etc. Thus, more and more countries carried out national body size survey and set up anthropometric databases. China National Institute of Standardization (CNIS) undertakes national body size survey in China. The first national adult body size survey in China is carried out during 1986–1987. The first national child body size survey is carried out during 2006–2008 by CNIS. As dramatic changes in Chinese body size have taken place during the past 30 years, with the support of the National R&D Infrastructure Development Program of China, CNIS conduct the new Chinese adult anthropometric survey during 2013–2018. With a vast territory, there are great differences in the human body dimensions among different regions in China. To collect more representative human body dimensions, the stratified multistage cluster sampling method is used in the nationwide anthropometric survey. According to the anthropologic studies, the nationwide survey area are divided into 6 regions, which are northeast and north China, central and western region, the lower reach of Yangtze river and the middle reach of Yangtze river, Two Guangs and Fujian, and Yun, Gui and Sichuan. In this anthropometric survey, the 3D body scanners are mainly used, together with Martin measuring instruments for manual anthropometric measurement partly. Up to Jan. 30, 2018, the nationwide field survey has been completed, which collected over 26,000 adults from 32 cities in 24 provinces. The new Chinese adult anthropometric data and statistic reports will be released soon.

**Keywords:** Anthropometry · Body size · National survey · China

## 1 Introduction

Anthropometric data that is consisted of data about human body properties such as physical anthropology and biomechanics provides a foundation for industrial design and ergonomics. The design of products or environments for human use shall be human-centered and accommodating the human physiologic properties and limitations. Therefore, anthropometric data provide a foundation not only for ergonomics but also for production and living, and promote human-centered technological innovations. Being increasingly aware of the importance of anthropometry, many developed


S. Bagnara et al. (Eds.): IEA 2018, AISC 826, pp. 434–441, 2019.
https://doi.org/10.1007/978-3-319-96065-4_48

countries have undertaken anthropometrical initiatives jointly or independently to develop their own anthropometrical databases [1].

In the context of rapid economic growth and technological progress, companies in China pay more attentions to innovation and user experience of products, which in turn make these companies pay more attentions to human-centered design and anthropometrical data. However, existing anthropometrical data of Chinese adult population are collected during 1986–1987 [2], and rapid progress in healthcare and living standards have led to great changes in Chinese body dimensions. These body data collected 30 years ago therefore cannot represent the body dimensions of the present Chinese generation. For this reason, China National Institute of Standardization (CNIS) launched a new nationwide body size survey during 2015–2018 under the support of the National R&D Infrastructure Development Program of China.

## 2 Brief Introduction to Previous Anthropometrical Survey in China

### 2.1 Nationwide Anthropometrical Survey of Adult Population

During 1986–1987, CNIS has conducted the first nationwide body size survey of adult population aged from 18–60 years in China, which stratified sampling method is used to collect body dimension data from over 20,000 persons in 16 provinces, municipalities or autonomous regions, and over 70 body dimension items are taken. According to the statistical results of nationwide anthropometrical survey, a set of national standards for human body dimensions such as GB/T 10000-1988 *Human dimensions of Chinese adults* are developed [2]. Having been widely accepted and adopted, these anthropometric data and standards contribute to economic growth and labour safety much.

### 2.2 Nationwide Anthropometrical Survey of Child Population

During 2006–2008, under the support of the National R&D Infrastructure Development Program of China, CNIS has conducted the first nationwide body size survey of child population aged from 4 to 17 years in China, which sampled over 20,000 persons from 10 provinces, municipalities or autonomous regions and 136 body dimension items are taken. This anthropometric measurement was done by 3D body scanners mainly, together with Martin measuring instruments for manual anthropometric measurement partly. According to the child anthropometric data, a set of Chinese national standards for child body dimensions such as GB/T 26158-2010 *Human dimensions of Chinese minors* and GB/T 26159-2010 *Hand sizing system of Chinese minors* were developed [3, 4], which provide statistical data for the design of school supplies and household items for child use and architecture- and transportation-related products and facilities.

# 3 Latest Nationwide Chinese Adult Anthropometrical Survey

## 3.1 Background

Existing anthropometric data of Chinese adult population were collected during 1986–1987. As rapid economic growth and technological progress in China, Chinese anthropometric data is a pressing need for design and innovations, CNIS initiated a new round of nationwide anthropometric survey of Chinese adult population under the support of the National R&D Infrastructure Development Program of China in 2013. After a survey of needs for human body data in industries, measure methods standardization and national sampling plan, a nationwide anthropometric survey has been started. Owing to limit time and funds, only data about body dimensions are collected nationwide while small sample sizes are taken for such human ergonomic characteristic data as biomechanical data, vision data, auditory data and tactile data.

## 3.2 Measurement Items

Through the survey of the need for anthropometric items in industries, the international and national standards about anthropometry [2–6] are referenced, according to the principle of economic efficiency and the correlation between anthropometric parameters, 169 anthropometric items are measured in this survey, which include 143 body dimensions (116 items for standing posture and 27 items for sitting posture), 12 hand dimensions, 5 foot dimensions and 9 head and face dimensions.

## 3.3 Measuring Methods

In this anthropometric survey, the 3D scanners are mainly used, together with Martin measuring instruments for manual measurement partly. 3D scanners are used for collecting data about whole body, head and foot dimensions. Whole body dimensions are measured in three postures (see Fig. 1). The two-dimensional scanners specially developed are used for measuring hand dimensions. The standard measuring instruments

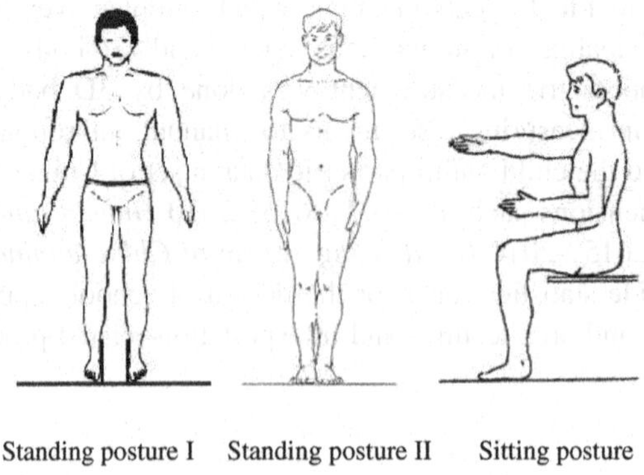


**Fig. 1.** Postures for 3D body size scanning

such as the anthropometer, sliding callipers and spreading callipers are used as manual measuring over 30 body dimensions such as head, face, hands and feet. In combination of interactive measurement and auto-calculation, the task of manual data-acquisition from 3D images is greatly reduced. Many measures are taken to guarantee body measure accuracy besides in terms of the international and national standards about anthropometry [7–9].

### 3.4 Nationwide Sampling Plan

Given the vast territory and great differences in body size among difference regions in China, a stratified multistage cluster sampling method is used.

**Stratified by Region.** Adults' body size varies greatly in different regions in China. To make persons selected more representative, the anthropometric survey area are divided into 6 regions according to the anthropologic studies, which are northeast and north China, central and western region, the lower reach of Yangtze river and the middle reach of Yangtze river, Two Guangs and Fujian, and Yun, Gui and Sichuan. In each region, the provinces included and the ratio of regional population to total population see Table 1. According to statistical results from China Population Census 2010, the sample sizes of the six regions are determined based on the ratio of regional population to total Chinese population.

**Table 1.** Stratified sampling by region

| No. | 1 | 2 | 3 | 4 | 5 | 6 |
|---|---|---|---|---|---|---|
| Region | Northeast and North China | Central and western region | The lower reach of Yangtze River | The middle reach of Yangtze River | Two Guangs and Fujian | Yun, Gui and Sichuan |
| Provinces, municipalities or autonomous regions included | Helongjiang, Jilin, Liaoning, Inner Mongolia, Hebei, Shandong, Beijing and Tianjin | Henan, Shanxi, Shannxi, Ningxia, Gansu, Xinjiang, Tibet and Qinghai | Jiangsu, Zhejiang, Anhui and Shanghai | Hubei, Hunan and Jiangxi | Guangdong, Guangxi, Hainan and Fujian | Yunnan, Guizhou and Sichuan |
| The ratio of regional population to total population | 21.57% | 24.19% | 15.52% | 12.90% | 12.30% | 13.51% |

**Sampling the Prefecture-Level Cities.** This survey took the population of prefecture-level cities (including the municipalities) as the first subpopulation. To take advantage of the partners' measuring instruments, the population of prefecture-level cities which the partners with measuring instruments (Beijing, Tianjin, Changchun, Dalian, Xi'an, Zhengzhou, Wuxi, Hangzhou, Jiaxing, Shanghai, Shenzhen, Foshan and Changsha) are in will be selected. More sample sizes are assigned to these cities with measuring instruments for cost effective purpose.

Considering the differences in body size among different regions, probability (of inclusion in the sample) proportional to size sampling (πps sampling) method is used, the sample of the subpopulation is in proportion to regional population. In each region, at least two prefecture-level cities is selected, the prefecture-level cities with higher population will be easily sampled. Generally, prefecture-level cities with higher population are too thickly populated areas in the region and thus representative anthropometric data shall be collected.

The prefecture-level cities in the same region are numbered according to their geographic position. According to sampling algorithm, a prefecture-level city is randomly selected and then the sample number of subsequent cities are determined at regular interval. In order to get discrete sample cities, the adjacent provinces are numbered successively and cities close to each other within a province are numbered as closely as possible. In the case those sampled cities that can't be surveyed anyway will be replaced by those with the closest number to the former in the same region.

The sample size for a prefecture-level city within a region will be determined according to the ratio of its population to the total population of the region. The average sample size of each selected city is about 800 persons.

**Sampling in a Prefecture-Level City.** Given the differences in body sizes and investigative efforts in between urban and rural area, the different sampling methods in between urban and rural populations are used in the sample prefecture-level city respectively, in urban area the organizations will be sampled, in rural area the villages will be sampled.

Considering that quite a few rural population migrate to urban areas for work and it is more difficult to make a survey in rural areas than in urban areas, the sample ratio of urban population to rural population in a prefecture-level city is set at 5a:1, where a refers to the ratio of urban population to rural population in the prefecture-level city.

*Sampling in Urban Areas.* In consideration of the differences in body sizes among persons with different occupations, the urban population are divided into 5 subgroups, i.e. colleges and universities, government agencies, companies in secondary industry (except construction), companies in tertiary industry (except catering or housekeeping), construction, catering and housekeeping. There are almost adults aged from 18–24 years in colleges and universities subgroup. In the case of without colleges and universities in the sample city, the sample size that should fall within the subgroup will be assigned to other four subgroups. The population in government agencies are identified as a single subgroup, because that government employees and retirees are well organized and vary greatly in age. Considering that there are potential variance in body sizes between blue-collar workers and non-blue-collar workers, the population in secondary industry with more blue-collar workers and those in tertiary industry with less blue-collar workers are identified as a subgroup separately. The secondary industry includes mining, manufacturing, electricity, gas, water production and supply. The population in construction, catering and housekeeping industries are identified as a single subgroup because there are a great many rural migrant workers.

Organizations are sampled with probability proportional to population in the sample city (PPS sampling). The sample size for the organization are determined according to the ratio of the type organization's population to the total population in the city.

The population of every type organization is random sampled, a sample size of about 50 persons with all age groups is taken within a sample organization.

*Sampling in Rural Areas.* Probability proportionate to size (PPS) sampling is used to sample towns within a sampled prefecture-level city. Villages in a sampled town are randomly sampled. The sample size of about 50 persons is taken in each sampled village.

**Replacement of Samples.** In the case those sampled organization, town or village that can't be surveyed anyway will be replaced by one of the same type with the closest number. If the latter still can't be surveyed, the one with the second closest number to the former will be selected, and so on. In the case those sampled person that can't be measured anyway will be replaced by the person of the same type in the adjacent subpopulation.

**The Proportion of Age in Sample.** In consideration of variance in body size between different age groups, the sample population is divided into 4 groups: 18–24 years old, 25–34 years old, 35–59 years old and 60–75 years old. In order to make samples relatively get together and facilitate the survey, the age of sample don't been considered during sampling subgroups, but the number of samples in different age groups is adjusted during the survey. The sample size for each age group in a region is calculated based on the total sample size of the region. The proportion of sample size of each age group is calculated based on statistic result of China Population Census 2010 (see Table 2). In the case that the sample size of an age group is exceeded the proportion set in the group, the sample of the age group are replaced by the sample of other age groups when the follow-up organizations and subpopulations are sampled.

**Table 2.** Proportion of each age group in the samples

| Age group | Percentage |
|---|---|
| 18–24 years old | 14% |
| 25–34 years old | 18% |
| 35–59 years old | 56% |
| 60–75 years old | 11% |

**Table 3.** Summary of sampled cities and sample size of Chinese adult population

| Region | No. | City | Sample size | Region | No. | City | Sample size |
|---|---|---|---|---|---|---|---|
| Northeast and North China | 1 | Changchun | 738 | Central and western region | 18 | Zhengzhou | 714 |
| | 2 | Xingtai | 596 | | 19 | Zhumadian | 1253 |
| | 3 | Jining | 650 | | 20 | Taiyuan | 891 |
| | 4 | Beijing | 1617 | | 21 | Xi'an | 1371 |
| | 5 | Tianjin | 331 | | 22 | Lanzhou | 558 |
| | 6 | Qingdao | 857 | | 23 | Xining | 401 |
| | 7 | Dalian | 525 | | 24 | Xinxiang | 976 |
| | 8 | Tangshan | 836 | | 25 | Jiaozuo | 620 |
| | | Subtotal | 6150 | | | Subtotal | 6784 |
| The lower reach of Yangtze River | 9 | Wuxi | 514 | The middle reach of Yangtze River | 26 | Wuhan | 1152 |
| | 10 | Shanghai | 1731 | | 27 | Hengyang | 878 |
| | 11 | Hefei | 539 | | 28 | Changsha | 858 |
| | 12 | Hangzhou | 736 | | 29 | Nanchang | 611 |
| | 13 | Jiaxing | 612 | | | Subtotal | 3499 |
| | | Subtotal | 4132 | | | | |
| Two Guangs and Fujian | 14 | Shenzhen | 1201 | Yun, Gui and Sichuan | 30 | Chengdu | 749 |
| | 15 | Foshan | 400 | | 31 | Chongqing | 1527 |
| | 16 | Guilin | 872 | | 32 | Kunming | 423 |
| | 17 | Fuzhou | 617 | | | Subtotal | 2699 |
| | | Subtotal | 3090 | | | Total | 26354 |

## 4 Conclusion

Up to Jan. 30, 2018, the nationwide anthropometrical survey has been completed, which collected body data from over 26,000 adults from 32 cities in 24 provinces. The sampled cities and sample size of Chinese adult population see Table 3.

Now the anthropometric data is drawn from the 3D images of over 26,000 Chinese adults. The new Chinese adult anthropometric data and statistic reports will be released soon.

**Acknowledgments.** This work are supported by National R&D Infrastructure Development Program of China (2013FY110200), National Key R&D Program of China (2017YFF0206602) and Quality Supervision, Inspection and Quarantine R&D Program of China (201510042).

## References

1. Zhao C (2013) Survey research on human body ergonomic data (in Chinese). Chin J Ergon 19 (1):76–79 (2013)
2. Standardization Administration of the People's Republic of China, GB/T 10000-1988 Human dimensions of Chinese adults (in Chinese)
3. Standardization Administration of the People's Republic of China, GB/T 26158-2010. Human dimensions of Chinese minors (in Chinese)

4. Standardization Administration of the People's Republic of China, GB/T 26159-2010 Hand sizing system of Chinese minors
5. International Organization for Standardization (ISO), ISO 7250-1:2008, Basic human body measurements for technological design – part 1: body measurement definitions and landmarks
6. Standardization Administration of the People's Republic of China, GB/T 16160-2008, Location and method of anthropometric surveys for garments (in Chinese)
7. Standardization Administration of the People's Republic of China, GB/T 5704-2008, Measuring instruments for anthropometry (in Chinese)
8. International Organization for Standardization (ISO) ISO 20685:2010, 3-D scanning methodologies for internationally compatible anthropometric databases
9. International Organization for Standardization (ISO) ISO 20685-2:2015, Ergonomics – 3-D scanning methodologies for internationally compatible anthropometric databases – part 2: evaluation protocol of surface shape and repeatability of relative landmark positions

# Use of Anthropometry and Fit Databases to Improve the Bottom-Line

Kathleen M. Robinette[1](✉) and Daisy Veitch[2]

[1] Fitmetrix LLC, 5428 Lytle Rd, Waynesville, OH 45068, USA
kathrobinette@gmail.com
[2] SHARP Dummies Pty Ltd., Belair, SA, Australia
daisy.veitch@gmail.com

**Abstract.** Many apparel companies do extensive market research to understand their customer base. This results in good information about the gender, ethnicity, age, income and other characteristics of the target market. In this paper we will show how this can be taken one step further with large anthropometric databases and small fit studies to better target the market, improve sales and reduce waste, through improved fit, faster product development and tighter inventory control. The improvement in population accommodation with good selection of the sizes will be quantified and illustrated using data from the WEAR Association database. An example of the creation of the "fit map" for an apparel item will be provided and compared against a priori assumptions about the range of fit. Then a comparison of raw data versus data weighted to the target market will be illustrated.

**Keywords:** Anthropometry · Sizing · Fit-mapping

## 1 Introduction

Fit problems continue to plague the apparel industry. Companies are producing sizes that end up on the sales rack because they don't fit anyone and needed sizes are not available, so companies are missing sales. Previous studies have demonstrated why published tables of body measurements, such as those found in sizing standards, or summary statistics, such as means and percentiles, do not provide good fitting products [1–3]. Perhaps the biggest limitation of published statistics is that they cannot be tailored to the product's user population (referred to as the target market). Raw body measurement and demographic data can be tailored, and this enables the product developer to limit the assortment of sizes to just those needed. This paper will describe and provide examples for the use of raw data from databases.

## 2 Representing the Target Market

### 2.1 Defining the Target Market

The target market is the population expected to purchase the product. The first task in defining the target market is deciding on the characteristics of the customers that are


S. Bagnara et al. (Eds.): IEA 2018, AISC 826, pp. 442–452, 2019.
https://doi.org/10.1007/978-3-319-96065-4_49

relevant to the product. Some characteristics that are important include: age, gender, country or region, occupation, income level, education level etc. This is done by executive decision sometimes informed by a marketing questionnaire survey.

The next task is to decide what measurements are important to the product fit. For example, for a jacket, shoulder width, chest circumference, waist-back length and arm length will be important. For a pair of pants waist circumference, hip circumference, and leg length will be important.

If unsure about what measurements might be available and important there are many sources for measurement descriptions. There are standards organizations that list these. For example, the International Standards Organization (ISO) has standard body measurement descriptions that can be purchased [4] as do ASTM International [5].

Measurement descriptions are also part of the documentation from past measurement surveys done by the US government and some of those can be downloaded for free. One example is the document of measurement descriptions from the CAESAR® project [6]. Another more recent example is a document from a survey completed by the U.S. Army in 2012 [7].

## 2.2 Finding, Adapting and Weighting the Sample

The third task is to find or collect raw data for a sample that includes people with the relevant characteristics and measurements. It must be raw data because it must be adjusted to represent the target market. It would be unusually rare to find a sample that is exactly the composition needed.

The raw data file or files must include people from all the target market demographic groups. It can contain other groups as well, but any groups not in the target market will be dropped. This is called sample segmentation. Once the unnecessary subsets are dropped it is recommended that at least 200 subjects remain. Also, if there are further subgroups, such as ethnicity groups, there should be at least 10 subjects in each of the remaining subgroups for weighting purposes.

For example, the data for North America from the CAESAR survey [8] includes males and females ages 18–71, but a target market for a pant might only include adult females ages 18–45. To adapt the CAESAR data to the market, males and subjects 45 and older are dropped. Figure 1 shows the waist and hip circumferences for the sample before and after segmentation. The full sample plot shows that the males had larger waists but smaller hips than the females. This is a proportioning difference that would substantially impact the sizes of the pant. There is quite a bit of overlap for males and females though. This might be an area where a unisex pant might work.

Figure 2 shows the affect of the age group segmentation on the female sample. The distribution of females is narrower for the 18–45 age group, with fewer women with the largest waists and largest hips. As a result, fewer sizes will be needed to accommodate this age group.

Next it is important to ensure that sub-categories are adequately represented. Table 1 show the distribution of ethnicity in the segmented sample. Each ethnic group has more than 30 subjects which is a good representation. However, it is anticipated

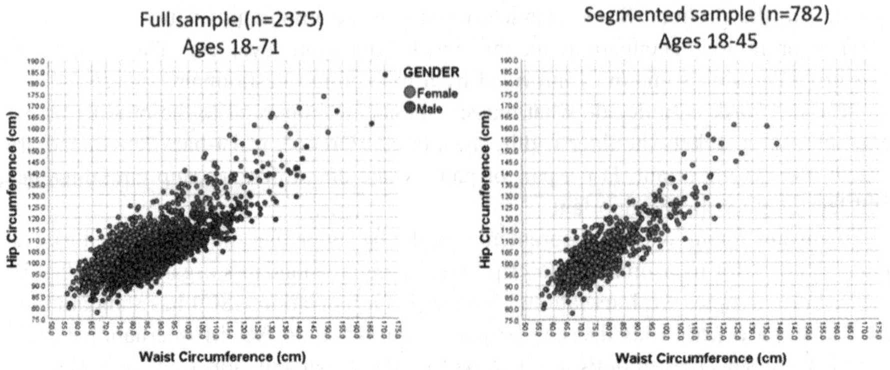


**Fig. 1.** Waist and hip circumferences for full and segmented samples.

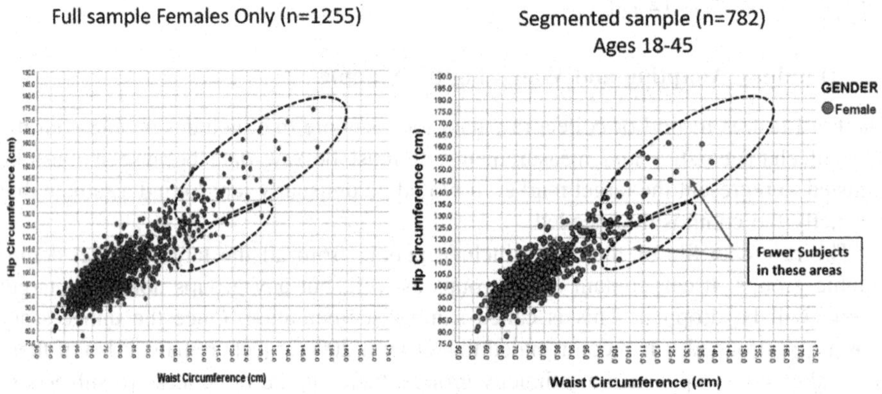


**Fig. 2.** Female sample before and after age group segmentation.

that the target market will be 20% Black and 10% Other, with 70% White. This requires an increase in the representation of Black subjects and a reduction in the representation of others. This is accomplished through weighting. It is also anticipated that 60% of the target market will be under age 30 and weighting for age group can be done at the same time.

**Table 1.** Ethnicity frequencies for the segmented sample.

| | | Ethnicity | | | |
|---|---|---|---|---|---|
| | | Frequency | Percent | Valid percent | Cumulative percent |
| Valid | Black | 109 | 13.9 | 13.9 | 13.9 |
| | Other | 114 | 14.6 | 14.6 | 28.5 |
| | White | 559 | 71.5 | 71.5 | 100.0 |
| | Total | 782 | 100.0 | 100.0 | |

An example of the target proportions for both age groups and ethnicity are shown in Table 2. This represents a sample with 60% below age 30, 20% Black, 10% Other and 70% White.

**Table 2.** Target market proportions in each cell

| Target population | Age 18–29 | Age 30–45 | Total |
|---|---|---|---|
| Black | .12 | .08 | .20 |
| Other | .06 | .04 | .10 |
| White | .42 | .28 | .70 |
| Total | .60 | .40 | 1.00 |

Sample weighting assigns a weight to each subject in the sample to adjust the sample to the desired proportions. The subject weights are calculated by dividing the target proportions, such as those shown in Table 2, by the sample proportions which for this example are shown in Table 3. The resulting cell weights, as shown in Table 4, are used to create a weight variable that is then applied to each subject for statistical calculations and representation of frequencies. The weight indicates how many subjects a particular subject will represent in the new weighted target market sample. Subjects in under-represented categories, such as the Black ethnicity, will count as more than one subject. Subjects in over-represented categories, such as the other ethnicity, will count as less than one subject. This provides a more accurate count when determining how many of each size to produce.

**Table 3.** Sample proportions in each cell

| Sample | Age 18–29 | Age 30–45 | Total |
|---|---|---|---|
| Black | 0.08 | 0.06 | 0.14 |
| Other | 0.07 | 0.07 | 0.15 |
| White | 0.24 | 0.48 | 0.71 |
| Total | 0.39 | 0.61 | 1.00 |

**Table 4.** Cell weights

| Category | Age 18–29 | Age 30–45 |
|---|---|---|
| Black | 1.54 | 1.30 |
| Other | 0.81 | 0.56 |
| White | 1.76 | 0.59 |

In the statistical software package SPSS® [9] the weight variable is created using the syntax feature. Figure 3 shows the code in the syntax feature for creating the weight variable. In this example the age categories were labelled 1 and 2 representing the 18–29 and 30–44 categories respectively. Once the weighting variable is available the software allows you to weight any analysis using it.

An example of the impact of weighting is shown in Fig. 4. In this figure Waist Circumference for the original segmented sample is compared to the Waist Circumference for the sample after the ethnicity and age group weighting variable was applied. The mean has shifted to the right (larger) slightly with the new sample composition.

Sample segmentation and weighting allows the creation of a more representative sample of the customers with a more accurate count of the number of people in each size category. It is important to start with a sample that is as close to the customer population as possible. While this example used a sample from the adult civilian population of North America, if the target market is in another area of the world or is a younger or older age group it will be important to find a sample closer to that population.

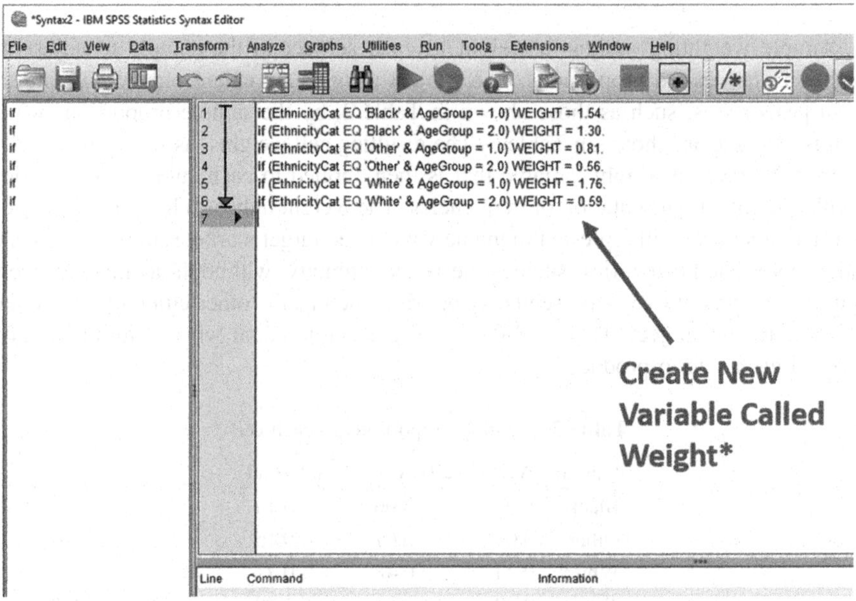


**Fig. 3.** Screenshot showing syntax for creating weight variable.

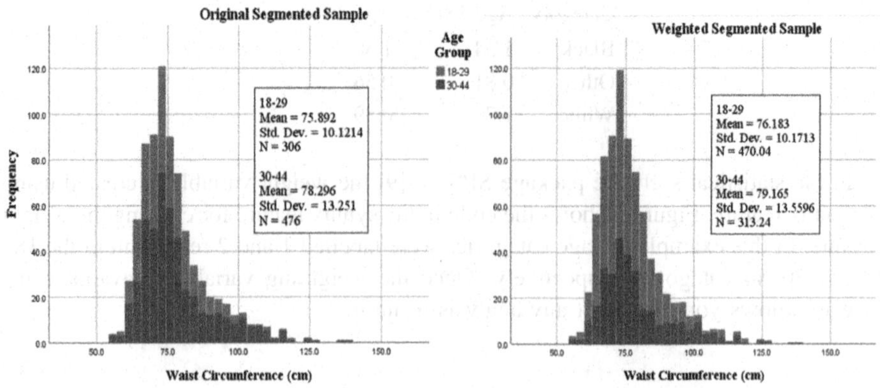


**Fig. 4.** Waist circumference before and after weighting.

Several organizations now have datasets available. For example, the WEAR Association [10] has data that is searchable, and downloadable by members that includes several different regions of the world. There is a visual index of the measurements to help determine if the measurements of interest are available. There is also an American and European raw data set of adults that can be purchased and includes 3-D scans in both standing and seated poses.

The National Center for Health Statistics [11] has data for U.S. populations of all ages that is continuously updated and can be downloaded for free, no membership required. The number and variety of measurements available are limited however.

If a suitable data set is not available, then a new one will need to be collected. This can be done as part of fit testing, provided a large enough sample is obtained.

## 3 Mapping Fit

### 3.1 Fit Data

The first part of fit mapping is establishing the fit database. This requires fit testing of mock-ups or prototypes of the product using people drawn from the target market. There are three types of data needed from these tests: (1) fit scores including size of best fit, (2) anthropometry, and (3) demographics. By relating the anthropometry and demographics to the fit scores it is possible to establish what range of body measurements are accommodated in a size.

There are several publications that describe fit mapping [2, 12, 13]. The Navy did an effective fit mapping study of their women's uniform [14]. The data sheet from that study is shown in Fig. 5. Six different items were tested. Two tops were tested, a shirt and a coat. Four bottoms were tested, two skirts and two pairs of pants. Fit was assessed using 14 sizes of each item on 906 Navy women. Different sizes were tried until the best fitting one for each garment type was determined. This one was evaluated by both an expert fitter and the subject. Fifteen body measurements were taken, and some demographics were collected.

One important discovery was that only 16% of the women had the best fit in the same size for all four bottoms. There were some women, (4%), whose best fit size was different in every bottom. More than half, (56%), had their best fit in two different sizes depending on the garment. This clearly demonstrated that fit and size assumptions can be very inaccurate and fit data must be collected for each new type of product to establish the range of fit.

Another important finding was that tops and bottoms needed to be separately issued. The important measurements related to good fit for the shirt were Bust Circumference and Neck Circumference, while the important measurements for the bottoms were Waist Circumference and Hip Circumference. The correlations between the body measurements for the tops and the body measurements for the bottoms were not strong enough. This meant that it was not possible to reliably predict what size pant or skirt would fit given the top size. This emphasizes the point that fit is product specific and at this time there is no body of knowledge about how and why things fit the way they do. The only way to know is to fit test.

SUBJECT #: ____________ DATE: ____________

SS#: ____________ DATA COLLECTOR: ____________

DATE OF BIRTH: ____________ DATA ENTERER: ____________

PLACE OF BIRTH: ____________ RANK/OCCUPATION: ____________

W☐ B☐ A☐ H☐

WEIGHT ______

HEIGHT ______

NECK CIRCUMFERENCE ______

SHOULDER CIRCUMFERENCE ______

CHEST CIRCUMFERENCE (at scye) ______

BUST CIRCUMFERENCE (at bustpoint) ______

CHEST CIRCUMFERENCE (below bust) ______

WAIST CIRCUMFERENCE ______

HIP CIRCUMFERENCE ______

WAIST BACK ______

SLEEVE INSEAM ______

SLEEVE OUTSEAM ______

SLEEVE LENGTH ______

WAIST HEIGHT (outseam) ______

CROTCH HEIGHT (inseam) ______

| ITEM | WHITE SHIRT | WHITE SKIRT | WHITE SLACKS | BLUE SKIRT | BLUE COAT | BLUE SLACKS |
|---|---|---|---|---|---|---|
| Navy Size | | | | | | |
| Commercial Size | | | | | | |
| Rating Fitter | | | | | | |
| Rating Subject | | | | | | |

FIT ASSESSMENT: 1 = Excellent 2 = Good 3 = Fair 4 = Poor

Comments (Fit Subject's):

Comments (Data Collector's):

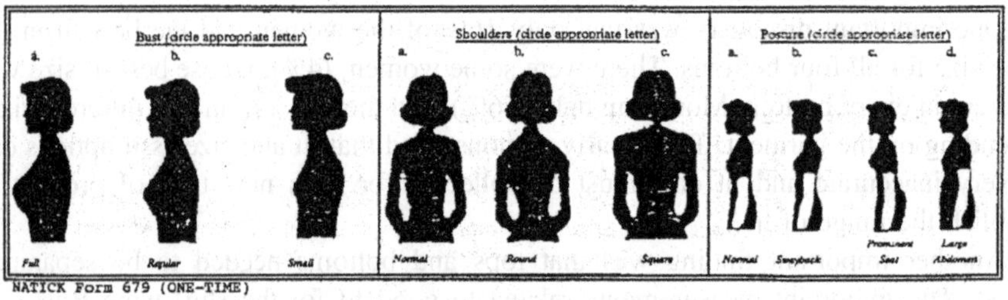


NATICK Form 679 (ONE-TIME)

**Fig. 5.** Photocopy of Navy Women's Uniform data sheet [14]

The Navy developed a bottom garment size range using the range of fit from the fit test results [15]. While the raw fit scores are no longer available, a size prediction chart was created, and for fit mapping demonstration purposes the range of fit for a single size can be derived from that chart.

Note, the size prediction chart uses body measurements not garment measurements. These are not the same thing. The range of fit is the range of people who fit in the pants during the fit test. The waist of the pant was smaller than the waist of the people who wore it (negative ease), for example, while the hip of the pant was larger than the hip of the people who wore it (positive ease).

## 3.2 Mapping Fit

Mapping the fit is simply applying the fit range for each size to the tailored sample representing the target market. For example, we can map how well the Navy garments would fit our target market by comparing the fit ranges within a size to our target market sample. The Navy is required to fit all personnel, so they have many sizes that a commercial manufacturer will not need. Luckily it is not necessary to use all the sizes for fit mapping. Four sizes from the Navy Women's Pant Size Prediction chart, (10M, 12M, 14M, and 16M) are centrally located, and have the same range of fit within a size so they appear to be accurate representations of the range of fit. The four selected sizes are mapped against our new target market in Fig. 6.

Even though the target market sample used here is a civilian population ages 18–45, these 4 Navy sizes seem to fit the middle well. This is a good thing because the population clusters around the center. There are fewer people within the same range as one moves away from the center.

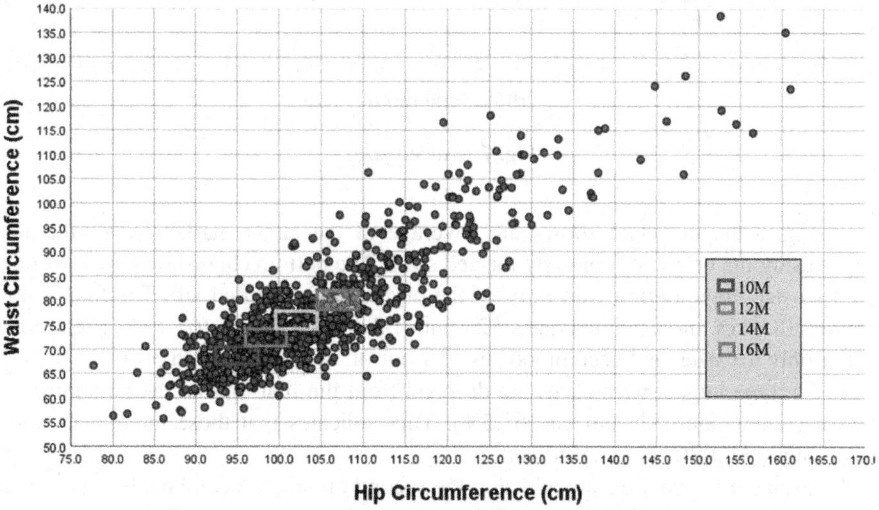


**Fig. 6.** Four Navy sizes mapped against new target market.

The range of fit for a single size can be used to determine the best assortment of sizes for the target market. For example, the original four navy sizes might be kept, and a few additional ones added, such as shown in Fig. 7. Two sizes were added to the four in the Misses (M) size range (grade), and five were added for an additional size range that has a larger hip for the same waist.

Because the raw data on the target market is available it is possible to count the number of people who fall in each size as well as counting the number accommodated by each size range. By moving the sizes around the proportion of the target market accommodated can be optimized not just for the number who will fit, but also considering the cost, manufacturability and other factors. For example, the use of two sizes ranges might be compared to the use of three size ranges in terms of the percentage of the target market accommodated. An illustration of this trade-off is shown in Fig. 8.

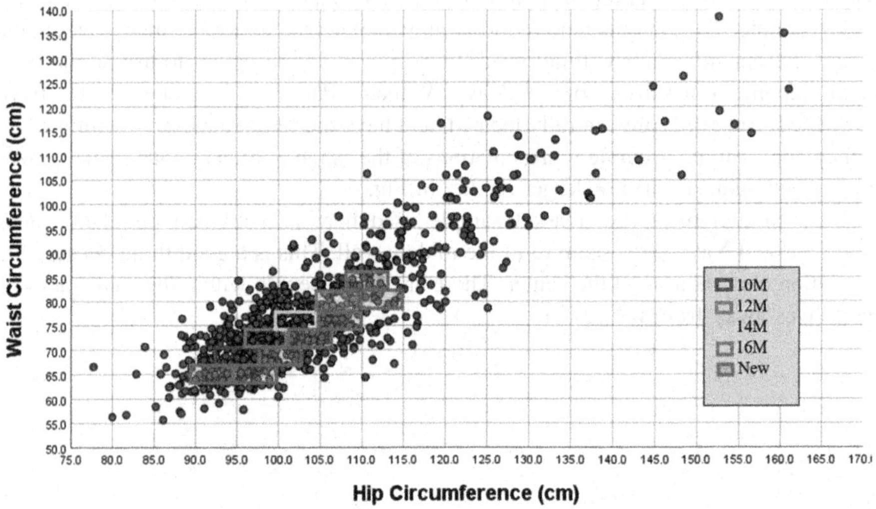


**Fig. 7.** Added sizes

In Fig. 8 the top chart shows the coverage of the target market with three size ranges using the original 4 centrally located sizes. The bottom chart shows the coverage of a two-size range option with three of the original sizes shifted left. The first option requires 18 sizes and accommodates 526 out of the 782 subjects (67.2%). The second option only 10 sizes and accommodates 354 out of 782 subjects (45.3%).

If the three largest waist sizes are dropped from the first option, the remaining 15 sizes accommodate 481 subjects (61.5%). This indicates that those three sizes combined only add 5.7% more customers.

If two larger waist sizes are added to the second option, the combined 12 sizes will accommodate 381 subjects (48.7%). This indicates that adding two sizes only increases customer accommodate by 3.4% in a two size range system.

These comparisons illustrate the kind of trade studies that can be done to determine the most cost-effective size assortment for a given market. The cost of a third size range can be weighed against the benefit of the additional potential customers.

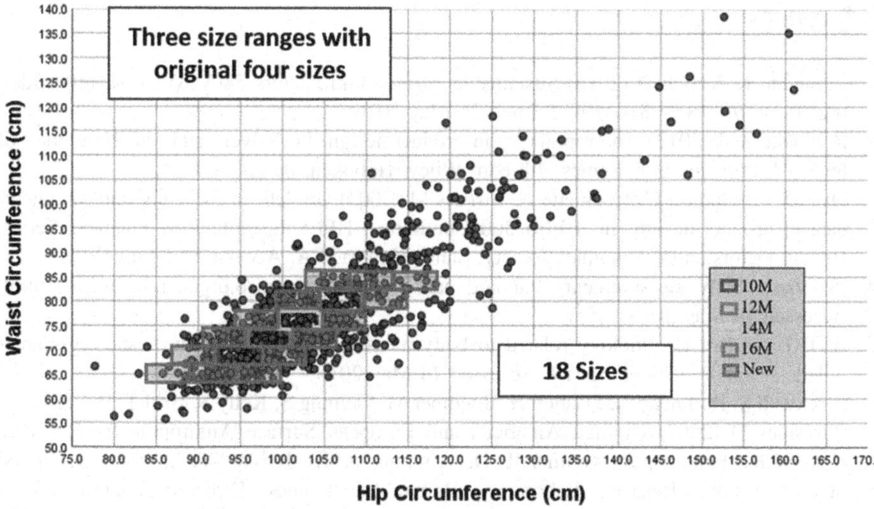


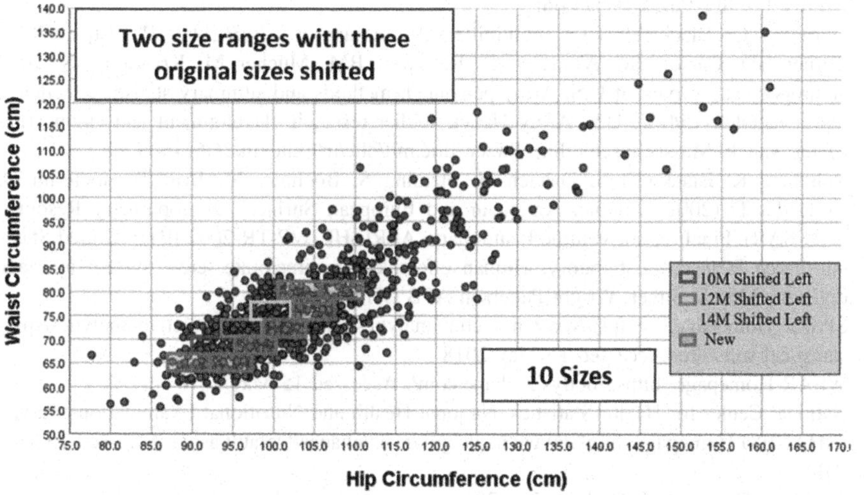


**Fig. 8.** Comparison of two size assortment options.

## 4 Summary

With the availability of raw data bases combined with fit test scores it is possible to compare sizing options for a given target market. Informed decisions can be made that, in addition to the percentage of the target market accommodated, takes cost, manufacturability, and other factors into account. The improved product development process leads to increased efficiency, reduced costs, tighter inventory control and improved customer satisfaction. In summary, the result is a better fit for the customers, no wasted sizes, and an improved bottom-line.

## References


1. Robinette K, Veitch D (2016) Sustainable sizing. Hum Factors 58(5):657–664. https://doi.org/10.1177/0018720816649091 Epub 26 May 2016
2. Robinette KM (2012) Anthropometry in product design. In: Salvendy G (ed) Handbook of human factors and ergonomics, 4th edn. Wiley, Hoboken, pp 330–346
3. Dainoff M, Gordon C, Robinette K, Strauss M (2003) Guidelines for using anthropometry data in product design. In: HFES 300 Committee, HFES Institute Best Practices Series, Human factors and ergonomics society, Santa Monica, CA. Accessed 14 May 2018
4. ISO basic body measurements standard part 1. https://www.iso.org/standard/65246.html. Accessed 14 May 2018
5. ASTM standard terminology related to body measurements. https://www.astm.org/search/fullsite-search.html?query=5219. Accessed 14 May 2018
6. Blackwell S, Robinette K, Daanen H, Boehmer M, Fleming S, Kelly S, Brill T, Hoeferlin D, Burnsides D (2002) Civilian American and European Surface Anthropometry Resource (CAESAR), Final report, volume II: Descriptions, AFRL-HE-WP-TR-2002-0173, United States Air Force Research Laboratory, Human Effectiveness Directorate, Crew System Interface Division, 2255 H Street, Wright-Patterson AFB OH 45433-7022. http://www.dtic.mil/dtic/tr/fulltext/u2/a408374.pdf
7. Gordon CC, Blackwell CL, Bradtmiller B, Parham JL, Barrientos P, Paquette SP, Corner BD, Carson JM, Venezia JC, Rockwell BM, Mucher M, Kristensen S: 2012 anthropometric survey of U.S. Army personnel: methods and summary statistics, technical report natick/tr-15/007, U.S. Army Natick soldier research, development and engineering center Natick, Massachusetts. http://www.dtic.mil/dtic/tr/fulltext/u2/a611869.pdf
8. Robinette K, Blackwell S, Daanen H, Fleming S, Boehmer M, Brill T, Hoeferlin D, Burnsides D (2002) Civilian American and European Surface Anthropometry Resource (CAESAR). Final report, volume I: summary, AFRL-HE-WP-TR-2002-0169, United States Air Force Research Laboratory, Human effectiveness directorate, crew system interface division, 2255 H Street, Wright-Patterson AFB OH 45433-7022
9. SPSS Homepage. https://www.ibm.com/analytics/data-science/predictive-analytics/spss-statistical-software. Accessed 16 May 2018
10. WEAR Homepage. https://bodysizeshape.com/. Accessed 15 May 2018
11. National Center for Health Statistics, National Health and Nutritional Examinations Survey (NHANES) Homepage. https://wwwn.cdc.gov/nchs/nhanes/Default.aspx. Accessed 15 May 2018
12. Choi HJ, Zehner GF, Hudson JA (2010) A manual for the performance of protective equipment fit-mapping. Technical report, AFRL-RH-WP-SR-2010-0005. http://www.dtic.mil/docs/citations/ADA519894. Accessed 16 May 2018
13. Whitestone JJ, Robinette KM (1997) Fitting to maximize performance of HMD systems. In: Melzer J, Moffitt K (eds) Head mounted displays, designing for the user, Chap 7. McGraw Hill Publishing, New York, pp 175–206
14. Mellian SA, Ervin C, Robinette KM (1990) Sizing evaluation of Navy women's uniforms. Technical report no. 182, Navy clothing and textile research facility, Natick MA and AL-TR-1991-0116, Armstrong Laboratory, Air Force Systems Command, Wright-Patterson AFB OH. http://www.dtic.mil/dtic/tr/fulltext/u2/a231463.pdf. Accessed 16 May 2018
15. Robinette KM, Mellian SA, Ervin CA (1990) Development of sizing systems for Navy Women's Uniforms. Technical report AL-TR-1991-0117. Armstrong Laboratory, Air Force Systems Command, Wright-Patterson AFB OH. http://www.dtic.mil/dtic/tr/fulltext/u2/a250071.pdf

# Importance of Human Anthropometry in the Interior Development of Autonomous Vehicles

Sibashis Parida[1,2](✉), Samuel Brock[1,2], Sylvester Abanteriba[2], and Mattias Franz[1]

[1] BMW Group, Knorrstraße 147, 80788 Munich, Germany
sibashis.parida@bmw.de
[2] RMIT University, 124 La Trobe Street, Melbourne 3000, Australia

**Abstract.** Currently autonomous driving is one of the dominant trends in the automotive industry. Cars that no longer need the driver's attention will be used in a completely new way than is currently the case. Fully autonomous driving would allow drivers to participate in the so called "non-driving secondary activities". In the future the car would be a place to work, a place to socialize, to relax, to meditate, spend quality time with the family, take a nap, and a whole lot more. The autonomous driving vehicle would be more of a living space, rather than just a mode of transportation.

In order to facilitate these non-driving activities, the interior of the vehicle and the vehicle seats would be reconfigured and redesigned differently, as compared to the conventional vehicle interiors until now. Given the fact that the space available within a classical vehicle is limited, a lot of importance needs to be given to human anthropometry. In order to offer optimal space and comfort, the interior of the vehicle needs to be flexible and needs to be adaptable and customizable to the anthropometry of the occupant.

The paper helps to understand the importance of human anthropometry in the design and development of the vehicle seating and seating configuration. Discussed in the paper are potential advantages, if human anthropometry is used as a parameter in the design and development of the interior of autonomous driving cars.

**Keywords:** Anthropometry · Autonomous vehicles · Vehicle automation Self-driving · Seating

## 1 Introduction

### 1.1 Autonomous Vehicles

Autonomy is a topic currently dominating discussions within the automotive industry. Increasing traffic congestion in urban and particularly city centers, legislative, social trends, expense and customer preference are factors affecting changes to personal mobility, while the potential revenue is motivating a large push from existing and new businesses. As described by the Society of Automotive Engineers (SAE), there are six


S. Bagnara et al. (Eds.): IEA 2018, AISC 826, pp. 453–463, 2019.
https://doi.org/10.1007/978-3-319-96065-4_50

levels of vehicle automation from Level 0 (fully manual control of driving task) and Level 1 (driver assistance systems and comfort functionality) to Level 4 (highly automated, but retains control input as there may be request for occupant to intervene). Level 5 is fully autonomous with no user input besides destination, and no driver controls are necessary [1].

The consensus is that there will be different models of ownership with the increase in adoption of sharing and on-demand services together with private ownership, implemented across both conventional and new concept vehicles. The sharing model is favored by some [2] and could result in a more fragmentary market with a range of vehicles and services at all price points. This is an important point because the potential users of each product will be different. It will be the use that will determine the vehicle [3] and interior concept.

### 1.2 Visions and Expectations

Most publicly shown concepts and general reporting on the topic seem to nearly universally feature passengers turning around to talk to one another or using mobile devices whilst looking supremely comfortable in open flexible spaces enabled by new energy drivetrains. The main vision presented common among concepts is that the vehicle users will be able to reclaim driving time for personal or business use when in an autonomous driving mode.

But with fully autonomous vehicles some time away, the focus is currently on vehicles with partial autonomy. In the interim, occupants may be allowed to participate in other activities in certain limited situations. However the driver may still be required to pay attention or take over the driving task at short notice. In this case it is envisioned that the workstation can be minimized with space-saving retractable controls. The space can then be used for relaxation, entertainment, socializing with other passengers or as a mobile office.

Many concepts are presented as a “third living space”, but are generally vague in description and highly speculative. It is not possible to allow for the wide range of all activities we do in our other spaces in the confines of a vehicle. This leads to current research in defining exactly which activities are likely to take place and how the needs and expectations of the vehicle occupants can be accommodated in a new vehicle concept.

## 2 Anthropometry

### 2.1 Physical Characteristics of the Driver

Potential customers come in all shapes and sizes described by anthropometry, the measurement of physical dimensions to study the variety, proportions and capability of the human body [4, 5]. Anthropometry is used for defining the vehicle package – the seat design, layout and interior space arrangement – for comfortable use by all occupants. There are a large number of dimensions of the human body that can be measured, but only a few are useful or relevant to vehicle design. Some are commonly used, the 16 joint and body landmark locations specified by the SAE in the H-point procedure for example. Some measures such as body mass index (BMI) are derived from several others.

There are broadly four body types with proportions consisting of body and legs being either long or short, with approximately 70% of the population covered by any combination in between [6]. Measurements and proportions are unique to the individual as is posture, body shape [7] and ability. Stature and BMI may be very similar between individuals, but each will have unique shape, proportion and posture. Anthropometry varies considerably with gender, age and region. Postures are not predicted by anthropometry but by individual preferences which varies considerably among individuals [8] and also with age and ailment.

### 2.2 Populations and Percentiles

Within a population of individuals, body dimensions follow a statistical distribution with most anthropometric variables conforming quite closely to the normal distribution [9]. Appropriate sampling of a population is necessary to make valid statements on the distribution of body measurements within that population [6]. The population and gender mix used for product design is determined from projected or historical sales, vehicle class, target demographic or region. Distributions change with gender, age and nationality mix, and also generally change over time.

Percentiles describe the ranking of a single measurement relative to the rest of the population distribution, the probability that an individual has a body dimension greater than a percentage of a population. Stature (standing height) is typically the primary variable used to describe body size, with auxiliary measures such as corpulence (waist circumference), proportion (leg length) [10] and recently BMI [7] used to relate accommodation in the vehicle package.

Most applications of percentiles in ergonomics aim to accommodate extreme small and large sizes of $5^{th}$ percentile female and $95^{th}$ percentile male (commonly abbreviated to the form F05 and M95 respectively), following the rule of thumb from Konz [11]: "Let the smallest woman reach. Let the largest man fit". This approach assumes that 90% of the population would automatically be accommodated but this is not true. By itself, a percentile does not capture the true multivariate nature of the human body [5] and in fact a much smaller percentage of the population is accommodated. Percentiles do not generally project to other dimensions in that a tall person may not necessarily be a wide person for example. The $50^{th}$ percentile describes the mean in the case of the normal distribution, but simply accommodating the $50^{th}$ percentile person is also flawed since roughly half of the population will find the product too small while the other half too large. Safety-relevant applications use 1st percentile female to 99th percentile male manikins [6]. Accommodating the entire population is not a feasible approach either, hence statistical methods are used. There will always be a percentage of the population that are excluded from the design either from anthropometry or preference.

### 2.3 Digital Human Models

Digital human models (DHMs) were introduced to product development initially as a digitization of traditional human anthropometry occupant templates for use in CAD packages. Individual manikins are constructed from simplified segments that are joined together through kinematic linkages intended to represent the mechanics of the human body. Scaling and mutability of individual body segments is incorporated through correlation to univariate stature and sitting height percentiles [12]. The main function of a DHM is to provide a tool to evaluate product ergonomics and usage. Anthropometry is critical as without it, a DHM is simply a human-shaped visualization tool with completely arbitrary size which has little value. RAMSIS is most common for automotive seating applications and allows evaluation of posture and functional analyses including vision and reach for a wide range of simplified but anthropometrically accurate body typologies. There is interest in the extreme sizes as well as special cases such as children.

## 3 Interior Development

### 3.1 Driving Task

Current interior development focuses heavily on the driver which is expected, since a driver is always present in a conventional vehicle. The driving task is very well described having remained fundamentally unchanged for decades. Current vehicle interiors in support of this task must have the seat, dashboard, displays and control inputs arranged around the driver, but also constrain the positioning [13]. This arrangement must be in consideration of the intended user population with accommodation traditionally provided by task-oriented models in the SAE Standards, which form the basis for many current processes [14].

Interactions that are not directly controlling the vehicle are secondary and tertiary tasks which can include operating lights and wipers, or climate and navigation respectively. These tasks have special postures and motions that are distinct from the driving task. Current non-driving secondary tasks might include reaching for a drink, making adjustments to the infotainment or using a mobile device. Passengers are currently accommodated after the driver.

### 3.2 Current Accommodation Practices

Current practices accommodate the driver as comfortably as possible given they have to perform the driving task. Vehicle interiors and seating in particular must deliver a balance between the sometimes conflicting functions and demands of being comfortable, supportive, practical, safe and attractive. Comfort is a main brand and product differentiator to the consumer [15]. Consideration of anthropometry is key to comfortable seating and is generally provided through adjustability. The greater the number of independently adjustable dimensions in the vehicle, the more likely it is that a comfortable posture and good fit can be achieved by the occupant [16] for their body dimensions and preferences. This is especially true at the higher end of the market, but is always balanced by cost.

The current process for interior design typically iterates a subjective evaluation methodology to find optimal solutions [17] for the physical arrangement and seat shape to suit a population of expected users within the boundaries of target functional values. The SAE Standard templates are used to set the reference points, design position and posture of the driver. The initial package arrangement is created around the driver using a variety of templates and carry-over parts to establish space and reach for primary and secondary controls, fields of view and other functions. Each parameterized design candidate is compared to target. Each manufacturer and supplier has their own internal procedures and guidelines built up with experience, benchmarking and customer feedback. This initial package is then refined to remove obvious flaws, evaluating with validated DHM data. Assessments for vision, reach, comfort and others are performed automatically for the range of different sized manikins to ensure that the intended range of occupants are accommodated within the simulated vehicle package. The point of using virtual tools like DHMs is that the occupant package can be developed and assessed early in the vehicle program to get as close as possible to the final design and reduce the need for expensive and time intensive [18] physical prototyping.

User trials are then conducted to highlight issues not picked up with the DHMs including long-term comfort, fatigue and other subtleties. Subjects used for assessment are selected such that the anthropometric distribution closely represents that of the intended user. Assessment can be either static sit-in or a ride-drive test that often includes several design candidates and competitor products for comparison, with feedback used to modify the design. Anthropometry is also considered for optimal positioning of secondary controls, displays, ingress/egress and restraint systems [8] in a similar manner.

Development is often more successful for the medium size male with occupants below this size sometimes substantially less comfortable [17] and certain dimensions such as hip breadth are much less accommodated, especially for females. It is common to use a group that has an equal distribution of small females to medium males to large males [19] though different distributions and weightings for assessment are sometimes used. It can be difficult to assemble and maintain a representative sample [15], as people at the tails of the distribution are comparatively rare.

# 4 Transition to Autonomy

## 4.1 Changing Customer Requirements

Automation provides the ultimate comfort in that it allows the driver to do something other than driving, making the journey more productive or enjoyable [20] and this experience will become a key user requirement and commercial differentiator. Responsibility of the driving task is removed from driver, and the possibility to engage in other activities is presented. The projected activities can be broadly categorized as relaxation, entertainment, socializing and mobile office. The interior space must be changed to allow these activities, since the benefits of automation may not be perceived unless users can actually engage in them.

Posture and subsequently comfort changes with the activity, so it is important to define the activities passengers want to do and how they prefer to sit whilst doing them. One of the ways this can be achieved is through analogies to other sectors such as mass transport. Kamp et al. [21] observed the activities and postures chosen by train passengers with socializing, relaxing (watching the scenery), and reading the top three activities. The most common posture for all activities including use of a mobile device was with the head free of support, trunk against backrest, arms free, and legs free with both feet on the floor. A flatter seat allows a greater range of sitting postures [17] and freedom of movement especially cruising with low lateral accelerations or low speed. Jung et al. [22] also examined train passengers and observed upright, relaxed, and extended posture subcategories. Activities with extended postures naturally required more longitudinal space or more precisely, less restriction. Most activities had relaxed posture, but this also changed over the travel duration to a slumped position. Having face-to-face seating requires more longitudinal space to accommodate the legs, plus slumping.

Eost and Flyte [23] examined people using the vehicle space as a mobile office and found problems include lack of space, storage and flat surfaces. Occupants adapt to prioritize the task at hand and most had creative ways of making the environment easier for them to work including resting work on steering wheel, placing laptops on passenger seat and knees and sliding the seats all the way back. Sang et al. [24] observed similar behaviors noting an increase in reported back pain where facilities were not available to support the activities performed in the vehicle.

Mobile devices such as smartphones and tablets have seen a rapid rise to ubiquity. Although Kamp et al. [21] observed that most passengers will simply relax, this is contradictory to forward-looking surveys where respondents say they would use their smartphone or other mobile device. Several studies such as that by Weston et al. [25] found that users reported discomfort in the torso, neck and arms, especially when unsupported. Asundi et al. [26] also found that laptop use on a desk or on lap gave high wrist, neck and shoulder extension which can cause discomfort. More upright postures promoted forward-flexed or slouched postures, but the postures were much different in chairs that had the ability to recline and adjust the hip rotation for comfort. This is important since it requires appropriate resting supports or fold-out tray tables that depend on anthropometry for positioning and clearances, as well as reclining functionality.

It will be the end use that determines the vehicle concept, particularly the interior. Shared and private vehicles will have different uses and requirements. Future vehicles will be more focused on passenger experience and understanding passenger requirements will become an important area of research with challenges from the wide variety of activities and potentially radical seating concepts. It must be remembered that these are still road vehicles and as such a simple translation of the office or living room to one with wheels [20] is likely to be unsuccessful. While most other modes of transport are characterized by low accelerations, road vehicle dynamics have relatively high and abrupt accelerations with high frequency vibrations from the road surface. The biomechanical response of the occupants to the vehicle dynamics may influence comfortable postures and even limit some tasks. Further, the vibration response of the human body does change with BMI and adjustment of the seat, particularly when reclined [27].

For the intermediate levels of automation, there is a need to account for the non-driving activities in addition to the traditional driving task. The operational difficulty as many have pointed out is the handover between the two modes, especially back to human control. This problem also applies to the design of the interior space in that the driving and other tasks have very different accommodation requirements. Further, it is possible that the activity may interfere with the driving task either physically such as laptop or through delayed response.

Customers are also using their vehicles more, though trip duration may not necessarily be longer. Fagnant [28] suggests that potentially people would travel more, and more people would travel with shared autonomous vehicles.

### 4.2 Changing Customers

For autonomous vehicles, anthropometry is important for consideration of future users with people on average getting bigger, older and more diverse. Width and circumference measurements have been increasing for several decades [29] with the rate outpacing height, which has been stable from about 1990. The net result is an observed increase in BMI. That is, future seated occupants are likely to be more corpulent than currently described with a shift in body proportions, reach and comfort impacting accommodation over the service life of the seat. Seated hip breadth is a design dimension observed to be increasing at a significant rate of between 2–3% per decade dependent on population. Trends in hip breadth were observed by Molenbroek et al. [29] and in large-scale anthropometric surveys such as sizeGermany and sizeNorthAmerica [30] finding male chest circumference also increased considerably. Clearances to other parts of the interior space such as tray tables are also affected as seen by airlines in particular.

Emerging markets are also experiencing rapid increases in stature. BRICS countries (Brazil, Russia, India, China and South Africa) are particularly difficult as their specific anthropometries are very different. Sitting height to stature proportion [31] is very different and whereas some have a much higher prevalence of high BMI individuals, others tend towards low BMI. Jung [22] also highlights the regional differences in preference for activities as well as anthropometry. The spread of the global automotive market and in particular the Chinese market has been a challenge typically addressed with unique but functionally equivalent seating [15], which is likely to continue.

Population age is also observed to be increasing on average with growing numbers of individuals aged 60 or older a particular concern. Driving postures and body shapes are significantly associated with age [32] and are of interest due to the vulnerability to injury in operation. Older drivers are likely to adopt a lower, slouched driving posture due to vision and joint discomfort, but were less sensitive to discomfort than younger occupants [33]. Older people or those with mobility issues are less able to adapt to an unsuitable design whereas younger people can readily adapt posture [34]. The extent to which impairments and other health conditions affect occupant posture and accommodation is not addressed. Anthropometry of people over 65 is very under-represented in sample populations and typically those with musculoskeletal impairments or very high BMI are explicitly excluded, despite the increasing prevalence.

One of the main benefits communicated for autonomous vehicles is increased mobility for the elderly and disabled, thus the potential intended user would have a disproportionately higher inclusion of such individuals. This is particularly important for shared mobility vehicles as it is in modes of public transport, where ingress/egress, handrail placement, and step height have special importance for anthropometry and biomechanics and are covered by accessibility regulations. The population diversity in public transport is much greater than that used for private vehicles.

The demographics of sharing users and private customers might also change. The average buyer age is now over 50 with 29% over 60 [35] while others suggest younger generations indicate preference for shared or on-demand vehicles [36] over private vehicles. Those with mobility difficulties may also turn to shared services due to increased flexibility.

### 4.3 Changing Development

The focus of interior development is moving towards overall occupant well-being. Perceived discomfort has been observed to be significantly different for each activity [37] indicating that each had very different postures. Driving has a very narrow range of postures and definitions, and there is a need to define the postures for new non-driving activities for incorporation into the vehicle design.

Currently, the only anthropometric inputs to development are through the use of percentile identifiers in the DHMs. Individual anthropometric dimensions are considered only indirectly through the scaling algorithm, though torso and legs are typically accommodated separately. The application of individual anthropometric dimensions as design variables was limited due to poor correlation with predictive outcomes, especially to posture which is an individual preference. Some however are inherited from the legacy templates, including buttock-popliteal length which defines the maximum seat cushion length, hip and shoulder breadths which set the minimum cushion and backrest widths, and sitting height which places the back of head on the head rest, as well as dimensions for reach and adjustment ranges.

Seating that can accommodate non-driving activities will require new functionality such as rotation and reclining for which adequate space and clearance has to be ensured within the narrow vehicle interior. Space and motion requirements are assessed with anthropometric manikins to determine the bounding envelope but with independently movable seat components and complex seat kinematics, specific anthropometric dimensions will need to be directly considered to avoid clashes. In particular, movement of the seat position may require temporary adjustment and potential discomfort in some occupants. A large man with very long legs in a slumped posture will require much more space for rotation than a small woman for example, and corpulent individuals will likely experience more discomfort if the seat requires returning to upright for rotational clearance.

Relevant dimensions might include the shoulder-knee diagonal which accounts for leg abduction but is posture- and activity-dependent, elbow rest height and shoulder-elbow length for resting supports while using a mobile device, sitting body depth which relates to waist circumference, or buttock-toe length or elbow-elbow breadth for bounding envelope consideration. These dimensions are currently not utilized because they do not relate to the driving task. Individual dimensions also adhere to a distribution, and although there are defined relationships they are not projected by the stature or BMI percentiles. BMI is a non-specific overall measure used for scaling adipose tissue of individual body segments. Evaluation will become a highly multivariate problem with complex correlations. Further, some individual dimensions such as waist circumference were observed to be increasing on average while others were stable.

The current technical difficulty is the integration of these new functions with the existing vehicle functionalities. Rotation is a particular challenge, as is occupant restraint [3] which is likely to remain so [20]. Increased virtualization [38] will aid development with automated analyses allowing evaluation of a much larger number of possibly quite radical seating configuration concepts, while the user's anthropometric changes can be explicitly specified in RAMSIS [35]. As automation competence and reliability increases, the emphasis on the driving task may diminish and the controls are expected to be minimized which will free up some interior space, though it is likely that the complexity of the activities will also increase as well as the amount of time spent doing them. And for this reason, the challenges in development are expected to remain.

## 5 Conclusion

Anthropometry is a critical consideration for physical accommodation of the driver. When the driving task is removed, the primary focus shifts to comfort and offering users the ability to more effectively utilize the interior space for new activities they value and expect from automation. User experience will become a commercial differentiator with shared and private vehicles and services anticipated to have very different uses, users and hence vehicle interiors. Consideration of the rapidly changing physical attributes, age and diversity of the future user will be of high importance. Current difficulty in integration with existing vehicle functions is likely to remain a significant challenge even as the emphasis on driving diminishes. Anthropometry may find new importance with direct input to development providing the ability to evaluate complex seating concepts enabling new activities and creating a vehicle that is more than the most comfortable compromise for driving.

## References

1. SAE International (2016) Surface vehicle recommended practice J3016 SEP2016: taxonomy and definitions for terms related to driving automation systems for on-road motor vehicles
2. Lavieri PS, Garikapati VM, Bhat CR et al (2017) Modeling individual preferences for ownership and sharing of autonomous vehicle technologies. Transp Res Rec J Transp Res Board 2665:1–10
3. Winner H, Wachenfeld W (2016) Effects of autonomous driving on the vehicle concept. In: Maurer M, Gerdes JC, Lenz B et al (eds) Autonomous driving: technical, legal and social aspects. Springer, Heidelberg, pp 255–275

4. Roebuck JA (1995) Anthropometric methods: designing to fit the human body. Human factors and ergonomics society
5. Ziolek SA, Wawrow P (2004) Beyond percentiles: an examination of occupant anthropometry and seat design. SAE technical paper (2004-01-0375). https://doi.org/10.4271/2004-01-0375
6. Grünen RE, Fabian G, Bubb H (2015) Anatomical and anthropometric characteristics of the driver. In: Bubb H, Bengler K, Grünen RE et al (eds) Automobilergonomie. Springer Fachmedien Wiesbaden, Wiesbaden
7. Reed MP (2013) Measuring and modeling human body shapes for vehicle design and assessment. http://mreed.umtri.umich.edu/mreed/pubs/Reed_2013_3D_Anthropometry.pdf. Accessed 15 Mar 2018
8. Reed MP, Manary MA, Flannagan CA et al (2000) Effects of vehicle interior geometry and anthropometric variables on automobile driving posture. Hum Factors 42(4):541–552. https://doi.org/10.1518/001872000779698006
9. Lämkull D, Berlin C, Örtengren R (2008) Digital human modeling: evaluation tools. In: Duffy VG (ed) Handbook of digital human modeling: research for applied ergonomics and human factors engineering. CRC Press, New York
10. Bubb H, Fritzsche F (2008) A scientific perspective of digital human models: past, present, and future. In: Duffy VG (ed) Handbook of digital human modeling: research for applied ergonomics and human factors engineering. CRC Press, New York
11. Konz SA (1983) Work design: industrial ergonomics, 2nd edn. Grid Pub, Columbus Ohio
12. Godil A, Ressler S (2008) Shape and size analysis and standards. In: Duffy VG (ed) Handbook of digital human modeling: research for applied ergonomics and human factors engineering. CRC Press
13. Reed MP, Manary MA, Flannagan CAC et al (2000) Comparison of methods for predicting automobile driver posture. SAE technical paper (2000-01-2180). https://doi.org/10.4271/2000-01-2180
14. Reed M, Roe RW, Manary MA et al (1999) New concepts in vehicle interior design using ASPECT. SAE technical paper (1999-01-0967). https://doi.org/10.4271/1999-01-0967
15. Kolich M (2008) A conceptual framework proposed to formalize the scientific investigation of automobile seat comfort. Appl Ergon 39(1):15–27. https://doi.org/10.1016/j.apergo.2007.01.003
16. Gyi D (2013) Driving posture and healthy design. In: Gkikas N (ed) Automotive ergonomics: driver-vehicle interaction. CRC Press, Boca Raton
17. Reynolds M (2012) Sitting posture in design position of automotive interiors. Int J Human Factors Model Simul 3(3–4):276. https://doi.org/10.1504/IJHFMS.2012.051554
18. Herriots P, Johnson P (2013) Are you sitting comfortably? A guide to occupant packaging in automotive design. In: Gkikas N (ed) Automotive ergonomics: driver-vehicle interaction. CRC Press, Boca Raton
19. Fazlollahtabar H (2010) A subjective framework for seat comfort based on a heuristic multi criteria decision making technique and anthropometry. Appl Ergon 42(1):16–28. https://doi.org/10.1016/j.apergo.2010.04.004
20. Diels C, Erol T, Kukova M et al (2017) Designing for comfort in shared and automated vehicles (SAV): a conceptual framework
21. Kamp I, Kilincsoy U, Vink P (2011) Chosen postures during specific sitting activities. Ergonomics 54(11):1029–1042. https://doi.org/10.1080/00140139.2011.618230
22. Jung E, Han S, Jung M et al (1998) Coach design for the Korean high-speed train. Appl Ergon 29(6):507–519. https://doi.org/10.1016/S0003-6870(97)00010-0
23. Eost C, Flyte MG (1998) An investigation into the use of the car as a mobile office. Appl Ergon 29(5):383–388. https://doi.org/10.1016/S0003-6870(98)00075-1

24. Sang K, Gyi D, Haslam C (2010) Musculoskeletal symptoms in pharmaceutical sales representatives. Occup Med (Lond) 60(2):108–114. https://doi.org/10.1093/occmed/kqp145
25. Weston E, Le P, Marras WS (2017) A biomechanical and physiological study of office seat and tablet device interaction. Appl Ergon 62:83–93. https://doi.org/10.1016/j.apergo.2017.02.013
26. Asundi K, Odell D, Luce A et al (2010) Notebook computer use on a desk, lap and lap support: effects on posture, performance and comfort. Ergonomics 53(1):74–82. https://doi.org/10.1080/00140130903389043
27. Paddan GS, Mansfield NJ, Arrowsmith CI et al (2012) The influence of seat backrest angle on perceived discomfort during exposure to vertical whole-body vibration. Ergonomics 55(8):923–936. https://doi.org/10.1080/00140139.2012.684889
28. Fagnant DJ, Kockelman K (2015) Preparing a nation for autonomous vehicles: opportunities, barriers and policy recommendations. Transp. Res. Part A Policy Pract 77:167–181. https://doi.org/10.1016/j.tra.2015.04.003
29. Molenbroek JFM, Albin TJ, Vink P (2017) Thirty years of anthropometric changes relevant to the width and depth of transportation seating spaces, present and future. Appl Ergon 65:130–138. https://doi.org/10.1016/j.apergo.2017.06.003
30. Seidl A, Trieb R, Wirsching H-J et al (2016) SizeNorthAmerica—the new North American anthropometric survey: conceptual design, implementation and results. In: Advances in physical ergonomics and human factors. Springer, pp 457–468
31. Schmidt S, Amereller M, Franz M et al (2013) A literature review on optimum and preferred joint angles in automotive sitting posture. Appl Ergon 45(2):247–260. https://doi.org/10.1016/j.apergo.2013.04.009
32. Park J, Ebert SM, Reed MP et al (2016) Statistical models for predicting automobile driving postures for men and women including effects of age. Hum Factors 58(2):261–278. https://doi.org/10.1177/0018720815610249
33. Kyung G, Nussbaum MA (2010) Assessment of aging effects on drivers' perceptual and behavioral responses using subjective ratings and pressure measures. Meas Behav 2010:318
34. Porter JM, Marshall R, Case K et al (2008) Inclusive design for the mobility impaired. In: Duffy VG (ed) Handbook of digital human modeling: research for applied ergonomics and human factors engineering. CRC Press
35. Bubb H, Grünen RE, Remlinger W (2015) Anthropometric vehicle design. In: Bubb H, Bengler K, Grünen RE et al (eds) Automobilergonomie. Springer Fachmedien Wiesbaden, Wiesbaden
36. Cyganski R (2015) Automated vehicles and automated driving from a demand modeling perspective. In: Bubb H, Bengler K, Grünen RE et al (eds) Automobilergonomie. Springer Fachmedien Wiesbaden, Wiesbaden
37. Kamp I (2012) The influence of car-seat design on its character experience. Appl Ergon 43(2):329–335. https://doi.org/10.1016/j.apergo.2011.06.008
38. Stadler S, Hirz M, Thum K et al (2013) Conceptual full-vehicle development supported by integrated computer-aided design methods. Comput-Aided Des Appl 10(1):159–172. https://doi.org/10.3722/cadaps.2013.159-172

# Anthropometric Implications of the Global Obesity Epidemic

Bruce Bradtmiller[1(✉)], Neal Wiggermann[2], and Monica L. H. Jones[3]

[1] Anthrotech, Inc., Yellow Springs, OH 45387, USA
bruce@anthrotech.net
[2] Hill-Rom, Inc., Batesville, IN 47006, USA
[3] University of Michigan Transportation Research Institute,
Ann Arbor, MI 48109, USA

**Abstract.** Makers of products that must accommodate anthropometric variability need to understand body size and shape of all people, including those who are obese. This is particularly important in safety critical applications such as medical devices and promoting the mobility of people with obesity in public spaces. To address the lack of data on this population segment we undertook a preliminary anthropometric study of individuals with high body mass index (BMI).

Several challenges, unique to this population segment, were encountered. 1. Measurement definitions/techniques needed, in some cases, to be altered to accommodate the larger size of the participant. 2. Definition of the boundaries of the sample (e.g., limiting by BMI or weight) is unknown, yet critical in determining the final statistics. 3. Issues of sample acquisition – how to obtain access to the population – materially affected the composition of the final sample. 4. The usual statistics used to report anthropometric data assume a generally normal distribution. In this case, the distribution is non-normal and affects the interpretation of the statistics.

**Keywords:** Obesity · Universal design · Bariatric

## 1 Background

An obesity epidemic has been ongoing in the United States for several decades and has been noted in other parts of the world more recently (Finucane et al. 2011; Hales et al. 2018; James 2008; Mitchell et al. 2011). Obesity carries with it increased risk of various negative health outcomes, such as heart disease and stroke, diabetes, arthritis and some cancers (Bhaskaran et al. 2014; US HHS 1988; US HHS 2013). Governments, employers and organizations promoting healthy living have begun a variety of campaigns to combat the epidemic and they may be showing some signs of success. However, complete success will be decades away at best, and in the interim, people with obesity need to carry out the activities of daily living in the home, in the workplace and in public spaces. Further, they need to be accommodated with clothing, protective equipment and workspaces that take into account a broader range of anthropometric variability than has traditionally been used in product or workspace design.


S. Bagnara et al. (Eds.): IEA 2018, AISC 826, pp. 464–471, 2019.
https://doi.org/10.1007/978-3-319-96065-4_51

Designers and product developers traditionally rely on publicly available anthropometric databases to create designs that accommodate the intended target audience. In traditional practice, the target is the central 90% or 95% of the distribution of a given population. More recently, proponents of universal design recommend that the composition of an environment should be such that "it can be accessed, understood and used to the greatest extent possible by all people, regardless of their age, size, ability or disability" (Center for Excellence in Universal Design 2014). At present, we are unaware of any publicly available anthropometric databases that include significant numbers of obese individuals. This lack of anthropometric data impairs the ability of well-intentioned designers and product developers to achieve universal design; specifically, it impedes the design of products that accommodate obese individuals.

To address this gap we conducted an anthropometric data collection focused on people with obesity. We adhered to the World Health Organization definition of obesity, which is having a Body Mass Index (BMI) of 30 or more. This initial effort is limited in scope but produced informative results. It has also raised several technical issues that need to be addressed prior to undertaking a large-scale anthropometric survey. This paper focuses primarily on the technical issues associated with survey design and definition of anthropometric measures unique to obese individuals. Preliminary results are also presented.

## 2 Survey Planning

Preparation for an anthropometric survey includes dimension selection and definition, determination of sample size, definition of the sample of interest, and planning for sample acquisition. This population segment required specific consideration of each of these factors.

### 2.1 Dimension Selection and Definition

The intent of this study was that the anthropometric dimensions selected would be widely applicable, and typical anthropometric surveys include a large number of dimensions. For example, in the US Army's 1987–1988 anthropometric survey (Gordon et al. 1989), over 130 dimensions were measured, and the resulting data were used for the design of clothing, protective equipment, head gear, airplane cockpit design, tank and automotive interior design and so on. However, the duration of time required to complete an extensive number of measurements is significant and individuals with high BMI often have difficulty standing for long periods and may experience fatigue. Bariatric medicine specialists at our data collection locations advised that the measurement protocol should not exceed 10 min for each individual. We therefore prioritized measurements, favoring those needed for hospital furniture design, clothing design and general body shape characteristics. A total of 30 dimensions were measured, but not all dimensions were measured at all of the four data collection locations. To maximize the size of the sample for analysis, 10 "common" dimensions were measured at all locations and the remaining 20 were divided across the locations. Table 1 lists the 30 dimensions included in this study.

**Table 1.** Complete dimension list

| | |
|---|---|
| Abdominal extension depth, supine[a] | Hip circumference, front view, standing |
| Axilla to cubital fold, sitting | Knee-knee breadth, sitting[a] |
| Bideltoid breadth, sitting | Knee-knee breadth, supine[a] |
| Buttock circumference, side view, standing | Navel height, omphalion, standing |
| Buttock-knee length, sitting | Neck circumference, sitting |
| Buttock-popliteal length, sitting | Neck width, sitting |
| Calf breadth, left and right, supine | Preferred head section angle, supine[a] |
| Calf circumference, sitting | Stance breadth |
| Chest circumference, standing | Stature[a] |
| Forearm-forearm breadth, sitting[a] | Thigh breadth, left and right, supine |
| Front chest midline-elbow breadth, supine[a] | Upper arm circumference, sitting |
| Hand breadth | Waist/hip breadth, sitting[a] |
| Head breadth | Waist/hip breadth, supine[a] |
| Head length | Waist circumference, standing |
| Heel to popliteal with foot extended, sitting | Weight[a] |

[a]Dimensions measured at all sites.

In anthropometry, there are standard dimension definitions used world-wide (ISO 7250-1 2017) in order to promote the cross-utilization of anthropometric data across industries and countries. In general, we attempted to follow those definitions, although some modifications were required. For example, if a dimension could be measured in either a standing or seated position, the seated position was chosen.

Body posture, always a part of a measurement definition, was also modified for specified dimensions due to the size of the participants. For example, Knee-to-Knee Breadth, Sitting, is typically measured with the knees pressed together. This is an uncomfortable position for individuals with large thighs. Taking measurements with participants in more natural postures has the added benefit of the data being more directly applicable to product design. In this instance, the dimension is used in chair design, and the modified posture is more typical of how people would typically sit in a chair. Similarly, when measuring stance breadth (the distance from the outer edge of one foot to the outer edge of the other foot), we defined the dimension as a natural stance, rather than one with the heels together. This modification again accommodated participants with larger thighs and/or balance impairments that are derived from co-morbidities associated with obesity. We used the same natural stance breath posture for measuring stature. Finally, designers of workspaces and furniture need to accommodate the width of the middle part of the body, but for people with obesity, the widest part is not necessarily at the hips or the waist – it could be either. Therefore, we defined maximal breadth at either the hip or waist, and the measurer indicated which location was the wider.

Several new dimensions were defined with supine body positions that are not typically used in anthropometry. For the supine position, we asked the participant to lie comfortably on the measurement table. This was necessary because on larger people the body contours are deformable and differ between postures (e.g. seated vs. supine).

Thigh and calf breadth were measured bilaterally from a supine posture, since fluid retention can differ bilaterally and may affect lower limb dimensions.

### 2.2 Sample Size

Sample size for anthropometric surveys is typically determined by three parameters: (1) variability across the population of the attribute being measured, (2) the level of confidence required in the resulting summary statistics, and (3) the level of precision needed in the resulting summary statistics (ISO 15535: 2012). The last two parameters are largely based on how the final data will be used and are ultimately a matter of judgment. The first parameter – variability – is data-based and is usually established by examining one or more similar populations. Given the lack of anthropometric data available for individuals with high BMI, it was not possible to estimate variability in advance. As a result, available resources and budget determined the sample size of the current study. The variation observed in the current data set will be extremely useful in determining sample sizes for larger studies of this population in the future.

### 2.3 Sample Acquisition

Mobility is one of the challenges of daily living for people of size. As a result, many limit the number and duration of excursions outside the home. To minimize inconvenience and to increase the participate rate, we elected to conduct data collection at bariatric weight loss clinics and one academic institution. Clinical staff aided in participant recruitment. This approach did create a sample bias, in that those people of high BMI who do not seek weight loss treatment are not represented in the study. For this study, the benefit of increased access to potential participants was a higher priority than randomness; this resulted in a convenience sample.

## 3 Survey Results

Results of this anthropometric study will be reported in a subsequent publication. The focus of this proceeding is to identify a statistical approach that will provide useful and practical guidance for designers of living spaces, workspaces, furniture and equipment.

Anthropometric data are typically presented in terms of mean, standard deviation and percentiles. Data from anthropometric dimensions in typical large-scale anthropometric surveys follow a normal distribution as a result of the sampling strategy and a large sample size. However, given that the inclusion criterion of the current study sample is defined as BMI $\geq$ 30, this population segment is not normally distributed. Percentiles are specifically used to characterize the sample because they are non-parametric—equally valid regardless of the data distribution. Mean and standard deviation are also still meaningful in the absence of normality; they are just no longer sufficient to characterize the distribution. The bigger issue is that this is a convenience sample and a population segment, which precludes using these data for population estimates.

Figure 1 illustrates the distribution of BMI and body weight for the present sample. Shapiro–Wilk goodness of fit tests were conducted to identify dimensional distributions that were non-normal. These tests were carried out at the 5% significance level. Figure 2 presents other design-critical dimensions also demonstrating non-normal distributions.

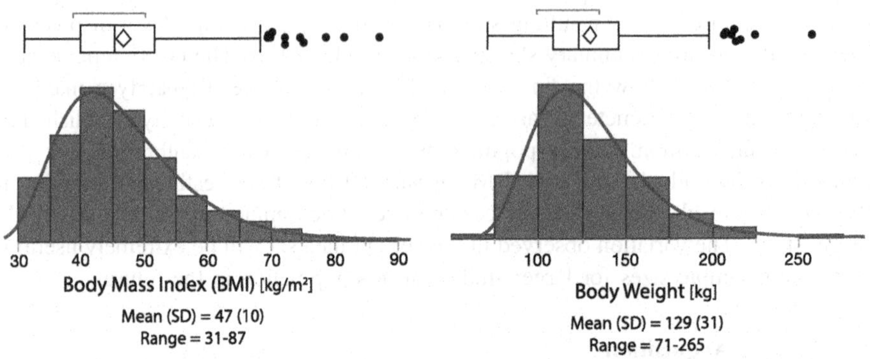


**Fig. 1.** Distribution of Body Mass Index (BMI) and weight.

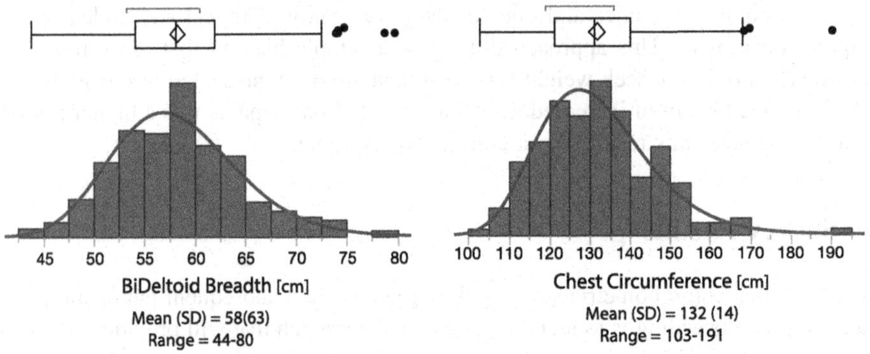


**Fig. 2.** Distribution of BiDeltoid breadth and chest circumference.

Table 2 presents the 95th percentile, 99th percentile and maximum for the current study sample compared to 95th and 99th percentile values from ISO 7250-3, a guidance standard for the design of products and workspaces to be marketed world-wide.

Selected dimensions were included based upon compatible definitions between the present study and the ISO standard. With the exception of stature, all values from the present bariatric sample were larger than the comparable design values from the ISO standard.

**Table 2.** Selected percentiles from present study and ISO 7250-3. (Values in mm and kg)

| | Present sample | Present sample | ISO 7250-3 | Present sample | ISO 7250-3 | Present sample |
|---|---|---|---|---|---|---|
| | n | P95 | P95 | P99 | P99 | Max |
| Stature | 115 | 1865 | 1959 | 1895 | 2054 | 1896 |
| Weight (kg) | 288 | 187 | 117 | 217 | 143 | 265 |
| Bideltoid breadth | 288 | 702 | 550 | 752 | 592 | 798 |
| Calf circ. | 187 | 599 | 422 | 684 | 464 | 710 |
| Chest circ. | 187 | 1543 | 1225 | 1724 | 1354 | 1905 |
| Neck circ. | 187 | 542 | 440 | 600 | 449 | 641 |

## 4 Discussion and Conclusion

Well-informed product and workspace design considers the anthropometric characteristics of the user population. Typically, the design range is the central 90% or 95% of the distribution of a given population. Because the population segment including people with obesity is typically at the extreme end of the anthropometric distribution of the overall population, this segment is likely to be systematically excluded from the design space. For various social and medical reasons, people with obesity are often not included in anthropometric databases that purport to represent the entire population, which obviously compounds the situation. In this study we address a portion of that problem by specifically recruiting and collecting anthropometric data from people with obesity.

Some of the postures which are standard for traditional anthropometric data collection are either not possible or uncomfortable for many people of size. In particular, those postures, which require that the heels or knees be placed together, are problematic. An alternative is a modified standard posture, with the heels, for example, 30 cm apart, or the knees 20 cm apart. However, given the large variability in this population segment, we determined that it would not be possible to achieve standard leg positioning that would work for all participants. Instead, we modified those postures to include a comfortable stance, and a comfortable sitting posture. The result of this change is that the data contain greater apparent dimensional variability due to posture variability. At the same time, that increased variability is reflective of the larger variability in this particular population segment and will be more relevant for designers than data from people forced into unnatural positions. These revised techniques and newly defined supine and maximal dimensions can be used in subsequent anthropometric surveys of people of size.

Engineers and designers often use anthropometric standards in creating new products. A commonly used standard is ISO 7250-3, "Basic human body measurements for technological design – Part 3: Worldwide and regional design ranges for use in product standards", which is useful for global applications as it takes into account worldwide anthropometric variability. A comparison of dimensional values within the current bariatric sample and the ISO standard shows that products designed to the international standard would disaccommodate people of size. Our sample was larger on all the non-stature dimensions compared in Table 2 for both the 95th and 99th percentiles.

Determination of sample size requires the establishment of an acceptable level of precision and a tolerable level of confidence. It also requires knowledge of the variability of the characteristics being measured. In our case, we were hampered by the lack of data from previous studies which would have allowed us to estimate variability of these anthropometric measures. One of the contributions of this effort was the development of a baseline data set whose dimensional variability can be used establish statistically based sample sizes for subsequent anthropometric surveys of people of size.

The data collected in this study are necessarily from a convenience sample and cannot be used to infer population estimates. Further they are from a population segment, rather than from a complete population. At the same time, comparison of these data with international design standards suggests that products designed to current standards are likely to be inadequate in accommodating people of size. These data may be used to identify potential flaws or shortcomings in product design. Future research should be directed at building a database with a larger sample size, and a targeted sample so it can be used to accurately represent this population segment.

## References

Bhaskaran K, Douglas I, Forbes H, dos-Santos-Silva I, Leon D, Smeeth L: Body-mass index and risk of 22 specific cancers: a population-based cohort study of 5.24 million UK adults. Lancet 384(9945):755–765 (2014)

Centre for Excellence in Universal Design Homepage (2014). http://universaldesign.ie/What-is-Universal-Design/. Accessed 17 Apr 2018

Finucane M, Stevens G, Cowan M et al (2011) National, regional, and global trends in body-mass index since 1980: systematic analysis of health examination surveys and epidemiological studies with 960 country-years and 9·1 million participants. Lancet 377:557–567

Hales C, Fryar C, Carroll M, Freedman D, Ogden C (2018) Trends in obesity and severe obesity prevalence in US youth and adults by sex and age, 2007–2008 to 2015–2016. J. Am. Med. Assoc. Published online and downloaded 26 Mar 2018

ISO 7250-1 (2017) Basic human body measurements for technological design – Part 1: Body measurement definitions and landmarks. International Organization for Standardization, Geneva

ISO 15535 (2012) General requirements for establishing anthropometric databases. International Organization for Standardization, Geneva

ISO 7250-3 (2015) Basic human body measurements for technological design – Part 3: Worldwide and regional design ranges for use in product standards. International Organization for Standardization, Geneva

James W (2008) WHO recognition of the global obesity epidemic. Int. J. Obesity (London) 32 (Suppl. 7):S120–S126

Mitchell N, Catenacci V, Wyatt H, Hill J (2011) Obesity: overview of an epidemic. Psychiatr Clin North Am 34(4):717–732

US Department of Health and Human Services, Public Health Service, National Institutes of Health, National Heart, Lung and Blood Institute: Clinical guidelines on the identification, evaluation, and treatment of overweight and obesity in adults NIH Publication No. 98-4083 (1988)
US Department of Health and Human Services, Public Health Service, National Institutes of Health, National Heart, Lung and Blood Institute: Managing overweight and obesity in adults: Systematic evidence review from the Obesity Expert Panel (2013). https://www.nhlbi.nih.gov/sites/default/files/media/docs/obesity-evidence-review.pdf. Accessed 7 May 2018

# SOOMA - Software for Acquisition and Storage of Anthropometric Data Automatically Extracted from 3D Digital Human Models

Flávia Cristine Hofstetter Pastura[1(✉)], Tales Fernandes Costa[1], Gabriel de Aguiar Mendonça[1], and Maria Cristina Palmer Lima Zamberlan[1,2]

[1] Laboratório de Ergonomia (LABER), Instituto Nacional de Tecnologia (INT), Av. Venezuela 82, Térreo, Anexo 4, Rio de Janeiro, Brazil
{flavia.pastura, cristina.zamberlan}@int.gov.br, talesfc@gmail.com, gabrieldeaguiarmendonca@gmail.com
[2] Universidade Estácio de Sá (UNESA), Av. Pres. Vargas 642, Rio de Janeiro, Brazil

**Abstract.** SOOMA is a software tool developed by Laboratório de Ergonomia (LABER) of the Instituto Nacional de Tecnologia (INT) of Brazil for use in projects involving anthropometric characterization of the human body. Its main objective is to automate the location of body landmarks and calculation of anthropometric measurements based on digital human models generated by 3D laser scanning. SOOMA was designed as a modular application, allowing implementation of algorithms for landmarks location and anthropometric measurements calculation independent of the graphical user interface responsible for controlling user interaction. Data and files are stored in a local database immediately after production. Optionally, stored data and files may be copied to a centralized document management system that provides long term preservation and decentralized access by registered users. The Control module is a *C#* windows application and the local database is an opensource database. Currently, the Script module is based on *R* but other algorithm execution tools can be used. Opensource software *LogicalDOC* is used as document management system and some specialized functions for visual verification of landmark location and 3D pdf generation depend on additional opensource tools (*CloudCompare*, *Meshlab* and *Miktex*).

**Keywords:** Anthropometry · Body landmark · Digital human model

## 1 Introduction

Conventional anthropometric measurement is a manual task performed by specialized technicians using direct measuring tools. Accuracy and consistency of results may be affected by faulty procedures and human errors. Manual direct measurement allows only for linear measures, as surface and volume measurements are not possible with


S. Bagnara et al. (Eds.): IEA 2018, AISC 826, pp. 472–481, 2019.
https://doi.org/10.1007/978-3-319-96065-4_52

conventional measurement tools. Also, as subjects are not permanently available for measurement tasks, only previously defined and planned measures are taken.

Modern optical technologies like 3D laser scanning make it possible to capture human body measurements independent of physical contact [1]. Laser light projected over the human body is reflected and captured by cameras and mapped as a large set of points (point cloud) in 3D coordinate system. These coordinates can be used to identify body landmarks and extract linear, surface and volume measurements with higher accuracy and consistency than is possible with direct manual measurements [2, 3].

Specialized software tools can be used to visualize the 3D digital model and manually extract anthropometric measurements. However, experience shows this is an arduous task and results may vary depending on user's ability with the tool. To obtain anthropometric measurements from even a small population based on manual extraction is a significant challenge. To assure good quality and productivity levels, standard procedures for extraction of anthropometric data are needed.

For this, INT's Ergonomics Laboratory (LABER) developed SOOMA as a supporting tool for the extraction of body landmarks and anthropometric measurements (1D, 2D and 3D). Following design guidelines were used:

- **Easy of use** – user friendly interface and functionalities for a qualified user to execute standard procedures.
- **Agility** – immediate results evaluation so that user may decide whether to accept or reject 3D model acquired.
- **Transparency** - promotion of results consistency and quality by preserving all data necessary to verification and reproduction of results at any time.
- **Flexibility** – algorithms are not embedded in compiled code and can be changed according to specific needs of research project or be updated with corrections and improvements. All registered data on subject model is available for use by the algorithms (e.g. sex, height, weight, posture, …).
- **Persistence** - all data on human model and software settings, whether entered, acquired or calculated is stored independent of user action.
- **Low dependency** – there is no dependency regarding scanner equipment brand or model, nor on local network or internet connection. For 3D model acquisition a digital file must be provided, but there is no hard dependency on proprietary file formats.

## 2 Description

**SOOMA** is composed by two modules: **Control** and **Script**. The first is responsible for user interface and local database storage. Second module is responsible for execution of algorithms specially designed for processing a digital human model file for the extraction of body landmarks and anthropometric measures calculation. Storage of 3D model files and data is managed in two stages: short-term storage of data and files is assured by an embedded database (open source software *SQLite*) and long-term storage is provided by copying data and files to a Content Management System (open source software *LogicalDOC*) (see Fig. 1).

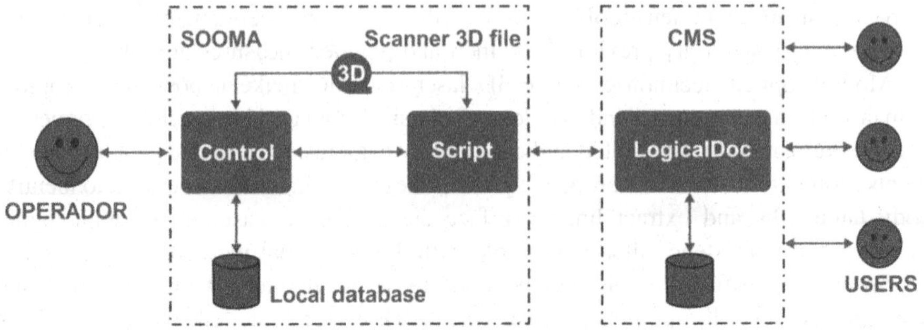


**Fig. 1.** SOOMA architecture.

Main concepts and artifacts considered on SOOMA development were:

- **Project Datasheet** – body postures, landmarks and anthropometric measures of interest are not fixed on code and may be defined according to specific requirements of each research project. Configuration data for each project is entered by a ".xlsx" file that must be imported by SOOMA. Specific project requirements are processed by the algorithms executed by the **Script** module.
- **Subject** – each person that contributes with digital models is identified by an alphanumeric code to which all model files and data refer to. For privacy considerations the person name and email are not exposed with other model data.
- **Subject Form** – for each subject a set of characterization data is registered, including anthropometric measures manually collected during scanning sessions. This dataset is available for use by the extraction and measurement algorithms;
- **Body Posture** – for each subject several digital models may be captured, where each model refers to a specific body posture of interest to the research project and has its own model file generated by the 3D scanner.
- **Body Landmarks** – for each 3D model file the **Script** module extracts a set of XYZ coordinates corresponding to the set of body landmarks defined for the project.
- **Anthropometric Measures** – based on extracted landmarks the **Script** module calculates a set of anthropometric measures defined for the project.
- **Session Dataset** – the Subject Form along with the digital model file Body Landmarks and Measures results compose the Session dataset. Subjects may be scanned multiple times along the project duration and for each contribution, new sessions datasets are created with updated subject data at the time of the session.

SOOMA is a standalone *C#* [4] application installed on Windows desktops or notebooks and for most operations may be used independent of local network or internet access. Software integrated with SOOMA (*SQLite* [5], *CloudCompare* [6], *Meshlab* [7] and *Miktex* [8]) must be installed on same equipment as SOOMA, except for *LogicalDOC* [9] which can be accessed on a dedicated server on local network or internet.

Module **Script** language and platform are not fixed and may be chosen according to each project requirements. However, as communication between modules is by command line and files exchange, the **Script** platform must be installed on same equipment as SOOMA. During SOOMA implementation, the open source software for statistical computing *R* [10] was used as module **Script** platform. So, our first algorithm for landmark extraction and anthropometric measurements was developed as an *R* script. In the future other algorithm execution platforms may be used.

## 3 Operation

Every day use of SOOMA may be resumed as a 4-step workflow (see Fig. 2).

**Fig. 2.** Simple SOOMA workflow.

1. **Fill Subject Form** – user identifies the subject on scanned 3D model files and fills a screen data form with subject classification information and some conventional manual measurements (ex.: gender, height, weight, body posture etc.);
2. **Process 3D model** – user supplies file containing subject's 3D digital model and executes **Script** to process it. Results are collected by **Control**, automatically stored on local database and presented to user on screen results grid;
3. **Verify results** – user verifies quality of landmarks location and anthropometric measurements results and decides whether proceed to next step or go back to step 2. To help users on this step, SOOMA provides integration with *CloudCompare* for visualization of landmarks together with digital model (see Fig. 3).
4. **Export to CMS** – digital model and results data are automatically saved on local database, but to assure long term storage independent of equipment used, an integration with *LogicalDOC* allows users to send files and data to a centralized document management system.

Step 1 is executed once for each Session Dataset, while Steps 2 & 3 are repeated for each body posture model file. Step 4 is optional and may be executed immediately following Step 3 or at a later time for all models on the Session Dataset.

## 4 Algorithm Development

SOOMA first algorithm for automated landmark extraction and anthropometric measurement was developed as an R script. Main concepts for this algorithm development, as well as tests procedures and results are described. For tests and validation, a total of 70 volunteers were scanned (35 men and 35 women) and a test procedure based on five

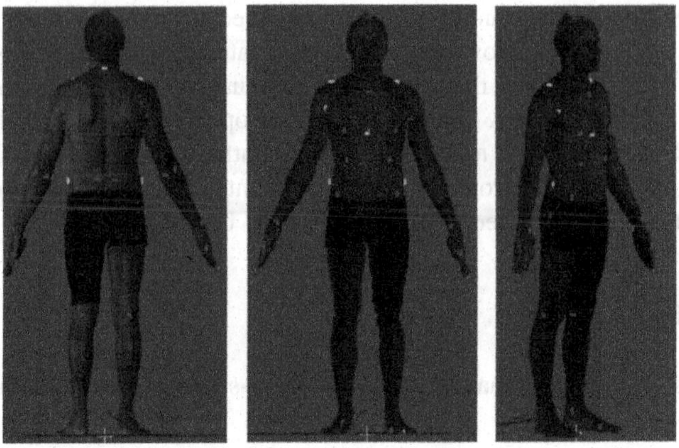

**Fig. 3.** Visualization of landmarks with *CloudCompare*. (Color figure online)

steps was defined and executed: (1) Subject scanning; (2) Model selection and cleaning; (3) Landmarks extraction; (4) Measurements Calculation and (5) Results validation.

## 4.1 Subject Scanning

Before scanning, subject signed a Terms of Agreement and personal data was collected. Then subject dressed a measurement attire and 70 colored markers were affixed to volunteer's body (61 blue circle-shaped stickers and 9 green elastic bands). Next, anthropometric measurements were manually taken based on positioned markers.

Finally, subject was scanned by a full body 3D scanner (*Cyberware WBX)* with laser vertical pitch of 2 mm, in a standing posture, conforming to *CAESAR* scanning research protocol [11]. Typical scanning period was about 36 s and each volunteer was scanned three times at same body posture.

To ensure resulting 3D model was adequate, a checking procedure was performed at each scanning session: equipment calibration, lighting control, and control of body posture and movements [12, 13].

Afterwards, scanner software *Cyscan* (*Cyberware* proprietary software) was used to generate a 3D triangular polygonal mesh file based on Stanford Triangle Format (".ply"). Typical 3D mesh generated contains about 700.000 vertices, where each vertex corresponds to a point of reflected laser light detected by the scanner cameras. Each point is defined by six variables: a XYZ coordinate and RGB based reflected light color.

## 4.2 Model Selection and Cleaning

Five quality criteria were used to select the best human model from each three scans set: (1) Non-distorted image; (2) Visualization of five essential landmarks (C7, left and right acromion and trochanterion) (3) No rotation of the trunk in top plane view;

(4) Elastic bands placed parallel to the ground; (5) Landmarks not hidden by elastic bands; (6) Elastic bands not hidden by measurement attire. Cleaning of digital model was performed by manually deleting environmental noise using *CloudCompare* software.

### 4.3 Markers Location

A built-in functionality is provided by the **Control** module for use with ".ply" file format, where a point cloud version is automatically generated with only the triangles vertices coordinates and RGB color. But any other open format may be converted to a point cloud model by the **Script** module implementation. The point cloud model is imported by the **Script** module and executed by statistical software package R to isolate the points that form the colored markers from all other points on the model. To improve isolation of colored markers, a DIFF variable is calculated for each point so that only points with DIFF > 5 are selected. Blue points are found by setting $DIF_B = (B - R) + (B - G)$ and green points by setting $DIF_G = (G - R) + (G - B)$.

For each blue marker (composed by about 50 blue points) a weighted arithmetic mean of XYZ coordinates is calculated as candidate to represent the location of the marked body landmark. To push this candidate towards the center of the marker the point's $DIF_B$ is used as weight, considering that points located at the center of the markers have a stronger blue color. A similar procedure is used to locate the centerline of the green elastic bands.

However, selecting a landmark candidate is not enough. It is necessary to associate the candidate with a specific body landmark, and this is not always a clear and easy task, especially when distance between markers is not large. Also, significant effort and time is dispended on fixation of the circular markers and elastic bands, not always with adequate results. So, there is interest in methods that allow for automated location and classification of landmarks independent of colored markers on the subject body.

### 4.4 Landmark Region Calculation

SOOMA first algorithm includes a method for automated classification of candidate landmarks that also allows for estimation of landmarks without markers. Method hypothesis is that each body landmark has a characteristic region for its possible location, defined by a range of values based on the landmark relative height and angle at a certain direction. The set of relative heights and angles that define these landmark regions depend on the population physical characteristics, and for its calculation the following procedure was used:

a. The marker center point for all blue markers on the 70 subjects were calculated and visually validated and associated with body landmarks using *CloudCompare*.
b. Variable $\boldsymbol{HPs_i}$ is defined as the ratio of height of marker center point corresponding to landmark $\boldsymbol{i}$ to the subject $\boldsymbol{j}$ stature. So, $\boldsymbol{HPs}$ vector is composed of 61 variables which correspond to the relative heights ($hps_{ij}$) of all calculated landmarks:

$$hps_{ij} = \frac{\text{height of marker center point calculated as landmark } i \text{ of subject } j}{\text{stature of subject } j} \quad (1)$$

c. For trunk and lower limbs, height of marker center point is obtained by value of its Z coordinate – which is the distance to scanner base. For upper limbs, it is necessary to consider that they are inclined in the project standard posture, and height calculation must discount the angle elevation relative to limbs positioned alongside the trunk. For this, some geometric manipulation is applied to get arms aligned with Z-axis and have relative heights calculated in relation to the tip of the middle finger.
d. For calculation of marked angle $AP_i$, all points contained within range of ±6 mm (marker size) from z coordinate are selected and location angle is determined by X and Y coordinates of three points: the marker itself, the center of the shape and a reference point (see Fig. 4).

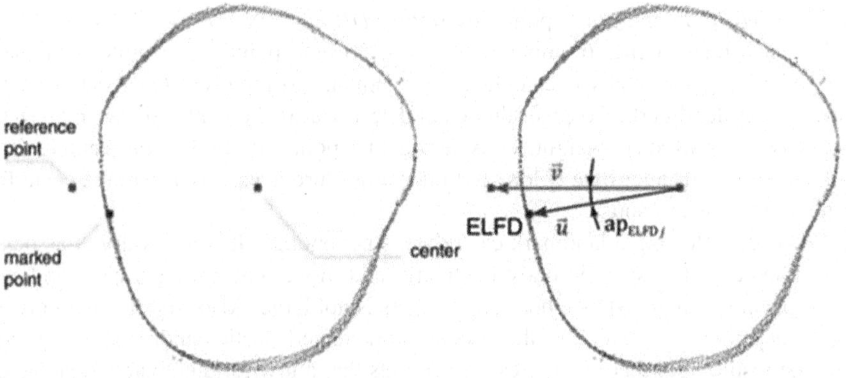


**Fig. 4.** (A) points to calculate angle of right lateral femoral epicondyle (B) apELFDj - angle of the lateral femoral epicondyle.

e. With $\boldsymbol{HPs_i}$ and $\boldsymbol{AP_i}$ defined, intervals with high probability of containing the population means were created. For each variable, normality assumptions were evaluated using the *Shapiro-Wilk* hypothesis test, suitable for small samples [14]. When normality assumption at a 5% significance level was accepted, 95% confidence intervals (CI) were defined for population means. When normality assumptions were not met, interquartile ranges (IQR) were defined, on which limits were first and third quartiles. This type of range has high probability of containing the central tendency, since they have 50% of central data.
f. Results of low/high relative heights and angles define the limits of a set of volume regions where landmarks are probable located, which we call Landmark Regions.

## 4.5 Landmark Estimation and Classification Using Landmark Regions

For each subject model, the Landmark Regions can be used to estimate body landmarks coordinates without using markers by following procedure:

a. identify the Landmark Region set of relative height and angle corresponding to the desired landmark.
b. convert the relative height range of Landmark Region to a height range in meters by multiplying the lower and upper limits height range by the subject's stature.
c. select all subject's cloud points contained between the lower and upper height limits and within the region angle range.
d. Estimated Landmark's coordinates are obtained by calculating the arithmetic means of selected cloud points coordinates.
e. for each subject model, calculate the Euclidean distance between the Estimated Landmark and all landmark candidates calculated from the blue marker points.
f. landmark candidate which is closest to an Estimated Landmark's coordinates, with a maximum calculated distance value of 10 cm, is classified as the corresponding Body Landmark and marked as "Located".
g. if no landmark candidate is close enough to an Estimated Landmark but the landmark region contains point clouds within it, the estimated landmark is classified as the corresponding Body Landmark and marked as "Estimated".
h. if no landmark candidate is close enough to an estimated landmark and the landmark region is empty of point clouds, the corresponding Body Landmark is set with invalid coordinates and marked as "Not found".

This method of estimation and classification considers landmarks as independent from each other. Each landmark is estimated considering only the expected information for that landmark.

## 4.6 Measurement Calculation

Lengths, widths and diameters are calculated with Euclidean distance methods from the coordinates of the Body Landmarks obtained by localization and estimation.

A smoothing method is used to reduce perimeter overestimation due to variations in the circumferences point coordinates. Cross-section surface area and volume of body segments are calculated by adaptations of the perimeter method. The body segment is split in several parts, measurement is made as for a simple geometry part and repeated for higher number of parts until differences in sequence of measurements falls below a certain value (see Fig. 5).

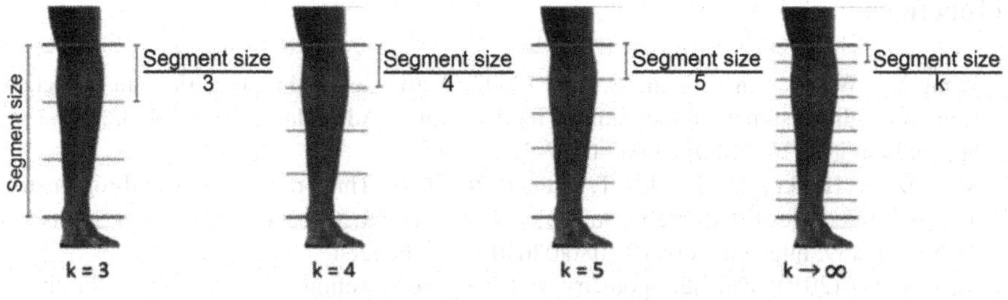


**Fig. 5.** Body segment division in equal parts for volume calculation.

## 5 Results Validation

In anthropometry, height ratio is a resource used to obtain body segment size estimates. In this work, sample means of the relative heights of landmarks were used to determine these segments due to strong correlation between stature and segment size. It was found that heights of the upper limb landmarks presented moderate correlation with stature, assuming values around 0.7, compared to trunk and lower limbs, where most of the correlations were above 0.8, not considering hands and feet.

Normality assumptions were met for almost all landmarks for both sexes for Shapiro-Wilk test at a significance level of 5%, allowing use of confidence intervals to estimate almost all landmarks. Obtained ranges presented low amplitudes with few cases above 1% of stature. This was observed even when normality assumption was not accepted and interquartile interval, less precise than confidence interval, was used. Considering average height of men (175.9 cm) and women (164.0 cm), height ranges presented amplitudes around 1.2 cm, size of marker used to identify the landmarks.

With height and angle ranges calculated from the test sample, mean difference between markers' centers and estimated landmarks (mean error) was 1.646 cm for men and 1.445 cm for women. Estimated circumference heights were about 1.5 cm distant from the elastic bands.

Considering the three measurement methods - traditional, with markers and with estimated landmarks and circumferences - the mean and median values remained very close to each other for all anthropometric measurements both for men and women.

In all we consider that experience has shown that SOOMA is a viable tool for anthropometric data extraction with standard procedures that results in good quality and productivity levels. Also, results achieved signals that body landmark extraction without marker fixation is a viable goal. However, additional work is needed to enlarge population used for building landmark regions dataset and for developing scripts able to process additional body postures.

**Acknowledgements.** We are grateful to Universidade Estácio de Sá (UNESA) through the UNESA Productivity Research Program 2018 and to Conselho Nacional de Desenvolvimento Científico e Tecnológico (CNPq) through the Institutional Capacity Program (PCI/MCTIC) for supporting this work.

## References

1. Wang MJ, Wu W, Lin K, Yang S, Lu J (2007) Automated anthropometric data collection from three-dimensional digital human models. Int J Adv Manuf Technol 32:109–115. https://doi.org/10.1007/s00170-005-0307-3
2. Schranz N, Tomkinson G, Olds T, Daniell N (2010) Three-dimensional anthropometric analysis: differences between elite Australian rowers and the general population. J Sports Sci 28(5):459–469. https://doi.org/10.1080/02640411003663284
3. Stewart AD (2010) Kinanthropometry and body composition: a natural home for three-dimensional photonic scanning. J Sports Sci 28(5):455–457. https://doi.org/10.1080/02640411003661304
4. C# Homepage. https://pt.wikipedia.org/wiki/C_Sharp. Accessed 25 May 2018

5. SQLite Homepage. https://www.sqlite.org. Accessed 25 May 2018
6. CloudCompare Homepage. http://www.danielgm.net/cc. Accessed 30 Mar 2017
7. Meshlab Homepage. http://www.meshlab.net. Accessed 25 May 2018
8. Miktex Homepage. https://miktex.org/. Accessed 25 May 2018
9. LogicalDOC Homepage. https://www.logicaldoc.com/download-logicaldoc-community. Accessed 25 May 2018
10. R Homepage. https://www.r-project.org. Accessed 27 Aug 2017
11. Robinette KM, Blackwell S, Hoeferlin D, Fleming S, Kelly S, Burnsides D (2002) Civilian American and European Surface Anthropometry Resource (CAESAR) Final Report Volume I: Summary, National Technical Information Service, 5285 Port Royal Road, Springfield, Virginia 22161
12. Brunsman MA, Daanen HAM, Robinette KM (1997) Optimal postures and positioning for human body scanning. In: International conference on recent advances in 3-D digital imaging and modeling proceedings, pp 266–273. (Cat. No. 97TB100134). https://doi.org/10.1109/im.1997.603875
13. Daanen HAM, Brunsman Ma, Robinette KM (1997) Reducing movement artifacts in whole body scanning. In: International conference on recent advances in 3-D digital imaging and modeling proceedings, pp 262–265. (Cat. No. 97TB100134). https://doi.org/10.1109/im.1997.603874
14. Razali NM, Wah YB (2011) Power comparisons of Shapiro-Wilk, Kolmogorov-Smirnov, Lilliefors and Anderson-Darling tests. J Stat Model Anal 2(1):21–33. https://doi.org/10.1515/bile-2015-0008

# Design for Plus Size People

J. F. M. Molenbroek[1,3(✉)], R. de Bruin[1,2,3], and T. Albin[2,3]

[1] Faculty of Industrial Design, Delft University of Technology (TU Delft), Landbergstraat 15, 2628 CE Delft, The Netherlands
j.f.m.molenbroek@tudelft.nl
[2] Erin Ergonomics and Industrial Design, Nijmegen, The Netherlands
[3] High Plains Engineering Services, Minneapolis, USA

**Abstract.** Obesity is a growing issue in western societies with consequences for the field of human centered design. Most anthropometric data sources assume the data follow the Gaussian distribution, with population data symmetrically distributed above and below the mean value. This assumption is often true in length measurements like body heights, but may not be true for measurements more sensitive to body mass, like body weight, hip width, elbow-to-elbow width, and body depth. While length measurements have remained relatively stable over time in western societies, mass related measurements are increasing.

The authors have experience in providing data via an interactive website DINED, which seeks to make anthropometry accessible without requiring expert knowledge about anatomy and statistics. Currently all DINED dimensions are assumed Gaussian, including those related to body mass. This might not work when designing for plus size people. Future additions in DINED will be about design for obesity and about how to implement 3D scanning into the design process in order to redress these defects.

**Keywords:** Anthropometrics · Ergonomics education · Product design Plus-size

## 1 Introduction

Obesity is a growing issue in western societies. According to the World Health Organization obesity has tripled from 1975–2016 worldwide. At the moment about 2 billion people have a body mass index (BMI) of more than 25 (overweight) and 650 million are even over 30 (obese) [1]. For the first time in history obesity has outnumbered underweight [2]. At lot of health issues are connected with a high BMI and there is a tendency to focus mainly on this aspect of plus sized people. Although prevention programs on healthy living, with enough exercise and healthy foods, are put into effort to tame the obesity-epidemic, fact is that there actually are vast amounts of people overweight. And overweight people tend to be larger people, not in height, but in width. From a product designer's point of view this presents new challenges. Especially when following the inclusive design ideology 'not to exclude people by design', which means in this case; taking care that the largest people are able to use products comfortably. But how large are the largest? Anthropometrics are needed so


S. Bagnara et al. (Eds.): IEA 2018, AISC 826, pp. 482–495, 2019.
https://doi.org/10.1007/978-3-319-96065-4_53

products and services can be designed that fit, or physically accommodate, users [3]. Designers need to be aware that not only the one-dimensional aspect of body circumference is affected. Increasing body mass also affects movement of the body like forward bending, field of view downwards, and hence leads to usability problems with many products, like toilets [4], surgical tables, chairs, wheelchairs, and public transport. However, data from overweight or obese people are rare in public databases. The group in itself is not easy to investigate. Being overweight makes travelling to a research site a challenge. In addition, experience teaches that when the purpose of measuring is not made absolutely clear or doesn't relate to an actual design problem, subjects are prone to feel embarrassed by the act of measuring. Some [5] have therefore chosen the method of self-reporting to tackle this problem, although this method clearly has its drawbacks in terms of reliability.

In this paper, we will explore the current methods available when designing for plus sized people.

## 2 DINED as a Tool for Designers and Researchers

### 2.1 Ergonomics and Anthropometry in Education

User-centered design has been taught at the faculty of Industrial Design Engineering (IDE) since its start in 1969. Ergonomics is one of the 4 pillars of IDE from the start, next to Aesthetics, Marketing and Engineering. Currently the scope is towards consumer products, services and social designs. Courses at IDE are aimed at giving insight in the optimization of the human-product interaction from the viewpoint of the consumer (ergonomics) and from the viewpoint of the producer (marketing).

One of the main topics is observational research. Students are taught to study the behavior of people (consumers) using products and services in order to find potential user problems, which can lead to clues for product innovation. One of the outcomes for example could be that the product and user do not fit 'physically'. Here anthropometrics come onto stage.

The best way to teach about anthropometry is to let students experience the impact of anthropometrics themselves. A design-assignment e.g. could be: 'Design a minimum shower cabin that fits all students but makes it possible to use the shower in a comfortable way'. In this way students have to think about relevant body dimensions, experience for themselves through repeated measurements what it takes to measure accurately and finally experience for themselves that nobody is average: not one body will score the 'P50' average on all body dimensions.

### 2.2 History of DINED.nl

To support the anthropometry and ergonomics education, and to explain related statistics and anatomy, a tool called DINED was gradually developed (www.dined.nl). Now DINED has become an anthropometric information system including data from several populations around the world and tools to make design decisions more easily.

It all started around 1980 as a table on 1 A4 cardboard showing the P5-P50-P95 percentiles of body dimensions frequently used in product design (see Fig. 1).

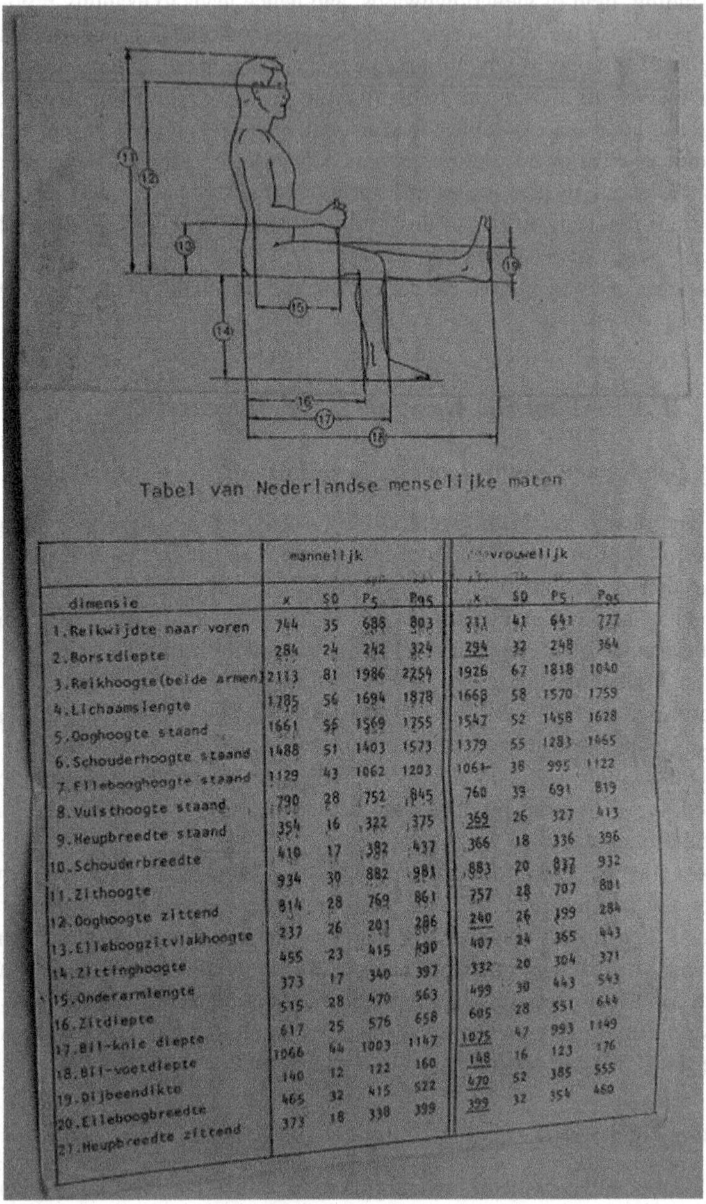

Tabel van Nederlandse menselijke maten

| | mannelijk | | | | vrouwelijk | | | |
|---|---|---|---|---|---|---|---|---|
| dimensie | x | SD | P5 | P95 | x | SD | P5 | P95 |
| 1. Reikwijdte naar voren | 744 | 35 | 688 | 803 | 711 | 41 | 641 | 777 |
| 2. Borstdiepte | 284 | 24 | 242 | 324 | 294 | 32 | 248 | 364 |
| 3. Reikhoogte (beide armen) | 2113 | 81 | 1986 | 2254 | 1926 | 67 | 1818 | 1040 |
| 4. Lichaamslengte | 1785 | 56 | 1694 | 1878 | 1668 | 58 | 1570 | 1759 |
| 5. Ooghoogte staand | 1661 | 56 | 1569 | 1755 | 1547 | 52 | 1458 | 1628 |
| 6. Schouderhoogte staand | 1488 | 51 | 1403 | 1573 | 1379 | 55 | 1283 | 1465 |
| 7. Elleboghoogte staand | 1129 | 43 | 1062 | 1203 | 1064 | 38 | 995 | 1122 |
| 8. Vuisthoogte staand | 790 | 28 | 752 | 845 | 760 | 39 | 691 | 819 |
| 9. Heupbreedte staand | 354 | 16 | 322 | 375 | 369 | 26 | 327 | 413 |
| 10. Schouderbreedte | 410 | 17 | 382 | 437 | 366 | 18 | 336 | 396 |
| 11. Zithoogte | 934 | 30 | 882 | 981 | 883 | 20 | 837 | 932 |
| 12. Ooghoogte zittend | 814 | 28 | 769 | 861 | 757 | 28 | 707 | 801 |
| 13. Elleboogzitvlakhoogte | 237 | 26 | 201 | 286 | 240 | 26 | 199 | 284 |
| 14. Zittinghoogte | 455 | 23 | 415 | 490 | 407 | 24 | 365 | 443 |
| 15. Onderarmlengte | 373 | 17 | 340 | 397 | 332 | 20 | 304 | 371 |
| 16. Zitdiepte | 515 | 28 | 470 | 563 | 499 | 30 | 443 | 543 |
| 17. Bil-knie diepte | 617 | 25 | 576 | 658 | 605 | 28 | 551 | 644 |
| 18. Bil-voetdiepte | 1066 | 64 | 1003 | 1147 | 1075 | 47 | 993 | 1149 |
| 19. Dijbeendikte | 140 | 12 | 122 | 160 | 148 | 16 | 123 | 176 |
| 20. Elleboogbreedte | 465 | 32 | 415 | 522 | 470 | 52 | 385 | 555 |
| 21. Heupbreedte zittend | 373 | 18 | 338 | 399 | 399 | 32 | 354 | 460 |

**Fig. 1.** DINED-table 1980

It consisted of an estimation of Dutch adults based on German DIN33402; the German dataset was adjusted according to predictions for the Dutch population based on the Dutch Growth Diagrams 1980, which resulted in 3,5% higher values for the Dutch male and 2% for the Dutch female, respectively. These data were later coupled with data collected from the Dutch population (amongst others students and elderly people). New possibilities of computer graphics led to an interactive DINED website with more visual tools for designers, thus making anthropometry accessible without requiring expert knowledge about anatomy and statistics [6].

In Table 1 an overview is given on the current collection of searchable anthropometric data made available via DINED. Preparations are currently going on to develop DINED-3D. These data will become available as statistical shape models.

**Table 1.** DINED-datasets in 2018

| 1D and 2D data | Size dataset | Variables | Age |
|---|---|---|---|
| Dutch students 1986 | n = 354 | v = 50 | |
| Dutch students 2014 | n = 400 | v = 40 | |
| Dutch elderly GDVV 1983 | n = 822 | v = 30 | |
| Dutch elderly GERON 1998 | n = 600 | | Age 50+ |
| Dutch adults GERON 1998 | n = 150 | | Age 20–30 |
| Dutch children KIMA 1993 | n = 2400 | | Age 0–12 |
| Chilean children 2012 | n = 3046 | | Age 6–18 |
| Chilean adults 2016 | n = 2946 | | Age 18–99 |

## 3 Current Usage of DINED

According to Google Analytics each year about 50k users use the DINED.nl website to find anthropometric data. Since a few years past, users need to set up a (free) user account, which enables one to see the professional status and background of DINED users. From being originally aimed at TU Delft students, the tool is now used by educational institutions all of the world, many companies involved in product development and even medical institutions. It would be interesting to analyze to what purpose the various offered tools are used, to improve the tool in general and to extend the platform, though this is yet a plan for the future. Current tools available at DINED are:

- Percentiles: Calculating percentiles assuming normal distribution (Fig. 2)
- Reach Envelopes: Developed especially to show large differences related to age (Fig. 3)
- Profiler: to show nobody is average (Fig. 4)
- Ellipse: to show correlation between 2 variables (Fig. 5)
- Other sources: a library of digital sources in anthropometry
- Raw data: excel tables with measured data.

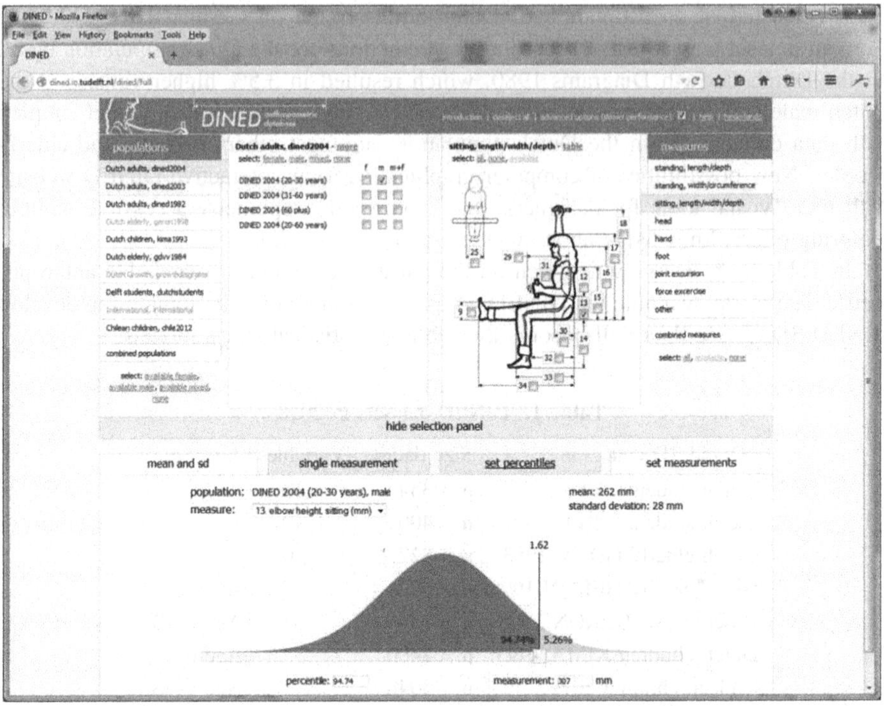


**Fig. 2.** Calculating percentiles assuming normal distribution using DINED percentiles

### 3.1 Percentiles

In this tool, several populations and body dimensions can be chosen. Thereafter the designer can choose how much percentage will be excluded from the design. The data is based on real measurements or on estimated data [3, 6–8].

### 3.2 Reach Envelope

This is an interactive tool that shows the designer immediately which region it is comfortable (green) or out-of-reach (red). The data is based on real measurements of 750 elderly that draw lines of comfort on a large white board [9].

### 3.3 Profiler

This is an interactive tool that allows you to enter a number of measurements of one's own and to compare them with one of the populations inside DINED. The outcome is a profile of percentiles of each body dimensions and will show in most cases that your dimensions are not average! A useful eye-opener for designers that follow the ego-design approach.

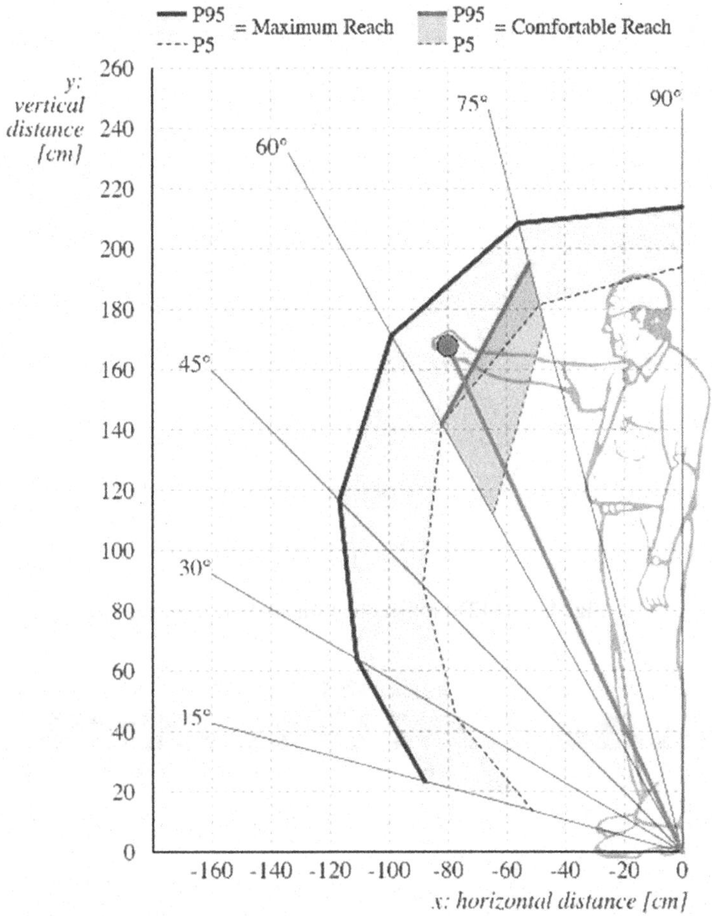


**Fig. 3.** DINED reach envelopes, developed especially to show large differences related to age (Color figure online)

### 3.4 Ellipse

This is an interactive tool that allows the designer to show the correlation between the raw data of 2 variables. If no raw data are available an ellipse can also be drawn if mean and SD values of both dimensions and the correlation coefficient, R, between them are known. The resulting ellipse, including scatterplot of the data, gives insight in the bi-variate distribution and it allows the drawing of 'rectangles' representing a size of a product like, for instance, the different sizes in a sizing system.

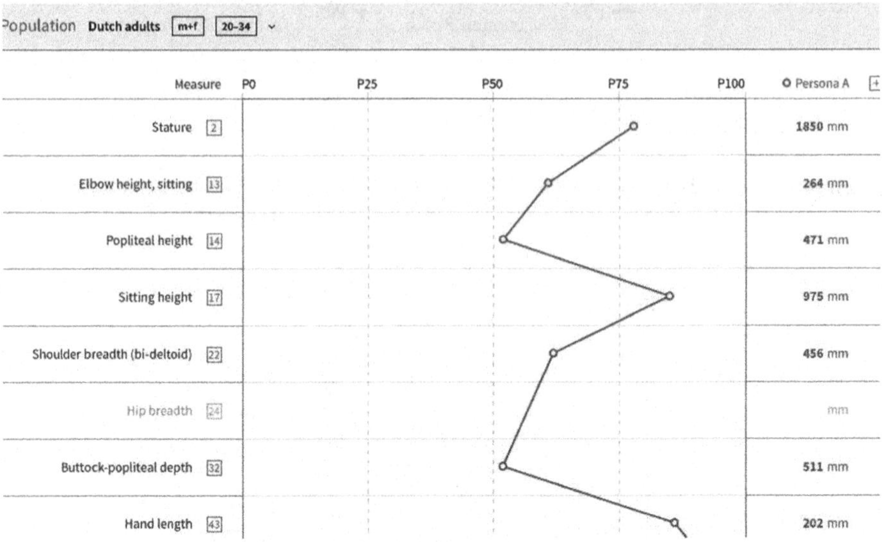


**Fig. 4.** DINED profiler shows nobody is average…

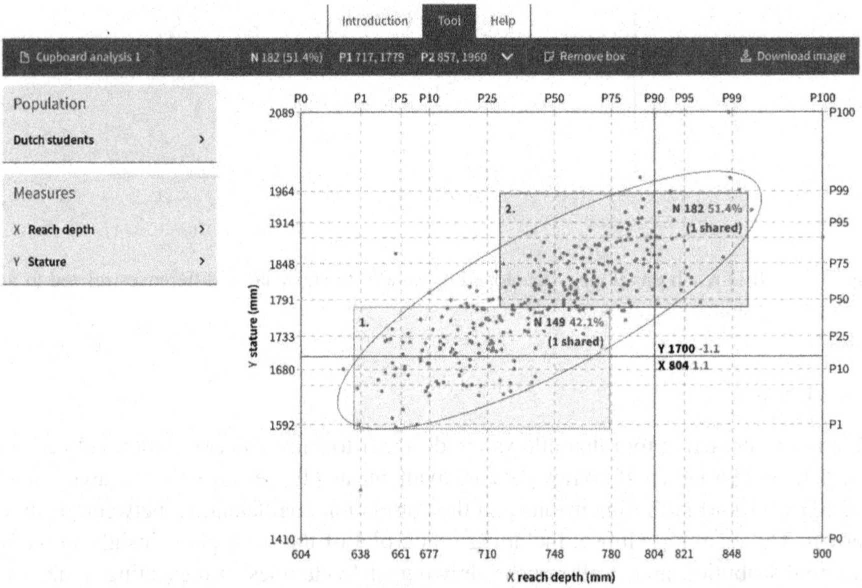


**Fig. 5.** DINED Ellipse showing correlation between two variables (stature and reach depth). Rectangles showing two sizing options and corresponding inclusion percentages (rectangles 1 and 2)

# 4 Future Additions to DINED

Several future additions to the DINED website are currently planned:

- Publishing and visualization of own anthropometric datasets
- Knowledge base for using 3D-scans in design
- Experience data of 4D anthropometry, involving the 'moving' human body.

## 4.1 Publish Your Own Data

Adding to the existing databases would be the possibility to use DINED as a platform to publish your own data. After verification of the quality of the real measured data, the population (sample) you measured data will be displayed with means, sd and percentiles if the data is (normal) Gaussian distributed.

## 4.2 Knowledge Base 3D-Anthropometry

This addition will include our knowledge, tools and experience about how to use 3D scans for designers [10]. One tool is able to integrate a scatterplot form Ellipse with 3D-scans from a person that will pop-up after a mouse-click on a dot in the scatterplot. Caesar data are available for TU Delft Campus use only. The first dataset that will become available are the 3D-scans from the project 'Ventilation mask' [11, 12].

## 4.3 Experience 4D-Anthropometry

In our Bodylab we are able to capture 4D data from humans in motion. After some experiments we learned we have to develop extra digital space and software to be able to manipulate these 4D data. Illustrative: a single experiment involving capturing motions of the shoulder joint took up nearly 0,5 terabyte in raw data. Sharing these and other experiences might help other researchers and developers.

# 5 Design for Plus-Size People

One group of people that has been growing in number over the last decades and has not been taken into consideration by all of the before mentioned anthropometric design tools, offered to designers to 'make products that fit all human beings'; people that have an above average weight. The absolute maxima to date are a man by the name John Brower Minoch (1941–1983) and a woman called Carol Yager (1960–1993) that weight at least 635 kg and 720 kg respectively [13]. Though design goals will not accommodate for the extreme, there is still lack of data on much larger group of people that are overweight (Fig. 6).

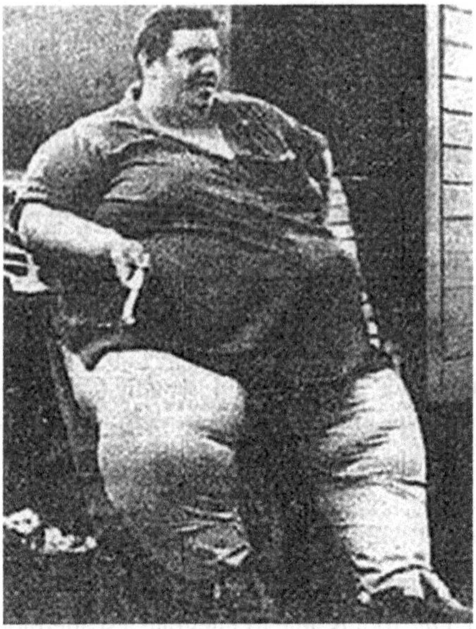

**Fig. 6.** John Brower Minoch in 1983, height 180 cm, weight 635 kg [22]

### 5.1 Own Data

A first start to fill the knowledge gap about anthropometric dimensions of 'Plus-size' people was made in measuring 21 'seriously overweight' people in 2003 and later in 2010 another 3 people in 2010. The results are shown in Table 2.

### 5.2 Other Methods to Estimate Sizes of Plus Size People

From the Gaussian distribution we know each percentile can be estimated or calculated by z * SD when z is a fixed number for each percentile P. P5 = mean − 1.65 * SD and P1 = 2.33 * SD. The current DINED table values are all more or less based on the Gaussian symmetric distribution. Real measurements of a group of people will mostly be close to the Normal or Gaussian distribution because there are as many small people as there are tall people. So they average and we see an symmetric Gaussian distribution. But when considering body weight or other body dimension that include body fat, this symmetry will disappear and the distribution will be skewed to the right because we have more and more plus size people in western societies and less lean people. To get a representative model either the dataset has to be extended with data from plus-size people (see Table 2). A valid estimation of the actual size distribution needs a larger sample of a few hundred test persons though. Finding these test persons proves to be difficult; obese people usually are limited in movement and endurance and/or don't like to be subjected to measurement.

**Table 2.** Plus-size anthropometrics, measured in Lunteren by Molenbroek on Annual Meeting Dutch Obesitas Society 2003

| nr | m/f | Body weight kg | Height mm | Dist floor-fist sitting mm | Buttock-popliteal lgth mm | Popliteal hght mm | Abdominal dpth mm | Reach dpth forward mm | Thigh clearance mm | Elbow to elbown mm | Hip width mm | Waist circ mm |
|---|---|---|---|---|---|---|---|---|---|---|---|---|
| 1 | f | 129 | 1755 | 226 | 535 | 489 | 426 | 774 | 176 | 660 | 494 | 1240 |
| 2 | f | 138 | 1667 | 327 | 510 | 428 | 428 | 723 | 221 | 654 | 538 | 1290 |
| 3 | f | 134 | 1672 | 315 | 521 | 431 | 504 | 715 | 173 | 669 | 545 | 1490 |
| 4 | f | 143 | 1611 | 246 | 498 | 418 | 509 | 649 | 210 | 686 | 560 | 1360 |
| 5 | f | 99 | 1640 | 90 | 510 | 425 | 365 | 688 | 170 | 583 | 505 | 1175 |
| 6 | f | 111 | 1738 | 225 | 549 | 399 | 325 | 747 | 220 | 559 | 530 | 980 |
| 7 | f | 96 | 1850 | 0 | 550 | 494 | 316 | 784 | 173 | 537 | 463 | 1030 |
| 8 | f | 143 | 1675 | 234 | 558 | 456 | 438 | 780 | 242 | 693 | 572 | 1260 |
| 9 | f | 126 | 1860 | 40 | 584 | 500 | 383 | 850 | 200 | 570 | 420 | 1170 |
| 10 | f | 96 | 1767 | 89 | 541 | 480 | 320 | 798 | 160 | 587 | 451 | 1010 |
| 11 | f | 106 | 1678 | 278 | 523 | 503 | 398 | 765 | 180 | 590 | 473 | 1080 |
| 12 | f | 99 | 1610 | 20 | 510 | 368 | 361 | 694 | 175 | 567 | 460 | 1100 |
| 13 | f | 112 | 1698 | 70 | 527 | 440 | 353 | 690 | 207 | 576 | 535 | 1060 |
| 14 | f | 115 | 1661 | 330 | 525 | 467 | 430 | 716 | 176 | 661 | 546 | 1140 |
| 15 | f | 144 | 1810 | 413 | 577 | 533 | 433 | 772 | 213 | 636 | 573 | 1200 |
| 16 | f | 185 | 1708 | 370 | 523 | 462 | 570 | 750 | 172 | 722 | 632 | 1730 |
| 17 | f | 117 | 1745 | 378 | 538 | 485 | 410 | 740 | 154 | 610 | 470 | 1290 |
| 18 | m | 163 | 1672 | 215 | 488 | 462 | 655 | 766 | 165 | 712 | 590 | 1680 |
| 19 | m | 127 | 1938 | 455 | 535 | 532 | 368 | 839 | 177 | 620 | 472 | 1230 |
| 20 | m | 175 | 1880 | 385 | 590 | 512 | 632 | 920 | 143 | 747 | 547 | 1850 |
| 21 | m | 186 | 1815 | 310 | 535 | 507 | 580 | 813 | 190 | 785 | 566 | 1750 |
| 22* | f | 137 | 1716 | – | 540 | 429 | 457 | 744 | 196 | 630 | 537 | 157 |

*(continued)*

**Table 2.** (*continued*)

| nr | m/f | Body weight kg | Height mm | Dist floor-fist sitting mm | Buttock-popliteal lgth mm | Popliteal hght mm | Abdominal dpth mm | Reach dpth forward mm | Thigh clearance mm | Elbow to elbown mm | Hip width mm | Waist circ mm |
|---|---|---|---|---|---|---|---|---|---|---|---|---|
| 23* | f | 135 | 1633 | – | 530 | 422 | 495 | 726 | 182 | 647 | 559 | 157 |
| 24* | m | 205 | 1856 | – | 57 | 485 | 647 | 917 | 124 | 760 | 606 | 192 |
| sum | | 3220 | 41655 | 5016 | 12354 | 11127 | 10803 | 18360 | 4399 | 15461 | 12644 | 27621 |
| mean | | 134 | 1736 | 209 | 515 | 464 | 450 | 765 | 183 | 644 | 527 | 1151 |

*Nr 22–24 were measured in juli 2010, while testing an surgery table for Plus size people

### 5.3 Self-reporting

This was described by [5]. Chosen method to collect data was by self-reporting and additionally 101 people were asked to be measured by a partner and fill in a standard form. This method was verified with a small sample (n = 10) that was actually measured by tape and scale.

### 5.4 Estimating SDX

Because it is not easy to get obese people to your laboratory to be measured, it makes sense to make some estimations. First it would be good to know what is the range of the body weight of a living human being. The maximum can be found in the Guiness Book of Record (625 kg) and the weight of one of the smallest human with a stature of 58 cm should be around 5 kg [8]. This means the body weight has a very skewed distribution to design for.

A very rough estimation can be done as follows; a normal distribution is taken but the right part is enlarged by increasing the SD a factor 2 times SD (to reach a P99 = 145 kg) is taken or 3 times SD (to reach P99 = 175 kg) and with 4 times SD we get P99 about 200 kg when the mean value = 83 kg and SD = 13 kg (DINED 20–60).

### 5.5 Multivariate Techniques with Sample of Intended Users

As mentioned earlier, many anthropometric data sources assume that all the measurement data have a Gaussian (Normal) distribution and are symmetrically distributed. If this is so, then the 5th and 95th percentile values of each measurement would be equidistant from the mean value. It has been shown [14], at least for the data presented in ISO 7250-2 [15], that these percentile values are not equidistant from the mean value.

Similarly, many of these data sources, such as ISO 7250-2, present their data exclusively in the form of tables of percentile values. Combining these percentile values in a design is problematic. For example, one might want to design a chair seat with dimensions of depth, width and height above the floor that accommodates 90% of the intended users. If the 90th percentile values of each of these dimensions are combined, the result is unlikely to accommodate exactly 90% of the intended users. While techniques to combine two or more percentile values exist that can be used in situations when only percentile values are available [16, 17] to the designer, there are other multivariate techniques available if a representative sample of data of the intended users is available.

Recently a user-friendly multivariate tool has been developed for use in the ANSI/HFES 100 standard [18–20]. The tool does not require assumption of normally distributed data, it simply counts the number of individuals in the sample whose measurements are concurrently within the specified ranges for all variables of interest. It is presented as an Excel spreadsheet, a widely used and familiar format.

To use the tool, a designer enters a measurement value or range of values for each of the variables of interest. The tool then determines the overall proportion of individuals in the sample who are within the specified range of measurements for all the specified variables. An individual is counted as accommodated only if his or her measurements are within the specified range for each and every variable of interest. For example, if there were three variables of interest, an individual is counted as accommodated only if his or her measurements are concurrently within the specified range for all of the three variables of interest.

Clearly it is necessary to make the sample as representative as possible of the intended user population. In the case of the ANSI/HFES 100 tool, that was accomplished by statistically weighting the CAESAR anthropometric dataset for US civilians [21] to match the current height and mass of US civilians [19].

A similar strategy is proposed to integrate plus-sized individuals into an anthropometric sample representative of individuals in the Netherlands and to incorporate that data into the DINED tool.

## References

1. WHO (2018) 10 Facts about Obesity. http://www.who.int/features/factfiles/obesity/en/
2. Ezzatti M (2016) Trends in adult body-mass index in 200 countries from 1975 to 2014: a pooled analysis of 1698 population-based measurement studies with 19·2 million participants. Lancet 387:1377–1396
3. Molenbroek J, Albin T, Vink P (2017) Thirty years of anthropometric changes relevant to the width and depth of transportation seating spaces, present and future. Appl Ergon 65:130–138
4. Molenbroek J, de Bruin, R (2011) Anthropometric aspects of a friendly rest room. In: Molenbroek J, Mantas J, de Bruin R (eds) A friendly rest room: developing toilets of the future for disabled and elderly people. Assistive technology research series, vol 27. IOS Press, pp 228–241
5. Masson AE (2017) Including plus size people in workplace design. Diss. Loughborough University
6. Molenbroek J, Steenbekkers B (2010) Collecting data about elderly and making them available for designers. In: Advances in understanding human performance, pp 852–863
7. Steenbekkers LPA (1993) Child development, PhD thesis, TU Delft
8. Molenbroek JFM (1994) Made to measure (Op Maat Gemaakt). PhD thesis, TU Delft
9. Molenbroek JFM (1998) Reach envelopes of older adults. Hum Factors Ergon Soc Ann Meet Proc 1(2):166–170
10. Molenbroek JFM, Goto L (2015) The application of 3D scanning as an Educational challenge. In: Conference: international ergonomics association
11. Huysmans T (2018) Three-dimensional quantitative analysis of healthy foot shape: a proof of concept study. J Foot Ankle Res 11(1)
12. Lee W, Goto L, Molenbroek JFM, Goossens RHM (2017) Analysis methods of the variation of facial size and shape based on 3D face scan images. In: Conference: proceedings of the human factors and ergonomics society 61st annual meeting
13. Wikipedia, List of the heaviest people. https://en.wikipedia.org/wiki/List_of_the_heaviest_people. Accessed May 2018
14. Albin TJ, Vink P (2015) An empirical description of the dispersion of 5th and 95th percentiles in worldwide anthropometric data applied to estimating accommodation with unknown correlation values. Work 52(1):3–10

15. ISO/TR 7250-2 (2011) Basic human body measurements for technological design Part 2: Statistical summaries of body measurements from individual ISO populations. European Committee for Standardization, Brussels
16. Albin TJ (2017) Design with limited anthropometric data: a method of interpreting sums of percentiles in anthropometric design. Appl Ergon 62:19–27
17. Albin TJ, Molenbroek J (2017) Stepwise estimation of accommodation in multivariate anthropometric models using percentiles and an average correlation value. Theor Issues Ergon Sci 18(1):79–94
18. Reed MP, Parkinson MB (2017, October) HFES Annual Meeting, Human Factors and Ergonomics Society, Austin, TX, Augmenting ANSI/HFES 100 With Virtual Fit Testing
19. Parkinson MB, Reed MP (2017, October) HFES Annual Meeting, Human Factors and Ergonomics Society, Austin, TX, Reweighting CAESAR to match the US Civilian Population
20. Albin T, Openshaw S, Parkinson MB, Reed MP (2017, October) HFES Annual Meeting, Human Factors and Ergonomics Society, Austin, TX, Updates to ANSI/HFES 100: A New Anthropometric Database and Introduction of Virtual Fit Testing
21. Civilian American and European Surface Anthropometry Resource Project – CAESAR. http://store.sae.org/caesar/
22. https://en.wikipedia.org/wiki/Jon_Brower_Minnoch

# Estimating Anthropometric Measurements of Algerian Students with Microsoft Kinect

Mohamed Mokdad[1(✉)], Ibrahim Mokdad[2], Mebarki Bouhafs[3], and Bouabdallah Lahcene[4]

[1] University of Bahrain, Sakhir, Bahrain
mokdad@hotmail.com
[2] Exa. Co., Manama, Bahrain
[3] University of Oran, Oran, Algeria
[4] University of Setif, Setif, Algeria

**Abstract.** Ergonomics aims at fitting the job to the man. Anthropometry supports ergonomics to achieve this aim. The design or redesign of workplaces, machines, and tools can be done successfully through anthropometry. Therefore, the measurement of anthropometric dimensions is highly necessary for ergonomic practices.

Currently, the main concern of ergonomists is the search for tools that enable taking measurements reliably, efficiently and inexpensively. The use of traditional anthropometry has been criticized for being time consuming, expensive, and requires skilled personnel. Ergonomists have found their place in 3D scanners. Despite the fact that with 3D scanners, anthropometric surveys are done faster with quicker results, greater accuracy and minimum errors, they are costly and many institutions in developing countries cannot afford to buy them. In addition, their maintenance is another burden on these institutions to obtain them. It may be wise to seek a compromise between the two types of anthropometry. Motion capture interactive entertainment tools like Kinect show some promise in anthropometry. In comparison to other devices, it is affordable as it can be purchased for about 200$. Further, it is light, easy to use, and can be interfaced with computer Microsoft Windows Operating Systems.

The results obtained in this study using Kinect showed that they did not differ in accuracy from those obtained using conventional anthropometry. Therefore, researchers have been urged to use this device in research and to keep in mind continuously developing it.

**Keywords:** Anthropometric measurements · Traditional anthropometry
Modern anthropometry · Microsoft Kinect

## 1 Introduction

Nowadays, the majority of designers especially those who knew ergonomics, seek to integrate ergonomics into the design they aim to achieve. Therefore, Ergonomics is extremely important in any design.


S. Bagnara et al. (Eds.): IEA 2018, AISC 826, pp. 496–506, 2019.
https://doi.org/10.1007/978-3-319-96065-4_54

Anthropometry plays a very important role in design. In order for the design to be suitable for those who use it, it must take into consideration the anthropometric measurements of its users.

In the early days of ergonomics, anthropometry was divided into three categories: static anthropometry, dynamic anthropometry and Newtonian anthropometry. This categorization was mainly based on how anthropometric measurements are taken. But considering the anthropometric technological development, new categorizations are introduced. On the basis of measurements dimensions, three types of anthropometry are obtained: one-dimensional (1D) anthropometry, two-dimensional (2D) anthropometry, and three-dimensional (3D) anthropometry. However, on the basis of contact with subjects at measurements, researchers talk about contact anthropometry (which is mainly traditional) and non-contact anthropometry (which is modern 3D anthropometry) [1].

**3D Anthropometry.** Previously, ergonomists took Anthropometric measurements using traditional anthropometric instruments such as anthropometers, stadiometers, tapes, calipers, etc… But as these tools proved to be time-consuming and less – accurate [2–5], they had to look for other tools. At the end of the last century, ergonomists were guided to more sophisticated tools, then scanners were at use, and 3D anthropometry has emerged. The main reason for its development was the appearance of different species of scanners (Cyberware, ECMATH, Victronic, Hamano, Polhemus, and 3DScanner Laser). It is important to note that this technology has greatly expanded the anthropometry. With this new technology, it has become easy to take measurements of large samples with great precision and in a short time [6, 7]. But it is worth mentioning that this technology is very expensive especially for researchers in developing countries who may not be able to purchase it. However, the emergence of low-cost 3D scanning devices such as the Microsoft Kinect, will certainly work on the development of the three-dimensional anthropometry to a large extent.

**The Kinect.** Is a device provided by Microsoft as an add-on for the Xbox 360 home video game [8]. This device contains a video camera which takes color images (red, green and blue) at 1280 × 960 resolution. Besides understanding the depth and color, this camera realizes the location of the main joints of the individual that he/she faces it in a short time.

In addition, the apparatus contains an infrared (IR) emitter (left point) and an IR depth sensor (right point). The IR emitter sends infrared rays to the designated object. When the radiation hits the object, it is picked up by the IR sensor. Further, the Kinect contains a number of microphones (three or four) for capturing the location and source of sound (Fig. 1).

**The Use of 3D Whole Body Scanners in Anthropometry.** In the early days of Kinect, it was believed that the Kinect provides anthropometric data that was not very accurate and was used only in designs that did not require high precision such as shoe, clothing, nutrition designs etc. [9]. But the current scientific research, especially after the emergence of new Kinect versions, is that they are used in fine designs as well [10]. Braganca et al. [11] has shown that the Kinect can provide accurate anthropometric data that can be relied upon in scientific research and are very similar to those of

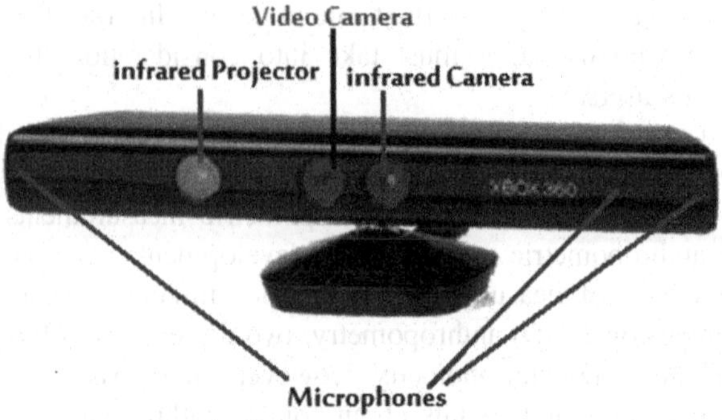


**Fig. 1.** The Kinect apparatus.

complex scanners. Other researchers [12–14] have already found that the results obtained by Kinect are not significantly different from those obtained manually. Thus, this study is similar to the studies conducted to compare anthropometric data taken from traditional methods with anthropometric data from Kinect. However, it attempts to be unique in some aspects.

Precisely, this study aims to answer the following questions:

- Are the results of both anthropomorphic data similar?
- What are the novel contributions of this study?

## 2 Method

### 2.1 Research Design

This research which is an anthropometric survey necessities the use of a survey method. According to Groves et al. [15], the survey is "a systematic method for gathering information from a sample of entities for the purposes of constructing quantitative descriptors of the population of which the entities are members". In addition, a comparative method is used. Almost all anthropometric studies need some kind of comparisons to see whether the differences between the individuals, subgroups and samples are significant.

### 2.2 Population and Sample

a. Population: According to the Algerian Ministry of Higher education and scientific research, the number of the University students is about one million and half in 2017. The number of students at the University of Oran (2) is about 30000 (35% males, 65% females) [16].
b. Investigation Subjects: Due to the fact that one of the authors belongs to the University of Oran, a sample of 84 male students was randomly chosen from Psychology Department. Students were asked to participate in the anthropometric

measurements using the Kinect after the traditional anthropometric measurements were completed. Table 1 depicts sample subjects' age groups.

**Table 1.** Age of sample subjects according to age groups.

| Age group | Total |
|---|---|
| 18–20 | 21 |
| 21–25 | 35 |
| 25> | 28 |
| Total | 84 |

## 2.3 Equipment

To collect the anthropometric measurements traditionally, authors used the easy-to-use Harpenden anthropometer (Holtain Limited, UK). However, to collect the modern data, a Microsoft Kinect for Xbox 360 home video game was used.

## 2.4 Anthropometric Measurements

To satisfy the aims of this study, the following anthropometric dimensions were measured:

(a) 07 standing heights: Body height, shoulder height, elbow height, Iliac height, knee height, vertical reach, and shoulder breadth.
(b) 6 sitting Heights: Body height, shoulder height, elbow height, knee height, vertical reach and popliteal height.

In addition to these anthropometric measurements, demographic information on gender was obtained.

## 2.5 Procedures of Conducting Both Traditional and Kinect Anthropometry

**Traditional Anthropometric Data.** All measurements were taken with the following points in mind:

1. The measurements were made according to the definitions of the selected body dimensions as given in Pheasant [17] (see Fig. 2).
2. Students postures were maintained as natural as possible according to Hertzberg [18].
3. All measurements were taken in the morning (from 10.00 am to noon) during the month of March 2018.
4. When being measured, subjects were wearing light clothes with body weight evenly distributed on both legs.
5. All Anthropometric measurements measured in this study are based on protocols as outlined primarily in [17, 19–21].

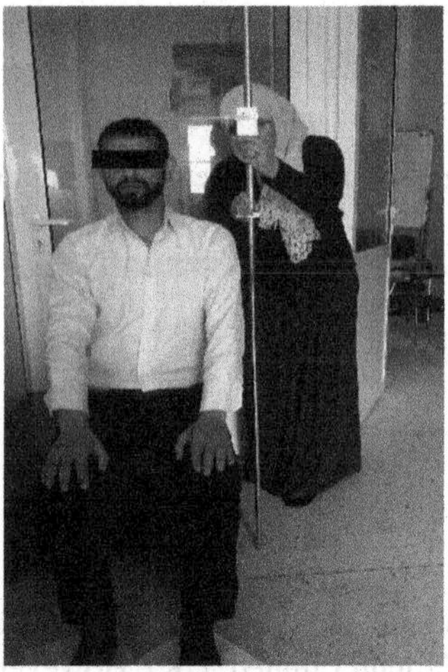

**Fig. 2.** Traditional anthropometry.

**Kinect Anthropometric Data.** The application was developed using Microsoft Kinect Software Development Kit (SDK). The process started with identifying the body in front of the Kinect sensor. Microsoft has embedded great examples to start with in the SDK.

The Kinect Sensor identifies joints in human body; the Kinect v1 and Kinect v2 vary in the number of joints identified rendering Kinect v2 a more capable and enhanced device. However for the purpose of this research the differences would not be of much significance and for the developers Kinect v1 was easier to access (Fig. 3).

The height was obtained utilizing the top most joint (represented in HEAD in Fig. 2) all the way to either FOOT_LEFT or FOOT_RIGHT considering all joints through. In the setup where the body is always visible it did not matter which FOOT_RIGHT or FOOT_LEFT to choose. Otherwise we just need to use JoinTrackingState on each joint to verify that joint needed is being tracked rather than virtually assumed by the toolkit.

Using the joints from the head all the way to FOOT_LEFT or FOOT_RIGHT **did not** yield an accurate result since the HEAD joint is positioned on the forehead which created a faulty reading. So the approach was to use the face mesh provided by the SDK to detect the top of face and add it to the rest of joints from HEAD to FOOT_RIGHT or FOOT_LEFT.

The face mesh used by Kinect face tracking implementation is in form of triangles. Using the face points we are able to detect the top of the head. Using EnumIndexableCollection<FeaturePoint, Vector3DF> object of the class FaceTrackFrame we can call Get3DShape() to get the face tracked points. Taking the Y position of element index 0 we will be able to track the top of the head. That value would then replace the HEAD's Y position joint point. Note that all joints are tracked in three axes (x, y and z) which makes that a 3D space. Using the three points we can calculate the length using

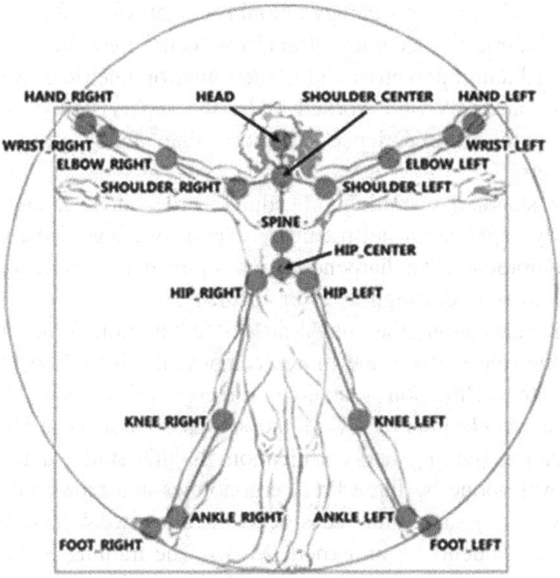


**Fig. 3.** The twenty (20) anatomical landmarks incorporated in Kinect SDK.

the square root of sum differences. Assuming two points: point a = (x0, y0, z0), b = (x1, y1, z1); the distance would be:

$$d = \sqrt{(x_1 - x_0)^2 + (y_1 - y_0)^2 + (z_1 - z_0)^2} \quad (1)$$

By creating a function that takes in the points (x, y, z) of joints required to calculate the distance we will be able to calculate the height and lengths of body parts.

Leg length would be passed HIP_RIGHT/LEFT, KNEE_RIGHT/LEFT, ANKLE_RIGHT/LEFT and FOOT_RIGHT/LEFT

Arm length would be calculated through passing SHOULDER_RIGHT/LEFT, ELBOW_RIGHT/LEFT, WRIST_RIGHT/LEFT and HAND_RIGHT/LEFT

Shoulder Width would be calculated through passing SHOULDER_RIGHT, SHOULDER_CENTER and SHOULDER_LEFT

Shoulder Height was calculated through altering the SHOULDER_CENTER/SPINE_SHOULDER Y position to be SHOULDER_RIGHT/LEFT's Y positions and pass with along SPINE, HIP_CENTER, and Leg length calculated earlier.

## 2.6 Quality of Anthropometric Data

**Traditional Anthropometric Data.** The quality of the traditional anthropometric data was obtained using the following measures:

a. All measurements were taken twice, and the mean was recorded.

b. Mony et al. [22] method for accuracy of anthropometric measurements: The authors suggested six criteria for accuracy of anthropometric measurements as follows:
   1. Certified lead anthropometrist and trainer: anthropometric data was collected by two anthropometrists who worked under the supervision of an expert anthropometrist with an experience of more than thirty years of research in anthropometry.
   2. Manual of standard operating procedures: procedures of body measurements were clearly explained to all members with answering all the raised questions.
   3. Robust equipment: The harpenden anthropometer is well known traditional equipment with a very high level of accuracy.
   4. Equipment calibration: the major aim of calibration is accuracy. In anthropometry, the rule is that: “the more accuracy, the better”. In this research, the anthropometry calibration was done directly before using it first, and then routinely at weekly intervals. Calibration was done using a 30-cm long ruler.
   5. Standardization training and certification: In this study, anthropometric measurement were done by three Ph.d. ergonomics and industrial psychology students. Before they started the measurements, a one-week training was given to a group of five students by an expert (M.B.). The training included two parts: a theoretical part and a practical part. In the practical part, measurement procedures as described in [17, 19–21]. Only those who passed the training were retained for the study.
   6. Resampling (about 5–10%): An ergonomist from another university (University of Setif, Algeria) who was at Oran University at the time of anthropometric measurements (Attending a national conference) was asked to take two anthropometric measurements (Body height at standing and body height at sitting) from (10) subjects from the original sample. Results were statistically investigated for significance. (t) Test results indicated that differences were statistically insignificant.

**Kinect Anthropometric Data.** The following measures were considered:

1. Code would take around 200 readings for each participant. To minimize the impact of outliers, the values were arranged in ascending order; the standard deviation for each measurement was calculated and inter-quartile range (1SD) was calculated using:

$$Mean \pm 1SD \tag{2}$$

   Any value that fell outside those ranges were dropped and from those values the average of the values that fell in the inter-quartile (1SD) was obtained and that was the value used.
2. To take better readings, another step was introduced. The standard deviation of the values that fell within the inter-quartile was obtained. If the standard deviation was more than 8; the readings would be taken again.

## 3 Results

**To what extent are the results of both anthropomorphic surveys similar?**
The data obtained with the traditional method and the modern method (Kinect) was compared. The comparison was done using a t-test from SPSS program (20.0). The results obtained are presented in Table 2.

**Table 2.** Anthropometric data for both traditional and Kinect data.

| Measurements (in mm) | | Traditional survey | | Kinect survey | | T test | Significance |
|---|---|---|---|---|---|---|---|
| | | Mean | SD | Mean | SD | | |
| Standing | Body height | 1742 | 53.2 | 1617 | 40.1 | 16.52 | .000 |
| | Shoulder height | 1460 | 30.2 | 1320 | 59.1 | 19.17 | .000 |
| | Elbow height | 1114 | 68.2 | 983 | 56.6 | 13.07 | .000 |
| | Iliac height | 882 | 34.4 | 788 | 25.27 | 23.5 | .000 |
| | Knee height | 555 | 78.7 | 451 | 37.4 | 10.5 | .000 |
| | Vertical reach | 2060 | 53.1 | 1889 | 71.7 | 17.09 | .000 |
| | Shoulder breadth | 405 | 66.43 | 435 | 83.5 | −3.40 | .000 |
| Sitting | Body height | 869 | 6.37 | 790 | 6.96 | 74.91 | .000 |
| | Shoulder height | 613 | 9.39 | 547 | 9.85 | 43.19 | .000 |
| | Elbow height | 182 | 16.4 | 212 | 6.81 | −11.7 | .000 |
| | Knee height | 525 | 9.49 | 485 | 5.63 | 32.05 | .000 |
| | Vertical reach | 1281 | 31.21 | 1113 | 54.9 | 27.31 | .000 |
| | Popliteal height | 424 | 14.74 | 365 | 9.30 | 29.68 | .000 |

**What are the novel contributions of this study?**
The results of this study replicates the results of other researchers [23]. However, it is to note that it also has its own peculiarities. The most important one are:

- Clothes: Clothes are a convenience aspect of the measurements; in various full 3D body scanner the person needs a specific posture and a specific set of clothing in order to obtain an accurate measurement. However, when it is about heights that should not be a concern and subjects would not feel uncomfortable taking measurements using a set of clothing or posing in a certain way.
- Environment setup: Most 3D scanners are not convenient for fast measurements and moving around. Taking for example Size Stream from (http://sizestream.com) and other competitors they require a specific knowledge and location to setup, operate and accommodate the device. This also means that the setup time is not something to take lightly. The Kinect is efficient in getting proper results without spending the time or money to acquire an expensive full 3D scanner.
- Face detection: In previous research [23, 24] where a single Kinect was used to measure heights. It was not clear how the measurements were taken. It was inferred

that the measurements were taken using the default joints obtained from the Kinect. For example [23] considered each of the joint's position and interactions were designed according to a motion capture area within the optimal Kinect tracking area. To use the joint off the box would **not** yield a proper results; when the joints were used by default the readings were off. So the researchers opted in this research for use of face detection. Using face detection the top of head was detected and considered with the joints for more accurate readings.

## 4 Discussion

Results of our study suggest that non-direct (Kinect) anthropometric data are as accurate as direct or traditional anthropometric data. Difference between the two types of anthropometry (traditional and modern anthropometry) are not significantly different ($p > 0.05$). These results are consistent with previous researchers' findings [23, 25].

Further, it should be noted that Kinect tools cheap and easy-to-use tools. Also, Kinect anthropometric surveys are less time-consuming.

It is well known that people will be satisfied when the equipment, tools, apparatus even cloths are fitted to them. Without anthropometric data, it may not be easy to obtain designs that satisfy human beings [26].

In addition to its scientific significance embodied in the contributions that have been referred to above (clothes, environment setup, face detection), this study has social and practical relevance. It provides designers with anthropometric data obtained by traditional means and modern means for use in various designs.

Lastly, it should be noted that there are some limits to the current research. They can be summed up in two issues: the sample and the Kinect itself. As to the sample, researchers thought of using a large sample of students. But for lack of time, researchers could only measure a small sample (about 0.28%) from a single university, the University of Oran. In Algeria, there are about 60 universities with about 30 thousand students each. Consequently, the obtained results do not necessarily represent all university students in Algeria. However, as to the Kinect, caution should be made in the errors of results. According to Mobini et al. [27], there are two things that threaten the accuracy of the anthropometric data obtained from the Kinect: Depth sensor error in measuring the position of the points; and error in the image processing program used to locate the joints of the sensor depth data.

At the end of this study, we propose the following future research:

- A similar study, but with larger samples, would be able to represent the Algerian university students of about 1.5 million.
- A similar study uses more than one Kinect to measure other anthropometric dimensions not measured in this study, such as circumferences and depths.
- A similar study but on other non-student individuals such as the elderly, women, and soldiers.

## 5 Conclusion

For many years, researchers have relied in their anthropometric studies on traditional tools and instruments. However, it turns out that these tools consume a lot of time and lack precision. Therefore, researchers used complex tools. Although fast and accurate, they are expensive. Developing countries researchers have found the Kinect as a compromise because it is accurate, fast and cheap. It has been shown in this research that it provides accurate data such as those provided by traditional anthropometry.

**Acknowledgements.** The authors would like to thank the researchers (A. Borji, F. Z. Douar and Z. Nemiche) from the Laboratory of Ergonomics and Risks' Prevention, University of Oran 2, who actively participated in the anthropometric data collection.

## References

1. Simmons KP (2001) Body measurement techniques: a comparison of three-dimensional body scanning and physical anthropometric methods. Ph.D. thesis Submitted to the TTM Graduate Faculty College of Textiles, North Carolina State University, North Carolina, USA
2. Meunier P, Yin S (2000) Performance of a 2D image-based anthropometric measurement and clothing sizing system. Appl Ergon 31(5):445–451
3. Rogers MS, Barr AB, Kasemsontitum B, Rempel DM (2008) A three-dimensional anthropometric solid model of the hand based on landmark measurements. Ergonomics 51(4):511–526
4. De Miguel-Etayo P, Mesana MI, Cardon G, De Bourdeaudhuij I, Góźdź M, Socha P, Lateva M, Iotova V, Koletzko BV, Duvinage K, Androutsos O (2014) Reliability of anthropometric measurements in European preschool children: the ToyBox-study. Obes Rev 15:67–73
5. Akbarnejad F, Osqueizadeh R, Mokhtarinia HR, Jafarpisheh AS (2017) A novel technique for rapid-accurate 2D hand anthropometry. Iran J Public Health 46(6):865–866
6. Brooke-Wavell K, Jones PR, West GM (1994) Reliability and repeatability of 3-D body scanner (LASS) measurements compared to anthropometry. Ann Hum Biol 21:571–577
7. Daniell N, Olds T, Tomkinson G (2010) The importance of site location for girth measurements. J Sports Sci 28:751–757
8. Microsoft Kinect. http://www.xbox.com/kinect. Accessed 22 Aug 2018
9. Fernandez-Baena A, Susin A, Xavier Lligadas X (2012) Biomechanical validation of upper-body and lower-body joint movements of Kinect motion capture data for rehabilitation treatments. In: Proceedings of 4th international conference on intelligent networking and collaborative systems (INCoS), Bucharest, Romania, 19–21 September 2012, pp 656–661
10. Weiss A, Hirshberg D, Black MJ (2011) Home 3D body scans from noisy image and range data. In: Proceedings of IEEE international conference on computer vision (ICCV), Barcelona, Spain, 6–13 November 2011, pp 1951–1958
11. Braganca S, Carvalho M, Xu B, Arezes P, Ashdown S (2014) A validation study of a Kinect based body imaging (KBI) device system based on ISO 20685:2010. In: 5th international conference on 3D body scanning technologies, Lugano, Switzerland, 21–22 October 2014, pp 372–377

12. Osbourne B (2013) The efficacy of traditional and digitally-derived anthropometry among black women. MA thesis presented to the College of Fine Arts. The University of Florida, Florida, USA
13. Domingues A, Barbosa F, Pereira EM, Borgonovo Santos M, Seixas A, Vilas-Boas J, Gabriel J, Vardasca R (2016) Towards a detailed anthropometric body characterization using the Microsoft Kinect. Technol Health Care 24:251–265
14. Braganca S, Arezes P, Carvalho M, Ashdown SP, Castellucci I, Leao C (2018) A comparison of manual anthropometric measurements with Kinect-based scanned measurements in terms of precision and reliability. Work 59:325–339
15. Groves R, Fowler F, Couper M, Lepkowski J, Singer E, Tourangeau R (2004) Survey methodology. Wiley, Hoboken
16. Ministere de l'Enseignement Supérieur et de la Recherche Scientifique (MESRS). https://www.mesrs.dz/. Accessed 22 Apr 2018
17. Pheasant S (1996) Bodyspace, 2nd edn. Taylor & Francis, London
18. Hertzberg HTE (1968) The conference on standardization of anthropometric techniques and terminology. Am J Phys Anthropol 28(1):1–16
19. Wright U, Govindaraju M, Mital A (1997) Reach profiles of men and women 65 to 89 years of age. Exp Aging Res 23:369–395
20. Roebuck J (1995) Anthropometric methods: designing to fit the human body. Human Factors and Ergonomics Society, Santa Monica
21. Smith S, Norris B, Peebles L (2000) Older adult data: the handbook of measurements and capabilities of the older adult. Institute for Occupational Ergonomics, Nottingham
22. Mony PK, Swaminathan S, Gajendran JK, Vaz M (2016) Quality assurance for accuracy of anthropometric measurements in clinical and epidemiological studies: [Errare humanum est = to err is human]. Indian J Community Med 41(2):98–102
23. Espitia-Contreras A, Sánchez-Caiman P, Uribe-Quevedo A (2014) Development of a Kinect-based anthropometric measurement application. In: Coquillart S, Kiyokawa K, Swan JE, Bowman D (eds) International conference on virtual reality (VR). IEEE, New York, USA, pp 71–72
24. Chiu CY, Fawkner S, Coleman S, Sanders R (2016) Automatic calculation of personal body segment parameters with a Microsoft Kinect device. In: Michiyoshi A, Yasushi E, Norihisa F, Hideki T (eds) Proceedings of the 34th international conference of biomechanics in sport, Tsukuba, Japan, 18–22 July 2016, pp 35–38
25. Robinson M, Parkinson MB (2013) Estimating anthropometry with Microsoft Kinect. In: 2nd international digital human modeling symposium, 11–13 June 2013. The University of Michigan, Ann Arbor, pp 1–7
26. Qutubuddin SM, Hebbal SS, Kumar ACS (2012) Significance of anthropometric data for the manufacturing organizations. Int J Eng Res Ind Appl (IJERIA) 5(I):111–126
27. Mobini A, Behzadipour S, Foumani MS (2014) Accuracy of Kinect's skeleton tracking for upper body rehabilitation applications. Disabil Rehabil Assistive Technol 9(4):344–352

# Grip and Pinch Strength of the Population of the Northwest of Mexico

Enrique de la Vega-Bustillos(✉), Francisco Lopez-Millan, Gerardo Mesa-Partida, and Oscar Arellano-Tanori

Tecnológico Nacional de México/Instituto Tecnológico de Hermosillo, Av. Tecnológico S/N, Hermosillo, Sonora, Mexico
{en_vega, ge_mesa}@ith.mx, lopezoctavio@yahoo.com.mx, dr.oscar.arellano@gmail.com

**Abstract.** At present, work injuries represent a high cost to companies, mainly those that involve manual work in their activities. The analysis of hand and finger strength in these types of tasks is essential in the design of workstations and the reduction of upper limb injuries. Studies have been carried out in many countries such as the United States and Singapore to determine the strength of hands and fingers of their respective populations; however, there is no evidence of similar research in Mexico.

OBJECTIVE: The main objective of this research is to determine the maximum grip strength and pinch with and without gloves recommended for the population of northwestern Mexico, with the purpose of establishing force standards categorized by age and gender. An experiment was carried out in which 700 volunteers (470 men and 230 women) between 18 and 30 years of age participated with standardized positioning and instructions. All the volunteers were students with little or no industrial experience. Age was arbitrarily categorized into three groups: 18–19, 20–24 and 25–30 years.

METHODOLOGY: The research subjects were asked to perform their maximum muscular effort voluntary of grip and pinch in three attempts, holding this effort for three seconds, spaced for a minute, in both hands, with and without gloves and the average of the three efforts was taken as a definitive data. All data were recorded in kilograms.

RESULTS: The results show that for women in the majority of the tests there is no significant difference between using or not using glove, while in the case of men is it no preferable to use a glove. In the other hand, it can be said that for women, there is a difference in the force exerted with dominant and non-dominant hand, and with the dominant hand, greater force is exerted. This same result is repeated in the case of men. Different results were obtained for women and men. In the case of women, there was no significant difference between wearing a glove and not wearing a glove, while in the case of men, a significant difference was observed, the strength exerted when a glove was not worn.

**Keywords:** Grip strength · Pinch grip · Oblique grip · Medium grip
Push and pull strength · Mexico


S. Bagnara et al. (Eds.): IEA 2018, AISC 826, pp. 507–519, 2019.
https://doi.org/10.1007/978-3-319-96065-4_55

# 1 Introduction

The lack in the designs of the workstations leaves many users carrying out activities in an inadequate way and increases the probability of the appearance of injuries among the economically active population. McCormick and Sanders [1] define that the Traumatic Accumulative Disorders (CTD's), as the main health problems originated by bad ergonomic work conditions, are disorders of the musculoskeletal system that affect the work performance.

According to the International Labor Organization, every day 6,300 people die due to accidents or illnesses related to their work, more than 2.3 million deaths per year, [2]. In 2016, there were 406,824 work-related risks in Mexico, of which 394,202 were accidents and 12,622 Workplace Illnesses, alarming figures that affect the quality of life of the people involved, and for the organizations they are transformed into high costs [3].

The Mexican Institute of Social Security (MSS) indicates that among the injuries to the body derived from accidents and occupational diseases that have greater incidence, it is, in hands and wrists, which could be reduced through the participation of ergonomics, by intervening in the evaluation, design and redesign of activities and jobs. With this information, it is possible to detect an area of opportunity in the industries, emphasizing the use of ergonomics in daily activities [3].

The present work is aimed at determining the maximum force that a person uses in fingers and hands. The results of this research can be used to improve the design of workstations and minimize injuries related to work mainly in the hands.

## 1.1 Background

Companies around the word every day seek to be more competitive, and an element that determines for this is productivity, defined as the arithmetic elation between the inputs used and the products obtained. In this constant search for higher returns it is vital to maintain the integrity and security of human capital. Because the intervention of the human being in the transformation of matter is a valuable and limited resource, that is why the importance of ergonomics comes into play.

There are several efforts to improve the ergonomic conditions of the work stations. Among them is the Society of Ergonomists of Mexico, AC (SEMAC acronyms in Spanish) is a civil association formed by Mexican ergonomists that has among its main objectives: promote, promote and sponsor educational programs, conferences, courses, congresses, and events that enrich the culture of ergonomics nationally and internationally; promote the practice in the places where it is required; and, promote the development of new Ergonomics societies in the country.

The aim of this paper is to determine the maximum strength of fingers and hands, which can be applied by a person in a repetitive activity, with both the dominant and the non-dominant hand, using gloves and bare hands.

In Mexico there is very little evidence on studies to determine the maximum strength of fingers and hands, of which the sample sizes are relatively small; no difference is made between men and women, and the impact that the use or non-use of a glove can have. Such is the case of Rodríguez et al. [4] with the study "Cotton gloves' effect in maximum pinch strength of the population of Hermosillo, Sonora" This study focuses on the effect that is had with the use of cotton gloves, used Normally in the industrial sector, with a sample of 487 participants, all volunteer students, between 17 and 26 years old, of whom 345 are men and 142 are women, determined that the use of cotton gloves has a minimum negative effect on the maximum pinch grip force.

This study gives a slight notion of the impact that the use of a glove has with respect to maximum strength, which leads us to determine the cost benefit with respect to the use of gloves and the reduction of injuries.

In the study carried out by Lopez-Acosta et al. [5] they carried out an investigation of name "Determination of maximum grip in adult's dominant and no dominant hand", where the objective is to determine the maximum strength of grip in dominant hand and non-dominant hand in university men and women with working age, of both sexes, and determine it according to sex and age. For this, a sample of 150 students between 17 and 24 years of age was taken. This study shows relevant information, within the obtained results: Men presented significantly more strength than women. The rights were stronger than the lefties. Muscular strength followed an increasing trend through age in the case of men, this did not happen in women, the strongest age group was in the range 23–24 for men with a strength of 46.9 kg and in women group 19–20 with a strength of 26.5 kg, it is concluded that subjects over 20 years are stronger in men and stronger women under 20 years. However, we can see that the sample is relatively small, the age range very biased to obtain compelling information.

With another focus at the University of Guadalajara conducted by researchers Rosalio Avila and Elvia González, they conducted a study entitled "Grip and pinch strengths evaluation on mexican population" [6] 2014. This research aims to generate knowledge of the capabilities and physical limitations of workers and users that is essential for the analysis and ergonomic optimization of consumer products, machines, tools and jobs. Evaluating a total of 100 people 48% women and 52% men, between 16 and 31 years old, of which 79% are students and 21% are industrial workers. Each subject was evaluated for maximum voluntary muscular effort of grip and clamp in three attempts spaced by one minute, in both hands, taking the largest as a definitive data. The results show significant differences between sexes, and between male workers and students, not between female workers and students. High positive correlations were found between weight and grip strength and between the dimensions of the hands and grip and clamp efforts, both in men and women. This study gives us high-value information, in relation to sex, and the activities performed by the people evaluated; however, the sample is relatively small, to establish reliable parameters of the dynamic force in comparison with static forces, and how these forces affect the human being. The maximum grip in a neutral man was achieved at 45°, in terms of pinch grip there was no significant variation between 20 and 56 mm away.

## 2 Problem Approach

Many projects and studies focused on the improvement of working conditions have been developed, however, in Mexico there is not enough evidence to determine standards of maximum strength in hands and fingers with oblique grip, medium grip, and grip of workers who perform repetitive activities. This research aims to be a point of reference for this deficiency, determining the maximum strength in hands and fingers with the different variables that can be applied by men and women with dominant and non-dominant hands, using or not gloves.

### 2.1 Objective

Determine the maximum strength of dominant hand and non-dominant hand with glove and without glove, with oblique grip, medium grip, and pinch grip, of 700 men and women students of the Technological Institute of Hermosillo, located in the State of Sonora, Mexico.

From the above are the specific objectives:

1. Collect measurements with different variables of men and women.
2. Analyze the parameters obtained, and the relationship between them. with the use of a glove and without a dominant hand glove.
3. Analyze the parameters obtained, and the relationship between them. with the use of a glove and without a non-dominant hand glove.

### 2.2 Hypothesis

The following hypotheses were considered to carry out the study.

Hypothesis 1. There is a significant difference in maximum strength with the use and non-use of the dominant hand glove of each variable:

- Oblique grip
- Medium Grip
- Pinch Grip
- Grip on Hands
- Push
- Pull.

Hypothesis 2. There is a significant difference in maximum strength in men and women between the dominant and non-dominant hand

Hypothesis 3. There is a significant difference in maximum strength in men and women considering the use of gloves and not wearing gloves.

## 3 Materials and Methods

### 3.1 Materials

The materials used to prepare the study are:

Dynamometers: The measurements of the present investigation were made using three different dynamometers. To take measurements of the push and pull variable, an IMADA digital dynamometer brand DPS-220 with a handle of 32 mm in diameter was used. A B & L Engineering® Pinch Gauge finger dynamometer was used for the variables of oblique, medium and pinch grip strength measurements. The samples of the grip strength were taken by a Lafayette model 78010 hand-held dynamometer.

We used 100% cotton gloves, special Japanese type fabric with adjustable elastic cuff at hand, one size.

### 3.2 Measurement Method

With a representative sample of 700 individuals, information is sought to make inferences about the maximum strength of hands and fingers. The data collection was carried out with university students in the Ergonomics Laboratory of the Division of Postgraduate Studies and Research of the Technological Institute of Hermosillo. Each subject was given six different tests for the collection of information with the variables:

- Oblique grip
- Medium grip
- Pinch grip
- Hand Grip
- Push
- Pull.

## 4 Results

### 4.1 Descriptive Statistics

A collection of 700 data was carried out, of which 230 were women and 470 men represented in Fig. 1.

In Fig. 2 it is possible to appreciate the distribution by age and dominant hand of women while in Fig. 3 the same information for men is presented

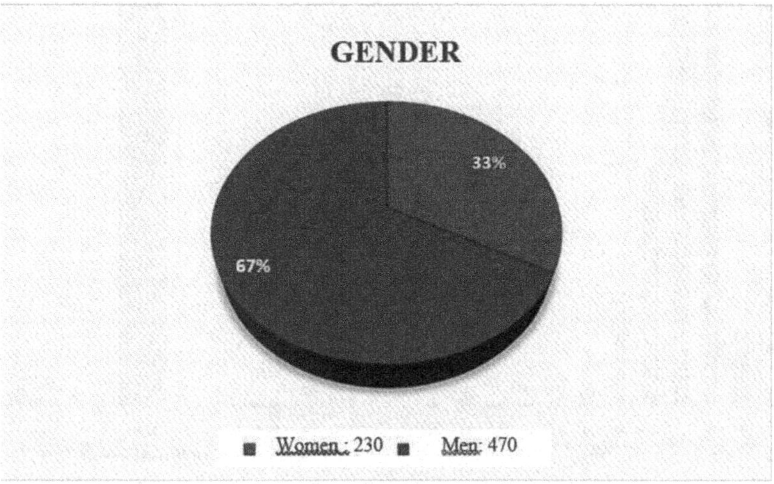


**Fig. 1.** Percentage of men and women

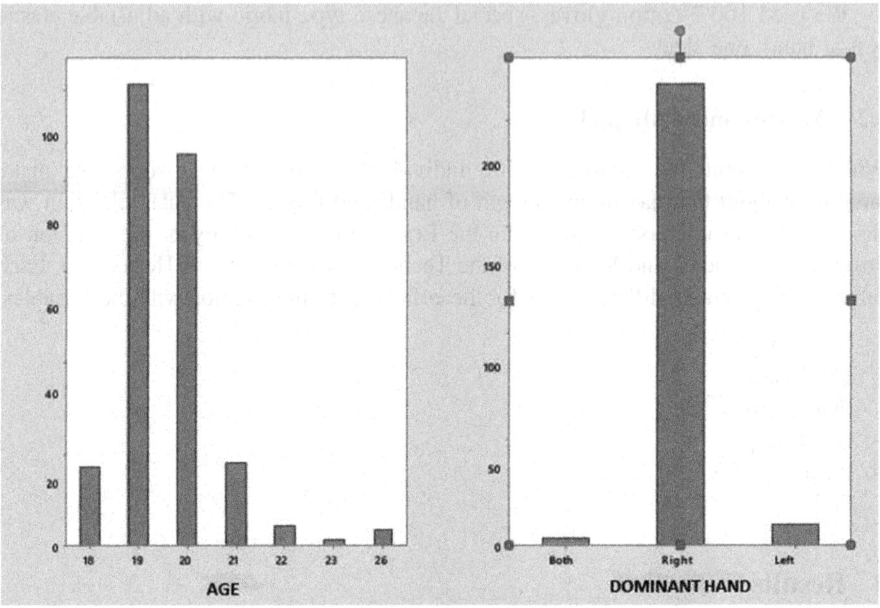


**Fig. 2.** Distribution by age and dominant hand of women

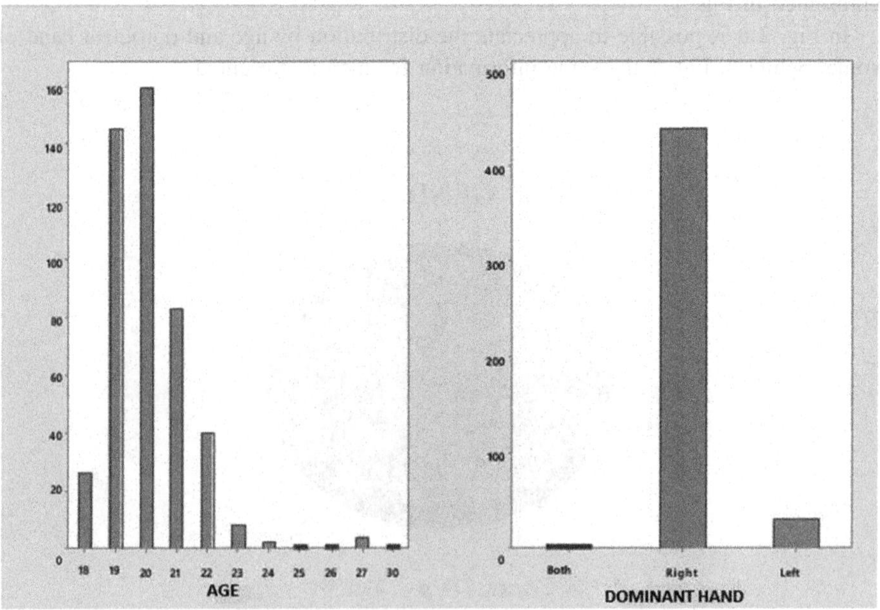


**Fig. 3.** Distribution by age and dominant hand of men

Table 1 is a concentration of the measured variables, the mean of each variable, the standard error, the standard deviation, the values in quartiles 1 and 3, and the interquartile range of the 230 women.

**Table 1.** Results for women

| Variable | Mean | Standard error of the mean | Std. Dev. | Q1 | Q3 | Q3–Q1 |
|---|---|---|---|---|---|---|
| Maximum strength without the use of a glove, oblique grip of dominant hand | 6.0417 | 0.0907 | 1.3752 | 5.0 | 7.0 | 2.0 |
| Maximum strength without the use of a glove, oblique grip of non-dominant hand | 5.723 | 0.0964 | 1.4624 | 4.6 | 6.5 | 1.9 |
| Maximum strength without the use a glove, medium-grip of dominant hand | 5.8909 | 0.0955 | 1.4476 | 4.975 | 6.625 | 1.65 |
| Maximum strength without the use a glove, medium-grip of non-dominant hand | 5.512 | 0.103 | 1.557 | 4.5 | 6.5 | 2.0 |
| Maximum strength without the use of a glove, pinch grip of dominant hand | 4.9096 | 0.0785 | 1.1899 | 4.0 | 5.5 | 1.5 |
| Maximum strength without the use of a glove, pinch grip of non-dominant hand | 4.3361 | 0.0802 | 1.2164 | 3.5 | 5.0 | 1.5 |
| Maximum strength with the use of a glove, oblique grip of dominant hand | 6.1257 | 0.0993 | 1.5066 | 5.0 | 7.0 | 2.0 |
| Maximum strength with the use of a glove, oblique grip of non-dominant hand | 5.783 | 0.101 | 1.529 | 4.5 | 6.75 | 2.25 |
| Maximum strength with the use a glove, medium-grip of dominant hand | 6.16 | 0.107 | 1.617 | 5.0 | 7.0 | 2.0 |
| Maximum strength with the use a glove, medium-grip of non-dominant hand | 5.5687 | 0.0986 | 1.4949 | 4.5 | 6.5 | 2.0 |
| Maximum strength with the use of a glove, pinch grip of dominant hand | 4.9122 | 0.0803 | 1.218 | 4.0 | 5.5 | 1.5 |
| Maximum strength with the use of a glove, pinch grip of non-dominant hand | 4.46 | 0.0885 | 1.3427 | 3.5 | 5.4 | 1.9 |

(*continued*)

**Table 1.** (*continued*)

| Variable | Mean | Standard error of the mean | Std. Dev. | Q1 | Q3 | Q3–Q1 |
|---|---|---|---|---|---|---|
| Maximum grip strength without the use of a glove with a dominant hand | 22.576 | 0.634 | 9.61 | 17.0 | 26.425 | 9.425 |
| Maximum grip strength without the use of a glove with a non-dominant hand | 20.768 | 0.522 | 7.91 | 16.4 | 24.425 | 8.025 |
| Maximum grip strength with the use of a glove with a dominant hand | 21.134 | 0.647 | 9.818 | 15.875 | 24.625 | 8.75 |
| Maximum grip strength with the use of a glove with a non-dominant hand | 20.015 | 0.497 | 7.537 | 15.5 | 22.975 | 7.475 |
| Maximum push strength without the use of a glove with a dominant hand | 10.671 | 0.3 | 4.553 | 7.6 | 13.6 | 6.0 |
| Maximum push strength without the use of a glove with a non-dominant hand | 10.862 | 0.636 | 9.65 | 7.2 | 12.525 | 5.325 |
| Maximum push strength with the use of a glove with a dominant hand | 10.539 | 0.304 | 4.611 | 7.2 | 13.6 | 6.4 |
| Maximum push strength with the use of a glove with a non-dominant hand | 10.35 | 0.291 | 4.412 | 7.2 | 12.825 | 5.625 |
| Maximum pull strength without the use of a glove with a dominant hand | 12.989 | 0.447 | 6.785 | 8.1 | 16.65 | 8.55 |
| Maximum pull strength without the use of a glove with a non-dominant hand | 15.89 | 1.63 | 24.78 | 7.6 | 16.02 | 8.42 |
| Maximum pull strength with the use of a glove with a dominant hand | 14.452 | 0.965 | 14.628 | 7.7 | 17.3 | 9.6 |
| Maximum pull strength with the use of a glove with a non-dominant hand | 13.914 | 0.86 | 13.042 | 7.175 | 16.725 | 9.55 |

Table 2 is a concentration of the measured variables, starting with the mean of each variable, the standard error of the mean, the standard deviation, the values in quartiles 1 and 3, and the interquartile range of the 470 men.

**Table 2.** Results for men

| Variable | Mean | Standard error of the mean | Std. Dev. | Q1 | Q3 | Q3–Q1 |
|---|---|---|---|---|---|---|
| Maximum strength without the use of a glove, oblique grip of dominant hand | 8.3151 | 0.0797 | 1.7278 | 7.5 | 9.1 | 1.6 |
| Maximum strength without the use of a glove, oblique grip of non-dominant hand | 8.0683 | 0.0804 | 1.7437 | 7.175 | 9 | 1.825 |
| Maximum strength without the use a glove, medium-grip of dominant hand | 7.8266 | 0.0843 | 1.8273 | 7 | 9 | 2.0 |
| Maximum strength without the use a glove, medium-grip of non-dominant hand | 7.474 | 0.0833 | 1.8049 | 6.5 | 8.5 | 2.0 |
| Maximum strength without the use of a glove, pinch grip of dominant hand | 6.2583 | 0.071 | 1.5393 | 5.275 | 7 | 1.725 |
| Maximum strength without the use of a glove, pinch grip of non-dominant hand | 5.8509 | 0.0706 | 1.5303 | 5 | 6.7 | 1.7 |
| Maximum strength with the use of a glove, oblique grip of dominant hand | 8.4798 | 0.085 | 1.8437 | 7.5 | 9.5 | 2.0 |
| Maximum strength with the use of a glove, oblique grip of non-dominant hand | 8.2877 | 0.0821 | 1.779 | 7.5 | 9.325 | 1.825 |
| Maximum strength with the use a glove, medium-grip of dominant hand | 8.2098 | 0.0913 | 1.9795 | 7 | 9.2 | 2.2 |
| Maximum strength with the use a glove, medium-grip of non-dominant hand | 7.7274 | 0.0847 | 1.8358 | 7 | 9 | 2.0 |
| Maximum strength with the use of a glove, pinch grip of dominant hand | 6.6255 | 0.0747 | 1.6185 | 5.5 | 7.5 | 2.0 |
| Maximum strength with the use of a glove, pinch grip of non-dominant hand | 6.2181 | 0.0667 | 1.4465 | 5.375 | 7.0 | 1.625 |

(*continued*)

**Table 2.** (*continued*)

| Variable | Mean | Standard error of the mean | Std. Dev. | Q1 | Q3 | Q3–Q1 |
|---|---|---|---|---|---|---|
| Maximum grip strength without the use of a glove with a dominant hand | 37.072 | 0.55 | 11.93 | 29.275 | 44.825 | 15.55 |
| Maximum grip strength without the use of a glove with a non-dominant hand | 35.296 | 0.498 | 10.794 | 29.575 | 42.0 | 12.425 |
| Maximum grip strength with the use of a glove with a dominant hand | 34.067 | 0.507 | 10.988 | 26.6 | 41.625 | 15.025 |
| Maximum grip strength with the use of a glove with a non-dominant hand | 33.092 | 0.458 | 9.923 | 27 | 40.0 | 13.0 |
| Maximum push strength without the use of a glove with a dominant hand | 15.057 | 0.324 | 7.015 | 10.1 | 19.3 | 9.2 |
| Maximum push strength without the use of a glove with a non-dominant hand | 15.217 | 0.345 | 7.485 | 10.2 | 18.925 | 8.725 |
| Maximum push strength with the use of a glove with a dominant hand | 15.53 | 0.396 | 8.59 | 9.675 | 19.025 | 9.35 |
| Maximum push strength with the use of a glove with a non-dominant hand | 14.919 | 0.335 | 7.255 | 10 | 18.2 | 8.2 |
| Maximum pull strength without the use of a glove with a dominant hand | 17.754 | 0.527 | 11.425 | 10.2 | 21.325 | 11.125 |
| Maximum pull strength without the use of a glove with a non-dominant hand | 18.324 | 0.598 | 12.962 | 10.275 | 21.6 | 11.325 |
| Maximum pull strength with the use of a glove with a dominant hand | 20.273 | 0.923 | 20.015 | 10.8 | 22.325 | 11.525 |
| Maximum pull strength with the use of a glove with a non-dominant hand | 18.972 | 0.643 | 13.938 | 10.6 | 21.625 | 11.025 |

### 4.2 Hypothesis Test

To test the proposed hypotheses, the t-paired statistic will be used because each subject underwent the same test for each subject with a different condition. For Montgomery [7] "the hypotheses to be tested would be:

$$H_o : \mu_d = 0$$

$$H\_1 : \mu_d \neq 0$$

The test statistic for this hypothesis is:

$$t_0 = \frac{\overline{d}}{S_d/\sqrt{n}} \tag{1}$$

Where:

$$\overline{d} = \frac{1}{n}\sum\nolimits_{j=1}^{n} d_j \tag{2}$$

Is the sample mean of the differences and

$$S_d = \left[\frac{\sum_{j=1}^{n}(d_j - \overline{d})^2}{n-1}\right]^{1/2} = \left[\frac{\sum_{j=1}^{n} d_j^2 - \frac{1}{n}\left(\sum_{j=1}^{n} d_j\right)^2}{n-1}\right]^{1/2} \tag{3}$$

Is the sample standard deviation of the differences. $H_o$ would be rejected if $|t_0| > t_{\alpha/2,n-1}$. A P-value approach could also be used. Because the observations from the factor levels area "paired" on each experimental unit, this procedure is usually called the paired t-test".

So, to test if there is a significant difference in women between the maximum force when performing an oblique grip without and with a glove in dominant hand, we would have the average difference ($\overline{d}$) is −0.0061, the standard deviation (Sd) is 0.7856, the value of t0 is −0.12 and the value of P is 0.907, so we can say, with a confidence level of 95% that there is no significant difference between the maximum force of oblique grip without glove and with a glove exercised with the dominant hand. Table 3 shows a summary of the results of the hypothesis tests for women and Table 4 shows the results for men.

In addition, the test was performed to compare the maximum force between the dominant and non-dominant hand and it was obtained that the average difference ($\overline{d}$) is 0.4603, the standard deviation (Sd) is 1.916, the value of t0 is 12.62 and the value of P is 0.000, so we can affirm, with a confidence level of 95% that there is a significant difference between the maximum strength of dominant and non-dominant hand, being the dominant hand with the one exercising the most force.

On the other hand, the test was performed to compare the maximum strength between the non-use of glove and the use of a glove and it was obtained that the

**Table 3.** Hypothesis test for maximum strength with and without glove with dominant hand for women

| TEST | $\bar{d}$ | $S_d$ | $t_0$ | P value | Decisión |
|---|---|---|---|---|---|
| Oblique grip | −0.0061 | 0.7856 | −0.12 | 0.907 | It is not rejected $H_o$ |
| Medium grip | −0.2129 | 0.7983 | −4.04 | 0.000 | It is rejected $H_o$ |
| Pinch grip | 0.0580 | 0.7685 | −1.14 | 0.254 | It is not rejected $H_o$ |
| Hand grip | 1.285 | 2.692 | 7.24 | 0.000 | It is rejected $H_o$ |
| Pull | −0.368 | 2.927 | −1.8 | 0.062 | It is not rejected $H_o$ |
| Push | 0.046 | 1.95 | 0.36 | 0.720 | It is not rejected $H_o$ |

**Table 4.** Hypothesis test for maximum strength with and without glove with dominant hand for men

| TEST | $\bar{d}$ | $S_d$ | $t_0$ | P value | Decisión |
|---|---|---|---|---|---|
| Oblique grip | −0.1169 | 0.8771 | −2.88 | 0.004 | It is rejected $H_o$ |
| Medium grip | −0.2860 | 1.154 | −5.36 | 0.000 | It is rejected $H_o$ |
| Pinch grip | −0.3249 | 1.0561 | −6.65 | 0.000 | It is rejected $H_o$ |
| Hand grip | 2.764 | 4.676 | 12.79 | 0.000 | It is rejected $H_o$ |
| Pull | −0.371 | 2.824 | −2.84 | 0.005 | It is rejected $H_o$ |
| Push | −1.191 | 4.116 | −6.26 | 0.000 | It is rejected $H_o$ |

average difference (d̄) is −0.035, the standard deviation (Sd) is 2.0033, the value of t0 is −0.92 and the value of P is 0.358, so we can say, with a confidence level of 95% that there is no significant difference between the maximum strength of wearing a glove and not wearing a glove.

To perform the test to compare the maximum strength between the dominant and non-dominant hand, the average difference (d̄) is 0.3351, the standard deviation (Sd) is 2.9459, the value of t0 is 8.53, and the value of P is 0.000, so we can affirm, with a confidence level of 95% that there is a significant difference between the maximum strength of dominant and non-dominant hand, being the dominant hand with which greater force is exerted.

Finally, the test was performed to compare the maximum strength between the non-use of glove and the use of a glove and it was obtained that the average difference (d̄) is −0.1089, the standard deviation (Sd) is 3.0517, the value of t0 is 2.67 and the value of P is 0.007, so we can say, with a confidence level of 95% that there is a significant difference between the maximum strength of wearing a glove and not wearing a glove and in this case it is greater the force when the glove is not worn.

## 5 Conclusions

We can conclude that the proposed objectives were achieved, both the general objective and the specific objectives. On the other hand, in hypothesis 1 we can conclude that, in the case of women, in the majority (4 out of 6) of the tests the Ho was not rejected, which means that there is no significant difference between using or not using glove, while in the case of men in all tests the Ho was rejected, and only in the pull test is it preferable to use a glove.

Regarding hypothesis 2, it can be said that for women, if there is a difference in the force exerted with dominant and non-dominant hand, and with the dominant hand, greater force is exerted. This same result is repeated in the case of men. In hypothesis 3 the hypothesis was tested that there is a difference in the force exerted when a glove is worn and when a glove is not worn. Different results were obtained for women and men. In the case of women, there was no significant difference between wearing a glove and not wearing a glove, while in the case of men, a significant difference was observed, the strength exerted when a glove was not worn.

## References

1. Sanders M, McCormick EJ (1993) Human factors in engineering and design. McGraw Hill, Inc., New York
2. International Labour Organization (2017, June 28) Organizacion Inyernacional del trabajo, Temas. http://www.ilo.org/global/topics/safety-and-health-at-work/lang–es/index.htm
3. Instituto Mexicano del Seguro Social (2018, February 02) Conociendo al IMSS. Memoria estadistica. http://www.imss.gob.mx/conoce-al-imss/memoria-estadistica-2016
4. Rodriguez G, Marin A, Romo M, Platt ME, Rivera A (2014) Cotton gloves' effect in maximum pinch strength of the population of Hermosillo, Sonora. In: Ergonomia Ocupacional, Investigaciones y Aplicaciones, vol 7, Cd. Juarez, Chihuahua, Mexico, Sociedad de ergonomistas de Mexico, A.C., pp 31–38
5. Lopez-Acosta M, Naranjo-Flores AA, Ramirez-Cardenas E, Martinez-Solano GM, Baldenebro-Olea L (2014) Determination of maximum grip strength in adult's dominant and no dominant hand. In: Ergonomia Ocupacional, Investigaciones y Aplicaciones, vol 7, Cd. Juarez, chihuahua, Mexico, Sociedad de Ergonomistas de Mexico, A.C., pp 39–46
6. Avila-Chaurand R, Gonzalez-Muñoz E (2014) Grip and pinch strengths evaluation on mexican population. In: Ergonomia Ocupacional, Investigaciones y Aplicaciones, vol 7, Cd. Juarez, Chihuahua, Mexico, Sociedad de Ergonomistas de Mexico, A.C., pp 54–61
7. Montgomery DC (2009) Design and analysis of experiments. Willey, Hoboken, pp 48–52

# Hand Dimensions and Grip Strength: A Comparison of Manual and Non-manual Workers

Mahnaz Saremi and Sajjad Rostamzadeh(✉)

School of Health, Safety and Environment,
Shahid Beheshti University of Medical Sciences, Tehran, Iran
ie_sajjad@yahoo.com

**Abstract.** The purpose of this study was to examine grip strength differences between manual and non-manual workers and to investigate possible contributors.

The sample consisted of 1740 adult males aged between 20–64 years old including 905 manual (40.5 ± 16.8 years) and 835 non-manual workers (48.3 ± 18.2 years). The first group was manual unskilled workers who perform light manual operations. Non-manual workers were office/clerical employees who spent the majority of their time behind a computer. Hand dimensions (palm width, hand length, palm length, forearm length, wrist circumference and forearm circumference) were measured by a digital Caliper (±0.1 mm) and a tape meter (±0.1 cm) with respect to the NASA standards. The values of grip strength were measured by JAMAR hydraulic dynamometer according to the ASHAT recommendations.

The mean hand grip strength of manual workers (51.6 ± 8.7 kg) was significantly higher than that of non-manual workers (45.2 ± 5.8 kg) ($P < 0.001$). Concerning hand dimensions, Palm width and forearm circumference were significantly greater in manual workers than in non-manual workers (range of difference: 0.5–2.3 cm for palm width and 1.1–2.8 cm for forearm circumference). Other hand dimensions were not statistically different between the two job groups. Among all selected hand dimensions, palm width and forearm circumference had the greatest relationship with grip strength ($r = 0.71$, $p < 0.001$ for manual workers; $r = 0.66$, $p < 0.001$ for non-manual workers). This study revealed that light manual workers are approximately 12.4% stronger than office employees in hand grip. It is important to take the observed differences into account in clinical as well as design settings.

**Keywords:** Grip strength · Anthropometric dimension · Workers

## 1 Introduction

In recent years, increased automation has reduced exposure to heavy physical activities in different workplaces. One of the main reasons behind this trend could be attributed to the increase in the use of ergonomic tools and equipment tailored to the needs of each occupational group [1]. An occupational group is a broad-based group of employees with comparable occupational responsibilities located at comparable levels of responsibility within an organization or industry [2–4].


S. Bagnara et al. (Eds.): IEA 2018, AISC 826, pp. 520–529, 2019.
https://doi.org/10.1007/978-3-319-96065-4_56

In many jobs, workers perform their tasks by applying force on tools' handles. Defined as an essential motor ability of humans, muscular strength is used by workforces to overcome external and internal resistance and to generate motion at the work time [5]. Stronger muscle strength in upper extremities is related to the quality and value of grip strength. It is well-known that stronger grip strength in later life is related to less disability, morbidity, and mortality [6–8]. In fact, the ability of a hand to adjust itself to the object that holds depends on the strength of its muscles [9] as well as skeletal muscle size of the hand [10, 11].

Grip strength is a simple, inexpensive and widely-used clinical measure of muscle strength [12], which has been considered as a possible predictor of overall body strength among both genders [13, 14]. It is generated by a combination of the intrinsic and extrinsic hand muscles [15]. Various factors can affect handgrip strength. For example, epidemiological studies have shown that handgrip strength (HGS) has a curvilinear relationship with age [16–18], increases linearly with height, weight, and BMI [19–21], and is significantly greater in males than females [17, 22, 23] as well as in dominant hand compared to the non-dominant one [16, 24]. Some hand anthropometric dimensions such as palm width, hand length and forearm circumference are also found to be related to grip strength [25–27]. A general assumption exists that hand grip strength increases with hand size [10, 11].

The occupation is another important factor influencing human handgrip ability. Some studies have previously shown differences between the grip strength of different occupational groups. They stated that workers who exerted forceful activities with their upper extremities in workplaces, particularly those engaged in material handlings (e.g. lifting and carrying) have a higher grip strength than other workers who perform activities such as writing, keyboarding, etc. [2, 28].

As a matter of fact, the anthropometric and HGS data of different occupational groups are essential for the safe and usable design of products, equipment, and hand tools [29–31]. The strength evaluations are also necessary to predict the capability of workers while performing an occupation requiring strength without incurring injurious strains [31, 32]. In this study, workers were divided into two occupational groups according to their manual demand at work (i.e. manual vs. non-manual workers). We hypothesized that manual workers has higher HGS than non-manual workers and this difference could be attributed to their greater values in some anthropometric dimensions.

Therefore, the current study was conducted with the following aims:

(1) To investigate the hypothesis that light manual workers have higher HGS compared to office/clerical employees as non-manual workers.
(2) To determine anthropometric differences between the two occupational groups.
(3) To determine demographic and anthropometric factors related to HGS in each occupational group.

## 2 Materials and Methods

### 2.1 Subjects

The present study was conducted among 1740 adult males aged from 20 to 64 years old. Subjects were divided into two jobs and nine five-year interval age-groups. To this end, we chose two occupational groups. The first group was manual unskilled workers who perform light manual tasks which generally require no special training to perform, i.e. groundskeepers and gardeners, laborers performing lifting, digging, mixing, loading and pulling operations, and housekeeping such as cleaning, dusting, emptying trashes, sweeping and washing the floor of office rooms and so on in their routine daily work. The second group was office/clerical workers including all clerical type of work regardless the level of difficulty, such as bookkeepers, collectors (bills and accounts), messengers and office helpers, office machine operators (including computer), shipping and receiving clerks, typists, secretaries, and legal assistants, etc. finally, the sample was composed of 905 manual and 835 non-manual workers.

In this study, subjects were selected using a stratified random sampling method from different organization and companies. Then, they were informed about the main objectives of the study. People with a history of any surgery and fracture in hand, neck or upper extremities; neurological disorders; and upper limb injury/deformities in the past 6 months were excluded from the study. Subjects with more than one year work experience were recruited. In the following, the procedure and duration of the tests were described to the participants and each subject completed an informed consent agreement. The study was approved by the local ethics committee.

### 2.2 Measurements Procedures

Selected hand anthropometric dimensions of both sides (hand length, palm length, palm width, forearm length, wrist circumference, and forearm circumference) were measured by a digital Caliper ($\pm$0.1 mm) and a tape meter ($\pm$0.1 cm), based on the NASA standards [33]. Calibration of all equipment was conducted prior to and during the data collection period. The age of participants was recorded from their identity cards. Using a stadiometer, height was nearly recorded to 0.1 cm and weight was measured to around 0.1 kg. Body mass index (BMI) was calculated in kg/m$^2$.

Handheld grip strength dynamometry (Hersteller/manufactures; SEHAN Corporation, Masan-Korea; Distributer Rehaforum Medical GmbH, Elmshorn-Germany) was utilized to measure the muscular force generated by flexor mechanism of the hand and forearm (HGS), because of its well-known validity and reliability [34–36]. Before starting the test, hand dominancy was determined by asking subject the following question: "Which hand do you write?" Then, grip strength test was performed for both hands according to proposed recommendations of the American Society of Hand Therapists (ASHT) [36]. Subjects were seated on a chair without armrests and rest on it. Their shoulders were adducted and neutrally rotated, elbow flexed to 90°, forearm and wrist in 0–15° of extension and 0–15° of ulnar deviation. Three tests of grip strength at an interval of 1 min were performed for both dominant and non-dominant hands of each participant. The mean was recorded as HGS value for each hand.

### 2.3 Statistical Analysis

Normality test was checked by the one-sample Kolmogorov-Smirnov (K-S) test. Two-sample t-test was performed to compare mean HGS between the occupational groups and sides. In this regard, an F-test was first performed to test the equality of the two population variances being compared in order to determine the appropriate t-test statistics to be used. Pearson correlation coefficient test was used to assess the relationship between handgrip strength with demographic and anthropometric variables. All statistical analysis was performed using the SPSS version 23 (IBM Corporation, New York, NY, United States). A value of $p < 0.05$ was accepted as statistically significant.

## 3 Results

Demographic characteristics of manual and non-manual workers are given in Table 1. Ninety-one percent of participants were right-handed dominants.

**Table 1.** Mean ± standard deviation (SD), minimum and maximum of age, weight, height, and BMI as a function of occupational group.

| Variable | Manual workers (n = 905) | | Non-manual workers (n = 835) | |
|---|---|---|---|---|
| | Mean ± SD | Min-Max | Mean ± SD | Min-Max |
| Age (years) | 40.5 ± 16.8 | 21–62 | 48.3 ± 18.2 | 20–64 |
| Weight (Kg) | 71.3 ± 9.3 | 64–96 | 78.1 ± 4.3 | 63–102 |
| Height (cm) | 173.2 ± 7.1 | 158–193 | 171.9 ± 3.1 | 166–201 |
| BMI ($Kg/m^2$) | 23.2 ± 2.8 | 18.6–31.7 | 25.8 ± 4.0 | 19.3–36.2 |

Table 2 represents the mean handgrip strength for different age groups according to hand dominnacy and job-group. The mean grip strength in light manual workers was significantly higher than that of office/clerical employees ($p < 0.05$); whatever the age group. More precisely, the average amount of difference was 12.4% in the case of dominant hand and 10.9% in the case of non-dominant hand; all ages included. Results also showed that dominant hand was stronger than non-dominant hand in both groups (11% for light manual workers and 9.5% for office/clerical employees).

It was observed that the age group of 35–39 had the highest mean grip strength both in light manual and office/clerical workforces. In addition, the relation between the mean grip strength and age was curvilinear, suggesting that the average of grip strength increased up to the 35–39 age range and then decreased gradually, in both manual and non-manual workers as well as in both sides.

**Table 2.** Grip strength values ± SD (kg) for dominant and non-dominant hands of manual (n = 905) and non-manual (n = 835) workers by age groups.

| Age group | Grip strength (kg) (Mean ± SD) | | | | | |
|---|---|---|---|---|---|---|
| | Dominant hand | | | Non-dominant hand | | |
| | Manual workers | Non-manual workers | d | Manual workers | Non-manual workers | d |
| 20–24 | 50.1 ± 8.7 | 46.6 ± 7.7 | 3.5 | 45.3 ± 7.5 | 40.2 ± 7.6 | 5.1 |
| 25–29 | 52.7 ± 7.9 | 46.8 ± 7.6 | 5.9 | 47.8 ± 8.2 | 41.9 ± 9.2 | 5.9 |
| 30–34 | 54.4 ± 8.2 | 47.7 ± 9.6 | 6.7 | 49.0 ± 7.3 | 45.0 ± 10.1 | 4.0 |
| 35–39 | 58.3 ± 9.4 | 51.0 ± 8.3 | 7.3 | 51.3 ± 9.7 | 47.3 ± 8.6 | 4.0 |
| 40–44 | 55.2 ± 9.2 | 49.0 ± 7.9 | 6.2 | 48.8 ± 10.2 | 42.6 ± 7.4 | 6.2 |
| 45–49 | 52.9 ± 8.3 | 46.3 ± 9.1 | 6.6 | 45.1 ± 9.3 | 41.9 ± 8.0 | 3.2 |
| 50–54 | 51.1 ± 10.2 | 43.2 ± 11.2 | 7.9 | 46.8 ± 8.2 | 40.3 ± 8.1 | 6.5 |
| 55–59 | 47.6 ± 8.4 | 40.4 ± 9.2 | 7.2 | 41.2 ± 6.8 | 35.9 ± 8.1 | 5.3 |
| 60–64 | 42.2 ± 9.1 | 35.9 ± 8.2 | 6.3 | 38.1 ± 7.2 | 33.2 ± 7.4 | 4.9 |
| Total | 51.6 ± 8.8 | 45.2 ± 8.8 | 6.4 | 45.9 ± 8.1 | 40.9 ± 8.7 | 5.0 |

d: $HGS_{manual\ worker} - HGS_{Non\text{-}manual\ worker}$

Table 3 represents the distribution of selected hand dimensions by hand side and occupational group. Palm width and forearm circumference were significantly different between the two occupational groups ($P < 0.05$). Light manual workers had greater palm width and forearm circumference than office/clerical employees, in both hands. Other hand dimensions were not statistically different between the two groups.

**Table 3.** Selected hand anthropometric dimensions (kg) for both hands of manual and non-manual workers.

| Variable | Mean ± SD | | d |
|---|---|---|---|
| | Manual workers | Non-manual workers | |
| *Dominant hand* | | | |
| Hand length | 20.05 ± 0.90 | 19.25 ± 0.75 | 0.80 |
| Palm length | 11.65 ± 0.63 | 10.80 ± 0.51 | 0.85 |
| Palm width | 11.82 ± 0.26 | 9.92 ± 0.40 | **1.90** |
| Forearm length | 27.45 ± 1.15 | 27.00 ± 0.98 | 0.45 |
| Wrist circumference | 19.65 ± 1.00 | 18.83 ± 0.90 | 0.82 |
| Forearm circumference | 29.35 ± 2.27 | 26.53 ± 0.54 | **2.82** |
| *Non-dominant hand* | | | |
| Hand length | 19.70 ± 0.85 | 19.00 ± 0.89 | 0.70 |
| Palm length | 11.45 ± 0.56 | 11.00 ± 0.61 | 0.45 |
| Palm width | 11.00 ± 0.41 | 9.30 ± 0.48 | **1.70** |
| Forearm length | 26.91 ± 1.25 | 26.12 ± 1.10 | 0.79 |
| Wrist circumference | 19.83 ± 0.98 | 19.00 ± 0.79 | 0.83 |
| Forearm circumference | 29.25 ± 2.03 | 26.95 ± 2.03 | **2.30** |

**d** = $(\text{hand dimension})_{Manual\ workers} - (\text{hand dimension})_{Non\text{-}Manual\ workers}$
$\mathbf{d_{Bold}}$ = Significant at 0.01 level.

Pearson correlation test showed that all of the demographic and anthropometric factors were significantly correlated with HGS (Table 4). More precisely, results showed a significant negative relationship between the age variable with grip strength in both sides and both job-groups. This means that as age rises, HGS will have a decreasing trend whatever the type of job. Among selected hand anthropometric dimensions, palm width had the most significant correlation with hand grip strength ($r = 0.710$ and $r = 0.689$, $p < 0.01$ for dominant and non-dominant hands of light manual workers, respectively; $r = 0.660$ and $r = 0.582$, $p < 0.01$ for dominant and non-dominant hands of office/clerical employees, respectively). Moreover, forearm circumference was also other distinctive variable when was matched alongside the other selected variables ($r = 0.645$ and $r = 0.631$, $p < 0.01$ for dominant and non-dominant hands of light manual workers, respectively; $r = 0.492$ and $r = 0.477$, $p < 0.01$ for dominant and non-dominant hands of office/clerical employees, respectively).

**Table 4.** Pearson's correlation coefficients between the hand grip strength of each hand with demographic and anthropometric factors.

| Variables | Manual workers (n = 905) | | Non-manual workers (n = 835) | |
|---|---|---|---|---|
| | Dominant | Non-dominant | Dominant | Non-dominant |
| Age | −0.301** | −0.275** | −0.277** | −0.266** |
| Weight | 0.368** | 0.344** | 0.261** | 0.291** |
| Height | 0.402** | 0.339** | 0.370** | 0.365** |
| BMI | 0.254** | 0.261** | 0.200* | 0.198* |
| Hand Length (HL) | 0.413** | 0.372** | 0.361** | 0.370** |
| Palm Width (PW) | 0.710** | 0.689** | 0.660** | 0.582** |
| Palm Length (PL) | 0.466** | 0.507** | 0.432** | 0.455** |
| Wrist Circumference (WC) | 0.240** | 0.222** | 0.200* | 0.295** |
| Forearm Circumference (FC) | 0.645** | 0.631** | 0.492** | 0.477** |
| Forearm Length (FL) | 0.242** | 0.233** | 0.185* | 0.195* |

*Correlation is significant at the 0.05 level (2-tailed).
**Correlation is significant at the 0.01 level (2-tailed).

## 4 Discussion

The present study was conducted to compare difference values of HGS and hand anthropometric dimensions between light manual workers and office/clerical employees. The contribution of anthropometric and demographic factors on handgrip strength was also investigated as a function of job type. Our results revealed a significant difference between the grip strength of two studied groups of workers; indicating that light manual workers had approximately 10% stronger hands than their office coworkers. This finding is in line with previous studies [2, 4, 5, 37]. In general, manual workers use their hands and arms more than office workers and exert more force in the course of frequent griping and grasping. This is mainly due to the fact that strong and repetitive movements increase the muscle strength, muscle mass and ultimately grip strength and arms strength in manual workers. In addition, higher HGS in manual workers can be due to the effects of posture, movement and hand load on shoulder muscles, as well as greater muscles mass [38–40].

Grip strength values for both light manual workers and office/clerical employees reached their peak between the ages of 35–39 and then decreased gradually with the increase of age. In other words, the relationship between age and grip strength is non-linear [10, 16, 24]. This finding is consistent with the results of other studies in which the highest grip strength is found between the ages of 25 and 50 years. This result can be due to the age related decrease in physical activity, changes in muscle fibers and overall muscle mass and chronic diseases [41, 42]. Owing the overall mean difference between two hands, our result do align with generally accepted "10% rule" suggesting that the dominant hand grip strength is approximately 10% greater than the non-dominant handgrip strength [43].

The value of grip strength is related to the hand size. Our finding is in good agreement with previous studies in which strong correlation has been found between handgrip strength and hand size [44–46]. Therefore, individuals with large hands were more likely to have higher grip strength than those with small hand sizes probably because of their greater muscle mass. However, it seems that some hand dimensions could be more determinative than the others, since palm width and forearm circumference were much strongly correlated with the HGS. It could be explained by the fact that light manual workers have massive muscle in upper extremities because they use strong and repetitive movements in order to carry out their job tasks, which could in turn results in greater hand grip strength.

## 5 Conclusions

In conclusion, light manual workers have higher grip strength than office employees, probably because of their larger palm width and forearm circumference. Grip strength data are essential for the design of ergonomic products for different age groups of youth, adults and elderly workforces. Ergonomic design of hand tools can reduce the pressure on a neuromuscular system of the hands and arms and greatly prevent contact stress. Therefore, paying attention to the required grip strength of workers to perform their tasks would be a valuable guideline for designers. However, it may be worthwhile to investigate the eventual contribution of more parameters such as other body dimensions, nutritional status, and lifestyle in further studies.

## References

1. Kuijt-Evers LFM, Groenesteijn L, De Looze MP, Vink P (2004) Identifying factors of comfort in using hand tools. Appl Ergon 35(5):453–458
2. Josty IC, Tyler MPH, Shewell PC, Roberts AHN (1997) Grip and pinch strength variations in different types of workers. J Hand Surg Am 22(2):266–269

3. Angst F, Drerup S, Werle S, Herren DB, Simmen BR, Goldhahn J (2010) Prediction of grip and key pinch strength in 978 healthy subjects. BMC Musculoskelet Disord 11(1):94
4. Ding H, Leino-Arjas P, Murtomaa H, Takala E-P, Solovieva S (2013) Variation in work tasks in relation to pinch grip strength among middle-aged female dentists. Appl Ergon 44(6): 977–981
5. Dahlke G, Butlewski M, Drzewiecka M (2016) Impact of the work process on power grip strength: a practical case. Occup Saf Hyg
6. Cooper R, Kuh D, Cooper C, Gale CR, Lawlor DA, Matthews F et al (2010) Objective measures of physical capability and subsequent health: a systematic review. Age Ageing 40(1): 14–23
7. Adamczyk JG, Hołun M, Boguszewski D, Siewierski M (2013) Evaluation of the effectiveness of hand grip strength training using a device Powerball®. The effectiveness of the force training with the help of Powerball®. Musculoskelet J Hosp Spec Surg 3(6):35–44
8. Volaklis KA, Halle M, Meisinger C (2015) Muscular strength as a strong predictor of mortality: a narrative review. Eur J Intern Med 26(5):303–310
9. Trzaskoma Z, Trzaskoma Ł (2001) The proportion between maximal torque of core muscles in male and female athletes. Acta Bioeng Biomech 3(Suppl 2):601–606
10. Peters MJH, van Nes SI, Vanhoutte EK, Bakkers M, van Doorn PA, Merkies ISJ et al (2011) Revised normative values for grip strength with the Jamar dynamometer. J Peripher Nerv Syst 16(1):47–50
11. Shahida MSN, Zawiah MDS, Case K (2015) The relationship between anthropometry and hand grip strength among elderly Malaysians. Int J Ind Ergon 50:17–25
12. Walker-Bone K, D'angelo S, Syddall HE, Palmer KT, Cooper C, Coggon D et al (2016) Heavy manual work throughout the working lifetime and muscle strength among men at retirement age. Occup Env Med oemed-2015
13. Tietjen-Smith T, Smith SW, Martin M, Henry R, Weeks S, Bryant A (2006) Grip strength in relation to overall strength and functional capacity in very old and oldest old females. Phys Occup Ther Geriatr 24(4):63–78
14. Barlow MJ, Findlay M, Gresty K, Cooke C (2014) Anthropometric variables and their relationship to performance and ability in male surfers. Eur J Sport Sci 14(sup1):S171–S177
15. Abe T, Counts BR, Barnett BE, Dankel SJ, Lee K, Loenneke JP (2015) Associations between handgrip strength and ultrasound-measured muscle thickness of the hand and forearm in young men and women. Ultrasound Med Biol 41(8):2125–2130
16. Werle S, Goldhahn J, Drerup S, Simmen BR, Sprott H, Herren DB (2009) Age-and gender-specific normative data of grip and pinch strength in a healthy adult Swiss population. J Hand Surg (Eur Vol)
17. Ekşioğlu M (2016) Normative static grip strength of population of Turkey, effects of various factors and a comparison with international norms. Appl Ergon 52:8–17
18. Mathiowetz V, Kashman N, Volland G, Weber K, Dowe M, Rogers S (1985) Grip and pinch strength: normative data for adults. Arch Phys Med Rehabil 66(2):69–74
19. Jürimäe T, Hurbo T, Jürimäe J (2009) Relationship of handgrip strength with anthropometric and body composition variables in prepubertal children. HOMO-J. Comput Hum Biol 60(3): 225–238
20. Shyamal K, Yadav KM (2009) An association of hand grip strength with some anthropometric variables in Indian cricket players. Facta Univ Phys Educ Sport 7(2):113–123
21. Wu S-W, Wu S-F, Liang H-W, Wu Z-T, Huang S (2009) Measuring factors affecting grip strength in a Taiwan Chinese population and a comparison with consolidated norms. Appl Ergon 40(4):811–815

22. Li K, Hewson DJ, Duchêne J, Hogrel J-Y (2010) Predicting maximal grip strength using hand circumference. Man Ther 15(6):579–585
23. Kamarul T, Ahmad TS, Loh WYC (2006) Hand grip strength in the adult Malaysian population. J Orthop Surg 14(2):172–177
24. Schlüssel MM, dos Anjos LA, de Vasconcellos MTL, Kac G (2008) Reference values of handgrip dynamometry of healthy adults: a population-based study. Clin Nutr 27(4):601–607
25. Günther CM, Bürger A, Rickert M, Crispin A, Schulz CU (2008) Grip strength in healthy caucasian adults: reference values. J Hand Surg Am 33(4):558–565
26. Anakwe RE, Huntley JS, McEachan JE (2007) Grip strength and forearm circumference in a healthy population. J Hand Surg (Eur Vol) 32(2):203–209
27. Shim JH, Roh SY, Kim JS, Lee DC, Ki SH, Yang JW et al (2013) Normative measurements of grip and pinch strengths of 21st century korean population. Arch Plast Surg 40(1):52–56
28. Chatterjee S, CHOWDHURI BJ (1991) Comparison of grip strength and isometric endurance between the right and left hands of men and their relationship with age and other physical parameters. J Hum Ergol (Tokyo) 20(1):41–50
29. Mittal A, Kumar S (1998) Human muscle strength definition, measurement and usage. Part guidelines for the practioners. Int J Ind Ergon 101–21
30. Eksioglu M (2011) Endurance time of grip-force as a function of grip-span, posture and anthropometric variables. Int J Ind Ergon 41(5):401–409
31. Ekşioğlu M, Recep Z (2013) Hand torque strength of female population of Turkey and the effects of various factors. In: Occupational safety and hygiene. CRC Press, pp 51–56
32. Eksioglu M (2004) Relative optimum grip span as a function of hand anthropometry. Int J Ind Ergon 34(1):1–12
33. Book NAS (1978) Volume II: a handbook of anthropometric data. NASA Reference: Publication 1024
34. Mathiowetz V, Weber K, Volland G, Kashman N (1984) Reliability and validity of grip and pinch strength evaluations. J Hand Surg Am 9(2):222–226
35. Svens B, Lee H (2005) Intra-and inter-instrument reliability of Grip-Strength Measurements: GripTrack™ and Jamar® hand dynamometers. Br J Hand Ther 10(2):47–55
36. Fess EE, Moran C (1981) American society of hand therapists: clinical assessment recommendations. Garner Soc
37. Nygård C-H, Eskelinen L, Suvanto S, Tuomi K, Ilmarinen J (1991) Associations between functional capacity and work ability among elderly municipal employees. Scand J Work Environ Health 122–127
38. Frontera WR, Hughes VA, Lutz KJ, Evans WJ (1991) A cross-sectional study of muscle strength and mass in 45-to 78-yr-old men and women. J Appl Physiol 71(2):644–650
39. Metter EJ, Lynch N, Conwit R, Lindle R, Tobin J, Hurley B (1999) Muscle quality and age: cross-sectional and longitudinal comparisons. J Gerontol Ser A Biomed Sci Med Sci 54(5): B207–B218
40. Häkkinen A, Kautiainen H, Hannonen P, Ylinen J, Mäkinen H, Sokka T (2006) Muscle strength, pain, and disease activity explain individual subdimensions of the Health Assessment Questionnaire disability index, especially in women with rheumatoid arthritis. Ann Rheum Dis 65(1):30–34
41. Adedoyin RA, Ogundapo FA, Mbada CE, Adekanla BA, Johnson OE, Onigbinde TA et al (2009) Reference values for handgrip strength among healthy adults in Nigeria. Hong Kong Physiother J 27(1):21–29
42. Sternäng O, Reynolds CA, Finkel D, Ernsth-Bravell M, Pedersen NL, Aslan AKD (2015) Factors associated with grip strength decline in older adults. Age Ageing 44(2):269–274
43. Bechtol CO (1954) Grip test. J Bone Jt Surg Am 36(4):820–832

44. Pieterse S, Manandhar M, Ismail S (2002) The association between nutritional status and handgrip strength in older Rwandan refugees. Eur J Clin Nutr 56(10):933
45. Lauretani F, Russo CR, Bandinelli S, Bartali B, Cavazzini C, Di Iorio A et al (2003) Age-associated changes in skeletal muscles and their effect on mobility: an operational diagnosis of sarcopenia. J Appl Physiol 95(5):1851–1860
46. Koley S, Kumaar SB (2012) The relation between handgrip strength and selected hand-anthropometric variables in Indian inter-university softball players. Facta Univ Phys Educ Sport 10(1):13–21
47. Patrick JM, Bassey EJ, Fentem PH (1982) Changes in body fat and muscle in manual workers at and after retirement. Eur J Appl Physiol Occup Physiol 49(2):187–196

# Estimation to Use the Stick Figure of Kinect® Version 2 for Digital Anthropometry

Sabine Wenzel(✉), Juliana Buchwald, and Hartmut Witte

Biomechatronics Group, Technische Universität Ilmenau,
Max-Planck-Ring 12, 98693 Ilmenau, Germany
sabine.wenzel@tu-ilmenau.de

**Abstract.** In this paper we examine the possibility to use the integrated stick figure and their joint positions of Microsoft® Kinect® version 2 for digital measurement of human body dimensions. Manual anthropometric measurement data and the indirect parameter calculation based on Kinect® joints were compared for selected human body dimensions. This paper takes a closer look at the length measurements of body height and arm span.

15 female participants took part in a subject test. Participants stood frontal to the Kinect® in two different stages (feet together and slightly apart; both with and without shoes) and in three different poses. Each parameter was analysed by three different algorithms based on the Kinect® joint data.

For determination of body height, the indirect parameter calculation seems to be a possible replacement of the direct manual anthropometric measurement, if the accuracy claim is low. We observed deviations between the values derived from Kinect® data and the direct anthropometric values up to 5 cm. Arm span data as well shows linear trends, but deviations are higher. Multivariate analyses seem to be necessary.

**Keywords:** Digital anthropometry · Stick figure · Microsoft® Kinect®

## 1 Introduction

Anthropometry plays an important role in industrial design, ergonomics and clothing design. Manual measurement of anthropometric data is time consuming. With techniques of digital anthropometry it will be easier to adjust workplaces to individual properties of workers. The workplace adaption to workers is as well important for a higher productivity as for health prevention.

Different methods of digital anthropometry exist. The approaches based on low-cost techniques (like Microsoft® Kinect®) analyses depth images and silhouettes (e.g. Annichini et al. 2013; Aslam et al. 2017; Robinson and Parkinson 2013) or automatically calculated stick figures of the Kinect® device (e.g. Chiu et al. 2016; Hamilton et al. 2014). Those studies used the first generation of Microsoft® Kinect®, based on structured light and stick figures with 20 joints. The depth sensing technology used in the second generation of Kinect® changed to the time of flight principle (Lun and Zhao 2015). With this the joint number (up to 25) and the joint position accuracy were


S. Bagnara et al. (Eds.): IEA 2018, AISC 826, pp. 530–543, 2019.
https://doi.org/10.1007/978-3-319-96065-4_57

increased (Wang et al. 2015). Thus, we prove whether the new technology may help to improve the quality of anthropometric data.

## 2 Materials and Methods

### 2.1 Manual Anthropometry

Based on 3D-coordinates of the joints, we can calculate 1D-data (length and angle). A calculation of circumference data or weight dimensions is not possible. Thus, in our subject test (see Sect. 2.3), we collected only length measurement data for a comparison between manual and digital anthropometry, based on stick figure data from Kinect®.

Human body dimensions were collected manually with anthropometric instruments (anthropometer and callipers). One experimenter took all measurement data of all participants. The manual measurement recorded anthropometric points according to DIN 33402:2005-12, and Flügel et al. (1986). The paper presented here takes a closer look on the length measurements of body height and arm span.

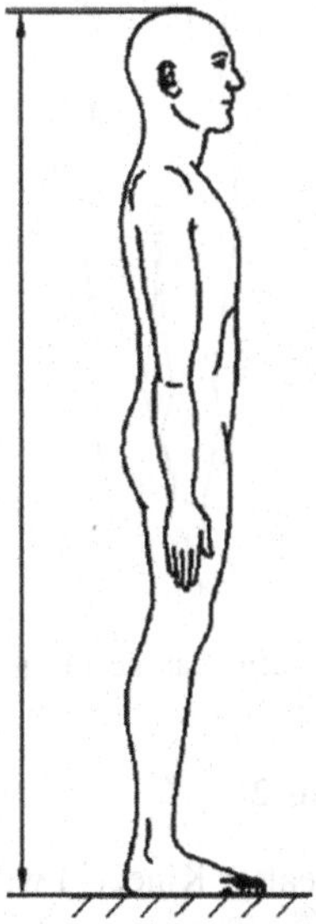

**Fig. 1.** Determination of the body height according to DIN 33402-2:2005-12. (DIN 33402-2 2005)

Figure 1 represents the way of measuring the parameter body height. It is measured as the vertical distance from ground to the highest point of the head (*vertex cranii*) in standing position. The feet are posed in a close parallel contact. The head is posed according to the 'Frankfurter Horizontale': the connection lines of the highest bony point of the outside end of the bony ear channel (*porion*) and the deepest point of *orbita* of both sides are posed in a horizontal plane– for practical means, taking into account the systematic error, the reference lines are drawn from the highest point of the outside end of the skinny ear channel to the outer corner of the eye (DIN 33402-2 2005).

The arm span (see Fig. 2) is defined by the maximum horizontal distance of the outer points of the middle fingertips. Arms and hands are stretched sideways in standing position with feet posed like described for the body height measurement. The measurement ensues as well at the human backside (Flügel et al. 1986).

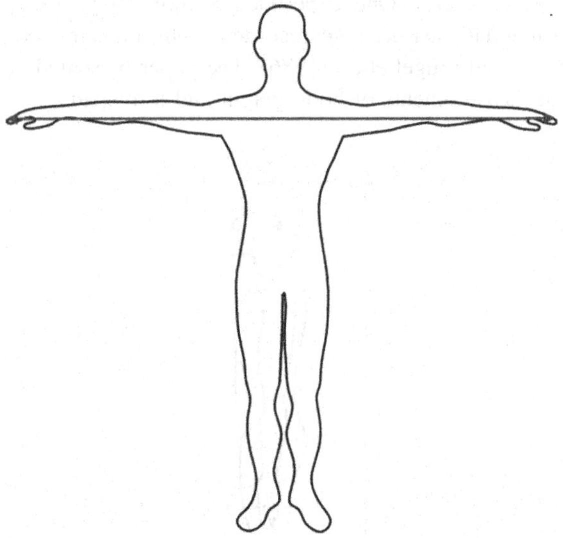

**Fig. 2.** Definition of the arm span according to Flügel et al. (1986).

## 2.2 Microsoft® Kinect® Version 2

Microsoft® Kinect® version 2 (hereafter Kinect®) was developed for gesture control of the gaming console Xbox One®. The appendant Microsoft® Kinect® Software Development Kit (SDK) allows using the raw data generated by Kinect®. Thus, using Kinect® for scientific research is possible.

Kinect® consists of a depth camera, a real image HD-camera and a microphone-array. The cameras own a frame rate of 30 fps (frames per seconds). With an internal Microsoft®-library, Kinect® generates based on the depth data stick figures up to six persons. One stick figure consists of 25 "joints" (relevant locations see Figs. 4 and 5). Each joints' data contains information on the spatial location (x-; y-; z-value) in relation to the depth camera. This information is the basis of a digital measurement of human body dimensions (see Sect. 3).

### 2.3 Subject Tests with Kinect®

We carried out a subject test with 15 female participants in the age of 18 to 36 years (average age = 24.1 years), to minimise the number of possible confounders. First the participants were measured manually by one experimenter. Thus, we got the comparison value (defined as the gold standard for this test) for the digital measurement.

After that we detected different static standing poses from frontal (see Fig. 3) for the digital anthropometry. Kinect® was located on a height of one meter and a distance to the participant of 2.5 m. This allows a depiction of the whole body in the middle of the field of view.

Figure 3 shows these standing poses measured with the Kinect®. All positions were detected frontal to the camera. In position 1 the participant stands straight with hanging arms. For position 2 the participant had to bend the arms in 90° sideway to the body. The participant stretched the arms and hands sideway for position 3. All measurements carried in static. The holding time was ten seconds. By a frame rate of 30 fps implies

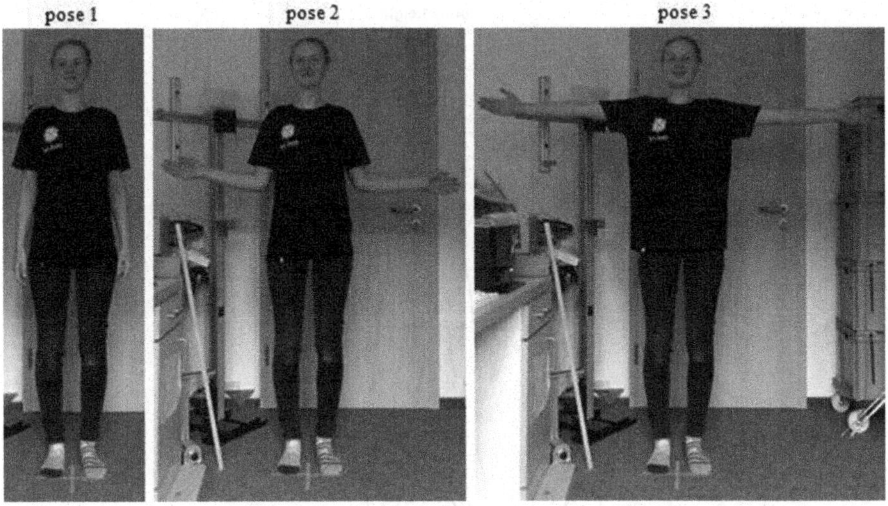


**Fig. 3.** Standing poses of the subject test for Kinect® measurements.

this 300 frames/stick figures for the analysis. These three poses were done by different stages:

- stage 1: feet together without shoes
- stage 2: feet slightly apart without shoes (see Fig. 3)
- stage 3: feet together with shoes on
- stage 4: feet slightly apart with shoes on.

## 2.4 Calculation of Anthropometric Data from Kinect® Data

Figure 4 shows the three different calculation abilities we used for the body height analysis: the direct distance between head and foot (ability 1 - Eq. 1), the direct distance between head and ankle (ability 2 - Eq. 2), and the addition of the distance between head and hip and of the distance between hip and ankle (ability 3 - Eq. 3). For the first calculation we differed in left and right side.

$$h_{body,1} = \left|\overrightarrow{head} - \overrightarrow{foot}\right| \tag{1}$$

$$h_{body,2} = \left|\overrightarrow{head} - \overrightarrow{ankle}\right| \tag{2}$$

$$h_{body,3} = \left|\overrightarrow{head} - \overrightarrow{hip}\right| + \left|\overrightarrow{hip} - \overrightarrow{ankle}\right| \tag{3}$$

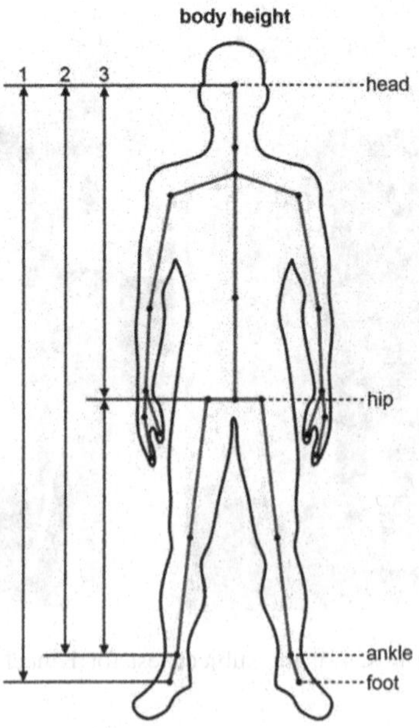


**Fig. 4.** Calculation abilities for the body height based on joint positions of the stick figure: ability 1 – foot to head, ability 2 – ankle to head, ability 3 – addition of ankle to hip and hip to head.

For arm span, we correspondingly analysed different calculation abilities (Fig. 5). The analysis depends on: the distance between hand tip and hand tip directly (ability 1 - Eq. 4), the direct distance between wrist and wrist (ability 2 - Eq. 5), and the addition of the distance between $wrist_{right}$ and $shoulder_{right}$, of the distance between shoulder and shoulder and of the distance between $shoulder_{left}$ and $wrist_{left}$ (ability 3 - Eq. 6). Only pose 3 were analysed.

$$l_{arm\,span,1} = \left|\overrightarrow{handtip_{right}} - \overrightarrow{handtip_{left}}\right| \tag{4}$$

$$l_{arm\,span,2} = \left|\overrightarrow{wrist_{right}} - \overrightarrow{wrist_{left}}\right| \tag{5}$$

$$\begin{aligned} l_{arm\,span,3} = & \left|\overrightarrow{wrist_{right}} - \overrightarrow{shoulder_{right}}\right| \\ & + \left|\overrightarrow{shoulder_{right}} - \overrightarrow{shoulder_{left}}\right| + \left|\overrightarrow{shoulder_{left}} - \overrightarrow{wrist_{left}}\right| \end{aligned} \tag{6}$$

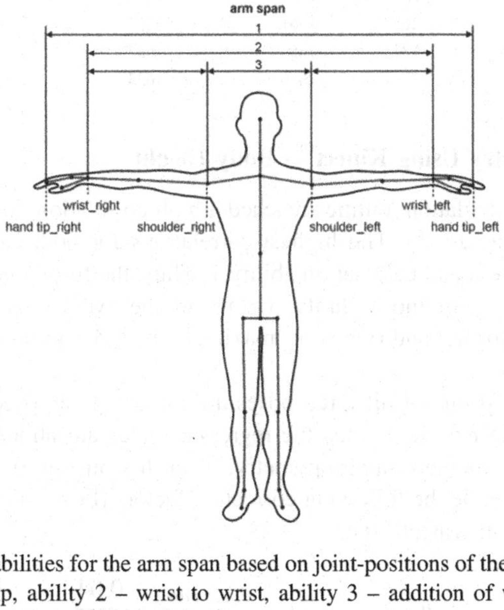


**Fig. 5.** Calculation abilities for the arm span based on joint-positions of the stick figure: ability 1 – hand tip to hand tip, ability 2 – wrist to wrist, ability 3 – addition of wrist to shoulder and shoulder to shoulder and shoulder to wrist.

## 3 Results

### 3.1 Manual Anthropometry

Table 1 reports the participants' ages and the values of their manually measured parameters.

**Table 1.** Manually acquired anthropometric data of participants (all female).

| Participant | Age [years] | Body height [mm] | Arm span [mm] |
|---|---|---|---|
| 1 | 20 | 1753 | 1720 |
| 2 | 36 | 1690 | 1708 |
| 3 | 24 | 1639 | 1600 |
| 4 | 24 | 1532 | 1526 |
| 5 | 24 | 1603 | 1579 |
| 6 | 28 | 1742 | 1751 |
| 7 | 20 | 1670 | 1650 |
| 8 | 25 | 1588 | 1583 |
| 9 | 20 | 1742 | 1722 |
| 10 | 27 | 1642 | 1654 |
| 11 | 26 | 1594 | 1531 |
| 12 | 23 | 1584 | 1534 |
| 13 | 26 | 1661 | 1636 |
| 14 | 18 | 1654 | 1635 |
| 15 | 27 | 1640 | 1633 |

### 3.2 Anthropometry Using Kinect® - Body Height

All stages and all calculation abilities reached a high correlation rate after Pearson and Spearman (see Appendix 1). The highest correlations for both can be found by the combination of stage 3 and calculation ability 1. Thus, the further considerations focus on this combination. For more clarity, we report the average values of each pose, because the values for left and right vary maximally by 0.5% in pose 1, 0.7% in pose 2 and 1.9% in pose 3.

The linear regressions of all three poses are almost superposed (see Fig. 6). The rises of the lines are near to 1, thus the regression lines are almost in parallel to the bisection line (anthropometry-anthropometry, black line in Fig. 6). Based on a linear regression we can define the following correction factors (Eqs. 7–9) for calculating the body height based on Kinect® data.

$$h_{body\_new,pose\,1} = \frac{h_{body,pose\,1} + 0.04}{0.94} \tag{7}$$

$$h_{body\_new,pose\,2} = \frac{h_{body,pose\,2} + 0.11}{0.99} \tag{8}$$

$$h_{body\_new,pose\,3} = \frac{h_{body,pose\,3} + 0.08}{0.97} \tag{9}$$

Appendix 2 lists all parameters of the body height beginning with the manual anthropometry above the Kinect® calculations to the new body height calculations based on the Kinect® data and their correction factors.

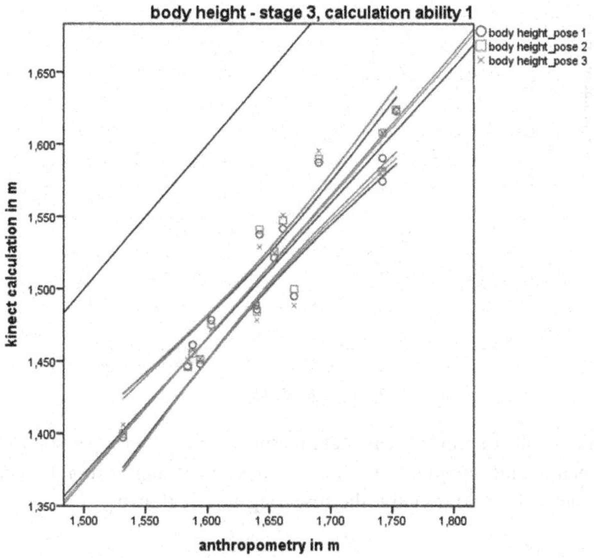


**Fig. 6.** Body height: results of Kinect® calculation ability 1 in comparison to the results of the manual measurement ("anthropometry"). Values for stage 3. Linear regression with 95% confidence intervals. Additionally, the bisecting line is drawn.

### 3.3 Anthropometry Using Kinect® - Arm Span

Figure 7 shows a trend for a linear regression. Thus, the determination of a correction factor is possible. Unfortunately, the lines are not in parallel to the bisection line (black line in Fig. 7). Based on the linear regression we can define the following correction factors (Eqs. 10–12) for calculating the new arm span.

$$l_{arm\,span\_new,1} = \frac{l_{arm\,span,1} - 0,61}{0,56} \tag{10}$$

$$l_{arm\,span\_new,2} = \frac{l_{arm\,span,2} - 0,54}{0,41} \tag{11}$$

$$l_{arm\,span\_new,3} = \frac{l_{arm\,span,3} - 0,57}{0,4} \tag{12}$$

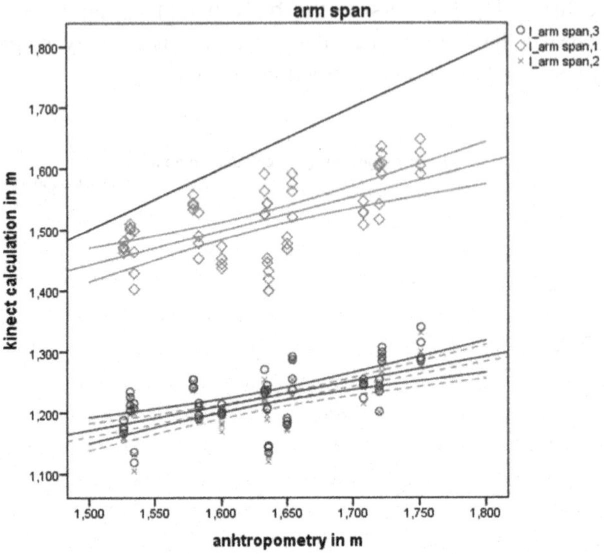


**Fig. 7.** Arm span: results of Kinect® calculation ability 1 to 3 in comparison to the results of the manual measurement ("anthropometry"). Values of pose 3 for stage 1 to 4. Linear regression with 95% confidence intervals. Additionally, the bisecting line is drawn.

Appendix 3 lists all parameters of the arm span, beginning by the manual anthropometry above the Kinect® calculations to the new arm span calculations based on the Kinect® data and their correction factors.

# 4 Discussion

Based on a subject test with 15 participants, we analysed different abilities for calculating body height and arm span with Kinect® joint data. We could show a linear regression for the selected parameters. Thus, a determination of correction factors was possible. Wang et al. (2015) find the mean offset of the joint accuracy in standing exercises decrease from Microsoft® Kinect® version 1 to version 2 from 7.6 cm–24.5 cm to 6.6 cm–14.6 cm. Thus, the quality of these factors has to be looked at critically, due to the given error rate of Kinect® data.

The size of the analysed sequence has to be checked, if it is necessary to hold the pose ten seconds or it is possible to reduce the measuring time. Maybe there is a possibility to smooth the sequence to reduce outliners.

By using the Kinect® system there are some major disadvantages, which still have to be corrected or eliminated, like the variability of joint positions related to the human body. Especially the finger tips are disjointed, and we sometimes had problems with the localization of the hand position. Such detecting problems from knees downwards to feet (during stage 3 and 4 "feet together") also could be observed.

The body height receive derivations of 0.2%–2.4% (pose 1); 0.1%–2.7% (pose 2) and 0%–3.2% (=5.3 cm; pose 3). The arm span has not delivered so good results. The derivations are between 0.1% and 11.8%. In all measurements one participant (participant 13) has an increased derivation between 11.4% and 16%. Compared to Hamilton et al. (2014) we decrease the derivation of the body height. The body height values of their female participants were described by deviations up to 8.6 cm. Robinson and Parkinson received (2013) a max. error of 3.6 cm to 7.1 cm for estimating stature and assumed a relatively accuracy. With multiple Kinect sensors they improved accuracy with a max. error from 2.3 cm to 4.1 cm. At the moment based on our results and the derivations we cannot define the best calculation ability (Hamilton et al. 2014; Robinson and Parkinson 2013).

## 5 Resume and Outlook

The quick and contactless measurement by Kinect® seems to be an alternative to manual anthropometry, if we only look at the advantages. Kinect® joint data can be used for selected body dimension calculation (for not so high-precision applications like workplace adaption), if the result accuracy may be limited to five percent. The analysis of the body height promises a first approach for a replacement of the manual anthropometric measurement. We found linear trends between Kinect® measurement data and manually acquired anthropometric data, and thus defined correction factors for exemplary parameters. The quality of these factors has to be improved. We have to analyse causes of occurring errors, and to improve our algorithms for the Kinect® data re-calculations. Therefore, we will take into account more parameters.

With this we will have the potentiality to derive digital anthropometric data with sufficient accuracy for subsequent processing in technical purposes.

# Appendix

Appendix 1: Pearson- and Spearman-Correlations of body height calculation

| body height | anthr. | body height,1_left | | | body height,1_right | | | body height,2_left | | | body height,2_right | | | body height,3_left | | | body height,3_right | | |
|---|---|---|---|---|---|---|---|---|---|---|---|---|---|---|---|---|---|---|---|
| | | pose 1 | pose 2 | pose 3 | pose 1 | pose 2 | pose 3 | pose 1 | pose 2 | pose 3 | pose 1 | pose 2 | pose 3 | pose 1 | pose 2 | pose 3 | pose 1 | pose 2 | pose 3 |
| pearson correlation stage 1 | 1 | ,853** | ,874** | ,800** | ,807** | ,851** | ,856** | ,906** | ,917** | ,853** | ,858** | ,828** | ,883** | ,908** | ,918** | ,851** | ,853** | ,826** | ,879** |
| significance (1-seitig) | | ,000 | ,000 | ,000 | ,000 | ,000 | ,000 | ,000 | ,000 | ,000 | ,000 | ,000 | ,000 | ,000 | ,000 | ,000 | ,000 | ,000 | ,000 |
| n | 15 | 15 | 15 | 15 | 15 | 15 | 15 | 15 | 15 | 15 | 15 | 15 | 15 | 15 | 15 | 15 | 15 | 15 | 15 |
| spearman correlation stage 1 | 1,000 | ,915** | ,935** | ,864** | ,791** | ,879** | ,863** | ,938** | ,963** | ,867** | ,844** | ,840** | ,890** | ,951** | ,948** | ,856** | ,844** | ,829** | ,880** |
| significance (1-seitig) | | ,000 | ,000 | ,000 | ,000 | ,000 | ,000 | ,000 | ,000 | ,000 | ,000 | ,000 | ,000 | ,000 | ,000 | ,000 | ,000 | ,000 | ,000 |
| n | 15 | 15 | 15 | 15 | 15 | 15 | 15 | 15 | 15 | 15 | 15 | 15 | 15 | 15 | 15 | 15 | 15 | 15 | 15 |
| pearson correlation stage 2 | 1 | ,895** | ,939** | ,928** | ,901** | ,931** | ,940** | ,920** | ,957** | ,945** | ,907** | ,949** | ,955** | ,925** | ,960** | ,945** | ,907** | ,946** | ,951** |
| significance (1-seitig) | | ,000 | ,000 | ,000 | ,000 | ,000 | ,000 | ,000 | ,000 | ,000 | ,000 | ,000 | ,000 | ,000 | ,000 | ,000 | ,000 | ,000 | ,000 |
| n | 15 | 15 | 15 | 15 | 15 | 15 | 15 | 15 | 15 | 15 | 15 | 15 | 15 | 15 | 15 | 15 | 15 | 15 | 15 |
| spearman correlation stage 2 | 1,000 | ,913** | ,938** | ,953** | ,891** | ,917** | ,948** | ,945** | ,971** | ,981** | ,933** | ,962** | ,978** | ,945** | ,974** | ,978** | ,940** | ,963** | ,971** |
| significance (1-seitig) | | ,000 | ,000 | ,000 | ,000 | ,000 | ,000 | ,000 | ,000 | ,000 | ,000 | ,000 | ,000 | ,000 | ,000 | ,000 | ,000 | ,000 | ,000 |
| n | 15 | 15 | 15 | 15 | 15 | 15 | 15 | 15 | 15 | 15 | 15 | 15 | 15 | 15 | 15 | 15 | 15 | 15 | 15 |
| pearson correlation stage 3 | 1 | ,949** | ,959** | ,942** | ,945** | ,947** | ,940** | ,924** | ,933** | ,934** | ,922** | ,928** | ,924** | ,926** | ,934** | ,934** | ,920** | ,926** | ,922** |
| significance (1-seitig) | | ,000 | ,000 | ,000 | ,000 | ,000 | ,000 | ,000 | ,000 | ,000 | ,000 | ,000 | ,000 | ,000 | ,000 | ,000 | ,000 | ,000 | ,000 |
| n | 15 | 15 | 15 | 15 | 15 | 15 | 15 | 15 | 15 | 15 | 15 | 15 | 15 | 15 | 15 | 15 | 15 | 15 | 15 |
| spearman correlation stage 3 | 1,000 | ,957** | ,962** | ,933** | ,954** | ,962** | ,947** | ,945** | ,931** | ,938** | ,892** | ,901** | ,945** | ,945** | ,942** | ,938** | ,894** | ,911** | ,945** |
| Signifikanz (1-seitig) | | ,000 | ,000 | ,000 | ,000 | ,000 | ,000 | ,000 | ,000 | ,000 | ,000 | ,000 | ,000 | ,000 | ,000 | ,000 | ,000 | ,000 | ,000 |
| n | 15 | 15 | 15 | 15 | 15 | 15 | 15 | 15 | 15 | 15 | 15 | 15 | 15 | 15 | 15 | 15 | 15 | 15 | 15 |
| pearson correlation stage 4 | 1 | ,900** | ,918** | ,867** | ,929** | ,937** | ,929** | ,916** | ,928** | ,932** | ,950** | ,948** | ,946** | ,919** | ,929** | ,932** | ,950** | ,948** | ,943** |
| significance (1-seitig) | | ,000 | ,000 | ,000 | ,000 | ,000 | ,000 | ,000 | ,000 | ,000 | ,000 | ,000 | ,000 | ,000 | ,000 | ,000 | ,000 | ,000 | ,000 |
| n | 15 | 15 | 15 | 15 | 15 | 15 | 15 | 15 | 15 | 15 | 15 | 15 | 15 | 15 | 15 | 15 | 15 | 15 | 15 |
| spearman correlation stage 4 | 1,000 | ,937** | ,937** | ,822** | ,916** | ,937** | ,915** | ,922** | ,927** | ,913** | ,961** | ,929** | ,942** | ,922** | ,927** | ,913** | ,958** | ,929** | ,942** |
| significance (1-seitig) | | ,000 | ,000 | ,000 | ,000 | ,000 | ,000 | ,000 | ,000 | ,000 | ,000 | ,000 | ,000 | ,000 | ,000 | ,000 | ,000 | ,000 | ,000 |
| n | 15 | 15 | 15 | 15 | 15 | 15 | 15 | 15 | 15 | 15 | 15 | 15 | 15 | 15 | 15 | 15 | 15 | 15 | 15 |

Appendix 2: List of parameters of body height: anthropometrical measurement and calculations

| participant | anthropometry [m] | body height - stage 3, ability 1 kinect calculation | | | | | | | | | new body height calculation based on kinect data and correction factor | | | | | |
|---|---|---|---|---|---|---|---|---|---|---|---|---|---|---|---|---|
| | | pose 1_left [m] | pose 1_right [m] | **body height_pose 1_average [m]** | pose 2_left [m] | pose 2_right [m] | **body height_pose 2_average [m]** | pose 3_left [m] | pose 3_right [m] | **body height_pose 3_average [m]** | body height_new,1 [m] | diff_1 [cm] | body height_new,2 [m] | diff_2 [cm] | body height_new,3 [m] | diff_3 [cm] |
| 1 | **1,753** | 1,620 | 1,626 | **1,623** | 1,623 | 1,625 | **1,624** | 1,622 | 1,623 | **1,623** | 1,769 | 1,615 | 1,752 | 0,148 | 1,755 | 0,215 |
| 2 | **1,690** | 1,589 | 1,586 | **1,588** | 1,591 | 1,589 | **1,590** | 1,594 | 1,597 | **1,596** | 1,731 | 4,138 | 1,717 | 2,717 | 1,727 | 3,732 |
| 3 | **1,639** | 1,493 | 1,485 | **1,489** | 1,492 | 1,485 | **1,489** | 1,492 | 1,487 | **1,490** | 1,627 | 1,240 | 1,615 | 2,435 | 1,618 | 2,096 |
| 4 | **1,532** | 1,399 | 1,396 | **1,398** | 1,400 | 1,400 | **1,400** | 1,406 | 1,406 | **1,406** | 1,529 | 0,274 | 1,525 | 0,675 | 1,532 | 0,004 |
| 5 | **1,603** | 1,477 | 1,480 | **1,479** | 1,470 | 1,481 | **1,476** | 1,464 | 1,480 | **1,472** | 1,615 | 1,243 | 1,602 | 0,148 | 1,600 | 0,300 |
| 6 | **1,742** | 1,589 | 1,592 | **1,591** | 1,606 | 1,610 | **1,608** | 1,605 | 1,611 | **1,608** | 1,735 | 0,743 | 1,735 | 0,665 | 1,740 | 0,179 |
| 7 | **1,670** | 1,496 | 1,494 | **1,495** | 1,505 | 1,495 | **1,500** | 1,479 | 1,498 | **1,489** | 1,633 | 3,702 | 1,626 | 4,374 | 1,617 | 5,299 |
| 8 | **1,588** | 1,464 | 1,459 | **1,462** | 1,460 | 1,451 | **1,456** | 1,459 | 1,457 | **1,458** | 1,597 | 0,934 | 1,581 | 0,669 | 1,586 | 0,243 |
| 9 | **1,742** | 1,575 | 1,574 | **1,575** | 1,584 | 1,579 | **1,582** | 1,585 | 1,581 | **1,583** | 1,718 | 2,445 | 1,709 | 3,341 | 1,714 | 2,757 |
| 10 | **1,642** | 1,538 | 1,537 | **1,538** | 1,540 | 1,542 | **1,541** | 1,531 | 1,527 | **1,529** | 1,678 | 3,619 | 1,668 | 2,568 | 1,659 | 1,676 |
| 11 | **1,594** | 1,451 | 1,445 | **1,448** | 1,453 | 1,450 | **1,452** | 1,454 | 1,450 | **1,452** | 1,583 | 1,102 | 1,577 | 1,673 | 1,579 | 1,462 |
| 12 | **1,584** | 1,446 | 1,447 | **1,447** | 1,445 | 1,447 | **1,446** | 1,451 | 1,452 | **1,452** | 1,581 | 0,262 | 1,572 | 1,228 | 1,579 | 0,513 |
| 13 | **1,661** | 1,538 | 1,545 | **1,542** | 1,546 | 1,549 | **1,548** | 1,551 | 1,551 | **1,551** | 1,682 | 2,145 | 1,674 | 1,324 | 1,681 | 2,044 |
| 14 | **1,654** | 1,525 | 1,518 | **1,522** | 1,528 | 1,524 | **1,526** | 1,527 | 1,523 | **1,525** | 1,661 | 0,717 | 1,653 | 0,147 | 1,655 | 0,064 |
| 15 | **1,640** | 1,489 | 1,483 | **1,486** | 1,486 | 1,483 | **1,485** | 1,493 | 1,464 | **1,479** | 1,623 | 1,660 | 1,611 | 2,939 | 1,607 | 3,330 |

## Appendix 3: List of parameters of arm span: anthropometrical measurement and calculations

| | | | | kinect calculation | | | new arm span calculation based on kinect data and correction factor | | | | | |
|---|---|---|---|---|---|---|---|---|---|---|---|---|
| stage | participant | arm span_anth [m] | arm span_anth [m] | arm span_1 [m] | arm span_2 [m] | arm span_3 [m] | arm span_new,1 [m] | diff_1 [cm] | arm span_new,2 [m] | diff_2 [cm] | arm span_new,3 [m] | diff_3 [cm] |
| 1 | 1 | 1720 | 1,72 | 1,603 | 1,25 | 1,254 | 1,773 | 5,321 | 1,732 | 1,171 | 1,710 | 1,000 |
| 1 | 2 | 1708 | 1,708 | 1,529 | 1,241 | 1,245 | 1,641 | 6,693 | 1,710 | 0,176 | 1,688 | 2,050 |
| 1 | 3 | 1600 | 1,6 | 1,473 | 1,2 | 1,214 | 1,541 | 5,893 | 1,610 | 0,976 | 1,610 | 1,000 |
| 1 | 4 | 1526 | 1,526 | 1,472 | 1,159 | 1,177 | 1,539 | 1,329 | 1,510 | 1,624 | 1,518 | 0,850 |
| 1 | 5 | 1579 | 1,579 | 1,558 | 1,237 | 1,241 | 1,693 | 11,386 | 1,700 | 12,100 | 1,678 | 9,850 |
| 1 | 6 | 1751 | 1,751 | 1,626 | 1,3 | 1,315 | 1,814 | 6,329 | 1,854 | 10,266 | 1,863 | 11,150 |
| 1 | 7 | 1650 | 1,65 | 1,468 | 1,174 | 1,186 | 1,532 | 11,786 | 1,546 | 10,366 | 1,540 | 11,000 |
| 1 | 8 | 1583 | 1,583 | 1,453 | 1,209 | 1,216 | 1,505 | 7,764 | 1,632 | 4,871 | 1,615 | 3,200 |
| 1 | 9 | 1722 | 1,722 | 1,591 | 1,267 | 1,284 | 1,752 | 2,979 | 1,773 | 5,117 | 1,785 | 6,300 |
| 1 | 10 | 1654 | 1,654 | 1,576 | 1,284 | 1,287 | 1,725 | 7,100 | 1,815 | 16,063 | 1,793 | 13,850 |
| 1 | 11 | 1531 | 1,531 | 1,501 | 1,225 | 1,235 | 1,591 | 6,007 | 1,671 | 13,973 | 1,663 | 13,150 |
| 1 | 12 | 1534 | 1,534 | 1,429 | 1,133 | 1,136 | 1,463 | 7,150 | 1,446 | 8,766 | 1,415 | 11,900 |
| 1 | 13 | 1636 | 1,636 | 1,4 | 1,131 | 1,143 | 1,411 | 22,529 | 1,441 | 19,454 | 1,433 | 20,350 |
| 1 | 14 | 1635 | 1,635 | 1,544 | 1,236 | 1,237 | 1,668 | 3,286 | 1,698 | 6,256 | 1,668 | 3,250 |
| 1 | 15 | 1633 | 1,633 | 1,592 | 1,254 | 1,271 | 1,754 | 12,057 | 1,741 | 10,846 | 1,753 | 11,950 |
| | | | | | | | | | | | | |
| 2 | 1 | 1720 | 1,72 | 1,607 | 1,241 | 1,243 | 1,780 | 6,036 | 1,710 | 1,024 | 1,683 | 3,750 |
| 2 | 2 | 1708 | 1,708 | 1,528 | 1,243 | 1,247 | 1,639 | 6,871 | 1,715 | 0,663 | 1,693 | 1,550 |
| 2 | 3 | 1600 | 1,6 | 1,437 | 1,169 | 1,198 | 1,477 | 12,321 | 1,534 | 6,585 | 1,570 | 3,000 |
| 2 | 4 | 1526 | 1,526 | 1,481 | 1,167 | 1,188 | 1,555 | 2,936 | 1,529 | 0,327 | 1,545 | 1,900 |
| 2 | 5 | 1579 | 1,579 | 1,541 | 1,251 | 1,254 | 1,663 | 8,350 | 1,734 | 15,515 | 1,710 | 13,100 |
| 2 | 6 | 1751 | 1,751 | 1,648 | 1,331 | 1,34 | 1,854 | 10,257 | 1,929 | 17,827 | 1,925 | 17,400 |
| 2 | 7 | 1650 | 1,65 | 1,488 | 1,179 | 1,191 | 1,568 | 8,214 | 1,559 | 9,146 | 1,553 | 9,750 |
| 2 | 8 | 1583 | 1,583 | 1,529 | 1,187 | 1,191 | 1,641 | 5,807 | 1,578 | 0,495 | 1,553 | 3,050 |
| 2 | 9 | 1722 | 1,722 | 1,609 | 1,276 | 1,291 | 1,784 | 6,193 | 1,795 | 7,312 | 1,803 | 8,050 |
| 2 | 10 | 1654 | 1,654 | 1,521 | 1,227 | 1,238 | 1,627 | 2,721 | 1,676 | 2,161 | 1,670 | 1,600 |
| 2 | 11 | 1531 | 1,531 | 1,504 | 1,22 | 1,226 | 1,596 | 6,543 | 1,659 | 12,754 | 1,640 | 10,900 |
| 2 | 12 | 1534 | 1,534 | 1,403 | 1,105 | 1,119 | 1,416 | 11,793 | 1,378 | 15,595 | 1,373 | 16,150 |
| 2 | 13 | 1636 | 1,636 | 1,4 | 1,12 | 1,145 | 1,411 | 22,529 | 1,415 | 22,137 | 1,438 | 19,850 |
| 2 | 14 | 1635 | 1,635 | 1,543 | 1,241 | 1,245 | 1,666 | 3,107 | 1,710 | 7,476 | 1,688 | 5,250 |
| 2 | 15 | 1633 | 1,633 | 1,564 | 1,221 | 1,237 | 1,704 | 7,057 | 1,661 | 2,798 | 1,668 | 3,450 |
| | | | | | | | | | | | | |
| 3 | 1 | 1720 | 1,72 | 1,517 | 1,201 | 1,202 | 1,620 | 10,036 | 1,612 | 10,780 | 1,580 | 14,000 |
| 3 | 2 | 1708 | 1,708 | 1,547 | 1,253 | 1,254 | 1,673 | 3,479 | 1,739 | 3,102 | 1,710 | 0,200 |
| 3 | 3 | 1600 | 1,6 | 1,444 | 1,18 | 1,199 | 1,489 | 11,071 | 1,561 | 3,902 | 1,573 | 2,750 |
| 3 | 4 | 1526 | 1,526 | 1,463 | 1,158 | 1,167 | 1,523 | 0,279 | 1,507 | 1,868 | 1,493 | 3,350 |
| 3 | 5 | 1579 | 1,579 | 1,544 | 1,25 | 1,254 | 1,668 | 8,886 | 1,732 | 15,271 | 1,710 | 13,100 |
| 3 | 6 | 1751 | 1,751 | 1,605 | 1,282 | 1,289 | 1,777 | 2,579 | 1,810 | 5,876 | 1,798 | 4,650 |
| 3 | 7 | 1650 | 1,65 | 1,477 | 1,179 | 1,181 | 1,548 | 10,179 | 1,559 | 9,146 | 1,528 | 12,250 |
| 3 | 8 | 1583 | 1,583 | 1,49 | 1,206 | 1,21 | 1,571 | 1,157 | 1,624 | 4,139 | 1,600 | 1,700 |
| 3 | 9 | 1722 | 1,722 | 1,636 | 1,294 | 1,299 | 1,832 | 11,014 | 1,839 | 11,702 | 1,823 | 10,050 |
| 3 | 10 | 1654 | 1,654 | 1,563 | 1,254 | 1,255 | 1,702 | 4,779 | 1,741 | 8,746 | 1,713 | 5,850 |
| 3 | 11 | 1531 | 1,531 | 1,491 | 1,198 | 1,204 | 1,573 | 4,221 | 1,605 | 7,388 | 1,585 | 5,400 |
| 3 | 12 | 1534 | 1,534 | 1,464 | 1,199 | 1,208 | 1,525 | 0,900 | 1,607 | 7,332 | 1,595 | 6,100 |
| 3 | 13 | 1636 | 1,636 | 1,419 | 1,127 | 1,135 | 1,445 | 19,136 | 1,432 | 20,429 | 1,413 | 22,350 |
| 3 | 14 | 1635 | 1,635 | 1,446 | 1,189 | 1,206 | 1,493 | 14,214 | 1,583 | 5,207 | 1,590 | 4,500 |
| 3 | 15 | 1633 | 1,633 | 1,526 | 1,223 | 1,233 | 1,636 | 0,271 | 1,666 | 3,285 | 1,658 | 2,450 |
| | | | | | | | | | | | | |
| 4 | 1 | 1720 | 1,72 | 1,542 | 1,232 | 1,235 | 1,664 | 5,571 | 1,688 | 3,220 | 1,663 | 5,750 |
| 4 | 2 | 1708 | 1,708 | 1,508 | 1,215 | 1,224 | 1,604 | 10,443 | 1,646 | 6,166 | 1,635 | 7,300 |
| 4 | 3 | 1600 | 1,6 | 1,453 | 1,183 | 1,202 | 1,505 | 9,464 | 1,568 | 3,171 | 1,580 | 2,000 |
| 4 | 4 | 1526 | 1,526 | 1,469 | 1,16 | 1,173 | 1,534 | 0,793 | 1,512 | 1,380 | 1,508 | 1,850 |
| 4 | 5 | 1579 | 1,579 | 1,534 | 1,241 | 1,242 | 1,650 | 7,100 | 1,710 | 13,076 | 1,680 | 10,100 |
| 4 | 6 | 1751 | 1,751 | 1,592 | 1,275 | 1,285 | 1,754 | 0,257 | 1,793 | 4,168 | 1,788 | 3,650 |
| 4 | 7 | 1650 | 1,65 | 1,469 | 1,171 | 1,181 | 1,534 | 11,607 | 1,539 | 11,098 | 1,528 | 12,250 |
| 4 | 8 | 1583 | 1,583 | 1,478 | 1,188 | 1,196 | 1,550 | 3,300 | 1,580 | 0,251 | 1,565 | 1,800 |
| 4 | 9 | 1722 | 1,722 | 1,624 | 1,284 | 1,307 | 1,811 | 8,871 | 1,815 | 9,263 | 1,843 | 12,050 |
| 4 | 10 | 1654 | 1,654 | 1,592 | 1,288 | 1,292 | 1,754 | 9,957 | 1,824 | 17,039 | 1,805 | 15,100 |
| 4 | 11 | 1531 | 1,531 | 1,51 | 1,209 | 1,213 | 1,607 | 7,614 | 1,632 | 10,071 | 1,608 | 7,650 |
| 4 | 12 | 1534 | 1,534 | 1,499 | 1,205 | 1,216 | 1,588 | 5,350 | 1,622 | 8,795 | 1,615 | 8,100 |
| 4 | 13 | 1636 | 1,636 | 1,432 | 1,131 | 1,146 | 1,468 | 16,814 | 1,441 | 19,454 | 1,440 | 19,600 |
| 4 | 14 | 1635 | 1,635 | 1,453 | 1,197 | 1,208 | 1,505 | 12,964 | 1,602 | 3,256 | 1,595 | 4,000 |
| 4 | 15 | 1633 | 1,633 | 1,526 | 1,225 | 1,237 | 1,636 | 0,271 | 1,671 | 3,773 | 1,668 | 3,450 |

## References

Annichini M, Arena R, Fanini M, Fattorel M, Pavei D, Tasson D, Garro V, Lovato C, Giachetti A (2013) Shape processing for digital anthropometry. In: Eurographics Workshop

Aslam M, Rajbdad F, Khattak S, Azmat S (2017) Automatic measurement of anthropometric dimensions using frontal and lateral silhouettes. IET Comput Vis 11(6, 9):434–447

Chiu C-Y, Fawkner S, Coleman S, Sanders R (2016) Automatic calculation of personal body segment parameters with a Microsoft Kinect device. In: Ae M, Enomoto Y, Fujii N, Takagi H (eds) 34 International Conference of Biomechanics in Sport 2016, pp 35–38. https://ojs.ub.uni-konstanz.de/cpa/issue/view/127

DIN 33402-2:2005-12 (2005) Ergonomics - Human body dimensions - Part 2: Values

Flügel B, Greil H, Sommer K (1986) Anthropologischer Atlas. Verlag Tribüne, Berlin

Hamilton MA, Quartuccio J, Mead P, Nunnally A, Lund R, Feild A (2014) Detecting key inter-joint distances and anthropometry effects for static gesture development using Microsoft Kinect. Proc Hum Factors Ergon Soc Ann Meet 58(1):2260–2264

Lun R, Zhao W (2015) A survey of applications and human recognition with Microsoft Kinect. Int J Pattern Recogn Artif Intell 29(05):1–48

Robinson M, Parker MB (2013) Estimating Anthropometry with Microsoft Kinect, pp 1–7. https://www.semanticscholar.org/paper/Estimating-Anthropometry-with-Microsoft-Kinect-Robinson/72db7b00e5c0209abf077aaedd5fb820d1a5bc54?tab=abstract

Wang Q, Kurillo G, Ofli F, Bajcsy R (2015) Evaluation of pose tracking accuracy in the first and second generations of Microsoft Kinect. In: Balakrishnan P (ed) IEEE International Conference on Healthcare Informatics (ICHI). IEEE Computer Society, Dallas, pp 380–389

# Assessment of Dimensional Needs for Designing Spaces for Wheelchair Users

Adrián Leal-Pérez[1(✉)], Libertad Rizo-Corona[1], John Rey-Galindo[1,2], Carlos Aceves-González[1,2], and Elvia González-Muñoz[1,2]

[1] University of Guadalajara, 44250 Guadalajara Jal, Mexico
adrian26leal@gmail.com, libertad.rizo.din@gmail.com, {john.rey, c.aceves, elvia.gmunoz}@academicos.udg.mx
[2] Ergonomics Research Center, 44250 Guadalajara Jal, Mexico

**Abstract.** The design of spaces must take in consideration among other things, the physical characteristics of the variety of users. Regarding buildings of public use this diversity is even more, being used to pay special attention to those users who are considered extremes in certain conditions. A segment among this variety is wheelchair users, who are particularly challenged by the built environment. In the context of this study, it is possible to find some handbooks with guidelines for designing spaces for people using a wheelchair. However, it is not clear where the dimensions come from since there is not a specific reference nor a database with dimensions of people in such condition in that context. The objective of this study was twofold: (1) to assess some design dimensions provided by two handbooks for designing spaces for people using a wheelchair; and (2) to identify and understand the problems that users of wheelchair face in performing some of their daily life activities. An anthropometric survey of 14 wheelchair users was performed and then their dimensions were contrasted against dimensions presented by the handbooks. After an interview about their daily life activities was made, to incorporate them in the formulation of the dimensional parameters for space design. The participants were asked to rate the difficulty level of some indoor and outdoor daily activities. The comparison of the results through coefficient of variation showed that the data used in the accessibility standards handbooks correlates with the mean of some of the dimensions taken in this study, for example, maximum height point near the knee (0.4%), height to the rim handle (9.2%), elbow to elbow width (1.2%). Overall the development of anthropometric data for special populations, in this case, wheelchair users contributes to the creation of a more inclusive society through the improvement of design guides, spaces and policies.

**Keywords:** Wheelchair users · Inclusive spaces · Design guides

## 1 Introduction

The present work examines the daily needs of wheelchair users while using the public spaces as well as public buildings from an ergonomics and human factors perspective. Many government and public associations have developed some guides and handbooks


S. Bagnara et al. (Eds.): IEA 2018, AISC 826, pp. 544–554, 2019.
https://doi.org/10.1007/978-3-319-96065-4_58

to design spaces for the multiple characteristics of the population, among them, wheelchair users, although the provenance of such criteria is unknown.

It is unknown as well, if the design recommendations from these handbooks have a correspondence with the characteristics and needs of wheelchair users. In consequence, the aim of the present project was to identify the correspondence between the guides and handbooks and the actual dimensions of an anthropometric sample and identify the difficulties and needs of this group of users while using the public space in the local area of Guadalajara, Mexico.

### 1.1 Background

**Social Context**

Health system in Mexico traditionally tend to classify people on wheelchair within the concept of disability; however, the new norms and laws recently published are in line with the tendency to differentiate between disability and limitation; for example, the Handbook of Technical Standards and Accessibility [1], or the Accessibility Law for Mexico City [2] distinguishes between people with disabilities and people with limited mobility according to temporality criteria:

"Person with disabilities: Any person who, due to congenital or acquired reasons, has one or more physical, psychosocial, intellectual or sensory deficiencies, whether permanent or temporary and which, when interacting with the barriers imposed by the social environment, may impede their full and effective inclusion, on equal terms with others. Person with limited mobility: a person who temporarily or permanently due to illness, age, accident, surgical operation, genetics or any other condition, makes a slow, difficult or unbalanced journey" [2].

In addition, the latter concept also includes "children and adults who travel with them, pregnant women, seniors, people with luggage or packages that prevent their proper transfers, as well as the person who accompanies them in said transfers" [2]. This indicates that it is recognized that there are very varied characteristics inherent to people that can hinder the use of services in relation to the population without limitations. Moreover, the United Nations through the WHO defines a person with disabilities as that who "is restricted in class or in the number of activities that can perform due to ordinary difficulties caused by a permanent physical or mental condition or more than six months".

According to data from the National Survey of Demographic Dynamics carried out by INEGI [3], the most frequent types of disability in Mexico are limitation to: walking, climbing or descending using the legs (64.1%) and seeing, even when wearing glasses (58.4%) and at the opposite end of the spectrum, difficulty in speaking or communicating (18%) [4]. Therefore, since people with limited mobility constitute the largest group, ensuring the integration of their mobility is considered a priority.

Equally important, according to the World Disability Report [5], physical barriers are an important aspect that demerits the provision of services, interferes with the participation of this segment of the population in education, social and working life, and causes inequality with respect to people without disabilities. The above-mentioned report makes nine recommendations to address these gaps, including strengthening and supporting research on disability, improving data collection on disability and enabling access to all conventional systems and services.

**The Use of Public Space**

The use of public space is essential for the development of activities of daily living, such as going to work, school or even recreation. The conditions of this space directly affect the way in which these activities are carried out and therefore the quality of life of those who use them [6]. The public space is not only the street or the outdoor walkways, but the space built for serving the general population, for the provision of services such as schools, hospitals and government offices, among others. Ideally, the public space should have features that make it safe and easy to use for the entire population. That is, it should consider the characteristics of the diversity of users and correspond to them by allowing easy access without any additional adaptation [7].

Within this diversity of the population, there are especially vulnerable groups such as the elderly, children, pregnant women and people with physical or cognitive disabilities [8, 9]. To respond to these characteristics, space must have materials and dimensions as well as implements that facilitate the safe movement of everybody.

In the particular case of wheelchair users, the dimensions of both the exterior areas and the built spaces for public use are crucial to providing safety and ease of use. In the Outside, for example, sidewalks must be free of obstacles and guarantee a minimum traffic space, and ramps must have an adequate inclination and a waiting area that can hold the person in a wheelchair without putting him or her at risk of sliding into vehicular traffic. Also, the interiors of public buildings must provide spaces with adequate heights such as attention counters, sinks and elevator controls, among other considerations.

People in wheelchairs have needs to be covered in public and private spaces as varied as people without limited mobility. There are currently some official documents, one made by a government health institute and the other by a City municipality as a technical recommendation, which include dimensions of people in wheelchairs that are taken as a reference for the design of public spaces interiors (bathrooms and corridors) to exteriors (sidewalks, entrances, parking lots and ramps).

However, there are two factors that could improve the application of these handbooks. The first is the consideration of the appropriate percentiles, in most cases the extremes, i.e. 5th and 95th [10], following a criterion for each dimension, since if we consider the average in all cases, a large part of the population is excluded. Secondly, it is necessary to validate the data with the Mexican population, since the procedure, origin and date of these measurements are not specified in both documents.

# 2 Method

## 2.1 Participants

The participants were eleven men and three women between the ages of 19 and 50, users of a wheelchair with a total or partial dexterity of the upper extremities. Eight of them were contacted through an association that develops autonomy of people in wheelchairs and the remaining six were contacted individually with the support of a local association and measured in their homes or workplaces.

Participants were required to be independent wheelchair users, i.e., that their transfers did not require the assistance of anyone else. Although all participants were able to move their upper extremities and propel their chair independently, some did not have the ability to flex their fingers, indicating that the minimum mobility profile for their participation in the study corresponds to that described for quadriplegia in cervical vertebra C7.

To be admitted in this study, participants had to be 18 years or older, agree to contribute in the research by signing the informed consent form, need a wheelchair for mobilization and to maintain their upper limbs dexterity in whole or in part. Only the data of those participants for whom sociodemographic and anthropometric data were collected, were taken into account for this study.

## 2.2 Instruments

An anthropometric protocol was designed based on Jarosz and Nowak [11, 12] considering 22 structural and functional measurements. To collect the anthropometric data, the following instruments were used:

- Martin anthropometer
- Vernier type gauge/calipers (400 and 700 mm)
- Handle/Grip cone

Additionally, a questionnaire was adapted with information from the Evaluation of Daily Activity Questionnaire [13] to assess the difficulties of the wheelchair user in everyday life. This questionnaire included a list of seven questions for indoor spaces, seven for outdoor spaces and seven open-ended questions for participants to give their opinions regarding limitations or suggestions for changes in the spaces where they carry out their work activities. This allowed complementing the quantitative approach.

To make the analysis of qualitative data Nvivo V12 software was used.

## 2.3 Procedure

The measures were taken over a period of four months, from November 2017 to February 2018. The activity was divided into three posts to know the reaches, dimensions of the "person-chair" system and some measures of the hand. The objective and procedure of the study were explained to each participant previously to start each stage of the study.

## 2.4 Data Analysis

Collected data, was divided into three categories, anthropometric nature, the perception of the difficulty of daily life activities and opinions regarding daily life and work activities.

The anthropometric data were analyzed using SPSS software, determining the kurtosis and asymmetry of its distribution to verify the normality of the data, as well as the arithmetic mean that would be used for subsequent comparison with the dimensions of the existing handbooks. This comparison was carried out in two simultaneous ways; by comparing the means of their indexes of variation (formulae 1) and by direct comparison of the values.

We started by determining the mean of each dimension and assumed that the measurements in the handbooks were the mean since no more information was available; the following formula was subsequently used:

$$Coefficient\ of\ variation = \left(\frac{data\ from\ observations}{data\ from\ the\ handbooks} * 100\right) - 100 \quad (1)$$

Thus obtaining the coefficient of variation, which is expressed as the percentage difference between the two dimensions.

The data on the perception of difficulty in activities of daily living were assessed on a scale with the values "no difficulty", "little difficulty", "medium difficulty", "high difficulty" and "unable to do so". For this data, the response frequency of the values was determined using the SPSS software. Opinions regarding activities of daily living and work were analyzed with NVivo V12 software using thematic analysis, where the frequency of terms was determined and shown in a graphical form. For this purpose, a transcript was made of all the answers given by the participants in the open-ended questions section on the workspaces and the additional comments provided in the section on the perception of difficulty. Finally, the 20 most relevant terms were maintained.

## 3 Results

The data obtained were compared with the handbooks #1 and #2 made by different government agencies for the design of spaces considering the wheelchair users. It is worth mentioning that it was not possible to compare all the measurements of the anthropometric protocol obtained from the sample, since not all the measurements were included in the handbooks, but only those contained in the protocol developed for the sample and present in the handbooks. It is important to mention that handbook #1 showed the information in a grid with intervals from which to infer the measurements, not explicitly stated. It also shows three views of the same image whose measurements do not correspond to each other, for the purposes of this comparison the measurements of the front and side views were taken.

The comparison of both handbooks was made in two ways: through the coefficient of variation, calculated from the averages of each of the dimensions and through the difference in centimetorthers between the measurement proposed in the handbook and the ideal measurement according to the anthropometric criteria indicated in the Table 1 for each case, obtained from the sample.

From the 17 dimensions compared from the handbooks #1 and #2, nine were found to have a variation coefficient of more than 10% between the sample and the handbooks. Of these nine dimensions, four have a coefficient of variation of more than 20%, and one of them reaching 50%. Although, differences of 10% may seem small, brought

**Table 1.** Comparison between the data from the sample and the handbook #1

| Dimension name | Handbook dimension | Dimension criteria | Sample dimension | Sample mean | Variation coefficient[a] | Gap between dimension (cm) |
|---|---|---|---|---|---|---|
| Lateral reach with straight back | 70 | 5th percentile back straight | 72.5 | 77.5 | 10.7% | 2.5 |
| Lateral reach men | 100 | 5th percentile back straight | 72.5 | 77.5 | 22.5% | 27.5 |
| Lateral reach women | 90 | 5th percentile back straight | 72.5 | 77.5 | 13.8% | 17.5 |
| Lateral reach bending the body | 80 | 5th percentile back straight | 72.5 | 77.5 | 3.1% | 7.5 |
| Frontal reach with straight back | 80 | 5th percentile back straight | 57.5 | 65.5 | 18.1% | 22.5 |
| Frontal reach men | 72 | 5th percentile back straight | 57.5 | 65.5 | 9% | 14.5 |
| Frontal reach women | 67 | 5th percentile back straight | 57.5 | 65.5 | 2.2% | 9.5 |
| Lateral reach bending the body | 85 | 5th percentile back straight | 57.5 | 65.5 | 22.9% | 27.5 |

Note: [a]Obtained from the mean of each dimension compared. For this purpose, it was assumed that dimensions used in the handbook where the mean. Dimensions expressed in centimetres.

to real life on space design applications, impact the everyday use of the space by wheelchair users, resulting in the possibility to enter or not to a place because of the dimensions of a door or corridor, or affecting the possibility to reach some facilities. This is particularly important for users with extreme dimensions (Table 2).

Regarding the perception of difficulty in activities of daily living, in most of the 14 items on the questionnaire, participants reported not using assistive devices neither needing someone else to do the activity for them, although in some cases it was reported that they received help from someone else to answer the phone, open the door, carry heavy objects in the house and do the shopping.

**Table 2.** Comparison between the data from the sample and the handbook #2

| Dimension name | Handbook dimension | Dimension criteria | Sample dimension | Sample mean | Variation coefficient[a] | Gap between dimension (cm) |
|---|---|---|---|---|---|---|
| Total wheelchair height | 91.5 | Average[b] | 79 | 79 | 13.6% | 12.5 |
| Total height to the Wheel handle | 58.5–65/mean 61.7 | 5th and 95th Percentile[b] | 53.5–59 | 56 | 9.2% | 5 y 6 respectively |
| Maximum height point near the knee | 58.5–65/media 61.7 | 5th and 95th Percentile[b] | 56.5–68 | 62 | 0.4% | 2 y 3 respectively |
| Height to the first toe | 20.5 | 95th Percentile | 24 | 17.5 | 14.6% | 3.5 |
| Elbow to elbow width | 80 | 95th Percentile | 92.5 | 81.5 | 1.2% | 12.5 |
| Lateral Reach | 61 | 5th Percentile | 72 | 77 | 26.2% | 11 |
| Frontal reach | 132 | 5th Percentile | 57.5 | 65.5 | 50.3% | 74.5 |

Note: [a]Obtained from the mean of each dimension compared. For this purpose, it was assumed that dimensions used on the handbook where the mean. Dimensions expressed in centimetres. [b]Criteria depends on the application, in this case, the design of spaces

Among the activities with less difficulty were found "opening doors in public buildings", "getting to the telephone in time to answer" although those who reported so, were unable to do it due to a lack of prehensile capacity in the hands. Moving around the house also proved to be of little or no difficulty for most people; it should be noted that during the interview, those subjects who mentioned living in two-storey houses said that they had adapted their space to be able to carry out their daily activities, according to their reasoning, as they did not need to go up to the second floor, it was not a limitation. This reasoning can be seen in Table 3, in relation to going from floor one to floor 2, however, those who had to overcome different obstacles recognized this problem.

The results from the frequency analysis are shown in Fig. 1. The graphic shows the 20 most mentioned topics, showing the main concerns of the wheelchair users while using the public space and buildings. The topics most recurrently mentioned are in order; ramps, spaces, work, and public.

One of the aspects that was frequently reported with some degree of difficulty was travelling by public transport (see Table 3), some participants mentioned preferring to pay more for a private transport service than travelling on city buses, while others highlighted the positive conditions for accessibility on alternative means of public transport such as the subway.

**Table 3.** The frequency of the difficulty reported by activity by the wheelchair users

| Answer | Change from floor 1 to floor 2 | Pick up objects from the floor | Reach objects above the head | Transport heavy objects | Move around in the streets | Use public transport | Front desk procedures | Carry the groceries |
|---|---|---|---|---|---|---|---|---|
| Not difficult | 5 | 10 | 4 | 4 | 2 | 2 | 3 | 3 |
| Little difficulty | 1 | 4 | 4 | 3 | 2 | 1 | 4 | 5 |
| Medium difficulty | – | – | 5 | 4 | 4 | 4 | 3 | 1 |
| High difficulty | 3 | – | 1 | 3 | 5 | 5 | 1 | 3 |
| Unable to do so | 4 | – | – | – | – | 2 | 2 | 2 |
| Lost data | 1 | – | – | – | 1 | – | 1 | – |
| Total | 14 | 14 | 14 | 14 | 14 | 14 | 14 | 14 |

Note: The questionnaire included 14 activities, those which were reported mostly with no difficulty are not shown.

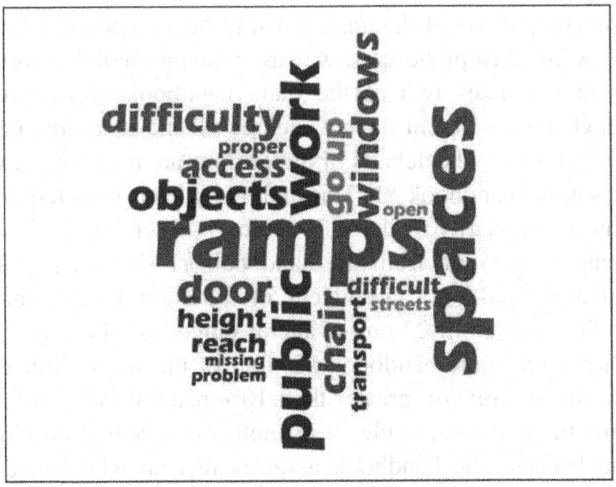


**Fig. 1.** Representation of frequencies of the 20 most mentioned terms in the interviews to wheelchair users (translation from the original in Spanish).

## 4 Discussion

This study allowed to know structural and functional anthropometrics dimensions of some wheelchair users and contrast such dimensions with those provided by two handbooks for designing spaces for people using a wheelchair; and to identify and understand the problems that users of wheelchair face in performing some of their daily life activities.

One of the main constraints of this study was the sample size, therefore, it was not possible to divide the data between men and women for a better comparison with the handbooks, as some of them show dimensions for each group. However, the goal of this research is to initiate the debate regarding the validity of the data available in the current handbooks and design guidelines.

One of the aspects that limited the comparison with the handbooks was the absence of measurement protocols that indicated how the measurements were taken, also the documents do not refer to the origin of the data or the criteria to determine each dimension.

For example, in both handbooks, the use of a single measure is predominant without specifying whether it corresponds to the 5th, 50th or 95th percentile. It is important to note that the use of the 5th and 9th percentile dimensions must be strategic, whether it is a matter of range measurements, where the 5th percentile should be used, or in traffic and clearance measurements, where the 95th percentile should be used to ensure the inclusion of the greatest amount of users. That data is essential to make a correct application of the proposed dimensions to the objectives of each project. Therefore, for the comparison of the data, it had to be assumed that the measurements in the handbooks were taken in the same way as in the protocol developed in this study.

This document compares two of the main handbooks considered as the most complete. Although both of them do not include all the measures necessary for the design spaces for people in wheelchairs or dimensions are not shown explicitly and are difficult to read as in the handbook #1. This ambiguity leads to different interpretations, so the completeness and clarity of the information is essential.

The coefficient of variation from the handbook #1 showed that six of the eight dimensions compared lead to a coefficient of variation greater than 10%, which translates into differences from 22 cm in lateral range, to 14 cm in frontal range for men. On the other hand in the handbook #2, four of the seven dimensions compared showed a coefficient of variation greater than 10% resulting in a difference of 12 cm between the elbow-to-elbow amplitude, or in another case a difference of 74 cm, which can be explained because the handbook assumes that all wheelchair users have the same capabilities and reach flexibility.

These discrepancies may be due to differences in the method of measurement, the indiscriminate use of average dimensions or the omission of the origin and date of the anthropometric data used in these guides and handbooks, which can be considered a methodological error that limits the validity of the data, making more difficult its application.

It was considered important to include the users' perception of their daily activities in order to know the users' perspective and to detect aspects that could not be detected by comparing dimensions.

It also was found that the type of wheelchair they use (active or orthopedic) have a strong impact on the users' self-perception, dimensions, and functionality. Some of the participants mentioned having no difficulty performing activities that others rated as difficult or impossible to do, such as climbing or descending stairs, reaching things over their heads, and so on. However, when doing a little digging during the interview, some of them mentioned the use of technical aids or the adaptation of their environment to reduce or eliminate this difficulty.

## 5 Conclusion

It is concluded that it is necessary to update these guidelines and to incorporate the following recommendations:

1. References on the origin and date of the data submitted.
2. Suitable criteria for determining each of the measures (5°, 95°, range, etc.), as using average data excludes the extremes of the population.
3. The subjective evaluation of users.

The first point ensures the validity of the application of these documents, the second allows the inclusion of population groups with characteristics that differ from the mean and the third point allows us to know aspects that are not evident only with the analysis of anthropometric measurements.

The assessment of the difficulty of daily activities showed that the mobility of people in wheelchairs is still exposed to various obstacles and limitations, so the application of principles contained in handbooks for the design of indoor and outdoor spaces to meet the mobility needs of this group is paramount.

**Acknowledgment.** We would like to thank the associations *Don Bosco Sobre Ruedas* and *COEDIS* for all the support given to contact the participants of this study. Both associations aimed at developing the autonomy of people in wheelchairs and raise awareness in society about this condition.

## References

1. Gobierno de la Ciudad de México (2016) Manual de Normas Técnicas de Accesibilidad, p 159
2. Mancera-Espinoza MA (2017) Ley de la Accesibilidad Para la Ciudad de México
3. INEGI (2015) Encuesta Nacional de la Dinámica Demográfica 2014
4. INEGI (2015) A Propósito Del Día Internacional De Las Personas Con Discapacidad, pp 1–17
5. World Health Organization (2011) World Report On Disability. World Health Organization, Malta
6. Lawton MP, Windley PG, Byerts TO (1982) Aging and the environment: theoretical approaches, 1st edn. Springer, Virginia

7. British Standards Institution (1999) Design management systems. British Standards Institution
8. Reamy BV, Unwin BK (2008) Hyperlipidemia management for primary care. Springer, New York
9. Casajús JA, Vicente-Rodríguez G (2011) Ejercicio Físico y Salud en Poblaciones Especiales, 1st edn. Consejo superior de deportes, Madrid
10. Pheasant S, Haslegrave CM (2005) Bodyspace: anthropometry, ergonomics, and the design of work, 3rd edn. Taylor & Francis, Boca Raton
11. Jarosz E (1996) Determination of the workspace of wheelchair users. Int J Ind Ergon 17:123–133
12. Nowak E (1996) The role of anthropometry in design of work and life environments of the disabled population. Int J Ind Ergon 17:113–121. https://doi.org/10.1016/0169-8141(95)00043-7
13. Hammond A, Tennant A, Tyson S, Nordenskiold U (2016) Evaluation of daily activity questionnaire: v1 (parts 1 and 2), outcome measure, Manchester

# Latin American Schoolchildren Anthropometry: Study of the Anthropometric Differences of the Rural and Urban Zones in Cotopaxi, Ecuador

T. Wendy L. Velasco(✉) and C. Cristina Camacho(✉)

San Francisco de Quito University, Quito 170901, Ecuador
wendy.velasco@estud.usfq.edu.ec, ccamacho@usfq.edu.ec

**Abstract.** Musculoskeletal disorders have been widely studied in the adult population, with a focus on workplaces, due to the high rates of absenteeism in the last decade (Punnett and Wegman 2004). However, there are a limited number of studies that focus on musculoskeletal disorders and child anthropometry. Studies have shown that the lack of proper fit of children to school furniture can generate anatomical and functional changes, and negatively affect the learning process (Castellucci et al. 2014). In Ecuador, as in several Latin American countries, statural growth is influenced by socioeconomic status (Castellucci et al. 2016). For this reason, the present study focuses on the creation of anthropometric tables of school children between the ages of 5 and 7 for the urban and rural zones in Cotopaxi, Ecuador. The data is based on 10 anthropometric measures of a sample size of 300 urban and 300 rural children. The anthropometric profile is then used to analyze the differences between urban and rural children, the potential causes for these differences, and to propose a design for school desks. As a result of the study, prototypes of school desks that comply with ergonomic standards and anthropometric measures of school children were created using 3D printers.

**Keywords:** Ergonomics · Anthropometric data · Latin american children
School furniture design · Musculoskeletal disorders

## 1 Introduction

"Anthropometry refers to the study of the measurements of the human body in terms of the dimensions of bone, muscle and tissue" [1]. Anthropometric data has been used as references in ergonomic studies that evaluate and establish furniture design criteria, so that they have a proper design that ergonomically accommodates a high percentage of the population [1].

Some anthropometric variables such as the weight and growth of the human body have been used as historical reference in order to register individuals' physical characteristics and incidence of diseases [2]. In this aspect, Gutiérrez and Apud [3] established that around 50% of the world population, at some period of their life, suffers from back pain.


S. Bagnara et al. (Eds.): IEA 2018, AISC 826, pp. 555–565, 2019.
https://doi.org/10.1007/978-3-319-96065-4_59

Musculoskeletal disorders have been widely studied in the adult population, with a focus on workplaces, due to the high rates of absenteeism in the last decade [4]. Such importance has been given to the physical well-being of workers, that the design of work furniture that takes into account the user's anthropometry, in order to create ergonomic designs, is a common practice [5].

However, there is a limited number of studies that focus on child musculoskeletal disorders and anthropometry. Al-Ansari and Mokdad [6] describe the importance of the role of anthropometry in the design of school furniture, given that "the quality of life of adulthood is defined by the early stages of life" [7] and "postural vices acquired at an early age are transformed into postural habits for adult life" [3]. Additionally, it has been documented, that "the various disorders that affect the adult can have their origin in childhood" [8]. In fact, in Europe, the 4th most reoccurring cause for insufficient development that has been detected in adolescents between 15 and 19 years old are musculoskeletal symptoms [3, 9].

Musculoskeletal disorders (MSD) are one of the diseases that affect children's statural evolution [9]. Schoolchildren spend a big part of their days sitting on chairs, and since they are in the process of growing up, deficiencies in the design of school furniture can generate bone deformation, among other problems [3]. Furthermore, in Ecuador, as in several Latin American countries, statural evolution is influenced by socioeconomic status [10]. Through previous studies, it has been observed that children of a high socioeconomic level are, on average, higher than those of a lower level socioeconomic status; being one of the main causes for this height difference, the influence of the quality of the alimentation and nutritional value of the population [11]. Considering that the prevalence of growth and MSD preferentially affects the most disadvantaged social classes, such as the ones living in rural areas, the present study focuses on gathering anthropometric measures of the rural and urban Zones in Cotopaxi, Ecuador, and analyzing the differences and potential causes for the measurements differences between the socioeconomic strata [8].

In Central America, the evaluation of socioeconomic levels was implemented through the Food and Nutritional Surveillance System (SISVAN) by using anthropometric indicators [12]. Anthropometric measurements indirectly allow the evaluation of, not only the child's development, but also the body composition and the effects of socioeconomic changes in a country or region [7].

Furthermore, studies have shown that the lack of accommodation of children to school furniture in the classroom, besides generating anatomical and functional changes, can affect the learning process [13]. This situation has generated worldwide concerns in the design of school furniture [2]. In fact, "in the last two decades there has been an adjustment of the standards in relation to tables and chairs for educational institutions" [13] in different countries, such as: Chile, Japan, and England [14–16].

Currently in Ecuador, anthropometric data for school children from 5 to 7 years old, only records weight, height, and body mass indexes (BMI) [17]. However, this information is insufficient to generate adequate standards for the design of school furniture [13].

The significant differences in the statural evolution and anthropometric characteristics of children of different social stratums have not been considered when designing school furniture [7]. In Ecuador, there are currently no anthropometric studies that corroborate the mentioned differences; therefore anthropometric tables from other countries are used [18]. However, it should be emphasized that international standards do not apply to the Ecuadorian population, due to the fact that these do not take into account the measurements of the most vulnerable population segments, which constitute a large percentage of the Ecuadorian population. In this way, the design of school furniture is not standardized for the anthropometric and ergonomic reality of Ecuadorian children. Given the information presented, the present study is extremely important, as it focuses on gathering anthropometric measures of the rural and urban Zones in Cotopaxi, Ecuador, and designing ergonomic school furniture that take into account these measures.

## 2 Methodology

The present study follows the method of "selection and use of people for the testing of anthropometric aspects of products and industrial designs" determined by the ISO 15537 standard [19], and the methodology for "physical design: anthropometry and biomechanics" of González et al. [20]. The following steps are established:

1. Determine the target population
2. Determine the sample size
3. Determine the design limits
4. Analysis of the data
5. Solve the design problems

### 2.1 Target Population

In the study conducted by Guerrero and Yajamín [18] with 400 children from 2 to 5 years old in the city of Quito - Ecuador, it was determined that there is no true accommodation to the highest percentile of the population. Also, in the study conducted by García-Acosta and Lange-Morales [21] carried out on 4611 children between the ages of 10 and 18 in the city of Bogotá - Colombia, there was a disagreement between the school furniture used and the required one. As it can be seen, several studies have focused on the evaluation of adequacy of school furniture; however there is a research gap in the population of 5 to 7 years old.

Due to the aforementioned information, the target population for the present study is children from 5 to 7 years of age, belonging to educational establishments in rural and urban areas of the province of Cotopaxi - Ecuador, with a total of 47,367 children. Among these, 26% (12,327 children) correspond to the urban area, and 74% (35,040 children) correspond to the rural area [22].

## 2.2 Sample Size

Given the limited time for a thorough study of the entire target population, and the limited access to legal representatives' authorization for the participation of children, it has been decided to apply a statistical sampling technique in order to calculate the sample size [23]. Following the sampling process of Nogales [24], the sample size was determined as follows:

The study uses the sample size formula for sampling by proportion of infinite populations, since the target population is greater than 10,000 units [23]. A confidence level of 95% and an acceptable sampling error (e) limit of 5% were used [23]. The proportions established correspond to the proportion of the population of children from urban and rural areas (corresponding to $p = 0.26$ for the urban area), and $1 - p$ as the expected prevalence of the parameter that is not estimated ($1 - p = 0.74$ for the rural area) [23]. The size of the sample calculated with the aforementioned equation is 600 children, 300 of the urban area and 300 of the rural area.

Two affixes are made to the sample, the first one corresponds to the allocation of children by area (urban/rural) for each district of the province of Cotopaxi, and the second one corresponds to the gender by area (urban/rural) for each district from the province of Cotopaxi.

For the affixation of the sample, a number of sample individuals proportional to the size of their population was assigned. The elements of the sample are distributed according to the size of strata 1 and 2, corresponding to the rural and urban areas respectively (Table 1). Regarding gender, information from INEC [25] is taken into account, where 48.5% of the population of the province of Cotopaxi belongs to the male gender, and 51.5% to the female gender (Table 2).

**Table 1.** List of the number of children per strata to be measured in each district. (Source. Own elaboration).

| District address | Rural | Urban |
|---|---|---|
| La Mana | 20 | 67 |
| Latacunga | 104 | 149 |
| Pangua | 22 | 5 |
| Pujilí | 67 | 26 |
| Salcedo | 41 | 29 |
| Saquisilí | 21 | 20 |
| Sigchos | 25 | 5 |
| Total | 300 | 300 |

**Table 2.** List of the number of boys and girls by area to be measured in each district. (Source. Own elaboration).

| District address | Boys rural zone | Girls rural zone | Boys urban zone | Girls urban zone |
|---|---|---|---|---|
| La Mana | 10 | 10 | 32 | 35 |
| Latacunga | 50 | 54 | 72 | 77 |
| Pangua | 11 | 11 | 2 | 3 |
| Pujilí | 32 | 35 | 13 | 13 |
| Salcedo | 20 | 21 | 14 | 15 |
| Saquisilí | 10 | 11 | 10 | 10 |
| Sigchos | 12 | 13 | 2 | 3 |
| Total | 145 | 155 | 145 | 155 |

### 2.3 Design Limits

The limits of anthropometric design are based on the measurements of the physical characteristics of the target population of the study, so that the accommodation of the characteristics can be projected in an ergonomic design of school furniture [20]. To understand the search for adjustment between human measurements and design limits, González et al. [20] establishes the following steps:

1. To establish the human physical characteristics necessary and sufficient for the study.
2. To select the appropriate population, representative for the design desired to make.
3. To determine the relevant statistical points, percentiles, to accommodate the established range of users.

Regarding step 1, in order to determine the human physical characteristics (anthropometric variables) required for the design and evaluation of school furniture, the dependence factor between the user's anthropometry and the dimensions of the object involved in the activity of the sitting posture was evaluated (in this case chairs and desks) [20]. Table 3 shows the anthropometric variables required for the design [20].

Regarding step 2 mentioned above, the target population was previously identified. Regarding step 3, the final determination of the design limits corresponds to establishing the appropriate percentiles to use when designing school furniture so that the appropriate range of the population is accommodated.

The statistical percentile is determined by the ranking of all the values of the sample data and the choice of the points from which the values are outside the scope of the design, either by excess or by default. Usually in ergonomics, the 95$^{th}$ percentile is used, corresponding to the maximum end of the design scope [20]. On the other hand, for the minimum end, it is common to use the ranks of the 1$^{st}$ to 5$^{th}$ percentiles [20]. Given this, and given that the design focuses on adjustable dimensions, González et al. [20] establishes that the 5–95$^{th}$ percentile rank must be met when designing school furniture.

**Table 3.** List of anthropometric variables considered for the design of school furniture. (Source. González et al. [20]).

| Code | Anthropometric dimension | Description | Application |
|---|---|---|---|
| E1 | Height | Vertical distance from the floor to the crown of the subject's head | Defines the relationship of body segments in relation to furniture objects |
| ES2 | Sitting height | Vertical distance from the seat surface to the crown of the subject's head (upright posture) | Defines the position of the body segments in the posture of interaction with the furniture |
| CA3 | Elbow height | Vertical distance from the seat surface to the elbow (arm bent at an angle of 90° in upright posture) | Reference to determine the height of the chair, and the height of the work surface of the table |
| HA4 | Shoulder height | Vertical distance from the seat surface to the acromium (lower and upper shoulder) | It has no direct relationship with the furniture. Used as a reference, the height of the backrest of the chair should not exceed this dimension |
| CM5 | Ulna-hand distance | Horizontal distance from the elbow to the tip of the third finger (middle). (Arm glued to the side, bent at an angle of 90°, extended hand) | It has no direct relationship with the furniture. Used as a reference, the depth of the seat of the chair should not exceed this dimension |
| MA6 | Thigh height | Vertical distance from the seat surface to the highest point of the thigh (sitting posture) | Determines the minimum height of the part below the working surface of the table. One must make a space of slack between the table and the thighs for mobility of the legs |
| AP7 | Knee height | Vertical distance from the sole of the foot to the knee (soles in a position parallel to the floor) | Defines the height of the seat of the chair |
| GP8 | Gluteal-patellar distance | Horizontal distance from the knee to the gluteus | It establishes the maximum depth that the seat of the chair can have, reference for the location of the backrest of the chair |
| AC9 | Hip width | Horizontal distance between the most lateral points of the hip | Determines the minimum width of the seat of the chair |
| AEC10 | Subscapular height | Horizontal distance between the most prominent points of the shoulder blades (sitting upright posture) | Defines the height of the edge of the backrest of the chair, to allow movement of the upper extremities |

## 2.4 Data Analysis

The data collected included: body specified dimensions, age, and gender. All tabulations were written on a single Excel® spreadsheet, obtaining the following parameters: fifth percentile (P5), fiftieth percentile (P50), ninety-fifth percentile (P95) and standard deviation for the rural and urban zones, respectively. For each dimension taken, the minimum value (P5) and the maximum value (P95) were identified, establishing the range that should be covered in future furniture designs. It is important to note, that "although other studies concluded that there are significant anthropometrical differences between genders" [26], a distinction between them was not made in the present study since all school groups in the Cotopaxi region have boys and girls together in the same classroom using the same furniture [21]. Therefore, García-Acosta and Lange-Morales [21] established that doing a gender distinction in the furniture design is not useful for practical purposes and effects.

The analysis of the data gathered in the present study showed as a result, the anthropometric tables for the rural and urban zones, which can be found in the following tables (measures are in centimeters) (Tables 4 and 5).

**Table 4.** Anthropometric table for the rural zone. (Source. Own elaboration).

| Parameter | E1 | ES2 | CA3 | HA4 | CM5 | MA6 | AP7 | GP8 | AC9 | AEC10 |
|---|---|---|---|---|---|---|---|---|---|---|
| $\mu$ | 112 | 90 | 15 | 38 | 29 | 9 | 34 | 38 | 22 | 31 |
| $\sigma$ | 6,2 | 6,8 | 1,9 | 2,6 | 1,9 | 1,1 | 2,1 | 2,5 | 5,2 | 7,9 |

**Table 5.** Anthropometric table for the urban zone. (Source. Own elaboration).

| Parameter | E1 | ES2 | CA3 | HA4 | CM5 | MA6 | AP7 | GP8 | AC9 | AEC10 |
|---|---|---|---|---|---|---|---|---|---|---|
| $\mu$ | 123 | 109 | 16 | 40 | 33 | 10 | 37 | 39 | 25 | 33 |
| $\sigma$ | 4,6 | 8,7 | 1,7 | 2,2 | 2,5 | 1,6 | 2,0 | 2,1 | 9,2 | 8,3 |

From the tables obtained, the difference between the dimensions of each of the populations can be evidenced. Within this aspect, it is important to emphasize the characteristics of height (E1) and width of the hip (AC9), since they present a difference greater than 10% between their maximum values. For the statue there are the maximum values of 120 cm for the rural area, and 131 cm for the urban area; for the width of the hip, the maximum values of 27 cm are for the rural area, and 38 cm for the urban area. The other dimensions are not considered because the difference is not relevant, being less than 10% between the two populations.

## 2.5 Solving the Design Problems

Through the dimensions established for the study populations, the next step was to come up with the design of the furniture. For the design, 3 of the 5 steps of the IDEO methodology were followed, corresponding to observation and research, brainstorming and prototyping [27]. In the first place, the design approach was established, which is an adjustable design with adjustable parts, as mentioned by Altaboli et al. [28], in his design proposal for a desk. In second place, the design of the positioning of the furniture is as two independent pieces, contrary to what Altaboli et al. [28] propose, but in line with what Castelluci et al. [13] propose, which is to design “the table and the chair separately due to the freedom of adaptation that can be given to children”.

For the design of the desk, the dimensions corresponding to the height of the knee and the width of the hip were used to determine the height and width of the desk, respectively. For the height, the ninety-fifth percentile (P95) of the measurement of the knee height (AR7) of the urban area was taken to achieve the maximum adjustment, and the fifth percentile (P5) of the measure of the knee height of the rural area was used (AR7) for the minimum scope. The percentiles mentioned above for the width of the desk were taken according to the width of the hip (AC9), and adding 15 cm needed for the adjustment of the seat to the chair (since this is a separate piece of furniture).

The final design of the desk is made up of 5 pieces: two pieces that are found as a floor base, two support pieces (with holes for adjustment that are located on the back), and the base of the desk (to which the pieces or legs of support are adjusted).

The following is the outline of the proposed design, and the original desk design currently used by the students (Fig. 1).

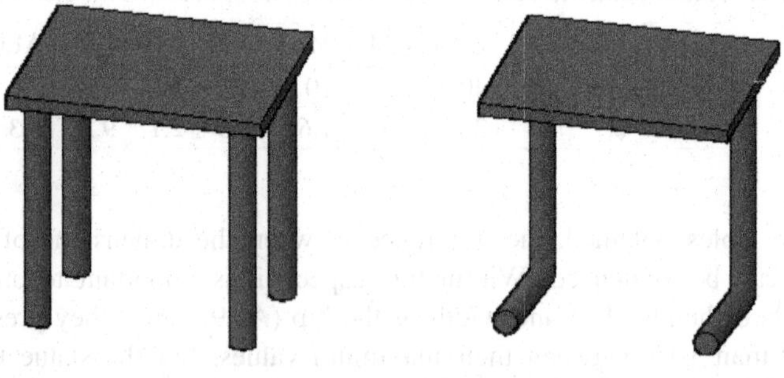

**Fig. 1.** (a) Actual desk design, (b) Proposed desk design. (Source. Own elaboration).

For the design of the chair, the dimensions corresponding to the sitting height, elbow height, shoulder height, thigh height, knee height, gluteal-popliteal distance, hip width and subscapular height were used to determine the height and width of the chair. For each dimension of the chair, the P95 is taken in order to achieve maximum adjustment, and the P5 is used for the minimum reach. The dimensions established for the width of the chair are fixed because of the requirement to fit the chair with the desk (since this a separate piece of furniture). Regarding to the dimensions of the height of the chair, three regulations are established at intervals of 5 cm that facilitate the adjustment of the population studied. The aforementioned regulations are located through pieces that are fitted and adjusted with screws.

The following is the outline of the proposed design, and the actual design of the chair currently used by the students (Fig. 2).

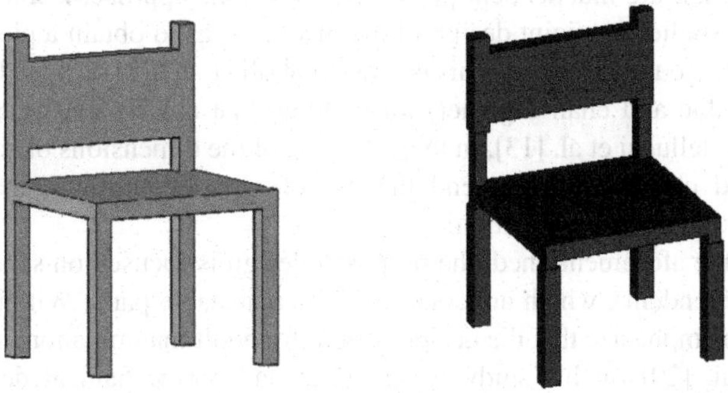

**Fig. 2.** (a) Actual chair design, (b) Proposed chair design. (Source. Own elaboration).

## 3 Conclusions

Several anthropometric studies conducted in the country have focused only on the collection of data and the creation of anthropometric tables, but mainly in adulthood, without taking into account that the causes of musculoskeletal disorders come from poor postural habits of people's childhood [3]. In Ecuador, there are no previous studies or anthropometric tables for the population of children in the age range of 5 to 7 years, as well as no studies related to the accommodation of children with regard to school furniture [18]. In several Latin American regions, research has been promoted within this field due to the need to improve the conditions to which children are exposed [12]. This establishes educational institutions as necessary study places, as these are the places where people spend most of their time during childhood [6]. Given the aforementioned, and as a result of the present research study, anthropometric tables have been obtained for the population determined in the study. This information allows applicability in several fields of study, not only within the country, but also in several neighboring regions that share similarities in their geography and socioeconomic factors to which the population is exposed.

Through the collected data, it has been possible to observe the existence of significant differences in the dimensions of the anthropometric characteristics of the two populations, rural and urban. Although an analysis of the possible causes is not carried out within this study, the hypothesis of nutritional differences is established as the premise of one of the predominant causes of these differences [12]. It is established as a recommendation, the realization of a future investigation focused in a nutritional study that aims at analyzing the causes of the existing anthropometric differences between the two populations defined throughout this study.

Additionally, through the information obtained, an ergonomic design proposal for school furniture has been made, specifically a table and a chair, which adjust to the real dimensions of both segments of the population. In this way, this proposal can be adjusted and regulated according to the height and width needs of each child for the determined age range. As already mentioned, the design proposal focuses on one main characteristic: comfort. Due to this factor, the furniture had a separate design, that is, the table and chair are independent pieces. However, the approach commonly given to these types of studies is a joint design of the pieces, so as to obtain a piece of furniture that is economic to manufacture. This is how Altaboli et al. [28] designed a single piece of furniture (table and chair together) for children that are 7 to 10 years old. On the other hand, Castellucci et al. [13], in their analysis of the dimensions of school furniture for 8 year old children, recommend the use of separate furniture that gives more freedom of mobility and adaptation.

Based on the aforementioned, the proposed design is focused on separate pieces of furniture (independent), which are constituted by adjustable parts. Within this aspect it is important to emphasize that the design is built for adult manipulation, as proposed by González et al. [20], in his study for physical and biomechanical design. In other words, the regulation of the furniture must be done by an adult, so it is guaranteed that the children cannot manipulate the furniture, preventing them from getting hurt or generating any damage to the furniture. Although, within the investigation, specific aspects of the design have not been taken into account, such as issues related to manufacturing costs and materials, a subsequent study is recommended for the benefit-cost analysis of the proposed design as well as feasibility of implementation by educational institutions.

## References

1. Lescay R, Becerra A, González A (2017) Antropometría. Análisis comparativo de las Tecnologías para la Captación de las Dimensiones Antropométricas. Revista EIA 13(26)
2. Maldonado-Macías A, Romero R, Zapata J, Martínez E, Noriega S (2015) Desarrollo de Datos Antropométricos para Niños con Discapacidad Motriz en la Ciudad Juárez. CULCyT (41)
3. Gutiérrez M, Apud E (1992) Estudio Antropométrico y Criterios Ergonómicos para la Evaluación y el Diseño de Mobiliario Escolar. Cuaderno Médico Social 33:72–80
4. Punnett L, Wegman D (2004) Work-related musculoskeletal disorders: the epidemiologic evidence and the debate. J Electromyogr Kinesiol 14:13–23
5. Harris C, Staker L, Pollock C, Trinidad S (2005) Musculo-skeletal outcomes in children using information technology – the need for a specific etiological model. Int J Ind Ergon 35:131–138
6. Al-Ansari M, Mokdad M (2009) Anthropometrics for the design of Bahraini school furniture. Int J Ind Ergon 39:728–735
7. Gracia B, Plata C, Rueda A, Pradilla A (2003) Antropometría por Edad, Género y Estrato Socioeconómico de la Población Escolarizada de la Zona Urbana de Cali. Colombia Médica 34(2):61–68

8. González A, Vila J, Guerra C, Quintero O, Dorta M, Pacheco J (2010) Nutritional condition of school age children. Clinic, anthropo-medical and alimentary assessment. Medisur 8 (2):15–22
9. Balague F, Troussier B, Salminen J (1999) Non-specific low back pain in children and adolescents: risk factors. Eur Spine J 6:429–438
10. Castellucci HJ, Catalán M, Arezes PM, Molenbroek JFM (2016) Evidence for the need to update the Chilean standard for school furniture dimension specifications. Int J Ind Ergon 56:181–188
11. Castellucci HJ, Arezes PM, Viviani CA (2010) Mismatch between classroom furniture and anthropometric measures in Chilean schools. Appl Ergon 41(4):563–568
12. Estrada G, Matienzo G, Apollinaire J, Martínez M, Gómez M, Carmouce H (2008) Perfil Antropométrico Comparado de Escolares Deportistas y no Deportistas. Medisur 5(2):37–45
13. Castellucci HJ, Arezes PM, Molenbroek JFM (2015) Analysis of the most relevant anthropometric dimensions for school furniture selection based on a study with students from one Chilean region. Appl Ergon 46:201–211
14. INN (Instituto Nacional de Normalización Chile) (2002) Norma Chilena 2566, Mobiliaria Escolar – Sillas y Mesas Escolares – Requisitos dimensionales. INN, Santiago de Chile
15. JIS (Japanese Industrial Standards) (2011) JIS S 1021 – School Furniture – Desks and Chairs for General Learning Space. JIS, Tokyo, Japan
16. BSI (British Standard Institution) (2006) BS EN 1729-1: 2006, Furniture – Chairs and Tables for Educational institutions – Part 1: Functional Dimensions. BSI, UK
17. Ministerio de Salud Pública del Ecuador (2011) Protocolo de Atención y Manual de Consejería para el Crecimiento del Niño y la Niña. Recuperado el 22 de octubre de 2017 desde. http://www.opsecu.org/manuales_nutricion/
18. Guerrero W, Yajamín M (2017) Comodidad en el Aula: Retos en la Aplicación de Antropometría en el Diseño de Mobiliario Preescolar. Tesis (Ingeniería Industrial), Universidad San Francisco de Quito, Colegio de Ciencias e Ingeniería, Quito, Ecuador
19. ISO (2009) Principios para la selección y empleo de personas en el ensayo de aspectos antropométricos de productos y diseños industriales (UNE-EN ISO 15537). ISO, Valencia
20. González P, Gómez M (2001) Temas de Ergonomía y Prevención. Universidad Politécnica de Catalunya, Barcelona
21. García-Acosta G, Lange-Morales K (2007) Definition of sizes for the design of school furniture for Bogotá schools based on anthropometric criteria. Ergonomics 50(10):1626–1642. https://doi.org/10.1080/00140130701587541
22. Ministerio de Educación (2015) Zonas, distritos y Circuitos Educativos. Recuperado el 12 de noviembre de 2017 desde. https://educacion.gob.ec/zonas-distritos-y-circuitos/
23. Montgomery DC, Runger GC (2002) Applied statistics and probability for engineers (2nd edn). Limusa Willey, México D.F.
24. Nogales A (2004) Investigación y técnicas de Mercado. ESIC Editorial, Madrid
25. INEC (2010) Fascículo Provincial Cotopaxi. Recuperado el 12 de noviembre de 2017 desde. http://www.ecuadorencifras.gob.ec/wp-content/descargas/Manu-lateral/Resultados-provinciales/cotopaxi.pdf
26. Jeong BY, Park KS (1990) Sex differences in anthropometry for school furniture design. Ergonomics 33(12):1511–1521
27. Hargadon A, Sutton RI (1997) Technology brokering and innovation in a product development firm. Adm Sci Q 42:716–749
28. Altaboli A, Belkhear M, Bosenina A, Elfsei N (2015) Anthropometric evaluation of the design of the classroom desk for the fourth and fifth grades of Benghazi primary schools. Procedia Manufact 3(1):5655–5662

# How to Assess Sitting (Dis)comfort? – An Analysis of Current Measurement Methods and Scales

Annika Ulherr(✉) and Klaus Bengler

Chair of Ergonomics, Technical University of Munich,
Boltzmannstr. 15, 85748 Garching, Germany
{annika.ulherr, bengler}@tum.de

**Abstract.** For at least 60 years, the research community of human factors and ergonomics investigates the topic of comfortable and optimized seating. Great amounts of scientific publications address the evaluation and often the objectification of the positive as well as negative sensations during sitting. The terms of comfort and discomforts are usually applied to determine the subjective feelings, regardless whether it is for aircraft, automotive or office seating. The great variety of scales being used and experiment designs are considered critical regarding the comparability of published results. After presenting the current state of the art, parameters affecting discomfort evaluations and a meta-analysis of already published studies are introduced. On these bases, the paper concludes with arguments for the necessity of more standardization in the area of sitting research, future research goals as well as important questions to be answered and discussed within the research community of sitting comfort/discomfort.

**Keywords:** Comfort · Discomfort · Assessment · Scales · Sitting

## 1 Introduction

Since the middle of the last century, modern humans developed a sedentary lifestyle due to drastic changes in transportation, communication, workplace, and entertainment [1]. According to a German study, office worker spent in total up to 9.6 h per day seated [2]. And even so the negative facts of prolonged sitting, e.g. problems with the metabolic health [1, 3], are common knowledge, people still sit whenever possible and want to do so comfortably.

Especially in the automotive context, the seat is considered crucial for the costumers and their purchase decision [4]. Customers expect the seat to fulfill the comfort requirements [5]. In addition, the importance of interior comfort will rise due to the shift to highly automated or even autonomous driving, since the driver will be able to focus more on comfort aspects, and conduct more often and more diverse secondary tasks in different seating positions [6].

The scientific community of human factors and ergonomics therefore addresses the topic of optimized seating already for a long time. To solely apply ergonomic criteria is not sufficient to ensure comfortable seats [7]. Therefore, subjective sensations need to


S. Bagnara et al. (Eds.): IEA 2018, AISC 826, pp. 566–575, 2019.
https://doi.org/10.1007/978-3-319-96065-4_60

be investigated, interpreted, and transferred into the seat development and seat construction. The identification of subjective sensations while sitting is normally based on the concepts of sitting comfort and discomfort.

Several questions regarding the comparability of published sitting (dis)comfort studies arise due to the vast variety of used terms, models, measurement tools and experiment designs [8]. Most of the published studies investigate similar research questions but with very different scales and methodologies. The authors, therefore, consider it critical to compare results without restriction [8].

The aim of this paper is to investigate parameters of and effects on sitting (dis)comfort experiments to emphasize the need of standardization in the area of sitting research. Additionally, the extent of comparability between different published experimental results will be quantified and presented. After the overview of the current state-of-the art, the core of this paper is a meta-analysis of published (dis)comfort studies aimed at statistically comparing their results. Also the collected data of a largescale discomfort study [9] will be analyzed to show influencing parameters on discomfort experiments.

The conclusion in combination with an extended discussion will present the authors view on future research questions and the necessity of agreed standards for experimental designs to enrich the scientific discourse.

## 2 Sitting Comfort and Discomfort - State of the Art

Hertzberg was one of the first who published a seat comfort/discomfort model, stating that comfort is defined as the absence of discomfort [10]. Further definitions in line with Hertzberg's followed, claiming comfort and discomfort being opposite entities [11, 12].

Zhang, Helander and Drury used a cluster analysis, which sorted the subjective sensations while sitting on an office chair into two separate groups, to define comfort and discomfort as independent assessments, which can occur – in contradiction to Hertzberg - simultaneously [13]. Their model states that 'discomfort' describes the subjective sensations associated with poor biomechanics and fatigue, whereas 'comfort' is dedicated for parameters of well-being [13]. Shortly after, Helander and Zhang verified their model with a field study and thereby underlined that comfort and discomfort assessments can co-occur, and that design only affects comfort [14].

The thereafter-published comfort/discomfort-models are all in principle based on theoretical considerations, and literature reviews instead of experimental investigations. The models by Looze et al., Moes, Vink and Hallbeck, and Naddeo et al. are in agreement with the approach by Zhang et al., and are consecutively built upon each other [15–18].

> *"There is still no consensus as to what type of scale is best for relating comfort/discomfort to pressure"* [19].

This statement by Kyung and Nussbaum from 2008 still describes quite accurately the current situation regarding the comfort and discomfort evaluation methods in general. As published by the authors before, there are no standards regarding the assessment procedures [8]. The used scales, sitting durations, secondary tasks, and number of measurements vary extensively.

## 3 Objectives

Sitting comfort and discomfort are still important and popular research topics. Between the 5th of April 2017 and the same day one year later, 86 new scientific documents were published fitting the keyword search term 'comfort AND sitting' on Scopus. The authors consider the comparability of the published findings critical based on the different used approaches to measure the subjective sensations while sitting [8].

The aim of this paper is to investigate further the quality and adequacy of currently used measurement methods and scales. Therefore, results of studies researching similar topics have to be statistically analyzed, and interpreted altogether [20]. In addition, effects of the experimental setup and the composition of the subject group on the discomfort evaluation are examined in order to gain a better understanding of the impact of the test design on the outcomes.

## 4 Methods

The core of this paper is a meta-analysis to assimilated and objectively evaluate published results of (dis)comfort studies. The results of the Meta-analysis will additionally be compared to the presented findings of a large-scale discomfort study on car seats [9].

### 4.1 Meta-analysis of published (Dis)comfort Experiments

The aim of a meta-analysis is to estimate "the size of an effect in the population by combining effect sizes from different studies that test the same hypothesis" [21]. The first step therefore is an extensive literature search for published experiments with similar, or better the same, research questions.

When the necessary information was available, the effect sizes of the publications were collected, or if needed, calculated or converted in a uniform way, in this case *Cohen's d* [22]. When mean values ($M_i$), standard deviations ($s_i$), and sample sizes ($n_i$) are provided, Cohen's d is calculated:

$$d = \frac{M_A - M_B}{\sqrt{\frac{(n_A - 1)s_A^2 + (n_B - 1)s_B^2}{n_A + n_B - 2}}}. \quad (1)$$

When the statistical analysis in the publications are t-test, and the t-values as well as the group sizes ($n_i$) were reported, the formula for Cohen's d is

$$d = t \times \sqrt{\frac{1}{n_A} + \frac{1}{n_B}}. \quad (2)$$

Some of the publications use an ANOVA (Analysis of Variances) as statistical test. Eta squared ($\eta^2$) is the estimated effect size for an ANOVA [23], which is defined as:

$$\eta^2 = \frac{SS_{effect}}{SS_{total}}, SS_i = Sum\,of\ \ squares. \tag{3}$$

Alternatively, as:

$$\eta^2 = \frac{[F(k-1)]}{[F(k-1)+(N-k)]}. \tag{4}$$

To convert $\eta^2$ into Cohen's d the used formula is:

$$d = 2 \times \sqrt{\frac{\eta^2}{1-\eta^2}}. \tag{5}$$

After collecting Cohen's d for all relevant studies, provided that the needed information was given in the publication, the effect sizes get combined to get a better idea about the true effect [20]. If enough publications were found and effect sizes available, at least 15, there are further investigation methods (random or fixed effect method) applicable [20].

## 4.2 Large-Scale Discomfort Experiment

Within the project UDASim ('Global discomfort assessment for vehicle passengers by simulation') [24], funded by the German ministry of education and research (BMBF), a large-scale discomfort experiment was conducted to collect data sets for an artificial neural network to predict sitting discomfort based on simulation results [9, 25].

The conducted experiment had a mixed design. The independent, with-in subject variables are the seats (three) and the car packages (three). The between-subjects factor are the subject groups F05, M50, and M95 (females with $5^{th}$ percentile body height; males with $50^{th}$ percentile body height; males with the $95^{th}$ percentile body height). The dependent variable is the discomfort, measured four times per configuration for 30 body parts as well as an overall rating using the CP-50 scale.

All further information about the experimental setup, design, procedure, and participants is provided in a previous publication by the authors [9].

The effect sizes for the main effects were calculated based on the contrasts using:

$$r = \sqrt{\frac{F(1, df_R)}{F(1, df_R) + df_R}} \tag{6}$$

and

$$d = \frac{2r}{\sqrt{1-r^2}} \tag{7}$$

## 5 Results

The data of the conducted experiment was analyzed using IBM SPSS Statistics 22. Effect sizes were calculated with Microsoft Excel 2013.

The literature research concluded in 23 relevant papers investigating seating (dis)comfort of road vehicles (22 using car seats and one using truck seats). After reviewing the content of the publications, ten studies [7, 26–34] provided the necessary information to calculate effect sizes (Cohen's d). The authors also see the already known fact, that there is a tendency to publish only significant results, during this research. If at all numerical values are given, these are usually only for statistical significant effects.

Table 1 provides an overview of the absolute values of the calculated effect sizes for the ten publications. If more than one effect size was computable, only the largest one was chosen for the table.

**Table 1.** Effect sizes of the publications included in the meta-analysis

| No. | Effect on | Effect of | Cohen's d | Effect classification [35] | Annotation |
|---|---|---|---|---|---|
| [7] | Comfort (body parts) | Different seats | 1.79 | Large | Same experiment as [27] |
| [26] | Comfort | Car class | 0.22 | Small | |
| | Discomfort | (Sedan, SUV) | 0.27 | Small | |
| [34] | Discomfort (overall) | Support of the front thigh | 2.52 | Very large | |
| [33] | Discomfort (body parts) | Time/sitting duration | 0.65 | Medium | |
| | | Different seats | 0.51 | Medium | |
| [32] | Discomfort (overall) | Time/sitting duration | 2.48 | Very large | |
| | | Vibration | 0.81 | Large | |
| | | Seat foam | 0.33 | Small | |
| [31] | Discomfort (body parts) | Time | 1.15 | Large | Same experimental set up as [30] with a different scale |
| | | Driving position | 0.55 | Medium | |
| [30] | Discomfort (body parts) | Time/sitting duration | 1.14 | Large | |
| | | Driving position | 1.15 | Large | |

*(continued)*

**Table 1.** *(continued)*

| No. | Effect on | Effect of | Cohen's d | Effect classification [35] | Annotation |
|---|---|---|---|---|---|
| [29] | Comfort (overall) | Lumbar support prominence | 0.63 | Medium | |
| [27] | Overall Comfort index | Different seats | 4.66 | Very large | Same experiment as [7] |
| [28] | Discomfort (overall) | Time/sitting duration | 6.24 | Very large | |

The analysis of the mixed design, large-scale discomfort experiment shows several interesting results. Only the main effects on the overall discomfort ratings are reported.

For the main effect of the seat the Greenhouse-Geisser estimate of the departure from sphericity was $\varepsilon = 0.93$. This main effect was not significant $F(1.86, 68.93) = 1.37$, $p = 0.26$. Contrasts revealed that there were no significant differences between the ratings of seat 1 and 2, $F(1, 37) = 1.36$, $p = 0.25$, $r = 0.19$ ($d = 0.38$), as well as no significant differences between the discomfort of seat 2 and 3, $F(1, 37) = 3.16$, $p = 0.08$, $r = 0.28$ ($d = 0.59$).

For the main effect of the package the Greenhouse-Geisser estimate of the departure from sphericity was $\varepsilon = 0.96$. This main effect was not significant $F(1.91, 70.73) = 2.51$, $p = 0.09$. Contrasts revealed that there were no significant differences between the ratings of package 1 and 2, $F(1, 37) = 1.83$, $p = 0.19$, $r = 0.22$ ($d = 0.44$), as well as no significant differences between the overall discomfort of package 2 and 3, $F(1, 37) = 1.04$, $p = 0.32$, $r = 0.17$ ($d = 0.33$).

For the main effect of the time the Greenhouse-Geisser estimate of the departure from sphericity was $\varepsilon = 0.43$. This main effect was significant $F(1.29, 47.76) = 156.40$, $p < 0.01$. Contrasts revealed that there were significant differences between the ratings at the beginning (0 min) and after 15 min, $F(1, 37) = 176.31$, $p < 0.01$, $r = 0.91$ ($d = 4.47$), between 15 and 30 min, $F(1, 37) = 99.72$, $p < 0.01$, $r = 0.85$ ($d = 3.28$), as well as between 30 and 45 min, $F(1, 37) = 64.81$, $p < 0.01$, $r = 0.80$ ($d = 2.65$).

There was no significant effect of the subject group, indicating that the discomfort ratings of females of the $5^{th}$ percentile body height, males of the $50^{th}$ percentile body height and males of the $95^{th}$ percentile body height do not differ, $F(2, 37) = 2.11$, $p = 0.14$, $r = 0.23$ ($d = 0.48$).

The effect sizes of the three main effects on the overall sitting discomfort can be interpreted as a medium effect of the seat, a small effect of the car package and a very large effect of the time. The subject group has a small (almost medium) effect on the overall discomfort ratings.

## 6 Discussion

The meta-analysis showed that although many sitting comfort or discomfort experiments are published, the reported results often contain no numerical values or decisive information such as effect sizes or additional values for calculating them are missing. Therefore, it is not possible now to apply any further statistical methods for a deeper understanding of influences of experimental parameters (e.g. sitting duration) on sitting comfort and discomfort.

The conducted meta-analysis, nevertheless, provided interesting findings especially combined with the additional analysis results of the large-scale discomfort experiment. A closer look at the independent variables of the authors' experiments reveals clear differences in the effects of these on the subjective (dis)comfort ratings. The effect (Cohen's d) of different seats varies between 0.38 (analyzed experiment by the authors) and 4.66 [27] depending on the experiment design. Two analyses of the same experiment have shown that the seat influences the comfort evaluation of individual body parts much less than the overall comfort evaluation [7, 27].

Different scales also seem to have a major influence on the measurements of effects of experimental parameters. Thus, the same underlying test design with the use of two different scales causes the driving position to have a medium [31] or a large [30] effect on the discomfort of individual body parts.

The effect of the sitting duration on the discomfort was already published [36, 37]. The conducted meta-analysis in combination with the statistical analysis of the discomfort experiment supports these findings. The calculated effect size (Cohen's d) of the time on subjective discomfort ratings varies between 0.51 and 4.47, which can be interpreted medium to very large effect of the time. The average effect size of the calculated effects of the sitting duration is:

$$\bar{d}_{time} = \frac{\sum_{i=1}^{k} d_{i,time}}{n} = \frac{4.47 + 0.51 + 2.48 + 1.15 + 1.14 + 6.24}{6} = 2.67 \quad (8)$$

So the average effect size of the sitting duration on the subjective feeling of discomfort can be seen as a very large effect [35].

The relatively low effect of the subject group on the discomfort, given the extreme differences in the body height are an interesting finding but also needs to be investigated further regarding interaction effects with other subjective measurements, for instance BMI or age.

## 7 Conclusion

The aim of this paper was to investigate currently used (dis)comfort measurement methods. To get a deeper understanding and more quantitative results, a meta-analysis of already published sitting comfort and discomfort experiments was conducted.

On the one hand, the meta-analysis shows that relevant information for the calculation of effect sizes is often missing in publications and if it is given, then usually only for significant results.

Nevertheless, the meta-analysis, together with the results of the authors' large-scale discomfort study, shows that the effect sizes vary greatly. The influence of different experiment parameters on the subjective ratings also seem to be affected by the chosen (dis)comfort scale.

The influence of the sitting time on discomfort can be assumed certain, but whether it has to be considered in isolation or interacting with other parameters needs further investigation.

Altogether, the variances of effect sizes in combination with the variances of experimental designs, procedures, and assessment scales, emphasizes the need for higher experimental standards to investigate sitting comfort and discomfort. The authors still recommend a methodical reinvestigation of the human seat interaction [8] to do justice to this important field of seating comfort and discomfort, especially for the important tasks ahead, as automation and digitalization.

**Acknowledgement.** The presented work was neither government-funded nor commissioned by any industrial stakeholder. The used data set was collected within the project UDASim, funded by the German Federal Ministry of Education and Research (BMBF).

## References

1. Owen N, Healy GN, Matthews CE, Dunstan DW (2010) Too much sitting: the population health science of sedentary behavior. Exerc Sport Sci Rev 38:105–113
2. Techniker Krankenkasse (2013) Beweg Dich, Deutschland! TK-Studie zum Bewegungsverhalten der Menschen in Deutschland. Techniker Krankenkasse Pressestelle, Hamburg
3. Healy GN, Dunstan DW, Salmon J, Cerin E, Shaw JE, Zimmet PZ, Owen N (2008) Breaks in sedentary time: beneficial associations with metabolic risk. Diab Care 31:661–666
4. Bubb H, Bengler K, Grünen RE, Vollrath M (2015) Automobilergonomie. Springer Fachmedien Wiesbaden, Wiesbaden
5. Kolich M (2008) A conceptual framework proposed to formalize the scientific investigation of automobile seat comfort. Appl Ergon 39:15–27
6. Kyriakidis M, Happee R, de Winter JCF (2015) Public opinion on automated driving. Results of an international questionnaire among 5000 respondents. Transp Res Part F: Traffic Psychol Behav 32:127–140
7. Kolich M (2003) Automobile seat comfort: occupant preferences vs. anthropometric accommodation. Appl Ergon 34:177–184
8. Ulherr A, Bengler K (2017) Seat assessment – a discussion of comfort and discomfort models and evaluation methods. In: Proceedings of 1st international comfort congress ICC 2017
9. Ulherr A, Zeller F, Bengler K (2018) Simulating seat discomfort. An experimental design for using digital human models. In: Cassenti DN (ed) Advances in Human Factors in Simulation and Modeling: Proceedings of the AHFE 2017 International Conference on Human Factors in Simulation and Modeling, 17–21 July 2017, The Westin Bonaventure Hotel, Los Angeles, California, USA. Springer International Publishing, Cham, pp 354–365
10. Hertzberg HTE (1958) Seat comfort. "Annotated bibliography of applied physical anthropology in human engineering," WADC Technical Report, pp 30–56
11. Wachsler RA, Learner DB (1960) An analysis of some factors influencing seat comfort. Ergonomics 3:315–320

12. Shackel B, Chidsey KD, Shipley P (1969) The assessment of chair comfort. Ergonomics 12:269–306
13. Zhang L, Helander MG, Drury CG (1996) Identifying factors of comfort and discomfort in sitting. Hum Fact J Hum Fact Ergon Soc 38:377–389
14. Helander MG, Zhang L (1997) Field studies of comfort and discomfort in sitting. Ergonomics 40:895–915
15. de Looze MP, Kuijt-Evers LFM, van Dieën J (2003) Sitting comfort and discomfort and the relationships with objective measures. Ergonomics 46:985–997
16. Moes NCCM (2005) Analysis of sitting discomfort, a review. In: Bust PD, McCabe PT (eds) Contemporary ergonomics 2005. Taylor & Francis, London, pp 200–204
17. Vink P, Hallbeck S (2012) Editorial: comfort and discomfort studies demonstrate the need for a new model. In: Vink P, Hallbeck S (eds) Applied ergonomics. special section on product comfort, vol 43, pp 271–276
18. Naddeo A, Cappetti N, Vallone M, Califano R (2014) New trend line of research about comfort evaluation: proposal of a framework for weighing and evaluating contributes coming from cognitive, postural and physiologic comfort perceptions. In: Ahram T, Karwowski W, Marek T (eds) Proceedings of the 5th international conference on applied human factors an ergonomics AHFE 2014
19. Kyung G, Nussbaum MA (2008) Driver sitting comfort and discomfort (part II): relationships with and prediction from interface pressure. Int J Ind Ergon 38:526–538
20. Field AP (2003) Can meta-analysis be trusted? Psychol 16:642–645
21. Field AP (2018) Discovering statistics using IBM SPSS statistics. SAGE Publications, London, Thousand Oaks, California
22. Field AP, Gillett R (2010) How to do a meta-analysis. Br J Math Stat Psychol 63:665–694
23. Tabachnick BG, Fidell LS (2009) Using multivariate statistics. Pearson Education, Boston
24. Ulherr A, Bengler K (2014) Global discomfort assessment for vehicle passengers by simulation (UDASim). In: Proceedings of 3rd international digital human modeling symposium DHM 2014
25. Ulherr A, Yang Y, Bengler K (2017) Implementation of an artificial neural network for global seat discomfort prediction by simulation. In: Proceedings of the 5th international digital human modeling symposium. Bundesanstalt für Arbeitsschutz und Arbeitsmedizin, Dortmund, pp 274–282
26. Kyung G, Nussbaum MA, Babski-Reeves K (2008) Driver sitting comfort and discomfort (part I): use of subjective ratings in discriminating car seats and correspondence among ratings. Int J Ind Ergon 38:516–525
27. Kolich M, Taboun SM (2004) Ergonomics modelling and evaluation of automobile seat comfort. Ergonomics 47:841–863
28. El Falou W, Duchêne J, Grabisch M, Hewson D, Langeron Y, Lino F (2003) Evaluation of driver discomfort during long-duration car driving. Appl Ergon 34:249–255
29. Lim S, Chung MK, Jung J, Na SH (2000) The effect of lumbar support prominence on driver's comfort and body pressure distribution. Proc Hum Fact Ergon Soc Ann Meet 44:308–311
30. Smith J, Mansfield N, Gyi D, Pagett M, Bateman B (2015) Driving performance and driver discomfort in an elevated and standard driving position during a driving simulation. Appl Ergon 49:25–33
31. Smith J, Mansfield N, Gyi D (2015) Long-term discomfort evaluation. Comparison of reported discomfort between a concept elevated driving posture and a conventional driving posture. Procedia Manufact 3:2387–2394
32. Mansfield N, Sammonds G, Nguyen L (2015) Driver discomfort in vehicle seats – Effect of changing road conditions and seat foam composition. Appl Ergon 50:153–159

33. Cardoso M, McKinnon C, Viggiani D, Johnson MJ, Callaghan JP, Albert WJ (2018) Biomechanical investigation of prolonged driving in an ergonomically designed truck seat prototype. Ergonomics 61:367–380
34. Mergl C (2006) Entwicklung eines Verfahrens zu Optimierung des Sitzkomforts auf Automobilsitzen. Garching bei München
35. Cohen J (1992) A power primer. Psychol Bull 112:155–159
36. Hiemstra-van Mastrigt S, Kamp I, van Veen SAT, Vink P, Bosch T (2015) The influence of active seating on car passengers' perceived comfort and activity levels. Appl Ergon 47:211–219
37. Na S, Lim S, Choi H-S, Chung MK (2005) Evaluation of driver's discomfort and postural change using dynamic body pressure distribution. Int J Ind Ergon 35:1085–1096

# Evaluation of an Adaptive Assistance System to Optimize Physical Stress in the Assembly

Katharina Rönick(✉), Thilo Kremer, and Jurij Wakula

Institute of Ergonomics and Human Factors, Technical University Darmstadt, Darmstadt, Germany

k.roenick@iad.tu-darmstadt.de

**Abstract.** In the project "Mittelstand 4.0 Competence Center Darmstadt", an assistance system was developed, which automatically adapts to the individual adjustment of working height, reach and the provision of tools and materials with the help of anthropometric employee data. The data is read in via RFID card. This assistance system has now been evaluated in a study. In addition to reviewing the change in physical loads.

For evaluation, 30 subjects performed a typical assembly activity with and without the support of the assistance system. Using motion capturing, based on the Captiv system of the company TEA ergonomics, the posture was determined without and with the adjustment of the assistance system. For this purpose, the movements of selected parts of the body were analyzed and evaluated. In addition, an interview was conducted to investigate how the users perceived the adjustment of the assistance system and where they saw room for improvement.

**Keywords:** Assistance system · Physical stress
Anthropometric employee data

## 1 Introduction

According to a survey of employed persons conducted by Wittig et al. [1], 54.4% of the 20,036 respondents stated that they often work standing up, about 17% said they worked frequently in a forced posture, i.e. bent, squatting, kneeling or lying down. These strains are a cause of musculoskeletal disorders, which are still one of the main causes of days of illness in employment [2].

The resulting physical limitations increase with increasing age [3]. Developments in production through Industry 4.0 offer new opportunities to optimize workplace stresses. One option for workplace design is the use of adaptive support systems. They can be used at standing workplaces to optimize loads caused by body posture. In the further course of this work, an adaptive assistance system for optimizing the physical loads at standing workplaces in assembly is presented. It processes the individual anthropometric data of the employee and adjusts individually to the body dimensions of the employee. The article describes the results of the evaluation of the assistance system according physical stress.


S. Bagnara et al. (Eds.): IEA 2018, AISC 826, pp. 576–584, 2019.
https://doi.org/10.1007/978-3-319-96065-4_61

## 2 Fundamentals

This chapter defines adaption assistance systems in more detail and introduces the evaluated system for a standing working station.

### 2.1 Definition of Adaption

In the software area, a distinction is made between static and dynamic adaptation. Dynamically adaptive systems adapt context-dependently to provide users with the best software tool in any environment. In static systems, the adaptation does not take place in real time. At certain intervals, adjustments of the program and the functions are carried out independently of the context [4]. Regarding assistance systems, adaptation means systems that adapt independently to a changed environment or to changing requirements by the user. This also includes the support system presented here, which serves as a test vehicle for the evaluation presented below.

### 2.2 Adaptive Assistance System for a Standing Work Station

Weidner et al. [5] define assistance systems as systems that support the user in activities without replacing him. The user retains control over the execution of the activity and the role of system operator. The aim of the systems is to relieve the employee from physical or mental stress.

In order to avoid critical postures, an adaptive assistance system was designed and implemented as part of the "Mittelstand 4.0 Competence Center Darmstadt" project, which is designed to enable ergonomic posture during work by adapting to the user's body measurements. The adaptive position is calculated from the user's body measurements and adjusted using actuators. In addition, a subsequent manual adjustment of the position is possible.

The standing workplace consists of an assembly surface, which is covered with a rubber coating. In addition, two slanted shelves are arranged vertically behind the mounting surface at a distance of 250 mm. The shelves provide space for the reception of containers equipped with components. As a unit, these areas are called material staging and can be moved in height and depth. The overall view of the standing workstation can be seen in Fig. 1.

Users store their individual data of body height, body depth and forward reach on an RFID card. This is done with the aid of a fitting sensor, which is attached to the structure of the assembly table. The data can be select by the system and the adaptive position can be calculated from it.

The adaptive position is defined as those settings of the work surface and material supply which correspond to the calculations for physical and workload size according to [6, 7]. After logging in via RFID card, the system automatically moves to the calculated position. This process is called system adaptation.

Interaction with the system takes place via tablet PC. This is mounted on the right side on a swiveling and inclinable holder. The software consists of individual interfaces, which appear depending on the action to be performed. Possible actions are logon and logoff procedures as well as the possibility to subsequently adjust the system

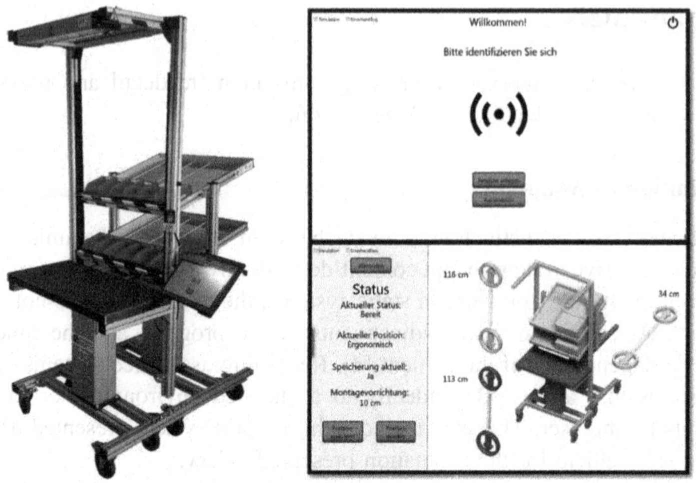


**Fig. 1.** Standing work station with material position area and mounting surface (left), examples of user interface (right)

to personal preferences. This increases the acceptance of the system. Examples for the user interface can be found in Fig. 1.

# 3 Methods

In this chapter the evaluation concept is described in detail. Also an overview of the assembly activity and the distribution of subjects is given. With the concept described above the following questions will be analyzed:

- How does the stress in neck, back and shoulder change with and without using the adaptive assistance system?
- Is there a difference of the postural discomfort with and without using the adaptive assistance system?

## 3.1 Test Procedure

The study to investigate the research questions was conducted as follows. All recorded data were collected anonymously.

At the beginning of the study, the relevant body measurements of the subjects were determined. For this purpose, body height, body depth and forward reach were measured in the subject according to [8] and stored on an RFID card. Subsequently, the test persons were equipped with the Captiv system of the company TEA ergonomics. The Captiv system is a motion capturing system that detects body movement by means of sensors that are attached to the positions of the individual body segments defined by the system. Using the corresponding software, postures of individual body segments can be evaluated. The study focused on the posture of the neck, shoulder and back. After applying the Captiv sensors, the test persons worked at the standing workplace for 30 min each in the adaptive setting and 30 min in a standard setting (working surface at a height of 1000 mm). To exclude possible side effects, the order of settings was randomized. Between the two settings there was a recovery time of 15 min for the subjects to avoid influences of physical fatigue.

To simulate an assembly task, the assembly of a Lego model was selected as the work task to simulate sensomotor operations. The central assembly processes of joining, handling, adjusting and checking are all included in the handling of Lego. The assembly steps were displayed on the interface of the work center during the activity. The parts required for assembling the Lego model were provided in gripper containers in the material supply area. One Lego model was mounted for each setting.

In order to obtain additional subjective assessments of posture in addition to objective measurement data, the subjects were asked to complete the questionnaire for postural discomfort according to Corlett and Bishop [9] at regular intervals. Three questionnaires were taken per setting - before the start of the activity, after half the working time and after the end of the activity. The questionnaires were filled out standing on the assembly table in order not to influence the physical loading process.

Afterwards, a short interview was conducted with the test persons to evaluate the adaptive adjustment and to take up suggestions and wishes for improvement.

### 3.2 Distribution of Subjects

A total of 22 male and 8 female subjects aged between 19 and 67 years (mean age 32.84 years) took part in the study. The physical size varied between 159 and 194 cm, the average size was 182 cm. The test persons had no background in assembly activities, but came from the university environment. Figure 2 shows the contribution of height and age for the test persons. As seen below 7 of the 30 test persons have a height below the 5% percentile or above the 95% percentile.

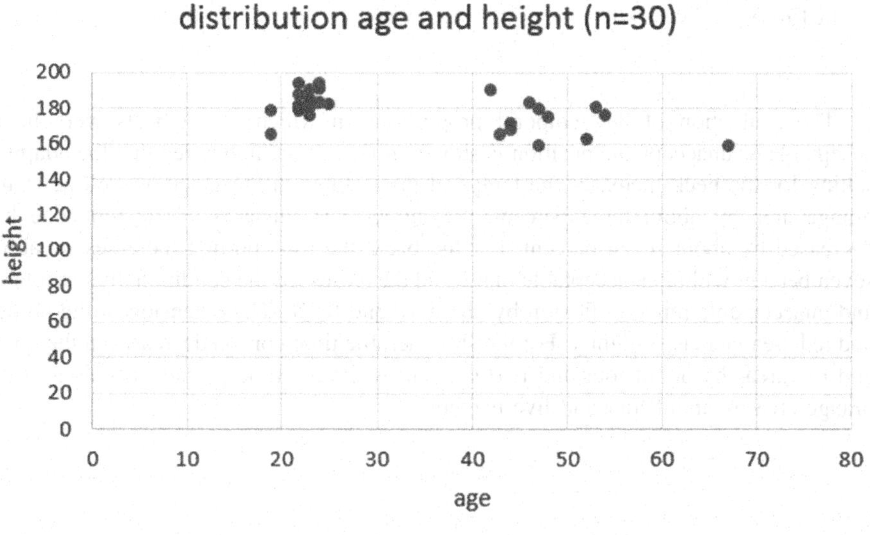


**Fig. 2.** Contribution of age and height of the subjects

# 4 Results

This chapter presents the results of the study. For the statistical evaluation, a t-test was carried out for each of the different questions. If the p value is less than 0.05, the test is significant. The results of the standard setting were compared with the results of the adaptive setting.

## 4.1 Posture-Related Stress

For the investigation of the question whether the stress caused by postures vary between the standard and the adaptive setting, the data of the Captiv system are evaluated. The guideline values formulated in [10] served as the basis for the evaluation of the data by the software-internal Captiv analysis tool. The geometric angles of certain body parts were calculated from the movement data of a recording. The rotation of the shoulder and the extension and flexion of the neck and back are of particular interest for recording the various settings of the adaptive assembly table. Using the analysis tool and in accordance with [10], certain postures are assigned to the zones green, orange and red on the basis of the measured angular relationships. The green zone represents an acceptable posture, the orange zone a conditionally acceptable posture and the red zone an unacceptable posture. The time components of these zones were then analyzed numerically and graphically. The classification of the zones is shown in Table 1. Due to technical problems with Captiv, only 20 data sets could be fully evaluated.

**Table 1.** Classification of the angles into the zones green, orange and red

| Body segment | Neck (extension/flexion) | Shoulder (rotation) | Back (extension/flexion) |
|---|---|---|---|
| Green zone | 0°–40° | 0°–20° | 0°–20° |
| Orange zone | | 20°–60° | 20°–60° |
| Red zone | <0°; >40° | <0°; >60° | <0°; >60° |

The evaluation of the temporal proportions in which the subjects were in an acceptable or unacceptable position is shown in Fig. 3. It can be seen that the adaptive setting for the neck increases the temporal proportions in the green as well as in the orange area by about three percent, the duration in the area of the red zone has decreased by about seven percent. For the back, the time portion increases by about seven percent within an acceptable range and decreases for the conditionally acceptable and unacceptable range of flexion by about 10 and 0,5%. The extension in the orange and red area increases slightly. For the shoulder, the time portion decreases in the green and red areas by about four and two percent, while the time portion increases in the orange area by about four and five percent.

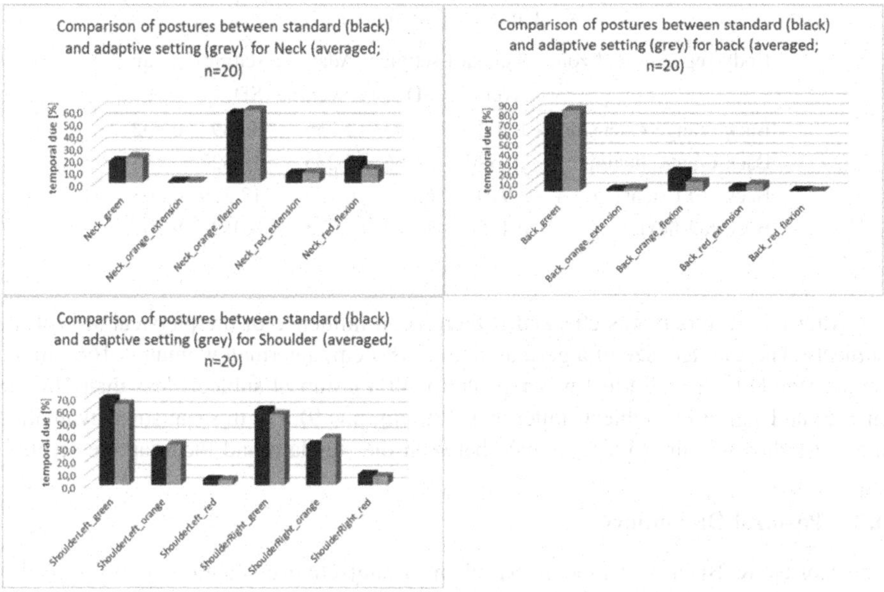


**Fig. 3.** Averaged temporal due postures and zones for neck, shoulder and back

Table 2 shows average, standard deviation (SD) and p-value. The t-test shows that the time components in standard and adaptive setting do not differ significantly from each other (cf. Table 2). Thus, there is no difference in neck, shoulder and back stress in standard or adaptive setting.

**Table 2.** Average, standard deviation (SD) and p-value according posture related stress in neck, shoulder and back (n = 20)

| Body segment and zone | Standard setting | | Adaptive setting | | p-value |
|---|---|---|---|---|---|
| | Average | SD | Average | SD | |
| Neck green | 17,505 | 13,962 | 20,625 | 12,965 | 0,468 |
| Neck orange extension | 0,730 | 0,783 | 0,755 | 0,643 | 0,913 |
| Neck orange flexion | 56,470 | 23,770 | 59,580 | 14,710 | 0,622 |
| Neck red extension | 7,385 | 9,211 | 7,920 | 6,544 | 0,833 |
| Neck red flexion | 17,895 | 20,763 | 11,120 | 15,162 | 0,246 |
| Shoulder Left green | 68,310 | 24,203 | 64,027 | 27,339 | 0,603 |
| Shoulder Left orange | 27,330 | 22,045 | 31,885 | 25,786 | 0,552 |
| Shoulder Left red | 4,360 | 4,750 | 4,085 | 2,864 | 0,826 |
| Shoulder Right green | 59,520 | 18,985 | 55,950 | 23,492 | 0,600 |
| Shoulder Right orange | 32,010 | 18,849 | 37,300 | 22,762 | 0,428 |
| Shoulder Right red | 8,470 | 4,091 | 6,755 | 3,377 | 0,156 |
| Back green | 74,325 | 18,836 | 81,280 | 20,752 | 0,274 |

*(continued)*

**Table 2.** (*continued*)

| Body segment and zone | Standard setting | | Adaptive setting | | p-value |
|---|---|---|---|---|---|
| | Average | SD | Average | SD | |
| Back orange extension | 1,700 | 2,593 | 2,735 | 3,589 | 0,302 |
| Back orange flexion | 19,590 | 21,665 | 9,235 | 18,827 | 0,115 |
| Back red extension | 3,910 | 6,004 | 6,732 | 12,319 | 0,363 |
| Back red flexion | 0,485 | 1,449 | 0,022 | 0,052 | 0,161 |

After this results it was checked if there is an influence of the physical size of the subjects. The average size of a german man is 180 cm, a german woman is 166 cm tall on average [11]. So the t-test was repeated with a group of subjects less than 166 cm (n = 5) and a group of subjects taller than 180 cm (n = 9). For this statistical evaluation the t-test shows again no significance between the standard and the adaptive setting.

### 4.2 Postural Discomfort

The Corlett & Bishop method is based on a subjective evaluation of postural discomfort in relation to defined body regions. According to Corlett and Bishop [9], the subjectively perceived discomfort related to different body segments is evaluated by the test persons on a scale of 0 (none) to 7 (very strong). The questionnaire was completed after 0, 15 and 30 min. A total of 30 data records are available. For the evaluation, only the body segments neck, shoulder and back are considered, as they have already been used for the evaluation of posture.

**Table 3.** Average, standard deviation (SD) and p-value according postural discomfort in neck, shoulder and back (n = 30)

| Body segment | Standard setting | | Adaptive setting | | p-value |
|---|---|---|---|---|---|
| | Average | SD | Average | SD | |
| Neck | 1,230 | 1,501 | 0,470 | 1,106 | 0,028 |
| Back | 1,200 | 1,606 | 0,900 | 1,447 | 0,450 |
| Shoulder left | 0,200 | 0,484 | 0,100 | 0,305 | 0,343 |
| Shoulder right | 0,170 | 0,461 | 0,070 | 0,365 | 0,356 |

The t-test shows a significant difference in the neck area (see Table 3). Back and shoulders are not significant. Thus there is a difference between the standard and the adaptive setting only in the area of the neck as far as subjective discomfort is concerned.

### 4.3 Interview

For the evaluation also an interview was conducted with the subjects to value the adaptive adjustment and obtain further design recommendations for the adaptive assistance system. The qualitative evaluation of the interview shows that 22 of 30 subjects prefer the adaptive attitude of the workplace. 20 of 30 test persons said that no further correction of the adaptive adjustment is necessary. As design recommendations for the assistance system, the test persons mentioned further settings regarding the distance and the angle setting of the material supply levels. Regarding the user interface, the control element could be adjusted even more haptically. 16 of the 30 test persons wanted feedback on the adaptive adjustment of the system.

## 5 Discussion

The evaluation has shown that the setting of the adaptive assistance system influences the subjectively perceived discomfort in the neck area. When evaluating the postures of the neck, shoulders and back the average showed differences between the standard and the adaptive setting of the standing workstation. For every body segment the temporal proportion in an unacceptable posture decreases. But for the t-test no significance could be determined. The t-test with groups of subjects less and taller than the average size of a german man or woman also shows no significance between the settings. According to Bubb [12] the comparison of technical design variants should include 30–50 test persons. Thus, the examination of the differences in posture between standard and adaptive posture should be repeated with more subjects of smaller or taller physical size in order to make a better statement.

The interview of the test persons showed that the adaptive assistance system was assessed as positive by the test persons. The output of feedback on the adaptive setting of the assistance system could further increase the acceptance of the assistance system.

**Acknowledgement.** Special thanks go to the Federal Ministry of Economics and Energy for supporting the project “Mittelstand 4.0 - Kompetenzzentrum Darmstadt”.

## References

1. Wittig P, Nöllenheidt C, Brenscheidt S (2013) Grundauswertung der BIBB/BAuA-Erwerbstätigenbefragung 2012: mit den Schwerpunkten Arbeitsbedingungen. Arbeitsbelastungen und gesundheitliche Beschwerden, Dortmund/Berlin/Dresden
2. Knieps F, Pfaff H (2017) BKK Gesundheitsreport: Vol 2017. Digitale Arbeit - digitale Gesundheit: Zahlen, Daten, Fakten; mit Gastbeiträgen aus Wissenschaft, Politik und Praxis. Medizinisch Wissenschaftliche Verlagsgesellschaft, Berlin
3. Walch D, Günthner WA, Neuberger M (2009) Auswirkungen der demographischen Entwicklung auf die Intralogistik. Ansätze zum Erhalt der Erwerbsfähigkeit von Logistikmitarbeitern. In: Industrie Management 2/2009 - Technologiegetriebene Veränderungen der Arbeitswelt, Bd. 02, 67ff. GITO-Verlag

4. Geihs K (2008) Selbst-adaptive Software. Informatik-Spektrum, vol 31, No 2. Springer, Berlin Heidelberg
5. Weidner R, Redlich T, Wulfsberg JP (2015) Technik, die die Menschen wollen – Unterstützungssysteme für Beruf und Alltag – Definition, Konzept und Einordnung. In: Weidner R, Redlich T, Wulfsberg JP (eds) Technische Unterstützungssysteme. Springer, Berlin Heidelberg
6. Deutsches Institut für Normung e.V. (1986) Körpermaße des Menschen – Teil 2: Werte. (DIN EN ISO 33402-2). Beuth Verlag, Berlin
7. Deutsches Institut für Normung e.V. (07.1986) (1988) Arbeitsmaße im Produktionsbereich; Begriffe, Arbeitsplatztypen, Arbeitsplatzmaße. (DIN EN ISO 33406). Beuth Verlag, Berlin
8. Deutsches Institut für Normung e.V. (2010) Wesentliche Maße des Körpers für die technische Gestaltung – Teil 1: Körpermaßdefinitionen und –messpunkte. (DIN EN ISO 7250-1). Beuth Verlag, Berlin
9. Corlett EN, Bishop RP (1976) A technique for assessing postural discomfort. Ergonomics 19(2):175–182
10. Deutsches Institut für Normung e.V. (2009) Sicherheit von Maschinen – Menschliche körperliche Leistung – Teil 4: Bewertung von Körperhaltungen und Bewegungen bei der Arbeit an Maschinen (DIN EN ISO 1005-4). Beuth Verlag, Berlin
11. Länderdaten Homepage. https://www.laenderdaten.info/durchschnittliche-koerpergroessen.php. Accessed 28 May 2018
12. Bubb H (2003) Wie viele Probanden braucht man für allgemeine Erkenntnisse aus Fahrversuchen? In: Winner H, Landau K (eds) mensch + fahrzeug. Darmstadt

# Anthropometry for Ergonomic Design of Workstations: The Influence of Age and Geographical Area on Workers Variability

M. Micheletti Cremasco[1], A. Giustetto[1](✉), R. Castellone[2], M. P. Cavatorta[2], and S. Spada[3]

[1] Department of Life Sciences and Systems Biology (DBIOS-ICxT), University of Torino, Via Accademia Albertina, 13, 10123 Turin, Italy
ambra.giustetto@unito.it

[2] Department of Mechanical and Aerospace Engineering, Politecnico di Torino, Corso Duca Degli Abruzzi, 24, 10129 Turin, Italy

[3] FCA, EMEA Region – Manufacturing Planning & Control – Direct Manpower Analysis & Ergonomics, Corso Settembrini 53, 10135 Turin, Italy

**Abstract.** The present paper reports some considerations about the results of the project "La Fabbrica si Misura", an anthropometric measurement campaign that involved a large sample of Fiat Chrysler Automobiles (FCA) workers of both genders throughout the Italian plants. The main purpose of our study is to investigate the anthropometric variability of the sample in terms of age and geographical area of the plant, in order to identify differences in workers' dimensions useful to design workstations adopting real users' as well as to choose Personal Protective Equipment or aids suitable in terms of different sizes and body proportions.

This paper shows the preliminary results from the analysis of the male sample anthropometric data collected during 2017. The analyses were conducted on more than 3000 male subjects aged between 18 and 65 years. In order to evaluate the differences between workers from different Italian areas and age classes, subjects from northern, central and southern plants were analysed separately and distinguished by age. Results highlighted differences referred to the variation of the percentiles of the whole sample and emphasize workers dimensional characteristics differences according to age and geographical provenance. Outcomes obtained from the analysis of the anthropometric features of interest resulted consistent with changes occurred during the history of the plant and its location, in terms of job demand, needs and workers' origin. Considerations on workers variability highlighted in the present study could give useful suggestions for the ergonomic design of workstations adapted on users' dimensions and needs.

**Keywords:** Anthropometry · Workstations design · Ergonomics


S. Bagnara et al. (Eds.): IEA 2018, AISC 826, pp. 585–595, 2019.
https://doi.org/10.1007/978-3-319-96065-4_62

# 1 Introduction

## 1.1 Background and Motivation

Anthropometry is the branch of the human sciences that deals with body measurements, particularly with measurements of body size, shape, strength and workload; it plays a significant role in the design of industrial environment, workplaces, machines and tools improving well-being, health, comfort, and safety [1]. A valid and reliable anthropometric database is important to design workplaces according to the ergonomic approach, where man is placed at the centre of the design with the aim of adapt workstations, tools, activities and working environments to human characteristics, needs and limits. Technical designers should refer to anthropometric dimensions collected from the users of interest, especially workers, whether it be "*design for all*" or "*ad hoc* design".

Database like "CAESAR" [2, 3], representative of a sample of North American and European populations, "NHANES" [4, 5] which provides data about the civilian U.S. population and "ANSUR" [6] representative of the U.S. military in the late 1980s are the most consistent and used references available for certain user populations. Some populations and their anthropometric characteristics are not included in these databases and there are difficulties in comparing different studies if they describe concise statistical informations like mean, standard deviation or different key percentiles of different body measurements about a variety of populations. The State Bureau of Technical Supervision [7] and Instituto Nacional de Tecnologia [8] provide anthropometric data, expressed in percentiles, for Chinese and Brazilian civilians, respectively. Barroso et al. [9] provide both summary statistics as well as percentile wise information of anthropometry for the Portuguese population. Many surveys have been directed at specific groups of users, including Philippines manufacturing industry workers [10], female maquiladora workers located along the Mexico-U.S. border [11], Taiwanese workers [12], Indian workers [13], Indonesian workers in the company of roof tiles [14] and male agricultural workers in India [15].

The collection of anthropometric data on samples of specific populations, such as workers, were found to be very useful for the design of workstations, safety equipment works, tools, systems and methods of work, furnishings, and user-friendly workspaces in order to reduce drudgery, increase efficiency, safety and comfort.

The current international standard on human body measurements for technological design ISO 7250-2:2010 [16] provides statistical summaries of body measurements together with database background information for working-age people in different countries: Austria, Germany, Italy, Japan, Kenya, Korea, the Netherlands, Thailand and United States. With regards to the Italian population, the reference in this standard is the anthropometric campaign "L'Italia si Misura" [17], that represents the general Italian adult population in 1990 and 1991.

Considering the normal dimensional variations of populations over time (secular trend) and the influence on populations dimensions caused by social factors such as immigration [18], it is relevant to collect updated and accurate anthropometric information of the population of interest for obtaining effective and reliable results [19]. For the design of workstations, it is fundamental to know the specific dimensions of the sample of workers of interest, to highlight the dimensional (anthropometric), functional and biomechanical subjects variability in relation to the characteristics of different people because of gender and constitution, but also of geographical area.

From the lack of a recent anthropometric database, updated, reliable and representative of the Italian working population, arose the need to carry out a new measurement campaign. "La Fabbrica si Misura" project (the Factory Measures itself), carried out between 2017 and 2018 collecting Fiat Chrysler Automobiles (FCA) workers' anthropometric data, represents an excellent opportunity to re-evaluate the anthropometric characteristics and variability of the Italian working population, in order to design workstations that match with real users scales [20].

### 1.2 Aims of the Study

In companies with plants in different regions, an anthropometric assessment of a large number of workers provides measurements useful to design workstation according to the ergonomic approach. In these valuations, the reference human sample can show strong dimensional (anthropometric), functional and biomechanical variability in relation to the characteristics of different people because of gender and constitution, but also of geographical area. The present paper reports some considerations about data processing from the results of "La Fabbrica si Misura" project, carried out in Italians FCA plants. Our main purpose was to study the anthropometric variability of the male sample considering age and geographical area, in order to identify differences in workers' dimensions useful to design workstations and instruments adopting real users' data. This assessment represents an important photograph of the present FCA manufacturing working context that can be used for several other considerations in order to pursue worker's health and well-being. Workers variability analysis can also contribute to create and update the anthropometric database to supply Personal Protective Equipment or aids suitable in respect of different sizes and body proportions, paying more attention to the worker's variability.

## 2 Materials and Methods

The anthropometric data analysed in the present study, derived from "La Fabbrica si Misura" project, an anthropometric campaign of FCA workers carried out by FCA, Polytechnic of Turin and the National Institute for Accident Insurance at Work (INAIL). The project consists in the acquisition of anthropometric data of at least 6000 subjects (3000 per gender), aged between 18 and 65, belonging to the working population of the FCA Italian production plants, located in different Italian regions from north to south: Piemonte, Emilia-Romagna, Abruzzo, Lazio, Molise, Campania and Basilicata. This project aims to provide support for the generation of virtual mannequins useful in assessing the ergonomic risk of workstations. For each subject anthropometric measurements of the whole body and body segments (depth, heights and widths) were chosen among those useful for virtual design applications in FCA.

The present study analysed the data of the male sample (n = 3174) with ages between 18 and 65 (average age of the whole sample = 43.1 years; dev.st = 10 years).

For each subject, in addition to the geographical reference of the place of work, subject's age and region of birth, as well as parents' region of birth, were derived from the database.

The sample was then divided into 5 age classes: the first class group individuals between the ages of 18 and 25, while the subsequent classes cover a 10-year interval as suggested by ISO 15535: 2012 [21].

Regarding the representativeness of the geographical areas, subjects were subdivided according to their work site in subjects from northern, central and southern plants, and are referred hereafter as north, centre and south, respectively.

In this preliminary analysis, we focused on three of the twelve measured anthropometric variables, choosing three main human body measurements, one for each dimension along the three body axes: body height (6.1.2), shoulder breadth (biacromial) (6.2.7) and body depth standing (6.1.10), measured in accordance with ISO 7250-1:2017 [22]. For each anthropometric variable, 5th, 50th, and 95th percentiles were calculated. These percentiles are indicated hereafter as P5, P50 and P95 respectively.

In order to evaluate differences between workers from different Italian areas, data were also analysed separately, distinguishing by area and age classes, and calculating the three reference percentiles.

# 3 Results

## 3.1 Distribution of the Values of the Three Percentiles Calculated for the Variables of Interest on the Whole Sample

The distribution of percentile values relating to the stature of the entire sample divided into the five age classes is shown in Fig. 1.

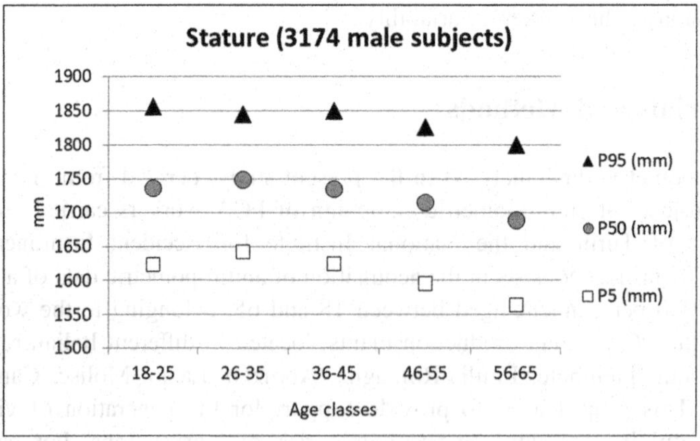


**Fig. 1.** Stature: distribution of the three percentiles of the entire male sample, calculated by age class.

Stature is greater in the first two age classes (18–29 years and 26–35 years) and tends to decrease towards the last age classes. The difference in stature, for all the three percentiles, is limited within the first three age classes while it becomes more important above the age of 45, with a decrease of 25–30 cm between one age group and the next.

In the last age class (56–65 years), given the same percentile, stature is about 50 cm lower than that of the first age class. The trend is rather constant for all the three percentiles, as it is the variation among the percentiles of the same age group. This distribution is due to phenomena such as the secular trend, which documents significant differences in body size between generations, and the physiological lowering of stature in ageing. This consideration is important in the definition of the heights of the workstations, as well as in the definition of their range of adjustment according to age.

With regard to the shoulder breadth, the values tend to increase moving from the lower age class to the greater age class of about 10–20 cm depending on the percentile (Fig. 2). The graph shows that there is less variation between the 5th and 50th percentile in the first age group (18–25 years), while there is a strong difference among the same percentiles in the third age class (36–45 years).

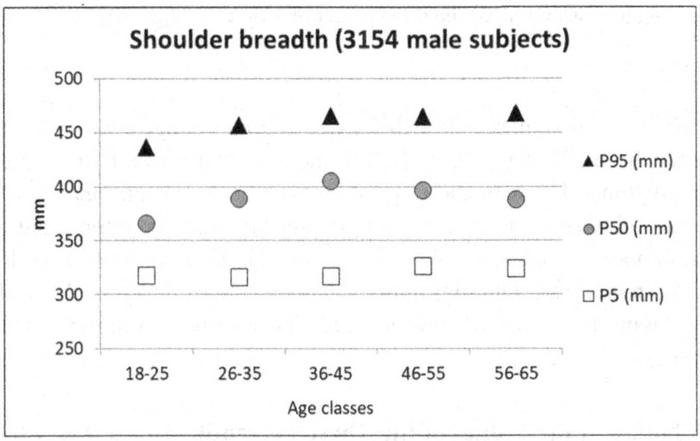


**Fig. 2.** Shoulder breadth: distribution of the three percentiles of the entire male sample, calculated by age class.

The difference among the percentiles of the other age classes is about 70 cm. This indicates the need to provide sufficient variability in the definition of the adaptability range of wearable devices and clothing sizes that involve shoulders and thorax. In general, despite the values of the first age group, the trend is fairly constant. This distribution reflects the normal trend of human bone growth because acromial width increases up to 25 years. The graph is therefore influenced by auxological causes and must be interpreted based on the physiological growth trends of the human bone structure.

Body depth shows a growing trend moving from the younger to the older age classes (Fig. 3). The trend is quite constant with a difference of about 20–60 cm between the value of the first class (18–25 years) and that of the last class (56–65 years) for each percentile.

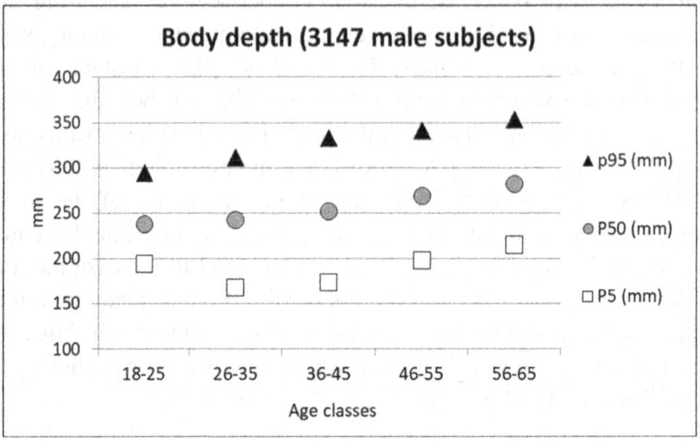


**Fig. 3.** Body depth: distribution of the three percentiles of the entire male sample, calculated by age class.

The graph shows that there is less difference among youngest workers and therefore less difference between the percentiles of the same age group. For the following classes, the difference between the percentiles is almost constant and about 60 cm.

The last age classes show greater body depth, this must be interpreted considering the increase in body thickness in old age in which an adipose deposit on the abdomen prevails. Also, in this case data highlight the importance of designing wearable devices and clothing taking into consideration not only the dimensional difference but also the different shape.

### 3.2 Distribution of the Values of the Three Percentiles of the Age Classes Divided by Geographical Area of the Plant

The sample has been divided according to the three Italian geographical areas (north, centre and south) of the plant and the distribution of data of each percentile for each age class in relation to the working plant location has been analysed. Regarding the stature (Fig. 4), the graph shows that the highest values for all the three percentiles and age classes are referred to the workers employed in plants located in the centre of Italy. The values of the southern sample are lower with an evident difference for age and slightly higher differences towards the older age classes. For each percentile, workers from north plants shows smaller values for stature in almost all age groups. In general, the distribution for age is quite similar between the data of the centre and the south, while the north has more differentiated values.

With regard to the shoulder breadth, the centre has greater values only for the P95 while subjects of the south have greater shoulder breadth values for the P50 and the P5. Data from the workers employed in the north have the most evident irregular trend

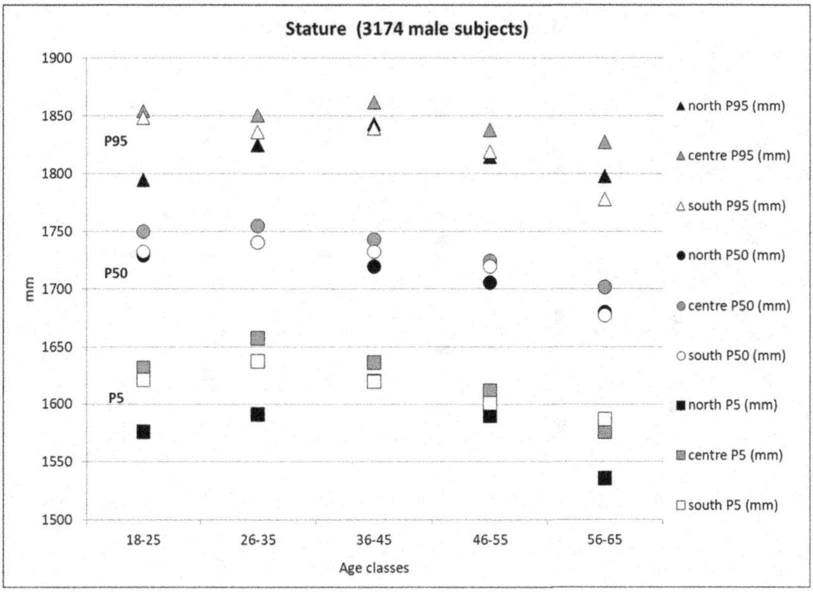


**Fig. 4.** Stature: distribution of the three percentiles of the male sample divided by age group and by geographical area of the plant

because of the 36–45 years class that has greatest values for all the three percentiles. In general, for all the percentiles an increasing tendency is shown: there is indeed a general tendency to a greater shoulder breadth in the older age groups. The lower values, especially in the first classes, may refer to the incomplete shoulders development (length of the clavicle and muscle development) that continue until the age of 25 (Fig. 5).

Regarding body depth, the graph shows a growing trend of values for all the percentiles and geographical areas as the age of the subjects increases (Fig. 6).

While for P50 and P95, central and southern subjects have a similar trend in each age classes, northern subjects show greater values in most cases.

For what concerns the P5, the trend of the subjects of the north and of the centre is more similar while the subjects of the south show much lower values.

The variation between subjects of different geographical areas is very limited in the first age group for the P50, and in the last age group for all three percentiles. In general, the graph shows larger and closer values in the last age group while more dispersed values in the younger age groups. This characteristic can be interpreted considering the difference in shape and constitution between the young and the elderly. The latter, in fact, generally have more similar physical characteristics, due to situations of weight increasing, frequent and consistent in ageing.

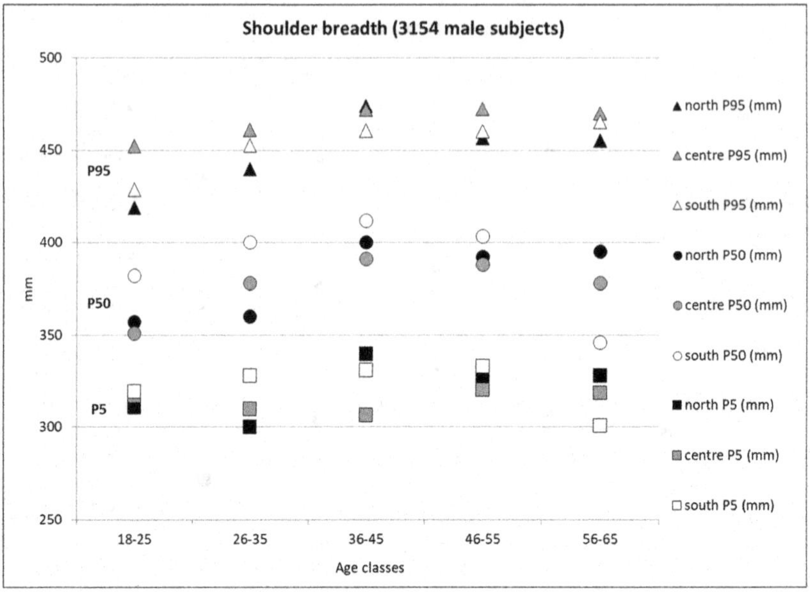


**Fig. 5.** Shoulder breadth: distribution of the three percentiles of the male sample divided by age group and by geographical area of the plant

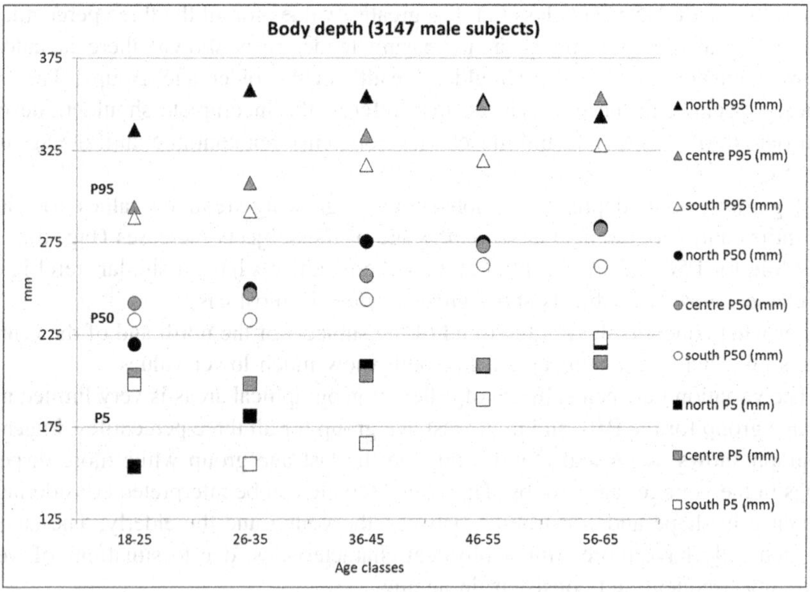


**Fig. 6.** Distribution of the values of the three percentiles relative to the body depth of the male sample divided by age group and geographical area of the plant

## 4 Discussion

With regard to the entire sample, it is evident that the new generations, up to the age of 45, have higher stature with a constant trend between P5, P50 and P95, while older subjects have smaller stature for all percentiles. The shoulder breadth increases between the first and second age classes and then increases more subtly in the following classes for all percentiles of the entire sample in a similar way. Even for the body depth, higher values are observed with increasing age, with a more evident increase after the age of 45. When evaluating the differences by geographical area of the plant, subjects of the centre show higher values for all percentiles considered and for all age classes, while subjects of the north show smallest values (Fig. 4). The south prevails for the P50 and P5 for almost all age classes, subjects of the north have higher values only for the last age class of P50 and P5, while subjects of the centre show greatest values for P95. The shoulder breadth is generally greater for older people, probably because of a more robust constitution, as also shown by the analysis of body depth values, which are greater and more coherent between geographical areas in the 56–66 years age class. The body depth (Fig. 6) has greater values for the north, indicating a higher tendency to overweight and obesity, and this is even more evident considering the relationship between greater body depth and lower stature.

To investigate the differences in characteristics between the subjects of the north and those of the centre and south, which result less different from each other, the composition in terms of the geographic origin of the northern sample has been explored in more detail, analysing the sample in terms of the subject's own and his parent's birth region (Fig. 7).

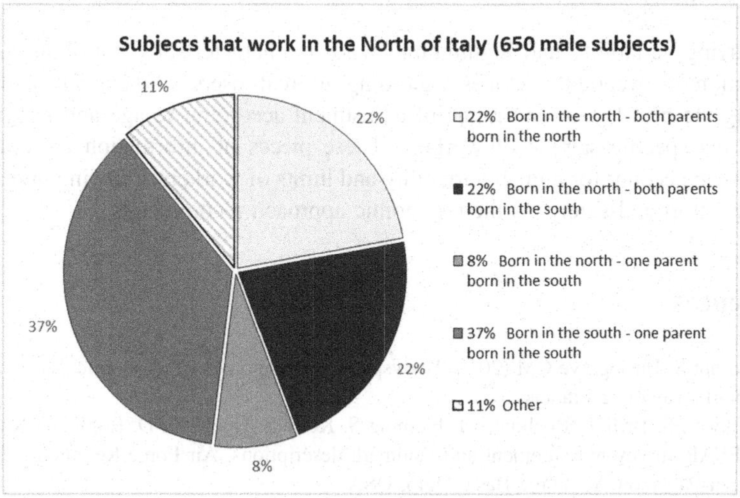


**Fig. 7.** Composition in terms of geographic origin of the sample employed in the north

The graph shows that most workers employed in the north (37%) were born in the south and have at least one parents born in the south. These subjects, added to those born in the north but with both or at least one parent born in the south (respectively 22% and 8%), represent 67% of the sample. Moreover, workers employed in the north are generally older than those from central and southern plants, this might be due to the composition of the northern plants where workers are of old employment, while in the centre and the south there are younger newly hired workers. The composition in terms of place of birth or parents origins, the older age and the stature of the northern sample must be interpreted considering the history of the northern sites: in these plants there is a clear influence of the great immigration of the 50–60's, so people show characteristics typical of the southern of those years. Anthropometric measurements are affected by both genetic and environmental characterization, which modulates the expression according to the conditions involved in growth. Young people or adults who have been employed for less time in the plants of the other two geographic areas are more similar in anthropometric characteristics and are the result of the secular trend and greater genetic mixture, as well as the best growth and lifestyle conditions.

## 5 Conclusion

In companies with plants in different regions, an anthropometric survey of a large number of workers provides useful measurements to design workstation based on the real dimensional, functional and biomechanical workers variability. According to the considerations above, FCA workers show a huge variability among them and data suggest that physical characteristics are related to specific factors like age and ageing, origin, geographical location as well as to socio-economic aspects and historical occurrence.

Studying variability and highlighting workers characteristics contribute to design workstation or wearable devices according to real users' dimensions and needs, choosing the best layout and range of adjustment according to age and geographical area of the specific sample of workers. These pieces of information are crucial for designers to account for human variability and limits of workers, allowing human well-being, comfort and safety, as the ergonomic approach recommends.

## References


1. Pheasant S, Haslegrave CM (2016) Bodyspace : anthropometry, ergonomics, and the design of work. Taylor & Francis
2. Blackwell S, Brill T, Boehmer M, Fleming S, Kelly S, Hoeferlin D, Robinette KM (2008) CAESAR survey measurement and landmark descriptions. Air Force Research Laboratory. Wright-Patterson Air Force Base, Ohio, USA
3. Robinette KM, Blackwell S, Daanen H, Boehmer M, Fleming S (2002) Civilian American and European surface anthropometry resource (CAESAR) final report, Volume I: Summary
4. National Conference on Health Statistics (2008) National health and nutrition examination survey (NHANES), U.S. Centers Disease Control Prevention

5. Calafat A (2006) National health and nutrition examination survey (NHANES) in USA. Epidemiology 17(6): S 76
6. Gordon CC, Churchill T, Clauser CE, Mcconville JT, Tebbetts I, Walker RA (1988) Anthropometric Survey of U.S. Army Personnel : Methods and Summary Statistics. Security, December 2014, p 640
7. The State Bureau of Technical Supervision (1989) Human dimensions of Chinese adults, GB 10000-88
8. Divisao de Desenho Industrial Instituto Nacional de Tecnologia (1998) ERGOKIT, Brazil
9. Barroso MP, Arezes PM, da Costa LG, Sérgio Miguel A (2005) Anthropometric study of Portuguese workers. Int J Ind Ergon 35(5):401–410
10. Del Prado-Lu JL (2007) Anthropometric measurement of Filipino manufacturing workers. Int J Ind Ergon 37(6):497–503
11. Liu WCV, Sanchez-Monroy D, Parga G (1999) Anthropometry of female maquiladora workers. Int J Ind Ergon 24(3):273–280
12. Wang EM-Y, Wang M-J, Yeh W-Y, Shih Y-C, Lin Y-C (1999) Development of anthropometric work environment for Taiwanese workers. Int J Ind Ergon 23:3–8
13. Gite L, Chatterjee D (1999) All India anthropometric survey of agricultural workers: proposed action plan, all India coordinated research project on human engineering and safety in agriculture, central institute of agricultural engineering, Bhopal, India
14. Sutalaksana IZ, Widyanti A (2016) Anthropometry approach in workplace redesign in Indonesian Sundanese roof tile industries. Int J Ind Ergon 53:299–305
15. Vyavahare RTT, Kallurkar SPP (2016) Anthropometry of male agricultural workers of western India for the design of tools and equipments. Int J Ind Ergon 53:80–85
16. International Organization for Standardization (ISO) (2010) ISO/TR 7250-2:2010 - Basic human body measurements for technological design – Part 2: Statistical summaries of body measurements from national populations
17. Masali M, Fubini E, Pierlorenzi G, Millevolte A, Anzil G, Ferrino M, Coniglio I, Fenoglio A, Salis N (2002) The anthropometric survey of Italian population "L'Italia si misura": a decade of research. In: De Waard B, Mooral, T (eds) Human Factors in Transportation, Communication, Health, and the Workplace. Shaker Publication, pp 409–420
18. de Vries C, Garneau CJ, Nadadur G, Parkinson MB (2010) Considering secular and demographic trends in designing for present and future populations. In: 36th design automation conference, Parts A and B, vol 1, pp 391–398
19. Rasmussen J, Waagepetersen RP, Rasmussen KP (2018) Projection of anthropometric correlation for virtual population modelling. Int J Hum Factors Model Simul 6(1):16–30
20. Spada S, Castellone R, Cavatorta MP (2018) "La Fabbrica si Misura": an anthropometric study of workers at FCA Italian plants. To be Present. IEA 2018 – 20th Congress International. Ergon. Assoc. Florence 26–30 August
21. UNI Ente italiano di normazione (2013) UNI EN ISO 15535:2013 - General requirements for establishing anthropometric databases
22. International Organization for Standardization (ISO) (2017) ISO 7250-1:2017 - Basic human body measurements for technological design – Part 1: Body measurement definitions and landmarks

# Modeling People Wearing Body Armor and Protective Equipment: Applications to Vehicle Design

Matthew P. Reed(✉), Monica L. H. Jones, and Byoung-keon Daniel Park

University of Michigan, Ann Arbor, MI 48109, USA
mreed@umich.edu

**Abstract.** Vehicle interiors are complex human-machine interfaces, posing substantial design challenges, particularly when the vehicle is also a workplace. These challenges are compounded by the wide variability in human size, shape, and preference. For law-enforcement officers, firefighters, soldiers, and other workers, specialized clothing or body borne gear can affect their accommodation, comfort, safety, and ability to perform. Digital human modeling has the potential to provide designers with accurate tools to represent human variability, but current software generally lacks the ability to represent accurate seated body shapes for occupants with body borne protective equipment. This paper presents an overview of research to develop body shape modeling tools for vehicles that incorporate body armor representations. Laser scan data drawn from a large-scale study of men in seated postures was used to develop a predictive model that generates body shape as a function of standard anthropometric dimensions and seat and workspace variables. The model is postured using a data-based approach that incorporates the effects of body armor and gear on posture. Importantly, the space claim for the body armor and body borne gear is validated by reference to laser scan data.

**Keywords:** Vehicles · Seats · Body armor · Human models

## 1 Introduction

Vehicles are important workplaces for a large number of people who work in law enforcement, fire-fighting, military, and other occupations. Digital human models (DHM) that represent human body size, shape, and posture in three-dimensional (3D) computer aided engineering (CAE) systems have been useful in improving accommodation and reducing worker stress in these environments. However, few commercial human modeling tools are capable of representing accurately a wide range of seated body shapes and modeling the effects of personal protective equipment (PPE) and body borne gear (BBG) remains a challenge.

This paper gives an overview of a new human modeling system for representing soldiers in vehicles. Data drawn from a large-scale 3D anthropometry survey were used to develop a parametric human body model capable of representing soldiers with a wide variety of size and shape. Posture prediction models, including the effects of body armor


S. Bagnara et al. (Eds.): IEA 2018, AISC 826, pp. 596–601, 2019.
https://doi.org/10.1007/978-3-319-96065-4_63

and BBG, was developed from landmark data obtained in the same study. Finally, the space claim for PPE and BBG is represented using an automatically configurable geometry model. The resulting space claim was validated using 3D scan data.

# 2 Methods

## 2.1 Data Sources

The data for the current work were obtained in the Seated Soldier Study, a large-scale investigation of soldier body shape and posture in vehicle environments [1]. In brief, 315 soldiers were recruited at 3 US Army posts. Using a VITUS XXL laser scanner, each soldier's body shape was captured minimally clad and in up to 24 posture conditions, including some seated and standing conditions with PPE and BBG.

Half of the participants were measured as they sat in a reconfigurable squad seat at a range of seat heights and back angles. Postures were gathered at three ensemble levels: uniform only, uniform + body armor (PPE), and PPE level + BBG. In each condition, a FARO Arm coordinate digitizer was used to record body landmarks, including points describing the position and orientation of the head, thorax, pelvis, and lower extremities.

## 2.2 Modeling Approach

**Body Shape.** Figure 1 shows the modeling approach schematically. Laser scan data from a range of seated postures were used to create a parametric, statistical body shape model that is capable of representing a wide range of male soldier sizes, shapes, and seated postures. The model was parameterized by stature, body weight, and erect sitting height. Because the underlying data included a range of lumbar spine flexions and recline angles, the torso posture could be parameterized based on the side-view eye location with respect to the hip joint centers. The modeling method has been described elsewhere [2]. In brief, a template model is fit to each scan using a two-step method involving radial-basis-function morphing to match landmark locations followed by an implicit surface method for fine fitting. A principal component analysis followed by linear regression is used to relate body shape to standard anthropometric variables.

**Posture.** Seated posture predictions for this squad (fixed-seat) application was based on regression models obtained from landmark data [1]. For the software implementation, two landmark locations were predicted: the position of the mean hip joint center relative to the seat reference point, and the location of the eyes relative to the hips. This effectively gives the position and posture within the seat. In addition to the anthropometric variables, the seat is characterized by seat height and seat back angle. Seat measurements are made using either the SAE J826 H-point manikin or the ISO 5353 seat index point tool [3].

Lower extremity postures can be varied to represent a range of seating configurations. A radial-basis-function morphing method [4] enables changes in thigh, leg, and foot orientation to produce a wide range of postures. The limb shapes in these postures are not validated, but any discrepancies are likely to be smaller than clothing effects.

**Body Armor.** Because body armor specifications are subject to limitations on public distribution, the research team created simplified geometry that approximates the space claim of a certain set of protective gear without including unnecessary details. Four sizes of body armor vest and helmet were created. The vest geometry was segmented into front, rear, and top components, each connected by reconfigurable geometric elements. In the software implementation, the correct size is first selected based on a sizing formula generated in actual fit tests. The vest is the applied to the human figure using methods designed to produce reliable overall dimensions – accurate representation of the human/armor interface was not an objective.

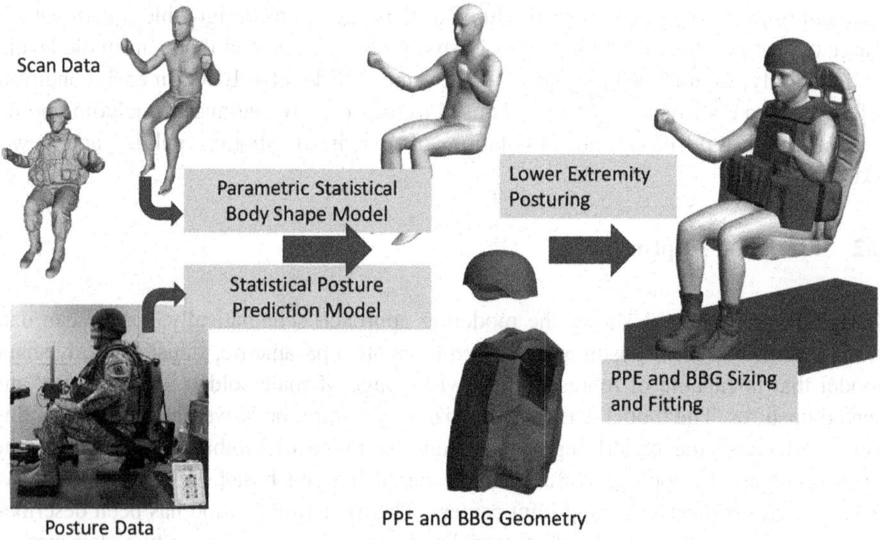


**Fig. 1.** Schematic of modeling approach.

# 3 Results

## 3.1 Representing Posture and Body Shape

Figure 2 shows a range of body shapes and postures. The figures were obtained by manipulating stature and body weight over a wide range, along with variation in seat height and seat back angle.

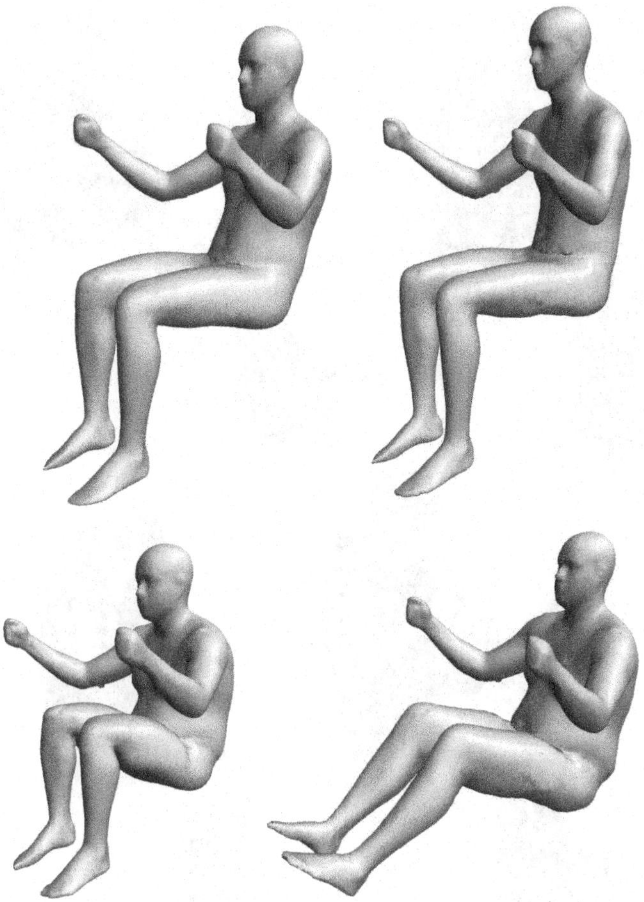

**Fig. 2.** Outputs from the body shape and posture models.

### 3.2 Body Armor and Gear

Figure 3 illustrates the simulation of body armor and BBG. The DHM is placed in a seat using the posture prediction model, demonstrating how the simulated figure scan be used to assessing accommodation by the seat and surrounding environment.

## 4 Discussion

### 4.1 Achievements

To our knowledge, this is the first parametric body shape model of seated adults based on an extensive database. The integrated posture-prediction capability is also the first to take into account the effects of PPE and BBG. Finally, the space claim of the simulated ensembles has been validated with reference to detailed scan data from over 200

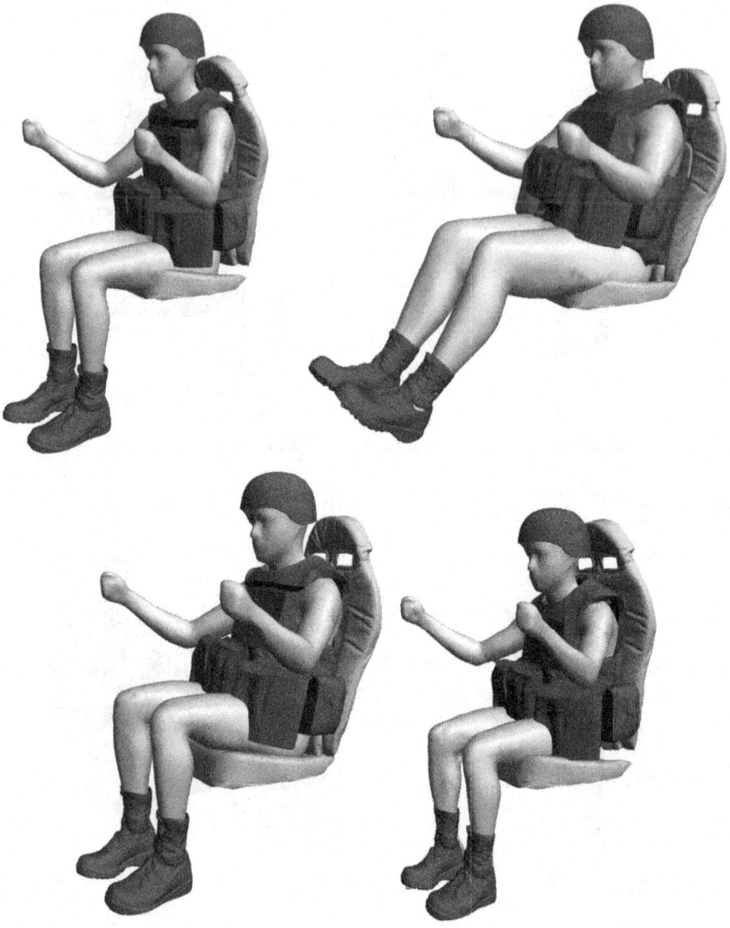

**Fig. 3.** Model outputs including body armor and body borne gear.

individuals. The models presented here are currently in use for seat design and assessment. A wider range of statistical body shape models for civilian applications is available online at http://humanshape.org/.

### 4.2 Limitations

The body shape and posture data were gathered from a diverse group of young soldiers, and hence the model may not be appropriate for other populations. Currently, only a male model is available. The posture manipulations of the limbs are not validated, although the data needed to do so are available. The simulated PPE and BBG represent only a small number of configurations. The posturing model is based solely on statistical relationships; the physical interaction between the sitter and seat is not modeled. This enables a highly efficient simulation that is validated in important ways (for example, reliable prediction of hip and head location). However, the effects of unusual seat configurations cannot be simulated by this method; additional data on posture and shape would be needed to generalize to other seats and PPE ensembles. The relationship between the body surface and the PPE and BBG geometry is not physically realistic. For example, the BBG may intersect the body surface, as visible in Fig. 3. Because the purpose of the simulations is to provide guidance for vehicle design, rather than for PPE design, only the overall positioning and space claim is validated. A more complex simulation is needed to apply the body shape model to PPE design and evaluation.

## References


1. Reed MP, Ebert SM (2013) The seated soldier study: posture and body shape in vehicle seats. Technical report UMTRI-2013-13. University of Michigan Transportation Research Institute, Ann Arbor, Michigan, USA
2. Park B-KD, Ebert S, Reed MP (2017) A parametric model of child body shape in seated postures. Traffic Inj Prev 18(5):533–536
3. Reed MP, Ebert SM (2016) Evaluation of the seat index point tool for military seats. SAE Technical Paper 2016-01-0309. SAE Int J Commercial Veh
4. Zhang K, Cao L, Fanta A, Reed MP, Neal M, Wang J-T, Line C-H, Hu J (2017) Automated method to morph finite element whole-body human models with a wide range of stature and body shape for both men and women. J Biomech 60:253–260

# Anthropometric Characteristics of Chilean University Students and Their Relation with the Dimensions of the Furniture of the Lecture Rooms

J. Freire(✉), E. Apud, F. Meyer, J. Espinoza, E. Oñate, and F. Maureira

Unit of Ergonomics, University of Concepción, Barrio Universitario s/n Concepción, Concepción, Chile
javierfreire@udec.cl

**Abstract.** The present study has the objective of associating the anthropometric dimensions of a sample of Chilean university students with the furniture used in classrooms. The ultimate goal is to propose suitable standards for Chilean universities. 17 body dimensions were measured in a sample of 176 students, using a Harpenden anthropometer Holtain. For the evaluation of the furniture, a metallic metric tape was used. To date, the furniture of 35 university classrooms has been evaluated. Mean values, standard deviations and percentile 5 and 95 were calculated for males and females and for the whole group. Average stature was 160.1 cm for women and 174.0 cm for men. The analysis of the relationship between body size of the students and the furniture they use, showed deficiencies not only in size of computer stations, chairs, and tables, but also in the design and maintenance making them even more uncomfortable, particularly for those students who spend a long time in sitting posture. Furthermore, all students use the same furniture without any consideration for differences in body size. In this respect, the discussion is centered on the need to take as reference the extremes of 5th and 95th percentile of the students' population, but more important is to establish the distribution of body dimensions, at least for a basic set of measurements for seats and tables design, also with inclusive criteria for populations with special needs.

**Keywords:** Anthropometric · Furniture · Students

## 1 Introduction

Tools, machines, workplaces, and clothing, among other implements, must be proportional to the dimensions of the users. Although this is a basic ergonomic concept, in practice, musculoskeletal symptoms are common among people, including students, workers, housewives, etc. For example, Mandal in 1981 [1], pointed out that, "more than half of the international population complains of having back pain." A classic research by Van Wely in 1970 [2], revealed a clear association between the workplace and the consultation for musculoskeletal problems in an industrial medical center.


S. Bagnara et al. (Eds.): IEA 2018, AISC 826, pp. 602–611, 2019.
https://doi.org/10.1007/978-3-319-96065-4_64

In Chile, the Ergonomics Unit of the Universidad de Concepción has carried out studies to detect the perception of musculoskeletal symptoms in sedentary people and also in those who perform dynamic physical activities. The results reveal that independent of the occupation, more than 60% of the people perceive at some stage of their lives, this type of symptoms [3].

As in many workplaces, one of the common problems in the classroom is the variable quality of seating and tables for students. Although ergonomics is more than the design of seats and tables, it is also true that many people, including workers and students, remain seated for a long time and it is usual that the furniture does not allow maintaining a good posture. Consequently, it is important to establish guidelines for discussion. The first thing that can be pointed out is that the ideal would be that these implements were adjustable, but this is generally not possible and the work surfaces have a fixed height. For this reason, these implements are usually designed for the largest people 95th percentile with the understanding that there is enough room to accommodate the smallest people. However, this is valid for desks, but not for seats. Whether the person can adapt to a desk will depend largely on the regulation of the seats.

Physical anthropometry in Ergonomics studies seeks to apply the knowledge of body size to the design of workplaces adapted to human characteristics. The standard, ISO 7250-1:2017 provides a description of anthropometric measurements which can be used as a basis for comparison of population groups and for the creation of anthropometric databases [4]. The basic list of measurements is intended to serve as a guide for ergonomists who are required to define population groups and apply their knowledge to the geometric design of the places where people work and live. It serves as a guide on how to take anthropometric measurements but also gives information to the ergonomist and designer on the anatomical and anthropometrical bases and principles of measurement which are applied in the solution of design tasks.

The present study, which is currently underway, has the objective of associating the anthropometric dimensions of a sample of Chilean university students with the furniture used in classrooms. The ultimate goal is to propose suitable standards for Chilean universities.

## 2 Methodology

Seventeen body dimensions were measured in a sample of 176 students, 75 female, and 101 male, using a Harpenden anthropometer (Holtain Ltd.U.K®). The 17 basic anthropometric dimensions are those recommended in the Chilean Standard 2639-2002, which is based on the international proposal of the ISO 7250 standard. For the evaluation of the furniture, a metallic metric tape was used. To date, the furniture of 35 university classrooms has been evaluated.

The 17 body measurements were taken according to the following definitions:

**Stature (Body Height):** It corresponds to the vertical distance from the highest point of the head called the vertex to the ground. For measurement, the person must be barefoot with their feet together, both arms hanging relaxed and the back straight with the head on the Frankfort plane.

**Eye Height:** It is the vertical distance from the external angle of the eye to the ground. The person stands completely upright with their feet together. The upper end of the anthropometer is located at eye level.

**Shoulder Height:** It is the vertical distance from the highest point of the lateral edge of the shoulder to the ground. The position of the person standing upright. The base of the anthropometer is located on the ground and the upper end at shoulder height.

**Elbow Height:** It is the vertical distance from the lowest point of the elbow to the ground. The person maintains the standing position mentioned above. The anthropometer is located on the side; the position marker arm is located at the lowest point of the elbow.

**Handheld Height:** It is the vertical distance from the axis of the handle at the height of the knuckles to the ground. The standing position is maintained, and a cylinder is held in your hand. The position marker arm is located at the height of the cylinder that is gripped.

**Sitting Height:** It is the vertical distance from the highest part of the head to the floor, with the person sitting and the back straight, Frankfort head in plane and knees at an angle of 90°.

**Shoulder Height (Sitting):** It is the vertical distance from the height of the shoulder to the floor with the person sitting. The anthropometer is positioned on one side parallel to the central axis of the body and the position marker arm is positioned at shoulder height.

**Elbow Height (Sitting):** It is the vertical distance from the lowest point of the elbow to the floor and the person is sitting. The position marker arm of the anthropometer is located at the lowest point of the elbow in 90° flexion.

**Thigh Clearance:** It is the vertical distance from the highest point of the thigh to the floor and the person is sitting. The anthropometer is located on one side of the limb to be evaluated parallel to it and the position marker arm is positioned at the height of the highest point of the thigh.

**Popliteal Height (Sitting):** It is the vertical distance from the lower superior of the immediate thigh behind the knee to the floor. The anthropometer is located parallel to the segment to be evaluated, the position of marker arm is in the posterior area of the knee.

**Buttock Popliteal Length (Seat Depth):** It is the horizontal distance from the most posterior point of the buttocks to the back of the knee. The seat is placed next to a wall in which the person approaches the maximum to this one. The base of the anthropometer is then positioned on the wall and, parallel to the axis of the extremity to be evaluated, the position marker arm is located at the posterior knee height.

**Buttock Knee Length:** It is the horizontal distance from the most posterior point of the buttocks to the anterior part of the knee. In the same position as the previous measurement, the position marker arm is located in the anterior area of the knee.

**Shoulder (Bideltoid) Breadth:** It is the distance between the most lateral points of the deltoid muscles below the level of the shoulders. The subject is positioned in a straight position with arms parallel to the axis of the trunk. The elbows position is at 90° angle.

**Hip Breadth (Sitting):** It is the horizontal distance between the points of both hips of greater diameter without pressing the tissues. The main bar of the anthropometer is located with both arms markers of position in the most prominent areas of the hips.

**Knuckle Wall Reach (Front Reach):** It is the horizontal distance from a vertical surface to the axis of the grip, while the subject supports both shoulder blades against the vertical surface. The subject is standing or sitting fully erect. With the shoulder blades and buttocks firmly supported against the vertical surface, the arm extended fully horizontally. The hand holds the measuring cylinder with the handle axis vertically.

**Upper Functional Scope (Sitting):** It is the vertical distance from the knuckle, with the arm extended vertically, to the ground. The person must be seating, with the arm stretched to the maximum upwards. The measurement is taken from the floor to the knuckle of the handheld.

**Forearm Range (Sitting):** It is horizontal distance from the back of the upper arm at the elbow to the hand grip, with the arm, bent at a right angle. The subject sits erect with upper arm hanging down, elbow at 90° with forearm and wrists in position.

## 3 Results

### 3.1 Age of the Students

The study includes the evaluation of anthropometric characteristics of undergraduate and postgraduate students. In this first stage, the evaluations were carried out by postgraduate students. For this reason, the age of the sample ranges from 20 to 60 years, which allows us to cover a broad age spectrum. As shown in Fig. 1, 50% of the students were under 30 years old.

### 3.2 Body Dimensions of the Students

Mean values, standard deviations and percentile 5 and 95 were calculated for males and females and for the whole group. Average stature was 160.1 cm for women and 174 for men. As average, male and female students were taller than reported in a study of

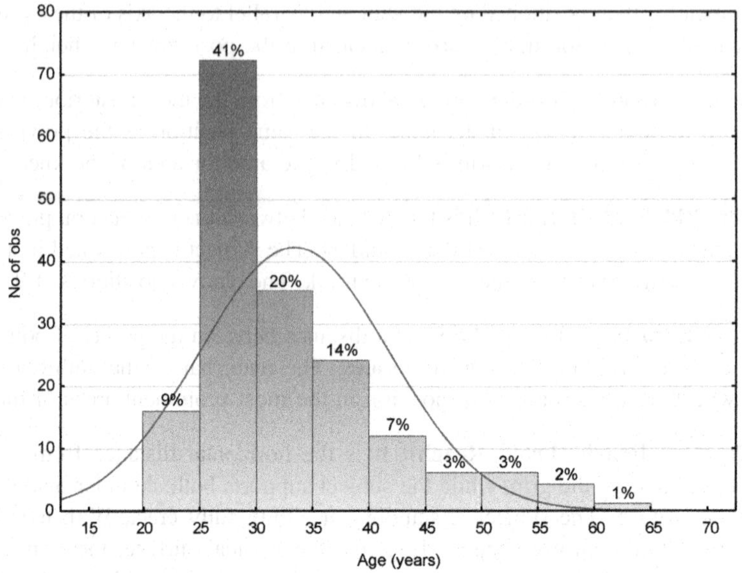


**Fig. 1.** Age distribution of the sample of male and female postgraduate students.

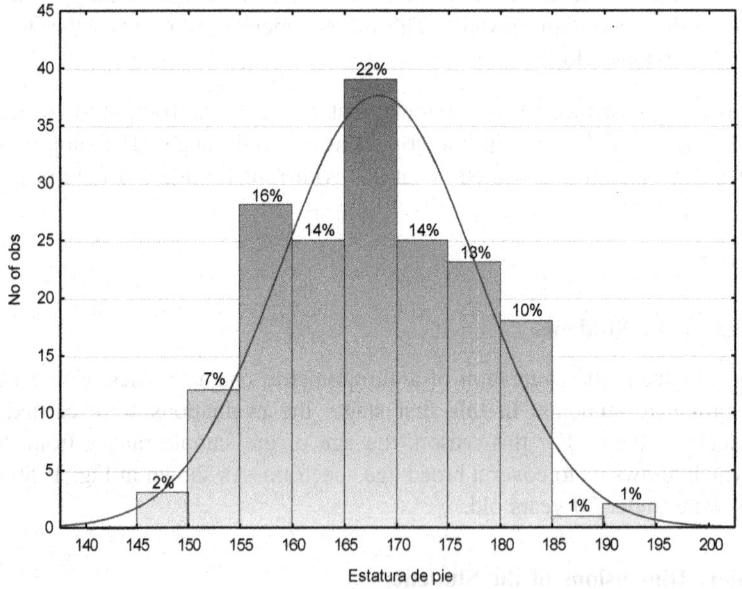


**Fig. 2.** Distribution of standing height of the students

anthropometric characteristics of Chilean population recently published by Castelluci et al. [5]. The trend for University students to be taller has also been demonstrated in Chile in previous studies, Apud et al. [6].

Looking at the whole group and considering the differences in body size between males and females, it is interesting to look closer at the distribution of stature which is shown in Fig. 2. As it can be seen the differences between extremes are 43 cm. When analyzing females and males separately (Table 1) the differences are reduced to 23 y 32 cm respectively. It is also interesting to see in Fig. 2, that 66% of the sample have statures between 155 and 175 cm. These references are important to define the sizes of furniture for lecture rooms.

Consequently, considering that furniture is used by both, male and female students, in Table 2, are shown the results obtained for the whole group of students.

**Table 1.** Stature of male and female students

| Gender | Means | Std. Dev. | Max | Min |
|---|---|---|---|---|
| Male | 174.0 | 6.8 | 160.0 | 192.0 |
| Female | 160.1 | 5.6 | 149.0 | 172.0 |
| **All groups** | **168.1** | **9.3** | **149.0** | **192.0** |

**Table 2.** Mean values, min, max and percentiles 5 and 95 for a sample of 176 students.

| Anthropometric characteristic (measurements are in centimeters) | Mean | Min | Max | P5 | P95 | Std. Dev. |
|---|---|---|---|---|---|---|
| Stature (body height) | 168,1 | 149,0 | 192,0 | 153,7 | 184,0 | 9,3 |
| Sitting height | 128,9 | 103,5 | 145,5 | 119,7 | 140,0 | 6,6 |
| Knuckle wall reach (front reach) | 72,6 | 55,6 | 88,0 | 63,0 | 83,0 | 6,1 |
| Shoulder height (sitting) | 101,3 | 84,3 | 131,0 | 90,8 | 112,0 | 6,7 |
| Elbow height (sitting) | 65,8 | 55,8 | 79,0 | 58,6 | 73,3 | 4,6 |
| Thigh clearance | 56,3 | 41,5 | 67,0 | 49,8 | 62,5 | 4,2 |
| Popliteal height (sitting) | 41,2 | 30,5 | 51,0 | 35,6 | 47,0 | 3,5 |
| Hip breadth (sitting) | 40,1 | 30,2 | 50,0 | 33,0 | 46,0 | 3,7 |
| Shoulder (bideltoid) breadth | 45,6 | 30,3 | 57,5 | 37,0 | 54,0 | 5,1 |
| Buttock popliteal length (seat depth) | 46,6 | 38,9 | 58,8 | 40,5 | 52,5 | 3,5 |
| Buttock knee length | 58,3 | 49,0 | 71,0 | 51,5 | 65,5 | 4,1 |
| Forearm (sitting) range | 42,3 | 24,7 | 84,0 | 32,2 | 53,0 | 7,2 |
| Upper functional scope (sitting) | 160,5 | 138,3 | 189,0 | 142,6 | 178,0 | 10,4 |

### 3.3 The Furniture in the Lecture Rooms

In Table 3 it is possible to see the differences between the different types of furniture in comparison to the anthropometric characteristics of the students, taking 5th and 95th percentiles as reference. It is important to note that furniture type I and II (see Fig. 3 and 4) were the most common to find, whose main features is that the table is attached to the chair without any possibility of alteration or accommodation.

**Table 3.** Characteristics of the furniture in relation to the anthropometric characteristics.

| Furniture feature | King of furniture | | | | P5 | P95 | Anthropometric characteristics |
|---|---|---|---|---|---|---|---|
| | I | II | III | IV | | | |
| Top of the table | 69 | 81 | 80 | 66 | 58.6 | 73.3 | Elbow height (sitting) |
| Table width | 24 | 30 | 480 | 250 | 37 | 54 | Shoulder (bideltoid) breadth |
| Depth of the table | 50 | 30 | 36 | 26 | 63 | 83 | Knuckle wall reach |
| Chair height | 43 | 45 | 43 | 43 | 35.6 | 47 | Popliteal height (sitting) |
| Chair width | 46 | 39 | 35 | 40 | 33 | 46 | Hip breadth (sitting) |
| Depth of the chair | 35 | 39 | 38 | 35 | 40.5 | 52.5 | Buttock popliteal length (seat depth) |
| Chair backrest | 74 | 83 | 83 | 70 | 90.8 | 112 | Shoulder height (sitting) |
| All Measurements are in centimeters | | | | | | | |

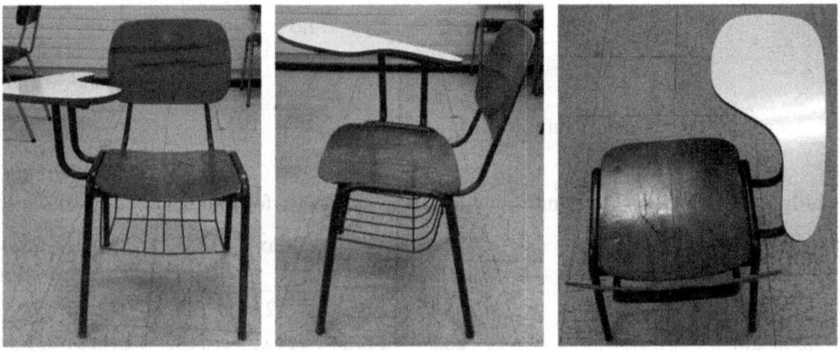

**Fig. 3.** Type I furniture used in the university.

**Fig. 4.** Type II furniture.

In relation to the type III of seat, it was found mainly in classrooms in which the tables were screwed to the floor without the possibility of moving them. The characteristics of these surfaces are mainly that they are long tables with a depth not greater than 36 cm. The number of seats per table will depend on the length, which in some cases reached 280 cm, so that they sat around 15 students per table. One of the problems of this situation is the little space for students to carry out their activities. Another negative aspect is that the chairs do not have armrests as seen in Fig. 5. The students have to rest their arms on the table whose height ranged between 80 and 85 cm, which is much higher than the height of the elbow to the floor in a sitting position that in the case of the 5th percentile corresponds to 58.6 cm and for the $95^{th}$ percentile to 73,3 cm (Table 4).

**Fig. 5.** Type III furniture.

**Table 4.** Recommendation of furniture.

| Furniture feature | King of furniture | | | |
|---|---|---|---|---|
| | Quartile I | Quartile II | Quartile III | Quartile IV |
| Top of the table | 62,5 | 65,8 | 68,5 | 73,3 |
| Table width | 55 | 55 | 55 | 55 |
| Depth of the table | 60 | 60 | 60 | 60 |
| Chair height | 38,5 | 41,2 | 43,4 | 47 |
| Chair width | 38 | 40,1 | 42 | 46 |
| Depth of the chair | 44,5 | 46,6 | 49 | 52,5 |
| Chair backrest | 97 | 101,3 | 105,3 | 112 |

As for type IV furniture, they are located in the oldest classrooms of the university and their main characteristic is that they are fixed seats that cannot be moved in any direction as seen in Fig. 6. Their height is fixed; the width is the same as its depth. Another important aspect to keep in mind is that it is not possible to accommodate the distance to the desk since the latter is also bolted to the floor. Undoubtedly, this type of furniture does not help students to maintain good posture.

Analyzing the anthropometric dimensions in relation to the characteristics of the furniture it was observed that many dimensions were not adequate. For example, the minimum popliteal height was 30.5 cm and the smallest seat height was 35 cm, while the minimum height of a table was 73 cm and the minimum elbow height to the floor in sitting posture was 55.6 cm. These examples confirm the poor proportionality between the student's size and the furniture.

**Fig. 6.** Type IV furniture.

## 4 Conclusion

As a conclusion, although the ideal will be to have furniture that could be regulated to the size of each individual, this is not realistic in developing countries. Therefore, a proposal will be presented to the authorities of the University for types and sizes of furniture to be used by students, based on the distribution of body size. This will be done when the sampling is completed. Meanwhile, the idea is to work with ranges calculated on the basis of subdivisions in quartiles as shown in figure n for the main dimensions for chairs and tables. It is important to consider that with 4 sizes of furniture the students could choose according to their body size the one that better suits their characteristics. It is also necessary to consider training in the use of furniture. Finally, the ultimate goal, at the end of the project, is to advance towards a standardization which allows students to develop their activities in comfortable conditions.

## References


1. Mandal AC (1981) The seated man (Homo Sedens) the seated work and practice. Appl Ergon 12(1):19–26
2. van Wely P (1970) Design and disease. Appl Ergon, 262–267
3. Apud E, Meyer F, Espinoza J, Oñate E, Freire J, Maureira F (2016) Ergonomics and labour in forestry. In: Pancel L, Köhl M (eds) Tropical Forestry Handbook. Springer, Berlin, Heidelberg
4. https://www.iso.org/standard/65246.html

5. Castellucci H, Arezes PM, Viviani CA (2010) Mismatch between classroom furniture and anthropometric measures in Chilean schools. Appl Ergon 41(4):563–568
6. Apud E, Gutiérrez M (1997) Diseño ergonómico y características antropométricas de mujeres y hombres adultos chilenos. Primeras Jornadas Iberoamericanas de Prevención de Riesgos Ocupacionales
7. Jalil N, Iman D, Mohamad A (2013) Student's Body Dimensions in Relation to Classroom Furniture Samira Baharampourl. Health Promot Perspect 3(2):165–174

# Anthropometric Factors in Seat Comfort Evaluation: Identification and Quantification of Body Dimensions Affecting Seating Comfort

Benjamin Heckler(✉), Manuel Wohlpart, and Klaus Bengler

Chair of Ergonomics, Technical University of Munich,
Boltzmannstr. 15, 85747 Garching, Germany
{Benjamin.Heckler,Manuel.Wohlpart,Bengler}@tum.de

**Abstract.** The objective of the presented study was the identification and quantification of anthropometric factors in seat comfort evaluation. Therefore, a "comfort-critical" and a "comfort-reference" seat were evaluated by 70 participants (38 men, 32 women) with a questionnaire consisting of 22 items. To identify anthropometric effects a certain requirement had to be fulfilled. The first analysis should show that the "comfort-reference" seat was rated better, compared to the basic "comfort-critical" seat, due to its additional adjustment tracks. The results showed that the "comfort-reference" seat was assessed better in 19 items. Based on these findings a second analysis was investigating, if the assumed anthropometric effects occurred more frequently on the worse rated "comfort-critical" seat. Therefore the participants were divided in three groups depending on their body dimensions. A statistical comparison of the three groups were performed for eight measured anthropometric variables. The number of significant differences between the body dimension groups were higher for the "comfort-critical" seat compared to the "comfort-reference" seat. The data show that anthropometric effects are existing in seat comfort evaluation and a deeper understanding of how body dimension affecting seat comfort needs to be researched.

**Keywords:** Anthropometry · Seat comfort · Evaluation method

## 1 Introduction

Comfortable seats are essential when offering passengers a pleasurable environment in modern car interiors. One of the main challenges in the construction process of car seats is the balance between a soft initial feeling during the first contact, a durable foam to maintain the required contour and avoid a negative posture during long-term driving and lateral support in order to hold the occupants in position by performing turning maneuvers [1].

The unique anthropometric proportion of each human is one of the most influencing factors on seating comfort, determining the shape, properties and adjustment tracks of car seats in the development process [2]. A study from Kolich et al. found anthropometric differences affecting the subjective comfort evaluation [3].


S. Bagnara et al. (Eds.): IEA 2018, AISC 826, pp. 612–622, 2019.
https://doi.org/10.1007/978-3-319-96065-4_65

Other studies investigated the effect of the anthropometry on different objective comfort parameters, such as posture [4] and interface pressure [5]. The anthropometric variables measured in mentioned seat comfort studies were stature, weight, gender, age, BMI, RPI, percentage of subcutaneous fat and ectomorphic index [2]. In some experiments the anthropometry correlated with several pressure parameters [6, 7].

Paul et al. researched the influence of specific body dimension in more detail. The authors investigated the correlation between pressure parameters and eight body dimensions in total. The anthropometric variables measured in the experiment were body mass, sitting height, bideltoid breadth, sitting acromial height, hip circumference, hip breadth feet apart, sitting knee height and buttock-to-knee length [5]. They could show, that some body dimensions correlated with different pressure variables and concluded that research is needed to evaluate if the pressure parameters correlate with a subjective comfort evaluation in order to simplify the prediction of seat pressure comfort, based on occupants' anthropometry. The main scope of the presented study was to analyze how several body dimensions influence the subjective assessment of seat comfort.

In order to demonstrate that anthropometric effects exist in current car seats, some requirements had to be fulfilled. One requirement was the correct selection of experimental seats. The tested seats needed to differ in their seat properties and adjustment ability, in order to research the assumed anthropometric effects occurring more frequently in more basic seats compared to fully adjustable seats. Therefore, a valid questionnaire enabling a differentiation of seat comfort between two diverse seats was required.

In the literature many seat comfort studies were found, presenting a great variation of questionnaires and measurement tools used for the subjective evaluation of seating comfort [8, 9]. The items and scales used in the questionnaires differ mostly. As a result, the diverse experimental designs and measurement tools of seat comfort studies lead to controversial and limited comparison between the respective findings. In this context, the further aim of the experiment was the verification of a questionnaire, which is able to differentiate the comfort properties of two diverse seats. The investigation of anthropometric effects in seat comfort evaluation was then based on those results.

## 2 Objectives

The scope of the study presented in this paper was the identification and quantification of body dimensions affecting the evaluation of seat comfort. For identifying the anthropometric factors certain requirements had to be fulfilled. The research approach was based on the assumption that the comfort rating of a seat with basic adjustment tracks is worse compared to a multi contour seat with additional adjustment functions. The verification of this hypothesis was the first aim of this study. If confirmed, the assumed anthropometric effects in seat comfort evaluation can potentially be detected. The first hypothesis concerned the possibility of a questionnaire to differentiate seat comfort between two seats:

1. The comfort rating of a multi contour seat is better compared to a basic seat.

If the first hypothesis is confirmed, anthropometric effects in seat comfort evaluation should be researchable by further analysis. The second hypotheses tested with the experiment, should demonstrate that the assumed anthropometric effects occur more frequently in the basic seat, because of it being less adjustability compared to a multi contour seat:

2. Anthropometric effects occur more frequently at a basic seat compared to a multi contour seat.

This analysis should demonstrate that anthropometric effects are existing in seat comfort evaluation. For this reason, the experimental evaluation proceeded in series-production vehicles on original seats.

# 3 Method

## 3.1 Experimental Seats

Two different serial-production cars were used for the experiment. One test vehicle was an Audi Q2 equipped with a sport seat that had four basic adjustment tracks (Fig. 1). The seat had a sportive design with prominent backrest bolster and stiff foam properties. Due to its characteristic and basic adjustment possibilities the seat was defined as a "comfort-critical" seat. The definition results from the assumption that the seat comfort rating of some passengers would be negatively affected by their anthropometry.

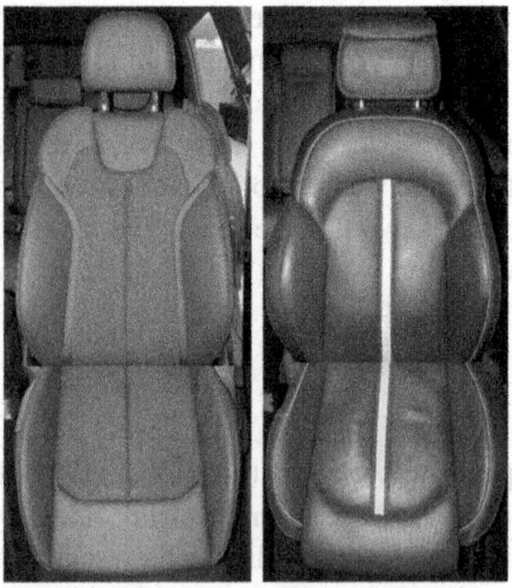

| | Seat adjustment tracks |
|---|---|
| 1 | Longitudinal adjustment |
| 2 | Height adjustment |
| 3 | Backrest inclination |
| 4 | Headrest height adjustment |
| *5* | *Cushion depth adjustment* |
| *6* | *Cushion tilt adjustment* |
| *7* | *Cushion bolster adjustment* |
| *8* | *Backrest head inclination* |
| *9* | *Lordosis height adjustment* |
| *10* | *Lordosis depth adjustment* |
| *11* | *Backrest bolster adjustment* |

**Fig. 1.** Left: Sport seat ("comfort-critical") of an Audi Q2 with basic adjustment tracks. Middle: Multi contour seat ("comfort-reference") of an Audi A8 with 11 adjustment tracks. Right: List of the four basic adjustments (1–4) of the sport seat and the additional *seven adjustment functions (5–11)* of the multi contour seat.

The second experimental car was an Audi A8 equipped with a multi contour driver seat, which had eleven adjustment function and a smooth foam composition to enable a comfortable seating position for each passenger. The seat was defined as a "comfort-reference" seat because of the assumption that all participants find a good to perfect seat position, independently of their unique body dimensions, resulting in a positive comfort sensation.

### 3.2 Measurement Tools

Eight body dimensions were measured of each participant before the test drives using the method described in the manual of ergonomics [10]. The measurement tools included an anthropometer, a tape measure and a scale.

The seat comfort was evaluated with a questionnaire consisting of 22 items (Table 2). The structure of the questionnaire based on a stepwise assessment of several comfort parameters. The accessibility of the adjustment elements and the adjustability of the seating position was evaluated with the first items. Then an assessment of the initial touch with the seat, the contour and functionality of different seat components as well as the lateral support of the cushion and backrest bolster followed. The rating criteria of the seat climate, the pressure distribution in eight body areas and the overall comfort rating completed the questionnaire. Each item was rated with a five-point ordinal scale. For each evaluation point a clear and plain statement was formulated. The worst rating was a 1 ("seat is unacceptable"), followed by a 2 ("seat needs to be improved"). The first positive rating was a 3 ("seat is fine/okay"), exceeded by a 4 ("seat is good to very good"). The best rating was a 5 ("seat is perfect"). The questionnaire was programmed on a tablet.

For the objective assessment of the seat comfort pressure data were collected with two pressure mats (XSensor Technology Corporation, PX100:40.40.02, PX100:40.64.02), however, the analysis of the pressure parameter is not focus of this present paper.

### 3.3 Experimental Design

The experimental design of the presented study was a mixed-model design. The first hypothesis was reviewed by a within-subject design. The independent variable (IV) was the respective seat and the dependent variable (DV) the subjective comfort rating consisting of 22 items. The statistical test used for the means comparison of the ordinal data was a Wilcoxon signed-rank test.

A between-subject design was used for the identification of anthropometric factors by testing the second hypothesis. The subjects were divided in three body dimension groups depending on their anthropometry (explained in Sect. 3.4 Participants). In that case the IV were the body dimension groups and the DV was the subjective assessment via the comfort questionnaire. The comfort rating of the three independent groups were tested with a one-way ANOVA on ranks (Kruskal-Wallis test).

### 3.4 Procedure and Setup

The acquired participants were instructed to wear casual clothes for the experiment. At first the anthropometry was determined by measuring eight body dimension. Besides the *stature* and *body weight* three body dimensions of the upper body (*shoulder breadth (bideltoid), sitting height* and *waist circumference*) and lower body (*seat depth, hip breadth* and *thigh circumference*) were measured.

After this procedure the participants were asked several questions about their demographics, driving behavior and their current mood. Then, the actual experiment started with the comfort evaluation of the first seat. To avoid order effects, the experimental procedure was randomized. Half of the subjects started with the Q2 sports seat and the other half with the A8 multi contour seat. Before the evaluation the participants were instructed to adjust the seat to their preferred driving position. The first comfort evaluation was executed in a static condition, followed by an approximately 25-min real test drive in public road traffic on a standardized route.

After the second comfort rating the subjects switches the car and the experiment continued on the second seat that was prepared with the pressure mats. The seat was adjusted by the test person, then the pressure data were recorded, followed by the comfort assessment. At the end of the evaluation a second pressure measurement was executed. Then, the mats were removed and placed on the other seat. The experiment continued with the comfort rating of the static condition without the pressure mats, followed by the real test drive. At the end the first seat was evaluated once again with the pressure mats placed on the seat. In total, both seats were evaluated in three conditions (static, real test drive and static with pressure mats).

The statistical analysis was conducted using the data of the static condition with pressure mats because each subject had spent nearly the same amount of time in the sitting position. Thereby, the time influencing the seat comfort was almost similar for all participants, which should cause for the results to be most comparable in that case.

### 3.5 Participants

Overall, 38 men (Ø 37.9 ± 9.7 years) and 32 women (Ø 29.5 ± 8.3 years) participated in the experiment. The subjects were acquired by their stature and body weight asked in the pre-questionnaire in order to receive a sample with a broad anthropometric spectrum. The variety of different typologies was a requirement for the investigation of anthropometric effects. For the statistical analysis the participants were divided in three groups for each anthropometric variable separately. The first group consisted of the 15 smallest percentiles of each body dimension, the middle group of the 15 subjects around the mean percentile and the third of the 15 tallest participants of each anthropometric variable (Table 1).

**Table 1.** Average values of three body dimension groups for eight anthropometric variables.

| Anthropometric variable | Small (n = 15) | Middle (n = 15) | Tall (n = 15) |
|---|---|---|---|
| *Stature* | Ø 162.2 cm (SD: 3.0 cm) | Ø 175.9 cm (SD: 2.2 cm) | Ø 188.2 cm (SD: 2.3 cm) |
| *Body weight* | Ø 57.6 kg (SD: 3.0 kg) | Ø 72.8 kg (SD: 2.2 kg) | Ø 95.6 kg (SD: 11.0 kg) |
| *Sitting height* | Ø 85.3 cm (SD: 2.0 cm) | Ø 91.3 cm (SD: 0.6 cm) | Ø 97.6 cm (SD: 2.0 cm) |
| *Shoulder breadth* | Ø 40.6 cm (SD: 0.9 cm) | Ø 45.5 cm (SD: 0.8 cm) | Ø 50.0 cm (SD: 2.4 cm) |
| *Waist circumference* | Ø 70.4 cm (SD: 3.2 cm) | Ø 81.7 cm (SD: 1.1 cm) | Ø 102.4 cm (SD: 7.8 cm) |
| *Seat depth* | Ø 85.3 cm (SD: 2.0 cm) | Ø 91.3 cm (SD: 0.6 cm) | Ø 97.6 cm (SD: 2.0 cm) |
| *Hip breadth* | Ø 35.6 cm (SD: 0.7 cm) | Ø 39.3 cm (SD: 0.6 cm) | Ø 42.6 cm (SD: 1.2 cm) |
| *Thigh circumference* | Ø 51.9 cm (SD: 1.5 cm) | Ø 57.8 cm (SD: 0.9 cm) | Ø 63.2 cm (SD: 1.8 cm) |

# 4 Results

## 4.1 Mean Comparison of the Comfort Questionnaire

To verify the first hypothesis, the 22 items of the questionnaire were evaluated by a mean comparison between the two seats with a Wilcoxon signed-rank test. Because of the one-tailed hypothesis the significance level was set on $\alpha = .025$ to analysis if the multi contour seat was assessed better. The results in Table 2 show that the "comfort-reference" seat (A8 seat) was rated significantly better in 19 of 22 Items compared to the "comfort-critical" seat (Q2 seat). Due to this result the first hypothesis can be confirmed.

**Table 2.** Results of the Wilcoxon signed-rank test of the comfort questionnaire between the "comfort-reference" seat (A8) seat and the "comfort-critical" seat (Q2).

| | Item | A8 seat (n = 70) Mean ± SD | Q2 seat (n = 70) Mean ± SD | Wilcoxon signed-rank test |
|---|---|---|---|---|
| 1 | Accessibility adjustment elements | 3.4 ± 1.2 | 3.5 ± 0.7 | z = −.37, p = .710 |
| 2 | Seat adjustment ranges | 4.6 ± 0.7 | 3.6 ± 0.9 | z = 5.93, p = .000 |
| 3 | Sitting position in the cockpit | 4.4 ± 0.6 | 4.2 ± 0.7 | z = 2.40, p = .016 |
| 4 | Cushion initial touch | 4.2 ± 0.6 | 3.7 ± 0.7 | z = 4.88, p = .000 |

*(continued)*

**Table 2.** (*continued*)

| | Item | A8 seat (n = 70) Mean ± SD | Q2 seat (n = 70) Mean ± SD | Wilcoxon signed-rank test |
|---|---|---|---|---|
| 5 | Backrest initial touch | 4.1 ± 0.6 | 3.7 ± 0.7 | z = 3.87, p = .000 |
| 6 | Cushion length | 4.4 ± 0.7 | 3.3 ± 0.9 | z = 6.13, p = .000 |
| 7 | Cushion inclination angle | 4.5 ± 0.6 | 3.4 ± 1.0 | z = 5.76, p = .000 |
| 8 | Distance to headrest (adjustment range) | 4.0 ± 1.0 | 3.3 ± 1.0 | z = 3.51, p = .000 |
| 9 | Headrest contact surface comfort | 3.3 ± 0.9 | 3.3 ± 0.8 | z = −.63, p = .530 |
| 10 | Location of lumbar support | 4.5 ± 0.6 | 3.4 ± 0.9 | z = 5.84, p = .000 |
| 11 | Cushion lateral support | 3.9 ± 0.7 | 3.5 ± 0.8 | z = 3.75, p = .000 |
| 12 | Backrest lateral support | 4.2 ± 0.7 | 3.5 ± 0.8 | z = 4.55, p = .000 |
| 13 | Seat climate (temperature) | 3.6 ± 0.8 | 3.5 ± 0.8 | z = 2.10, p = .036 |
| 14 | Pressure distribution transmission tunnel | 4.1 ± 0.7 | 3.8 ± 0.9 | z = 2.42, p = .015 |
| 15 | Pressure distribution thigh | 4.2 ± 0.6 | 3.6 ± 0.9 | z = 4.45, p = .000 |
| 16 | Pressure distribution thigh bolster | 4.0 ± 0.6 | 3.6 ± 0.9 | z = 2.77, p = .006 |
| 17 | Pressure distribution buttocks | 4.2 ± 0.6 | 3.9 ± 0.7 | z = 2.62, p = .009 |
| 18 | Pressure distribution buttocks bolster | 4.2 ± 0.6 | 3.8 ± 0.7 | z = 3.25, p = .001 |
| 19 | Pressure distribution shoulder | 4.0 ± 0.7 | 3.6 ± 0.7 | z = 3.34, p = .001 |
| 20 | Pressure distribution lordosis | 4.4 ± 0.6 | 3.6 ± 0.9 | z = 5.40, p = .000 |
| 21 | Pressure distribution backrest bolster | 4.3 ± 0.7 | 3.6 ± 0.9 | z = 4.46, p = .000 |
| 22 | Overall seat comfort | 4.1 ± 0.4 | 3.5 ± 0.6 | z = 5.35, p = .000 |

## 4.2 Comparison Between the Body Dimension Groups

For the identification of anthropometric effects, the distribution between the three groups of each body dimension were tested with a non-parametric Kruskal-Wallis test for each seat separately. The significance level was set on $\alpha = .05$ due to the two-tailed hypothesis. The analysis of the "comfort-critical" seat show differences between the three groups for all eight anthropometric variables. The body dimensions *stature, body weight, shoulder breadth* and *thigh circumference* had significant differences between the three groups in 24 items (Table 3). For the variable *waist circumference*, the subjective evaluation of the items location of lumbar support ($\chi^2 = 8.26$, $p = .016$), backrest lateral support ($\chi^2 = 9.82$, $p = .007$), pressure distribution transmission tunnel ($\chi^2 = 8.93$, $p = .012$) and the pressure distribution backrest bolster ($\chi^2 = 7.21$, $p = .027$) show significant differences. The accessibility of the adjustment elements ($\chi^2 = 9.37$. $p = .009$), backrest lateral support ($\chi^2 = 8.00$, $p = .018$), pressure distribution backrest bolster ($\chi^2 = 7.27$, $p = .026$) differ significantly for the *sitting height*. The *seat depth* had

**Table 3.** Results of the Kruskal-Wallis test between three body dimension groups of four anthropometric variables for the "comfort-critical" seat (Q2) with items that differ significantly.

| Anthropometric variable | Item | Kruskal-Wallis test |
|---|---|---|
| *Stature* | Accessibility adjustment elements | $\chi^2(2) = 6.68$, $p = .035$ |
| | Seat adjustment ranges | $\chi^2(2) = 6.27$, $p = .043$ |
| | Backrest initial touch | $\chi^2(2) = 6.11$, $p = .047$ |
| | Cushion length | $\chi^2(2) = 8.87$, $p = .012$ |
| | Cushion inclination angle | $\chi^2(2) = 9.12$, $p = .035$ |
| | Pressure distribution transmission tunnel | $\chi^2(2) = 8.43$, $p = .015$ |
| | Pressure distribution thigh | $\chi^2(2) = 8.92$, $p = .012$ |
| | Pressure distribution thigh bolster | $\chi^2(2) = 7.79$, $p = .020$ |
| | Pressure distribution buttocks bolster | $\chi^2(2) = 6.25$, $p = .044$ |
| *Body weight* | Cushion inclination angle | $\chi^2(2) = 7.87$, $p = .020$ |
| | Location of lumbar support | $\chi^2(2) = 7.34$, $p = .026$ |
| | Backrest lateral support | $\chi^2(2) = 7.54$, $p = .023$ |
| | Pressure distribution transmission tunnel | $\chi^2(2) = 6.62$, $p = .036$ |
| | Pressure distribution lordosis | $\chi^2(2) = 8.74$, $p = .013$ |
| | Pressure distribution backrest bolster | $\chi^2(2) = 11.77$, $p = .003$ |
| *Shoulder breadth* | Cushion length | $\chi^2(2) = 6.93$, $p = .020$ |
| | Cushion inclination angle | $\chi^2(2) = 10.34$, $p = .006$ |
| | Backrest lateral support | $\chi^2(2) = 10.18$, $p = .006$ |
| | Pressure distribution transmission tunnel | $\chi^2(2) = 8.00$, $p = .018$ |
| | Pressure distribution shoulder | $\chi^2(2) = 6.28$, $p = .043$ |
| | Pressure distribution backrest bolster | $\chi^2(2) = 9.34$, $p = .009$ |
| *Thigh circumference* | Backrest lateral support | $\chi^2(2) = 6.95$, $p = .031$ |
| | Pressure distribution thigh bolster | $\chi^2(2) = 8.41$, $p = .015$ |
| | Pressure distribution backrest bolster | $\chi^2(2) = 6.00$, $p = .050$ |

two items that show significant differences (cushion inclination angle ($\chi^2$ = 8.39, p = .015), distance to headrest ($\chi^2$ = 8.54, p = .014)). The seat climate ($\chi^2$ = 8.80, p = .012) was the only item that differs significantly between the anthropometric variable *hip breadth.*

The analysis of the A8 multi contour seat show that the subjective assessment differs significantly in only six of eight measured body dimension. For the variable *stature* the item pressure distribution transmission tunnel ($\chi^2$ = 9.00, p = .011) and for the *waist circumference* the cushion length ($\chi^2$ = 6.271, p = 0.043), pressure distribution thigh ($\chi^2$ = 6.44, p = .040) and pressure distribution thigh bolster ($\chi^2$ = 7.65, p = .022) had significant differences. Seat adjustment ranges ($\chi^2$ = 8.64, p = .013) was the only item that differed between the *sitting height* groups and the accessibility of the adjustment elements ($\chi^2$ = 6.22. p = .045) showed significant differences for the three body dimension groups of the *thigh circumference*. For the groups *seat depth* and *hip breadth,* the item seat climate ($\chi^2$ = 9.91, p = .007, $\chi^2$ = 7.61, p = .022) was significantly different.

In summary, for the "comfort-critical" seat 34 items were significantly different between the three groups of the eight anthropometric variables. In comparison, the "comfort-reference" seat showed statistical differences only in eight items overall.

## 5 Discussion

The aim of the presented study was divided in two parts. The first objective was to verify the subjective evaluation method by a comparison of two seats with diverse seat properties and a different number of adjustment tracks. This analysis was defined as a basic requirement for investigating anthropometric effects in seat comfort evaluation.

The "comfort-reference" seat was rated better in 19 of 22 items compared to the "comfort-critical" seat. Due to this result it can be assumed that a differentiation between two seats was possible with the questionnaire and the requirement for exploring anthropometric factors was fulfilled. The fact that both seats did not differ in three items (accessibility adjustment elements, headrest contact surface comfort, seat climate (temperature)) can be regarded as the potential of the questionnaire for a selective comfort assessment. For example, the foam characteristic of the head rest was nearly similar for both seats. Because both seat hat very stiff headrests, it seems obvious that the participants rated the item "headrest contact surface comfort" for both seats only with an average score, regardless of other comfort related characteristics. Also the accessibility of the adjustment elements did not differ much between both seats, resulting in a nearly equal assessment score. Therefore, it can be assumed that a selective evaluation of various seat components and comfort influencing factors is possible.

More research is needed in order to verify if the questionnaire can be used as a subjective measurement tool for seat comfort evaluation. An experiment with a larger number of different seats could show if the questionnaire has the ability to classify seats according to their comfort characteristics.

The anthropometric effects in seat comfort evaluation were analyzed with a comparison of three groups for eight anthropometric variable. The results for the two diverse seats showed a ratio of 34 to eight items that had significant differences between the three body dimension groups. The fact that more anthropometric effects occurred on the "comfort-critical" seat indicates that the anthropometry affects the comfort evaluation. The most influencing body dimensions were *stature* and *body weight*. On the upper body the *shoulder breadth* was the most affecting variable and for the lower body the *thigh circumference*. The experiment showed that the anthropometry is an influencing factor that needs to be considered, even in highly engineered premium cars.

The anthropometric effects need to be investigated further with more basic seats from volume produced cars in order to measure the impact of anthropometric variables in a wider context. The presented method can be used for evaluating the shape and material characteristic of seats as well as the need for additional adjustment tracks, to make a seat more suitable for a greater anthropometric spectrum.

## 6 Conclusion

With the subjective assessment method applied in this experiment, the participants were able to differentiate the seat comfort between two diverse seats. Additionally, the results show that a selective evaluation of different comfort parameters should be possible, by using the 22 items of the comfort questionnaire.

The statistical analysis of the three body dimension groups for the eight anthropometric variables displayed that anthropometric effects are existing even in premium car seats. The identification of anthropometric factors in seat comfort evaluation was demonstrated by the comparison of a "comfort-critical" seat with a "comfort-reference" seat. More anthropometric effects were detected at the "comfort-critical" seat, due to its sportive contour and basic adjustment tracks. Based on these findings a deeper understanding of how body dimensions influencing seat comfort is essential in order to adapt seat characteristics and adjustment possibilities in future seat concepts.

**Acknowledgement.** The authors want to thank the AUDI AG for supporting the experiment, by supplying the test cars and measurement tools (tablet, pressure measuring system, facilities).

## References

1. Kolich M (2004) Predicting automobile seat comfort using a neural network. Int J Ind Ergon 33(4):285–293
2. Hiemstra-van Mastrigt S, Groenesteijn L, Vink P, Kuijt-Evers LF (2017) Predicting passenger seat comfort and discomfort on the basis of human, context and seat characteristics: a literature review. Ergonomics 60(7):889–911
3. Kolich M (2003) Automobile seat comfort: occupant preferences vs. anthropometric accommodation. Appl Ergon 34(2):177–184
4. Park SJ, Kim CB, Kim CJ, Lee JW (2000) Comfortable driving postures for Koreans. Int J Ind Ergon 26(4):489–497

5. Paul G, Daniell N, Fraysse F (2012) Patterns of correlation between vehicle occupant seat pressure and anthropometry. Work 41(Supplement 1):2226–2231
6. Kyung G, Nussbaum MA (2008) Driver sitting comfort and discomfort (part II): relationships with and prediction from interface pressure. Int J Ind Ergon 38(5–6):526–538
7. Kyung G, Nussbaum MA (2013) Age-related difference in perceptual responses and interface pressure requirements for driver seat design. Ergonomics 56(12):1795–1805
8. Ulherr A, Bengler K (2017) Bewertung von Sitzen. Zeitschrift für Arbeitswissenschaft, 1–7
9. Kolich M (1999) Reliability and validity of an automobile seat comfort survey (No. 1999-01-3232). SAE Technical Paper
10. Schmidtke H, Groner P (1989) Handbuch der ergonomie: mit ergonomischen Konstruktionsrichtlinien und Methoden

# 3D Body Modelling and Applications

S. Alemany(✉), A. Ballester, E. Parrilla, A. Pierola, J. Uriel, B. Nacher, A. Remon, A. Ruescas, J. V. Durá, P. Piqueras, and C. Solves

Instituto de Biomecánica de Valencia, Universitat Politècnica de Valencia, 46022 Valencia, Spain
sandra.alemany@ibv.org

**Abstract.** Human body metrics have become a significant source of product innovation to industries where consumer fit, comfort and ergonomic considerations are key factors. This is especially the case for fashion (e.g. footwear or apparel), health (e.g. orthotics or prosthetics), transport and aerospace (e.g. seats or human-machine interfaces), and safety (e.g. protective equipment or workstations) among others. Large-scale databases of 3D body scans are today a research tool for most of the leading companies of those sectors.

In the last few years, new emerging businesses using 3D body data (e.g. garment and footwear customization, size recommendation, health monitoring) are increasing the number and size of 3D body scan repositories. 3D body databases are growing very fast and the development of 3D modelling tools is leveraging the practical application and exploitation of these data.

This paper presents three applications of 3D body modelling methods based on Principal Component Analysis (PCA): (1) shape analysis applied to the ergonomic sizing and design of products, (2) creation of 3D avatars from body measurements, and (3) serial 3D creation of harmonised watertight meshes acquired with any type of 3D body scanner.

**Keywords:** 3D human models · Body avatars · 3D mannequins Anthropometry

## 1 Introduction

The ergonomic design of products and environments with a high interaction with humans requires accurate and detailed knowledge about body shape, dimensions and their variability among the population [1, 2]. Anthropometric tables of percentiles and digital human models (DHM) are the main tools for designers to manage anthropometric summaries of the population. The anthropometric tables used in ergonomics are statistical reports, mainly mean, standard deviation and percentiles, describing the variability of human body dimensions. DHM are 3D representations of the human body with articulated parts that can be posed and scaled introducing limited anthropometric dimensions [3]. These DHM are widely used to analyse reaches modifying the pose and considering the range of movement of different joints of the body. Although DHM can be adapted to some anthropometric measurements, the lack of realistic body shapes is a limitation for many applications such as wearables, clothing or protective equipment.


S. Bagnara et al. (Eds.): IEA 2018, AISC 826, pp. 623–636, 2019.
https://doi.org/10.1007/978-3-319-96065-4_66

3D scanners have made it possible to capture the whole body shape at once and have become mainstream technology employed to perform anthropometric surveys. Since the CAESAR project [4], the 3D body scanning survey of reference, more than 20 similar large-scale studies have been performed around the world [5] to characterize the body size and shape of different populations. In particular, the CAESAR database's accessibility has boosted the development of data driven models to create 3D body avatars. Despite the existing 3D body databases that geographically cover the globe, the generation of multi-nationality models combining training databases of different surveys is still a challenge. The reasons for this are the restricted accessibility conditions of the data and the lack of harmonisation among body scans [6]. For instance, the variability of the posture of the arms and hands (Fig. 1) requires several adaptations to processing algorithms to standardise the body scan.

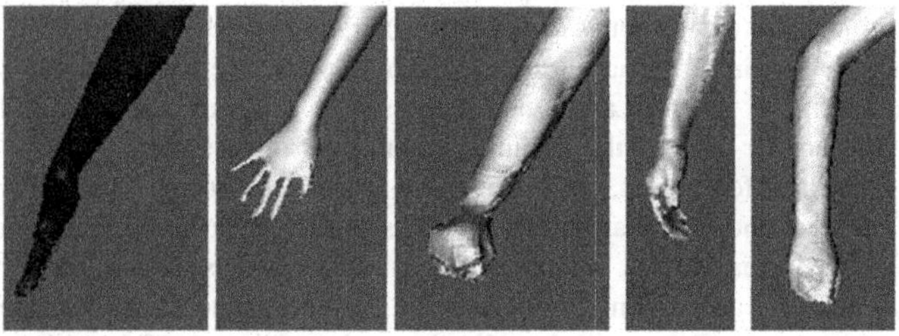

**Fig. 1.** Variability of the hand posture in different sizing surveys.

Allen et al. [7] proposed a new method to generate a parametric model, learned from a set of registered high-resolution 3D scans in a canonical pose applying principal component analysis (PCA). In this work, the parametric model of the 3D body was used to create 3D body avatars by editing multiple components correlated with body attributes (such as height and weight). Later, many authors extended the data driven approach based on PCA to create 3D body avatars using different types of partially correlated data such as 2D images or 3D body scans. The 3D body avatar is a clean watertight mesh of the body that can represent a statistical shape (3D aggregated data) or an individual body shape.

Statistical 3D body avatars can be created directly with a combination of components from the PCA model or from aggregated tables of standard anthropometry that represent percentiles, sizes or morphotypes. The resulting statistical avatars using these methods are more realistic than DHM used by current CAD software [3]. Along this line, Reed et al. compared two methods for generating boundary mannequins using a PCA model [8]. One method introduces the standard anthropometry of the boundary mannequins and the other is a direct mapping of the first three PC of the PCA. While the mapping of the shape space does not require a pre-selection of measurements, the number of mannequins and distribution in the body shape space is arbitrary. Although it has been demonstrated that it is possible to achieve more realistic DHM using a PCA model of the body, it is important to consider that potential training databases are limited to a canonical position, while the use of the DHM for ergonomics requires a multi-pose configuration, with the sitting position being the main pose.

Individual 3D body shapes can be created on the basis of the PCA model approach from a set of anthropometric measurements (1D–3D) [9–13] or using more detailed data of the person, mainly 2D images (2D–3D) [14–19] or 3D body scans obtained with low cost depth sensors (3D–3D) [20–23]. In the latter two cases, apart from fitting the 3D shape, it is also necessary to deal with the pose variability of the user and the influence in body shape. Allen et al. [24] and Hasler et al. [25] developed a unified model that describes both human pose and body shape. In these models, pose and body shape were stored in a single model and correlations between them were automatically exploited to induce realistic muscle bulging and fat deformation during animation. Anguelov et al. [26] presented in 2005 the SCAPE method (Shape Completion and Animation for PEople)—a data-driven method for building a human shape model that includes variations in both subject shape and pose. In this method, pose and shape are treated independently. An improved version of the SCAPE method has been proposed by Hirshberg et al. [27] to achieve smoother joints and a highly realistic articulated body model. In these methods, both models, sizes and shapes are fitted and optimised simultaneously.

The applications and uses of individual avatars are mainly focused on visualisation (e.g. the virtual trying on of clothing, animation) or anthropometry feature extraction for retail (e.g. bespoke clothing, size recommendation) [28, 29] or health (e.g. obesity monitoring) [30]. With body dimensions and 2D images, it is possible to approximate the 3D body shape of a person to represent the morphotype and size of the body; however, detailed data of the face in order to experience a good self-identification of a 3D body avatar requires more information than that provided by 3D scan data (Fig. 2).

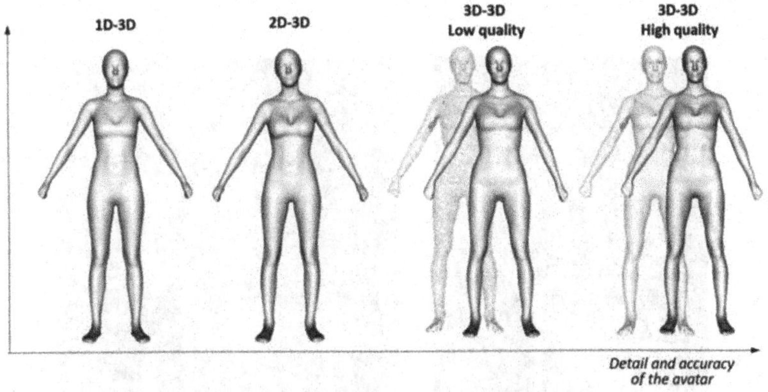


**Fig. 2.** Individual avatar of the same subject created using four methods: 1D–3D created from age, weight, stature bust girth, waist girth and hip girth; 2D–3D obtained from a frontal and a lateral photo [19]; 3D–3D of a low quality raw scan; and 3D–3D of a high quality raw scan.

In this paper, three applications of 3D body avatar creation based on a PCA model of the Spanish population are presented: (1) statistical 3D body avatars applied to the ergonomic design using PCA mapping; (2) creation of statistical 3D body avatars from aggregated measurements; and (3) serial 3D creation of harmonised watertight meshes acquired with any type of 3D body scanner.

## 2 The Dataset

The databases of 3D body scans used to model the shape space of adult bodies were acquired on two anthropometric surveys conducted by IBV for the apparel industry. Both studies used the 3D laser-based body scanner (Vitus Smart XXL) to acquire the surface of the body with a resolution of $\pm 1$ mm in accordance with the international DIN EN ISO 20685 standard.

The study of the female population finished in 2009 and measured 9,600 females with ages from 12 to 70 years old including regular and big sizes [31]. The stratification of the sample covered ten age groups enabling the analysis of specific market segments. Two standing postures were captured for the whole sample to determine the set of measurements for apparel application. The posture used to develop the shape space model of adults was standing with the feet parallel and hips wide apart, their arms extended and open at $\sim 60^{\circ}$, and fists closed with the hand dorsum pointing outwards (Fig. 3). A set of 15 markers were placed on the body to locate anatomical positions related to reference points required for the calculation of body measurements described in the ISO 7250 or ISO 9559 standard. A standard attire was provided to each subject according to the ISO 20685 standard (Fig. 3). The measuring protocol was complemented with a demographic questionnaire including information about clothing shopping habits and problems in finding well-fitting clothes.

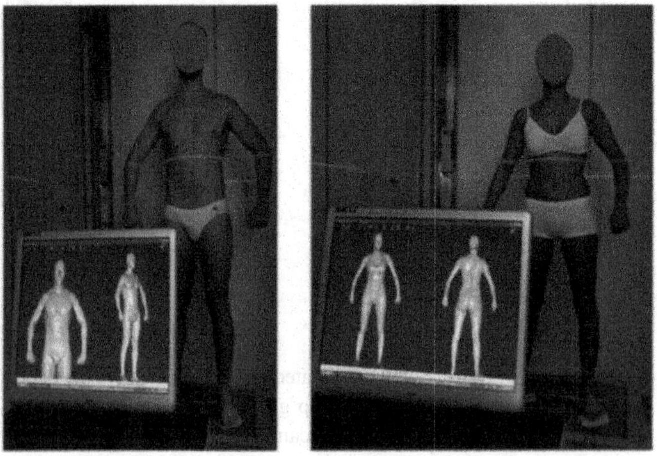

**Fig. 3.** Scanning protocol used in this survey.

The anthropometric survey of the male population was conducted between 2014 and 2015 in cooperation with the Spanish clothing sector [5]. The sample was stratified into four age groups according to the market groups defined by the clothing industry associations. 1,800 males were scanned in standing and sitting postures. The standing posture was the same as that described for the study of females (Fig. 3). Subjects also answered a questionnaire about clothing shopping habits and fitting problems. A standard attire was provided to each subject according to the ISO 20685 standard (Fig. 3).

## 3 Shape Space of Full Bodies

The raw 3D body scans of the two databases were registered and harmonised using an adaptation of different template-based methods [7, 32, 33]. In this case, we used a high resolution template body mesh of 150 K vertices, 99 K triangles, a 17-bone skeleton and a set of 35 landmarks as described by Ballester et al. [6] (Fig. 4). This process provided a database of individual body dimensions and a database of individual 3D homologous avatars with anatomical one-to-one vertex correspondence among them with a harmonised canonical pose.

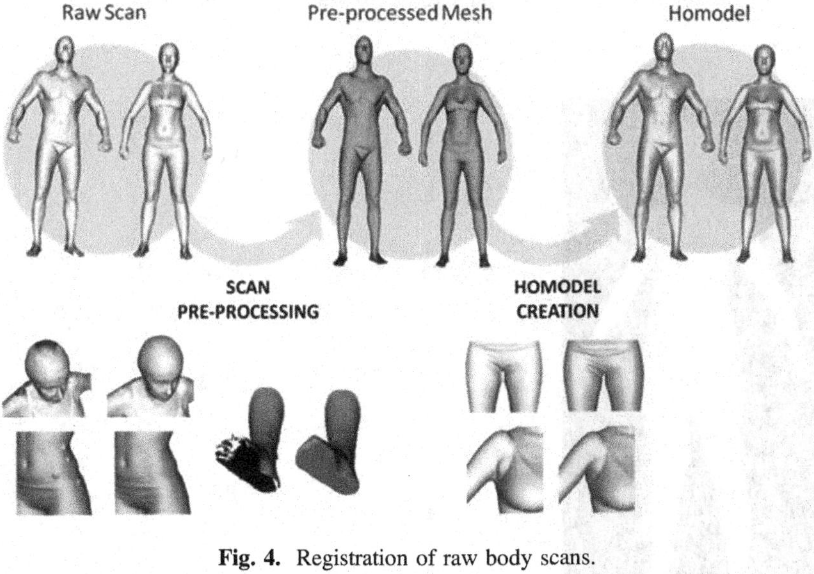


**Fig. 4.** Registration of raw body scans.

Database parametrisation was achieved by conducting a Rigid Procrustes Alignment and Principal Component Analysis (PCA) on the homologous data [34]. The posture variability was harmonized to create different shape spaces for males and females.

# 4 Applications

## 4.1 Statistical 3D Body Avatar Creation from the PCA Mapping Applied to Sizing Analysis

The parametric model obtained from the PCA enables the generation of statistical shapes mapping the shape space of bodies. This approach was proposed by Zehner [35] and Robinette [36, 37] for anthropometric body dimensions to select boundary mannequins as an alternative to the percentile method. The PCA method maximises the variability of the data, with the first components being a useful exploration tool to conceive the sizing strategy to optimize the fitting of products [38, 39]. This problem has been applied for sizing clothing by Veitch et al. using two principal components of body measurements to define 36 body categories: twelve sizes and three shapes in each size [40].

In this paper, the approach is extended to the 3D using the PCA model of the Spanish population of females. The resulting components were explored to define a set of statistical shapes representing the main variability of the body shape for sizing (Fig. 5).

The first component shows the variability in size combining height and girth variations (Fig. 5). The second component shows the variability of girths. The third component is mainly related to the length of the limbs. The fourth component shows a variability related to the posture of the back, shoulders and neck. The fifth component shows the variation of the body shape from ‘triangle’ to ‘rounded’.

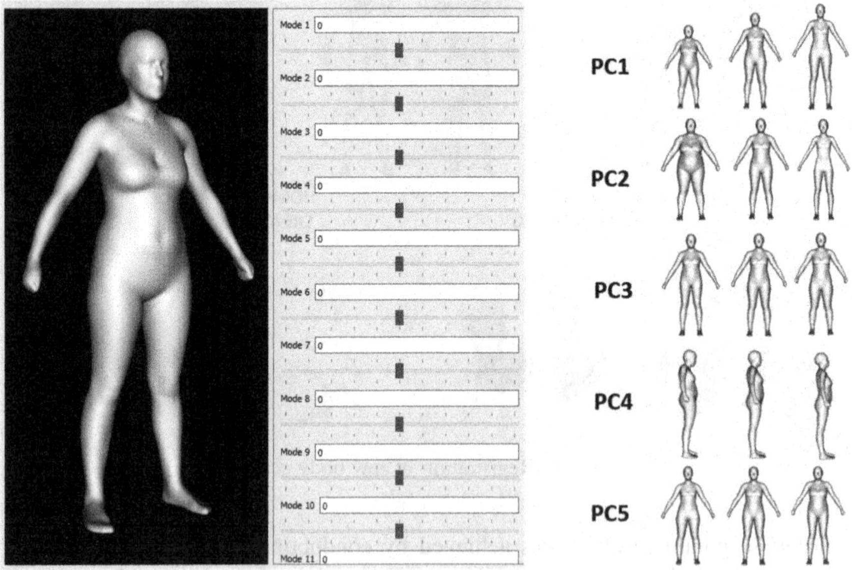


**Fig. 5.** Left: Interface to explore the PCA components. Right: Mean body shape and ± 3SD of the first five components.

The first two components were chosen to create a set of statistical avatars that covers the main size variability of stature and girths. A combination of ±1SD, ±2SD and ±3SD was used to map the avatars over the bi-variant distribution of PC1 and PC2 (Fig. 6). As a result, 22 statistical avatars were obtained. A summary of the digital body dimensions of the avatar set is reported in Table 1.

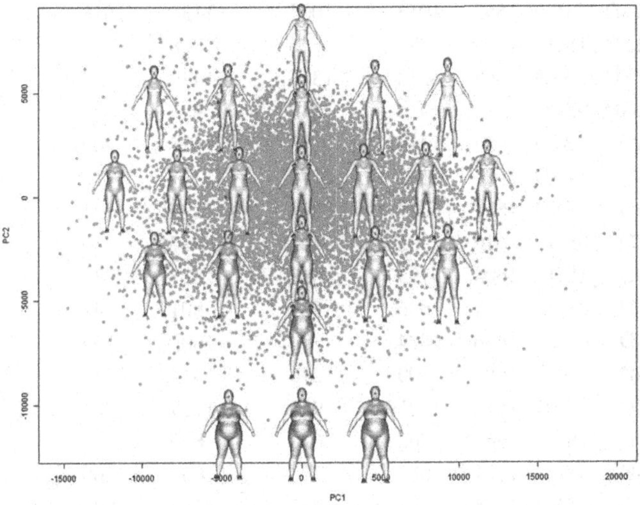


**Fig. 6.** Distribution of avatars mapping the first two components.

Although PCA mapping is an interesting tool to explore body shape variability and conceptualise a sizing system, for many applications, and especially to relate the body with the product scalability and dimensions, it is necessary to apply systematic sizing criteria based on dimensions instead of shape. Furthermore, the relationship of body-product dimensions is required to recommend the selection of the size. An alternative method is the creation of statistical body avatars from representative body dimensions.

### 4.2 Statistical 3D Body Avatar Creation from Measurements Applied to Reference Bodies for the Clothing Industry (1D–3D)

The design process of the clothing sector to fit a product to a body shape uses as a reference a collection of sizing tables that are mainly a summary of standard body dimensions for each size considering different types of garments (e.g. dresses, shirts, trousers). For the following application, IBV conducted an analysis of the Spanish database of dimensions aiming at determining the distribution of intervals combining pairs of primary and secondary dimensions for different types of garments. For each interval, 37 body measurements were considered according to EN 13402-1. Sizing tables were used to report the mean anthropometric values of each population interval. The 1D–3D method was used to generate the statistical 3D body models representing each interval of the sizing tables using as input the 37 body dimensions (Fig. 7).

**Table 1.** Digital body dimensions of body avatars generated by the combination of PC.

| PC1 | PC2 | Height | Bust girth | Waist girth | Hip girth | Crotch height |
|---|---|---|---|---|---|---|
| −3SD | 0SD | 1445 | 937 | 874 | 989 | 625 |
| −2SD | −1SD | 1488 | 1068 | 1028 | 1136 | 637 |
| −2SD | 1SD | 1473 | 798 | 691 | 860 | 657 |
| −2SD | 0SD | 1500 | 926 | 853 | 990 | 660 |
| −1SD | −1SD | 1553 | 1053 | 1005 | 1130 | 678 |
| −1SD | 1SD | 1538 | 785 | 666 | 868 | 699 |
| −1SD | −3SD | 1569 | 1338 | 1316 | 1439 | 659 |
| −1SD | 0SD | 1555 | 915 | 832 | 992 | 695 |
| 0SD | −3SD | 1634 | 1319 | 1297 | 1425 | 700 |
| 0SD | −2SD | 1624 | 1137 | 1098 | 1231 | 712 |
| 0SD | −1SD | 1617 | 1019 | 955 | 1107 | 721 |
| 0SD | 0SD | 1610 | 905 | 811 | 997 | 730 |
| 0SD | 1SD | 1604 | 793 | 668 | 895 | 738 |
| 0SD | 2SD | 1599 | 686 | 529 | 801 | 747 |
| 1SD | 0SD | 1666 | 894 | 790 | 1002 | 765 |
| 1SD | −3SD | 1698 | 1300 | 1276 | 1414 | 741 |
| 1SD | −1SD | 1683 | 1026 | 957 | 1129 | 761 |
| 1SD | 1SD | 1669 | 762 | 620 | 890 | 781 |
| 2SD | 0SD | 1721 | 883 | 770 | 1009 | 800 |
| 2SD | −1SD | 1748 | 1013 | 932 | 1132 | 802 |
| 2SD | 1SD | 1735 | 751 | 597 | 900 | 823 |
| 3SD | 0SD | 1777 | 873 | 750 | 1017 | 835 |

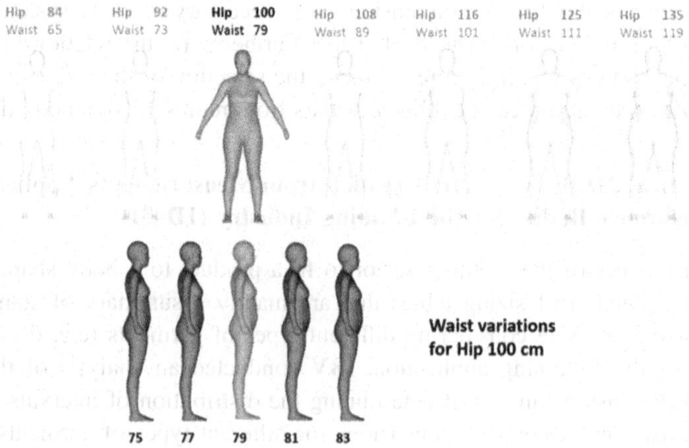


**Fig. 7.** Statistical avatars representing each interval of females represented by a combination of two primary measurements.

For the 1D–3D method, a regression model was created relating a set of body measurements with the combination of the PCA scores that match them. The creation of the avatar was based on iteratively resolving a minimisation problem that finds the combination of shape parameters (i.e. principal component scores) that best matches the body measurements entered. Additionally, the resulting body shape is optimized by iterative measurement of the target measurements. The performance of this method in our implementation achieves maximum error values in the order of 0.01 cm [6].

### 4.3 Avatar Creation from Individual 3D Body Scans (3D–3D)

Raw body scans are dense and unorganised meshes made up of hundreds of thousands of points containing many undesirable artefacts, especially in areas that are not reachable by 3D scan sensors. Typical artefacts include “holes”, surface noise and additional material (e.g. bumps, bridges, etc.).

The purpose of a method to create 3D body avatars from raw scans (3D–3D) is to facilitate the automatic conversion of raw body scans into simulation-ready avatars to be used as a basis for design development [41]. In order to be used in simulation environments, raw body scans need to be converted into smooth, watertight meshes, free of noise, artefacts and occlusion areas.

The current technology used for the serial processing of large body scan databases gathered in national sizing surveys is optimal for the high resolution scanners that provide raw scans with a high dense mesh and controlled and repeatable artefacts focussed on armpits, crotch, top of the head and shoulders. The mass use of low cost depth sensor devices, most of which are operated manually, has boosted the use of 3D body scans with coarse surfaces, missing parts and separated layers due to the movements of the body during the scanning process.

Adaptations of the technology presented by Trieb et al. [6] were aimed at making the technology robust and versatile enough to process any type of raw scan in standing

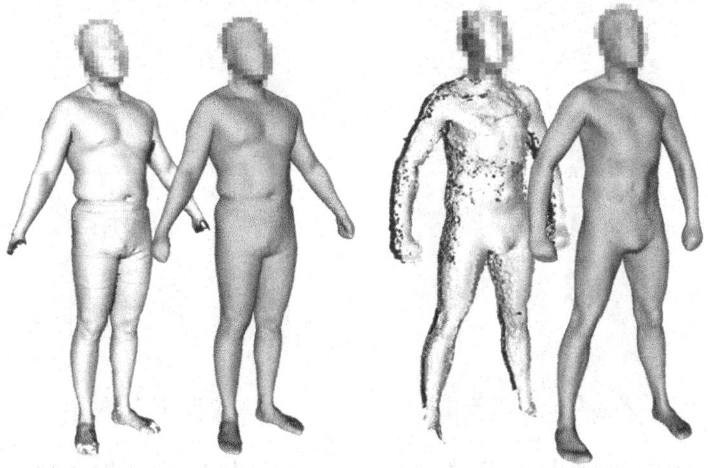

**Fig. 8.** Generation of avatars. Left: High quality raw scan. Right: Low quality raw scan.

pose using a non-rigid registration approach (Fig. 8). In order to seamlessly integrate 3D–3D technology into current design software, a web service that can be reached via API was developed. Different types of scans were tested in order to validate the performance of the registration method. Figure 9 shows the raw scans obtained using this technology with a 3D body scan captured with the Vitus Smart high-resolution scanner and the same person scanned with 8 views of Kinect v2. The resulting individual avatar is represented in Fig. 10. The visual appearance of the avatar is quite similar considering the significant difference of the surface quality of the original data.

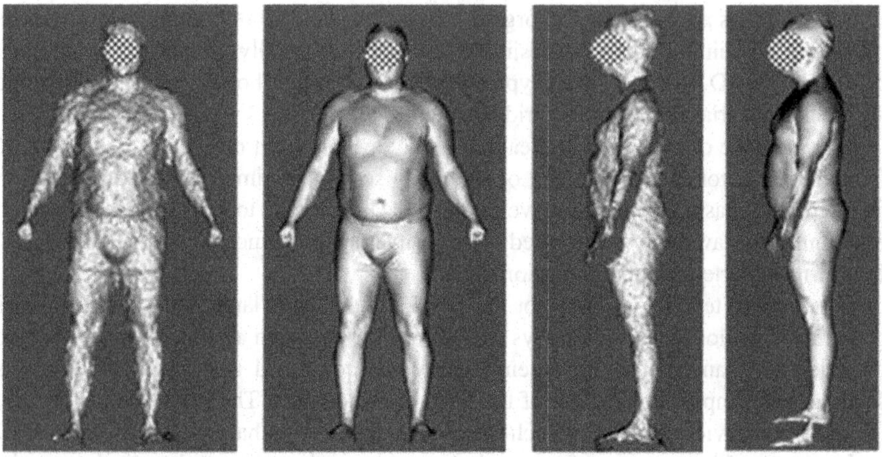

**Fig. 9.** Frontal and lateral views of a subject scanned with Kinectv2 (left) and Vitus Smart (right).

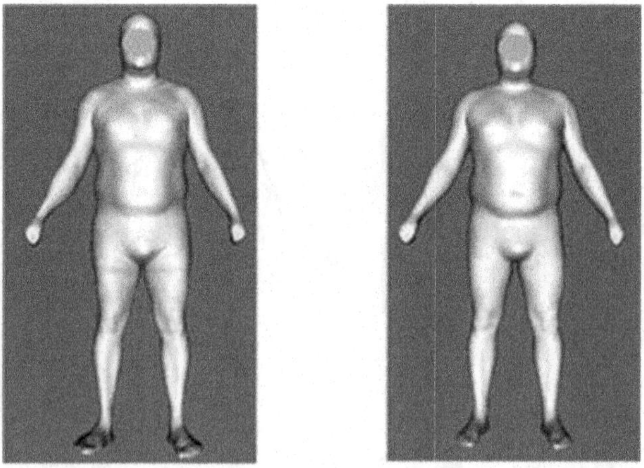

**Fig. 10.** Avatar generated from the Kinectv2 scan (left) and from the Vitus Smart scan (right).

More research is required to quantify the differences in 3D body measurements that could be relevant for the adoption of this technology in different applications.

## 5 Conclusions

In this paper, the methods of 3D body avatar creation based on a data-driven approach were reviewed considering those more suitable for the generation of aggregated avatars, representing a target population, or individual avatars representing an individual person.

The contribution of the PCA body models to these methods has been demonstrated as an efficient way to generate a synthesis of the shape space of 3D bodies. However, the generation of this model requires the availability of 3D body databases. Although many 3D anthropometric surveys were done in the past, access to the data is in general restricted preventing the possibility of building a global model representing different geographic regions. Furthermore, the combination of current 3D body databases to achieve more global models requires more research related to compatibility in particular, how aspects related to the scanning posture or clothing may affect the PCA model and alter the shape variability explained by the components.

3D anthropometric surveys are very costly. The sample sizes measured were quite big since this is a requirement to achieve low errors estimating mean anthropometric values and percentiles. However, the estimation of the optimal sample size for 3D body shape modelling purposes has not been analysed. A relevant question of research would be to study if it is possible to reduce the sample size for 3D body modelling introducing a new way of planning 3D body scanning surveys.

Finally, the creation of avatars that are in the borders or out of the parametric shape space described by the database of the PCA model may induce higher errors compared to the creation of avatars with body parameters within the range of the database. For data driven method applications, it is important to define the conditions of validity of the input data and additional work is required in order to quantify the 3D body reconstruction errors within the shape space.

For the 1D–3D method, it is possible to use different configurations of body dimensions as input data, from a short set of measurements to a complete set describing the body anthropometry in detail. More insights about the optimal combination of measurements used as input data in relation to accuracy and visualisation are necessary to guide the avatar creation process. This is also important to avoid breaking the statistical rules about how many predictors can be derived from data when doing regression analysis.

Nevertheless, it is important to consider the inconsistency of the 1D–3D method. This inconsistency is related to the type of input measurements, typically manual anthropometric measurements, and the type of measurements used to control the achievement of the target measurements during the creation process, that are necessary digital body measurements calculated over the 3D avatar. Biases between manual anthropometric measurements and digital measurements have been reported by many authors [42, 43]. The definition of body measurements based on anatomical points used on manual anthropometry cannot be reproduced in the digital measurements that require geometric features to be defined. 1D–3D data is a tool that can be used to generate both individual and aggregated avatars and to re-evaluate existing databases addressing these questions.

In the particular case of using data driven models to enhance current DHM, the articulated structures and body meshes of such models are still not compatible with the compacted parametric type of files used by the DHM. The seamless integration of both body modelling techniques is an important challenge for computer graphics.

Data driven approaches offer a variety of methods to create realistic avatars for different applications, profiting from the shape information of the body provided by existing databases. In the last few years, new emerging businesses employing 3D body data (e.g. garment and footwear customization, size recommendation, health monitoring) have increased the number and size of 3D body scan repositories. 3D body databases are growing very fast and the development of 3D modelling tools is crucial to leverage the practical application and exploitation of these data. However, for the moment, the access to most of the existing and future 3D data is not and will not be open preventing the exploitation and use by means of the data driving methods presented.

**Acknowledgments.** The authors thank the European Commission, the Instituto Valenciano de Competitividad Empresarial (IVACE) and the Agencia Estatal de Investigación del Ministerio de Economía, Industria y Competiti-vidad (MINECO) for the financial support of this research though the following projects: In-Kreate (funded by the European Union's Horizon 2020 Research and Innovation programme under Grant Agreement no. 731885), BodyPass (funded by the European Union's Horizon 2020 Research and Innovation programme under Grant Agreement no. 779780), 3DBODY_HUB (submitted to IVACE with a funding of Generalitat Valenciana and the European Regional Development Fund and the proposal nº IMDEEA/2018/49) and Torres Quevedo (funded by MINECO under the program Torres Quevedo 2016).

## References

1. Pheasant S (1991) Ergonomics, work and health. Palgrave, Basingstoke
2. Duffy VG (2016) Handbook of digital human modeling: research for applied ergonomics and human factors engineering. CRC Press, Boca Raton
3. Reed MP et al (2014) Developing and implementing parametric human body shape models in ergonomics software. In: Proceedings of the 3rd international digital human modeling conference, Tokyo
4. Robinette, KM, Daanen H, Paquet E (1999) The CAESAR project: a 3-D surface anthropometry survey. In: Proceedings of the second international conference on 3-D digital imaging and modeling, IEEE (1999)
5. Ballester A et al (2015) 3D body databases of the spanish population and its application to the apparel industry. In: Proceedings of 6th international conference on 3D body scanning technologies, Lugano, Switzerland
6. Trieb R et al (2013) EUROFIT—integration, homogenisation and extension of the scope of large 3D anthropometric data pools for product development. In: 4th International conference and exhibition on 3D body scanning technologies, Long Beach, CA, USA (2013)
7. Allen B, Curless B, Popović Z (2003) The space of human body shapes: reconstruction and parameterization from range scans. ACM Trans Graphics (TOG) 22(3):587–594

8. Reed MP, Park BKD (2017) Comparison of boundary manikin generation methods. In: 5th International digital human modeling symposium
9. Zeng Y, Fu J, Chao H (2017) 3D human body reshaping with anthropometric modeling. In: International conference on internet multimedia computing and service. Springer, Singapore
10. Reed MP et al (2014) Developing and implementing parametric human body shape models in ergonomics software. In: Proceedings of the 3rd international digital human modeling conference, Tokyo
11. Wuhrer S, Shu C (2013) Estimating 3D human shapes from measurements. Mach Vis Appl 24(6):1133–1147
12. Koo B-Y et al (2015) Example-based statistical framework for parametric modeling of human body shapes. Comput Ind 73:23–38
13. Baek S-Y, Lee K (2012) Parametric human body shape modeling framework for human-centered product design. Comput Aided Des 44(1):56–67
14. Seo H, Yeo YI, Wohn K (2006) 3D body reconstruction from photos based on range scan. In: International conference on technologies for e-learning and digital entertainment. Springer, Heidelberg (2006)
15. Xi P, Lee W-S, Shu C (2007) A data-driven approach to human-body cloning using a segmented body database. In: 15th pacific conference on computer graphics and applications. PG 2007. IEEE
16. Zhu S, Mok PY, Kwok YL (2013) An efficient human model customization method based on orthogonal-view monocular photos. Comput Aided Des 45(11):1314–1332
17. Saito S et al (2014) Model-based 3D human shape estimation from silhouettes for virtual fitting. In: Three-dimensional image processing, measurement (3DIPM), and applications 2014, vol 9013. International Society for Optics and Photonics
18. Mok PY, Zhu S (2018) Precise shape estimation of dressed subjects from two-view image sets. In: Applications of computer vision in fashion and textiles, pp 273–292
19. Ballester A et al (2016) Data-driven three-dimensional reconstruction of human bodies using a mobile phone app. Int J Digital Hum 1(4):361–388
20. Weiss, A, Hirshberg D, Black MJ (2011) Home 3D body scans from noisy image and range data. In: 2011 IEEE international conference on computer vision (ICCV). IEEE
21. Lu Y et al. Accurate nonrigid 3D human body surface reconstruction using commodity depth sensors. Comput Animation Virt Worlds e1807
22. Park B-K, Reed MP (2014) Rapid generation of custom avatars using depth cameras. In: Proceedings of the 3rd international digital human modeling conference
23. Tong J et al (2012) Scanning 3D full human bodies using kinects. IEEE Trans Vis Comput Graphics 18(4):643–650
24. Allen B et al (2006) Learning a correlated model of identity and pose-dependent body shape variation for real-time synthesis. In: Proceedings of the 2006 ACM SIGGRAPH/Eurographics symposium on computer animation. Eurographics Association
25. Anguelov D et al (2005) SCAPE: shape completion and animation of people. In: ACM transactions on graphics (TOG), vol 24, no 3. ACM (2005)
26. Hasler N et al (2009) A statistical model of human pose and body shape. In: Computer graphics forum, vol 28, no 2. Blackwell Publishing Ltd
27. Hirshberg DA et al (2012) Coregistration: simultaneous alignment and modeling of articulated 3D shape. In: European conference on computer vision. Springer, Heidelberg
28. Istook CL, Hwang S-J (2001) 3D body scanning systems with application to the apparel industry. J Fashion Mark Manag Int J 5(2):120–132
29. D'Apuzzo N, Gruen A (2009) Recent advances in 3D full body scanning with applications to fashion and apparel. Optical 3-D measurement techniques IX (2009)

30. Treleaven P, Wells J (2007) 3D body scanning and healthcare applications. Computer 40 (7):28–34
31. Alemany S, González JC, Nácher B, Soriano C, Arnáiz C, Heras H (2010) Anthropometric survey of the Spanish female population aimed at the apparel industry. In: Proceedings of the 2010 international conference on 3D body scanning technologies. Lugano, Switzerland
32. Amberg B, Romdhani S, Vetter T (2007 June) Optimal step nonrigid ICP algorithms for surface registration. In: IEEE conference on computer vision and pattern recognition, 2007. CVPR 2007. IEEE, pp 1–8
33. Sumner RW, Popović J (2004, August). Deformation transfer for triangle meshes. In: ACM Transactions on graphics (TOG), vol 23, no 3, pp 399–405. ACM
34. Gower JC (1975) Generalized procrustes analysis. Psychometrika 40(1):33–51
35. Zehner GF, Meindl RS, Hudson JA (1993) A multivariate anthropometric method for crew station design. Kent State University oH
36. Robinette KM, McConville JT (1981) An alternative to percentile models (No 810217). SAE technical paper
37. Robinette KM (1998) Multivariate methods in engineering anthropometry. In: Proceedings of the human factors and ergonomics society annual meeting vol 42, no 10. SAGE Publications, Sage, Los Angeles
38. Lacko D et al (2017) Product sizing with 3D anthropometry and k-medoids clustering. Comput Aided Des 91:60–74
39. Lee W et al (2016) Application of massive 3D head and facial scan datasets in ergonomic head-product design. Int J Digital Hum 1(4):344–360
40. Veitch D, Veitch L, Henneberg M (2007) Sizing for the clothing industry using principal component analysis—an Australian example. J ASTM Int 4(3):1–12
41. Durá JV, Caprara G, Ballester A, Pierola A, Kozomara Z (2018) Preliminary results of the InKreate Project. Revista de Biomecánica 65
42. Han H, Nam Y, Choi K (2010) Comparative analysis of 3D body scan measurements and manual measurements of size Korea adult females. Int J Ind Ergon 40(5):530–540
43. Markiewicz Ł et al (2017) 3D anthropometric algorithms for the estimation of measurements required for specialized garment design. Expert Syst Appl 85:366–385

# Ergonomics for Children and Educational Environments

# Differences in Visual Attention Performance Between Action Game Playing and Non-playing Children

Min-Sheng Chen(✉), Tien-Sheng Chiu, and Wei-Ru Chen

National Yunlin University of Science and Technology, Douliou, Taiwan, Republic of China
chens@yuntech.edu.tw, bokas3kas3@gmail.com, D10321001@yuntech.org.tw

**Abstract.** Games often act as a teaching tool, but prolonged video game playing may have effects on the cognitive ability of schoolchildren. The present research aims to investigate such effects. The study recruited schoolchildren from grades 1 to 6 as participants to examine differences in attention performance between action game players and non-players. The experiment used the modified UFOV (useful field of view) operated with such factors as distance and clues. The results revealed that the players are significantly superior to the non-players in reaction speed and accuracy, suggesting that the players have better attention in the space and selection realms. In addition, distance also had a significant effect on the participants: increase in distance significantly lowered the accuracy of the non-players, whereas that of the players changed little. The results suggest that video games can strengthen the visual attention of children. It is recommended that the research findings be considered in the design of teaching tools related to attention training. In addition, the task characteristics of the action game content can be incorporated in educational materials to improve the effectiveness of training and assistance.

**Keywords:** Schoolchildren · Action games · Game players · Visual attention

## 1 Introduction

With the continuous progress in science and technology, the number of people who use electronic products is increasing annually. Consumers of electronic products are becoming younger; many children have started to use these products for learning and social activities, as well as for playing video games.

Video games are everywhere. In addition to many potential negative effects, could prolonged video game playing influence children's cognitive ability? Dye et al. [6] found that different types of games produce significant effects on humans' cognitive abilities, such as visual attention, spatial ability, and memory. Green and Bavelier [8] stated that in completing game tasks, action game players use their visual search ability and visual attention intensively, thus making their visual attention performance superior to that of non-game players. Oei and Patterson [11] used various experiments, such as the flanker, Go/No-go, and task switching tasks to investigate the cognitive execution


S. Bagnara et al. (Eds.): IEA 2018, AISC 826, pp. 639–648, 2019.
https://doi.org/10.1007/978-3-319-96065-4_67

performance of players of different types of games, including action, strategy, development based, and recreational types. They found that players of development-based games trained for 20 h are significantly superior to players of other types of games in execution ability, namely, task switching and execution ability. Oei and Patterson [11] suggested that developmental game players use various tasks continuously, such as planning, strategy, and reconstruction, during the process of game training, which helps improve their cognitive execution ability. Dobrowolski et al. [4] investigated the performance of strategy and action game players and non-players in task switching and multiple object tracking. They found the performance of the strategy game players to be superior to that of the action game players and non-players. Based on the research above, different types of games classified by their design subject matters and tasks can help improve different specific cognitive abilities. Thus, investigation of the effects of different factors on game players' cognitive ability is needed for future design of relevant game subject matters that can strengthen schoolchildren's motivation and cognitive ability in learning based on the specific ability to be improved. This is also one of the research objectives.

Among various types of games, action games may be the most familiar to the public and have the highest market share in the video game market. Action games are action-oriented and can train players in hand–eye coordination and reaction capacity, in combination with the game content. The types of action games include the following: shooter, fighting, action adventure, racing, action role playing, and music rhythm games. Green and Bavelier [8] used flanker compatibility, useful field of view, and multiple object tracking tasks to investigate the visual selective attention of game players and non-players. They found that action game players are superior to their non-gaming counterparts in comparability, spatiality, and temporary loss of attention, suggesting that action game players are superior to non-players in visual selective attention. Similarly, Dye and Bavelier [5] used useful field of view tasks, attentional blink tasks, and multiple object tracking to investigate the visual attention of schoolchildren who were action game players. They reported that action game players are superior to non-players in visual selective attention and sustained attention. When performing object tracking, action game players can track more objects compared with non-game players. Buckley et al. [1] used Goldmann perimetry to investigate the field of vision of action game players and non-players. In the test, a light source would randomly occur in a 360-degree semicircular instrument. The participant was required to gaze at the light source. The gaze was verified by an inspector and recorded on a Goldmann graph. Their study showed that action game players have a wider central and peripheral field of vision than non-players, suggesting that game players have a wider field of vision.

The above research results indicate that action games can benefit players' visual attention. Thus, some scholars have used games as the subject matter for cognitive training and one of the tools for improving cognitive ability. Oei and Patterson [10] used different types of games as training materials for investigating the effects of games on humans' cognitive ability. In their experiment, the participants were divided into experimental and control groups, with the former further subdivided into five groups that were trained for different types of games for one month. The types of games for training included action games (Modern Combat: Sandstorm), spatial memory games

(Tvishi Technologies), matching games (Bejeweled 2; PopCap Games), hiding games (Hidden Expedition-Everest; Big Fish Games), and simulated life games (The Sims 3). Oei and Patterson used a series of tasks (e.g., attentional blink task, filter task, visual search/spatial working memory task, and complex span task) to investigate the performance of visual attention, visual search, visual, and visual spatial memory after completing the training. Their results indicated that only the action game group exhibited good performance after being trained, suggesting that action games can strengthen cognitive ability, particularly in the performance of visual attention. The performance of the action game players improved significantly after training.

The current work aims to investigate the differences in performance of visual attention between action game players and non-players. The research primarily uses the spatial attention experiment and utilizes the concept of distractibility in the useful field of view task for the experiment. The experiment requires the participants to judge the objects and their spatial positions in a short time. Different distances and clues were used in the experiment for investigating the differences in response to field of vision and other stimuli among the participants in different groups. These research achievements are expected to act as reference for relevant subject matters, such as game training or education of schoolchildren for increasing learning appeal.

## 2 Research Methods

The experiment investigated the differences in visual attention performance between action game players and non-players. The experiment was a spatial test using a modified useful field of view task. The accuracies and response times of the participants were collected for subsequent assessment of attention performance.

### 2.1 Experimental Objective

To investigate the visual attention of action game playing and non-playing children, the participants were required to employ divided attention in the spatial test. It included central and peripheral identification. In addition, they were asked to identify the central and peripheral target objects and their positions. The focus was the distance between the central and peripheral target stimuli, as well as the presence or absence of clues. Finally, the research aimed to investigate the effect of the stimulation distance and presence or absence of clues on the action game playing and non-game playing schoolchildren.

### 2.2 Participants

The research involved 24 children aged 7 to 12 years, including 12 action game players and 12 non-players. The game players played an action game for at least four hours each week for six consecutive months. The non-game players played a video game for less than one hour a week or did not play at all. The vision of each participant was normal (1.0) or above 0.8 after correction.

## 2.3 Experimental Design and Materials

The experiment followed a three-factor design: 2 (group) × 3 (distance) × 2 (clue). The inter-group factors were the two groups (action game players and non-players). The intra-group factors were distance and clue. The distance was between the peripheral target objects and the center of a circle. The distance between the target and the center of the circle was considered long when it was 2.5 cm, moderate when it was 5 cm, and long when it was 7.5 cm. The peripheral target objects were scattered in eight directions at a 45° angle. There were 24 positions in total (as shown in Fig. 1).

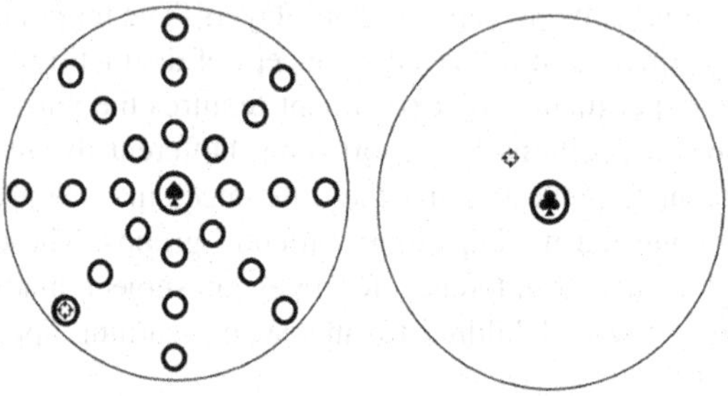

**Fig. 1.** Experimental stimulation (left: long distance, presence of clue; right: short distance, absence of clue)

The clue types were divided into presence and absence. Clue presence was characterized by the use of a circular diagram to guide the participant to identify the position of the target object. The dependent variables were accuracy rate and response time. The diagrams with possible club or spade for the central target objects and the diagram with the peripheral target objects as targets were 4 cm long and 4 cm wide. The radius of the circular diagram was 8 cm.

## 2.4 Experimental Equipment

The experiment used an ASUS A52 J 15.6-inch laptop running on Window 7 for displaying visual stimulation, equipped with a KINYO KBX03 external digital keyboard and E-prime 2.0 for creating and gathering relevant data. Statistical analysis software SPSS 22.0 was employed for data analysis.

## 2.5 Experimental Procedure

The experiment was a spatial test for visual attention. The experimental procedures are shown in Fig. 2. The participants were informed of the research objectives, experimental procedures, and notes before the experiment. Each participant signed a letter of consent and underwent one training session. The experiment started formally after we verified that the participants were familiar with operation of the reaction keys.

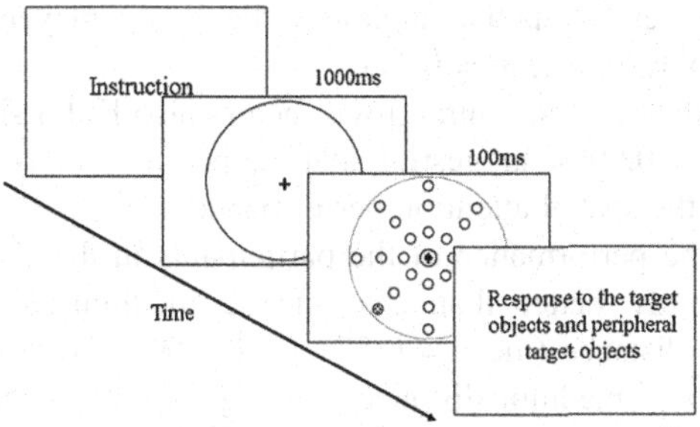


**Fig. 2.** Flowchart for testing spatial visual attention

At the beginning of the experiment, the participant would see a "+" at the center of the circle and gaze at it for one second. Subsequently, the target objects at the circle center and peripheral target objects would appear simultaneously. The display time was 100 ms. The target object at the circle center would be displayed as a spade or a club. Each participant was first required to identify the target object at the circle center as a spade or club. The participant was required to respond by pressing Key Z when the central stimulus was a spade, and Key X when the central stimulus was a spade. The participants were asked to respond to the peripheral target objects after identifying the target objects at the circle center, by pressing key 1 (lower left), 2 (lower middle), 3 (lower right), 4 (left), 6 (right), 7 (upper left), 8 (upper middle), and 9 (lower right) according to the numerical positions using an external numbers keyboard. The response was primarily designed for compliance with the spatial compatibility and required no additional memory. Each participant was required to respond correctly and quickly upon the display of a picture. The response was recorded as correct only after the participant correctly responded to the target objects and peripheral target objects. The accuracy rate and response time of the participant was collected after the experiment.

# 3 Research Results and Analysis

The accuracy scores and response times of the participants were collected for analysis. The accuracy was the ratio of the correct answers to all answers. Response time was the time required for a participant to press the key after he/she discovered the central and peripheral response stimuli. All data collected in the research were subject to repeated measure with SPSS Statistics 22. The least significant difference method (LSD) was used for post hoc tests and further understanding of the differences among various factors.

## 3.1 Accuracy

The accuracy collected were subject to 2 (groups: game players and non-game players) × 3 (distance: short, medium, and long) × 2 (clue: presence or absence) ANOVA variance analysis. The ANOVA results indicated that a significant main effect between the groups ($F$ (1,22) = 152.016, $p < 0.001$), suggesting that the game players and non-players have differences in spatial attention: the game players (mean = 0.85) are superior to the non-players (mean = 0.37).

However, the distance of the intra-group factors also had a significant main effect ($F$ (2,44) = 7.901, $p$ = 0.001), suggesting that the peripheral target objects at different distances influenced the spatial attention performance of the participants. Based on the LSD post hoc test, the performance of the participants in the short distance (mean = 0.65) was significantly better than that in the medium distance (mean = 0.61, $p$ = 0.038) and long distance (mean = 0.57, $p$ = 0.003). The spatial attention performance in the case of a medium distance was significantly better than that in long distance ($p$ = 0.029). The presence or absence of clues had no significant effect on the spatial attention performance of the participants ($p$ = 0.578).

The post hoc test on the effects of distance on the two groups revealed that the action game players had similar performance in the cases of short and medium distance ($p$ = 0.912), performing better in the case of a medium distance compared with a long distance ($p$ = 0.047). The non-players had different accuracies in the case of different distances. They non-players had a better accuracy in the case of a short distance compared with a medium distance ($p$ = 0.013). They also had a better accuracy in the case of a medium distance than a long distance ($p$ = 0.007), as shown in Fig. 3.

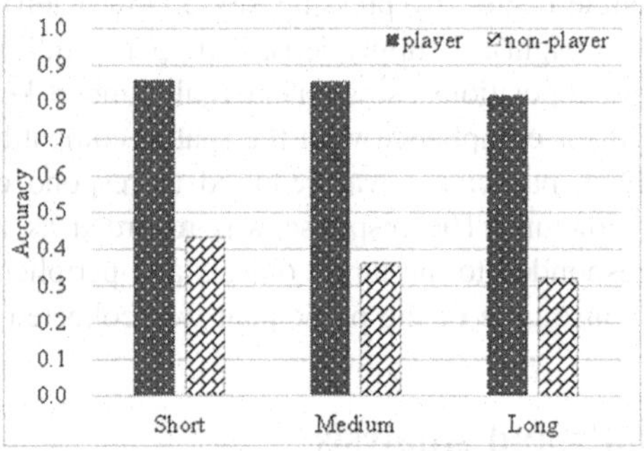


**Fig. 3.** Accuracy performance of the players and non-players in the case of different distances

## 3.2 Response Time

Similarly, the collected response times were subject to 2 (groups: game players and non-players) × 3 (distance: short, medium, or long) × 2 (clues: present or absent) ANOVA with the SPSS Statistics 22.

The analysis results indicated that the distance of the intra-group factors also had significant main effects ($F$ (2,44) = 7.23, $p$ = 0.002), suggesting the effects of the peripheral target objects at different distances on the response time of spatial attention of the participants. The LSD post hoc test showed that the performance of the participants in the case of short distance (mean = 1337.556) is significantly better than that for medium (mean = 1779.378, $p$ = 0.001) and long distances (mean = 1963.451, $p$ = 0.009). However, no significant differences were seen in the spatial attention between the participants in the cases of the medium and the long distance ($p = 0.265$). Similar to the accuracy, neither presence nor absence of clues had a significant effect on spatial attention ($p = 0.122$).

A post hoc test to investigate the effects of distance showed that the action game players had different accuracies in the case of different distances. The players were superior in the short distance than the medium distance ($p = 0.002$). The players were superior players in the medium distance than the long distance ($p = 0.011$). Meanwhile, the non-players were superior in the short distance than the medium distance ($p = 0.012$), as shown in Fig. 4.

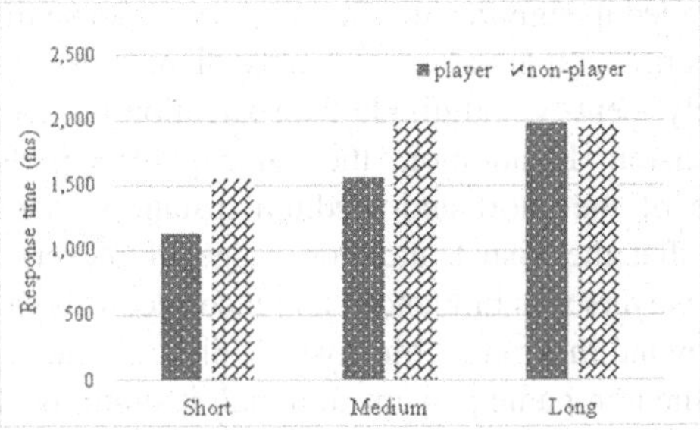


**Fig. 4.** Response time of players and non-players in the case of different distances

## 4 Discussion

The research investigated the differences in accuracy and response speed in a spatial test between the action game-playing and non-playing children using useful field of view task. The game players were found to outperform the non-players in the spatial test. In particular, the two groups have significant differences in the accuracy. Green and Bavelier [8] also found that action-game players outperform non-players in visual attention and spatial distribution of attention in both interference tasks and multiple target objects. Thus, prolonged action game training will help increase visual attention capacity and distribution of spatial attention. Castel et al. [2] investigated the performance of game players in a visual search task using experiments involving multiple interference objects and load capacity, and found that game players outperform non-players under the conditions of multiple interference objects and high load. They concluded that the improved visual attention and high capacity to inhibit unimportant interference objects of the game players enable them to search out the target objects in a quick and accurate manner.

However, such action game training is not only limited to young adult players but can also benefit schoolchildren in terms of improving attention. Dye et al. [7] investigated changes in visual attention in children and adults aged 7 to 22 years after they were subject to action game training; their study found that game training has benefits for the different age groups. They noted more significant effects in younger children have. Their research findings coincide with the present results. Action game training can increase children's visual attention capacity, cultivate their attention screening and selection strategies, and further improve their performance in attention-related tasks.

In addition, the research also finds that distance has a significant effect on attention, particularly on the accuracy. The research findings indicate that both the response time and accuracy of the participants in the case of short distance are better than those in the other two cases, suggesting that the peripheral target objects closer to the central target objects are more likely to be accurately and quickly detected by the participants. This result is similar to that obtained by Green and Bavelier [8] and Dye and Bavelier [5]. With the increase in distance, the accuracy of the response will also decrease gradually. The main reason lies in the limitations of the field of vision and the differences caused by the spatial distribution capacity of the divided attention. In the current results, differences in distance led to significant differences in response time in the action game players, but their accuracy remained considerably high, suggesting that game-playing children can accurately identify and divide their attention but they require a long time before making a decision. In addition, the game players performed excellently in accuracy in the case of the short and medium distances, compared with the long distance, suggesting that the visual attention range of players is 5 to 7.5 cm. The attention of the child players was influenced and incorrect responses occurred when the target objects were beyond this range. Meanwhile, distance had no significant effect on the response time of the non-game players, although it similarly affected their accuracy significantly. This finding indicates the difficulty faced by the non-players in making correct judgments and correctly dividing attention, leading to incorrect responses in the case of the long distance. Moreover, it indicates that child game players have a wider range of field of vision for attention and higher ability to divide attention than non-players [12]. Chisholm et al. [3] also proposed that rich game experience helps improve participants' attention control.

The presence or absence of clues had no significant effect on performance, suggesting that the participants were not influenced by clues with respect to spatial attention identification and task division. This result is inconsistent with that in previous research [7–9]. The reason may be that the clues in the research are not directly associated or compatible with the target objects. Thus, it is impossible for the participants to improve their performance in tasks using these clues. These clues may have only been an interference or background and may not have contributed to attention distribution and performance. Another possible factor is the absence of definite spatial guidance information of the clues in the research. Previous studies [7] used definite spatial information, such as arrows, to guide the spatial distribution of the participants' attention. Thus, it is suggested that clues of high relevance and directionality be used in the future design of clues relevant to games or spatial attention learning for facilitating the distribution of spatial attention.

## 5 Conclusions and Suggestions

The research mainly investigated the differences in visual spatial attention between action game playing and non-playing children. The results showed that the players significantly outperformed the non-players in terms of accuracy and response time.

Thus, game experience may improve children's ability to control their attention, including divided and selective attention and range of field of vision, and further enable them to complete relevant tasks quickly and accurately. Therefore, the characteristics and tasks of action games can be incorporated in attention training to increase the appeal of the training and improve selective and divided attention.

Significant differences were noted in the response time of game-playing children. Their slower responses for longer distance targets may indicate increased time taken for making decisions. Nonetheless, the distance had no effect on their accuracy, which was consistently high. In the case of the non-players, they reported a decreasing accuracy as the distance increased. This result suggests that the increase in distance produces an excessively high attention load on the non-players such that they cannot accurately make arrangements and perform control. Meanwhile, the clue messages showed no significant effects on both the players and non-players, possibly because the clues may have not effectively attracted or guided the participants. Hence, in investigating the effects of clues on players and non-players, focus should be given to different clue types or methods for understanding clue characteristics capable of attracting the attention of players and non-players.

## References


1. Buckley D, Codina C, Bhardwaj P, Pascalis O (2010) Action video game players and deaf observers have larger Goldmann visual fields. Vision Res 50(5):548–556. https://doi.org/10.1016/j.visres.2009.11.018
2. Castel AD, Pratt J, Drummond E (2005) The effects of action video game experience on the time course of inhibition of return and the efficiency of visual search. Acta Physiol (Oxf) 119 (2):217–230. https://doi.org/10.1016/j.actpsy.2005.02.004
3. Chisholm JD, Hickey C, Theeuwes J, Kingstone A (2010) Reduced attentional capture in action video game players. Atten Percept Psychophys 72(3):667–671. https://doi.org/10.3758/APP.72.3.667
4. Dobrowolski P, Hanusz K, Sobczyk B, Skorko M, Wiatrow A (2015) Cognitive enhancement in video game players: the role of video game genre. Comput Hum Behav 44:59–63. https://doi.org/10.1016/j.chb.2014.11.051
5. Dye MW, Bavelier D (2010) Differential development of visual attention skills in school-age children. Vision Res 50(4):452–459. https://doi.org/10.1016/j.visres.2009.10.010
6. Dye MW, Green CS, Bavelier D (2009) Increasing speed of processing with action video games. Curr Dir Psychol Sci 18(6):321–326. https://doi.org/10.1111/j.1467-8721.2009.01660.x
7. Dye MW, Green CS, Bavelier D (2009) The development of attention skills in action video game players. Neuropsychologia 47(8):1780–1789. https://doi.org/10.1016/j.neuropsychologia.2009.02.002
8. Green CS, Bavelier D (2003) Action video game modifies visual selective attention. Nature 423(6939):534–537. https://doi.org/10.1038/nature01647
9. Hubert-Wallander B, Green CS, Sugarman M, Bavelier D (2011) Changes in search rate but not in the dynamics of exogenous attention in action videogame players. Atten Percept Psychophys 73(8):2399–2412. https://doi.org/10.3758/s13414-011-0194-7

10. Oei AC, Patterson MD (2013) Enhancing cognition with video games: a multiple game training study. PLoS ONE 8(3):e58546. https://doi.org/10.1371/journal.pone.0058546
11. Oei AC, Patterson MD (2014) Playing a puzzle video game with changing requirements improves executive functions. Comput Hum Behav 37:216–228. https://doi.org/10.1016/j.chb.2014.04.046
12. Spence I, Feng J (2010) Video games and spatial cognition. Rev Gener Psychol 14(2):92–104. https://doi.org/10.1037/a0019491

# Relationship Between Educational Furniture Design and Cognitive Error

Ali Jafari, Shirazeh Arghami(✉), Koorosh Kamali, and Saeedeh Zenozian

Zanjan University of Medical Sciences, Zanjan 4515786349, Iran
arghami@zums.ac.ir

**Abstract. Introduction:** Learners' cognitive error plays a significant role in teaching-learning processes. This study is aimed to investigate the relationship between educational furniture design and cognitive error.
**Methods:** Thirty 18–22 years old students participated in the experiment. Four educational furniture, with different ergonomic characteristics were chosen. The furniture included two types of arm table student chairs (type 1 & 2), one set of library chair and desk (type 3) and one set of adjustable chair and desk (type 4). Each participant spent 90 min on each type of furniture while reading a book and making some notes. A before-after experiment designed to assess cognitive errors by Stroop Test. Paired T-test analysis was used for statistical comparisons at the .05 confidence level in SPSS software.
**Findings:** Comparison before-after errors showed that chairs of type 1 and type 2 could increase errors respectively .86 (P-value = .034) and .63 (P-value = .039). The increasing errors by furniture of type 3 was .40 (P-value = .184). Furniture of type 4 made errors reduction up to .16 (P-value = .517). Comparison between arm table (group 1) and separated table (group 2) showed that group 1 significantly increased errors up to .75 (P-value = .003). Group 2 has insignificantly increased error to .11.
**Conclusion:** The findings revealed a relationship between the ergonomic characteristics of the educational furniture and the number of cognitive errors, as the more ergonomics characteristics of the furniture, the less error. There is also an error percentage reduction using separated chair and desk.

**Keywords:** Educational furniture design · Cognitive error · Stroop test SCWT

## 1 Introduction

Cognitive error is drawing more and more attention in the ergonomics and the teaching-learning literature, as well as, the lay press [1]. This is why the term of human (cognitive) error is applied in both professional and public conversation [2]. In ergonomics, human error has been regarding as the main cause of many incidents and system failures [3]. As a result, the ergonomic redesign of work environment has already paved the way for error control [4]. In spite of it has been intuitively believed that well-designed furniture brings into being better cognition process and fewer errors, there are few studies in sedentary mental work.


S. Bagnara et al. (Eds.): IEA 2018, AISC 826, pp. 649–656, 2019.
https://doi.org/10.1007/978-3-319-96065-4_68

For more than decades, the relationship between the physical structure of furniture and musculoskeletal disorders has been discussing, and ergonomics improvement has been introducing of furniture structure [5–11]. Yet, it seems the relationship between ergonomic designed furniture and cognitive error has not held the attention of researchers. One can find, however, limited studies that imply ergonomic designed furniture improved the outcomes of cognitive task such as laparoscopic surgery [9] or innovation in startup companies [12]. Yet, the paucity of scientific discussion about the potential impact of educational classroom furniture on learners' cognitive error is intelligible.

The educational setting, as an influential environment, encompasses the childhood and youth of almost all people [13]. Accordingly, classrooms are one of the main physical structures in society [14]. For this reason, some studies have concentrated on the effect of this setting. Gilavand (2016) showed that educational furniture impacts on the score of Hermance's achievement motivation questionnaire in elementary schools [13]. In a review article, Lewinski (2015) described the effects of classrooms' architecture (acoustics, light, color, temperature, and seat arrangement) on students' academic performance in classrooms [15].

In recent years, academic teaching-learning process emphasizes on critical thinking, as a higher cognitive skill. In Iran, more than 20% of the population is academic student [16], which a considerable portion of them is studying health. Health professionals are the front line providers of health-care services at the community level. Thus, critical thinking is a prerequisite for effective management and judgment for their tasks. It has been argued that the absence of critical thinking can be caused by cognitive error in classrooms [1]. Therefore, preventing of cognitive error is crucial for health students in the classrooms. Since health students should attend classes for a remarkable period of time in a sedentary position, it is necessary to control negative effects of classroom furniture on cognition processes. This study is aimed to investigate the relationship between educational furniture design and cognitive error of academic students in School of Public Health at Zanjan University of Medical Sciences (Iran).

## 2 Methodology

This study was approved by research ethics committee of Zanjan University of medical sciences [ZUMS.REC.1396.191].

### 2.1 Sampling

Sampling
Due to the lack of similar studies, a pilot study was performed to determine the sample size. The pilot study included ten students (5 female and 5 male), which were randomly selected. Relaying on the results of the pilot study, thirty 18–22 years old students (15 female, and 15 male) who had no physical and mental disorders or drug consumption were entered into the experiment. The participants were asked to have enough sleep, as well as, to avoid using substances containing caffeine or tea for 10 h before tests. Prior to the study, the participants were acquainted with the trend of the experiment. Also, a verbal informed consent was obtained.

## 2.2 The Experiment

A before-after experiment was performed in the ergonomics lab in a Latin square pattern, based on the alteration of participants' cognitive errors in Stroop Color and Word Test (SCWT). SCWT is a common neuropsychological test to assess cognitive interference of a specific external stimulus feature impedes the processing of the other concurrent stimulus attribute [17]. All tests were carried out at 8–10 am, before the participants doing any tedious work. Each participant spent 90 min on each type of furniture while reading a book and making some notes. Then, the investigators checked the notes to ensure the participants' engagement. Paired T-test analysis was used for statistical comparisons at the .05 confidence level in SPSS software.

There are four types of educational furniture, with different ergonomic characteristics, at the School of Public Health. Therefore, each participant attended the lab four times. In each time, the participant used one of the furniture. The ergonomic characteristics of the furniture are described as followed by Figs. 1, 2, 3 and 4 and Table 1:

1. Type 1: An arm table student chair (Fig. 1), which has thick and soft seat and backrest. The backrest can supports neck to low back. Also, a lumber support is included. The table is made of wood.
2. Type 2: An arm table student chair (Fig. 2), which is to some extent similar to type 1, however, with smaller dimensions and slimmer seat and backrest. The curved backrest supports low back and only a part of upper back. Also, the table is made of hard plastic.

**Fig. 1.** Arm table student chair (Type 1)

**Fig. 2.** Arm table student chair (Type 2)

3. Type 3: A set of wooden library furniture with no soft material (Fig. 3). This set is used in the library and included a large separated desk. The chair has no armrest. In addition, the design of the backrest is not adapted to the curvature of the human body.
4. Type 4: An adjustable chair accompanied by a desk (Fig. 4). This type of furniture consists of an office wheelchair and a separated simple desk. Seat height adjustment, backrest tilt and curvature of the lumbar area make this chair more ergonomic. Also, an adjustable head-neck support is provided. The seat and backrest made of soft and elastic fiber with the possibility of air penetration.

**Fig. 3.** A set of wooden library furniture (type 3)

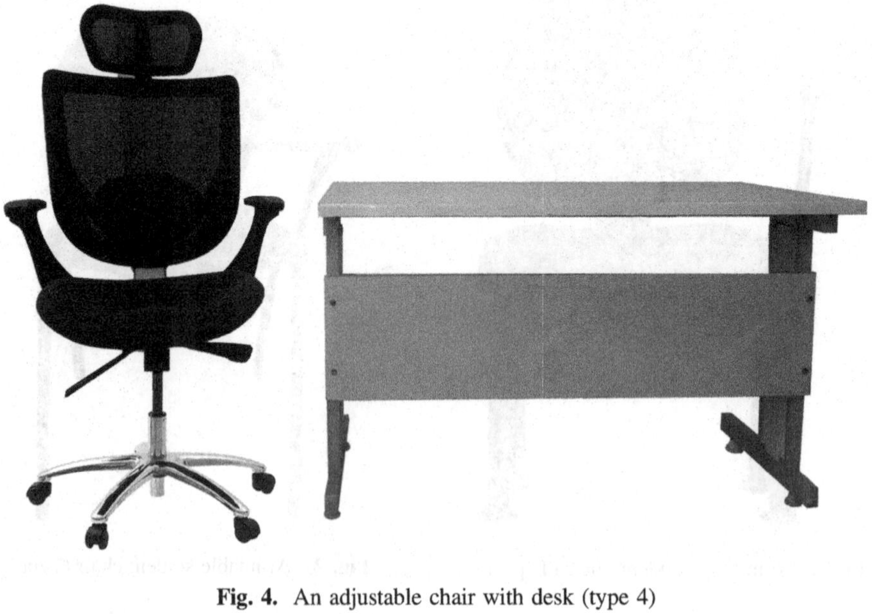

**Fig. 4.** An adjustable chair with desk (type 4)

**Table 1.** The furniture dimensions

| Dimensions | Type 1 | Type 2 | Type 3 | Type 4 |
|---|---|---|---|---|
| Seat height (cm) | 47 | 45.5 | 46.5 | 42–50 |
| Seat depth (cm) | 46.5 | 40.5 | 40 | 52.5 |
| Seat width (cm) | 48.5 | 41 | 43.2 | 50 |
| Backrest height (cm) | 57.5 | 38 | 45.5 | 56 |
| Backrest width (cm) | 43.5 | 48 | 38 | 48.5 |
| Armrest height (cm) | 27 | 19 | – | 21.5 |
| Seat-table height (cm) | 27 | 19 | 30.5 | 32.5–24.5 |
| Backrest-table distance (cm) | 35 | 34 | adjustable (separated table) | adjustable (separated table) |

# 3 Results

In this experimental study, we measured students' cognitive error induced by different design of educational furniture based on SCWT. The comparison of before-after SCWT results showed that chairs of type 1 and type 2 could significantly increase errors respectively .86 (P-value = .034) and .63 (P-value = .039). The increasing errors by furniture of type 3 was .40, however, not significant (P-value = .184). Surprisingly, furniture of type 4 made errors reduction up to the .16, however, the difference was not significant (P-value = .517). The more statistical analysis is presented in Table 2.

**Table 2.** Paired T-test analysis results between 4 furniture

| Furniture | Mean | | Number | Std. deviation | Std. error mean | P-value |
|---|---|---|---|---|---|---|
| Type 1 | Before | 1.34 | 30 | 1.47 | .26 | .034 |
| | After | 2.20 | 30 | 2.04 | .37 | |
| Type 2 | Before | 1.40 | 30 | 1.42 | .26 | .039 |
| | After | 2.03 | 30 | 1.65 | .30 | |
| Type 3 | Before | 1.26 | 30 | 1.41 | .25 | .184 |
| | After | 1.66 | 30 | 1.58 | .28 | |
| Type 4 | Before | 1.13 | 30 | 1.13 | .20 | .517 |
| | After | .97 | 30 | 1.37 | .25 | |

Another comparison carried out by dividing the furniture into two groups. Group 1 consisted of chairs without desk (type 1 + type 2) and the group 2 included chairs with desk (type 3 + type 4). Comparing the new data showed that both of these groups could increase errors. However, group 1 significantly increased errors up to .75 (P-value = .003). While, group 2 insignificantly increased error to .11 (P-value = .554) (Table 3).

**Table 3.** Paired T-test analysis results between group 1 and 2

| Furniture | Mean | | Number | Std. deviation | Std. error mean | P-value |
|---|---|---|---|---|---|---|
| Group 1 | Before | 1.36 | 60 | 1.43 | .18 | .003 |
| | After | 2.11 | 60 | 1.84 | .23 | |
| Group 2 | Before | 1.2 | 60 | 1.27 | .16 | .554 |
| | After | 1.31 | 60 | 1.51 | .19 | |

## 4 Discussion and Conclusion

In this study, we tried to investigate the relationship between the design of educational furniture and students' cognitive error based on SCWT. As mentioned before, there are remarkable studies which have discussed the impact of furniture design on physical stresses or posture, as well as, comfort and satisfaction [18–20]. However, it seems it is the first time that the relationship between the educational furniture and cognitive error is scientifically argued.

The results showed that Type 1, Type 2, Type 3 could increase errors, which the increases were significant for Type 1 (P-value = .034), and Type 2 (P-value = .039). Type 4, surprisingly, made an insignificant reduction of error. Also, dichotomization the furniture based on having a separated table or not, resulted in insignificant fewer errors for furniture with separated table (P-value = .554).

In contrast to type 1 and 2, furniture type 3 and 4 have separated desks. The main difference between furniture type 3 and 4, however, was related to the chairs. The chair type 4 had the advantage of seat height adjustment, backrest tilt, lumbar support, and adjustable head-neck support. Whereas, the chair of type 3 was quite wooden and suffering from lack of ergonomic features such as elbow support, height adjustability and etc.

Although furniture type 3 caused a rise of .4 errors on SCWT, it was fewer than the errors were induced by furniture type 1 and type 2. In point of view of learners' concentration, therefore, one may infer that a separated chair and table might be more effective than ergonomic features of the chair.

In addition, the results showed that type 4 did not increase the number of errors, rather could make a reduction to .16. Therefore, it can be concluded that a learning workstation in the classroom including a chair with ergonomic features which is aggregated to a separated desk may generate more positive effects on educational processes.

It should be noted that in this experimental study, each participant spent 90 min on each furniture type. Since Gunzelmann et al. (2010) found that the more task time, the more cognitive fatigue. Then, we logically expect further effects in the real world, as students in classroom spend longer time in a sedentary position [21].

## 5 Limitations

In this study, we examined the one of the cognitive effects of four types of furniture on 30 participants of school of public health for a 90-min reading task based on SCWT, which is not enough to achieve a rigorous conclusion. Therefore, we recommend further studies based on other types of seating furniture in the classroom, more varied participants spending longer time on the other cognitive tasks, and using other cognitive assessment methods.

**Acknowledgement.** The authors would like to express their highest appreciation to Zanjan University of Medical Science which sponsored the study [Grant Number: A-12-56-51].

## References

1. Huang GC, Newman LR, Schwartzstein RM (2014) Critical thinking in health professions education: summary and consensus statements of the millennium conference 2011. Teach Learn Med 26(1):95–102
2. Hansen FD (2006) Human error: a concept analysis. J Air Transp 11(3):61–77
3. Stewart M (1992) Simulation of human error in reinforced concrete design. Res Eng Des 4 (1):51–60
4. Rasmussen J, Vicente KJ (1989) Coping with human errors through system design: implications for ecological interface design. Int J Man Mach Stud 31(5):517–534
5. Boampong E, Effah B, Dadzie PK, Asibey O (2015) Ergonomic functionality of classroom furniture in senior high schools in Ghana. Int J Adv Sci Technol 2(1):6–11
6. Ward J, Coats J (2017) Comparison of the backjoy sitsmart relief and spine buddy LT1 H/C ergonomic chair supports on short-term neck and back pain. J Manipulative Physiol Ther 40 (1):41–49
7. Haller M, Richter C, Brandl P, Gross S, Schossleitner G, Schrempf A et al (eds) (2011) Finding the right way for interrupting people improving their sitting posture. In: IFIP conference on human-computer interaction, Springer, Heidelberg
8. Mououdi MA, Hosseini M (2018) The determination of the static anthropometric characteristics for the computer users from the monitoring room of one of the industries in the mazandaran province for designing an ergonomic chair. J Ergon 5(3):22–28
9. Rassweiler JJ, Klein J, Tschada A, Gözen AS (2017) Laparoscopic retroperitoneal partial nephrectomy using an ergonomic chair: demonstration of technique and matched-pair analysis. BJU Int 119(2):349–357
10. Dianat I, Karimi MA, Hashemi AA, Bahrampour S (2013) Classroom furniture and anthropometric characteristics of Iranian high school students: proposed dimensions based on anthropometric data. Appl Ergon 44(1):101–108
11. Jawalkar C (2014) Ergonomic based design and survey of elementary school furniture. i-Manag J Sch Edu Technol 9(4):27–31
12. Lee YS (2016) Creative workplace characteristics and innovative start-up companies. Facilities 34(7/8):413–432
13. Gilavand A (2016) The impact of educational furniture of schools on learning and academic achievement of students at elementary level. Int J Med Res Health Sci. 5(7S):343–348
14. Douglas D, Gifford R (2001) Evaluation of the physical classroom by students and professors: a lens model approach. Edu Res 43(3):295–309

15. Lewinski P (2015) Effects of classrooms' architecture on academic performance in view of telic versus paratelic motivation: a review. Front Psychol 6(746):1–5
16. Statistical Centre of Iran (2015) Iran in the mirror image, p. 133. [In Persian]. https: //www.amar.org.ir/Portals/0/Files/fulltext/1394/n_idaa_no.35_94.pdf
17. Scarpina F, Tagini S (2017) The stroop color and word test. Front Psychol 8(557):1–8
18. Lyons JB (2001) Do school facilities really impact a child's education? IssueTrak: a CEFPI brief on educational facility issues. In: Council of educational facility planners I, Scottsdale. https://files.eric.ed.gov/fulltext/ED458791.pdf
19. Odunaiya NA, Owonuwa DD, Oguntibeju OO (2014) Ergonomic suitability of educational furniture and possible health implications in a university setting. Adv Med Edu Pract 5:1–14
20. Savanur C, Altekar C, De A (2007) Lack of conformity between Indian classroom furniture and student dimensions: proposed future seat/table dimensions. Ergonomics 50(10):1612–1625
21. Gunzelmann G, Moore LR, Gluck KA, Van Dongen HP, Dinges DF (2010) Fatigue in sustained attention: generalizing mechanisms for time awake to time on task. In: Ackerman PL (ed) cognitive fatigue: multidisciplinary perspectives on current research and future applications. American Psychological Association, Washington, pp 93–94

# The Differences Between Bowing as Among Adult Members of Society in Japan and the Bowing of Japanese Students

Kohei Okado[1,2](✉), Hiroyuki Hamada[2], Noriyuki Kida[2], Tamotsu Matsuda[1], Rie Ohashi[1], Kazuki Kitamura[1], Andy Smith[1], David Todisco[1], Luke Jackson[1], and Tatsuya Ogimoto[1]

[1] Vories Gakuen Omi Brotherhood Junior High School and High School, Shiga, Japan
cowhey@gmail.com
[2] Kyoto Institute of Technology, Kyoto, Japan

**Abstract.** Regarding bowing in Japan, the bowing technique of students was compared with the bowing of adults. The bowing data of the students was analyzed. The subjects were junior high school first year and second years students (N = 12). For the measurement, three infrared cameras and TEMA, a high-performance moving image analysis software, was used. In the experiment, motion capture observation points were attached to the head, neck, waist and ankle of the subject. The recorded data was analyzed with TEMA, and the numerical values for six points of body movement were calculated. Moreover a questionnaire was conducted to survey students about their impressions of bowing. By comparing the bowing of adults and students, the following points were identified. First, it was observed that the variance in bowing between students differs greatly. The second observation is the difference in angle of the neck. The third observation is the difference in angle of the waist. The fourth observation is the difference in awareness of bowing. Through this study, it is demonstrated that there are many differences between the bowing of students and adults. Also, it was found that there are many differences between subjects.

**Keywords:** Japanese bowing · Junior high school · TEMA · Bowing of adults
Bowing of students

## 1 Introduction and Aims

How do the Japanese develop the ability to perform a correct bow according to customs? Although research on Japanese bowing has been done, research on bowing by students, especially junior high school students is rare. In this research, we analyze the difference between the bowing of adults and the bowing of Japanese junior high school students. Regarding the bowing of adults, there is a paper by Takeda et al. (2017), which defines the definition of bowing for adults. In comparison and analysis, this definition is used for the bowing of an adult. This research aims to further clarify the process by which Japanese acquire proper bowing.


S. Bagnara et al. (Eds.): IEA 2018, AISC 826, pp. 657–662, 2019.
https://doi.org/10.1007/978-3-319-96065-4_69

## 2 Methods

### 2.1 Participants

The survey was conducted at a junior high school in Japan. Participants (N = 12) included 1st graders (N = 10), and 2nd graders (N = 2). After clearly indicating the experimental method in writing and obtaining consent of the parents, the experiment was carried out. Moreover a questionnaire was conducted to survey students about their impressions of bowing.

### 2.2 Measuring Method

For the measurement, three infrared cameras and TEMA, a high-performance moving image analysis software, were used. In the experiment, motion capture observation points were attached to the head, neck, waist, and ankle of the subject. The recorded data was analyzed with TEMA, and the numerical values for the following six points were calculated.

1. *Time from start to end of bowing.*
2. *Time and speed from start of bowing to the bottom.*
3. *Time and stopped holding position at bottom.*
4. *Time and speed from the bottom of the bowing to the starting position.*
5. *Change in neck angle during bowing motion.*
6. *Change in waist angle during bowing motion.*

### 2.3 Targeted Bowing

In order to reflect the individual understanding of bowing, the targeted bowing was initiated only by the instruction of "Please bow".

### 2.4 Questionnaire

About bowing, questionnaires of seven items were prepared and were conducted for participants (10 people). The questionnaire is presented in Fig. 1.

1. What are the after-school activities you have experienced?
2. Did you have opportunities for bowing?
3. In what scenes did you do the bowing?
4. What kind of person was your target for bowing?
5. Have you had opportunities to receive a bow?
6. In what scenes have people bowed to you?
7. What kind of person bowed to you?

**Fig. 1.** Contents of questionnaire

# 3 Results

## 3.1 About Angle of Bowing

Takeda et al. (2017) indicated that the angle of the neck had hardly any changes or had changes in a negative (downward) direction. In the bowing of students, the angle of the neck was AV.18.7, SD.15.13. There were 4 students with a value less than 10°, one of them being −6.8°.

Takeda et al. (2017) indicated that there was a variation from 26.67° to 49.06° (average value 38.16 ± 7.92) in the angle of the waist. Regarding the angle of waist, it was shown that there was not a fixed angle, but it was relative to the target, situation, and participant's position.

In the bowing of students, the angle of the waist was AV. 50.6, SD.15.21. The bowing of students showed the bend is deeper at the waist compared with the adults. 6 students were over 49.06°.

## 3.2 About Time of Bowing

Takeda et al. (2017) indicated that the time from start to end of bowing was AV. 4.30 SD. 1.17. In the bowing of students, the time from start to end of bowing was AV.3.0, SD.0.63. The bowing of students showed a shorter time than adults.

In the time of bowing, Takeda et al. (2017) show that the extension time tends to be longer than the flexion time as a characteristic of adults. In the bowing of students, 4 students had longer extension time than flexion time, and 8 students tended to have more flexion time than xtension time.

## 3.3 About Time of Bowing

Regarding the speed of bowing among adults, Takeda et al. (2017) show that the speed of the extension was equal to or slower than the speed of the flexion.

In the bowing of students, there was 1 student whose speed of extension was faster than the speed of flexion, and there were 11 students whose speed of extension was lower than the speed of flexion.

## 3.4 Survey Questionnaire

Table 1 shows the survey results from the questionnaire.

Seven students answered they often bow. Five students answered they would have been bowed to. About situations of bowing, they can be classified into three groups from the questionnaire survey.

i. **Master-servant relationship**
ii. **Trust relationship**
iii. **Religious ritual**

Bowing in the master-servant relationship is a bow as a greeting to a higher-status person (adult, teacher, senior). Bowing in the trusted relationship is a bow to the

**Table 1.** The survey results from the questionnaire

| No | After-school activities | Opportunity of bowing | Scenes of bowing | Target for bowing | Opportunity to receive a bow | Scenes of receiving a bow | Person bowed to you |
|---|---|---|---|---|---|---|---|
| 1 | Tennis | Yes | Greeting, Expression of Gratitude | Teacher, Neighbors | No | | |
| 2 | Swimming | Yes | Greeting, Expression of Gratitude | Teacher, Parents | No | | Neighbors |
| 3 | Volleyball, table tennis, soccer, Badminton | Yes | Greeting, Sports | Techer, Senior, Opponent in Sports | Yes | Greeting | Junior |
| 4 | Piano, Swimming, Judo, Tennis | Yes | Judo | Opponent in Sports | Yes | Judo | |
| 5 | Swimming, Japanese Drum | Yes | Greeting | Student | Yes | Greeting | Junior |
| 6 | Swimming, Piano, Dance, Japanese Calligraphy | Yes | Greeting Sports, Greeting, | Senior, Shrine, Temple | No | | |
| 7 | Swimming, Piano, Violin, Basketball, English, Japanese Calligraphy | Yes | Recital, Religious Ceremony, Visit to a Shrine | Teacher, Friend | Yes | Visit to a Shrine, Funeral | Senior, Friend |
| 8 | English, Hyakunin Isumi | No | Greeting, Apology | Adult | Yes | Funeral | salesperson |
| 9 | Tennis, Swimming | No | Expression of Gratititude | Adult | No | Tradeoff for an act | target for an act |

opponent in sports, or a bow to an audience at a recital etc. A bow as a religious ceremony is the bowing at a funeral, visits to a shrine, visits to temples etc.

About receiving a bow, it could be classified into two groups from questionnaire survey.

- Master-servant relationship
- Interests relationship

Receiving a bow in the master-servant relationship category is receiving a bow from a lower-status person (younger, junior). Be bowed to in an interests relationship is exemplified as a tradeoff for an act.

From the questionnaire survey, there was no element of bowing as hospitality, nor bowing as manners, as students do not have this experience.

# 4 Discussion

## 4.1 Bowing of Students

By comparing the bowing of adults and students, the following points were identified. First, it was observed that the variance in bowing between students differs greatly. Students are only provided opportunities to practice "master-servant" style relationship bowing, and do not get to learn proper social bowing manners. It can be thought that the bowing behavior of students is strongly influenced by school teacher relationships.

The second observation is the difference in angle of the neck. In social bowing, it is important that the angle is hardly changed. However, in students' bowing it was found that most subjects were bending the neck around 30°.

The third observation is the difference in angle of the waist. In social bowing, waist angle is said to be suitable from 15° to 30°. However, in students' bowing it was found that most subjects were bending the waist over 30°.

The fourth observation is the difference in awareness of bowing. As an adult, bowing is a behavior of courtesy. It is commonly thought that proper bowing is necessary for mutual communication, and it is important for building trust in business relationships. On the other hand, the bowing of students is mainly an act of submission in the master servant relationship, and doesn't allow for a sense of equality.

## 4.2 Conclusions and Suggestions

Through this study, it is demonstrated that there are many differences between the bowing of students and adults. Also, it was found that there are many differences between subjects. It is necessary to better clarify the definitions of social bowing in order to reduce students' bowing variance. Additional clarification would allow for the development and implementation of educational programs designed to encourage students to become aware of the social benefits and improved of communication as results of proper social bowing.

We have already developed the easy-to-use programs of proper social bowing in Japan. It is useful for not only students in Japan, but also adults, for foreigners, exchange students who want to adopt proper social bowing. We want to contribute to inform proper social bowing in Japan to dozens of countries.

## Reference

Takeda T, Yamashiro K, Lu X, Kawakatsu S, Ota T (2017) Evaluation of Japanese bowing of non-experts by experts. In: Duffy V (eds) Digital human modeling. applications in health, safety, ergonomics, and risk management: ergonomics and design. DHM 2017. Lecture notes in computer science, vol 10286. Springer, Cham, pp 169–178

# Using Natural Gesture Interactions Leads to Higher Usability and Presence in a Computer Lesson

Shannon K. T. Bailey[1(✉)], Cheryl I. Johnson[1], and Valerie K. Sims[2]

[1] Naval Air Warfare Center Training Systems Division, Orlando, FL, USA
{shannon.bailey, cheryl.i.johnson}@navy.mil
[2] University of Central Florida, Orlando, FL, USA
valerie.sims@ucf.edu

**Abstract.** In the last few years, motion-tracking technology has become much more accurate and cost effective, opening the door to advances in natural user interfaces (NUIs), such as gesture-based human-computer interactions. Gesture-based interactions involve using the body to input a command in a computer system. The degree to which NUIs are easy to use should be addressed by research as these technologies become more widespread. Before the usefulness of NUIs can be assessed and optimal interfaces refined, the extent to which NUIs are "natural" or "intuitive" should be determined. The purpose of the present study was to determine how natural gestural computer interactions are perceived by users in a computer-based science lesson. An experiment was conducted comparing natural gestures to arbitrary gestures in a computer lesson to answer the research question of how natural gestures are perceived by users in terms of system usability and presence. Perceived usability of the computer lesson was measured using the System Usability Scale. Natural gestures were rated almost a full point higher in usability (on a 5-point scale) than arbitrary gestures. Presence (i.e., the feeling of "being there" in a virtual environment) was also measured after the computer lesson using the Presence Questionnaire. Natural gesture interactions were seen as inducing a higher sense of control in the computer lesson, more immersion, and better interface quality than arbitrary gesture interactions. The finding that natural gestures are perceived as more usable and contribute to higher feelings of presence should encourage instructional designers and researchers to keep the user in mind when developing gesture-based interactions.

**Keywords:** Gesture · Human-computer interaction · Natural user interface Presence · Usability

Disclaimer: This research was funded by the RADM Fred Lewis Interservice/Industry Training, Simulation & Education Conference (I/ITSEC) Scholarship and performed while S.B. was a graduate student at the University of Central Florida. Presentation of this material does not constitute or imply its endorsement, recommendation, or favoring by the U.S. Navy or Department of Defense (DoD). The opinions of the authors expressed herein do not necessarily state or reflect those of the U.S. Navy or DoD.


S. Bagnara et al. (Eds.): IEA 2018, AISC 826, pp. 663–671, 2019.
https://doi.org/10.1007/978-3-319-96065-4_70

## 1 Introduction

Motion-tracking technology has become much more accurate and cost effective in recent years, allowing for advances in gesture-based human-computer interactions, referred to as "natural user interfaces" (NUIs). Gesture-based interactions involve using the body to input a command in a computer system. These gesture interactions can be classified as more "natural" or "arbitrary" depending on how closely they correspond with the actions in the computer lesson [1]. For example, a NUI in which the user moves her hand to the left to move an on-screen object left may be perceived as a natural interaction because the interface directly corresponds with the computer action. A gesture that does not correspond to the desired computer action could be considered an arbitrary interaction, such as making a "double-tap" gesture with the index finger to move an object left.

As gesture-based interfaces become more technologically feasible to implement in computer systems, the ease with which users interact with these new interfaces is important to evaluate so that the technology is not imposing unnecessary mental or physical demand on the user from poor usability [2]. Natural gestures may be perceived as more usable and may increase the feeling of presence (e.g., immersion, system fidelity) the participant feels when interacting with the computer system. Alternatively, NUIs that are not designed from a user-centered approach may cause unnecessary effort to be exerted if the gestures are not perceived as user-friendly. In other words, if the interaction method has poor usability or decreases presence, the NUI may detract from the task itself. For example, if users are spending mental effort to recall arbitrary gestures that are not easy to remember or perform, they may spend less mental effort on the computer task itself. Gesture-based interactions are often developed based on what a computer system can recognize [3], and developers should not assume that any gesture-based interaction has acceptable usability. To ascertain whether NUIs are indeed perceived as usable by users, we compared natural to arbitrary gesture interactions in a computer lesson.

An experiment was conducted comparing natural gestures to arbitrary gestures in a computer lesson to answer the research question of how natural gestures are perceived by users. The natural and arbitrary gestures were developed in two previous studies [4]: First, participants performed gestures for a list of actions that they thought were natural, and then a second group of participants rated those gestures to determine which gestures were rated as most natural and arbitrary for each action. The current experiment sought to evaluate empirically the perception of these gestures when implemented in a computer lesson. Additionally, the gestures were taught using either video- or text-based instructions to answer research questions not addressed in the current proceedings paper due to page limitations. Instruction type (Video or Text instructions) was included as an independent variable in addition to gesture type (Natural or Arbitrary), but because instruction type was not a significant predictor of the dependent variables described below, it will not be discussed further. For extensive discussion of the theory, hypotheses, and null results regarding the Instruction variable that did not fit within the space limitations of the current proceedings, see [4]. Participants rated their perception of the gesture-based interactions on their usability and factors related to feelings of presence in a virtual environment. By asking participants to rate the gesture interactions on these factors, we can determine how natural and usable users feel NUIs are for completing a computer lesson.

# 2 Methodology

## 2.1 Design

The experiment utilized a 2 × 2 between-subjects design with two independent variables: Gesture Type (Natural or Arbitrary interaction) and Instruction Type (Video or Text instructions). The Natural Video condition used natural gestures to complete the computer lesson, and the gestures were taught with video-based instructions. The Natural Text condition also used natural gestures, but the gestures were taught using text-based instructions. The Arbitrary Video condition used arbitrary gestures to interact with the computer lesson, and the arbitrary gestures were taught using video-based instructions. The Arbitrary Text condition used arbitrary gestures taught via text-based instructions. The dependent variables were the participant ratings of usability on the System Usability Scale (SUS) and feelings of presence on the Presence Questionnaire (PQ). Because gestures in the Natural Gesture conditions corresponded with the actions in the computer lesson, it was hypothesized that participants in the Natural Gesture conditions would rate the interactions as higher in usability and feel more presence in the computer lesson than participants in the Arbitrary Gesture conditions.

## 2.2 Equipment

**Testbed.** The experimental testbed was a computer lesson developed using the Unity 3D game engine that utilized the Microsoft Kinect V1 infrared motion-tracker to capture gesture-based commands. The computer lesson was presented on a 30" LCD 1080p television screen. During the experiment, the participants stood nine feet from the screen, facing the motion-tracker, and used gestures specific to their condition to complete the computer lesson. The lesson presented a conceptual description of the Hubble space telescope and explained how lenses and mirrors refract and reflect light. Figure 1 is a screenshot of the computer lesson with which participants interacted using gestures. Participants advanced through the computer lesson by performing gesture-based commands that were either arbitrary or natural (Table 1). For example, in the interactive slide on refraction of lenses (Fig. 1), participants moved the lens into the beam of light by using the "select" gesture to highlight the lens, then the "down" gesture to move the lens into the beam of light. Next, participants used the "enlarge" gesture to increase the size of the lens. When the lens was moved into the beam of light, the light refracted, illustrating the conceptual information of refraction. The same type of interaction was completed for mirrors, illustrating the concept of reflection.

**Table 1.** Natural and arbitrary gesture-based interactions for the computer lesson

| Action | Natural gesture condition | Arbitrary gesture condition |
|---|---|---|
| 1. Up | Hand moves vertically from waist to above the head | Both fists move inward horizontally to center of chest |
| 2. Down | Hand moves vertically from above the head to the waist | Both fists move outward horizontally from center of chest |
| 3. Left | Hand moves right-to-left horizontally at chest height | Hand moves vertically from waist to above the head |
| 4. Right | Hand moves left-to-right horizontally at chest height | Hand moves vertically from above the head to the waist |
| 5. Clockwise | Hand moves in clockwise circle from chest to waist | Hand closes into a fist at chest height |
| 6. Counter-Clockwise | Hand moves in counter-clockwise circle from chest to waist | Arm moves straight out to the right side, parallel to the floor |
| 7. Select | Index finger points forward at chest height | Both hands with palms facing in toward each other move outward horizontally from center of chest |
| 8. Enlarge | Both hands with palms facing in toward each other move outward horizontally from center of chest | Closed fist moves horizontally right-to-left at chest height |
| 9. Shrink | Both hands with palms facing in toward each other move inward horizontally to center of chest | Closed fist moves diagonally from waist to opposite shoulder |

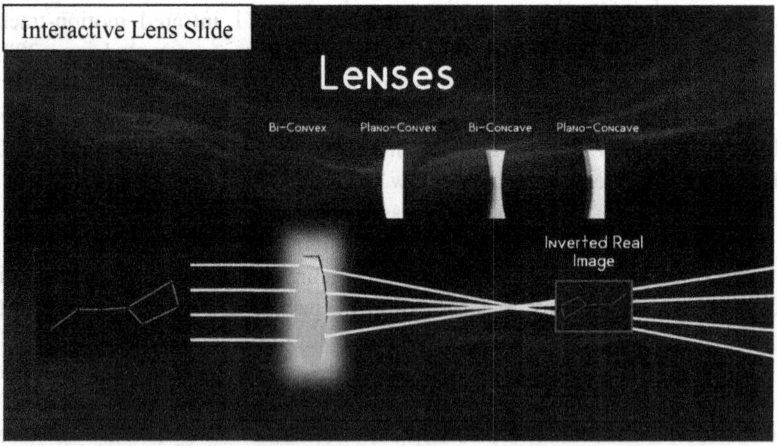


**Fig. 1.** Screenshot of the interactive section for lenses depicting refraction.

### 2.3 Participants

Participants ($n$ = 102) were recruited from a university. There were 70 self-identified females and 32 males. Participants were at least 18 years old ($M$ = 18.7 years, $SD$ = 1.5 years). Only predominately right-handed individuals were included in the study.

### 2.4 Materials

The measures included in this proceedings paper are a subset of the materials used in the experiment. The current analyses were conducted to determine empirically that the natural gesture interactions were in fact rated by users as more usable and supported stronger feelings of presence than arbitrary gestures when implemented in a conceptual computer lesson.

**System usability scale (SUS).** The SUS [5] is a 10-item measurement of perceived usability in which participants rate their agreement with statements on a scale of 1 ("Strongly disagree") to 5 ("Strongly agree"). Below the SUS, participants were asked to write any additional comments they had regarding the computer system in the blank space provided.

**Presence questionnaire (PQ).** The PQ [6] contains 19 items that corresponding with four subscales related to feelings of presence felt in a virtual environment: involvement, sensory fidelity, adaptation/immersion, and interface quality. Participants responded on a scale of 1 ("Not at All") to 7 ("Completely") to each item.

### 2.5 Procedure

Participants agreed to the informed consent and were assigned randomly to one of the four conditions (Natural Video, Natural Text, Arbitrary Video, or Arbitrary Text). Once assigned to a condition, participants completed a tutorial (either Video or Text instructions) on how to perform the gestures (either Natural or Arbitrary) depending on their respective conditions and practiced the gestures three times so that the gesture interactions could be performed during the experiment. Then, each participant completed the computer lesson on optics concepts using the gestures to move through the lesson. The computer lesson took approximately 10–12 min to complete. Following the computer lesson, participants completed the SUS, PQ, written reactions to the lesson, and demographic questionnaires followed by a debriefing.

## 3 Results

### 3.1 Usability

Participant ratings on the SUS items were averaged to determine the perceived usability of the computer lesson in each condition. The scores on the SUS ($\alpha$ = .86) were distributed normally ($Skew$ = −2.08, $Kurtosis$ = −1.31) around the mean of 3.728 ($SD$ = 0.761). The range of average SUS scores was between 1.80 and 5.00 on the 5-point Likert-type scale, with higher scores indicating more perceived usability. Means and standard deviations for each condition are presented in Table 2.

To determine whether SUS ratings differed by condition, a 2X2 between-subjects ANOVA was conducted with SUS scores as the dependent variable, and two levels of each independent variable: Gesture type (Natural or Arbitrary) and Instruction type (Video or Text). Levene's Test of Equality of Error Variance was significant ($F[3, 98] = 3.627$, $p = .016$); however, the *FMax* ratio did not exceed 5 [7], so no adjustments were made ($\sigma^2_{NatureText} = 0.241, \sigma^2_{ArbitaryText} = 0.743$; $FMax = 3.08$). There was a main effect for type of Gesture ($F[1, 98] = 35.714$, $p < .001$, $\eta p^2 = .267$), where those interacting with Natural gestures ($M = 4.087$, $SD = 0.544$) rated the system higher in usability than those using Arbitrary gestures ($M = 3.306$, $SD = 0.767$). There was not a main effect for Instruction ($F[1, 98] = 0.028$, $p = .868$, $\eta p^2 < .001$), nor an interaction between Gesture and Instruction ($F[1, 98] = 0.832$, $p = .364$, $\eta p^2 = .008$).

**Table 2.** System usability scale means and standard deviations by condition

| Condition | *n* | SUS Ratings | |
|---|---|---|---|
| | | *M* | *SD* |
| Arbitrary text | 23 | 3.257 | 0.862 |
| Arbitrary video | 24 | 3.354 | 0.678 |
| Natural text | 27 | 4.159 | 0.491 |
| Natural video | 28 | 4.018 | 0.592 |

### 3.2 Presence

The PQ consists of four subscales measuring different dimensions of sense of presence felt by participants during the computer lesson: involvement ($\alpha = .84$), sensory fidelity ($\alpha = .67$), adaptation/immersion ($\alpha = .70$), and interface quality ($\alpha = .70$). Provided in Table 3 are the means and standard deviations for each subscale on a 7-point scale, with higher scores on the involvement, sensory fidelity, or immersion subscales indicating more presence as related to each dimension. Interface quality differs from the other subscales in that a higher score indicates less presence because it measures the extent to which the computer system distracts from performance in the virtual environment – that is, a higher score indicates lower interface quality.

First, an ANOVA was conducted for each subscale to determine whether that dimension of presence differed as a function of the conditions. In each 2 × 2 ANOVA, the PQ subscale was the outcome variable with Gesture interaction (Natural or Arbitrary) and Instruction type (Video or Text) as independent variables. All of the PQ subscales had normally distributed residuals (*Shapiro-Wilk ps* > .05).

For the involvement subscale, which measures the degree to which the participant feels the control of the computer interface is natural, there was a main effect for Gesture type ($F[1, 98] = 8.124$, $p = .005$, $\eta p^2 = .077$), indicating that participants in the

**Table 3.** Means and standard deviations for the presence questionnaire subscales

| Condition | $n$ | Involvement | | Sensory Fidelity | | Adaptation/ Immersion | | Interface Quality[a] | |
|---|---|---|---|---|---|---|---|---|---|
| | | $M$ | $SD$ | $M$ | $SD$ | $M$ | $SD$ | $M$ | $SD$ |
| Total | 102 | 4.68 | 0.99 | 5.43 | 1.13 | 4.96 | 0.93 | 3.07 | 1.18 |
| Arbitrary Text | 23 | 4.55 | 1.01 | 5.43 | 0.91 | 4.66 | 0.91 | 3.39 | 1.44 |
| Arbitrary Video | 24 | 4.23 | 0.94 | 5.17 | 1.34 | 4.65 | 1.06 | 3.39 | 0.94 |
| Natural Text | 27 | 5.00 | 0.87 | 5.43 | 1.11 | 5.37 | 0.85 | 2.99 | 1.27 |
| Natural Video | 28 | 4.87 | 1.02 | 5.64 | 1.14 | 5.07 | 0.75 | 2.62 | 0.87 |

*Note.* [a] Higher scores indicate *worse* interface quality. The subscale of interface quality differs from the other subscales in that a higher score reflects less presence.

Natural Gesture conditions rated the control of the computer interface as more natural than the Arbitrary Gesture conditions. Instruction type ($F$[1, 98] = 1.378, $p$ = .243, $\eta p^2$ = .014) and the interaction of Gesture and Instruction ($F$[1, 98] = 0.252, $p$ = .617, $\eta p^2$ = .003) were not significant predictors of the involvement subscale.

The sensory fidelity subscale, which measures the extent to which the senses are engaged with the computer system, was not predicted by either Gesture type ($F$[1, 98] = 1.106, $p$ = .295, $\eta p^2$ = .011), Instruction type ($F$[1, 98] = 0.016, $p$ = .900, $\eta p^2$ < .001), or their interaction ($F$[1, 98] = 1.126, $p$ = .291, $\eta p^2$ = .011).

In the adaptation/immersion subscale that measures the participants' perceived ability to concentrate on or be immersed in the computer task, there was a main effect for Gesture ($F$[1, 98] = 10.223, $p$ = .002, $\eta p^2$ = .094). Those in the Natural Gesture conditions rated their sense of immersion higher than those in the Arbitrary Gesture conditions, although all conditions were above the midpoint (i.e., 3) on the scale, indicating high immersion overall. There was not a main effect for Instruction type ($F$ [1, 98] = 0.767, $p$ = .383, $\eta p^2$ = .008), nor was there an interaction effect ($F$[1, 98] = 0.663, $p$ = .417, $\eta p^2$ = .007).

Finally, the interface quality subscale was compared by condition, which is the extent to which the interaction with the computer task distracts from or otherwise hinders performance in the virtual environment. This was the only subscale in which a lower score indicated higher presence in terms of better interface quality. Levene's Test was significant ($F$[3, 98] = 3.80, $p$ = .013), but the $FMax$ ratio was below the acceptable value of 5 ($\sigma^2_{NatureVideo} = 0.755, \sigma^2_{ArbitaryText} = 2.74$; $FMax$ = 2.75), so no adjustments were made. Results of the ANOVA indicated there was a main effect for Gesture type ($F$[1, 98] = 6.672, $p$ = .011, $\eta p^2$ = .064), such that those in the Natural Gesture conditions felt the interface quality was better than those in the Arbitrary Gesture conditions. Neither Instruction type ($F$[1, 98] = 0.666, $p$ = .417, $\eta p^2$ = .007) nor the interaction between Gesture and Instruction ($F$[1, 98] = 0.647, $p$ = .423, $\eta p^2$ = .007) were significant predictors of interface quality.

### 3.3 Participant Reactions

Participants responded to an open-ended question following the SUS asking them to write any additional comments related to the computer lesson. In general, participants wrote positively about their experience using the computer system, even in conditions associated with lower usability ratings. Table 4 lists representative responses for each condition, and these responses reflect the finding that usability was high overall (i.e., all conditions were rated above the mid-point on the usability scale).

**Table 4.** Participant perceptions of the computer system

| Condition | Perception of computer system |
|---|---|
| Arbitrary text | "I really enjoyed using gesture commands to walk through this experiment. Would 100% use again" |
| Arbitrary video | "Enjoyed this study!" |
| Natural text | "Very fun to do, feels very futuristic" |
| Natural video | "You did an amazing job designing the system. Well done!" |

## 4 Discussion

The purpose of this experiment was to investigate how natural gesture interactions are perceived by users in a computer lesson in order to inform how NUIs should be developed. An analysis of usability scores was conducted with natural and arbitrary gestures, as well as how the gestures were instructed (e.g., video- or text-based instructions). Although the participants' written reactions to performing gestures in the lesson (regardless of condition) were overwhelmingly positive, natural gestures were rated almost a full point higher in usability (on a 5-point scale) than arbitrary gestures. This finding indicates that natural gestures were perceived as having 15% higher usability than arbitrary gestures.

Presence, or the feeling of "being there" in a virtual environment, was measured after the computer lesson using the PQ with four subscales: involvement, adaptation/immersion, sensory fidelity, and interface quality. For all of the subscales except sensory fidelity, there was an effect for type of gesture interaction, such that natural gestures had higher "presence" ratings than arbitrary gestures. Natural gesture interactions were seen as inducing a higher sense of control in the computer lesson, more immersive, and better interface quality than arbitrary gestures.

This research has empirical implications for the study of NUIs. Previous research on gesture-based interactions has focused on what gestures the computer can recognize [3], and gesture design does not typically begin with a user-centered approach that takes into consideration what the user perceives is natural. The issue with studying natural gesturing from this perspective is that the researcher must first confirm that the gestures are usable and do not detract from feelings of presence (i.e., sensory fidelity, involvement, interface quality, and immersion) in a computer lesson. If arbitrary gestures are perceived as less usable than natural gestures and detract from presence during the computer lesson, then the interface may distract from the computer task.

## 5 Conclusion

An experiment was conducted to determine how natural gesture-based computer interactions are perceived by users in a computer lesson compared to arbitrary gestures. As expected, gestures that more closely align with the intended computer action are rated higher on usability and presence scales than arbitrary gestures that do not correspond with the computer action. Although reactions to the gesture-based NUI were still largely positive even for those using arbitrary gestures (e.g., above-midpoint SUS scores and positive participant written comments), natural gestures were perceived as better than arbitrary gestures in terms of usability. Furthermore, natural gesture interactions induced a higher sense of control in the computer lesson, were more immersive, and were seen to have better interface quality. If instructional designers and researchers want to maximize NUI usability, the evidence suggests that natural gestures should be implemented over arbitrary gestures.

## References


1. Schwartz RN, Plass JL (2014) Click versus drag: User-performed tasks and the enactment effect in an interactive multimedia environment. Comput Hum Behav 33:242–255
2. Grandhi SA, Joue G, Mittelberg I (2010) Understanding naturalness and intuitiveness in gesture production: insights for touchless gestural interfaces. In: Proceedings of the SIGCHI conference on human factors in computing systems, pp 821–824. ACM
3. Shiratuddin MF, Wong KW (2011) Non-contact multi-hand gestures interaction techniques for architectural design in a virtual environment. In: Proceedings of international conference on information technology and multimedia
4. Bailey SKT (2017) Getting the upper hand: Natural gesture interactions improve instructional efficiency on a conceptual computer lesson [Doctoral Dissertation]. University of Central Florida
5. Brooke J (1996) SUS: a "quick and dirty" usability scale. In: Jordan PW, Thomas B, Weerdmeester BA, McClelland IL (eds.) Usability evaluation in industry, pp 189–194. Taylor & Francis, London
6. Witmer BG, Jerome CJ, Singer MJ (2005) The factor structure of the presence questionnaire. Presence-Teleop Virt 14:298–312
7. Field A (2013) Discovering statistics using IBM SPSS Statistics, p 876

# Ergonomics Teaching Concept at Technical Universities on the Basis of Warsaw University of Technology

Ewa Górska and Aneta Kossobudzka-Górska(✉)

Faculty of Management, Warsaw University of Technology,
Narbutta 85, 02-524 Warsaw, Poland
{ewa.gorska,aneta.gorska}@pw.edu.pl

**Abstract.** Ergonomic products and ergonomic working conditions are one of the most crucial determinants of market existence for enterprises and condition to hire and keep the most valuable employees. That is why, employers are searching for qualified ergonomics professionals, whose competences are conformed with university degree or certification from acknowledged institutes (e.g. BCPE or CREE). The goal of this paper is to present current state and possible development directions to improve the teaching in the area of ergonomics at Warsaw University of Technology, so that students interested in human-centric approach could get specialized degree and in the future become certified. Preparation of a student to become ergonomic technology designer and to appreciate and experience ergonomic values, will allow him or her to become a user of comfortable products and to work in comfortable working environment. This requires adequate teaching that would trigger vast changes in values, knowledge, abilities and habits, development of cognitive abilities as well as preferences and attitudes.

**Keywords:** Ergonomics · Ergonomist competences · Euro Ergonomist

## 1 Introduction

### 1.1 A Subsection Sample

Ergonomic products [4] and ergonomic working conditions [3] are one of the most crucial determinants of market existence for enterprises and condition to hire and keep the most valuable employees. That is why, employers are searching for qualified ergonomics professionals, whose competences are conformed with university degree or certification from acknowledged institutes (e.g. BCPE or CREE). The goal of this paper is to present current state and possible development directions to improve the teaching in the area of ergonomics at Warsaw University of Technology, so that students interested in human-centric approach could get specialized degree and in the future become certified.


S. Bagnara et al. (Eds.): IEA 2018, AISC 826, pp. 672–682, 2019.
https://doi.org/10.1007/978-3-319-96065-4_71

Paper will present the following problems:

- Analysis of teaching in the area of ergonomics and corresponding subjects on the basis of WUT faculties that offer such,
- Analysis of ergonomics development trends related to digital transformation that is currently happening,
- Requirements towards ergonomists in the light of national legal regulations,
- Knowledge necessary to apply for the Euro Ergonomist certification,
- Proposal of complimentary subjects for teaching programs at WUT faculties, trainings, post-graduate studies, projects, workshops, student research groups, that would cover required knowledge and skills necessary for future WUT graduates to become ergonomics and industrial design specialists, Euro Ergonomists or to establish international cooperation in Professional Ergonomics,

Authors of this paper believe that if modern technology and IT solutions are to serve people, they need to be compatible with the knowledge about our needs and expectations. In the age of digitization and efficiency improvements, psycho-physiological needs and limitations of human beings, are rarely taken into consideration. Numerous publications point out that ergonomics knowledge is one of the triggers for innovation and should be popularized by inclusion in teaching programs of primary schools and consecutive learning stages. Teaching in the area of occupational safety and ergonomics in Poland is realized mainly through universities.

Preparation of a student to become ergonomic technology designer and to appreciate and experience ergonomic values, will allow him or her to become a user of comfortable products and to work in comfortable working environment. This requires adequate teaching that would trigger vast changes in values, knowledge, abilities and habits, development of cognitive abilities as well as preferences and attitudes.

## 2 Analysis of Ergonomics Development Trends and Applications

Ergonomics teaching programs need to have up-to-date content due to progress in development of knowledge in this area. Therefore, a review of ergonomics-themed scientific conferences that are planned for 2018 in USA, Europe and Poland, was made.

General conclusion is that they are focusing on presentation and discussion about modern approach, design tools, methods, techniques and solutions used for integration of human beings, smart technologies, human-computer solutions, automation and artificial cognitive systems in all areas of human activity:

- Humans and artificial cognitive systems,
- Intelligence, technology and analytics,
- Computational simulation and modeling
- Complexity of human and artificial systems,
- Smart materials and human interaction systems
- Human-autonomy teaming.

Essence of these scientific elaborations are relations that take place between human factor and:

- Artificial intelligence and social technologies,
- Game design and virtual environment,
- Cybers-security
- Ageing process and gerontology
- System interaction,
- Roots and unmanned systems,
- Software and system design,
- Support service engineering,
- Training, education and teaching,
- Cross-cultural decision making.

Upcoming international seminar held in Poznan, Poland will cover: new possible applications of ergonomics and ergonomic issues in Industry 4.0. Also the following topics will be discussed: ergonomics in product and manufacturing processes design, ergonomic awareness and its shaping, education and training in ergonomics and occupational safety, ergonomics for elderly and disabled people, locomotor system load in work processes, ergonomic aspects in occupational risk assessment.

Ergonomics Committee at PAU in 2014 presented the directions of ergonomics application in: development of smart cities, modeling of functional-spatial aspects of objects, rooms and furniture, latest approach in food manufacturing, vertical farming in cities, remote and e-learning as well as chrono-ergonomics. Federation of European Ergonomics Societies - FEES (2016) see the future of ergonomics application in areas of: Marketing, HCI, Safety science (social ergonomics), LEAN (workplace ergonomics), Occupational safety, Inclusive design (design for special needs), User experience expert.

## 3 Analysis of Teaching of Ergonomics and Related Subjects at Warsaw University of Technology

Executive Board of Polish Ergonomics Society adopted a resolution in 2017 to perform the analysis of current state of ergonomics teaching in national public and private universities. Findings of the analysis and the report is to be presented to Labor Protection Council at Polish Parliament. Similar research was conducted by LPC in 2006. Goal of the research, ordered by PES Executive Board, is to determine trends in ergonomics teaching processes with relation to the 2006 research results. That is why a questionnaire was elaborated that covers the following information:

1. Name of the university, name of unit (faculty, department, chamber, laboratory), specialization.
2. Do teaching programs in particular university include topics from the area of ergonomics?
3. Name of the subject in particular specialization that includes ergonomic content.
4. Do the post-graduate teaching programs include topics from the area of ergonomics?

5. Do course teaching programs include topics from the area of ergonomics?
6. Do the Ph.D. teaching programs include topics from the area of ergonomics?

Polish Ergonomics Society distributed the questionnaire among national public and private universities and asked university officials to present different activities in the area of ergonomic teaching. Apart from the review conducted by PES, the authors collected information from the point of view of a student searching for online information about ergonomics teaching at WUT websites. Data was collected between October 2017 and March 2018. Results of the ergonomics teaching programs review are collected in Table 1.

**Table 1.** Collected subjects on ergonomics and related topics (self-study).

| Subject | Faculty |
|---|---|
| *Mandatory basic* | |
| Intellectual Property Protection and Labor Law | Faculty of Material Engineering |
| Foundations of Social Psychology<br>Office Work Methodology | Faculty of Administration and Social Sciences |
| Effective Presentation Skills | Faculty of Mathematics and Information Science |
| *Mandatory directional* | |
| Ergonomics and Occupational Safety | Faculty of Transport |
| Ergonomics in Product and System Design<br>Industrial Ergonomics<br>Foundations of Ergonomics | Faculty of Management |
| Fire Safety Requirements | Faculty of Civil Engineering |
| Fire Safety Requirements for Buildings | Faculty of Civil Engineering, Mechanics and Petrochemistry in Plock |
| Occupational Safety in Construction Industry<br>Technical Safety | The Faculty of Civil Engineering, Mechanics and Petrochemistry in Plock |
| Occupational Safety | Faculty of Chemistry |
| Ergonomics and Occupational Safety<br>Occupational Safety of Manufacturing Processes | Faculty of Chemistry<br>Faculty of Building Services, Hydro and Environmental Engineering |
| Vibration | Faculty of Power and Aeronautical Engineering |
| Mechanical Vibration | Faculty of Automotive and Construction Machinery Engineering |
| Elements of Design Process | Faculty of Architecture |
| Visual Communication and Effective Presentation Skills | Faculty of Civil Engineering, Mechanics and Petrochemistry in Plock |

(*continued*)

**Table 1.** (*continued*)

| Subject | Faculty |
|---|---|
| Fire Safety Requirements | Faculty of Civil Engineering, Mechanics and Petrochemistry in Plock |
| Noise Protection Requirements | Faculty of Building Services, Hydro and Environmental Engineering |
| Air-conditioning and Ventilation | Faculty of Civil Engineering, Mechanics and Petrochemistry in Plock |
| | Faculty of Building Services, Hydro and Environmental Engineering |
| Labor Law and Civil Law<br>Office Work Methodology | Faculty of Administration and Social Sciences |
| Initial Designing<br>Designing in Theory and Practice | Faculty of Architecture |
| *Profile subjects* | |
| Digital Analysis of Signals<br>Diagnosis and Monitoring of vibroacoustic<br>Reduction of Machine Noise and Vibration<br>Basics of Machine Vibroacoustic<br>Legal Aspects of Noise and Vibration Protection<br>Active Reduction of Vibration in Mechanical Systems Active Noise and Vibration Reduction Methods<br>Vehicle Acoustics | Faculty of Automotive and Construction Machinery Engineering |
| OHSE and Ergonomics in Occupational Safety Management<br>Communication and Public Speeches | Faculty of Management |
| Safety of Vehicles and Road Traffic<br>Safety of Traffic Control Systems<br>Humans in Transport Systems<br>Safety Engineering<br>Air Law and Air Traffic Safety<br>Reconstruction of Road Accidents<br>Threats and Accidents in Road Traffic | Faculty of Transport |
| Traffic Safety and Management<br>Fire Safety Requirements<br>Fire Safety Requirements for Bridges and Tunnels<br>Quality, Environment and Safety Management | Faculty of Civil Engineering |
| Air-conditioning and Ventilation | The Faculty of Civil Engineering, Mechanics and Petrochemistry in Plock |
| *Ph.D. course of study* | |
| Ergonomics in Working Environment Modeling | Faculty of Management |
| Human Factors Methods | Faculty of Transport |

Warsaw University of Technology offers also extra-curricular subjects selected by students and HES available at particular faculties for their students and WUT students from other faculties. Review of such subjects that are related to ergonomics is collected in Table 2.

**Table 2.** Extra-curricular and HES subjects offered for faculty students and WUT students from other faculties.

| Extra-curricular subjects at HES level | Extra-curricular subjects at Faculty level |
|---|---|
| Psychological Aspects in Construction | Psychological Aspects in Construction |
| Ergonomics | Ergonomics |
| Ergonomics in Practice | Analysis and modeling of physiological processes |
| Design Thinking | Design Thinking |
| Working environment protection | Working Environment Protection |
| Intellectual property protection and labor law | Fire safety requirements |
| Labor law basics | Art of thinking and learning |
| Art of thinking and learning | Effective Presentation Skills |
| Memory training techniques | Effective audio-visual and written presentation skills with elements of copyrights |
| Effective Presentation Skills | Industrial Ventilation |
| Effective audio-visual and written presentation skills with elements of copyrights | Influence of vibration and noise on operators |
| Public speeches – how to be heard | Selected problems of lighting in transportation |
| Occupational Safety Management | Selected aspects in rail transport |
| | Occupational Safety Management |

Warsaw University of Technology is currently running 240 h post-graduate studies from the Occupational Safety – "Safety and Protection of Humans in Working Processes" at the Faculty of Automotive and Construction Machinery Engineering with cooperation of Central Institute of Labor Protection – National Research Institute. Study program includes subjects like Ergonomics (18 h), Occupational safety and risk management, Legal labor protection; Psycho-physiological Challenges of Human Beings in Working Environment; Dangerous, Harmful and Disruptive Factors in Working Environment; Internal Transport; Industrial Disasters; Risk of Fire and Explosion; Personal Protective Equipment; Information Technology in Occupational Safety; Medical Service and First Aid Systems in Enterprises; Organization and Methodology of Occupational Safety Trainings.

## 4 Requirements Towards Ergonomics and Industrial Design Professionals

Ways of formal competence assurance are as follows [3]:

- European Qualification Framework (EQF) – EU initiative that combines national education systems aimed at mutual understanding of competences in different countries. The goal of EQF is to promote mobility and enable learning throughout whole life.
- National Qualification Framework (NQF) – is a teaching quality improvement tool – describes mutual relations between competences, integrating various national competence sub-systems.
- National Standards of Professional Competences (KSKZ) – is a norm describing professional competences necessary to perform given profession. Currently the database holds 253 professional qualification standards and 300 professional competence standards.

Ergonomics and industrial design professional (Code: 214924) is included in national standards database [10]. Standard describes necessary collection of knowledge, skills, social competences that are crucial to perform this profession.

Professional Ergonomists need to fulfill the following requirements:

- General and specialized knowledge and skills,
- psychophysical features,
- attitudes and interests,
- professional and life experience,
- core and general competences,
- social competences (good communicative skills, teamwork, change enthusiast, empathy, negotiation, ability to influence employees behavior and attitudes, ability to talk to people, ability to create and maintain good relations with employer and employees).

Core competences for ergonomic professionals are:

- diagnose and elaborate analysis of ergonomic working conditions improvement at workstations;
- design various technical objects that are adjusted to the dimensions and shape of human body, are functional, reliable, safe and comfortable as well as esthetic in shape and color;
- use methods and software that aids ergonomic designing.

## 5 Knowledge Necessary to Apply for Eur.Erg. Certificate

Certificate is a formal assurance of qualification to perform a jab o task, based on throughout evaluation of knowledge, experience and competences. Main reason for certification of ergonomists is market demand for ergonomic professionals, who become a part of market competitive advantage of enterprises. Certification is performed by authorized certification body. The institutions is are Center of Registration for European Ergonomists – CREE in Europe and Board of Certification in Professional Ergonomics – BCPE in USA. Application for certificate issued by CREE requires documented achievements in 3 areas: education, supervised training, professional experience.

According to CREE professional ergonomist is a person who [1]:

- Performs ergonomic design,
- Properly analyzes and interprets ergonomic research results,
- Can document results and conclusions from ergonomic analyzes,
- Can determine the conformity of human abilities with planned or existing requirements,
- Can elaborate plans of ergonomic design and diagnosis,
- Can formulate and implement recommendations regarding ergonomic improvements,
- Evaluates results of implemented ergonomic solutions,
- Acts professional and works within his or hers competence area.

Candidates need to have broad knowledge about ergonomics gained during lectures, seminars and laboratory exercises.

CREE requires evidence of a basic education across the following Areas of Knowledge:

A. Principles of Ergonomics
B. Populations and General Human Characteristics
C. Design of technical systems
D. Research, evaluation and investigative techniques
E. Professional issues
F. Ergonomics: Activity and/or Work Analysis
G. Ergonomic Interventions
H. Ergonomics: physiological and physical aspects
I. Ergonomics: psychological and cognitive aspects
J. Ergonomics: social and organizational aspects
K. Project work

This includes: A-J – theory min. 20 ECTS, F-J – practice min. 48 ECTS, laboratory exercises and project work – min 6 weeks.

Professional ergonomists are also supported by Foundation for Professional Ergonomics – FPE located in USA [12]. Its mission is to provide leadership in evolving and growing the ergonomics profession; bridge gaps between research, education and practice; and promote professionalism in ergonomics practice; all for the benefit of the public.

FPE goals are:

- Establish scholarships for deserving students in the area of Ergonomics and Human Factors
- Dieter W. Jahns Student Practitioner Award for graduate students in Ergonomics and Ergonomics-related programs.

- Develop courses on professional issues and technical topics. FPE, in conjunction with the BCPE and the Human Factors and Ergonomics Society (HFES), will contact and collaborate with the established leaders of ergonomics and related professions to develop quality content and provide professional delivery.
- Continue collaboration with Ergonomists Without Borders to provide ergonomics expertise and resources to agriculturally and industrially developing communities, globally.

FPE elected international Ambassadors, who are leaders in the area of ergonomics around the World. Their main task is to inform about FPE goals and programs in their countries and identification of potential cooperation areas between countries.

## 6 Didactic Offer of Warsaw University of Technology Versus CREE Requirements

Conducted research and analysis lead to a conclusion that the highest number of subjects necessary to gather knowledge to become an ergonomic professional is offered by Faculty of Management and Faculty of Transport. Complete collection of offered subjects is collected in Table 3.

**Table 3.** Collection of corresponding subjects at WUT and CREE (self-study).

| Subjects realized at Warsaw University of Technology | Faculty HES | Min. CREE requirements |
|---|---|---|
| Principles of Ergonomics | FoM | Principles of Ergonomics, |
| Ergonomics in Products and System Design | FoM | Populations and General Human |
| Industrial Ergonomics | FoM | Characteristics, |
| Ergonomics in Working Environment Modeling | FoM | Design of technical systems, |
| OHSE and Ergonomics in Occupational Safety | FoM | Research, evaluation and investigative |
| Management | FoM | techniques, |
| Communication and Public Speeches | FoM (HES) | Professional issues, |
| Ergonomics in Practice | FoT | Ergonomics: Activity and/ or Work |
| Ergonomics and Occupational Safety | FoT | Analysis, |
| Human Factors Methods | FoT | Ergonomic Interventions, |
| Humans in Transport Systems | FoCh | Ergonomics: physiological and |
| Occupational Safety and Ergonomics | FoCE | physical aspects, |
| Quality, Environment and Safety Management | (HES) | Ergonomics: psychological and |
| Ergonomics | (HES) | cognitive aspects, |
| Psychological Aspects in Construction | (HES) | Ergonomics: social and organizational |
| Working Environment Protection | (HES) | aspects, |
| Occupational Safety Management | (HES) | |
| Design Thinking | FoA | |
| Design in Theory and Practice | FoA | |
| Initial Designing | FoA | |
| Elements of Design Process | FoMaIS, | |
| Effective Presentation Skills | FoCEMaP in Płock, FoM (HES) | |

Most important subjects for future professional ergonomist, run at the post-graduate studies of Faculty of Automotive and Construction Machinery Engineering, are: Ergonomics; Occupational Safety and Risk Management; Psycho-physiological Challenges of Human Beings in Working Environment; Dangerous, Harmful and Disruptive Factors in Working Environment; Information Technology in Occupational Safety.

## 7 Summary

In order to allow the students of Warsaw University of Technology to expand their competences gained in course of selected study profile with ergonomics related subjects, one must encourage university officials to include topics like research, evaluation and analysis methodology; task and work process analysis; laboratory exercises; physiology, psychology, cognitive, social and organizational aspects in non-curricular or HES subjects. It is also important to support ergonomics with computer-aided methods, diagnostic methods and techniques as well as designing with the use of ergonomic requirements. Regardless of the possibility to undertake individual study course, including ergonomics related subjects, students should be able to take directional courses and trainings. Student research groups can also organize such workshops and trainings that in the end would grant a certificate of participation. It is also important to review current subjects and check if they are up-to-date to the current changes and trends in ergonomics. Ergonomics related knowledge is sought also by students and graduates from other universities, whose experience and education is not sufficient to become an ergonomics professional or Eur.Erg. certified specialist. Current ergonomics of working systems in our economy is improving bur still in many industrial areas it is unsatisfactory. This is connected with economical condition of many Polish enterprises, which lack resources to invest in modern technologies, have ineffective occupational safety services that struggle with execution of ergonomic standards and legal workstation requirements, low ergonomic awareness of management and employees [7]. This current state is confirmed by research conducted by Poznan University of Technology. Basic knowledge of ergonomics and ergonomic awareness is insufficient. Students more easily learn about ergonomics and see its application in everyday life, elder people have difficulty to understand the concept [5]. This is an additional argument to raise the importance on this knowledge area and introduce Principles of Ergonomics as an obligatory subject in all faculties of Warsaw University of Technology.

## References


1. Centre for Registration of European Ergonomists Homepage. The Eur.Erg. title. www.eurerg.eu/. Accessed 15 Apr 2018
2. Labor Market Department - Ministry of Family, Labour and Social Policy. Classification of jobs and specializations. http://psz.praca.gov.pl/rynek-pracy/bazy-danych/klasyfikacja-zawodow-i-specjalnosci/wyszukiwarka-opisow-zawodow//-/klasyfikacja_zawodow/zawod/214924. Accessed 01 May 2018

3. Górska E (2015) Ergonomics: Design, Diagnosis, Experiments. OWPW, Warsaw
4. Górska E, Tytyk E (2017) Ergonomic recommendation for consumer products. Poznan University of Technology Reseaerch Papers Organization and Management No. 73/2017. Poznan University of Technology Publishing, Poznan
5. Jasiak AE (2017) Ergonomic awareness and its shaping. The Małopolska School of Economics in Tarnów Research Papers Collection, vol 36, no 4, pp 111–128, Tarnów
6. Marcinkowski JS (2006) Diagnosis of education in the area of occupational safety, labor protection and ergonomics in Polish Universities. Faculty of Engineering Management at Poznan University of Technology Publishing, Poznan
7. Muszyński Z (2016) Development of ergonomics in Poland and worldwide. The Małopolska School of Economics in Tarnów Research Papers Collection, vol 29, no 1, Tarnów
8. Warsaw University of Technology. ECTS Catalogue. https://ects.coi.pw.edu.pl/menu2/programy. Accessed 15 Apr 2018
9. Public Job Service. www.praca.gov.pl/eurzad/index.eup#/panelOgolny. Accessed 15 Apr 2018
10. Ordinance of Ministry of Labor and Social Poicy from 07/08/2014 about job and specialization classification for job market and scope of application (Dz.U. 2018 pos. 227)
11. Statue from 20/04/2004 about hiring promotion and jaob market institutions, art. 4 stat. 7 (Dz. U. 2008, No 69 pos. 415)
12. Foundation for Professional Ergonomists. https://www.ergofoundation.org/. Accessed 02 May 2018

# Children's Creativity at School: Learning to Produce Multimedia Stories

Laura Anastasio Forcisi(✉) and Françoise Decortis

Laboratoire Paragraphe, Equipe C3U, Université Saint-Denis, 2 rue de la Liberté, 93526 Saint-Denis Cedex, France
forcisilaura@gmail.com, f.decortis@gmail.com

**Abstract.** The study presented here has two objectives: on the one hand, we seek to understand how a resource system is structured to serve creativity, and on the other hand we try to identify the patterns that are inherent in the cycle of the creative activity of imagination. The reference situation, which is based on the creation of a collective tale following a school trip, is analyzed through the prism of the instrumental approach to provide an anthropocentric point of view on the technique (Rabardel 1995). We use models NAM (Decortis 2008) and MARO (Bationo-Tillon 2013) to take into account the diversity of the instruments (instructions of the teacher, interactions of the pupils, resource sheet, text, illustration, etc.) in posing classes of situation to map the creative and narrative activity.

**Keywords:** Children · Narration · Text · Reader · Author
Creativity · Scheme · Instrumental approach

## 1 Introduction

We don't create from nothing, children either. For us, creativity is a dynamic activity, located, finalized, taking into account the point of view of the subject. It is part of a cycle of porous phases promoting iterations and allowing the subject to actively draw on the real, in his lived experience tinged with emotion but also in the collective experience [9]. The aim of this article is to understand the groundings of creativity in narrative construction. From the analysis of an educational situation, we try to answer the following questions: how is a structured a system of resources at the service of creativity? What are the schemes inherent in the cycle of the creative activity of the imagination? We look at how children develop a creative cycle supported by a system of varied resources (teacher's instructions, pupil interactions, resource sheet, text, illustrations, etc.) [4]. This study is included in my PhD in Ergonomics on narrative activities, where we support that creativity plays a key role in human learning and development [8] and it needs to be feed by various instruments.

We speak about narrative activities, which are situated, finalized, mediated and creative activities. In our study, pupils have to invent a collective tale after a classroom trip in the Massif Central. The entirety of the class writes the contents of the story, part by part, draw important scenes of the narrative and tape their voice on tablet to produce a multimedia tale with a several languages [10]. We look this multi-instrumented


S. Bagnara et al. (Eds.): IEA 2018, AISC 826, pp. 683–692, 2019.
https://doi.org/10.1007/978-3-319-96065-4_72

situation through the instrumental approach [11] to understand the means of activity with an anthropocentric point of vue on technology.

Complementary we use the Narrative Activity Model [5] which is based on Vygotsky's theories for who creativity is present in each person and throughout the life [13]. This model helps us to apprehend this creative situation with systematicity in four phases namely exploration, inspiration, production and sharing.

At the heart of the NAM model, the text occupies an important place. Here we propose an adaptation of the model of the activity of encountering a work of art (MARO) [2, 3] by replacing the work of art with the text. The child, who is here a creative subject, turns out to be a reader as well as an author.

## 2 Theoretical Background

### 2.1 Instrumental Approach

Children activity is analyzed according to the instrumental approach that are activities with instruments [11]. This approach is useful to understand the means of activity with an anthropocentric point of vue on technology. The concept of instrument, composed by schemes and artefact, is in the heart of this approach. It is really important for our study because this classroom has the specificity to be equipped by traditional artefacts, like paper and pen, but also by tactile tablets and a digital white board.

**Activity as an Object of Understanding**

This research in ergonomics focuses on the notion of activity, by activity we mean what is done, what is put into play by the subject to achieve the objective that is fixed in a given situation. The action is always oriented towards an objective, which is influenced by a multitude of characteristics of the situation (environmental, social, material, etc.) [12] as well as the subject. The activity is finalized, located, singular and mediatized.

**The Instrument as Mediator**

Instrumental approach is based on the concept of instrument, which integrates a character of mixed. It consists of one side of an artifact (the artifact can be material, cognitive, psychological or semiotic), and on the other side schemata that is a structured set of generalizable characters of the action that make it possible to repeat the same action, they are organizers of the action of the subjects [11].

### 2.2 The Contribution of Vygotski: The Cycle of the Creative Activity of the Imagination

In Vygotsky's cycle [13], reality and imagination are not dissociated but linked through different relations that form the basis of the creative process. The first relation shows that imagination is constructed from materials taken from the reality and inserted into the experience of the subject, imagination depends to the richness and the variety of the experiences. The second relationship is based on the historical and social experience of others, it is the experience of the child that feeds on the imagination by connecting the product of the imagination with other aspects of reality. The third relationship is about emotions, where feelings can influence the imagination, and conversely the imagination can influence emotions. In the last relationship, the imagination becomes reality and is embodied in something.

The imagination is based on a complex process, the products of the imagination are the fruit of a circular re-elaboration of reality. This process is structured in two stages: first, the accumulation of material by the child, then its re-elaboration through phases of dissociation and association. During the dissociation, the impressions coming from the reality change aspect like a phenomenon of mutation, the association represents the meeting of the dissociated and trans-figured elements. Finally, the imagination is doomed to be concretized, it must be incarnated, crystallized in external elements.

### 2.3 A Model of Narrative and Creative Activity: The NAM Model

The model of activity NAM [5, 6] is based on the cycle of creative imagination of Vygotski and offers to look activity steps with artefacts used. It is interesting to mobilize this model to apprehend globally a situation with systematicity with the four phases of NAM: exploration, inspiration, production and sharing. Exploration represents all the activities that come from the experience: everything seen, heard, touched …, it is a collection phase where elements are still perceived as an indivisible whole. Inspiration is the moment when the choice of ideas is made, we give importance to certain elements. It is a reflexive practice where children dissociate these complex wholes. During the production takes place the association of the dissociated elements and the expression. We rework elements chosen to produce new content. Sharing is the moment when the story is told by a child and shared with others by having an effect on them.

## 3 Reference Situation

In our study, based on a teaching situation in a class of CM1, we observe pupils creating a collective tale following a trip to the Massif Central. To initiate the project, childrens discover the narrative structure through famous tales by identifying the stages: initial situation, disruptive element, events, outcome, final situation. Pupils then write in group their version of the initial situation, the denouement, the final situation, etc. Each version is then submitted to the vote of the class. Pupils have great freedom for the content of their tale but they had to situate their story in the places visited during their school trip. When the text is finished, pupils divide the scenes in the story to draw them and then tell them through the BookCreator tablet app.

## 4 Methodology

To analyze the activity of the pupils and the teacher, the data collection methods mobilized were the observation of the sequence with audio recording, photo and video on the entirety of the sessions. They were completed by self-confrontations and semi-structured interviews with the teacher and several interviews with groups of pupils.

These methods of qualitative and subjective interviewing were essential in order to capture the intrinsic point of view of the child subject.

## 5 Application of the Model of Activity NAM

The NAM model helps to identify the effects of tablet on narrative and creative activity, and also to understand how the creative imagination cycle is changing. It repeats here four times: for the writing of each step of the narrative structure and during activity on tablet. Each session is organized in this way: time for individual reflection by taking notes on a paper support, oral communication with the group without writing, writing the text collectively, rereading, sharing with the class and collective vote.

The NAM model developed by Decortis meets evolutions at three moments: the invention of the first stage of the story, the invention of the following stages of the story, and the realization of the multimedia version of the story in BookCreator application.

For the development of the first stage of the story, during the exploration the pupils feed on various resources: in their memories of the school trip reactivated by the viewing of a slide show of the trip, in their imagination fueled by the known stories, and for some in a tool sheet with ideas provided by the teacher. The choices are then made individually influenced by the constraint of respect for places and immediately materialized on the draft notebook of each pupil. This first individual cycle ends with an oral sharing with the rest of the group (see Fig. 1).

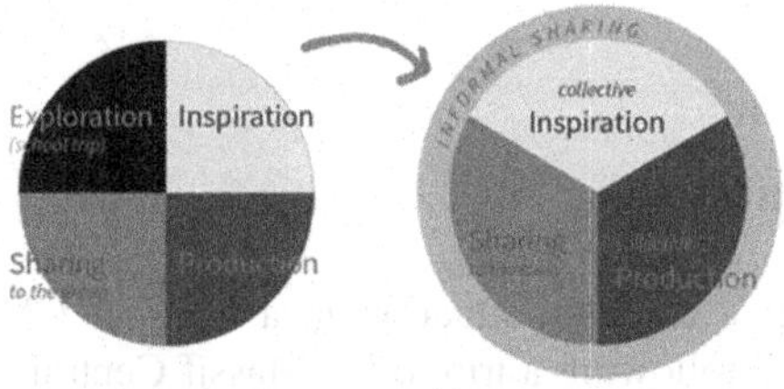


**Fig. 1.** NAM during writing of the first stage of the story.

There is a gap between the prescribed and the real. During the exploration, the teacher asked pupils to write only words such as note-taking, and the pupils individually composed micro-stories. This aspect influences the informal sharing in the followed collective cycle because pupils read their personal narrative one by one and then choose one as the basis to work.

This collective cycle is organized around an informal sharing between the different pupils of the group during all the pooling. Choices are the subject of negotiation into the group. Production is also collective, a pupil writes under the dictation of others. Finally, a pupil of the group comes to share their fragment of the story to the class to vote for the best fragment.

For the invention of the following stages of the tale, the preceding cycle encounters an evolution. The time of exploration where pupils will draw on their experiences disappears in favor of a time that we call understanding. This is indeed a time when pupils ensure consistency and continuity with the previous part of the story to invent the rest. The tool sheet distributed at the previous session to find ideas is totally abandoned. However, we note the emergence of supports created by the teacher to support understanding as debriefs to highlight the elements not yet mentioned or those to be discussed again. With the exception of the exploration phase which disappears in favor of the comprehension phase, the cycle is structured in a similar way (see Fig. 2).

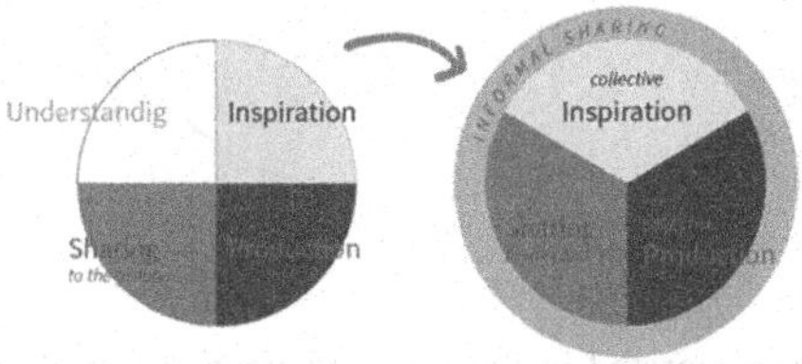


**Fig. 2.** NAM during writing of the following stages of the tale.

The textual tale is then transformed into a multimedia tale thanks to the Book-Creator application. After a fragmentation of the text, the activity is organized around a tablet on which groups of pupils pass in turn. Note that before the activity on tablet, an important preparatory work is necessary: illustration of the scenes of the tale, division of the text in units corresponding to the illustrated scenes, training to the reading of the tale. There are many situations of knowledge sharing between groups of pupils in difficulty and resource pupils, associated with informal exchanges between them. Production, central during the tablet activity, is supported by these different forms of sharing. The exploration and inspiration phases are evacuated because the children are exclusively in the realization of all that has been prepared before (see Fig. 3).

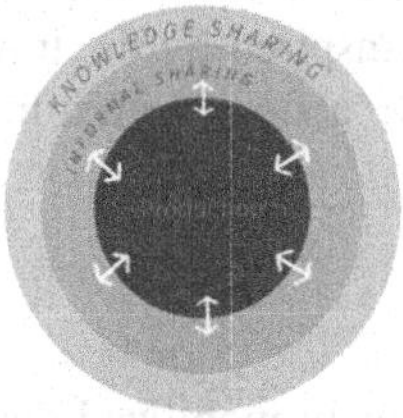


**Fig. 3.** NAM during activity on tablets.

## 6 Specific Scheme to the Narrative and Creative Activity

From located observations and interviews, we succeed to identify various schemes structuring the activity of narrative creation. These schemes fit in the different stages of the cycle of creative activity of imagination (see Fig. 4).

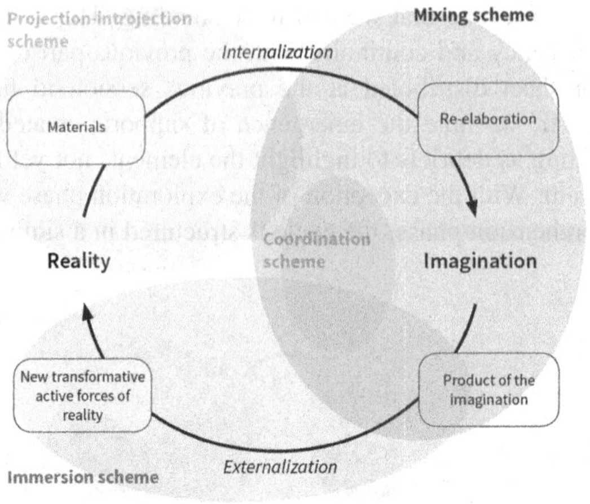


**Fig. 4.** Schemes integrated in the cycle of creative activity of imagination.

## 6.1 Projection-Introjection Scheme

Projection-introjection scheme, favored here by the imposition of the narrative constraint on the use of the places of the school trip, comes to create links between the lived experience and the content of the story. In this scheme, children imagine the story in their head, relate the story to their lived experience or slip into the skin of the characters of the story. There is for some children an identification with heroes of the tale: "I remember I had an idea except that it did not work out, it was the class of CM1 who was leaving in discovery class, adults they had just gone out to drink a fruit juice or something, there was no adult inside. And there was a blizzard that rage, so we were all alone inside, we had to manage ourselves and to find food, we didn't know when we could go out. The class of CM1 was us …" (Paul). The presentation of the slideshow awakens memories among pupils, which revives moments or places visited that can be remobilized for the content of the story. The children project their experience of history during the exploration, a pupil is inspired by the travel by train: "I would have loved that it happens in the present time, a story with a train that derails to go to Estables" (Céline).

## 6.2 Mixing Scheme

Mixing scheme is a scheme extremely mobilized during the phases of inspiration and production. These are strategies put in place by pupils to constitute their personal ideas into a coherent whole. Acceptation of the idea of others is important: "I'm trying to find compromises, my idea is my idea, I do not want to let go. To be accepted I try to break it a little my idea, but we both have to take a step if we want to agree in the group" (Paul). The metaphor of a cooking recipe is used by pupils who explain remixing elements. We find this idea of mixing and meeting several entities: "it's a bit like a kind of cooking. Each ingredient counts and needs all the ingredients to make a perfect recipe. I had a starting dough, I had several ingredients that is the ideas of others,

I choose a few, I put everything in a bowl and I mix and that's an idea" (Celine). This mixing scheme, however, is based on a common base, as seen earlier during the exploration, the pupils already produce personal micro-stories: "For example, Lou had read first, she had told her story. After everyone has read. Then, for example, I took bits of history and I added them to my story, it made a story. Then Lou did a little bit like that but with hers. Then in the end we agreed" (Maya).

### 6.3 Coordination Scheme

A coordination scheme is deployed in parallel with the mixing scheme, like a condition of realization. This scheme is not acquired by all pupils and raises many difficulties as the teacher notes: "Now this is the hardest part, you have to be okay together. The idea is not everyone gives an idea and then we make a micmac of all that, the idea is not either we take the ideas of Gertrude because it is all right and we will not break the head. The idea is that we really want to build a real adventure together. I'm sure you'll find something in common, that you'll be able to add things, enrich what your classmate has found." It is a scheme in which children must choose which ideas associate in the most harmonious way possible and minimize the feeling of disappropriation ("I preferred alone because you have your own ideas, in a group sometimes we withdraw your ideas," Louis). It is a social scheme of use that involves several subjects belonging to the class community [11].

### 6.4 Immersion Scheme

Immersion scheme occurs in reading and in writing the story. When the child immerses himself, he reads or writes with all his attention identifying each detail. Narrative details have a real importance in reading, it is a criterion of what makes a good story for pupils: "I that's why I voted because I liked the idea of the witch and also that she sings the song Source de la Loire tell me where is my dad, Source of the Loire tell me" (Mathias). Pupils also feel the need to dive into the text when constructing the story: "I wrote sentences, I wrote a story directly. The teacher said it was necessary to write nominal groups, for example small cook. I prefer to write sentences, it gives more inspiration than words" (Pauline).

## 7 From the Artwork to the Text: Analysis of the Effect of the Prescription to the Creative Activity Through the Model of Activity MART

The creativity of the children is built in the co-activity with the teacher, which opens or restricts the field of possibilities of the pupils. After having noted the instructions given by the teacher during the activity of the creation of the tale, we note an alternation of movements of opening and tightening of the creative pupil activity. We are comparing these movements of openness and tightening of pupils' creative activity in a model of the activity of encountering a text: the MART model, inspired by the MARO model developed by Bationo-Tillon [2, 3]. This model initially intended for visitor activity analysis articulates a series of situation classes according to two orientations: the sensory activity and the analytic activity. The sensory activity is based on the sensations and impressions expressed by the subject: immersion, impregnation and imagination. The analytical activity integrates three other classes of situation: detailed description of the work, emergence of questions and distancing where the work is related to areas of knowledge. We replace here the work with the text to study the oscillations between different models of the subject: the subject reader and the subject author (see Fig. 5).

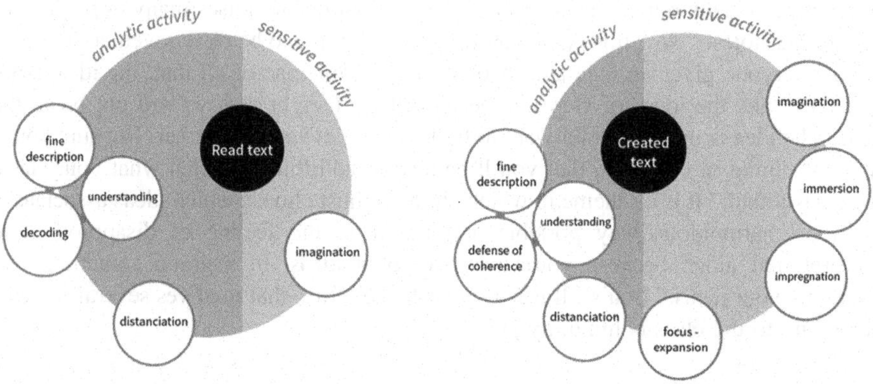


**Fig. 5.** Model of the activity of encountering a text for the subject reader (right) and the subject author (left).

## 7.1 When the Child Became a Reader

For the reader subject, analytic activity prevails. From the MARO model, we find the fine description in the encounter with the text, the subject engages in an active exploration where pupils fragment all the components of the text by naming and objectifying them. Associated with this class of situation, two other classes of situation emerge: decoding and understanding. Decoding is a deciphering of the text and the different words on which pupils stumble, it is an activity more detailed than the fine description. The understanding offers a more macro point of view, pupils try to understand the text as a whole without stopping on the complicated words: the goal is to grasp the meaning and logic of the text. The distanciation is still present but remains minimal, subjects create links between elements of the text and literary knowledge (literary genre of the tale). The emergence of question situation class disappears, we don't observe a shift to another space as in the MARO model. With this new situation classes, there is a desire to dig deeper into the text, whereas with a painting the subject moves elsewhere.

Sensory activity, based on the manifestation of sensations and impressions, is minimized. The imagination is maintained, there is an extension of the text by the reader who makes a transition from readable to visible. The resonance with personal experiences in contact with the work of art disappears here. Immersion is obstructed by the staccado rhythm of reading and the various elements to be taken into account (meaning of the text, new words of vocabulary, etc.), the emergence of the feeling is never explicited during activity. It seems that pupils are not fully immersed in the text even though there is increased concentration.

### 7.2 When the Child Became an Author

When the child is in the position of author, we observe a return sensitive classes of situation: immersion, impregnation and imagination. The imagination takes the form of an elaboration of the text that crystallizes each pupil's ideas through writing. The productions produced will be extended by readers beyond the spatio-temporal framework provided. The impregnation, absent in reading, is favored when the children rely with their experience of the school trip realized a few weeks before (for example a child wishes that the pupils are the heroes of the tale). Impregnation can take on a collective dimension with materials shared by all the class. Immersion is also maximized, pupils are invested in writing and complement their text with many details. This immersion in the text seems to be a need for pupils, who during the individual research phase of ideas, express the need to immerse themselves in the text by long sentences and not just words as they were asked in teacher's instructions.

We identify a class of situation straddling between sensory and analytic activity: the focus-expansion. Located between immersion and fine description, this is a moment where pupils detail, dig, deepen what they are writing.

The analytical activity, evacuated by the children, is regularly brought back by the teacher during the writing. The distanciation is reinforced by the respect of the corresponding narrative schema in literary genre of the tale. We find the fine description with its fragmentation of text components especially during debriefs before writing a new narrative stage. We also find understanding, where here the pupils are at the origin of the meaning and logic of the text produced. This trio of situation classes undergoes a major modification with the replacement of the decoding by a new class of situation: the defense of coherence. The teacher invites the pupils to free themselves from the problematic words but ensure the continuity of the story and the transitions from one part to the other.

## 8 Conclusion

Creativity in the context of the class is built in a double co-activity: in the co-activity with the teacher who acts as a bandmaster and co-activity with classmates. Language is a pivot instrument in this situation of creating a collective tale. Pedagogical choices made by the teacher highlighted the exploration and the link to be made with the school trip in the Massif Central realized a few weeks earlier. We were able to highlight these aspects thanks to the NAM model which gave us visibility of the activity over a long time, allowing us to take into account the collective and individual dimensions of the activity.

Our attention was also focused on the schemes structuring the narrative and creative activity. The projection-introjection scheme, the mixing scheme, and the immersion scheme turn out to be central to this type of activity. The coordination scheme varies depending if it is a collective or individual creation. These schemes, which are part of the creative activity cycle of Vygotski, are stabilized because we find them in other studies also in pedagogical situation realized within the framework of the thesis (creation of a comic multimedia, invention of micro-narratives with StoryCubes game, biographical stories, etc.) [1].

These schemes develop essentially during the sensory activity, indeed when the child being author this one prevails over the analytic which never totally disappears. Thanks to the MART model, we highlight oscillating movements between analytical activity and sensory activity during narrative creation if the subject is author or reader.

## References

1. Anastasio L, Decortis F (2018) Analyse de l'activité des enfants avec une interface tangible et narrative (Lunii – La fabrique à histoires): un instrument médiateur de la créativité?. Internet des objets, p 3
2. Bationo-Tillon A (2013) Ergonomie et domaine muséal. Activités, pp 10–12
3. Bationo-Tillon A, Decortis F (2015) Understanding museum activity to contribute to the design of tools for cultural mediation: new dimensions of activity? Le travail humain 79(1)
4. Bourmaud G (2013) De l'analyse des usages à la conception des artefacts: le développement des instruments. In: Falzon P (ed) Ergonomie constructive. PUF, Paris, pp 161–173
5. Decortis F (2013) L'activité narrative dans ses dimensions multi instrumentée et créative en situation pédagogique. Activités 10(1):3–30
6. Decortis F (2013) L'ergonomie orientée enfant. Concevoir pour le développement. PUF, Paris
7. Decortis F, Bationo-Tillon A (2014) "Once upon a time, there was a fairy who walked in paradise": the child finalised, mediated and creative narrative activity. Int J Arts Technol 7 (1):17–37
8. Decortis F, Bationo-Tillon A, Cuvelier L (2016) Penser et concevoir pour le développement du sujet tout au long de la vie: de l'enfant dans sa vie quotidienne à l'adulte en situation de travail. Activités, 13-2
9. Lahoual D, Decortis F, Bationo-Tillon A (2013) From analysis to modelling: a dynamic and situated approach to creativity. In: Workshop IDC 2013 – Interaction design and children, New York, USA
10. Malaguzzi L (1987) I cento linguaggi dei bambini: Narrativa del possibile. Proposte di bambini delle scuole dell'infanzia di Reggio Emilia, Italie
11. Rabardel P (1995) Les hommes et les technologies. Armand Colin, Paris
12. Suchman LA (1987) Plans and situated actions: the problem of human-machine communication. Cambridge University Press, Cambridge
13. Vygotski LS (1930/1983) Immaginazione e creatività nell'età infantile. Editori Riuniti, Paideia, Italy

# "Ergonomics on the Ground": A Case Study of Service Learning in Ergonomics Education

Jonathan Davy[1(✉)], Kim Weaver[2], Andrew Todd[1], and Sharli Paphitis[3]

[1] Department of Human Kinetics and Ergonomics, Rhodes University, Grahamstown, South Africa
j.davy@ru.ac.za
[2] Centre for Biological Control, Rhodes University, Grahamstown, South Africa
[3] Rhodes Community Engagement Department, Grahamstown, South Africa

**Abstract.** Service-learning combines academic learning with community-based service, is a pedagogical method that bridges the gap between traditional teaching approaches and prepares students to practice. In partnership with the Centre for Biological Control (CBC), we initiated a service-learning course aimed at providing a practical component for an Ergonomics Honours course. We were interested in how the course influenced student understanding of Human Factors and Ergonomics and Wilson's (2014) six notions, which were used as the theoretical basis for the course. Sixteen students worked alongside a group of people with disabilities (PWD) employed at the CBC to mass rear and harvest insects used to control invasive plant species. The students spent four hours a week for four weeks at the CBC mass rearing facility. Thereafter, students responded to questions aimed at exploring: how the SL course impacted their understanding of Wilson's notions; what they found challenging and enjoyable and what could be improved upon in the course. Student responses were analysed thematically. The experience enhanced the students' discipline-specific knowledge, particularly how they could apply the theory to the context, but it was evident that the students needed more time and more exposure to the system to derive a more detailed, nuanced understanding of the theory and how it related to practice. The SL course appears to have enhanced the students' understanding of HFE and its importance in contexts such as the CBC. The feedback from the students has also influenced the structure, aims and intended outcomes of this year's course to ensure that the course is mutually beneficial.

**Keywords:** Human Factors and Ergonomics · Education · Service-learning

## 1 Background

Human Factors and Ergonomics (HFE) advocates for the importance of adopting a systems approach to understanding complex sociotechnical systems [1]. In so doing, the different components of the whole (overall) system in focus are identified and the interactions between these components mapped and understood. In short, HFE purports taking a human centered approach to system optimization, with the dual outcomes of increasing worker safety and health and productivity [1]. An important consideration to


S. Bagnara et al. (Eds.): IEA 2018, AISC 826, pp. 693–702, 2019.
https://doi.org/10.1007/978-3-319-96065-4_73

achieve the abovementioned competencies is the design of both undergraduate and postgraduate HFE programs.

In South Africa, undergraduate and postgraduate degrees in HFE have historically only been offered by the Department of Human Kinetics and Ergonomics at Rhodes University. While this is changing, the focus of this paper is an aspect of the educational programs at the HKE department at Rhodes University.

### 1.1 The Case Study: Department of Human Kinetics and Ergonomics at Rhodes University

In the undergraduate Human Kinetics and Ergonomics courses, students walk away with a sound understanding of the various human systems and sub-systems. This is achieved by dividing the different human systems into three, broad domains; the biophysical, physiological and psychological and then integrating these domains into a holistic understanding of the human body and its functionality in context. This approach was founded on the work by Charteris et al. [2]. Although this gives the students an excellent background into these human systems, there has been an increased focus on understanding human in context, specifically through adopting a systems approach.

Aligned with this, recent literature has advocated for providing students\practitioners with an understanding of the broader systems within which humans operate; this is the case for both Human Factors and Ergonomics (HFE) [2] and more recently Sports Science [3, 4]. These and other authors suggest that humans perform activities of daily living, work and sport within a complex systems framework and consequently, these activities can only be understood within the context of that system [1, 5–7].

Wilson [1] contends that in order to be effective at understanding any human orientated system it is vital to adopt the following six notions as a guiding framework; systems focus, context, interactions, holism, emergence and embedding. In an educational setting, it is very difficult to give students an understanding of these notions without them actually experiencing what they mean. In essence, these notions can only be properly understood through experiential learning, in a context that Wilson would call "*in the wild*". It is therefore contended that in HFE (and Sport Science) there is a need to ensure that students are not only gaining an understanding of human systems, the theory through which we can understand them in context, but also the actual broader systems within which they may operate.

Learning about ergonomics is and can be achieved through many 'classroom' methods such as lectures, presentations, demonstrations, simulations and field trips [8, 9]. These approaches enable students to access HFE principles, theories and methods theoretically, but provide limited opportunities to gain insights in how to "do ergonomics on the ground". In order to prepare students to practice ergonomics, it is important that they experience real work systems alongside "real" workers [8, 10]. Furthermore, in the context of the Rhodes University and the broader [11] South African higher education system, this requires embedding in a system, an idea supported by service-learning [12, 13]. This stems from the need to instill a sense of democratic awareness, social responsibility and limit a tertiary institution's isolation from their immediate community [12].

**Service-Learning**

Service-learning, an approach that combines discipline-specific academic learning with community-based service has been identified as a pedagogical method that could bridge the gap between traditional teaching approaches and preparing students to practice ergonomics [9, 12]. Learning occurs through the service experiences of students where they apply their course content to communities that they work with and the problems they have [14]. Part of this experience involves constant reflection: Eyler [15] highlights the importance of reflection in academic service-learning courses, as it enhances the learning of those involved by reflecting on their experiences. With respect to ergonomics education, Page and Stanley [9] argued that the integration of an ergonomics service-learning component to any ergonomics course provides opportunities for students to confront real world problems that they are likely to encounter after graduating. Students in such programs develop insights into how their theoretical knowledge and skills with respect to ergonomics are applicable to the real world, while also highlighting the gaps in their skills set that would be necessary after university [9].

More broadly, participation in service-learning is associated with students being able to put theory into practice; understand course content better; improve their writing and thinking skills; improve their appreciation of diversity; develop relationships with people across cultural, racial and economic barriers; develop good communication skills; breaking down stereotypes of people; resolving conflict; time management and managing fear and stress [16, 17]. In spite of the benefits associated with service-learning implementation, constraining factors exist. These can include limited financial support; poor planning and hurried implementation; partnership difficulties; heavy workloads of students; political tensions in communities; logistical issues such as transport and timetabling; lack of participation in partnership activities; cultural and language differences amongst students; class sizes and lack of departmental commitment to service-learning [16]. In order to curb these constraints, some criteria on making service-learning a reality of an academic programme have been provided by Bender and Jordaan [18, p. 636] and include; service with the community must be meaningful and relevant to all stakeholders; enhanced academic learning must take place while serving the community; structured opportunities for reflection should be provided to students; purposeful civic learning must intentionally prepare students for active community participation in a diverse and democratic society; a scholarship of engagement should be promoted.

Given the increasing emphasis for students to have practical experience around how to practice ergonomics and the role that service-learning can play in facilitating this learning, we, in conjunction with the Science faculty's community engagement representative and the Centre for Biological Control's Waainek staff developed and implemented a service-learning component to an Ergonomics Honours module. We were interested in how the experience influenced student understanding of HFE and Wilson's six notions (their discipline specific knowledge), how it influenced their personal growth and development and how the course could be improved upon in the future. This paper focuses specifically on the discipline specific knowledge and how the feedback from the students on the course, aligned with our own assessment, should be integrated into improving the course going forward.

## 2 Overview of the Implementation of the Service-Learning Component

The service-learning course was introduced as part of an Ergonomics Honours course entitled Systems, Biomechanics and Chronobiology. The goals of the course were multifaceted. Firstly, we aimed to provide students with a deeper understanding of the biomechanical and chronobiological considerations associated with human performance and health within both a work and sporting context. Secondly, we wanted students to integrate their knowledge from these two supposedly separate domains. Thirdly, we intended to push students to understand the broader systems context that both of these sub-disciplines form part of. Lastly, we promoted taking systems approach and applying it within a real-world context so that they could gain a sound understanding of the six notions as outlined by Wilson [1] through experience. This paper deals with the last aim.

A service-learning component was introduced in the 8$^{th}$ week of the 12 week course. The purpose of this component was to provide students with an opportunity to embed themselves within in a real work environment and to apply their knowledge to and about the complex nature of the work systems within which humans operate. As part of the student's coursework was the seminal paper by Wilson [1] which explores the fundamental principles notions of Ergonomics, which highlights embedding as one of the notions. Wilson's paper essentially presents a conceptual framework for Ergonomics and assists specialists to understand how to go about doing Ergonomics.

### 2.1 The Work Context: An Overview

We partnered with the Centre for Biological Control (CBC), which is based in the Department of Zoology and Entomology at Rhodes University. The CBC has a facility, referred to as the Waainek Facility, that mass-rears insects known as biological control agents to assist in controlling alien invasive plants. This facility is supported by the Department of Environmental Affairs and supplies these insects to land owners and managers all over the country to control problematic weeds. The CBC trains and employs abled-bodied people and people with disabilities at this facility. Waainek is situated just off the University's campus and was therefore in close proximity for the students to visit and engage with the employees.

**Pre service-Learning Course Preparations**

Prior to sending the students to Waainek, ethical clearance was obtained from the Department of Human Kinetics and Ergonomics ethics committee. The students were briefed by HKE and community engagement staff on the concept of community engagement, service-learning and reflection. The students were encouraged to keep reflective journals of their experiences at the facility, which could be used to develop a better understanding of the Waainek work system following the end of the course. The students were required to spend a minimum of four hours per week at the facility, where they worked alongside the employees, most of whom are physically disabled. The service-learning module lasted for four weeks. The work mainly involved harvesting insects off invasive plants species while seated around small, plastic pools that were housed in plastic tunnel green houses.

Following the completion of the service-learning component, students completed a set of guiding questions (Table 1) to gain insights into their experience. These reflections were used to assess the service-learning component of the course and whether the learning goals were met. They were also used as an evaluation tool for the service-learning component of the course. These insights from the student's reflections can often highlight the growth and development from participating in service-learning activities, Eyler et al. [19, p. 62] classifies this along four dimensions: "sense of civic responsibility; respect for diversity; development of individual skills such as knowledge and academic concentration; and knowledge of self".

**Table 1.** Reflective questions for the service-learning course at the Waainek facility

| | |
|---|---|
| Discipline-specific knowledge | Describe the six fundamental principles notions of Ergonomics (as discussed by Wilson, 2014) and highlight what you have learned about each one through this service-learning experience? |
| Course evaluation | In what ways do you think the Biological Control Research Group in general and more specifically the Waainek Team benefited from this service-learning course, and do you have any suggestions based on your experience for improving the course? |

### 2.2 Data Analysis

Twelve from the sixteen students consented to have their reflection pieces included for analysis. Student written responses, specifically those related to how the experience influenced the students' understanding of Wilson's 6 notions, were analyzed thematically by three of the researchers using the following approach: familiarization with data, which included reading independently through the text, noting important and recurring content. Secondly, comparisons were made between the different observations of each researcher, with differences being discussed and resolved. Thereafter, the researchers finalized the important text or codes for each response, before finally establishing an appropriate theme for the codes observed. The abovementioned analyses of the student responses to the each of Wilson's notions were primarily deductive in nature in that we used the six notions as the theoretical framework.

## 3 Results and Discussion

### 3.1 Discipline Specific Knowledge

With respect to the discipline specific knowledge, we explored how the experience enhanced the students' understanding of Wilson's six notions. The first of notions is the importance of adopting a **systems focus**. In the student responses, there was a very strong focus on the importance placed on, firstly, understanding the *components of the work system* (the human, task/activity, technology/tools, the environment and the organizational factors. This was achieved through the application of Carayon's [20] work systems model, which served as an excellent model through which to map the Waainek work system. Secondly, the students were able to use their experience at Waainek to further their understanding of the *system of systems*, a concept introduced by Thatcher and Yeow [21] that was integrated into the course as a means of appreciating the complexity of a system.

A vital component of taking an effective ergonomics approach is understanding the importance of **context**, Wilson's second notion. Through their experiences, the students reported on several very important contextual factors; in particular the importance of understanding the human contextual background relating to the *socio-economic background*. Due to the large inequality prevalent in South Africa, the result of a variety of structural and systematic exclusionary practices during Apartheid, this observation is very important, as it affords the largely privileged group of students' insights into the daily challenges faced by those from a lower socioeconomic status in the context of South Africa. However, a deeper interrogation of the student responses revealed that while most were able to clearly articulate what context was (the work space in which they worked, the town of Grahamstown which is part of rural Eastern Cape, the poorest province in the country), very few were able to show how context would impact the outcomes of the system as a whole.

The third notion emphasizes the need to not only understand that there are various elements in the system, but to focus on the **interactions** that occur between these elements. A very important observation made by the students was that the exposure to a "real" working environment allowed them to see how the work place can only be optimized through an *understanding and explanation of interactions*, rather than just looking at the each component in isolation. However, as with the second notion, students were often able to articulate what interactions are important, but not on the quality of the interactions within the system. Students understood that there are various interactions within the system but were unable to identify any *sub-optimal interactions,* which in essence would highlight areas in the work system that perhaps needed to be addressed. In both cases, this may reflect the inexperience of the students in how to ask the right questions of those who work in the system. This would be important so as to fully grasp the context (why the mass rearing facility exists, how it came into being, why it employs people with disabilities) and the various interactions (between different elements of the system, and between the different levels of stakeholders in the system).

With respect to the notion of **holism**, the students articulated the importance of *understanding stakeholders* in the system, while also focusing on understanding the *worker holistically*. The last emergent theme was that the students could not *access the whole system.*

With respect to stakeholders, the students emphasized the importance of understanding who the relevant stakeholders are within the system, in essence who the system decision makers, actors, influencers and experts are, categories introduced by Dul et al. [22]. The stakeholders were limited to those with whom the students had interacted – the director of the CBC, the community engagement officer, the manager of the facility and the workers at the facility. However, it would been important to understand other stakeholders such as those in government who fund the project, the regulators of biological control in the country, the funders and the recipients of the insects – in the course, the Risk Management Framework created by Rasmussen [23] offered an excellent model for this, but the students did not apply it. A second broad aspect that emerged around understanding the *worker holistically*. This stressed the importance of appreciating that workers have a life outside of work, and that the home and work contexts, while geographically dispersed, still have an impact on the worker. This insight demonstrates an important perspective often neglected by managers – that the work-home life interaction can affect performance and functionality.

While the above examples highlight some understanding of the notion of holism, some students reported that they actually could not access the notion because they did not see the whole system. While this may have been the product of limited time at the mass-rearing facility and a poor appreciation of the scale of the CBC in the broader context, it could also reflect a lack of assertiveness on the part of the students to look for the "whole" as opposed to passively sit and work just with the workers. As lecturers, we recognized the importance of now emphasizing this in subsequent service-learning courses that involve this work system and encouraging the students to investigate further into the whole system.

Emergence, the fifth notion proposed by Wilson, speaks to how, in the real world, systems, tools, equipment, the environment are in ways that are not expected or predicted by the designers, what Wilson refers to as the **emergent properties**. Three overarching themes emerged under this notion. Firstly, the participants commented on the concepts of *flexibility and adaptability*, as emergent properties of the Waainek work system. This focused on how the workers, most of whom have some physical impairment, were able to overcome challenges they faced during their work through innovative, emergent uses of technologies. An example of this relates to how the workers used small paintbrushes to harvest the smaller insects that are damaged if harvested by hand.

There was also evidence that some students did *not have access to or notice any emergent properties*. This may have been the result of a lack of understanding of the notion, as others gave examples or, that there wasn't enough time to understand how the work is done and how this may change over time or when challenges arise. The course also occurred over the start of the winter period, which meant that the need for the insects and the speed at which they were reared and harvested, slowed down, which, in and of itself, is an emergent characteristic of the work system.

The final and arguably most important notion purported by Wilson, is the importance of **embedding** in a system to gain insights into the way things are done (as opposed to how they are imagined to be done). Three important themes emerged from the data analysis and centered on the *necessity of embedding* to understand the system, the link between *embedding and sustainability* and lastly, that to properly embed, *takes time*.

In terms of the first theme, which stressed the necessity of embedding for successful HFE practice, two subthemes were identified. The participants reported that spending time at the mass rearing facility, alongside the workers facilitated a better understanding of the components of the work system and their interactions, a common theme that has emerged throughout all six notions. The second, more prominent theme revolved around feeling and seeing how work is done, which the students argued was key for two important processes: the first related to understanding the difficulties experienced by the workers at the facility, which in turn, the students argued, would be important for the optimization of the work environment.

A second, important theme that arose revolved around the importance of taking time to understand the system, to form relationships in order to have a meaningful, *sustainable impact*. While only noted by one student, this comment highlights a vitally important aspect of any community engagement and service-learning work that has evidently not come to fruition in previous and the current course – that the relationship must be mutually beneficial, clearly defined and sustainable [12]. While the service-learning aspect of this course was exploratory, the lessons learned from it have indeed made us, as teachers, fundamentally more aware of how service-learning could and should be enhanced to ensure that there is not just a benefit to the students, but also to the workers and the overall system.

The last major theme centered on the fact that embedding *takes time* and that perhaps the time period that the students had at the mass rearing facility was inadequate, which one student referred to as "*partial embedding*". It became clear through the responses that this limited time provided a limited scope of the system, which could account for why some students did not understand or emphasize their understanding of the other notions, such as context, interaction, holism and emergence. A final and quite important barrier to effective embedding in the context that the students was the *language barrier*; most of the Honours students were white, English-speaking South Africans, while the workers were all black, isiXhosa speaking. Indeed, understanding each other linguistically and culturally is an important step towards understanding how the relationship can be established and be mutually beneficial. How this relates to the theme of time is that in order to learn the language would take time.

### 3.2 Course Evaluation – Student and Lecturer Perspectives

While it was evident that the service-learning course component enhanced the ability of the students to apply the theory practically to some extent, there were some suggestions from the students around how to make the programme more inclusive and beneficial to all involved. The students were all aware that they were there to learn from the workplace experience and from the Waainek team but they were not aware of what exactly they were going to '*give back*'. Said differently, it became evident that intended outcomes of the course were not as clear as originally thought. We, as the facilitators should have made it explicit at the beginning that their recommendations of how the CBC can improve the working situation of the employees would be taken into consideration by the management team. Additionally, we should have asked the students before they went into the workplace to think of innovative cost effective changes the CBC can make to improve the work environment that could have been co-constructed and the feasibility discussed. This has been included in the course at the time of writing this paper. This should satisfy the criterion for the service-learning course to be mutually beneficial to both the students and the community partner.

Another course critique from the students was that many them felt that they should have spent *more time* at Waainek with the team. Furthermore, there was consensus amongst the students that they would have liked to have been exposed to other parts of the CBC team to see the other parts of the system and how it all works together. We as facilitators intend on extending the course to explore the whole CBC and not just the mass-rearing facility.

## 4 Conclusions

This paper captures our initial attempts at integrating a service-learning module into an Ergonomics Honours module aimed largely at increasing the students' practical understanding of fundamental defining features of HFE, as explained by Wilson (2014). Analyses of student responses revealed some practical application and enhanced understanding of the notions following the experience at the Waainek mass rearing facility during the service-learning. However, the short time frame and limited exposure to and experience in the mass rearing facility and the broader CBC resulted in a reduced opportunity to understand these notions and apply them to the whole system. Furthermore, one of the outcomes of service-learning is that it should be mutually beneficial. This iteration of the course did not satisfy this criterion. Future iterations of the course will ensure that there are tangible, data driven outputs from the course that can be used to co-construct context-specific, low-cost no cost interventions to the mass rearing facility (and the broader CBC) should there be evidence of this necessity.

## References


1. Wilson JR (2014) Fundamentals of systems ergonomics/human factors. Appl Ergon 45(1):5–13
2. Charteris J, Cooper LA, Bruce JR (1976) Human kinetics: a conceptual model for studying human movement. J Human Mov Stud 2:233–238
3. Balagué N, Torrents C, Hristovski R, Kelso JAS (2017) Sport science integration: an evolutionary synthesis. Eur J Sports Sci 17(1):51–62
4. Bittencourt NFN, Meeuwisse WH, Mendonça LD, Nettel-Aguirre A, Ocarino JM, Fonseca ST (2016) Complex systems approach for sports injuries: moving from risk factor identification to injury pattern recognition—narrative review and new concept. Br J Sports Med 50(21):1309–1314
5. Salmon PM, Goode N, Lenné MG, Finch CF, Cassell E (2014) Injury causation in the great outdoors: a systems analysis of led outdoor activity injury incidents. Accid Anal Prev 63:111–120
6. Hulme A, Finch CF (2015) From monocausality to systems thinking: a complementary and alternative conceptual approach for better understanding the development and prevention of sports injury. Inj Epidemiol 2(1):31
7. Hulme A, Salmon PM, Nielsen RO, Read GJ, Finch CF (2017) Closing Pandora's Box: adapting a systems ergonomics methodology for better understanding the ecological complexity underpinning the development and prevention of running-related injury. Theor Issues Ergon Sci 18(4):338–359
8. Stone NJ (2008) Human factors and education: evolution and contributions. Hum Factors 50 (3):534–539
9. Page LT, Stanley LM (2014) Ergonomics service learning project: implementing an alternative educational method in an industrial engineering undergraduate ergonomics course. Human Factors Ergon Manuf Serv Ind 24(5):544–556

10. Jones DR (1999) Hands-on human factors and ergonomics education. In: Proceedings of the human factors and ergonomics society annual meeting, vol 43, no 7. SAGE Publications, Los Angeles, pp 535–538
11. Rhodes University (2016) What is service-learning? https://www.ru.ac.za/community engagement/servicelearning/whatisservice-learning/
12. Osman R, Castle J (2006) Theorising service learning in higher education in South Africa. Perspect Educ 24(3):63–70
13. Matthews S (2017) Privilege, poverty, and pedagogy: reflections on the introduction of a service-learning component into a postgraduate political studies course. Educ Res Soc Change 6(2):45–59
14. Eyler J, Giles Jr DE (1999) Where's the learning in service-learning? Jossey-Bass, San Francisco
15. Eyler J (2002) Reflection: linking service and learning—linking students and communities. J Soc Issues 58(3):517–534. https://doi.org/10.1111/1540-4560.00274
16. Lazarus J (2007) Embedding service learning in South African higher education: the catalytic role of the CHESP initiative. Educ Change 11(3):91–108
17. Lazarus J, Erasmus M, Hendricks D, Nduna J, Slamat J (2008) Embedding community engagement in South African higher education. Educ Citizsh Soc Justice 3(1):57–83
18. Bender G, Jordaan R (2007) Student perceptions and attitudes about community service-learning in the teacher training curriculum. S Afr J Educ 27(4):631–654
19. Eyler JS, Giles DE, Stenson CM, Gray CJ (1993–2000) At a glance: what we know about the effects of service-learning on college students, faculty, institutions and communities, 1993– 2000, 3rd edn. Vanderbilt University
20. Carayon P, Hundt AS, Karsh BT, Gurses AP, Alvarado CJ, Smith M, Brennan PF (2006) Work system design for patient safety: the SEIPS model. BMJ Qual Saf 15(suppl 1):i50–i58
21. Thatcher A, Yeow PH (2016) A sustainable system of systems approach: a new HFE paradigm. Ergonomics 59(2):167–178
22. Dul J, Bruder R, Buckle P, Carayon P, Falzon P, Marras WS, van der Doelen B (2012) A strategy for human factors/ergonomics: developing the discipline and profession. Ergonomics 55(4):377–395
23. Rasmussen J (1997) Risk management in a dynamic society: a modelling problem. Saf Sci 27(2–3):183–213

# A Quantitative Content Analysis of L1 and L2 English Writings Using a Text Mining Approach

Ikuyo Kaneko(✉)

Juntendo University, 2-5-1 Takasu, Urayasu, Chiba 270-0023, Japan
ikaneko@juntendo.ac.jp

**Abstract.** This study aimed to explore personal letters written in English as the first language (L1) and the second language (L2) using a text mining approach. The purpose of this study is to examine the usage of emotional expressions in English letters written by native English writers and Japanese learners of English. Among the emotions, affection and condolence are focused on. This study is also interested in confirming whether quantitative content analysis can obtain the same results as those shown by an analysis using a different method. Production experiments were carried out to compare the usage of English emotional expressions by Japanese university students with that of native English writers. All the subjects were instructed to write two kinds of personal letters; a love letter and a letter of condolence, under certain conditions. The contents of the letters were analyzed by means of the "text mining" method. Words frequently used in the letters were obtained, and a co-occurrence network of major words was created to investigate the relationships among words, which helped to understand how the writers expressed their emotions in English. Different linguistic features and patterns were observed in the letters written by the native English writers and those by the Japanese learners of English, respectively. It was also shown that quantitative content analysis supported the results obtained in another study using a different method. The results of this study showed quantitative content analysis is reliable method to better understand texts.

**Keywords:** Emotional expression · Text mining approach
Quantitative content analysis

## 1 Introduction

Human emotions are of great interest in such various academic fields as philosophy, sociology, psychology, ethnology, and linguistics. However, the domain of emotions has not been fully examined because of the complex character of emotions [1].

Expression of delicate and complex feelings and emotions is considered to be a weakness of English communicative competence common to Japanese learners of English. Senses, feelings, and emotions are basics for us, but how to express them is not systematically taught in the English classroom in Japan. Therefore, even those who have a fine command of English often have difficulty in expressing their senses, feelings, and emotions [2].


S. Bagnara et al. (Eds.): IEA 2018, AISC 826, pp. 703–713, 2019.
https://doi.org/10.1007/978-3-319-96065-4_74

The purpose of this study is to examine the usage of emotional expressions used in two kinds of English letters written by native English writers and Japanese learners of English. Among the emotions, this study focuses on "affection" and "condolence" which have quite different sociolinguistic background but can be conceived as important emotions for L2 learners. This study is also interested in confirming whether quantitative content analysis proposed by Higuchi [3, 4] can obtain the same results as those shown by an analysis using a different method.

## 2 Previous Research

In second language (L2) writing research, to the author's knowledge, there are few studies on the use of emotional expression. Kaneko and Sudo [5], Kaneko [6] observed the similarities and differences in English emotional expressions written by Japanese learners of English and native English writers. Focusing on Japanese learners of English, they also investigated the correlation between the proficiency level regarding emotional expressions and the English proficiency measured by the Test of English for International Communication (TOEIC) [7]. TOEIC is an English language test designed specifically to measure the everyday English skills of people working in an international environment. The concept of T-Unit (minimal terminable unit), introduced by Hunt [8], and the ESL Composition Profile developed by Jacobs et al. [9], were used to evaluate writing proficiency. The results of the both studies showed correlations between the TOEIC scores and the ESL Composition Profile scores. Different features were also observed in the patterns of the letters written by the native English speakers, and those by the Japanese college students respectively.

Kaneko [10], Kaneko and Mizusawa [11] examined the usage of emotional linguistic features used in English letters written by native English writers and Japanese learners of English using Appraisal theory (Martin and White [12]). Appraisal theory is a linguistic framework developed within Systemic Functional Linguistics (Halliday [13–15]). Systemic Functional Linguistics is a linguistic approach which regards language as a social semiotic system, not as a set of rules. It identifies three modes of meaning which operate simultaneously in all utterances—textual, ideational, and the interpersonal. In appraisal theory, interpersonal meaning is highlighted. Appraisal theory concerns the way speakers/writers convey attitudinal meaning during spoken/written communication. It also deals with how the writer's emotions were expressed in the communication. The results of these two studies showed particular linguistic patterns of love letters and letters of condolence. It also provided different features in the patterns of the letters written by the native English writers and those by the Japanese learners of English, respectively.

## 3 Methodology

A text mining approach was adopted as an analytical approach to this study, since the goal of this study was to examine the usage of emotional expressions. Text mining is a process of data mining of sequences of characters. It is construed as a method for analyzing text data which derives useful and valuable information from voluminous text data by analyzing the connections between phrases, grouping words and phrases, making rules to classify phrases and sentences, and extracting potential factors for the distribution of phrases and sentences. Text mining enables us to analyze text quantitatively [16].

Among several computational tools, KH Coder [17] was selected for this study. Content analysis has been adopted to examine patterns in communication in a replicable and systematic manner qualitative data, mainly in the field of social sciences. By systematically labeling the content of text, patterns of content can be analyzed quantitatively. In contrast, qualitative methods have been used to analyze meanings of content within text. KH Coder is a free software program which can analyze contents of text quantitatively. Concretely, it extracts words automatically from data and statistically analyzes them to obtain a whole picture and explore the features of the data while avoiding the prejudices of the researcher [3].

## 4 Methods

### 4.1 Subjects

The subjects in this experiment were 16 native English writers (seven males and nine females) and 12 Japanese college students (five males and seven females). All of the Japanese participants were undergraduate students at a co-educational university located in Chiba, Japan. They were selected by their English abilities, as measured through TOEIC. Their scores were higher than 500 (Highest score: 650, Lowest score: 540, Average score: 592, Highest score in Reading Section: 300, Lowest score in Reading Section: 235, Average score in Reading Section: 267). The group of native English writers consisted of five British, five Australian, three Canadians, and three Americans. The youngest participant was 24 years old while the oldest was 58 years old.

### 4.2 Procedures

Production experiments were carried out to investigate the usage of emotional linguistic features by the native English writers and the Japanese college students. All the participants were instructed to write two kinds of personal letters: a love letter and a letter of condolence. Tables 1 and 2 show the prompts given in English to the native English writers and in English and Japanese to the Japanese participants.

**Table 1.** Directions for Task 1 (Love Letter).

| |
|---|
| Suppose you have a boyfriend or a girlfriend with whom you have been going out for 3 years. Today is your third anniversary. Please write a love letter to her/him and express your current feelings. The format and length of the letter are not restricted. Feel free to express how you are feeling. You can take notes and use a dictionary if necessary. Don't refer to how-to books on letter writing, please. Thank you very much for your cooperation. |

**Table 2.** Directions for Task 2 (Letter of Condolence).

| |
|---|
| Suppose you have a teacher who had been taking good care of you until you entered the university. You have just heard that the teacher suddenly passed away. You cannot attend the funeral because of your schedule. Please write a letter of condolence to the teacher's family and express your current feelings. The format and length of the letter are not restricted. Feel free to express how you are feeling. You can take notes and use a dictionary if necessary. Don't refer to how-to books on letter writing, please. Thank you very much for your cooperation. |

When the participants wrote the letters, the format and length of the letter were not restricted, nor was time for writing. The participants were free to express how they were feeling. They could take notes and use a dictionary if necessary. However, they were not allowed to refer to how-to books on letter writing.

Following Higuchi's procedure, the sample letters were analyzed by means of the "text mining" method. Using a free software program called KH Coder, words frequently used in the letters were obtained. In linguistics, words can be divided into two classes: content words and function words. Content words are words which refer to a thing, quality, state or action and which have lexical meaning when the words are used alone. Nouns, verbs, adjectives, and adverbs are mainly classified as content words. Function words, on the other hand, are words which have little meaning on their own but which show grammatical relationships in and between sentences. Conjunctions, prepositions, pronouns, articles (i.e., "a," "an," and "the") are function words [18]. In the analysis of word frequency, only content words are focused on since this study is interested in the usage of emotional expressions, which consist of content words.

In addition to word frequency, co-occurrence network of major words was also created to investigate the relationships among words, which helped to understand how the writers expressed their emotions in English. The co-occurrence network has been traditionally adopted in content analysis to statistically express the data [19, 20]. In this procedure, the co-occurrence structure in data can be visualized by drawing a network connecting words which tend to be used together. Since it is a network, words are connected by lines. There is not much meaning in the positions of words. Even if two words are nearby, it does not mean that the degree of co-occurrence is strong, unless those words are connected by a line [3].

## 5 Results

### 5.1 Frequent Words and Their Contexts

Tables 3 and 4 show the content words which are used more than five times in love letters written by the native English writers (NE) and the Japanese learners of English (JP).

The word most frequently used in love letters is "love," as both Tables show. It is used 29 times in love letters written by native English writers, while it appears 30 times in those written by Japanese learners of English. When we look at the top five words in the Tables, it can be observed that both groups of writers frequently use almost the same

**Table 3.** The content words used more than five times in love letters (NE).

| Rank | Word | Freq. | Rank | Word | Freq. | Rank | Word | Freq. |
|---|---|---|---|---|---|---|---|---|
| 1 | Love | 29 | 10 | Spend | 9 | | Mean | 6 |
| 2 | Year | 19 | 11 | Look | 8 | | Thank | 6 |
| 3 | Know | 18 | | Special | 8 | | Wonderful | 6 |
| 4 | Life | 17 | | Think | 8 | 14 | Amazing | 5 |
| | Time | 17 | 12 | Happy | 7 | | Believe | 5 |
| 5 | Make | 16 | | Just | 7 | | Come | 5 |
| 6 | Meet | 15 | | Let | 7 | | Person | 5 |
| 7 | Want | 14 | | World | 7 | | Remember | 5 |
| 8 | Feel | 13 | 13 | Beautiful | 6 | | Smile | 5 |
| 9 | Anniversary | 10 | | Forward | 6 | | | |
| | Day | 10 | | Hope | 6 | | | |

**Table 4.** The content words used more than five times in love letters (JP).

| Rank | Word | Freq. | Rank | Word | Freq. | Rank | Word | Freq. |
|---|---|---|---|---|---|---|---|---|
| 1 | Love | 30 | 11 | Say | 10 | | Talk | 7 |
| 2 | time | 24 | | Talk | 10 | 15 | Lot | 6 |
| 3 | Want | 20 | | Feel/Feeling | 10 | | Pass | 6 |
| 4 | Tear | 19 | 12 | Dear | 9 | | Smile | 6 |
| 5 | Know | 18 | | Tell | 9 | 16 | Anytime | 5 |
| 6 | Make | 17 | 13 | Good | 8 | | Imagine | 5 |
| 7 | Think | 16 | 14 | Anniversary | 7 | | Long | 5 |
| 8 | Happy | 15 | | Friend | 7 | | Marry | 5 |
| | Meet | 15 | | Hope | 7 | | Promise | 5 |
| 9 | Thank | 13 | | Look | 7 | | Special | 5 |
| 10 | Day | 11 | | Point | 7 | | Spend | 5 |
| | Just | 11 | | Really | 7 | | Thing | 5 |
| | Today | 11 | | Remember | 7 | | Way | 5 |

words: love, year, know, and time. Although the rank of "make" is slightly different between both subject groups, it is used on a fairly frequent basis in love letters (NE: 16 times, JP: 17 times). The word "want" is ranked third in Table 4; the list of frequently appearing words of Japanese writers, while it is ranked seventh in Table 3; the list of frequently appearing words of native English writers. What is to be noted here is that a remarkable difference is observed in the usage of the word "life" between the native English writers and the Japanese learners of English. It is frequently used in love letters written by the native writers while it is hardly used in those written by the Japanese learners of English; it is used only twice, thus not on the list. Instead, such words as "happy," or "thank" are used more frequently by the Japanese writers than the native English writers. Regarding adjectives, the native English writers frequently use "special," "happy," "beautiful," "wonderful, and "amazing," while the Japanese learners of English use "happy," "dear," "good," "long," and "special" five times or more.

Tables 5 and 6 show the content words which are used more than six times in letters of condolence written by the native English writers (NE) and the Japanese learners of English (JP).

**Table 5.** The content words used more than six times in letters of condolence (NE).

| Rank | Word | Freq. | Rank | Word | Freq. | Rank | Word | Freq. |
|---|---|---|---|---|---|---|---|---|
| 1 | Student | 19 | 8 | High | 10 | 11 | Best | 6 |
| 2 | Teacher | 18 | | Dear | 10 | | Difficult | 6 |
| 3 | Know | 17 | 9 | Attend | 9 | | Funeral | 6 |
| 4 | Year | 15 | | Make | 9 | | Loss | 6 |
| 5 | Think/Thought | 15 | | University | 9 | | Pass | 6 |
| | School | 15 | | Able | 7 | | Person | 6 |
| 6 | Family | 13 | 10 | Class | 7 | | Sorry | 6 |
| | Time | 13 | | Great | 7 | | Want | 6 |
| 7 | Condolence | 12 | | Hear | 7 | | Wish | 6 |
| | Help | 12 | | Lesson | 7 | | | |
| | Live/Life | 12 | | Need | 7 | | | |

In these Tables, we can observe that the word "teacher" is frequently used in letters of condolence. One of the remarkable features shown in letters of condolence written by native English writers is that they use the word "condolence" more frequently than the Japanese learners of English. This word is ranked seventh in Table 5, while it is not found in Table 6. Instead of using the word "condolence," the Japanese writers frequently use the word "sorry" which is ranked third in Table 6. They also use the words "funeral" and "attend" more frequently than the native English writers. The word "funeral" is ranked third, and the word "attend" is ranked fourth. These words are ranked ninth and 11th respectively in Table 5, the list of frequently appearing words of native English writers. Regarding adjectives, the native English writers frequently use "dear," "able," "great," "best," "difficult, and "sorry," while the Japanese learners of English use "sorry," "great," "sad," "dear," and "good" more than six times.

**Table 6.** The content words used more than six times in letters of condolence (JP).

| Rank | Word | Freq. | Rank | Word | Freq. | Rank | Word | Freq. |
|---|---|---|---|---|---|---|---|---|
| 1 | Teacher | 28 | | Know | 9 | | Live | 7 |
| 2 | Think | 17 | | Life | 9 | 9 | Accept | 6 |
| 3 | Sorry | 14 | | Sad | 9 | | Advice | 6 |
| | Funeral | 14 | 7 | Class | 8 | | Letter | 6 |
| 4 | Attend | 11 | | Dear | 8 | | Love | 6 |
| | Meet | 11 | | News | 8 | | Make | 6 |

*(continued)*

**Table 6.** (*continued*)

| Rank | Word | Freq. | Rank | Word | Freq. | Rank | Word | Freq. |
|---|---|---|---|---|---|---|---|---|
| | Time | 11 | | Say | 8 | | Respect | 6 |
| | Want | 11 | | School | 8 | | Sport | 6 |
| 5 | Family | 10 | | Sincerely | 8 | | Study | 6 |
| | Great | 10 | | Student | 8 | | Teach | 6 |
| | Really | 10 | | Tell | 8 | | University | 6 |
| 6 | Hear | 9 | 8 | Good | 7 | | | |

## 5.2 Co-occurrence Network of Words

Co-occurrence networks of major words can tell us what words are used together frequently. Figures 1 and 2 show a co-occurrence network of words used more than five times in love letters written by the native English writers (NE) and the Japanese learners of English (JP). In love letters written by the native English writers, 39 words that appear more than five times are designated as the scope of analysis, while 55 words that appear more than five times are designated as the scope of the analysis in love letters written by the Japanese learners of English. KH Coder generates a network by connecting 60 pairs of the most strongly co-occurring words by lines. Words not connected by lines are removed from the result diagram.

In Fig. 1, we can observe that "happy" and "anniversary," "look" and "forward," and "special" and "person" are strongly connected, which indicates the writers use such

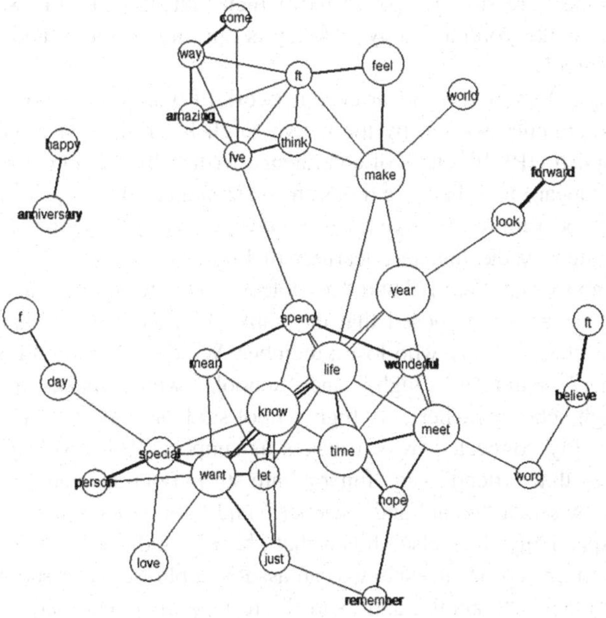


**Fig. 1.** Co-occurrence network of words used more than five times in love letters (NE).

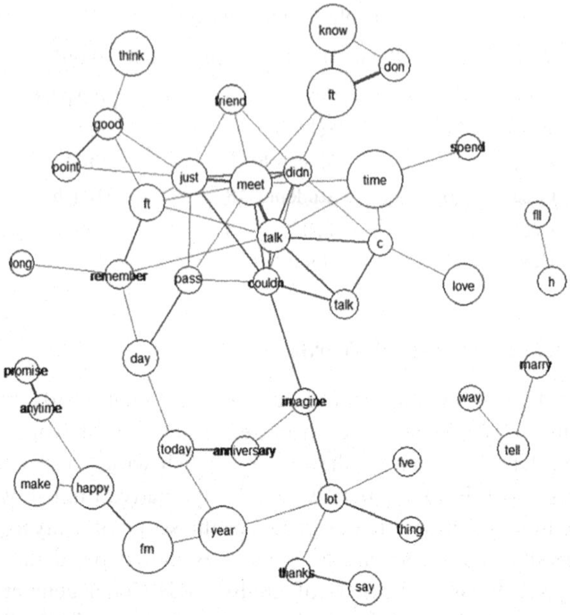


**Fig. 2.** Co-occurrence network of words used more than five times in love letters (JP).

phrases as "happy anniversary," "look forward," and "special person," respectively. In Fig. 2, it can be observed that "make" and "happy," "today" and "anniversary," and "say" and "thanks" are strongly connected, which indicates that the writers use such expressions as "make you/me happy," "today is our anniversary," and "I want to say thanks," respectively.

Figures 3 and 4 show a co-occurrence network of words used more than six times in letters of condolence written by the native English writers (NE) and the Japanese learners of English (JP). In letters of condolence written by the native English writers, 39 words that appear more than five times are designated as the scope of analysis, while 41 words that appear more than five times are designated as the scope of the analysis in love letters written by the Japanese learners of English.

Figure 3 shows that "family" and "condolence" are strongly connected. From this co-occurrence, it can be supposed that the native English writers express their condolence to a teacher's family who lost a member. We can observe strong connections among "teacher," "student," "high," and "school," which makes us assume such phrases as "high school teacher" or "high school student." Also, "sorry," "hear," and "loss," are strongly connected, which makes us suppose "sorry to hear your loss." It can be observed that "attend" and "funeral" are strongly connected as well.

Figure 4 shows that "great" and "teacher," and "good" and "advice" are strongly connected, respectively. It is also shown that "hear," "sad," and "news" are strongly connected. From this co-occurrence, we can assume a phrase "hear sad news." We can observe that "attend" "funeral," and "sorry" are strongly connected, which suggests that the Japanese writers are sorry that they are not able to attend a funeral.

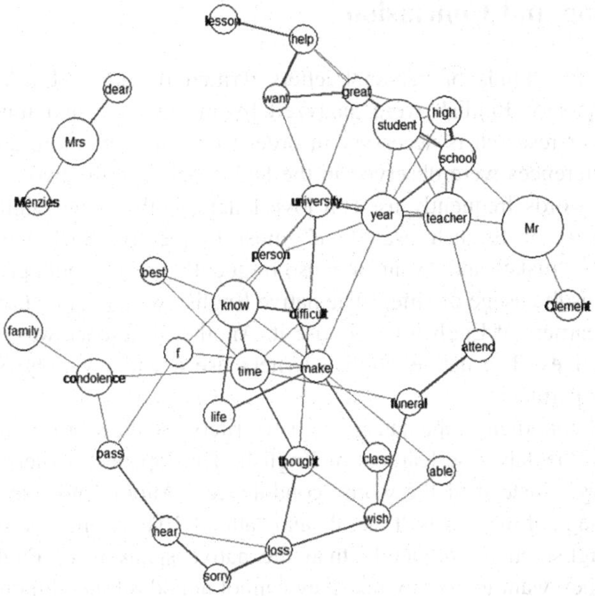


**Fig. 3.** Co-occurrence network of words used more than six times in letters of condolence (NE)

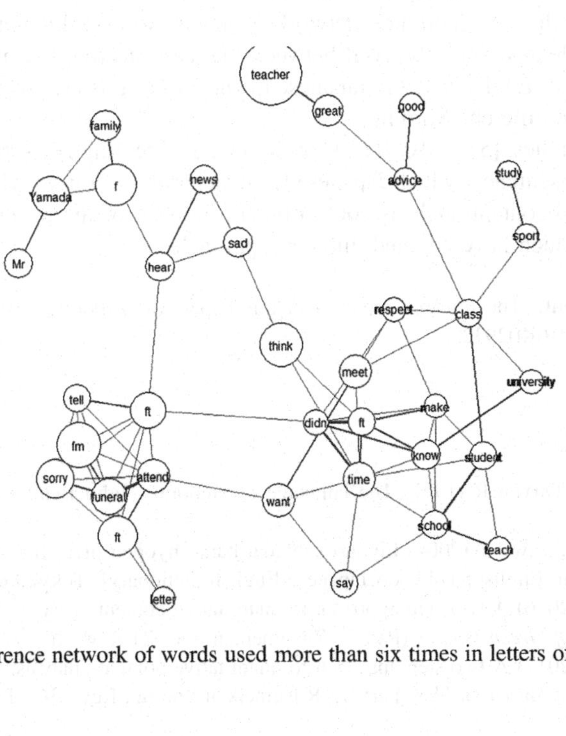


**Fig. 4.** Co-occurrence network of words used more than six times in letters of condolence (JP)

## 6 Discussion and Conclusion

In this study, two kinds of personal letters written by native English writers and Japanese learners of English were analyzed by means of a text mining approach. Although further research is necessary in order to accumulate more data, some similarities and differences were observed in the letters between the two subject groups.

Regarding words frequently used in love letters, both native English writers and Japanese learners of English use "love" most frequently. Such words as "year," "know," "time," "make" and "want" are also frequently used by both groups of writers. One exception is the usage of "life." The native English writers often use this word, but the Japanese learners of English hardly use it. Another difference was observed in the usage of adjectives. The native English writers use a wider variety of adjectives to modify his/her partner.

In letters of condolence, the native English writers use the word "condolence" more frequently than the Japanese learners of English. The Japanese writers frequently use the word "sorry," instead of the word "condolence." Another notable difference was found in the usage of the words "funeral" and "attend." These are used by the Japanese learners of English more frequently than the native speakers of English. It can be assumed that they want to convey that they cannot attend a funeral more strongly than the native English writers. Regarding the usage of adjectives, both groups of writers frequently use "sorry," "great," and "dear."

Turning to the co-occurrence network of major words, different links between words in both letters were observed between the subject groups. It indicates that the native speakers of English and the Japanese learners of English expressed their feelings and emotions in different wording.

Previous studies [5, 6, 10, 11] showed that native English speakers showed a tendency to use sentences which Japanese learners hardly ever used. This study showed that quantitative content analysis could obtain a portion of the results shown in previous studies which have adopted different approaches.

**Acknowledgement.** The research reported in this paper was supported by JSPS KAKENHI Grant Number JP16K02932.

## References

1. Niemeier S, Dirven R (1997) The language of emotions. John Benjamin Publishing Co., Amsterdam
2. Uechi Y, Tanizawa Y (2004) Eigono kankaku kanjo hyogen jiten [Sensory and emotional expressions in English: Tokyodo Japanese-English dictionary]. Tokyodo, Tokyo
3. Higuchi K (2016) A two-step approach to quantitative content analysis: KH Coder tutorial using *Anne of Green Gables* (Part I). Ritsumeikan Soc Sci Rev 52(3):77–91
4. Higuchi K (2017) A two-step approach to quantitative content analysis: KH Coder tutorial using *Anne of Green Gables* (Part II). Ritsumeikan Soc Sci Rev 53(1):137–147

5. Kaneko I, Sudo M (2008) Emotional expressions in L1 and L2 English writing. In: Bradford Watts K, Muller T, Swanson M (eds) JALT2007 conference proceedings, JALT, pp 949–956
6. Kaneko I (2008) Emotional expressions in English writing by Japanese college students. Theor Inf Cult 8:84–95
7. TOEIC Homepage. http://www.iibc-global.org/english/toeic.html. Accessed 26 May 2018
8. Hunt, K.: Grammatical structures written at three grade levels (Research report no. 3). National Council of Teachers of English, Urbana (1965)
9. Jacobs H, Zingraf S, Wormuth D, Hartfiel V, Hughey J (1981) Testing ESL composition: a practical approach. Newbury House, Rowley
10. Kaneko I (2012) Appraisal analysis of emotional expressions in English letters by first and second language writers. Theor Inf Cult 10:84–95
11. Kaneko, I., Mizusawa, Y.: An appraisal analysis of emotional expressions in first and second language writings by Japanese learners of English. In: Program and abstracts for the 2013 ASFLA national conference, p 50
12. Martin J, White P (2005) The language of evaluation: appraisal in English. Palgrave Macmillan, New York
13. Halliday M (1978) Language as social semiotic: the social interpretation of language and meaning. Edward Arnold, London
14. Halliday M (1985) An introduction to functional grammar, 1st edn. Edward Arnold, London
15. Halliday M (1994) An introduction to functional grammar, 2nd edn. Edward Arnold, London
16. Lee J (2017) Bunsho no keiryoteki bunseki [Quantitative analysis of text]. In: Lee J (ed) The science of measurement and evaluation of text. Hituji Shobo, Tokyo
17. KH Coder Homepage. http://khcoder.net/en/. Accessed 26 May 2018
18. Richards JC, Platt J, Platt H (1992) Longman dictionary of language teaching and applied linguistics, 2nd edn. Longman, Essex
19. Osgood CE (1959) The representational model and relevant research methods. In: Pool IDS (ed) Trends in content analysis. University of Illinois Press, Urbana
20. Danowski JA (1993) Network analysis of message content. In: Richards WD, Barnett GA (eds) Progress in communication sciences IV:197–221

# Effect of Redesigning School Furniture Based on Students' Anthropometry in North-West Nigeria

Ademola James Adeyemi[1](✉), Paul Ojile[2], Muyideen Abdulkadir[1], and Olusegun Isa Lasisi[1]

[1] Department of Mechanical Engineering, Waziri Umaru Federal Polytechnic, Birnin Kebbi, Nigeria
folashademola@gmail.com
[2] Department of Architecture, Waziri Umaru Federal Polytechnic, Birnin Kebbi, Nigeria

**Abstract.** Mismatch between classroom furniture and the students' anthropometry have been identified as a major cause of musculoskeletal disorders (MSDs) among schoolchildren. This paper is aimed at investigating the prevalence of MSDs and the effect of furniture intervention among students of tertiary institutions in Northwest Nigeria. The Cornell's MSD questionnaire was used to investigate the prevalence of MSDs in different body regions of 174 students aged between 17 and 25 years. Eleven students' anthropometric measures, relevant to furniture design, were gathered using appropriate equipment. Standardized regression equations for furniture design were used to compute the ergonomically compliant furniture dimensions for the participants. A new set of ergonomically designed furniture using the students' anthropometry was introduced and the Cornell's MSD questionnaire was used to investigate if the intervention made was significant or not. The first findings based on the questionnaire shows a high prevalent rate of MSDs especially at the back, neck and upper back among the students. Comparison of the obtained dimensions from the students' anthropometry with that of the existing furniture shows significant level of mismatch. The most significant variations were identified in the elbow height/Table height, and popliteal/seat height. New furniture compactible with the students' anthropometry was introduced in one of the institutions and a retest using the Cornell's MSD shows a reduction in the prevalence rate among the students. However, there is still need for ergonomic enlightenment on safe study posture to further reduce the occurrence of MSDs.

**Keywords:** Anthropometry · Musculoskeletal disorders · Furniture mismatch School ergonomics

## 1 Introduction

The provision of safe and comfortable furniture is vital towards achieving the learning objective of students in schools. It can also affect the health status of students in schools as poorly designed furniture may result in bad postural habits, which may lead to the


S. Bagnara et al. (Eds.): IEA 2018, AISC 826, pp. 714–722, 2019.
https://doi.org/10.1007/978-3-319-96065-4_75

development of musculoskeletal disorders (MSDs) in users [1]. For students that are still experiencing changes in physical structures [1], furniture provides support for the spine during the period of learning. However, a mismatch between individual anthropometric characteristics and the furniture dimensions will have negative effects on the body [2, 3]. Globally, this mismatch between school furniture dimensions and students' anthropometry has been identified as a major source of MSDs. These disorders cover different body regions such as the back, neck, shoulder, leg and many others. The effect of this mismatch is not limited to MSDs but may cause distraction, which could be at the detriment of students' ability to learn and can also lead to indiscipline among them (Woodcock *et al.* 2009). It may also lead to absenteeism and have psychosocial effect on both the students and their parents [4]. This serves as justification for the emphasis that both developing and developed nations are placing on standardization of school furniture [5, 6].

African countries, including Nigeria, have not shown enough commitment to this need as most African countries are yet to possess anthropometric database that is needed for such standardization [7, 8]. Information used for designing school facilities are obtained from foreign countries [8]. Researches on anthropometry have been carried out mainly by individuals and such researches remain uncoordinated and findings are independent of one another. In Nigeria, most of the researches on MSDs in schools are carried out in the southern part and the application of findings from such studies may not be applicable to the general population because of the differences in culture and standard of living of both regions [9]. Also, these studies failed to report the effect of any intervention or recommendation to reduce MSDS among the students. This study is therefore aimed at investigating the prevalence of MSDs and the effect of furniture intervention among students of tertiary institutions in the Northwest region of Nigeria.

## 2 Methodology

### 2.1 Sample Population

The study was part on an ongoing research to determine the anthropometric characteristics of students in tertiary institutions in Kebbi state, which is located in the Northwestern part of Nigeria. Four tertiary institutions in the state consented to participate in the research. Preliminary survey was conducted in the second and third quarter of 2017. A total of 250 students gave their consent to participate in the research. The students were aged between 17 and 25 years. However, only 174 of the students completed the entire process, which includes filling of questionnaire and measurement of anthropometric data. A follow up study carried out in the First quarter of 2018 involved only 80 of the students who participated in the previous exercise. The population reduced because some of the students in the previous exercise had graduated and the follow up survey was limited to classes that have benefited from the provision of new improved furniture.

## 2.2 Instruments

The study involved measuring the prevalence of MSDs and the anthropometry of the students. The self-administered Cornell's MSD questionnaire with a body map was used to measure MSDs in twelve body regions. The questionnaire measures the frequency of pain occurrence, the level of discomfort and level of interference of pain with students' academic activities. Eleven anthropometric body dimensions that relate to furniture measurements were measured using standardized standiometer, anthropometer and bio-impedance scale. The anthropometric dimensions were as defined in [10, 11].

Three sets of school furniture were identified and measured. Two of the selected furniture had long seats while the third set has separated chairs for three users. All the furniture had metallic frames with synthetic leather covering on the chairs and the writing surfaces. The measured dimensions were compared with the expected dimensions, which were obtained using the standardized regression equations [12–16].

Statistical analysis was carried out using SPSS 18 at a significant level of 0.05.

# 3 Results and Discussion

Table 1 shows the dimension of the selected furniture in the school. Furniture sets 1 and 2 were not demarcated for individual users. Hence, the seat width is approximated for 3 users.

**Table 1.** Dimensions of the existing furniture selected for the study.

| Dimension | Chair 1 (cm) | Chair 2 (cm) | Chair 3 (collapsible) |
|---|---|---|---|
| Seat height | 43 | 46 | 40 |
| Seat depth | 24 | 33 | 43 |
| Seat width | 152/3 | 152/3 | 45 |
| Backrest height | 67–79 | 63–75 | 40–90 |
| | Table 1 | Table 2 | Table 3 |
| Height | 77 | 78 | 76 |
| Width | 27 | 27 | 30 |
| Knee clearance | 75 | 73 | 30 |

Table 2 shows the range of acceptable dimensions based on the students' anthropometry and the standardized regression equations.

Figure 1 shows that the chairs of the three sets are not suitable for the entire population but could only accommodate certain percentiles in the students' population. The chairs of furniture 1 is only suitable for the 5$^{th}$ and 50$^{th}$ percentiles but not for the 95$^{th}$ percentile. The chairs of furniture 2 will only accommodate the 50$^{th}$ and 95$^{th}$ percentiles of the population while the chairs of furniture 3 were only suitable for the 5$^{th}$ percentile. Figure 2 however reveals that all the tables exceeded the safe level for the students and will therefore cause discomfort.

**Table 2.** Maximum and minimum recommended furniture dimensions in centimeter.

| Dimension | Equations | $5^{th}$ centile | $50^{th}$ Centile | $95^{th}$ Centile |
|---|---|---|---|---|
| Seat height (SH) | (P + 2)cos $30^0 \leq$ SH $\leq$ (P + 2)cos $5^0$ | 37.46–43.1 | 41.79–48.07 | 45.9–52.8 |
| Seat depth SD | 0.8 PB $\leq$ SD $\leq$ 0.99 PB | 33.32–41.23 | 37.8–46.78 | 41.08–50.84 |
| Seat width SW | 1HB $\leq$ SW $\leq$ 1.3HB | 25–32.5 | 29.3–38.09 | 33.83–43.98 |
| Table height (TH) | EH $\leq$ TH $\leq$ EH + 5 | 53.38–58.38 | 59.25–64.25 | 67.33–72.33 |

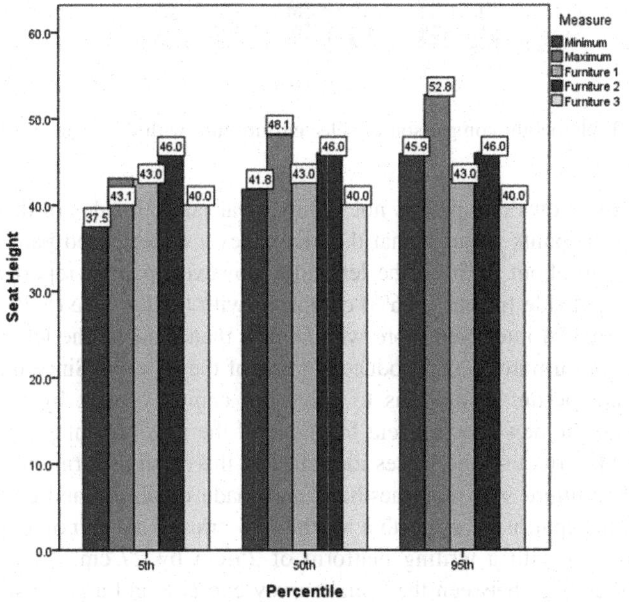


**Fig. 1.** Chair height comparison of selected furniture with the required dimensions

The response from the questionnaire demonstrated high reliability with Cronbach alpha of 0.964, intraclass correlation coefficient (ICC) of 0.958 and a confidence interval of 0.948-0.966.

Table 3 summarises the students' responses to the MSDs questionnaire. The percentage of students that have experience MSDs in any body region ranges from 14 to 67.3% of the participants. Lower back pain has the highest prevalence of 67.3%, followed by neck pain (59.4%) and upper back pain (43.7%). Moreover, pain is more severe at the lower back (19.4%), upper back (12.7%), knee (11.5%) and hip (10.3%) regions of the body. The regions where MSDs significantly interfered with students' studies are lower back (15.8%), hips (14.5%), neck (12.1%) and knee (10.3%).

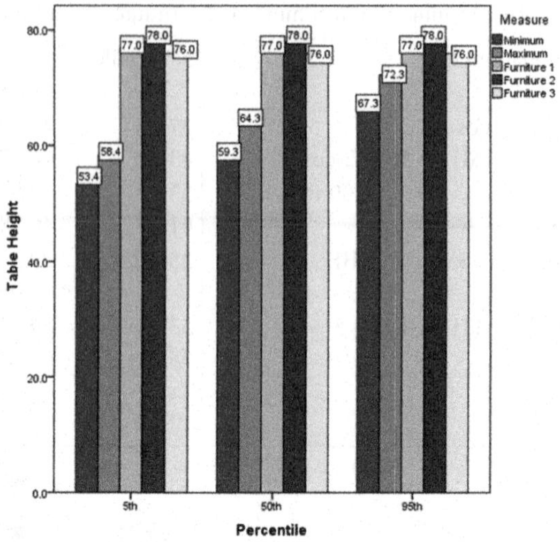


**Fig. 2.** Table height comparison of selected furniture with the required dimension

Table 3 also shows that pain is not symmetrical on both sides of the body. About 49% of the participants reported that they have never experienced pain on their right side compared to about 45% on the left side. However, pain is reported to be more severe in the right side (mean = 6.6%) compared with the left side (5.2%) and pain on the right side (6.3%) interfered more with studies than pain on the left side (4.9%).

A new set of furniture was introduced in one of the schools. Since the government policy does not permit institutions to carry out capital project by themselves, the institution made the new procurement from one of the government approved suppliers with the aim of correcting the lapses identified in the existing furniture in the school. The supplied furniture was more aesthetic and made of plastic and cotton materials. The height of the chair is 44 cm with a width of 45 cm and a depth of 47 cm. The table height was 60 cm with a writing platform of 25 cm by 27 cm. Table 4 shows the percentage differences between the initial survey exercise and a retest survey exercise carried out six months after the furniture intervention in the following year. Positive value indicates increase while negative value indicates reduction. There were reported increases in the percentage of students that never felt pain in all body regions except at the upper arm and right side of the forearm. Although there was increase in the number of students that stopped experiencing pain at the neck (4.5%), the number of cases occurring multiple times during the day also increased significantly (13.9%). Increases in the number of students who experienced pain multiple times in a day were also observed on the thigh and the fingers. However, these increases in frequency of occurrence neither increased severity nor interference with the students' studies.

The findings from Table 4 are indications that redesigning the tables and chairs in the classrooms alone may not be sufficient to eliminate or reduce the occurrence of MSDs among students. Rather, it has further highlighted the complexity and heterogeneous nature of pain prevention [17, 18]. It will also be difficult to provide furniture

**Table 3.** Summary of students responses in the Cornell's MSD questionnaire

| Body Region | Side | Frequency of pain occurrence | | | | | How uncomfortable | | | Interference with study | | |
|---|---|---|---|---|---|---|---|---|---|---|---|---|
| | | Never | 1–2 | 3–4 | Once daily | Multiple times | Slight | Moderate | Severe | None | Slight | Substantial |
| Neck | | 35.8 | 23.0 | 6.7 | 8.5 | 21.2 | 40.6 | 15.2 | 8.5 | 29.7 | 26.1 | 12.1 |
| Shoulder | Right | 49.1 | 17 | 12.1 | 3 | 4.6 | 26.1 | 15.2 | 4.8 | 21.2 | 24.8 | 4.2 |
| | Left | 38.2 | 9.7 | 3.6 | 7.3 | 4.8 | 24.2 | 15.8 | 3.0 | 20 | 23.6 | 3.6 |
| Upper back | | 37.4 | 17.6 | 9.7 | 7.9 | 8.5 | 23.6 | 14.5 | 12.7 | 21.8 | 24.2 | 9.1 |
| Upper arm | Right | 43.6 | 15.8 | 15.8 | 7.3 | 3.0 | 29.7 | 15.2 | 4.2 | 29.7 | 17.6 | 4.8 |
| | Left | 43.0 | 7.9 | 1.8 | 4.8 | 3.0 | 25.5 | 8.5 | 3.6 | 22.4 | 18.8 | 4.2 |
| Lower back | | 23.6 | 24.2 | 6.7 | 21.2 | 15.2 | 17.6 | 18.8 | 19.4 | 13.3 | 30.9 | 15.8 |
| Forearm | Right | 53.3 | 10.9 | 5.5 | 6.1 | 9.1 | 26.7 | 10.3 | 4.2 | 27.9 | 17.6 | 3.6 |
| | Left | 49.1 | 7.3 | 1.8 | 3.6 | 2.4 | 24.8 | 7.3 | 3.6 | 27.3 | 13.9 | 2.4 |
| Wrist | Right | 53.3 | 12.7 | 6.7 | 9.1 | 3.6 | 30.3 | 9.1 | 6.7 | 26.1 | 18.8 | 8.5 |
| | Left | 50.9 | 8.5 | 5.5 | 8.5 | 3.0 | 29.1 | 7.3 | 4.2 | 21.8 | 14.5 | 6.1 |
| Fingers | Right | 58.2 | 8.5 | 6.1 | 8.5 | 4.8 | 27.3 | 8.5 | 8.5 | 27.3 | 20 | 4.8 |
| | Left | 57.6 | 7.9 | 3.0 | 4.8 | 3.0 | 23.6 | 7.3 | 4.8 | 24.2 | 13.9 | 3.0 |
| Hip | | 44.8 | 18.2 | 6.7 | 9.7 | 12.1 | 28.5 | 18.2 | 10.3 | 23.0 | 23.0 | 14.5 |
| Thigh | Right | 45.5 | 14.5 | 9.1 | 5.5 | 15.8 | 23.0 | 20.6 | 6.7 | 26.7 | 21.8 | 7.3 |
| | Left | 37.0 | 12.1 | 7.9 | 3.6 | 7.9 | 20.0 | 15.2 | 8.5 | 25.5 | 17.6 | 6.7 |
| Knee | Right | 40.6 | 14.5 | 7.3 | 15.2 | 8.5 | 27.3 | 16.4 | 11.5 | 26.1 | 20.6 | 10.3 |
| | Left | 40.6 | 12.1 | 5.5 | 9.1 | 4.8 | 25.5 | 16.4 | 6.7 | 26.1 | 18.8 | 6.1 |
| Lower leg | Right | 49.1 | 9.7 | 3.6 | 10.9 | 7.3 | 29.1 | 12.7 | 6.1 | 26.7 | 19.4 | 6.7 |
| | Left | 46.7 | 10.9 | 1.8 | 7.9 | 15.8 | 28.5 | 10.9 | 7.3 | 25.5 | 18.2 | 6.7 |

**Table 4.** Difference in reported MSD before and after the intervention

| Body Region | Side | Frequency of pain occurrence | | | | | How uncomfortable | | | Interference with study | | |
|---|---|---|---|---|---|---|---|---|---|---|---|---|
| | | Never | 1–2 | 3–4 | Once daily | Multiple times | Slight | Moderate | Severe | None | Slight | Substantial |
| Neck | | +4.5 | −15.2 | −2.8 | +1.9 | +13.9 | +6.2 | −0.9 | −2.0 | +8.0 | 0 | −6.9 |
| Shoulder | Right | +13.2 | −4.0 | +6.1 | +0.9 | −2.0 | +6.4 | +5.6 | +1.7 | +8.7 | +9.0 | −2.9 |
| | Left | +0.8 | −0.6 | +1.6 | +4.4 | −2.2 | +9.6 | +5.0 | −0.9 | +8.6 | +8.9 | +0.3 |
| Upper back | | +6.8 | −0.1 | +1.4 | −0.1 | +1.9 | +10.2 | +0.2 | −1.0 | +5.5 | +4.4 | −2.6 |
| Upper arm | Right | −3.3 | +5.0 | +5.5 | +5.7 | +0.9 | +9.3 | +9.5 | +2.3 | +4.1 | +7.1 | +1.7 |
| | Left | −0.1 | +3.8 | −0.5 | +0.7 | +2.2 | +13.5 | +7.1 | +1.6 | +3.6 | +11.1 | +2.3 |
| Lower back | | +2.4 | −4.7 | +2.4 | +8.7 | −3.5 | +7.1 | +0.7 | +0.4 | +6.2 | +2.9 | +1.1 |
| Forearm | Right | −0.1 | +2.1 | +2.3 | +3.0 | +7.8 | +17.5 | +1.0 | +1.0 | +2.0 | +14.9 | −1.0 |
| | Left | +0.7 | +3.1 | +0.8 | +4.2 | +2.8 | +16.8 | +1.8 | +2.9 | +6.5 | +11.1 | +1.5 |
| Wrist | Right | +12.9 | −3.6 | −2.8 | +3.9 | −2.3 | +7.4 | +3.9 | −0.2 | +1.2 | +4.6 | +0,6 |
| | Left | +14.0 | +0.6 | −0.3 | +1.9 | −0.4 | +15.1 | +1.8 | −0.3 | +3.2 | +6.3 | +1.7 |
| Fingers | Right | +8 | +0.6 | −0.9 | +0.6 | −0.9 | +11.7 | +1.9 | +0.6 | +3.9 | +12.5 | −2.2 |
| | Left | +16.4 | −0.1 | −0.4 | +3.0 | −0.4 | +14.1 | +3.1 | +3.0 | +9.6 | +5.6 | +0.9 |
| Hip | | +8.4 | +3.9 | −2.8 | +3.3 | −6.9 | +5.9 | +3.9 | +0.1 | +18.2 | +1.7 | −2.8 |
| Thigh | Right | 0 | +4.0 | −1.3 | +3.6 | +3.7 | +5.6 | +5.4 | +1.1 | +5.8 | +1.5 | −0.8 |
| | Left | +0.6 | +4.8 | +1.5 | +1.6 | +1.2 | +0.8 | +4.3 | +5.8 | +7.0 | +3.2 | +1.1 |
| Knee | Right | +2.3 | +2.4 | +0.5 | +8.2 | −3.3 | +14.3 | +0.5 | +4.1 | +9.0 | +6.7 | −2.5 |
| | Left | +6.2 | +2.2 | −0.3 | +3.9 | −2.2 | +12.2 | −0.8 | +1.1 | +9.0 | +4.6 | +0.4 |
| Lower leg | Right | +5.4 | −0.6 | −1.0 | +3.4 | −3.4 | +17.7 | −2.3 | −0.9 | +8.8 | +4.0 | −0.2 |
| | Left | +6.5 | −1.8 | +9.9 | −7.9 | −13.2 | +17.0 | −0.5 | +0.5 | +7.0 | +5.2 | +1.1 |

that will fit the entire population because of anthropometric variation in people. Hence, holistic redesigning of the entire classrooms by considering the positing of boards, illumination and acoustic characteristics of the classroom should be critical inputs. These factors have previously been reported to be poorly factored into the design of school buildings in Nigeria [19]. There is also need to carry out enlightenment programs on the safe use of furniture among students [20].

## 4 Conclusion

The solution to the prevalence of MSDs among students lies in the provision of holistic interventions involving all stake holders. A major setback for provision of safe furniture for students use remains the non-availability of anthropometric data and standards in the country. This has continued to affect the quality of furniture available in schools since institutions cannot provide furniture directly but must rely on government approved suppliers who do not have data to work with. In addition to designing furniture based on students' anthropometry, there is also need to carry out proper sensitization on safe study postures and ergonomic practices that promote safety and comfort among the students.

## References


1. Carneiro V, Gomes A, Rangel B (2017) Proposal for a universal measurement system for school chairs and desks for children from 6 to 10 years old. Appl Ergon 58:372–385
2. Castellucci HI, Arezes PM, Molenbroek JFM, Viviani C (2014) The effect of secular trends in the classroom furniture mismatch: support for continuous update of school furniture standards. Ergonomics 58:1–11
3. Castellucci HI, Arezes PM, Molenbroek JFM (2014) Applying different equations to evaluate the level of mismatch between students and school furniture. Appl Ergon 45:1123–1132
4. James AA, Rohani JM, Rani MRA (2012) Development of a holistic backpack-back pain model for school children. In: 2012 Southeast Asian Network of Ergonomics Societies Conference, pp 1–5. IEEE
5. Molenbroek JFM, Kroon-Ramaekers YMT, Snijders CJ (2003) Revision of the design of a standard for the dimensions of school furniture. Ergonomics 46:681–694
6. BS EN (2006) Furniture-chairs and tables for educational institutions. Br Stand EN 1729-12006 2006:1–38
7. John K, Adeyemi AJ (2015) Anthropometric data for Tanzania's primary school furniture design. ARPN J Eng Appl Sci 10:890–895
8. Adeyemi AJ, Rohani JM, Akanbi G, Rani MRA (2014) Anthropometric data reduction using confirmatory factor analysis. Work J Prev Assess Rehabil 47:173–181
9. Cheng S, Foster R, Hester N, Huang CY (2003) A qualitative inquiry of Taiwanese children's pain experiences. J Nurs Res 11:241–250
10. Castellucci HI, Arezes PM, Viviani CA (2010) Mismatch between classroom furniture and anthropometric measures in Chilean schools. Appl Ergon 41:563–568

11. Dianat I, Javadivala Z, Asghari-Jafarabadi M, Asl Hashemi A, Haslegrave CM (2013) The use of schoolbags and musculoskeletal symptoms among primary school children: are the recommended weight limits adequate? Ergonomics 56:79–89
12. Agha SR (2010) School furniture match to students' anthropometry in the Gaza Strip. Ergonomics 53:344–354
13. García-Acosta G, Lange-Morales K (2007) Definition of sizes for the design of school furniture for Bogotá schools based on anthropometric criteria. Ergonomics 50:1626–1642
14. Gouvali MK, Boudolos K (2006) Match between school furniture dimensions and children's anthropometry. Appl Ergon 37:765–773
15. Musa AI, Ismaila SO (2011) Student anthropometric data and furniture mismatches in selected institutions in Abeokuta, Ogun State, Nigeria. Theor Issues Ergon Sci 15:1–9
16. Yanto ESH, Siringoringo H, Deros BM (2008) Mismatch between school furniture dimensions and student' s anthropometry (A Cross-Sectional Study in an Elementary). In: 9th Asia Pacific Industrial Engineering & Management Systems Conference, pp 656–665
17. Adeyemi AJ, Rohani JM, Rebi MRA (2013) A multifactorial model based on self-reported back pain among nigerian schoolchildren and the associated risk factors. World Appl Sci J 21:812–818
18. Adeyemi AJ, Rohani JM, Abdul Rani MR (2017) Backpack-back pain complexity and the need for multifactorial safe weight recommendation. Appl Ergon 58:573–582
19. Adeyemi AJ, Yusuf SA, Ezekiel OB (2017) Environmental considerations toward the provision of conducive learning environments in Nigerian schools. Arid Zone J Eng Technol Environ 13:449–457
20. Wilson I, Desai DA (2016) Anthropometric measurements for ergonomic design of students' furniture in India. Eng Sci Technol Int J 20:232–239

# Digital Discrimination: An Ergonomic Approach to Emotional Education for the Prevention of Cyberbullying

Margherita Bracci, Alison Margaret Duguid, Enrica Marchigiani, Paola Palmitesta, and Oronzo Parlangeli(✉)

Department of Social, Political and Social Sciences, University of Siena, Siena, Italy
{margherita.bracci,alison.duguid,enrica.marchigiani, paola.palmitesta,oronzo.parlangeli}@unisi.it

**Abstract.** The aim of this study is to shed some light on the complex relationship between cognitive, socio-affective and contextual (i.e. the technology and the way in which it is used) factors, which intervene in the context of prosocial and antisocial behavior, both in the real and virtual world. Results coming from a survey with a sample of 264 subjects show that those who perform victimization on the Internet are more likely to be more dependent on the use of the Internet and stay on social networks for more time in a day. In addition, while the aggressive behavior and the disengagement seem to be more correlated with a detachment towards the victim and therefore to a lower level of affective empathy, the helping behavior seems to be characterized by a greater cognitive capacity and a greater understanding of the other.

**Keywords:** Cyberbullying · Moral disengagement · Empathy
Prevention · Digital wellbeing · Communication technologies · Social network
Aggressive behavior

## 1 Introduction

Virtual aggressive behaviors and real world aggressive behaviors, such as bullying and cyberbullying, may have causes in common. There are however differences between real and virtual acts of aggression which can be ascribed to the fact that the latter, given the distance in terms of physical and sometimes temporal separation between the aggressor and the victim, make the harmful behavior more subtle, insidious and easier to carry out. The ease of Internet use, the anonymity it provides and the absence of real and direct confrontation with the victim can trigger a lowering of empathy and an absence of remorse around the aggressive behavior [1–3].

### 1.1 Empathy on the Net

Various studies [4–7] claim that empathy is fundamental in the moral development of the individual as it can inhibit the manifestation of aggressive behavior. However according to recent studies, in order to understand moral behavior it is necessary to


S. Bagnara et al. (Eds.): IEA 2018, AISC 826, pp. 723–731, 2019.
https://doi.org/10.1007/978-3-319-96065-4_76

associate the study of empathy and its components, both affective [8] and cognitive [9–11], to the reasoning and the motivation which gets individuals to modulate their behavior in a variety of contexts [12–14]. Human beings are sensitive to social signals such as facial expression, gaze direction and posture but also to the characteristics of the environment in which they find themselves. Not everyone however is capable of understanding others' feelings and intentions in the same way; these are abilities related to the concept of empathy, which encourages one to pay attention to, and to elaborate cognitively both the affective and the mental states of other individuals [15]. In the case of virtual social environments, it would seem that there is a certain coherence between real world behavior and virtual world behavior. For example, it appears that the more sociable have a wider social life in virtual environments [16, 17]. However, it also appears that the cognitive component of empathic abilities can be inhibited by a number of diverse factors while on the net. For example recent studies in the field of neuroscience have shown that the so-called moral or social emotions, for example admiration and compassion, involve a slower form information processing, due to a more demanding request for reflection and cognitive appraisal. Contrary to this need, digital communication is very fast and often it does not give the individual time to process information carefully [18].

## 1.2 Moral Emotions and Disengagement

Virtual environments facilitate the possibility of hiding one's identity. This raises the level of moral disengagement and provides easy recourse to thought mechanisms which are tilted towards justifying bad behavior and silencing the conscience [1]. This can lead to ever easier repetitions of the immoral behavior which over time can become increasingly serious [19–21]. The mechanisms of moral disengagement are nothing more than ways of preserving a positive self image, even when faced with offensive behaviors towards others [22, 23] allowing us to mitigate moral emotions such as feeling of guilt. It is, indeed, in many cases thanks to our anticipation of such moral emotions – negative in the case of regret, shame, guilt, or positive when we speak of pride and satisfaction – that we avoid committing actions which are considered ethically deplorable.

Reflexive cognitive processes which imply an evaluation of the self, encourage us to evaluate our behavior according to an internal standard (our moral identity) or to an external one (the perception that others have of us). In other terms they provide support for moral behavior or justification for violations [5, 24].

The consequences of an excessive, distorted or improper use of the net, which overrides the reflective capacity, constitute the beginning of a series of social problems for the present generation of young adolescents.

The aims of this work are to see if and how 'helping intentions', that is to say pro-social behavior, and individual factors linked to empathic capacity and moral disengagement are correlated, and to identify the way in which they intervene in the real world and in the virtual (e.g. Facebook, Whatsapp, Instagram). From an awareness of these indications it should be possible to promote a mindful use of the internet in order to augment the capacity in the young to become 'virtual social beings' with the same conception as in the real world, of self-respect and respect for others in an understanding of diversity.

## 2 The Study

The aim of this study is to shed some light on the complex relationship between cognitive, socio-affective and contextual (i.e. technology and the way in which it is used) factors which are likely to have a role in the context of pro-social and antisocial behavior, in both the real and the virtual world.

### 2.1 Method

**Sample and Data Collection.** The study was conducted through a self-reported questionnaire. Participants (N = 264, 114 male, 43.2%, and 150 female, 56.8%) were adolescents attending different high schools in the Tuscany region, from first to fifth year classes. They were invited to take part in the study, on a voluntary basis, by e-mails sent to the headteachers of their schools. In the e-mail, information about the aim of the study was provided, along with a link to a Google Forms webpage where participants could complete the survey online. Anonymity of respondents was assured and no recompense was offered for participation.

40.5% of the participant students attend the Scientific High, 25.8% Technical Institutes, 18.9% the Classical High, 10.6% the Human Sciences High and 8.3% the Linguistic High School.

### 2.2 Materials

The questionnaire was structured in 73 items and in 5 sections. The first section concerned personal data such as age, gender, type of school attended. The second section aimed at investigating the use of social networks, that is which type, for what reason etc. and to obtain information on specific relational issues such as "I feel on edge if I can't use internet" or "Sometimes I use one of my profiles to post content which cannot be seen by other users". The third section was aimed at measuring empathy through the Basic Empathy Scale (BES), a two factor model scale of 20 items designed to measure both cognitive empathy and affective empathy [25–27]. Answers were collected on a 5 point Likert Scale where 1 was "strongly disagree" and 5 was "strongly agree". For this study the Italian version of Albiero [27] was adopted. The fourth section aimed at evaluating moral disengagement through the Moral Disengagement Scale which comprises 32 items and 8 different mechanisms of moral disengagement, 4 items for each of them: moral justification, euphemistic language, advantageous comparison, displacement of responsibility, diffusion of responsibility, distorting consequences, attribution of blame, dehumanization [23]. This questionnaire collects answers via a 5 point Likert agreement scale where 1 corresponds to "strongly disagree" and 5 to "strongly agree". The last section of the questionnaire investigates the relationship between enacted or suffered anti-social or pro-social conduct and the moral emotions associated with them.

1. Antisocial behavior enacted, for example "Have you ever published photos, denigrating images, offensive terms etc.)";
2. Prosocial conduct enacted "Have you ever intervened in defense of someone who had been offended, denigrated, or publicly embarrassed on social networks?"

## 2.3 Results

Two hundred and ninety-seven surveys were filled in. We filtered out data from respondents who were upper rather than high school students to get a more homogeneous sample. The resulting sample included 264 participants, 56.8% Female and 43.2% Male. Socio-demographic and social network usage information is reported in Table 1. Most participants (68.6%) reported having more than 200 friends but 43.6% consider only from 10 up to 50 to be true friends. Most of the sample (86.7%) has only one profile on the social network and only 20.8% say they use one of his/her profiles to post content that cannot be shared by others net-users.

The mean values for moral disengagement differ significantly (*Mann–Whitney U* test, $p < .001$) between males (mean = 2.22) and females (mean = 1.90) and for all the factors considered ($p < .001$) but "distortion of consequences" and "advantageous comparison" (see Table 2). Results from a Mann-Whitney U test also indicate, in agreement with former scientific evidence, that mean empathy scores (Total, Affective and Cognitive) differ significantly between males and females when a global value is considered (mean male = 3.53 and female = 3.76; $p < .05$) and for Affective Empathy (mean male = 3.33; female = 3.66; $p < .01$) (see Table 2).

**Correlations Between Empathy and Moral Disengagement.** The results highlight a negative correlation between total empathy and the moral disengagement scale. ($r = .179$; $p < .006$). In relation to moral disengagement we find a negative correlation in particular with five factors: euphemistic labeling ($r = -.155$; $p < .005$); diffusion of responsibility ($r = -.253$; $p < .000$); advantageous comparison ($r = -.127$; $p < .005$); dehumanization of the victim ($r = -.169$; $p < .008$) and moral justification ($r = -.116$; $p < .068$).

*Affective Empathy.* Results highlight a negative correlation between affective empathy and the moral disengagement scale ($r = -.213$; $p < .001$). In this case too, with respect to affective empathy and the moral disengagement scale we find a negative correlation with the same five factors: euphemistic labeling ($r = -.206$; $p < .001$); diffusion of responsibility ($r = -.28$; $p < .000$); advantageous comparison ($r = -.155$; $p < .005$); dehumanization of the victim ($r = -.248$; $p < .000$); moral justification ($r = -.196$; $p < .000$).

*Cognitive Empathy.* No correlations were found between cognitive empathy and the moral disengagement scale.

*Antisocial Behavior and Prosocial Behavior on the Net.* With regard to the realization of aggressive and offensive behavior, 23.1% (no 76.9%; N = 260) say they have posted, at least once, embarrassing photos, or offensive words of others (Have you ever published embarrassing photos, or offensive words of others for everyone to see?).

**Table 1.** Main results for socio-demographic variables and social network usage

| Variable | N | % |
|---|---|---|
| Gender | | |
| Male | 114 | 43.2 |
| Female | 150 | 56.8 |
| Age | | |
| 14–15 | 36 | 13.6 |
| 16–17 | 100 | 37.9 |
| 18–20 | 114 | 43.2 |
| 21–25 | 14 | 5.3 |
| Started using internet from when I was | | |
| <6 years | 6 | 2.3 |
| 6–10 years | 65 | 24.6 |
| 11–14 years | 176 | 66.7 |
| >14 years | 17 | 6.4 |
| Using Facebook or other social networks | | |
| Never | 3 | 1.1 |
| t < 3 hours a day | 199 | 75.4 |
| t > 3 hours a day | 62 | 23.5 |
| If I cannot use the net I feel nervous (1 strongly disagree – 5 strongly agree) | | |
| 1 | 21 | 7.9 |
| 2 | 54 | 20.5 |
| 3 | 94 | 35.6 |
| 4 | 53 | 20.1 |
| 5 | 42 | 15.9 |

While 77% out of 260 say that they happened to intervene in aid of someone who had been offended on social networks. Those who carry out offensive behavior on line, posting photos or offensive comments have a higher score for disengagement.

No significant difference emerged however for empathy (total, cognitive or affective) between those who carry out offensive behavior and those who do not. What does emerge is a positive correlation between offensive behavior and moral disengagement ($r = .354$; $p < .000$) while helping behavior on the other hand correlates with cognitive empathy ($r = .226$; $p < .002$) and thus with total empathy ($r = 221$; $p < .001$). A significant difference emerges for offensive behavior related to net dependency (If I can't use internet I get agitated; 1-strongly agree, 5-strongly disagree, mean 3.16) among those who admit to having been offensive on line ($N = 60$; mean = 2.87) and those who don't ($N = 199$; mean = 3.27) (Mann-Whitney U test, $p < .016$). It would seem then that we can claim that higher dependency on the net is related to increased enacting of offensive behaviors.

**Table 2.** Average participants' scores on measures related to Empathy (Affective and Cognitive) and Moral Disengagement (MD)

| Variable | Mean | Sd | M | Sd | F | Sd |
|---|---|---|---|---|---|---|
| Empathy tot. | 3.66 | .50298 | 3.53 | .48728 | 3.76 | .49314 |
| Affective empathy | 3.52 | .62438 | 3.33 | .59516 | 3.66 | .61042 |
| Cognitive empathy | 3.84 | .56171 | 3.78 | .56270 | 3.87 | .55969 |
| MD | 2.04 | .50716 | 2.22 | .49351 | 1.90 | .47262 |
| Moral justification | 2.46 | .90476 | 2.89 | .84347 | 2.15 | .81631 |
| Euphemistic labeling | 2.03 | .70101 | 2.30 | .68480 | 1.83 | .64414 |
| Dehumanization | 2.92 | 1.0798 | 3.25 | 1.09570 | 2.66 | .99832 |
| Attribution of blame | 1.89 | .59620 | 2.07 | .60630 | 1.76 | .55311 |
| Distortion of consequences | 2.73 | .81558 | 2.75 | .80578 | 2.72 | .82526 |
| Diffusion of responsibility | 1.91 | .62722 | 2.11 | .63754 | 1.76 | .57859 |
| Displacement of responsibility | 1.95 | .71320 | 2.08 | .76311 | 1.86 | .65828 |
| Advantageous comparison | 1.34 | .53537 | 1.39 | .53632 | 1.30 | .53318 |

Where helping behavior is concerned on the other hand ("have you ever intervened in defense of a person who has been offended or denigrated publicly on social networks?") the results highlight the way in which those who intervene in favor of others had an average score which was significantly higher for cognitive empathy (N = 183; mean = 3.89) compared with those who tend not to intervene to defend those who have been attacked (N = 54; mean = 3.71), (Mann-Whitney U test, $p < 0.016$.). The results also highlight the way in which offensive behavior on the net is related to the number of hours spent on line (t < 3 hours a day; t > 3 hours a day). A higher number of hours a day spent on the networks corresponds to a greater ease in offensive behaviors on the net. The analysis shows a significant difference between the mean scores related to offending and the number of hours spent daily on the net (t < 3 h; N = 124; mean = 2.82) and more than 3 h (N = 60; mean = 3.02) (Kruskall Wallis test, $p < .002$).

There are no significant differences between those who spend a lot of time on social networks as far as helping behavior is concerned.

## 3 Discussion and Conclusions

The results highlight relationships between moral disengagement, empathy and aggressive or prosocial behavior which are worth further study. Firstly it appears that moral disengagement is related to affective empathy which leads to a hypothesis that the justification for disengagement, whether it is a case of euphemistic labeling, diffusion of responsibility or dehumanization of the victim, to give a few examples, such justification stems from a detachment from what the victim is feeling. This detachment is still emotive rather than cognitive: the reasons of another person can be understood but not his/her feelings. Behavior which victimizes does not seem to implicate, neither more nor less, cognitive involvement.

The opposite seems to hold if one thinks that when someone who offers to help, who intervenes because they can understand the victim's point of view, can see the world through his/her eyes; in fact helping behaviors do seem to be related to cognitive empathic capacity.

All of this, differently from other studies [28] needs to be put in the context of social relationships on the Internet. It has to be underlined that in this study we find that those who are more likely to perform victimizing acts on the Internet tend to be also more dependent on the use of the internet and to frequent social networks for more hours in the day. The scope of this research opens up perspectives which perhaps identify the possibility of moderating internet use and online presence as a key to containing such unacceptable behaviors as cyberbullying.

The first stage of an educational process should start through a series of steps, not in the virtual world, but in the real environments giving way to a strong social network and a close collaboration between family, school and society [29]. All stakeholders should be included in training courses on the use of technology and on social media. Above all, the school should favor maximum social cohesion among students within the classroom, aiming at reducing antisocial behaviors among peers [30] by providing emotional literacy that can help teenagers to talk about and recognize their own and others' emotions and how to regulate them. This learning path should be juxtaposed to a parental discipline and monitoring practices being able to reduce the time spent on the net by adolescents.

The second stage of this process, on the virtual side, should encourage design for reflection [31, 32] in order to educate young users to spend time enough to think and make moral behavior emerge, which need a longer time to be elaborated [18]. Furthermore "design for empathy" [31] should be specifically addressed in order to reduce affordances to a bully behavior on the net, working, on one side, on the affective aspects of empathy to promote a pro-social behavior and on the other, on the cognitive one to promote a helping attitude, which can encourage a mature and aware use of the network.

## References

1. Cross D, Barnes A, Papageorgiou A, Hadwen K, Hearn L, Lester L (2015) A social-ecological framework for understanding and reducing cyberbullying behaviours. Aggress Viol Behav 23:109–117
2. Sourander A, Klomek AB, Ikonen M, Lindroos J, Luntamo T, Koskelainen M, Ristkari T, Helenius H (2010) Psychosocial risk factors associated with cyberbullying among adolescents: a population-based study. Arch Gen Psychiatry 67(7):720–728
3. Tokunaga RS (2010) Following you home from school: a critical review and synthesis of research on cyberbullying victimization. Comput Hum Behav 26(3):277–287
4. Blair RJR (2005) Responding to the emotions of others: dissociating forms of empathy through the study of typical and psychiatric populations. Conscious Cogn 14(4):698–718
5. Tangney JP, Stuewig J, Mashek DJ (2007) Moral emotions and moral behavior. Ann Rev Psychol 58:345–372
6. Marshall LE, Marshal WL (2011) Empathy and antisocial behavior. J Forensic Psychiatry Psychol 22(5):742–759

7. Mehrabian A (1997) Relations among personality scales of aggression, violence, and empathy: validational evidence bearing on the Risk of Eruptive Violence Scale. Aggress Behav 23(6):433–445
8. Edgar JL, Nicol CJ, Clark CCA, Paul ES (2012) Measuring empathic responses in animals. Appl Anim Behav Sci 138(3–4):182–193
9. Batson CD (2012) The empathy-altruism hypothesis: issues and implications. In: Decety J (ed) Empathy: from bench to bedside. MIT Press, pp 41–54
10. Galinsky AD, Moskowitz GB (2000) Perspective-taking: decreasing stereotype expression, stereotype accessibility, and in-group favoritism. J Pers Soc Psychol 78(4):708–724
11. Kidd DC, Castano E (2013) Reading literary fiction improves theory of mind. Science 342 (6156):377–380
12. Decety J (2015) The neural pathways, development and functions of empathy. Curr Opin Behav Sci 3:1–6
13. Davidson RJ, Putnam KM, Larson CL (2000) Dysfunction in the neural circuitry of emotion regulation – a possible prelude to violence. Science 289(5479):591–594
14. Herpertz SC, Sass H (2000) Emotional deficiency and psychopathy. Behav Sci Law 18 (5):567–580
15. Carré A, Stefaniak N, D'Ambrosio F, Bensalah L, Besche-Richard C (2013) The Basic Empathy Scale in adults (BES-A): factor structure of a revised form. Psychol Assess 25 (3):679–691
16. Icevic Z, Ambady N (2013) Face to (face)book: the two faces of social behavior. J Pers 81 (3):290–301
17. Rosen L (2012) iDisorder: understanding our obsession with technology and overcoming its hold on us. Palgrave Macmillan, New York
18. Immordino-Yang MH, McColl A, Damasio H, Damasio A (2009) Neural correlates of admiration and compassion. PNAS 106(19):8021–8026
19. Ashktorab Z (2016) A study of cyberbullying detection and mitigation on Instagram. In: CSCW 2016 companion proceedings. ACM, New York, pp 126–130
20. Kowalski RM, Limber SE, Agatston PW (2012) Cyberbullying: bullying in the digital age, 2nd edn. Malden, Wiley-Blackwell
21. Patchin JW, Hinduja S (2012) Cyberbullying: an update and synthesis of the research. In: Patchin JW, Hinduja S (eds) Cyber-bullying prevention and response: expert perspectives. Routledge, New York, pp 13–36
22. Bandura A (1986) Social foundations of thought and action: a social cognitive theory. Prentice-Hall, Englewood Cliffs
23. Bandura A, Barbaranelli C, Caprara GV, Pastorelli C (1996) Mechanisms of moral disengagement in the exercise of moral agency. J Pers Soc Psychol 71(2):364–374
24. Haidt J (2003) The moral emotions. In: Davidson RJ, Scherer KR, Goldsmith HH (eds) Handbook of affective sciences. Oxford University Press, Oxford, pp 852–870
25. Joliffe D, Farrington DP (2006) Development and validation of the Basic Empathy scale. J Adolesc 29(4):589–611
26. Albiero P, Matricardi G, Speltri D, Toso D (2009) The assessment of empathy in adolescence: a contribution to the Italian validation of the Basic Empathy Scale. J Adolesc 32(2):393–408
27. Albiero P, Matricardi G, Toso D (2010) La Basic Empathy Scale, uno strumento per la misura della responsivita' empatica negli adolescenti: Un contributo alla validazione Italiana. Psicologia clinica dello sviluppo 3:693–706
28. Haddock AD, Shimerson RS (2017) An examination of differences in moral disengagement and empathy among bullying participant groups. J Relat Res 8:e15

29. Cohen-Almagor R (2018) Social responsibility on the Internet: addressing the challenge of cyberbullying. Aggress Viol Behav 39:42–52
30. van den Bos W, Crone EA, Meuwese R, Guroğlu B (2018) Social network cohesion in school classes promotes prosocial behavior. PLoS ONE 13(4):1–16
31. Bowler L, Mattern E, Knobel C (2014) Developing design interventions for cyberbullying: a narrative-based participatory approach. In: Kindling M, Greifeneder E (ed) iConference 2014 proceedings. iSchools, Illinois, pp 153–162
32. Parlangeli O, Mengoni G, Guidi S (2011) The effect of system usability and multitasking activities in distance learning. In: Proceedings of the CHItaly conference, 13–16 September. ACM Library, Alghero, pp 59–64

# Playful Learning for Kids with Special Educational Needs

Elisabetta Cianfanelli[1(✉)], Pierluigi Crescenzi[2(✉)], Gabriele Goretti[3(✉)], and Benedetta Terenzi[1(✉)]

[1] DIDA Department, University of Florence, Florence, Italy
elisabetta.cianfanelli@unifi.it,
benedetta.terenzi@gmail.com
[2] Department of Mathematics and Informatics, University of Florence, Florence, Italy
pierluigi.crescenzi@unifi.it
[3] School of Arts, Nanjing University, Nanjing, China
gabriele.goretti@qq.com

**Abstract.** It has been verified that today in the Italian schools there are about 15–20% of students presenting different kinds of "Special Educational Needs". In order to respond to BES, the schools follow the Customized Learning Plan (PDP), a useful tool for designing operational models, strategies, systems and learning criteria for each student. For the various operators involved in the treatment of the disorders and also for the parents, it is of great importance to have adequate, efficient and flexible support tools that respect the desires and the enthusiasm of children of different age, without creating additional psychological and social discomfort. This research presents a particular relevance focus on the issue of management of entertainment and playful learning through digital systems offering a 'playful interaction'. New concepts of 'game' and 'device for interactive activities' offer screen-based compounds to scenarios on the relation between user and device, including physical activities or integrated physical gestures, which relate the observation of screen to the physical reality.

**Keywords:** Interaction design · Product design · Playful experience

## 1 Learning Disorders and Technology

The Developmental Disorders Specific of Learning (DSA) are an area of clinical interest in which over the last thirty years an important advancement of knowledge has been achieved thanks to the numerous contributions derived from scientific research and the refinement of diagnostic investigation techniques. This allows today to be able to share the definition and classification of DSAs also among professionals and/or specialists of different backgrounds (psychologists, neuropsychiatrists, speech therapists, pedagogists), allow to make a diagnosis accurately and plan a targeted treatments.

Hammill defines the general characteristics of Learning Disability (LD) as a heterogeneous group of disorders manifested by significant difficulties in the acquisition and use of listening skills, oral expression, reading, reasoning and mathematics,


S. Bagnara et al. (Eds.): IEA 2018, AISC 826, pp. 732–742, 2019.
https://doi.org/10.1007/978-3-319-96065-4_77

presumably due at dysfunctions of the central nervous system [1]. In short, it deals with a diversified range of problems in cognitive development and scholastic learning, not primarily due to factors of severe mental handicap and definable on the basis of failure to reach expected learning criteria (for which there is a broad consensus) to the general potential of the subject [2].

These are usually difficulties that occur in the child from the early stages of his learning, when he must acquire new skills such as reading, writing and calculating, starting from a neuropsychological structure that does not favor the automatic learning of these specific skills. The evolution of these disorders, in fact, is favored by the precocity and adequacy of the intervention, as well as by the compensatory measures taken in the context of the scholastic pathway to encourage learning and for a favorable prognosis regarding social development and of the personality of those who present these problems.

These considerations, in recent years have also led Italy to focus on prevention, developing programs to strengthen the prerequisites of basic school learning to be used starting from kindergarten.

### 1.1 Relationship Between Technology and Teaching

Technology has already revolutionized our lives: from the way we work, to the way we shop or travel, just to name a few. It can therefore certainly change the way we study and acquire new knowledge. Human computer interaction is widely considered an integral component of many education and training systems at various levels of technology access [3]. An example is the Edugames: these in fact allow students to learn more actively than traditional educational methods and allow them to have fun while learning something new.

To date there are many studies that support the positive role of video games in the future of learning, thanks to their infinite potential. Of the many education and training systems developed in recent years, one of the most intensively studied is interactive computer game-based learning systems, which have been developed and applied in many teaching-learning activities, especially for children and adolescent learners [4, 5] because children in these age groups are intrinsically motivated to play games and often lack interest or motivation in their courses [6]. Studies show that the key factors in user acceptance of these systems are perceived ease of use, perceived usefulness, self-efficacy, and satisfaction [7].

## 2 Design of '1,2,3 Stella!' App

One of the main objectives of this project was to propose a valid alternative to what is already present on the market, in the field of apps for speech-language use. The multidisciplinary team has in fact conducted a careful preliminary analysis on the products currently on the market, evaluating different aspects of the applications. The study was supported by the NuovaMente Children's Diagnostic Center team, which includes child neuropsychiatrists, neuropsychologists, psychotherapists, speech therapists, pedagogists, learning tutors and educators.

The apps chosen for the analysis have been divided into three categories according to the type of use: specific apps for the treatment of speech therapy, non-specific apps, but which are used for some speech therapy treatments, apps with only a game goal. The app evaluation criteria were:

- Usability of children;
- Enjoyment (ability of the app to entertain and enjoy the child);
- Feedback for adults;
- Adaptability (set activities according to the different needs of the child);
- Interface design;
- Therapeutic suitability.

From this analysis it is possible to draw different conclusions. First of all, the specific apps for speech therapy are without an adequate graphic interface, even if they are valid for the speech therapy treatment to be adopted.

It has also been noted that many app for speech therapy treatments there are no the possibility of adapting the exercises according to different needs.

The game app, finally, provide a well designed and edited graphics, aimed at the fun of the child, of course, however, are not effective for treatment.

### 2.1 Difficulty in Learning and Mathematical Prerequisites

The work concerns the development of an app to aid the speech therapy treatment of children with difficulties in the prerequisites of the logical mathematical area[1]. This app, therefore, was created with the aim of enhancing the prerequisites of the logical-mathematical area that are weak in preschool children and to prevent the presence of a future DSA related to the world of computing, namely the Discalculia [8].

The prerequisites are a series of basic knowledge for school learning that the child should develop in order to easily continue on an educational path that reflects the normality [9]. For the kids is important to have and exercise this knowledge from the last year of the Infant School, so as to facilitate entry to the Primary School. The target is in fact represented by children aged between 4 and 6 years.

Instead, by numerical intelligence we mean "the ability to understand, interpret and reason through the complex cognitive system of numbers and quantities", that is to say, to be intelligent through quantity [10]. Some psychological studies have established that it is an innate ability in human beings (and not only), which are then born predisposed to numerical intelligence as to the verbal [9]. However, only some numerical skills are innate (ability to "see" a quantity in the right way), the others are cultural (ability to associate the quantity to the correct name) the ability to estimate [13].

---

[1] The project stems from the collaboration between Benedetta Terenzi, PhD and a contract professor at the Design Campus, scientific advisor and coordinator of the team, with Pierluigi Crescenzi, full professor of the Department of Mathematics and Computer Science 'Ulisse Dini' of the University of Florence, with the Children's Diagnostic Center Nuovamente (Firenze) which includes the speech terapist Elisa Cangialeoni, Ilaria Tilli, Giulia Filippi. The app is developed with the help of graduate in Computer Science Nicolae Puica [11] and the graduate in Industrial Design Martina Denti [12].

According to 2015 data from the International Academy for Research in Learning Disabilities (IARLD) 2.5% of the global school population presents difficulties in mathematics in comorbidity with other disorders, while 2 children in 1000 have severe Dyscalculia. The MIUR in 2015 calculated that in Italian schools, between state and non-state, of every order and degree, there are 186,803 students with learning disabilities and of these, 41,819 suffer from Discalculia.

### 2.2 The GUI Design

For the design were followed the heuristic principles of usability described in the decalogue of Jacob Nielsen, deriving from the application of factorial analysis techniques on usability problems. The app has been designed with a friendly graphic user interface (GUI), with the aim to have an instrument very simple and intuitive, for all the users. In fact, the app has provided different levels of use for the three different users. A coherent graphic layout was maintained in the different sections and within the different activities. In this way the different users do not feel disoriented in changing the sections provided by the app.

For the use by adults, have been inserted the recognizable commands now universally recognized used in other apps or websites, such as registration procedure 'new user' or 'log in', etc. For the icons and the bottom functions have been chosen very intuitive and easily understandable specifications.

The adult user has control of the information content and move freely between the various topics. The child user, on the other hand, deliberately follows only guided path. The registration procedures and the choice of the avatar can be skipped and resumed at any time. On every screen of children's games there are the commands 'exit' and 'help' have control over the activity.

Parents and children have access only to certain functions, chosen from time to time by the speech therapist. The children can only perform the activities activated by the speech therapist. During the activity by the children, in addition to the sound feedback there are some positive vocal feedback ("Good job!", " Keep it up!") Or negative ("Come on, try again!", "You're almost there!"), which serve to stimulate the child. The parent can monitor the activities performed by the child.

Speech therapists have a great deal of opportunity in choosing parameter settings for the child's activities, depending on the needs. They have also mission diaries and activity reports with very precise data indications, which can be used to produce accurate patient improvement graphs. All informations are saved in a cloud storage, accessible and manageable by the speech therapists.

## 3 IT Development of the App

The proposed app has been developed within the Corona SDK framework [14], which is a development environment that allows programmers to develop multi-platform apps by using LUA, a language supporting procedural, object-oriented, and functional programming language [15].

This framework is quite easy to be used and, most importantly, it allows the deployment of the app for different mobile operating systems, such as Android and iOS. Moreover, Corona SDK is widely used for the development of 2D games, and it supports several features which turned out to be very useful while developing our app (such as efficient management of animations, audio, and images).

This app makes heavy use of images at high resolution and the Corona SDK facilitates the management of such images by supporting the so-called *image sheets* which allow the programmer to load several different images from a single larger image file. To this aim it is sufficient to specify the width and the height of the global image (called *sheet*), the number of images contained in the sheet, and the width and the height of these images.

Corona SDK also allows the programmer to easily deal with *JSON* (JavaScript Object Notation) files. JSON is a simple standard for formatting and exchanging data, independently of the used programming language [16]. In our app, we used JSON in order to store all the local settings of a game, such as the ones we are going to describe in this section.

Finally, Corona SDK easily allows the programmer to manage a SQLLite database. SQLLite is "an in-process library that implements a self-contained, serverless, zero-configuration, transactional SQL database engine" [17]. In our app, we make use of a SQLLite database in order to globally store the activities of the children: the contents of this database are successively elaborated in order to evaluate the performances of the children themselves.

The app starts from the home page shown in the following figure, in which planets and stars rotate and a background music is played (from now on, we will avoid to specify that the app includes animations and sounds).

The ACCEDI button allow the children to access a centralized system, in order to record the activities and the results of the children themselves. The IMPOSTAZIONI button allows the parents or the therapist to change the settings of the game according to the following criteria.

1. How many exercises the children have to solve.
2. Focusing to a specific range of numbers.
3. Focusing on a specific subset of numbers.
4. Allowing the possibility of hearing an audio communicating the number to the children.
5. Selecting which kind of association the children have to perform: from decimal numbers to dot numbers, from dot numbers to decimal numbers, or both.

These settings are stored within the app and remains valid until the parents or the therapists decide to change them.

The GIOCA button of the home screen starts a new game, which is a sequence of association exercises in which the children have either to choose, among the three possibilities shown in the lower part of the screen, the decimal number corresponding to the dot number shown in the middle of the screen, or to choose, among the three possibilities shown in the lower part of the screen, the dot number corresponding to the decimal number shown in the middle of the screen (Fig. 1).

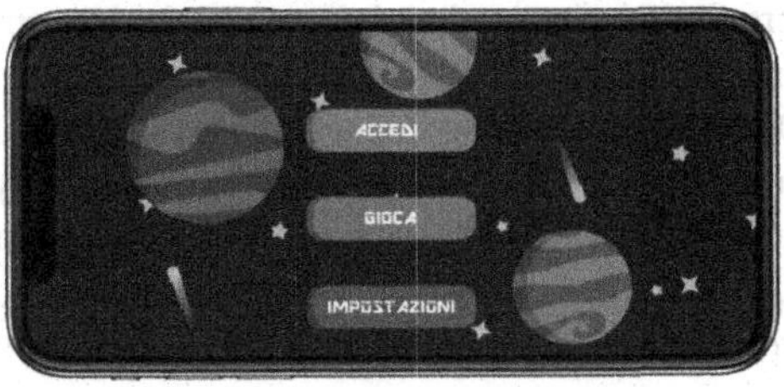


**Fig. 1.** App Home page

This second kind of exercise can be easily adapted to the integration of the app within a product-based screen-based interaction environment. Indeed, a prototype of such an integration has been realized by substituting the act of clicking on the correct button by the act taking a cube and facing the right face of the cube towards the screen (see the end of the next section).

As we already said, Corona SDK facilitates the task of including audio and animations in the app. Indeed, we used these features both for playing some background/feedback music and instruction audio files and for animating some drop-down menus (Figs. 2 and 3).

**Fig. 2.** App setting page

**Fig. 3.** App gaming page

## 4 Playful Interaction: Screen-Based E Product-Based Design

The year 2007 stands the launch of first IPhone [18]: this innovative product-system is working through innovative interface, avoiding traditional typing system and using innovative software able to transfer into the mobile system a real user-centred device. Moreover, the use of internet becomes continuous, not only a occasional need. Them the digital devices invade the user daily life, generating screen-addicted behaviours and consequent disorientation.

In parallel, within the international research context, we highlight a new interest about interaction design topics on product design and linked services aiming at avoiding monothematic researches on screen-oriented solutions. In fact, these studies develop product-based solutions connected to screen-based interaction design frameworks [19]. Then, following this design process, we have information and communication systems, sharing tools and interaction platforms not based only on a screen interface, but mostly supported from the performances of a 3D artefact.

Following this perspective, the product becomes a device aiming at improving the physical wellbeing and the quality of life, in a functional and aesthetic sense. Within the REI lab [20] of Design Campus-University of Florence set Care Device workshops programs, the research sessions focus on innovation done by embedding micro-technologies to the artefact, adding new steps of the product values chain. In addition the Care Device Workshops define a clear design framework on interaction design research connecting medical matters and the focus on shapes or psychological topics to semiotics.

According to Universal Design criteria [21, 22], the research process aims at connecting the design disciplines to the medical/paramedical care. In particular, this bridge is established through technologies embedded into the product. These applications evolve the artefacts to product systems including sensors, identification technologies, smart materials, becoming an auxiliary device to improve the user psycho-physical status.

The workshop set design-systems through user-friendly microcontrollers, composed by a hardware component (digital card) and specific software. The user-friendly sensors allow the researchers to run a proper prototyping and a fast learning about electronics and programming foundation. The digital sensors are high-precision tools that aims at changing a physical phenomenon (from environmental factors or from the contact with the user body) into a numeric value. Through a proper script on the microcontroller on the digital card, the input numerical values could be processes and interpreted to have a final output (digital or analogical). So, we define "learning by prototyping" process, in which we can experiment design solutions to implement realistic interaction frameworks in between user and device.

## 4.1 Design Methodology: Designing Devices Including Screen-Based and Product-Based Interaction

The workshop path aims at defining differences and contact points in between design disciplines and the medical/paramedical care, implementing the instances emerging from the medical research through the product-system performance. So the design project is a tool to express the medical/paramedical therapies and advices, on the other hand the medical research is a highly significant support for the interaction design performance. Then, the curative approach on specific disease or user functional problems stands in relation to design topics relating to the morphology, technology and interactive aspects of the product system.

The workshop program presents a design methodology according to specific user scenarios, following steps:

**(Phase 1)**

(A) Selection of the pilot subject according to medical/paramedical requests;

(B) Benchmarking analysis about other product or product-systems already existing in the market; highlighting on possible technology and design transfer from already existing product (even not in the same product area);

(C) Defining through a scenario-based design [23, 24] process the design concept and the interaction storyboard;
(D) Selecting the proper technologies to be embedded into the artefact.

**(Phase 2)**

(A) Design project development of the device and product-system;
(B) Prototype development.

**(Phase 3)** Evaluation session

(A) Interdisciplinary focus group to evaluate the design results and related interactive model [25];
(B) Prototype test, by simulating the user scenarios, supported by videos about the interactive performance [26, 27].

### 4.2 Screen/Product Based Interaction: Sensea [28] Microcontroller

Sensea microcontroller is an interactive platform that allows not expert users to create automations in a easy and fast way, avoiding complex trainings on electronics and programming systems.

Seansea is composed by smart electronic circuits able to communicate through new radio waives technologies, getting a high good level of communication without a high energy consumption. Microcontroller's power supply is provided by rechargeable batteries, working through Wireless Power Transfer [29] technology.

**Case History (1): MooMi project**

Project by Sara Foriglio/supervisors Gabriele Goretti (interaction design process), Michele Tittarelli (Max/MSP)/Scientific supervisor Elisabetta Cianfanelli

Internet addiction disorder is a clinical problem caused by the non-control of impulses. IAD could rise in different ways: cyber-sex addiction, social network addition, compulsive online shopping, videogame addiction or information overload stress.

Moomi project is composed by four toys-devices and by a dedicated screen-based app. These devices interact to the user in different ways: "Ted" fills with stains, "Ned" moves it self as if it melted, "Bill" stretches as if to fall asleep, "Carl" bends his head to one side. These visual feedbacks are activated progressively according to the time spent in front of the computer. Each Moomi toy embeds a proximity sensor aiming at counting the minutes spent in front of the screen. These playful tools provide a self-analysis support as for IAD users as for professionals who spend most of their time in front of a monitor.

**Case History (2): Usignolo project**

Project by Ilaria Forzoni/supervisors Gabriele Goretti (interaction design process), Michele Tittarelli (Max/MSP)/Scientific supervisor Elisabetta Cianfanelli

The voice may experience disabling vocal difficulties such as dysphonia (more or less severe vocal changes) that can result in organic lesions if correct vocal behaviours (vocal hygiene) are not used.

Usignolo project is an interactive device that allows the user to adjust the volume and tone of his voice in an optimal way: the values coming from the internal microphone are tested by the software that adjusts the intensity of the feedback based on the volume expressed, allowing to correct the volume itself (depending on the program set). Though a dedicated web site and app, it is possible to review the daily records, that could be evaluated by specialists of pathologies related to the voice.

## 5 1,2,3 Stella! Product-Based Prototype, Final Considerations

1,2,3 Stella! App and Care Device workshop program represented a very significant opportunity to implement a direct connection in between screen-based user experience and product-based interaction. As described in the Care Design tool as in 1,2,3 Stella! App as in Care Device devices (i.e. Usignolo case study) we establish a communication in between design disciplines and medical/paramedical studies (Fig. 4).

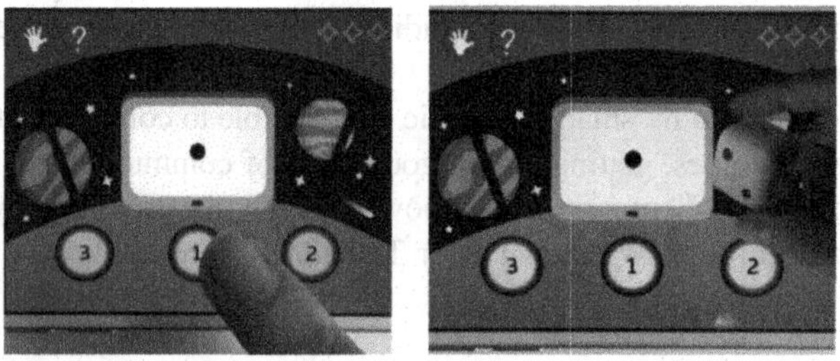


**Fig. 4.** Developing interactive scenario for 1,2,3 Stella! product-based interaction.

The prototype we are developing integrates with the existing app (described in Sect. 1.1) in a quite natural way. Since the activity requested to the children is to select either a decimal number or a dot number, we allow the children to perform this activity by taking a dice from the table on whose faces is represented the decimal number or the dot number. In other words, the prototype uses twenty little dices, one for each decimal number between 1 and 10 and one for each dot number between 1 and 10. Actually, the developed prototype currently includes only six dices representing the numbers between 1 and 3: this implies that in the settings of the game, the parents or the therapists have to tell the app to focus on the subset of numbers between 1 and 3. Whenever a dice is taken by the children, the micro-controller hidden within the dice will communicate to the app which decimal or dot number has been selected by the children. The app will respond to this interaction as if the children had clicked on the corresponding decimal/dot number on the screen.

## References


1. Hammill DD (1990) On defining learning disabilities: an emerging consensus. J Learn Disabil 23:74–84
2. Cornoldi C (2007) (a cura di) Difficoltà e Disturbi dell'Apprendimento. Il Mulino, Bologna
3. Lin HC et al (2017) Continued use of an interactive computer game-based visual perception learning system in children with developmental delay. Int J Med Inform 107:76–87
4. Kim B, Park H, Baek Y (2009) Not just fun, but serious strategies: using meta-cognitive strategies in game-based learning. Comput Educ 52:800–810
5. Hwang GJ, Wu PH (2012) Advancements and trends in digital game-based learning research: a review of publications in selected journals from 2001 to 2010. Br J Educ Technol 43:E6–E10
6. Papastergiou M (2009) Digital game-based learning in high school computer science education: impact on educational effectiveness and student motivation. Comput Educ 52:1–12
7. Prensky M (2003) Digital game-based learning. ACM Comput. Entertain 1:1–4
8. Wynn K (1995) The origins of numerical knowledge. Math Cogn 1:35–60
9. Lucangeli D (2017) La discalculia e le difficoltà in aritmetica: Guida con workbook. Giunti EDU, Firenze
10. Molin A, Poli S, Lucangeli D (2007) BIN 4-6 Batteria per la valutazione dell'intelligenza numerica. Trento, Erickson
11. Gelman R, Gallistel CR (1978) The child's understanding of number. Harvard University Press, Cambridge
12. Puica N (2018) App and specific Leaning Disorders. Design, Develop and Experimentation of an App for the Discalculia, Degree thesis, University of Florence
13. Denti M (2018) 1, 2, 3, Stella! App di ausilio al trattamento logopedico di bambini con difficoltà nei prerequisiti dell'area logico-matematica, Degree thesis, University of Florence
14. https://coronalabs.com/
15. https://www.lua.org/
16. https://www.json.org/
17. https://www.sqlite.org/
18. Mossberg W (2007) Testing Out the iPhone. Wall Street J
19. https://www.interaction-design.org/literature/book/the-encyclopedia-of-human-computer-interaction-2nd-ed/industrial-design
20. REI lab- Academic laboratory at Design Campus University of Florence. https://www.dida.unifi.it/vp-234-laboratorio-reverse-engineering-and-interaction-design.html
21. Mace RL (1998) Designing for the 21st century: an international conference on universal design. Hofstra University, Hempstead, New York
22. Bettye Rose Connell BR et al (1997) The principles of universal design, NC State University, The Center for Universal Design
23. Carroll JM (1999) Five reasons for scenario-based design. In: 32nd Hawaii international conference on system sciences
24. Jordan PW (1997) Designing pleasurable products: an introduction to the new human factors. Taylor & Francis, London
25. Britton TA, LaSalle D, Cheng W (2003) Priceless: turning ordinary, the products of the extraordinary experiences. Zhongxin Press, Beijing
26. ShiGuo L (2008) Experience and the challenge: product interaction design. Jiangsu Arts Press, Nanjing

27. Preece J, Rogers Y, Sharp H (2002) Human computer interaction would be beyond interactive. John Wiley & Sons, Inc
28. https://www.senseame.com
29. https://www.allaboutcircuits.com/technical-articles/introduction-to-wireless-power-transfer-wpt/

# Influence of Varying Backpack Loading and Velotypes on the Spatiotemporal Parameters of Gait and Energy Cost of Ambulation Among Primary School Children in Nigeria

## Short Title: Influence of Backpack Loading and Velotypes on School Children's Gait Performance

Echezona Nelson Dominic Ekechukwu[1,2(✉)], Callistus Chukwuwendu Okigbo[1], Adaobi Justina Okemuo[1,2], and Chioma N. Ikele[1]

[1] Department of Medical Rehabilitation, Faculty of Health Sciences and Technology, College of Medicine, University of Nigeria, Enugu, Nigeria
nelson.ekechukwu@unn.edu.ng

[2] Ergonomic Society of Nigeria (ESN), Enugu Region, Enugu, Nigeria
http://www.unn.edu.ng/

**Abstract. Objective:** To determine the effects of backpack loading and velotypes on Gait Parameters (GP) and Energy Cost of Ambulation (ECA) among Primary School Children (PSC) in Nigeria.

**Method:** A self-controlled cross-over study design, participants walked a 10 m distance at normal, slow and fast velotypes without backpack, carrying a backpack of 10%, 15%, and 20% of their Body Weights (BW). Gait parameters (stride length, stride duration, gait velocity, swing duration and Double support duration) were assessed using validated equations while ECA was assessed using the Ralstron's equation. Data was analysed using descriptive statistics and repeated measure ANOVA at $\alpha = 0.05$.

**Results:** A total of 69 PSC participated in this study. Over 25% of them carried backpack >10% of their BW. At normal velotype, most of the GP and ECA at the different backpack loadings had no significant difference ($P > 0.05$). At slower and faster velotypes, all the parameters were significantly different ($p < 0.05$) when compared across the different backpack categories with the exception of double support duration at slower velotype ($F = 3.191$, $p = 0.056$). Post hoc revealed that the most profound change in most of the parameters occurred at the 20% backpack loading phase. The ECA was significantly higher ($p < 0.05$) in slow velotype than the other velotypes in all categories of backpack loading.

**Conclusion:** Carrying a backpack load up to 20% body weight significantly affects energy cost of ambulation and gait parameters of primary school children. Irrespective of the backpack weights, energy cost of ambulation is highest at slow velotype.


S. Bagnara et al. (Eds.): IEA 2018, AISC 826, pp. 743–757, 2019.
https://doi.org/10.1007/978-3-319-96065-4_78

**Keywords:** Backpack loading and velotypes
Gait parameters and energy cost of ambulation · Primary school children

# 1 Introduction

Movement is a basic characteristic of life and an indispensable component of everyday living. Walking is a basic form of displacement for humans and essential for activities of daily living and social integration [1]. Children, the custodians of tomorrow require proper grooming and education [2]. Backpack carriage is the most common way children transport many materials such as books, water bottles and edibles to school [3]. Thus, carrying a backpack is a daily activity for most school children [4].

In India, more than 2.5 million elementary school children carry backpacks on their shoulders for 5 days in a week, [2] this practice is also not uncommon among Nigerian children [5–7]. Many adverse effects have been associated with carrying backpacks in children such as postural discomforts and fatigue, [8, 9] altered gait variables, [10, 11] plantar pressure, [12, 13] musculoskeletal pain, [14, 15] etc. Studies have shown that backpack design, [16, 17] strap pattern, [18, 19] duration of carriage, [20] backpack weight in relation to the child's body weight, [21, 22] and level of load placement, [23] are factors that influence safe backpack usage among school children.

Several studies [10, 11, 23–25] have reported the effects of backpack usage on several gait parameters. However, the effect of backpack on the energy cost of ambulation appears unknown. Also, while there are very scanty studies on the effects of varying backpack loads on gait parameters, there appear to be a dearth of published reports on the interactive effects of gait pace (velotype) and varying backpack loads on the spatiotemporal gait parameters and energy cost of ambulation. This study therefore assessed the influence of varying backpack weights and velotypes on the spatiotemporal gait parameters and energy cost of ambulation of primary school children in Nigeria.

# 2 Method

## 2.1 Participants

A total of 69 primary school children participated in this study with a self-controlled cross-over study design and repeated measures of their gait variables. Only school children in primary five and six recruited from three randomly selected primary schools in Enugu, Nigeria participated in this study. Children with any possible cofounders on gait performance such as neurological or cognitive impairment as well as any deformity that might dispose them to greater risk of carrying weighted backpacks were excluded from this study. A simple random sampling technique was used to select the participants from the screened pool. The minimum sample size was calculated using the sample size determination for finite population according to Slovin's formula [26] as shown in Eqs. 1 and 2:

$$n = \frac{N}{[1 + N\ (e^2)]} \quad (1)$$

where N = population size, n = sample size, e = error margin. Thus,

$$n = \frac{103}{[1 + 103(0.07^2)]} = 67 \quad (2)$$

Therefore, a minimum of 67 participants was projected for the study.

## 2.2 Procedure/Outcome Assessment

Ethical approval was sought and obtained from the Medical Research and Ethics Committee of University of Nigeria Teaching Hospital, Ituku-Ozalla. A written informed consent was obtained from the parents/guardians of the participants before they took part in this study. Demographic details such as age, weight, height and sex were assessed. A weighing scale (Harson, England) was used to measure the body-weight of participants; the precision of the weighing scale was 0.5 kg. A Hanging scale (Fazini, USA) was used to measure the backpack weights; the precision of the scale was 0.1 kg. A tape rule (Goldfish, China), a stop watch (SKMEI, China) and a digital camera (Pentax, USA) were used to measure the spatial parameters of gait, the temporal parameters of gait and to record motion pictures of the gait trials respectively.

Each participant performed a 10 m Walk Test (10MWT) without a backpack, and while carrying backpacks of 10%, 15% and 20% of their body weight at normal, slow and fast velotypes. On the first day, the participants performed the test without a backpack at the three velotypes, while they performed the test with a backpack weighing 10% of their body weight at the three velotypes on the second day. On the third and fourth days, they performed the 10MWT bearing backpacks of 15% and 20% of their body weights respectively at normal, slow and fast velotypes. The number of steps taken to walk the 10 m distance was counted by an assessor while another assessor used the stop watch to measure the duration of each test in the nearest 0.01 s. Possible error in count was controlled through video analysis of the motion that was captured by a camera placed lateral to the participant and perpendicular to the plane of movement. The gait parameters and energy cost of ambulation for each test were derived using the following Eqs. 3–10 [27–31]:

$$\text{No.of strides} = \frac{\text{no.of steps}}{2} \quad (3)$$

$$\text{Stride length (SL)} = \frac{\text{distance}}{\text{no.of strides}} \quad (4)$$

$$\text{Stride frequency (SF)} = \frac{\text{no.of strides}}{\text{time(s)}} \quad (5)$$

$$\text{Gait velocity (V)} = \frac{\text{distance}}{\text{time(s)}} = \text{SL} \times \text{SF} \quad (6)$$

$$\text{Stance duration (ST)} = -0.17 + 0.72\left(\frac{1}{\text{SF}}\right) \quad (7)$$

$$\text{Swing duration (SW)} = 0.18 + 0.27\left(\frac{1}{\text{SF}}\right) \quad (8)$$

$$\text{Double support duration (DS)} = 0.18 + 0.24\left(\frac{1}{\text{SF}}\right) \quad (9)$$

$$\text{Energy cost of ambulation (E)} = \frac{29}{\text{u}} + 0.0053\text{u} \quad (\textit{where u} = \textit{velocity in meters/minute}) \quad (10)$$

### 2.3 Data Analysis

Data obtained was cleaned and analyzed using SPSS version 21.0. Descriptive statistics of frequency, percentages, mean, and standard deviation were used to summarize the anthropometric and gait related variables. Repeated measure ANOVA was used to analyze the difference in gait parameters and energy cost of ambulation for the various groups of backpack loading at the different veloytypes. Bonferroni pairwise comparison was used for the post hoc test whenever the repeated measure ANOVA test was significant. The level of significance was set at $\alpha = 0.05$.

## 3 Results

### 3.1 Summary of Participants Variables

A total of 69 primary school children participated in this study, with an approximate male to female ratio of 1:2 (27 males and 42 females). The age range of the participants was between 9 and 15 years with a mean age of 11.42 ± 1.35years. The weight of the participants ranged from 28-55 kg with a mean weight of 37.4 ± 6.31 kg while the mean height of the participants was 1.47 ± 0.07 m, the height is ranged between 1.30-1.69 m. The mean body mass index of the participants was 17.16 ± 1.85 kg/m$^2$. The mean weight of the backpacks the participants carried to school was 2.92 ± 1.38 kg ranging from 0.5-7 kg of backpack weight. More than a quarter of the participants (26.1%) carried a backpack greater than 10% of their body weight as shown in Table 1.

**Table 1.** Summary of participants' demographic variables (N = 69)

| Variable | Category | Mean | S.D | Min | Max | Frequency | % |
|---|---|---|---|---|---|---|---|
| Sex | Male | | | | | 27 | 39.1 |
| | Female | | | | | 42 | 60.9 |
| BP-%W | ≤ 10% | | | | | 51 | 73.9 |
| | >10% <15% | | | | | 16 | 23.3% |
| | 15–20% | | | | | 1 | 1.4 |
| | >20% | | | | | 1 | 1.4 |
| AGE | | 11.48 | 1.35 | 9.00 | 15.00 | | |
| WEIGHT (kg) | | 37.40 | 6.31 | 28.00 | 55.00 | | |
| HEIGHT (m) | | 1.47 | 0.07 | 1.30 | 1.69 | | |
| BMI (kg/m$^2$) | | 17.16 | 1.85 | 13.79 | 22.73 | | |
| BPW (kg) | | 2.92 | 1.38 | 0.50 | 7.00 | | |
| RHR (bpm) | | 69.54 | 9.66 | 56.00 | 108.00 | | |

*Key: S.D = Standard Deviation, BP = Backpack Weight, BP-%W = Backpack weight in percentage of Body Weight, RHR = Resting Heart Rate in beats per minute.*

### 3.2 Comparison of Selected Participants Variables Across the Different Backpack Loading

At self-selected normal velotype (pace), only gait velocity was found to be significantly different when compared across the different backpack loads ($F = 7.19$, $p = 0.001$). The post-hoc comparison revealed that gait velocity was significantly higher at 20% backpack load than each of 15% (MD = 0.1, $p < 0.05$) and 10% (0.09, $p < 0.05$). However, there was no significant difference in the rest of the gait parameters and energy expenditure ($p > 0.05$) across the groups of backpack load at normal velotype as shown in Table 2.

At slow velotype, there was a significant difference in energy cost of ambulation across the groups of backpack loading ($F = 9.058$, $p < 0.001$); the post-hoc revealed that the energy expenditure at 20% backpack was significantly lower than those of other backpack loadings ($p < 0.05$). There was also a significant difference in each of stride frequency ($F = 12.36$, $p < 0.001$), and swing duration ($F = 11.149$, $p < 0.001$) across the different groups of backpack load. Post-hoc correction showed that stride frequency was significantly lower at backpack load of 10% when compared with other groups ($p < 0.05$) but had swing duration significantly higher at 10% backpack load than other groups ($p < 0.05$) as shown in Table 3. Also in Table 3, there was a significant difference in each of stride length ($F = 10.419$, $p < 0.001$) and velocity ($F = 13.614$, $p < 0.001$) across the groups of backpack loading. Post hoc revealed that at 20% loading, both variables were significantly higher than other loading groups ($P < 0.05$). In the same vein the significant difference in stance duration ($F = 9.526$, $p < 0.001$) was found to be between 10% backpack (0.83 ± 0.19 s) loading and no

**Table 2.** Comparison of participants variables across different backpack load at normal velotype (N = 69).

| Variable | BP | BP1 | BP2 | BP3 | F | P-value |
|---|---|---|---|---|---|---|
| Energy (cal/m/kg) | 0.83 ± 0.07 | 0.83 ± 0.04 | 0.84 ± 0.07 | 0.82 ± 0.05 | 1.797 | 0.170 |
| SL (m) | 1.24 ± 0.18 | 1.25 ± 0.18 | 1.23 ± 0.21 | 1.30 ± 0.24 | 2.450 | 0.075 |
| SF | 0.79 ± 0.13 | 0.76 ± 0.09 | 0.76 ± 0.10 | 0.80 ± 0.13 | 2.532 | 0.078 |
| Gait Velocity ($ms^{-2}$) | 0.97 ± 0.10[abcd] | 0.95 ± 0.11[abc] | 0.94 ± 0.19[abc] | 1.04 ± 0.24[ad] | 7.191 | 0.001* |
| St. Dur. (s) | 0.78 ± .0.20 | 0.79 ± 0.11 | 0.80 ± 0.15 | 0.75 ± 0.17 | 1.605 | 0.203 |
| Sw. Dur. (s) | 0.56 ± −.16 | 0.54 ± 0.43 | 0.54 ± 0.06 | 0.53 ± 0.07 | 1.995 | 0.157 |
| Ds. Dur. (s) | 0.17 ± 0.21 | 0.16 ± 0.15 | 0.14 ± 0.05 | 0.13 ± 0.07 | 1.163 | 0.311 |

*Key: SL-stride length. SF-stride frequency. St. Dur.-stance duration. Sw Dur-swing duration. Ds. Dur-double support duration. BP- no backpack condition. BP1- backpack with 10% body weight. BP2- backpack with 15% body weight BP3- backpack with 20% body weight *-significant ($p < 0.05$) ($X^a$–$X^b$) - significant difference in pair-wise post-hoc comparison.*

backpack loading (0.93 ± 0.19 s) groups ($p < 0.05$). There was no significant difference in double support duration across the different backpack loading groups at self-selected slow velotype ($F = 3.191$, $p = 0.056$) as shown in Table 3.

**Table 3.** Comparison of participants variables across different backpack load at slow velotype (N = 63)

| Variable | BP | BP1 | BP2 | BP3 | F | P-value |
|---|---|---|---|---|---|---|
| Energy (cal/m/kg) | 0.94 ± 0.15[abc] | 0.96 ± 0.19[abc] | 0.94 ± 0.16[abc] | 0.88 ± 0.11[d] | 9.258 | <0.001* |
| SL (m) | 0.98 ± 0.16[a] | 1.04 ± 0.15[bc] | 1.04 ± 0.20[bc] | 1.10 ± 0.19[d] | 10.419 | <0.001* |
| SF | 0.76 ± 0.16[acd] | 0.67 ± 0.10[b] | 0.72 ± 0.17[acd] | 0.75 ± 0.12[acd] | 12.316 | <0.001* |
| Gait Velocity ($ms^{-2}$) | 0.74 ± 0.19[abc] | 0.70 ± 0.16[abc] | 0.74 ± 0.18[abc] | 0.82 ± 0.18[d] | 13.614 | <0.001* |
| St. Dur. (s) | 0.83 ± 0.19[a] | 0.93 ± 0.19[b] | 0.89 ± 0.24[ac] | 0.82 ± 0.160[acd] | 9.526 | <0.001* |
| Sw. Dur. (s) | 0.56 ± 0.07[acd] | 0.59 ± 0.07[b] | 0.57 ± 0.08[ac] | 0.55 ± 0.06[ad] | 11.149 | <0.001* |
| Ds Dur. (s) | 0.15 ± 0.07 | 0.22 ± 0.22 | 0.19 ± 0.15 | 0.17 ± 0.11 | 3.191 | 0.056 |

*Key: SL-stride length. SF-stride frequency. St. Dur.-stance duration. Sw Dur-swing duration. Ds. Dur-double support duration. BP- no backpack condition. BP1- backpack with 10% body weight. BP2- backpack with 15% body weight BP3- backpack with 20% body weight *-significant ($p < 0.05$) ($X^a$–$X^b$) - significant difference in pair-wise post-hoc comparison.*

At fast walking pace, Post hoc for the significant difference in energy expenditure across the backpack loading groups ($F = 32.096$, $p < 0.001$) showed that energy expenditure at 20% backpack loading was significantly higher than other loadings ($p < 0.05$). Similar trend was also found for stride length ($F = 56.364$, $p < 0.001$), stride frequency ($F = 20.239$, $p < 0.001$), gait velocity ($F = 57.485$, $p < 0.001$). However, in stance duration ($F = 8.954$, $p < 0.001$), swing duration ($F = 11.333$, $p < 0.001$), and double support duration ($F = 12.655$, $p < 0.001$), these parameters were lower at 20% backpack loading than other groups ($p < 0.05$) as shown in Table 4.

**Table 4.** Comparison of Participants' variables across different backpack load at fast velotype (N = 69)

| Variables | BP | BP1 | BP2 | BP3 | F | P-value |
|---|---|---|---|---|---|---|
| Energy (cal/m/kg) | $0.80 \pm 0.02^{ab}$ | $0.80 \pm 0.02^{ab}$ | $0.82 \pm 0.04^{c}$ | $0.85 \pm 0.06^{d}$ | 32.096 | <0.001* |
| SL (m) | $1.38 \pm 0.16^{a}$ | $1.51 \pm 0.26^{b}$ | $1.59 \pm 0.24^{c}$ | $1.67 \pm 2.34^{d}$ | 56.364 | <0.001* |
| SF | $0.95 \pm 0.14^{abc}$ | $0.93 \pm 0.08^{ab}$ | $0.98 \pm 0.14^{ac}$ | $1.06 \pm 0.15^{d}$ | 20.239 | <0.001* |
| Gait Velocity ($ms^{-2}$) | $1.30 \pm 0.25^{ab}$ | $1.40 \pm 0.28^{ab}$ | $1.54 \pm 0.35^{c}$ | $1.75 \pm 0.38^{d}$ | 57.483 | <0.001* |
| St. Dur. (s) | $0.60 \pm 0.12^{abc}$ | $0.60 \pm 0.09^{abc}$ | $0.62 \pm 0.17^{abc}$ | $0.53 \pm 0.10^{d}$ | 8.954 | <0.001* |
| Sw. Dur. (s) | $0.47 \pm 0.05^{abc}$ | $0.47 \pm 0.03^{abc}$ | $0.47 \pm 0.05^{abc}$ | $0.44 \pm 0.04^{d}$ | 11.353 | <0.001* |
| Ds Dur. (s) | $0.08 \pm 0.04^{abc}$ | $0.08 \pm 0.03^{abc}$ | $0.08 \pm 0.04^{abc}$ | $0.05 \pm 0.04^{d}$ | 12.655 | <0.001* |

*Key: SL-stride length. SF-stride frequency. St. Dur.-stance duration. Sw Dur-swing duration. Ds. Dur-double support duration. BP- no backpack condition. BP1- backpack with 10% body weight. BP2- backpack with 15% body weight BP3- backpack with 20% body weight *-significant (p < 0.05) ($X^a$–$X^b$) - significant difference in pair-wise post-hoc comparison.*

### 3.3 Comparison of Participants Variables Across the Different Velotypes (Normal, Slow and Fast)

In ambulation without a backpack, there was a significant difference in energy expenditure when compared among the three gait velotypes ($F = 42.576$, $p < 0.001$), with a post hoc that showed that energy expenditure was significantly higher at slow velotype than other velotypes ($p < 0.05$). On the other hand, there was a significant difference in each of stride length ($F = 122.93$, $p < 0.001$), stride frequency ($F = 37.476$, $p < 0.001$), and gait velocity ($F = 151.659$, $p < 0.001$) across the different velotypes. Post hoc showed that these parameters were highest at fast velotype than the other Velotypes ($p < 0.05$). Also, there was a significant difference in each of stance duration ($F = 34.947$, $p < 0.001$), swing duration ($F = 16.522$, $p < 0.001$), and double support duration ($F = 10.598$, $p < 0.001$); their post-hoc revealed that walking at fast velotype recording significantly lower values in these parameters when compared with other velotypes ($p < 0.05$) as shown in Table 5.

**Table 5.** Comparison of Participants Variables across Different Velotypes without a Backpack (N = 63)

| Variable | Velotype1 | Velotype2 | Velotype3 | F | p-value |
|---|---|---|---|---|---|
| Energy (cal/m/kg) | $0.83 \pm 0.07^{a}$ | $0.94 \pm 0.15^{b}$ | $0.80 \pm 0.02^{c}$ | 42.516 | <0.001* |
| SL (m) | $1.24 \pm 0.18^{a}$ | $0.98 \pm 0.16^{b}$ | $1.38 \pm 0.16^{c}$ | 122.934 | <0.001* |
| SF | $0.79 \pm 0.13^{ab}$ | $0.76 \pm 0.16^{ab}$ | $0.95 \pm 0.14^{c}$ | 37.476 | <0.001* |
| Gait Velocity ($ms^{-2}$) | $0.97 \pm 0.20^{a}$ | $0.74 \pm 0.19^{b}$ | $1.30 \pm 0.25^{c}$ | 151.659 | <0.001* |
| St. Dur. (s) | $0.78 \pm 0.20^{ab}$ | $0.83 \pm 0.19^{ab}$ | $0.60 \pm 0.12^{c}$ | 34.947 | <0.001* |
| Sw. Dur. (s) | $0.56 \pm 0.16^{ab}$ | $0.56 \pm 0.07^{ab}$ | $0.47 \pm 0.46^{c}$ | 16.522 | <0.001* |
| Ds Dur. (s) | $0.17 \pm 0.21^{ab}$ | $0.15 \pm 0.06^{ab}$ | $0.08 \pm 0.04^{c}$ | 10.598 | 0.001* |

*Key: SL = Stride Length. SF = StrideFfrequency. St. Dur. = Stance Duration, Sw Dur = Swing Duration, Ds. Dur = Double support Duration, Velotype 1 = Normal walking pace, Velotype 2 = Slow walking pace, Velotype 3 = Fast walking pace,* * *= Significant,* ($X^{a}$–$X^{b}$) *= significant difference in pair-wise post-hoc comparison.*

When the participants carried backpacks weighing 10% of their body weights, there was a significant difference in energy expenditure when compared across the three velotypes ($F = 41.973$, $p < 0.001$) and the post hoc showed that energy expenditure was significantly higher at slow velotype than other velotypes ($p < 0.05$). On the other hand, there was a significant difference in each of stride length ($F = 143.881$, $p < 0.001$), stride frequency ($F = 178.708$, $p < 0.001$), and gait velocity ($F = 265.783$, $p < 0.001$) across the different velotypes. Post hoc showed that these parameters were highest at fast velotype than the other velotypes ($p < 0.05$). However, the significant difference in stance duration ($F = 115.542$, $p < 0.001$), swing duration ($F = 115.818$, $p < 0.001$), and double support duration ($F = 16.261$, $p < 0.001$) had a post-hoc that revealed that fast velotype category had significantly lower values in these parameters when compared with other velotypes ($p < 0.05$) as shown in Table 6.

**Table 6.** Comparison of Participants Variables across Different Velotypes with a Backpack of 10% Bodyweight (N = 63)

| Variable | Velotype1 | Velotype2 | Velotype3 | F | p-value |
|---|---|---|---|---|---|
| Energy (cal/m/kg) | $0.82 \pm 0.04^{a}$ | $0.96 \pm 0.19^{b}$ | $0.80 \pm 0.23^{c}$ | 41.973 | <0.001* |
| SL (m) | $1.25 \pm 0.18^{a}$ | $1.04 \pm 0.15^{b}$ | $1.51 \pm 0.26^{c}$ | 143.881 | <0.001* |
| SF | $0.76 \pm 0.09^{a}$ | $0.67 \pm 0.10^{b}$ | $0.93 \pm 0.08^{c}$ | 178.708 | <0.001* |
| Gait Velocity ($ms^{-2}$) | $0.95 \pm 0.16^{a}$ | $0.70 \pm 0.16^{b}$ | $1.40 \pm 0.28^{c}$ | 265.783 | <0.001* |
| St. Dur. (s) | $0.79 \pm 0.11^{a}$ | $0.93 \pm 0.19^{b}$ | $0.60 \pm 0.09^{c}$ | 115.542 | <0.001* |
| Sw. Dur. (s) | $0.54 \pm -0.04^{a}$ | $0.59 \pm 0.07^{b}$ | $0.47 \pm 0.03^{c}$ | 115.818 | <0.001* |
| Ds Dur. (s) | $0.16 \pm 0.14^{ab}$ | $0.22 \pm 0.22^{ab}$ | $0.8 \pm 0.03^{c}$ | 16.261 | <0.001* |

*Key: SL = Stride Length. SF = StrideFfrequency. St. Dur. = Stance Duration, Sw Dur = Swing Duration, Ds. Dur = Double support Duration, Velotype 1 = Normal walking pace, Velotype 2 = Slow walking pace, Velotype 3 = Fast walking pace,* * *= Significant,* ($X^{a}$–$X^{b}$) *= significant difference in pair-wise post-hoc comparison.*

When participants ambulated with backpacks weighing 15% of their body weight, there was a significant difference in energy expenditure (F = 27.331, p < 0.001) across the three velotypes. The post hoc analysis showed that energy expenditure was significantly higher at slow velotype than other velotypes (p < 0.05). On the other hand, there was a significant difference in each of stride length (F = 177.811, p < 0.001), stride frequency (F = 76.840, p < 0.001), and gait velocity (F = 224.36, p < 0.001) across the different velotypes. Their post-hoc analysis revealed that these parameters were higher at fast velotype than the other velotypes (p < 0.05). However, the significant difference in stance duration (F = 42.755, p < 0.001), swing duration (F = 63.585, p < 0.001), and double support duration (F = 25.421, p < 0.001) had a post-hoc in which fast velotype category recorded significantly lower values in these parameters when compared with other velotypes (p < 0.05) as shown in Table 7.

**Table 7.** Comparison of participant's variables across different velotypes with Backpack of 15% Body Weight (N = 69)

| Variable | Velotype1 | Velotype2 | Velotype3 | F | p-value |
|---|---|---|---|---|---|
| Energy (cal/m/kg) | $0.84 \pm 0.07^{ac}$ | $0.94 \pm 0.16^{b}$ | $0.82 \pm 0.04^{ac}$ | 27.331 | <0.001* |
| SL (m) | $1.23 \pm 0.21^{a}$ | $1.04 \pm 0.20^{b}$ | $1.59 \pm 0.24^{c}$ | 177.811 | <0.001* |
| SF | $0.76 \pm 0.10^{ab}$ | $0.72 \pm 0.17^{ab}$ | $0.98 \pm 0.14^{c}$ | 76.840 | <0.001* |
| Gait Velocity ($ms^{-2}$) | $0.94 \pm 0.19^{a}$ | $0.74 \pm 0.18^{b}$ | $1.54 \pm 0.35^{c}$ | 224.360 | <0.001* |
| St. Dur. (s) | $0.80 \pm 0.15^{a}$ | $0.89 \pm 0.24^{b}$ | $0.62 \pm 0.17^{c}$ | 42.753 | <0.001* |
| Sw. Dur. (s) | $0.54 \pm 0.06^{a}$ | $0.57 \pm 0.08^{b}$ | $0.47 \pm 0.05^{c}$ | 63.585 | <0.001* |
| **Ds Dur. (s)** | $0.14 \pm 0.05^{ab}$ | $0.19 \pm 0.15^{ab}$ | $0.07 \pm 0.04^{c}$ | 25.421 | <0.001* |

*Key: SL = Stride Length. SF = StrideFfrequency. St. Dur. = Stance Duration, Sw Dur = Swing Duration, Ds. Dur = Double support Duration, Velotype 1 = Normal walking pace, Velotype 2 = Slow walking pace, Velotype 3 = Fast walking pace, * = Significant, ($X^a$–$X^b$) = significant difference in pair-wise post-hoc comparison.*

Finally, during ambulation with a backpack of 20% of participants' body weight, there was a significant difference in energy expenditure (F = 9.777, p < 0.001). The post-hoc analysis showed that energy expenditure was significantly higher at slow velotype than the other velotypes (p < 0.05). On the other hand, there was a significant difference in each of stride length (F = 132.774, p < 0.001), stride frequency (F = 118.618, p < 0.001), and gait velocity (F = 223.547, p < 0.001) across the three Velotypes. Post hoc showed that these parameters were higher at fast velotype than the other Velotypes (p < 0.05). However, the significant difference in stance duration (F = 84.168, p < 0.001), swing duration (F = 86.424, p < 0.001), and double support duration (F = 26.062, p < 0.001) had a post-hoc that revealed that these parameters were significantly lower in the fast velotype category than the other velotypes (p < 0.05) as shown in Table 8.

**Table 8.** Comparison of Participants Variables across Different Velotypes with Backpack of 20% Body Weight (N = 63)

| Variable | Velotype1 | Velotype2 | Velotype3 | F | p-value |
|---|---|---|---|---|---|
| Energy (cal/m/kg) | $0.82 \pm 0.05^{a}$ | $0.88 \pm 0.11^{b}$ | $0.85 \pm 0.06^{c}$ | 9.777 | <0.001* |
| SL (m) | $1.30 \pm 0.24^{a}$ | $1.10 \pm 0.19^{b}$ | $1.67 \pm 0.23^{c}$ | 132.774 | <0.001* |
| SF | $0.80 \pm 0.13^{a}$ | $0.75 \pm 0.12^{b}$ | $1.06 \pm 0.15^{c}$ | 118.618 | <0.001* |
| Gait Velocity ($ms^{-2}$) | $1.04 \pm 0.24^{a}$ | $0.82 \pm 0.18^{b}$ | $1.75 \pm 0.38^{c}$ | 223.547 | <0.001* |
| St. Dur. (s) | $0.75 \pm 0.17^{a}$ | $0.82 \pm 0.16^{b}$ | $0.51 \pm 0.10^{c}$ | 84.168 | <0.001* |
| Sw. Dur. (s) | $0.53 \pm 0.06^{ab}$ | $0.55 \pm 0.06^{ab}$ | $0.44 \pm 0.04^{c}$ | 86.424 | <0.001* |
| Ds Dur. (s) | $0.13 \pm 0.07^{ab}$ | $0.17 \pm 0.16^{ab}$ | $0.50 \pm 0.04^{c}$ | 26.062 | <0.001* |

*Key: SL = Stride Length. SF = StrideFfrequency. St. Dur. = Stance Duration, Sw Dur = Swing Duration, Ds. Dur = Double support Duration, Velotype 1 = Normal walking pace, Velotype 2 = Slow walking pace, Velotype 3 = Fast walking pace, * = Significant, ($X^a$–$X^b$) = significant difference in pair-wise post-hoc comparison.*

## 4 Discussion

Most of the children in this study carried backpacks less than 10% of their body weights. This result differs from the report of Pau et al. [32] that reported the backpack to bodyweight ratio of school children in Italy as greater than 15% of their body weight. Also, Al-Saleem et al. [33] in their work among Saudi school children found that 72.2% of the children carried school bags greater than 15% of their body weight. The observed differences in the backpack to bodyweight ratio between this study and other studies may be attributed to socioeconomic and cultural differences in these environments. The comparing studies were conducted in developed countries and among children from higher socioeconomic status unlike in the present study that was done in Nigeria, a developing nation with lots of people living in poverty, [34] who may not afford backpacks and certain learning items like mini laptops, tablets etc. for their children. It is however, still worrisome that greater than a quarter of children in this study carried a backpack greater than 10% of their body weight. Nigerians therefore need to be enlightened about the rsik of heavy backpacks among school aged children so as to forestall the likelihood of increased musculoskeletal disorders.

At self-selected normal velotype, there was no significant difference in the measured spatio-temporal parameters (with the exception of gait velocity) when the participants ambulated without a backpack compared to when they ambulated with different loads of backpack. This may imply that at normal, comfortable walking pace, the body is able to regulate these gait parameters especially when the externally exerted weight does not exceed 20% of the body weight. This may have some applications in clinic, sports and ergonomics where the effects of increased loading may be compensated for by walking at self-selected normal speed. This result is in consonance with the reports of Pau et al. [32] that found no significant differences in stride and spatio-temporal gait parameters between loaded and unloaded conditions. However, it is in contrast with the findings of Abaraogu et al. [35] who reported significant effects of backpack load on cadence at normal walking speed as well as with the report of Hong and Bruggemann, [36] who found significant increase in the double support duration and decrease in the swing time during ambulation of their participants while bearing 20% of their body weights. The differences in the findings could be attributed to differences in participants' characteristics, study design and method of gait analysis. While the study by Abaraogu et al. [35] involved young adults, the present study was among school aged children. It has been established that children and adults have different strategies of countering gait destabilization. [37] Also, the study by Hong and Brueggmann, [36] recruited only male children and analysed gait using treadmill unlike in this present study that had both male and female children and analysed gait using video and observational techniques.

Also, there was no significant difference in energy expenditure during ambulation at normal walking pace during loaded and unloaded conditions. This indicates that at normal walking speed the system is able to compensate for any destabilization while maintaining the minimum energy requirement necessary for gait. While there appear to be no similar comparable study on the influence or effects of different backpack loading on gait energy expenditure, this result may have some clinical relevance. This knowledge may be applied to certain neurological conditions such as stroke where there is a drastic drop in gait energy efficiency due to heaviness of the paretic limbs. Therefore energy may be maximally conserved during neurological rehabilitation of stroke survivors by ambulating them at their self-selected normal speed.

At self-selected slow velotype, energy cost of ambulation decreased significantly at 20% body weight backpack load when compared to backpack-free ambulation. It has been previously established that energy expenditure increases at slower walking speed, [38] because energy expended during gait is a function of gait speed. Thus, in order to counter the expected increase in energy expenditure especially in a state of increased body loading, the body may compensate by selecting a more comfortable pace (velotype) so as to decrease energy expenditure. Alternatively, the discomfort imposed by the increased loading may cause the body to compensate by increasing the gait velocity so as to achieve gait purpose in a shorter time and thus decreasing the energy cost of ambulation in the process. Literature appears silent on the effects of loading on the spatiotemporal gait parameters and energy expenditure for children at slow walking velotype, more studies are therefore recommended. Similar study is also recommended in conditions where these compensatory strategies may not be readily available such as in children with neuro-developmental disorders compared with apparently healthy children.

All the measured spatio-temporal variables of gait were significantly different across the groups of backpack load at fast walking pace. This finding is in agreement with those of Abaraogu et al. [35] that reported a significant difference in the cadence of young adults while walking fast compared to their normal gait pace. Also, energy expenditure increased significantly while carrying backpacks of 15% and 20% body weight when compared with the unloaded state. This appears to suggest that there is an interactive effect of load and velotype (or velocity) on gait energy expenditure. With increase in loading, the participants tend to transit to a higher velotype by increasing their gait velocity as shown in the results above. Since gait energy expenditure is a direct function of gait velocity, [27] this increased velocity results in lower energy expenditure. However, during fast velotype, carrying a backpack of 20% body weight recorded the highest energy expenditure. Possible explanation could be that the body had attained its highest velotype for walking, and so further transitioning induced by increased loading may result in running. Therefore the body becomes least efficient when forced to continue walking fast with increased loading.

When variables were compared across the velotypes, energy cost of ambulation was highest in slow velotype both in the unloaded and loaded states whereas the normal velotype was most energy efficient when children carry a load up to 20% of their body weight. This implies that slow walking irrespective of the external load, is the least energy efficient pace. Also, while carrying at load greater than 20% of a child's body weight, walking at self selected normal pace conserves the most energy. This is also in line with the assertion by Schrack et al. [38] that energy expenditure increases at decreasing gait speed and also buttresses the initial position that the body becomes less efficient in conserving energy during ambulation when subjected to fast pace with increased loading (20% of body weight). However, more studies comparing the effects of higher loads (greater than 20% of body weight) on gait energy expenditure of both children and adults are recommended.

Stride length, stride frequency and gait velocity were consistently highest at fast velotype while stance duration, swing duration and double support duration were consistently lowest at fast velotype irrespective of the backpack loads. To walk at a fast velotype or pace or rate is to increase gait velocity. Gait velocity has been shown to be a product of cadence and step length [39]. Cadence is defined as the number of steps made per minute, [40] and therefore determines stride frequency, while stride length is approximately half the value of step length [30, 31]. It is therefore understandable that stride length and stride frequency were higher at fast velotype than other velotypes irrespective of the weights of the backpack. In the same vein, increase in gait speed is accompanied by a decrease in time spent on stance, swing and double support. This explains the findings that stride, swing and double support durations were lowest at fast velotypes when external load is not greater than 20% of a child's body weight. However, further studies on the effects of higher loading (beyond 20% of body weight) on these gait parameters at different velotypes among children and adults are desired.

While it is believed that this study may have thrown some light on some shrouded areas in literature especially on the effects varying backpack weights and velotypes on energy expenditure and spatio-temporal gait parameters of school children, the findings of this study should be interpreted with caution because the subjectivity of using only equations to estimate gait energy expenditure.

## 5 Conclusion

Based on the results of this study, the following conclusions were made: First, that more than a quarter of Nigerian school children carry backpacks greater than ten percent of their body weights. Secondly, that when children carry backpacks up to 20% of their body weight, their spatiotemporal gait parameters are significantly altered irrespective of their walking pace. Finally, that slow pace of walking is the least energy efficient pace irrespective of the weight of the child's backpack.

# References


1. Schmidt RA, Lee T, Winstein C, Wulf G, Zelaznik H (2018) Motor Control and Learning, 6th edn. Human kinetics, Champaign
2. Rai A, Agarwal S (2014) Assessing the effect of postural discomfort on school going children due to heavy backpacks. J Ergonomics S4:S4–011. https://doi.org/10.4172/2165-7556
3. Lasota A (2014) Schoolbag weight carriage by primary school pupils. Work: J Prev Assess Rehabil 48:21–26 http://www.ncbi.nlm.nih.gov/pubmed/23531575. Accessed Apr 2018
4. Dockrell S, Simms C, Blake C (2015) Schoolbag carriage and schoolbag-related musculoskeletal discomfort among primary school children. Appl Ergon 51:281–290
5. Hamzat TK, Abdulkareem TA, Akinyinka OO, Fatoye FA (2014) Backpack-related musculoskeletal symptoms among Nigerian secondary school students. Rheumatol Int 34 (9):1267–1273
6. Olatunya OS, Isinkaye AO, Agaja OT, Omoniyi E, Oluwadiya KS (2017) Backpack use and associated problems among primary school children in Nigeria: a call to action by stakeholders. Niger J Paediatr 44(3):157–162
7. Yamato TP, Maher CG, Traeger AC, Wiliams CM, Kamper SJ (2018) Do schoolbags cause back pain in children and adolescents? A systematic review. Br J Sports Med, pp.bjsports-2017
8. Janakiraman B, Ravichandran H, Demeke S, Fasika S (2017) Reported influences of backpack loads on postural deviation among school children: a systematic review. J Educ Health Promot 6:41
9. Rai A, Agarwal S (2014) Physical stress among school children due to heavy backpacks. Int J Emerg Trends Eng Dev 3:500–506
10. Pau M, Mandaresu S, Leban B, Nussbaum MA (2015) Short-term effects of backpack carriage on plantar pressure and gait in schoolchildren. J Electromyogr Kinesiol 25(2):406–412
11. Jorge JG, Faria AND, Furtado DA, Pereira AA, Carvalho EMD, Dionísio VC (2018) Kinematic and electromyographic analysis of school children gait with and without load in the backpack. Research on Biomedical Engineering, (AHEAD), pp 0–0
12. Castro MP, Figueiredo MC, Abreu S, Sousa H, Machado L, Santos R, Vilas-Boas JP (2015) The influence of gait cadence on the ground reaction forces and plantar pressures during load carriage of young adults. Appl Ergon 49:41–46
13. Pau M, Leban B, Corona F, Gioi S, Nussbaum MA (2016) School-based screening of plantar pressures during level walking with a backpack among overweight and obese schoolchildren. Ergonomics 59(5):697–703
14. Dockrell S, Simms C, Blake C (2015) Schoolbag carriage and schoolbag-related musculoskeletal discomfort among primary school children. Appl Ergon 51:281–290
15. Ogana SO, Osero JO, Wachira LJ (2017) Musculoskeletal pain and backpack usage among school children in Nairobi County. Kenya East Afr Med J 94(6):413–419
16. Gupta I, Kalra P, Iqbal R (2017) Physiological effects of backpack packing, wearing and carrying on school going children. In: International Conference on Research into Design (pp. 813–822). Springer, Singapore, January 2017
17. Hadid A, Gozes G, Atoon A, Gefen A, Epstein Y (2018) Effects of an improved biomechanical backpack strap design on load transfer to the shoulder soft tissues. J Biomech 76:45–52
18. Mallakzadeh M, Javidi M, Azimi S, Monshizadeh H (2016) Analyzing the potential benefits of using a backpack with non-flexible straps. Work 54(1):11–20

19. Golriz S, Hebert JJ, Bo Foreman K, Walker BF (2017) The effect of shoulder strap width and load placement on shoulder-backpack interface pressure. Work 58(4):455–461
20. Poursadeghiyan M, Azrah K, Biglari H, Ebrahimi MH, Yarmohammadi H, Baneshi MM, Hami M, Khammar A (2017) The effects of the manner of carrying the bags on musculoskeletal symptoms in school students in the city of Ilam. Iran Annal Trop Med Public Health 10(3):600
21. Rai A, Agarwal S (2015) Effect of weight of backpack and physiological stress among school going children. Int J Res 2(1):288–292
22. Brzęk A, Dworrak T, Strauss M, Sanchis-Gomar F, Sabbah I, Dworrak B, Leischik R (2017) The weight of pupils' schoolbags in early school age and its influence on body posture. BMC Musculoskelet Disord 18(1):117
23. Rodrigues S, Domingues G, Ferreira I, Faria L Seixas A (2017) March Influence of different backpack loading conditions on neck and lumbar muscles activity of elementary school children. In: Occupational Safety and Hygiene V: Selected papers from the International Symposium on Occupational Safety and Hygiene (SHO 2017), Guimarães, Portugal, p 485. CRC Press, 10–11 April 2017
24. Song Q, Yu B, Zhang C, Sun W, Mao D (2014) Effects of backpack weight on posture, gait patterns and ground reaction forces of male children with obesity during stair descent. Res Sports Med 22(2):172–184
25. Orantes-Gonzalez E, Heredia-Jimenez J, Soto-Hermoso VM (2015) The effect of school trolley load on spatiotemporal gait parameters of children. Gait Posture 42(3):390–393
26. Tejada JJ, Punzalan JRB (2012) On the misuse of Slovin's formula. Philipp Stat 61(1):129–136
27. Ralston HJ (1958) Energy-speed relation and optimal speed during level walking. Internationale Zeitschrift für Angewandte Physiologie Einschliesslich Arbeitsphysiologie 17(4):277–283
28. Öberg T, Karsznia A, Öberg K (1993) Basic gait parameters: Reference data for normal subjects, 10–79 years of age. J Rehabil Resour Dev 30(2):210–223
29. Wall JC, Charteris J, Turnbull GI (1987) Two steps equals one stride equals what? The applicability of normal gait nomenclature to abnormal walking patterns. Clin Biomech 2:119–125
30. Ibeneme SC (2008) Predicting the Critical Point for the Onset of Diabetic Foot Ulcer Using the Velocity Field Diagram. Doctoral Degree Thesis Submitted to The Faculty Of Health Sciences and Technology, College of Medicine. University of Nigeria, Enugu Campus, Enugu
31. Eke-Okoro ST (1989) Velocity field diagram of human gait. Clin Biomech (Bristol, Avon) 2:92–96. https://doi.org/10.1016/0268-0033(89)90045-4
32. Pau M, Mandaresu S, Leban B, Nussbaum MA (2015) Short-term effects of backpack carriage on plantar pressure and gait in schoolchildren. J Electromyogr Kinesiol 25(2):406–412
33. Al-Saleem SA, Ali A, Ali SI, Alshamrani AA, Almulhem AM, Al-Hashem MH (2016) A study of school bag weight and back pain among primary school children in Al-Ahsa. Saudi Arabia. Epidemiology (Sunnyvale, Calif.) 6(1):222
34. Dauda RS (2017) Poverty and economic growth in Nigeria: Issues and policies. J Poverty 21 (1):61–79
35. Abaraogu UO, Ugwa WO, Nnodim O, Ezenwankwo EF (2016) Effect of backpack strap patterns on gait parameters in young adults at self-selected normal and fast walking speeds. PM RJ 9:676–682. https://doi.org/10.1016/j.pmrj.2016.10.010
36. Hong Y, Brueggemann G (2000) Changes in gait patterns in 10-yearold boys with increasing loads when walking on a treadmill. Gait Posture 11(3):254

37. Curtis-Vinegra L (2015) Differences in balance and muscle activation strategies during gait initiation at different speeds between young and middle-aged adults (Doctoral dissertation, Seton Hall University)
38. Schrack JA, Simonsick EM, Ferrucci L (2013) The relationship of the energetic cost of slow walking and peak energy expenditure to gait speed in mid-to-late life. Am J Phys Med Rehabil/Ass Acad Physiatr 92(1):28
39. Loitongbam SS Joy AK (2018) Partial Body Weight Supported Training in Patients With Stroke: Evaluation of Changes in Certain Gait Parameters. Pre and Post Training. Paripex-Indian J Res 5(10)
40. Tudor-Locke C, Schuna JM, Han H, Aguiar EJ, Larrivee S, Hsia DS, Ducharme SW, Barreira TV, Johnson WD (2018) Cadence (steps/min) and intensity during ambulation in 6–20 year olds: the CADENCE-kids study. Int J Behav Nutr Phys Act 15(1):20

# Text Neck, More Technology, Less Health?

Sandra Zamira Genez Tarrifa(✉) and Ricardo De la Hoz Lara

Universidad Libre, Seccional Barranquilla,
Sede Km. 7 Vía Antigua, Puerto, Colombia
sandragenez@hotmail.com

**Abstract.** The term text neck or neck of text is a term introduced by Dr Dean Firhsman [1] and that has taken force little by little, and is described as the result of the excessive use of the mobile, typical of those who keep their heads too long downward and forward to see your cell phone. The repetition and constancy of this position can generate a permanent injury, whose main cause is the weight of the head.

According to the study "Evaluation of tensions in the cervical spine and posture caused by the position of the head" published by the surgeon Kenneth Hansraj, our head weighs between 4.5 and 5.5 kg. However, depending on the angle you are in, the effective weight increases. That is, if the head is at 15 °, the effective weight, which falls on the neck reaches 12.25 kg, this is the case when handling a mobile device (Smartphone, Tablet, among others). Typical symptoms are headache, pain in shoulders and neck [2].

Likewise. In Colombia in 2015, a survey conducted by Deloitte was applied, where issues related to the types of devices that are possessed, connectivity and usage trends were addressed. It was answered by a thousand people, men and women between the ages of 16 and 44, including people from the rural and urban sectors. With this in mind, we set out to investigate the effects of the excessive use of these mobile devices on young people.

**Keywords:** Technology · Text neck · Pain · Mobile devices
Symptomatology · Frequency

## 1 Background

In an experimental study among young adults conducted by the University of Gothenburg, Sweden and led by the occupational therapist Ewa Gustafsson, we found differences in posture, typing style and muscle activity among those young people with and without musculoskeletal symptoms in the neck and the upper extremities (Gustafsson et al. 2010, 2011). In the group with symptoms, almost all individuals had their necks flexed forward and without support. This causes static muscle changes, load on the neck and shoulders. In addition, they kept the phone with one hand and used only one thumb. This distinguished them from the group without symptoms, in which it was more frequent to sit with a straight neck, support the forearm, hold the phone with both hands and use both thumbs.


S. Bagnara et al. (Eds.): IEA 2018, AISC 826, pp. 758–767, 2019.
https://doi.org/10.1007/978-3-319-96065-4_79

Another study looking at the posture and typing style of university students writing on mobile devices found that almost all subjects had a flexed neck; almost half of them wrote with both thumbs and a third with a thumb [3].

Also in a comparative observational study in two groups conducted at the Australian Catholic University in Sydney, all participants were asked to stay 4 h per day in unsupported electronic devices. Twenty-two volunteers (13 women, 9 men) with neck pain (group "Text Neck") and 22 asymptomatic controls (15 women, 7 men) were evaluated by means of the head repositioning precision test to measure proprioception during the treatment. cervical movements (flexion, extension). A questionnaire was completed detailing the age, gender and average time per day using mobile electronic devices. Participants in the group with symptoms that indicated the intensity of cervical cortical pain. After analyzing the results obtained during this study, it was concluded that the participants with 'Text Neck' revealed an alteration of proprioception during the cervical spine compared to the asymptomatic controls [3].

This is why, the intention of this study is to identify if, the excessive use of mobile devices generates harmful effects at the neck level.

## 2 Methodology

The present study was carried out with adolescents and young adults from 17 to 36 years old, of which 84% correspond to women and 16% to men, this population is in stratification 1 and 2 mostly, with predominance of stratum 1 with 49.4%. In order to identify painful musculoskeletal symptoms, the Nordic Questionnaire was used, in the same way the global survey of mobile consumers was applied, in order to identify the exposure time, the frequency and duration of daily use of these devices, besides analyzing the weight and dimension of each of the mobile devices.

Likewise, goniometry was used in the evaluation of the range of movements of flexion, extension and inclination of the head in individuals when using mobile devices.

In addition, to evaluate the function of spinal segmental levels during neck posture to the exposure of mobile devices, deep reflexes were explored using the percussion hammer and indicating to the participants that they should be relaxed, with their eyes closed, for exploration. of the bicipital reflex The arm of the individual must be partially flexed and with the palm facing down. The evaluator should place his thumb or finger firmly on the biceps tendon (antecubital fossa). Then hit on your finger with the reflex hammer. You must feel the response, there is a flexion of the elbow even if you do not see it. For the triceps reflex the evaluator supports the arm and allows the participant's forearm to hang freely. The evaluator then strikes the tendon of the triceps above the elbow with the broad part of the hammer. For the reflex of the long supinator, the participant will have the forearm resting on the abdomen or in the lap. (bicipital, tricipital, radial style, pronator cube). It was used for the analysis of the variables under study, the statistical package of multivariate analysis statgraphics centurion [4].

## 3 Results

When the Nordic questionnaire [5] was applied to the study population, it was evidenced that the population presented painful musculoskeletal symptoms. First of all, pain was found in the neck with 57.8%, secondly in the dorsal or lumbar region of 55.3% and third in the hand. doll with 31.9%. As shown in graph No 1.

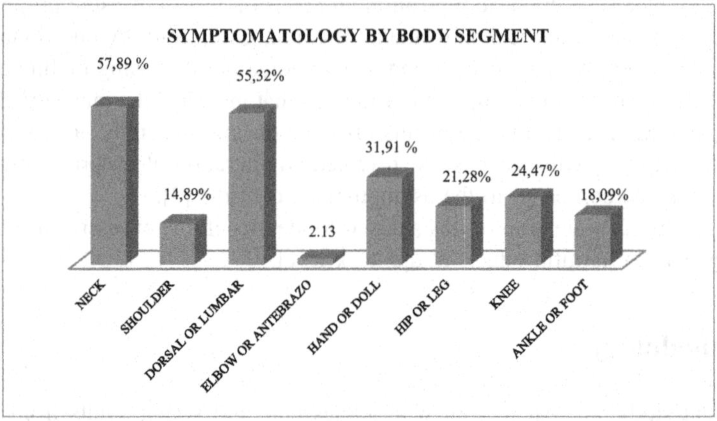


**Graph 1** .

When describing the severity of the neck pain, the respondents chose with option 45.4% which corresponds to the moderate level of pain, second option 2 and third option 4. As shown in the graph No 2.

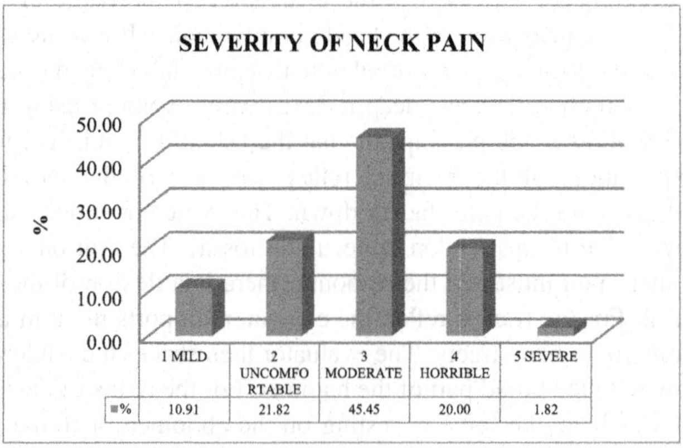


**Graph 2** .

When asked how long has been discomfort the last 12 months respondents answered that, neck pain occurs from 1 to 7 days in 25% of the population, 8–30 days in 19.2%, 2–4 months in 19.2%. As shown in graph No 3.

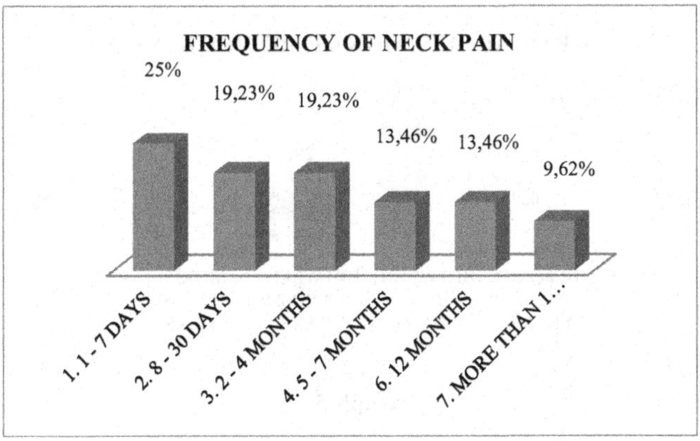


**Graph 3** .

In turn, the painful neck event occurs in 38.9% for less than 1 h, 31.9% with a duration of 1 to 24 h and 20.8% with a duration of 1 to 7 days (Graph 4).

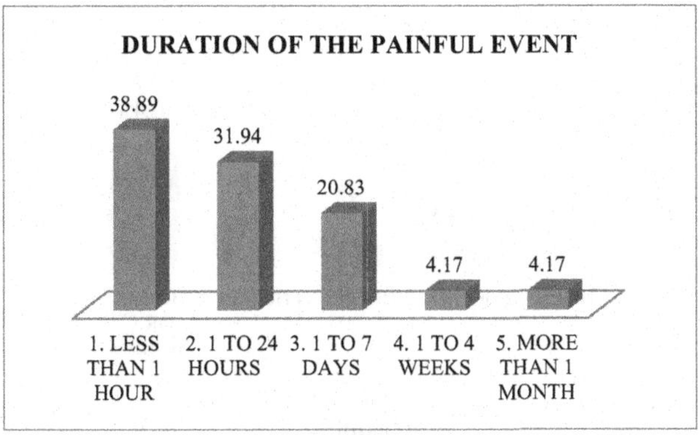


**Graph 4** .

During the application of the survey on the use of mobile devices, it can be seen that 60% (58% of neck symptoms) of the target population acquired a mobile device between 10–15 years of age, 29.5% among the 16 and 20 years. As shown in the following graphic 5.

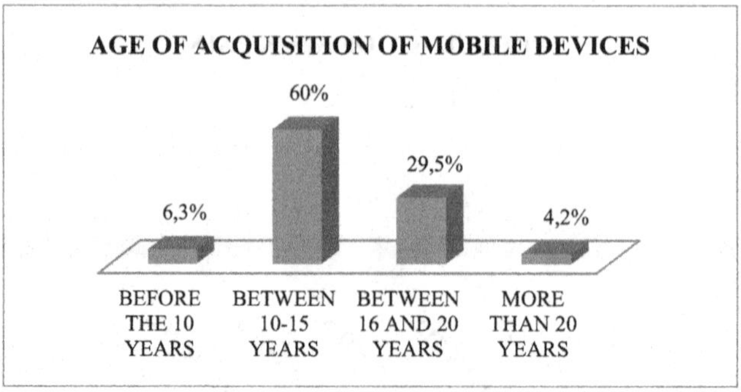


**Graph 5** .

Likewise, the target population uses their mobile device daily for 5 h or more for 63.4% and 4 h for 21.5%. As shown in graph No 6.

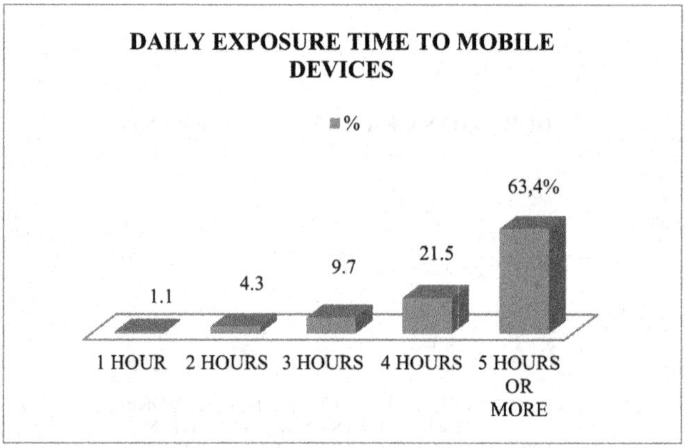


**Graph 6** .

As for the individuals who presented painful symptoms in the neck, when considering the weight and dimensions of the mobiles used, the following was found (Graph 7).

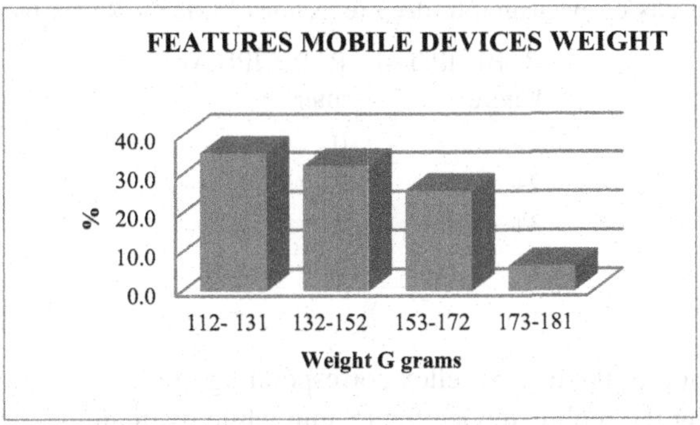


**Graph 7** .

Taking into account that the devices most used by the participating individuals have a weight of 112–131 grams, participants were instructed to sit on a chair with a backrest and use their mobile device for 15 min; When analyzing the accumulated weight during this time, it was determined that this equals 1,950 grams. It was also found that the models of the mobile devices most used by the young participants of the study are the Samsung J2 and J5 respectively.

After having identified the population with moderate symptoms of neck pain, participants were instructed to sit in a chair with a backrest, the deep reflexes of the upper limb (bicipital, triceps, style radial and pronator cube) were evaluated. I asked the participants to use their mobile device for 15 min they took again the deep reflexes of the upper limb (biceps, triceps, styleradial and pronator cube), having as reference for the evaluation of the deep reflexes the Scale of Siedel [4], presented to continuation. Table 1.

**Table 1.** The Scale of Siedel

| Degree type of response | Degree type of response |
|---|---|
| 0 | Unanswered |
| 1+ | Slow or diminished response |
| 2+ | Normal response |
| 3+ | Light response increase |
| 4+ | Abrupt increase in response |

Finding what:

The exploration of the bicipital reflex after having been exposed to the mobile device while maintaining the posture of neck flexion, 50% of the participating population showed a decrease in the response to the stimulus. The results are shown below in the following (Table 2).

**Table 2.** Results of evaluation of deep reflections MMSS - Reflex bicipital C5–C6

| R. BICIPITAL before | R. BICIPITAL after |
|---|---|
| 2+ | 0 |
| 2+ | 1+ |
| 3+ | 2+ |
| 1+ | 0 |

On examination of the triceps reflex corresponding to the nerve root C6, C7, young people, after using the mobile device for 15 min while maintaining the posture of neck flexion, 60% of the population showed a decrease in the response to the stimulus. Next, the results are shown in the following Table 3:

**Table 3.** Results of evaluation of deep reflections MMSS - Reflex Tricipital C6–C7.

| R. TRICIPITAL before | R. TRICIPITAL after |
|---|---|
| 3+ | 1+ |
| 3+ | 2+ |
| 2+ | 1+ |
| 2+ | 0 |
| 1+ | 0 |

To the exploration of the radial style reflex corresponding to the nerve root C5, C6, young people, after the use of the mobile device for 15 min maintaining the posture of neck flexion, 53% of the participating population showed a decrease in the response to the stimulus. Next, the results are shown in the following Table 4:

**Table 4.** Results of evaluation of deep reflections MMSS - Reflex Radial style C5-C6.

| R. TRICIPITAL before | R. TRICIPITAL after |
|---|---|
| 3+ | 2+ |
| 3+ | 1+ |
| 2+ | 1+ |
| 2+ | 0 |
| 1+ | 0 |

Taking into account that the motor reflex is the functional basis of muscle tone and in turn allows to assess the level of injury of the nervous system in humans. When evaluating the deep motor reflexes of MMSS in individuals it was possible to demonstrate that:

After being exposed to the use of mobile devices, the bicipital, tricipital and long supinator reflexes were explored, finding changes such as decreased intensity and/or absence: hyporeflexia and areflexia.

When taking the measurement of the angles of mobility adopted by the study population, it could be shown that 15.3% use their mobile device with neck flexion of 16°– 20°, 23.5% use their mobile device with neck flexion. 26°– 30°, 17.6% used their mobile device with neck flexion of 31°– 35°, 17.6% of 36°– 40°, sustained way. As shown in graph No 8 (Fig. 1).

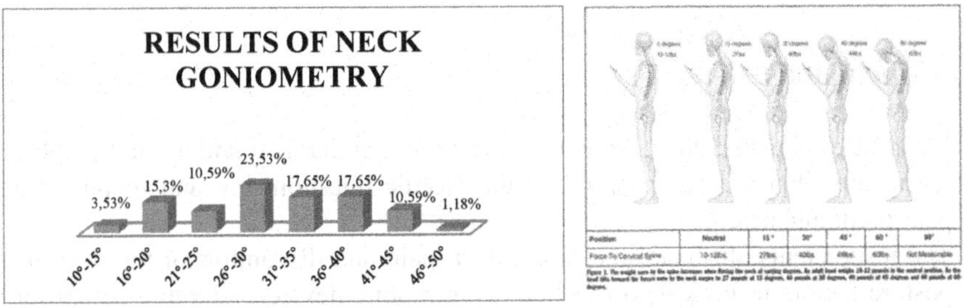


**Fig. 1.** Taken from the article "Evaluation of tensions in the cervical spine and posture caused by the position of the head" Dr Kennet Hansraj.

After having obtained the data of the studied variables, a correlation analysis is performed, by means of the Pearson correlation coefficient made in Statgrapich centurion, finding the following (6). Table 5:

**Table 5.** Variable analysis results

| Variables of analysis | | Results |
|---|---|---|
| POSTURE- NECK PAIN | Coef. Pearson | **0,3702** |
| | Sig. Bilateral | 0,0404 |
| POSTURE- DORSOR PAIN OR LUMBAR | Coef. Pearson | **0,3908** |
| | Sig. Bilateral | 0,0297 |
| POSTURE - HAND OR DOLL PAIN | Coef. Pearson | **−0,1922** |
| | Sig. Bilateral | 0,3002 |
| POSTURE - WEIGHT | Coef. Pearson | **0,9182** |
| | Sig. Bilateral | 0 |

Likewise, a multivariate analysis was carried out in order to know the relationship between the maintained neck flexion posture variables, painful neck symptomatology and deep reflexes of the upper limbs (MMSS) (Table 6):

**Table 6.** Results of multivariate analysis

| Variables of analysis | | Results |
|---|---|---|
| POSTURE- NECK PAIN-REFLEX BICIPITAL | Coef. Pearson | **0,4417** |
| | Sig. Bilateral | 0,0129 |
| POSTURE- NECK PAIN-REFLEX TRICIPITAL | Coef. Pearson | **0,4417** |
| | Sig. Bilateral | 0,0129 |
| POSTURE- NECK PAIN - REFLECTION OF STYLORADIAL | Coef. Pearson | **0,4417** |
| | Sig. Bilateral | 0,0129 |

# 4 Conclusions

- It can be established that neck pain is the symptom that is found in the first place in the results obtained when applying the Nordic questionnaire and moderate neck pain predominates.
- The respondents attributed the discomfort, pain and discomfort of the neck to the posture factors in neck flexion when using mobile devices, adopting inappropriate postures during sleep, stress, among others.
- There is a direct relationship between the maintained neck flexion posture and the painful neck symptomatology that the study population presents.
- There is a direct relationship, although a low prevalence between the maintained neck flexion posture and the painful symptomatology of the dorsal or lumbar spine segment presented by the study population.
- There is no direct relationship between sustained neck flexion and hand or wrist pain.
- There is a very strong relationship between the posture of neck flexion maintained and the weight perceived by the spine at the cervical level, it is important to emphasize that this favors the loss of the natural curve of the spine and leads to increases in tension around of the cervical spine.
- To the exploration of the deep reflexes of the individuals that manifested moderate or severe painful neck symptoms, it was found that:
  Although there is a direct relationship between the posture of maintained neck flexion and the painful symptomatology of the neck and the decrease in the intensity in the response to the stimulation of the bicipital, triceps and styleradial reflexes, this relationship is very low. It is important to keep this relationship in mind, even if it is low, since the motor reflex is the functional basis of muscle tone and in turn allows to evaluate the level of injury of the system Nervous in the human being.

- When analyzing the accumulated time of use of mobile devices daily and projecting it to 1 year i.e. 365 days, this is 365 to 1825 h per year of excess stress seen on the cervical spine.
- The conditions of use of mobile devices must be followed up in order to carry out actions that minimize the aforementioned risk factors.

## References


1. Neupane S, Ifthikar Ali UT (2017) Text Neck Syndrome -Systematic Review. Imp J Interdiscip Res 3(7):2454–1362. http://www.onlinejournal.in
2. Hansraj KK (2014) Assessment of stresses in the cervical spine caused by posture and position of the head. Surg Technol Int 25:277–279. http://www.ncbi.nlm.nih.gov/pubmed/25393825
3. Reid S, Portelli A (2015) The effects of 'text neck' on head repositioning accuracy: a two group comparative study. Physiotherapy 101:e1270. http://linkinghub.elsevier.com/retrieve/pii/S0031940615012092
4. Generales C (2002) complementarias en Atención Primaria. 28(10):573–582
5. Ramada-Rodilla JM, Serra-Pujadas C, Delclós-Clanchet GL (2013) Adaptación cultural y validación de cuestionarios de salud: Revisión y recomendaciones metodológicas. Salud Publica Mex 55(1):57–66

# Incidence of the Psychosocial Risk at Work: A Case Study of Technical Teachers at City of Morón

Concepción Nicolás Hernán[1,2(✉)]

[1] Tres de Febrero University, Buenos Aires, Argentina
nconcepcion@abc.gob.ar
[2] Dirección General de Cultura y Educación, Buenos Aires, Argentina

**Abstract.** The teaching exercise environment sets the workplace where pedagogic processes are carried out. The negative impact that is the result of the things that happen to the teacher or his educational environment configure the way people interact with each other. Considering every classroom of every school with every course is different, the intention is to show which are the dimensions that could impact most in their psyche; psychosocial factors are the influence of the issues inherent in the human condition, on the psyche of people; are those characteristics of working conditions and, above all, of their organization that affect the health of people through psychological and physiological mechanisms. The aim of the work is to carry out a brief analysis of the psychosocial dimensions that affect the technical school teachers and contribute to promote the improvement of working conditions in order to protect life, preserve and maintain the psychophysical integrity of workers. Taking as a case study the public technical schools of the city of Morón in the suburbs of Buenos Aires, it was decided to use the CoPsoQ that is an international instrument for the research, evaluation and prevention of psychosocial risks in its Spanish short version ISTAS 21. According to the results, we proceed to develop a table of priorities in function of a matrix to reach an approach of the dimensions that affect the most education workers. The results show that the dimension that most affect these workers are the pace of work and the predictability respect to the future and the most favorable are the double presence and the possibility of development.

**Keywords:** Ergonomics · Education · Environments · Psychosocial risk
Health and safety at work · Argentina

## 1 Introduction

When I teach and share many hours a week with other colleagues I understand from the inside that there is a problem in our profession that needs to be corrected. Having the possibility and authorization of the principals of the schools where I usually work, I intend through this study to turn visible and socialize the risk factors that can alter the pedagogic continuity of the students. The access to the data collection was possible because I am part of the staff of teachers of the technical School in Automotive,


S. Bagnara et al. (Eds.): IEA 2018, AISC 826, pp. 768–777, 2019.
https://doi.org/10.1007/978-3-319-96065-4_80

Electromechanics, Electronics and Electricity of the curricular offer of the General Directorate of Culture and Education.

This work was presented as work of graduation of the Degree in Hygiene and Safety at Work from the National University of Tres de Febrero. This project is part of a larger idea in progress that seeks to demonstrate how psychosocial risk factors impact on pedagogical processes and change it afterwards through teachers working situation improvements.

Workplaces, in a significant proportion, have potential dangers capable of reducing psychic and physical health; it is a right of workers to work in work spaces where conditions are subject to legislation on Hygiene and Safety not only in terms of infrastructure or technical issues but also in adequate environments that do not affect the psychophysical balance of workers.

The substantial body of knowledge that links the factors of the work organization with health problems of workers motivates the present work; according to the Superintendency of Workplace Risks (hereinafter referred to as SRT) there is a need for a local level approach to support public policies on: diagnosis, treatment and prevention [1].

The Argentine Republic, was the first country in Latin America to carry out the 1st National Survey on the Environmental Conditions of Work, directed by the guidelines of the Ibero-American Social Security Organization (OISS). It includes teaching activity within the category of Professional, Social and Community Services [2] and presents data related to the workplace risk system in force in the country along and elucidated aspects of working conditions in the sector compared to the total population studied in all sectors with references on training and preventive resources. From the results it was observed that social service workers sometimes suffer pressure and/or aggression from their partners, likewise almost one of four workers in the sector has suffered some kind of pressure and/or aggression from patients, students, public, others. Another result was that more than two thirds of workers in the sector feel affected by what refers to control (order, method and work pace). Likewise, a quarter of the population dedicated to social services sometimes works exceeded, seven percent work always exceeded, the most exposed population, as it can lead to different pathologies, such as stress or burnout.

Between 2008 and 2011, a team of ergonomists carried out at the request of the Department of Occupational Health of the General Directorate of Schools of the Province of Buenos Aires, Argentina, an investigation that allowed to know the reason or causes of the high degree of medical licenses of the staff, about the staff of 24 elementary schools [3]; in this case, the work is at the local level to better understand the problems that are discussed in everyday life, immersed as part of the educational community and within the possibilities and with limited resources.

## 2 Aims and Objectives

Our present study aims to (1) obtain results of the awareness survey to determine in which order the psychosocial risk dimensions affect or favor the psychophysiological integrity of this collective work; (2) detect opportunities for improvement and propose ideas for the improvement of risk indicators; (3) take real dimensions of the psychosocial risk of teachers and establish a basis for future research.

## 3 Limits and Scope of the Project

There are 369 public technical schools in the province of Buenos Aires, 127 in the Greater Buenos Aires Parties and 242 in the interior of the province. Morón is part of region 2 together with Ituzaingó City and Merlo City and there are no studies at local level on the subject. There are 4 public technical schools in the district; the present work was carried out on two of the four establishments: "EEST No. 6" and "EEST No. 8". The "EEST No. 4" shares a location with the Argentine Air Force and the specified areas teachers take possession of the classroom by internal competition and not by public act as in the other schools; so it would be out of this analysis and "EEST No. 1" has what is called 30% "rurality" which would imply a unfavorable conditions (worse conditions) to the schools analyzed, so it was decided to leave them for a subsequent analysis. The selection of the surveyed workers, made according to the availability of the teachers, consists of 23 teachers, a greater number of men than women, most of them full professors, provisional or substitute of specified areas.

## 4 Theoretical Framework

Occupational health refers not only to improvements in working conditions, but also to the degree of freedom that people have to organize it, individually or collectively, enabling the full development of their skills [4]. According to health definition, it is not something that is possessed as a good, but a way of functioning in harmony with its environment (work, leisure, social life, lifestyle in general). "Not only does it mean being free from pain or illness, but also having the freedom to develop and maintain your functional abilities (…) As the working environment constitutes an important part of the total environment in which man lives, health depends on working conditions." [5]. Psychosocial factors are the influence of the issues inherent in the human condition, on the psyche of people; are those characteristics of working conditions and, above all, their organization that affect the health of people through psychological and physiological mechanisms that we also call stress. According to the European Agency for Safety and Health at Work, workers feel stress when the demands of their work are greater than their ability to cope. In addition to mental health problems, workers subjected to prolonged stress periods can develop serious physical health problems, such as cardiovascular disease or musculoskeletal problems [6].

The teaching exercise involves cognitive theories, related to individual activities, as well as socio-interactionist, the collective activities are carried out in the social coexistence; in that process of activation of the brain, in a field of tension between the individual and the social [7]. The solidarity with which a teacher faces the different singularities with which individuals come to him, imply submerging in the daily life of human relationships. The impact of the negative things that happen to him and/or his environment, configure the ways in which people communicate and interrelate with each other. In each workplace the conditions and the environment will contribute or hinder the attempts of each teacher to achieve the internalization of culture in each of the participants of the educational space. Therefore, to think of a classroom as a space isolated from the rest of the educational community would be like thinking of a living cell outside the body, our approach in this case study is therefore compatible and takes into account the ergonomic systemic vision for address the problems of teachers. The risk factors in the teaching exercise, not only impact on the quality of teaching but also affect the psychophysical health of the agents. Achieving empathy and being able to recognize ourselves in the other, would also imply a contribution to achieving an environment conducive to sharing knowledge, since it would reduce the levels of violence and interferences that interrupt the communication that clearly affects the culture; life in the classroom must be considered in a systemic way since it is an open social system, of communication and exchange, with its multidimensionality, simultaneities and with a certain degree of immediacy, unpredictability and with history". [8]

## 5 Lines of Action and Applied Methodology

For the collection of the data, manifest authorization was requested to the managers of the establishments. This evaluation study is not presented as an end in itself, but as a tool for preventive action and change, which allows identifying and locating the risks of working conditions to which workers are subject to propose resolution measures that favor health of the workers. To do this, version 2 of the CoPsoQ-istas21 methodology was used (it is a second version of the Adaptation of the Copenhagen Psychosocial Questionnaire) to raise awareness of psychosocial risks at work, what is a public domain instrument that can be used freely and free of charge by all agents and professionals of occupational risk prevention:

> *"The answer to this questionnaire will help you to identify those working conditions derived from the organization of work, which may represent a risk to your health and which we technically call "psychosocial risks". The results, treated individually, will help you to know the situation in which you find yourself in your work. If you also share them with your colleagues, they can be useful for you to make proposals for preventive measures, that is, changes in working conditions to eliminate or reduce these risks".* [9]

They were presented to be completed by the person surveyed, so that they can observe their results; if the summation of each pair of answers points to a score within the green color, it would indicate that it is a favorable dimension for the development of the activities, if the sum of each pair of answers points to a score within the yellow color, say that there is a level of intermediate psychosocial exposure to health; If you indicate a score within the red color, it means that the exposure is unfavorable to health. In the following image (Fig. 1), the results of the whole of the sensitization surveys are reflected; based on the individual results, a matrix was constructed with the indicators obtained from all the respondents. Each row represents a Psychosocial dimension listed with a letter from A to P and each of the numbered columns represents a person surveyed, and in each of the cells of the double entry matrix, the color representative of the sum of the responses of each psychosocial dimension was placed.

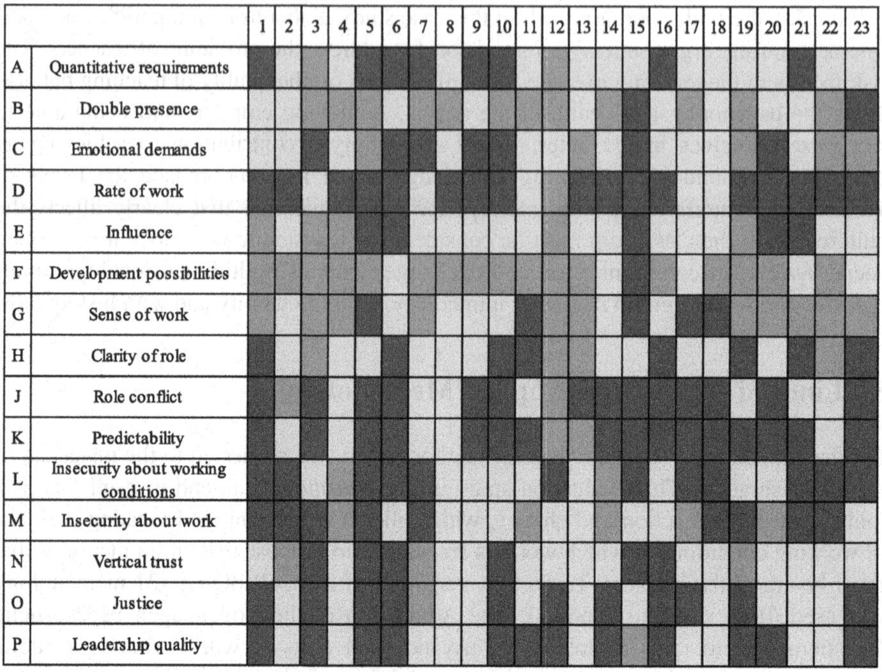


**Fig. 1.** Matrix of survey results (Color figure online)

# 6 Discussion

In a first impression some heterogeneity is observed in the results obtained, since the perception of risk is subjective and individual; to establish an order to estimate the risk, the dimensions that are perceived as the most harmful to health were first ordered,

| | Rate of work | Predictability | Role conflict | Leadership quality | Insecurity about working conditions | Justice | Insecurity about work | Quantitative requirements | Emotional demands | Clarity of role | Sense of work | Vertical trust | Influence | Double presence | Development possibilities |
|---|---|---|---|---|---|---|---|---|---|---|---|---|---|---|---|
| | D | K | J | P | L | O | M | A | C | H | G | N | E | B | F |
| Low | 4,35 | 8,70 | 8,70 | 4,35 | 8,70 | 4,35 | 21,74 | 8,70 | 17,39 | 17,39 | 47,83 | 13,04 | 21,74 | 52,17 | 73,91 |
| Medium | 21,74 | 21,74 | 30,43 | 43,48 | 39,13 | 47,83 | 39,13 | 56,52 | 47,83 | 47,83 | 17,39 | 60,87 | 56,52 | 39,13 | 21,74 |
| High | 73,91 | 69,57 | 60,87 | 52,17 | 52,17 | 47,83 | 39,13 | 34,78 | 34,78 | 34,78 | 34,78 | 26,09 | 21,74 | 8,70 | 4,35 |
| Total | 100 | 100 | 100 | 100 | 100 | 100 | 100 | 100 | 100 | 100 | 100 | 100 | 100 | 100 | 100 |

**Fig. 2.** Percents of survey in risk function

secondly, those dimensions that have an intermediate impact on health and, finally, the aspects that are favorable, in order of percent (Fig. 2).

The next graph (Fig. 3) shows the results in order of risk, it can be seen that the area in red is greater than the area in green and even that the area in yellow.

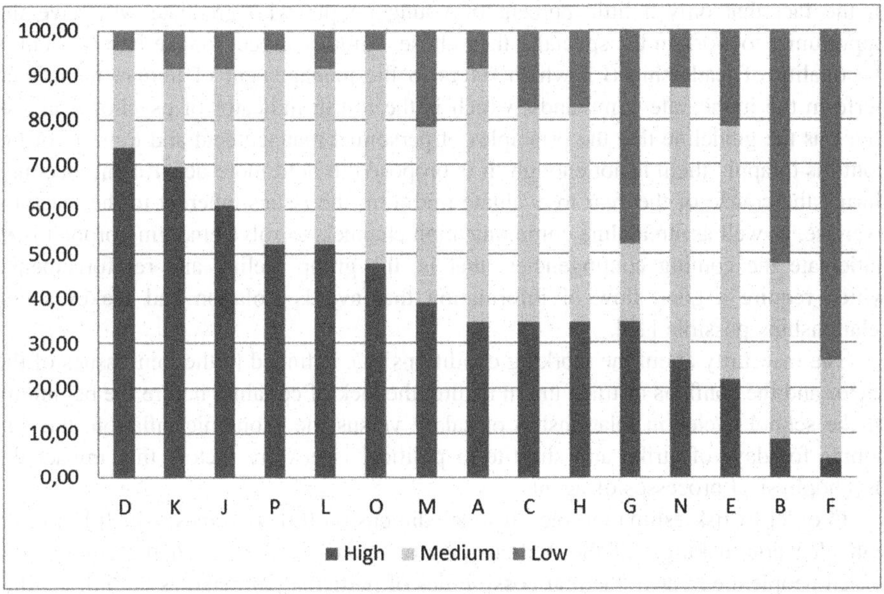


**Fig. 3.** Bar chart with the three risk levels ordered by estimate (Color figure online)

The work rhythm (D) is the dimension that most affects workers, we are aware that these dimensions are perceived as the ones that are most affecting health. One of the possible causes is related to work days lost due to different situations and contingencies, such as disinfection, bomb threats, water cuts and infrastructural dangers, which require teachers to increase the work pace to complete the curriculum and achieve to teach all preset contents. The different conjunctural situations to the school organization contribute to affect the planning, which leads to increasing demands during the school year; times are reduced and tasks become complex, since deviations occur on the planned. On the other hand the predictability (K), is followed by the idea of future restructuring and changes in the education system worries the agents, since the movements of hours would imply an adaptation to the new circumstances. The lack of adequate information, sufficient and timely to perform the work correctly and to adapt to changes implies an internal conflict of rejection. Here an opportunity for improvement is presented; it is proposed to work on the practices of information and communication management, trying to avoid superfluous questions and putting into debate the issues of work such as the administration of material resources or the improvement of channels of communication, which allow teachers in advance to find solutions to emerging concerns.

The third dimension in order of risk estimation is the role conflict (J), which refers to the contradictory demands that may arise at work, this high risk indicator could be related to the recognition, the lack of quantifiable objectives (unlike other industries such as recognition by production) and the belief that only the teacher is expected to deliver the students assessment scores in a timely manner. The technical school teacher aims to be part of the training of professionals in different areas, the contradiction lies in the fact that only a little percent of young people who graduate will have the opportunity to work in the specialty they chose. Another aspect to take into account is the quality of leadership (P), which refers to the management of human teams that perform the immediate commands, which is the fourth indicator of psychosocial risk gives us the guideline that the principles of personnel management and training of the controls to apply them is not enough. It is proposed to hold more department meetings during the course of the year to evaluate mid-term the issues inherent to the teaching exercise, as well as streamline communication channels to solve emerging problems or anticipate the coming contingencies, that is, the group feeling and reinforcements, which require a good flow of information that avoids isolation and makes social relationships possible [10].

The insecurity about the working conditions (L), is linked to the joint issues of the sector and the conflicts of trade union nature: the lack of certainty before the beginning of the school cycles in relationship of salary versus the economic inflation, the discounts for days of strike and short-term political issues are factors that impact the psychophysical processes of agents.

In order of risk estimation the "Justice" dimension (O) continues, which has to do with decision-making and the participation of agents in them; and the equity with which people are treated, the real possibilities of participation make us rethink to what extent each worker can influence work procedures, the resolution of conflicts or the distribution of tasks, so that it can be The commitment on the part of the higher-ups to create spaces for participation is imminent, where the voice of all of them re-signifies the idea of equality and parity, to reverse this indicator and that there are channels of action to solve the situations of injustice, such as discussion forums and mediations.

What refers to insecurity about employment (M) is the concern for the future in relation to employment, the occupation of the positions is performed by public act, "taking" the hours those teachers who are part of the official list (being the only authorized to title) many of the teachers in the technical area are alternates and provisional (professionals or future professionals in lists alternative or emergency) even without the pedagogical complementation that allows access to the official list to have the possibility of securing without the uncertainly of being displaced in teaching movements. Similar situation happens with teachers who are retired but still exercise, they lose priority in public events, in many cases teachers take substitutions that are not sure how long they will last, this will depend on the licenses taken by the professors and their return or not to the assigned curricular space if the license is extended or the teacher is re-qualified to passive activities (without students in classroom).

The results of quantitative psychological demands (A) derived from the amount of work are directly linked to the pace of work and time from the sources of quantity and distribution, in many cases the variable of salary and inflation of the economy, force to more work time or with inadequate tools, materials or processes, which may involve extra time for the completion of tasks within the allotted time represent a greater exposure.

Emotional psychological demands (C) are the efforts not to get involved in the emotional situation derived from the interpersonal relationships involved in work, especially in those occupations in which services are provided to people and which is intended to induce changes in them, such as the acquisition of skills and knowledge [9]. It is the main cause of emotional fatigue that requires long periods of rest; it requires specific skills that can and should be acquired to know what to do in each particular case without getting involved for health. Unlike what is observed in the work of the ergonomics team on primary schools, in this modality of secondary school emotional demands are not presented as an unfavorable risk factor for the psychophysical integrity of teachers, probably because the working groups they are more reduced (no more than 15 students) and that besides being already adolescents, implies that they can stand on their own in many cases as they are not small children.

Regarding the role Clarity dimension (H), the results show us that it is not a risk dimension high since the knowledge on the part of the workers is determined by the incumbencies at the moment of the assignment of the course on the part of the authority, the corresponded tasks are usually well assigned, the responsibility is clear, when the classroom is taken over.

The influence (E) is the following psychosocial dimensions, which refers to the levels of participation that teachers have on the fundamental decisions of their daily work, in the teaching exercise, the classroom is under possession of the teacher who is responsible for what happens in her, therefore the way of doing the work is his decision, not so in decisions that refer to structural changes or decisions outside the classroom.

The results reflected in the vertical trust dimension (N) have to do with the level of reliability of the information that flows from the management to the workers and the level with which the teachers can express their opinion, the origin of the lack of trust has a lot to do with the previous experience of organizational justice [9], one possible way to improve this indicator is to create less vertical communication channels, discussion panels, with the possibility of devolution and feedbacks that mobilize agents and take their opinions into account, contributing positively to this way to the organizational culture. Generate spaces for reflection so that teachers can talk about the meaning of their work, their role trainer and impediments, would be a possible clue to enrich some of the data collected in this investigation [3]

Referred to the sense of work (G), that more than half of the teachers surveyed believe that their tasks are of social utility and are able to visualize their contribution to the teaching service, they know that they are part of the training of young people, the significance of work is essential to achieve motivation in the agents, the acknowledgments and the expressions of affection, motivate the educating agents, accompanying the recognition on the part of the students and the educational community. The favorable factors are those indicators in the analysis of the results that a large part of the respondents feel that they are not harming their health.

The results of the dimension "possibility of development" (F) implies applying the skills and knowledge in the workplace, this is the positive strong point of the analysis, since in the technical areas, the specific contents and the constant need of updates go behind global progress, technologies change and imply that teachers have to update themselves so as not to become obsolete in terms of techniques and technologies of specific training fields, in a continuous manner the workshop teacher applies and teaches skills, there is a great offer of training sessions, conferences and congresses of the different technical areas to which agents can access.

Another dimension to consider is the "Double Presence" (B), which is the need to be in two places at the same time, it is intended to refer to the inability to reconcile family and work tasks even when there is a regulation for the reconciliation of work and family life. When teachers take possession of the courses by public act, the agent decides on their own or not, preferably by scoring, the schedules that they consider convenient, so that they organize their times, in many cases there is time to go and return at home between assigned curricular spaces, unlike what has been observed in the work of the ergonomics team on primary schools, exist differences between the elementary and this modality of secondary school; there is no double presence as an unfavorable risk factor for the psychophysical integrity of teachers [3].

## 7 Conclusions

The incidence of psychosocial factors on the health of technical teachers of technical schools was analyzed through an evaluation method that made teachers aware of the way in which psychosocial factors influence their daily tasks, unlike of another type of occupational hazards, the ways of determining the levels of exposure are not quantifiable, however there are tools such as the one used in this work to make the problem visible and exercise the right to know workplace risks in the workplace. Likewise, providing workers with a tool for the evaluation of work conditions and environment implies the socialization of the difficulties that all workers who share and form part of an educational community face. In terms of results, the pace of work and predictability are the most unfavorable dimensions for the health of technical teachers, since they have been the most homogeneous patterns and those with the greatest negative impact on the health of teachers. In contrast, the possibilities of development in the workplace can serve as a starting point to apply the proposed preventive actions on the other dimensions. Being a starting point to start thinking about an educational culture where continuous improvement is promoted in aspects related to the impact of psychosocial factors on health, it is necessary to reach the levels established by the regulations, on the one hand to guarantee the right to work in a space that does not affect the psychophysical integrity of teachers and in turn to encourage the right to participate in preventive processes and from there (and not least) generate new spaces for democratic participation, so that young people are formed in a knowledge culture in a preventive way.

## References

1. Superintendence of Work Risks (2013) Research Area in Occupational Health. Psychosocial risk factors in the Workplace in Argentina: Theoretical Framework. http://publicaciones.srt.gob.ar/Publicaciones//2013/3_Marco.pdf
2. Superintendence of Work Risks (2009) 1ra Encuesta Nacional a Trabajadores sobre Empleo, Trabajo, Condiciones y Medio Ambiente Laboral. Argentina. https://www.srt.gob.ar/wpcontent/uploads/2014/12/images_Publicaciones_Informes_investigacion_LIBRO_FINAL_Corregido_2014.pdf
3. Cuenca G (2011) Factores de riesgos psicosociales presentes en la actividad de los docentes de escuelas primarias en la Pcia. de Buenos Aires
4. Dejours C (2015) Psychopathology of Work, Clinical Observations. Karnac Books Ltd., London
5. Epelman M, Fontana D, Neffa JC (1990) Effects of new computerized technologies on the health of workers (ed) Humanitas, Buenos Aires
6. European Agency for Safety and Health at Work Homepage. https://osha.europa.eu/es/themes/psychosocial-risks-and-stress,last. Accessed 28 May 2018
7. Vygotsky L (1978) Mind and Society. Harvard University Press, Cambridge
8. J Sacristán (1991) EI curriculum: una reflexión sobre la práctica. Morata editiones, Madrid
9. Moncada S, Llorens C, Andrés R et al (2015) CoPsoQ-istas21 (versión 2) para la sensibilización sobre los riesgos psicosociales en el trabajo. http://www.copsoq.istas21.net/web/abreenlace.asp?idenlace=10071
10. Ortega Domínguez S, Sánchez Díaz C (2008) Revista electrónica de investigación Docencia Creativa, vol 2, pp 174–180

# Maximum Acceptable Weight of Lift in Adolescence Aged 15 to Less Than 18 Years Old

Naris Charoenporn[1(✉)], Amata Outama[2], Teeraphan Kaewdok[3], Poramate Earde[4], and Patchree Kooncumchoo[4]

[1] Department of Industrial Engineering, Faculty of Engineering, Thammasat University, Bangkok, Thailand
cnaris@engr.tu.ac.th
[2] Medical Engineering Program for the Graduate Studies, Faculty of Engineering, Thammasat University, Bangkok, Thailand
[3] Faculty of Public Health, Thammasat University, Bangkok, Thailand
[4] Department of Physical Therapy, Faculty of Allied Health Science, Thammasat University, Bangkok, Thailand

**Abstract.** By Convention No. 138 specifies the minimum age for working children which it must not be lower than the age of completing compulsory education and lower than 15 years old. The allowable work for working children must not be harmful to children's health, must be safe and not against good morals. Manual material handling is common tasks can be found in many conditions of working children aged 15 to less than 18 years old, especially lifting tasks. The lifting tasks may exceed the children's capacity and may lead them to problems of accumulative fatigue, musculoskeletal disorders and other health effects in the future. This study was conducted not only to determine what maximum weight is accepted by the children but also to compare its' results with NIOSH lifting equation whether suitable for working children or not. This research experiment get permission and approval from Thammasat University Research Ethics Subcommittee 3. The study of lifting tasks by simulating works in laboratory and applying psychophysical criteria shows that the maximum acceptable weight is significantly different between boys and girls. Increasing lifting frequency can significantly affect to the MAWL. The MAWL based on psychophysical criterion seems not to be different when lifting frequencies more than 1 lift/minute. The RWL (at origin) calculated from NISOH lifting equation based on the load constant recommended by ISO11295:2014 are higher than the MAWL of this study. It is possible using NIOSH lifting equation to assess risk of manual lifting at origin may be underestimate when applying to working children situation.

**Keywords:** Manual material handling · Lifting · Working children


S. Bagnara et al. (Eds.): IEA 2018, AISC 826, pp. 778–788, 2019.
https://doi.org/10.1007/978-3-319-96065-4_81

# 1 Introduction

Manual material handlings, MMH, (lifting-lowering, pushing-pulling, and carrying) are considered as risky tasks and able to result in Musculoskeletal Disorders (MSDs) effecting on injuries of several part of the such as hands, wrists, shoulders, especially lower back. Musculoskeletal disorders can affect muscles, joints and tendons in all parts of the body. Most WRMSDs develop over time. They can be episodic or chronic in duration and can also result from injury sustained in a work related accident. Additionally they can progress from mild to severe disorders. These disorders are not only life threatening but they also impair the quality of life of a large population. Lifting of high weight exceeding the body capacity with improper posture is an aspect of MMH can increase risk of the injuries including lifting of low weight but repetitively performs the task for continuously prolonged period.

Presently, there are still evidences showing that works related to MSDs (WRMSDs) have been found in MMH significantly. In Great Britain, HES, UK reported that the total number of WRMSDs cases (prevalence) in 2014/15 was 553,000 out of a total of 1,243,000 for all work related illnesses, 44% of the total [1]. The number of new cases of WRMSDs (incidence) in 2014/15 was 169,000, an incidence rate of 530 cases per 100,000 people. An estimated 9.5 million working days were lost due to WRMSDs, an average of 17 days lost for each case. WRMSDs represent 40% of all days lost due to work related ill health in Great Britain in 2014/15. Form the statistics on WRMSDs as reported by Safe Work Australia in 2016 also shows more serious MSE claims (360,180 cases) over the five year period between 2009 and 2014 [2]. The percentage of serious claims that involved MSDs remained stable at between 59 and 61 per cent. Even the frequency rate of serious MSD claims declined during 2009-2014 but the median amount of compensation for all serious claims increased continuously. All of the serious MSD claim resulted from muscular stress while lifting, carrying, or putting down objects. In the United State, 2015, musculoskeletal disorders (MSDs), such as sprains or strains resulting from overexertion in lifting, accounted for 31 percent (356,910 cases) of the total cases for all workers [3]. This resulted in an incidence rate of 298 cases per 100,000 full-time workers in 2015. Private industry workers who sustained an MSD required a median of 12 days to recuperate before returning to work in 2015. In Thailand, according to the 2014 Annual Report (Office of the Compensation Fund, 2014), there were 9,132,756 workers as registered to the compensation fund [4]. It was found that 845 workers suffered from MSD injuries in lifting tasks. That are considered about 9.3 cases per 100,000 workers.

According to the study of Water et al. (1994), the maximum of recommended weight limit (RWL) was 23 kg [5]. The RWL would be reduced when lifting task are dealing with higher risk. Six risk factors (horizontal distance, vertical height, moving distance in vertical, asymmetric angle, griping quality, repetitive frequency) including duration of lift are used for determining a RWL as known in the name of NIOSH lifting equation. The equation is widely accepted for considering maximum weight of lift for safe and presently adapted to international standards, ISO 11228-1: 2003 Ergonomics - Manual handling - Part 1: Lifting and carrying [6]. The practical details for applying the ISO 11228 are explained more in ISO/TR11295:2014 [7]. The maximum of recommended weight limit are recommended depending on age and genders, 20 kg for boys under 18 years old, and girls 15 kg. These maximum of RWL might not be suitable for repetitive and prolonged works and might increase risk of MSD injuries. For Thai regulation the maximum weight of lift is limited at 25 kg for boys and 20 kg for girls. This is also not suitable for preventing MSD injuries. The main objective of this research is to study how the maximum of RWL should be considered as the limit for working children to perform lifting tasks. The results of MAWL in this study would be also used to discuss about the suitable condition of the load constant as referenced in ISO/TC11295.

## 2 Methods

### 2.1 Psychophysical Criterion

To find out the maximum acceptable weight of lift in this research, the psychophysical criterion was applied. Psychophysical method is a randomized experiment based on personal perception of their lifting capability. The maximum acceptable weight of lift (MAWL) is the amount of weight which a person chooses to lift under given conditions for a defined period. This research set up the lifting period of 20 min for each experimental condition. The method is designed to quantify the subjective tolerance of people to the stresses of manual materials handling [5]. Determining the maximum weight limit for safe is based on the percentage of population acceptance, at least 75% of female group, not less than 99% of male, and 90% of all workers' acceptation which the population consist of 50% male and 50% female [8, 9].

### 2.2 Participants

Sixty volunteer (twenty-nine males, thirty-one females) were recruited from student population of a primary school. Their ages were between 15 and less than 18 years. They were healthy and did not had any neurological and musculoskeletal disorders. They were not athletes or weight lifters. They were investigated by Physical Activity Readiness (PAR-Q & YOU) test to ensure that they were ready to do physical activities [10]. The participants could lifted and moved objects with both hands. This research experiment was approved from Thammasat University Research Ethics Subcommittee 3. Participant information sheets were provided for all participants and their parents. All participants had to get permission from their parents to join the experiment. The written informed consent from parents and assent from students were obtained before data collection. Table 1 were the subjects' characteristics.

**Table 1.** The participants' characteristics in this study.

| Variable | Male (n = 29) | | | | Females (n = 31) | | | |
|---|---|---|---|---|---|---|---|---|
| | Min | Max | Mean | SD | Min | Max | Mean | SD |
| Body weight (kg) | 47.0 | 91.0 | 62.3 | 9.9 | 38.0 | 73.0 | 54.0 | 9.5 |
| Stature (cm) | 157.0 | 186.0 | 171.7 | 7.09 | 150.0 | 168.0 | 160.0 | 4.10 |
| BMI (kg/m$^2$) | 17.1 | 27.5 | 21.01 | 2.61 | 15.4 | 29.2 | 21.4 | 3.7 |

## 2.3 Lifting Conditions

The lifting experimental workstation was shown in Fig. 1(a). The lifting workstation was adjustable in height of both lifting (origin) and lowering (destination) levels to match the participants' anthropometries, knuckle and elbow height. The participants were assigned to lift and lower the plastic basket with the weight inside adjustable (subtract or add), Fig. 1(b). The hand location while grasping the basket and staring to lift was set at a vertical level bit higher than the knuckle height of individual participant so that the participants could grip the handle of basket without bending the trunk. The baskets with the modified handles optimal to grip were used as lifting objects. The basket was weighted by metal balls packed in plastic bags, Fig. 1(b). Each metal ball bag was 0.5 kg. The participant was asked to standing as close as the station while lifting. The destination to lower the basket was higher than the lifting origin so that the hand location while lowering the basket was around the standing elbow height. After the participants lowered the basket at destination, the baskets were returned automatically by the earth gravity (Karakuri design). The participants were asked to lift and lower the basket repetitively within 20 min at five different lifting frequencies. The environmental climate of the laboratory while running the experiment was 25–27 °C (dry temperature), 55%–65% (humidity), and air velocity less than 0.2 m/s.

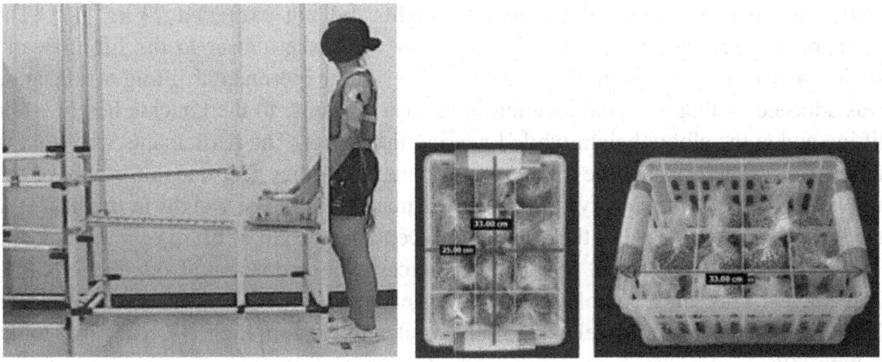

(a) (b)

**Fig. 1.** (a) The adjustable lifting workstation in the experiment, and (b) the plastic basket with modified handles that are weighted by 0.5 kg metal ball bags.

## 2.4 Experimental Design

There were five conditions in the experiment, depending on lifting frequency. Five experimental conditions were set in the experiment based on different lifting frequencies (1, 2, 4, 6, 12 lift/min). Randomized block design was used in this experiment, blocking on participants. When a participant arrived the laboratory, each condition of lifting frequency was assigned to the participant by random.

Each participant arrived the laboratory in the morning around 9–9.30 am on different days. The participants were instructed to know more details of the experiment procedure. The blood pressure was measured (Omron Model HEM-7130) while the participant sitting in relax. A heart rate monitor (Polar Model M400) was used to record heart rate of the participants before starting the experiment, while liftings, and resting after finishing each lifting condition. The average of the HR in the five minute sitting relax was used as the referenced resting HR. The experimental protocol was presented in Table 2.

**Table 2.** The experimental protocol for each participants in one day

| Lifting frequency | Run | Sitting & Resting | Lifting | Resting (min) |
|---|---|---|---|---|
| 5 lifting frequency conditions (1, 2, 4, 6, 12 lift/min) were selected randomly. | 1 | 5–10 min | 20 min | After finishing each lifting experiment the participants sat and rested until the HR recovers to the resting plus before starting next condition |
| | 2 | 5–10 min | 20 min | |
| | 3 | 5–10 min | 20 min | |
| | 4 | 5–10 min | 20 min | |
| | 5 | 5–10 min | 20 min | |

The initial weights in the basket of each lifting condition were different and shown in Table 3. This initial weights were considered from the study of Pinder and Boocock (2014) but the maximum of the initial weights did not exceeded 23 kg [11]. The participants were taught how to lift and lower the basket closer to the lifting workstation and to stand comfortably during lifting. The experimental lifting workstation was adjusted so that the hand location in vertical was near to the knuckle height while lifting and at the elbow height while lowering the basket. The participants were told to take off the weight inside the basket at any time when they feel uncomfortably or too heavy for lifting continuously 8 h from their imagination. The weight in the basket at the last decision was used as the maximum acceptable weight of lift of each participant. The participants were asked to rate their perceived exertion based on Borg's CR10 immediately after they completed the lifting experiment of each frequency condition [12]. They were allowed to relax themselves, sit and rest until their HR recovered to the resting HR.

**Table 3.** Initial weight used in each condition of five different lifting frequencies separated by gender [11]

| Lifting frequency (lifts/min) | Initial weight (kg) for male | Initial weight (kg)for female |
|---|---|---|
| 12 | 12.5 | 8.5 |
| 6 | 19.0 | 8.5 |
| 4 | 17.0 | 9.0 |
| 2 | 19.0 | 6.5 |
| 1 | 23.0 | 13.0 |

## 3 Results

The minimum, maximum, mean and standard deviations of the MAWL in each lifting frequency condition were summarized in Table 4 separated by gender. Increasing of lifting frequency seemed to affect decreasing of the MAWL, Fig. 2. The experimental data were statistically analyzed using analysis of variance (ANOVA). The means of MAWL in each lifting frequency condition between males and females were different significantly (P-value < 0.001). The MAWL of males in all lifting conditions were gather than females.

**Table 4.** The minimum, maximum, mean, and standard deviations of MAWL (kg) at different lifting frequencies, separated by gender.

| Lifting frequency (lifts/min) | The maximum acceptable weight of lift (kg) | | | | | | | |
|---|---|---|---|---|---|---|---|---|
| | Male (n = 29) | | | | Female (n = 31) | | | |
| | Min | Max | Mean* | SD | Min | Max | Mean* | SD |
| 12 | 5.5 | 15.0 | 11.55 | 2.28 | 5.0 | 11.5 | 7.56 | 1.50 |
| 6 | 6.0 | 20.0 | 16.20 | 3.37 | 5.0 | 13.0 | 8.35 | 1.59 |
| 4 | 7.5 | 18.5 | 15.50 | 2.60 | 5.0 | 13.5 | 9.08 | 1.56 |
| 2 | 9.0 | 23.0 | 18.46 | 2.88 | 5.5 | 12.0 | 7.29 | 1.87 |
| 1 | 14.0 | 25.0 | 21.34 | 2.57 | 7.5 | 16.0 | 11.72 | 1.72 |

* The mean of MAWL between male and female are different significantly (p-value < 0.001)

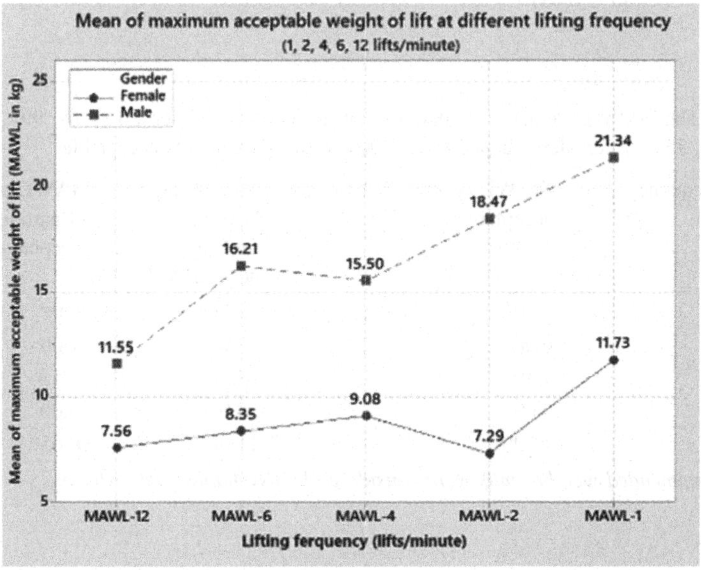


**Fig. 2.** The mean of MAWL at different lifting frequencies, separated by gender.

The multiple mean comparisons of MAWL in five different lifting frequencies were investigated in statistical by Tukey's method. In the male group, the lowest mean of MAWL was found at the lifting frequency of 12 lifts/min. It was significant different from the other lifting frequency conditions, Table 5. In the female group, it was not clearly seen the MAWL at the lifting frequency of 12 lifts/min significantly differed from the other lifting frequencies. Based on psychophysical criterion the MAWL at 99% of males acceptable, 75% of females acceptable, and 90% of all participants acceptable, the MAWL under these criterion were presented in Table 6.

**Table 5.** The result of multiple mean comparisons of the MAWL in five different lifting frequencies using Tukey's method.

| Lifting frequency | Female grouping* | | | |
|---|---|---|---|---|
| MAWL-60 s (1 lifts/min) | A | | | |
| MAWL-15 s (4 lifts/min) | | B | | |
| MAWL-10 s (6 lifts/min) | | B | C | |
| MAWL-5 s (12 lifts/min) | | | C | |
| MAWL-30 s (2 lifts/min) | | | C | |
| **Lifting frequency** | **Male grouping*** | | | |
| MAWL-60 s (1 lifts/min) | A | | | |
| MAWL-30 s (2 lifts/min) | | B | | |
| MAWL-10 s (6 lifts/min) | | | C | |
| MAWL-15 s (4 lifts/min) | | | C | |
| MAWL-5 s (12 lifts/min) | | | | D |

**Means that do not share a letter are significantly different.*

*($\alpha = 0.05$)*

**Table 6.** The MAWL under consideration of psychophysical criterion by 99% of males acceptable, 75% of females acceptable, and 90% of all participants acceptable.

| Lifting frequency (lifts/min) | MAWL at 99% of males acceptable | MAWL at 75% of females acceptable | MAWL at 90% of all participants acceptable |
|---|---|---|---|
| 12 | 5.50* | 6.50 | 6.00 |
| 6 | 6.00* | 8.00 | 7.00 |
| 4 | 7.50* | 8.50 | 7.55 |
| 2 | 9.00 | 6.00* | 6.00 |
| 1 | 14.00 | 11.00 | 10.05* |

** The recommended weight limits were considered by psychophysical criterion.*

Rating of the perceived exertion (RPE, Borg CR10) were obtained immediately when the participants finished lifting session of each frequency condition. The minimum, maximum, mean, and standard deviation of RPE were illustrated in Table 7. Increments of lifting frequencies seemed affect to decrements of the RPE. The highest mean of RPE was found when participants performed the lifting task at 12 lifts/min, Table 8. That was significantly different from the other lifting frequency conditions ($\alpha = 0.05$), both in the male and the female groups. Figure 3 shows the comparison bar chart of the averaged REP at five different lifting frequencies, between the males and the females.

**Table 7.** The minimum, maximum, mean, and standard deviations of perceived exertion score (Borg's CR10) at different lifting frequencies, separated by gender.

| Lifting frequency (lifts/min) | The perceived exertion score (Borg's CR10) | | | | | | | |
|---|---|---|---|---|---|---|---|---|
| | Male (n = 29) | | | | Female (n = 31) | | | |
| | Min | Max | Mean* | SD | Min | Max | Mean* | SD |
| 12 | 1 | 7 | 4.26 | 1.72 | 0 | 7 | 4.48 | 1.63 |
| 6 | 1 | 7 | 3.76 | 1.22 | 0 | 6 | 2.93 | 1.31 |
| 4 | 0 | 7 | 2.95 | 1.54 | 0 | 5 | 2.61 | 1.40 |
| 2 | 0 | 5 | 3.10 | 1.44 | 0 | 6 | 1.98 | 1.56 |
| 1 | 0 | 7 | 2.80 | 1.39 | 0 | 5 | 2.27 | 1.29 |

**Table 8.** The mean comparisons of the RPE between each pair of lifting conditions using Tukey's method.

| Lifting frequency | N | Mean | Grouping* | | Lifting frequency | N | Mean | Grouping* | |
|---|---|---|---|---|---|---|---|---|---|
| RPE5 s | 29 | 4.207 | A | | RPE5 s | 31 | 4.484 | A | |
| RPE10 s | 29 | 3.724 | A | B | RPE10 s | 31 | 2.935 | | B |
| RPE30 s | 29 | 3.034 | | B | RPE15 s | 31 | 2.613 | | B |
| RPE15 s | 29 | 2.810 | | B | RPE60 s | 31 | 2.274 | | B |
| RPE60 s | 29 | 2.793 | | B | RPE30 s | 31 | 1.984 | | B |

**Means that do not share a letter are significantly different. ($\alpha = 0.05$)*

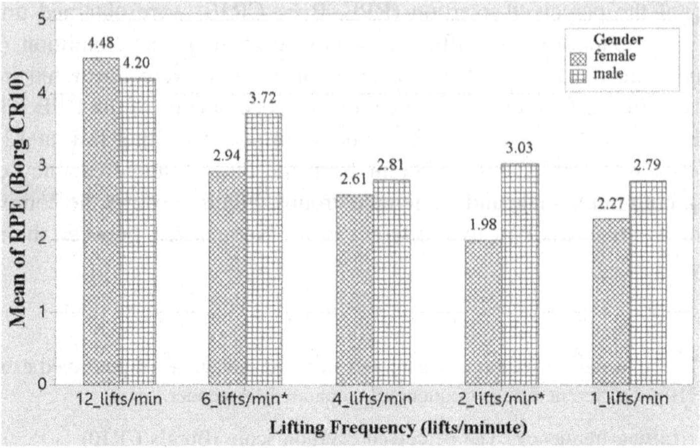


**Fig. 3.** The mean of RPE (Borg CR10) after finishing lifting tasks at five different frequencies. [*mean of RPE are different significantly between male and female. ($\alpha = 0.05$)]

## 4 Discussion

Many countries have allowed their children to work when their age reach to 15 years old or more. The allowable work for working children must not be harmful to their health, must be safe, and not against good morals. Manual lifting is a general activity that can be found in most of works, and widely accepted that it has high relationship with the problems of musculoskeletal disorders (MSDs). Because the manual lifting often results in over exertions with awkward postures. Repeatedly lifting in a prolonged period will also consume more energy from the body. This can lead to accumulative fatigue.

As mentioned earlier, the main purpose of this study was to examine the MAWL of children aged between 15 and less than 18 years. This is a starting point to investigate whether the recommended weight limit (RWL) suggested by NIOSH lifting equation whether will be suitable for children, and acceptable by them or not. Table 9 shows the

**Table 9.** Recommended weight limit (RWL in kg) calculated from NIOSH lifting equation 1991 at the origin (lifting) and destination (lowering) of male and female, at 5 different frequencies.

| Lifting frequency (lifts/min) | RWL (kg) of Male* | | RWL(kg) of Female* | | Psychophysical criterion of this study (kg) |
|---|---|---|---|---|---|
| | Lifting (Origin) | Lowering (Destination) | Lifting (Origin) | Lowering (Destination) | |
| 12 | 5.4 | 3.4 | 4.2 | 2.5 | 5.5 |
| 6 | 10.9 | 6.8 | 9.2 | 5.2 | 6.0 |
| 4 | 12.2 | 7.6 | 10.3 | 5.8 | 7.5 |
| 2 | 13.3 | 8.3 | 11.1 | 6.3 | 6.0 |
| 1 | 13.7 | 8.6 | 11.5 | 6.5 | 10.05 |

**Use load constants of 20 kg for male and 15 kg for female to estimate the RWL.*

RWL as calculated by NIOSH lifting equation. The load constant (LC) recommended in ISO/TR11295:2014 were applied, 15 kg for female and 20 kg for male. The RWL at origin with 12 lifts per minutes is probably same as the MAWL of this study. At the lower frequencies, the RWL at origin are more than the MAWL of the study, in both male and female. This seems the risk assessment using NIOSH lifting equation is underestimated when it is applied to working children. In contrast, at destination (lowering) the RWL are less than the MAWL of the study.

## 5 Conclusion

Increasing lifting frequency can significantly affect to the MAWL. The MAWL based on psychophysical criterion seems not to be different when lifting frequencies more than 1 lift/minute. The RWL (at origin) calculated from NISOH lifting equation based on the load constant recommended by ISO11295:2014 are higher than the MAWL of this study. It is possible using NIOSH lifting equation to assess risk of manual lifting at origin may be underestimate when applying to working children situation. For further studies based on other criterions, biomechanics and physiology are still required for considering load constant used in NIOSH lifting equation.

**Acknowledgement.** This research work was supported by Department of Labour Protection and Welfare, Ministry of Labour, Thailand.

## References

1. Health and Safety Executive (2015) TheWork-related Musculoskeletal Disorder (WRMSDs) Statistics, Great Britain
2. Safe Work Australia (2016) Statistics on Work-Related Musculoskeletal Disorders. Canberra, Australia
3. Bureau of Labor Statistics (2016) News Release: Nonfatal Occupational Injuries and Illnesses Requiring Days Away from Work 2015, U.S, 10 November 2016
4. Office of the Compensation Fund (2014) Annual Report of the Social Security Office. Ministry of Labour, Thailand
5. Water TR., Putz-Anderson V, Garg A (1994) Applications manual for the revised NIOSH lifting equation. U.S. Department of Health and Human Services, Public Health Service, Center for Disease Control and Prevention, National Institute for Occupational Safety and Health (NIOSH)
6. ISO 11228-1 (2003) Ergonomics–Manual handling Part I: Lifting and carrying. International Organization fot Standardization, TC159
7. ISO/TR 12295 (2014) Ergonomics - Application document for International Standards on manual handling (ISO 11228-1, ISO 11228-2 and ISO 11228-3) and evaluation of static working postures (ISO 11226), TC159
8. Snook SH (1978) The design of manual handling tasks. Ergonomics 21:963–985
9. Herrin GD, Jariedi M, Anderson CK (1986) Prediction of overexertion injuries using biomechanical and psychophysical models. Am Ind Hyg Assoc J 47:322–330
10. Canadian Society for Exercise Physiology (2002) Physical Activity Readiness Questionnaire PAR-Q and You: A Questionnaire for People Aged 15 to 69), Health Canada

11. Pinder ADJ, Boocock MG (2014) Prediction of the maximum acceptable weight of lift from the frequency of lift. Int J Ind Ergon 44:13
12. Borg G (1998) Borg's Perceived Exertion and Pain Scales, 2nd edn. Human Kinetics, Champaign
13. Author, F.: Contribution title. In: 9th International Proceedings on Proceedings, pp 1–2. Publisher, Location (2010)
14. http://www.springer.com/lncs. Accessed 21 Nov 2016

# Interdisciplinary Adaptation and Extension of the User Experience Questionnaire for Videos in Learning Environments

Aline Lohse(✉), Alexander Aust, Janina Röder, and Angelika C. Bullinger

Technical University of Chemnitz, Straße der Nationen 62, 09111 Chemnitz, Germany
aline.lohse@mb.tu-chemnitz.de

**Abstract.** A digital transformed or perpetual disruptive working environment will change working processes just as on-the-job trainings at work. Using video sequences in diverse scenarios is actually growing like for instruction, training on the job and information brokerage in case of fault reporting. The utilization of mobile devices with their highly intuitive software makes it possible. To ensure a user centered employment of such video sequences adaption of the method for evaluation is necessary. The evaluation method User Experience Questionnaire (UEQ DIN ISO 9241-210) as an approach to explore the perceived quality of a product means its appeal and usability in form of its objective and subjective effects. An adaption of the method UEQ is necessary for video sequences because of additional items like music, cuts and flashes. The adaption and extension process of UEQ includes the disciplines of media psychology and analyses of film realized in a mixed method approach. An analyses of central documents of those disciplines leads to a first extension of the UEQ. Afterwards an interdisciplinary focus group of lecturers formulates criteria independently according to pragmatic and hedonic quality. Based on a qualitative analyses of this focus group a second extension of the UEQ followed. Finally this version was tested interdisciplinary again with tutorials about technics for ideation processes by students of product-and process management, ergonomics and wood design.

**Keywords:** User experience · Evaluation method · Educational video

## 1 Introduction

### 1.1 Digitization in Learning

The transition from an analog to a digital world is increasingly influencing our everyday lives as well as the world of work [18, 30]. This development comes into the lecture halls through the students' lifeworld, as well as through research and new technologies, and opens up new design possibilities for teachers and learners to impart knowledge and competences [30]. Media formats of various kinds as well as technologies such as virtual reality and augmented reality are becoming increasingly


S. Bagnara et al. (Eds.): IEA 2018, AISC 826, pp. 789–798, 2019.
https://doi.org/10.1007/978-3-319-96065-4_82

important in the transfer of knowledge. The existing design guidelines and evaluation criteria for teaching materials and concepts necessary for the development and evaluation therefore require adaptation in order to be able to ensure the effective, efficient and secure development of products and processes, also in the course of the digital transformation. This applies to universities as well as the training departments in companies [13].

A central component of current teaching-learning formats is the video [21]. The informatics discipline, which can be found at the front of the application of digital teaching-learning formats, is an example of the frequency of learning media shown in Fig. 1. In the self-study phases of Inverted Classrooms [6], blended learning, and MOOCs videos can be embedded. Likewise, it enriches the presence phases and can be combined with group work, quizzes or discussion rounds for student activation [2, 6, 28]. The benefits of the instructional video are differentiated in a study by the Hochschulforum Digitization [4]. In the context of job-related further education and mobile teaching-learning formats, the instructional video is given further impetus [13, 28].

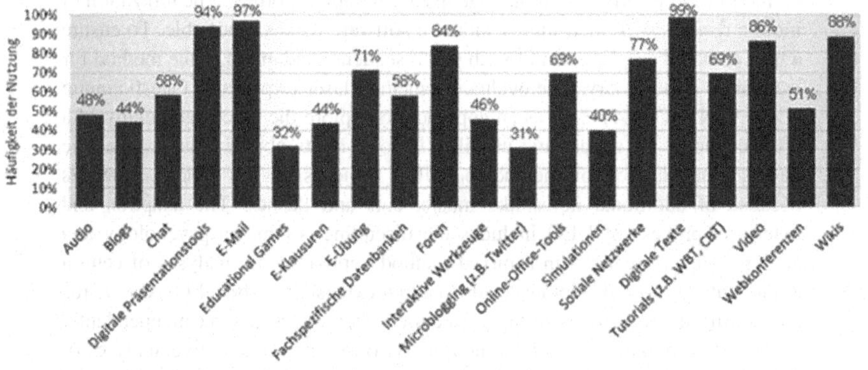


**Fig. 1.** Learning media and their incidence in the department of computer science [21]

### 1.2 Design and Evaluation of Educational Videos

For the design of educational videos, there are checklists that take into account factors such as the structure, readability of the font, duration, learning objectives, camera perspective, target groups, sound as well as the use of graphics and images and the factual and scientific correctness [6, 23]. The checklists remain very general and often emphasize only the importance of the factors without going down to the level of production and giving specific recommendations for action. The question of the effect of elements such as animations, pronunciation, music or editing remains completely unanswered. For motivation, emotions and thus music or animations play a crucial role in the subsequent learning success [12, 25]. From the user-centered research, usability engineering, it is known that the benefits of a product can be increased by systematically guided action, and the acceptance increases accordingly [19, 22]. In addition, in the field of user experience, design is focused in the sense of design and described in

ISO 9241-210 [15]. To measure the user experience, the User Experience Questionnaire is used, which includes both hedonic and pragmatic quality characteristics. These are used to assess emotions and the attachment to a product, as well as its usability itself. Basically, the UEQ is broadly suited to products and processes for software applications as well as physical products. However, this means that no or only very limited motion picture products can be evaluated. However, in order to be able to evaluate instructional videos on their effectiveness with regard to their learning success with the user, it is necessary to adapt or further develop the UEQ.

A research on the evaluation of educational videos in the context of higher education revealed that videos are usually collected through summative evaluation at the level of teaching events. "Did you use the video for learning?" Or "Did the video help you better understand the learning content?" are just a few examples. The contribution of the educational video to the learning success is not determined. In order to use such videos purposefully for the transfer of knowledge, an evaluation of learning success, which can be traced back to the design of an educational video, is necessary. The following describes the goal and the research design.

## 2 Research Objective and Research Design

In this chapter, the goal of this research contribution should first be mentioned and explained. Subsequently, the procedure of the project will be described.

### 2.1 Development of UEQ with Design Science Research

It is the aim of the present research contribution to further develop the UEQ and thus to make it usable for evaluating user-centered design of educational videos in order to measure the effectiveness of the learning success of a teaching video.

The evaluation of educational videos requires a complex concept, which combines the disciplines film industry, media psychology and the labor sciences [25]. The UEQ represents a very good basis for further development through the validated hedonic and pragmatic quality characteristics [9, 10]. An extension to central elements of the film such as music, cuts and impressions is afforded.

In the sense of Design Science Research [10, 28] a revision and evaluation of the UEQ was planned in three steps. The first revision was done after a systematic literature search, the second revision after the first evaluation with students and the third revision after an evaluation with a group of experts [7, 20] from interdisciplinary teaching-learning designers (Fig. 2).

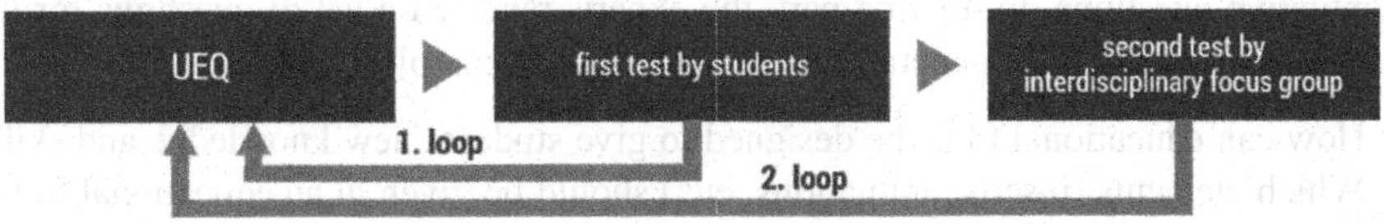


**Fig. 2.** Research design of the UEQ developement

### 2.2 Setting and Methods

On the basis of a systematic literature research in the disciplines of media design with a focus on film, media psychology and occupational science, the existing UEQ was shortened or supplemented by hedonic and pragmatic quality features. In order to fully refer to the user-centered approach, videos for learning creativity techniques (duration 12 months) were created within the teaching-learning project INNO Design [27]. In order to maximize the professional distance between the students who created the videos and the ones they were to use, a strong interdisciplinary group was chosen. Students of industrial science (n = 16) at the Faculty of Machines developed and produced the videos for students (n = 8) of the Department of Wood Design (HG) at the University of Applied Arts. According to the usability engineering process shown in Fig. 3, the videos were created according to the user group, applied by them and evaluated using the UEQ in a first revised version. At the same time, the students of the TUC evaluated the videos of their fellow students.

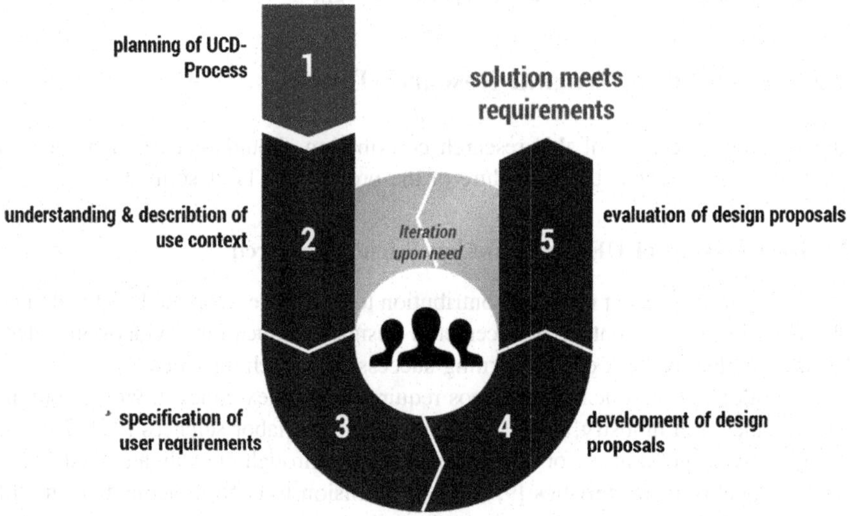


**Fig. 3.** UE process for the development of videos

After the first evaluation by students of both universities, the UEQ was revised. In order to be able to accept and apply the questionnaire for the assessment of instructional videos in other disciplines, a second evaluation was carried out with instructional learning designers from 8 in various disciplines.

The expert group was presented with a mixed questionnaire of qualitative and quantitative questions. In the first part, the experts received a list of questions for free answering in the field of general design guidelines. Examples were:

1. How can educational films be designed to give students new knowledge and skills?
2. Which elements (inserts, animations, etc.) should be given in an educational film in order to convey new facts to students?
3. What would your concept for the realization of an educational film look like?

In a second part of the survey, the following points were evaluated quantitatively:

Music, impressions, animations, choice of technique (interview, laying technique, etc.), cuts, pronunciation, emotions, suspense, clarity as well as the effect of educational videos. The answer options from "completely true" to "not at all" were given on a five-point scale.

**Results and further research**

In this chapter, the adjustments of the UEQ are presented in detail, followed by the results of the evaluation with the students and the expert group.

## 2.3 Adaptations to the User Experience Questionnaire

The quality characteristics of the UEQ include criteria that are described with adjectives [4, 9, 14]. For the pragmatic quality characteristics, adjective pairs, e.g. confusing - clear or unintelligible - understandable, used. The hedonic features with e.g. drowsing - activating or monotonous - varied characterized. Adjective pairs were removed from the existing UEQ, which made no sense in terms of the evaluation of educational videos. For the quantitative evaluation of the product, the ordinal scale is used as shown in Fig. 4. A value of 1 means that the "negative" adjective (e.g. monotonous) is completely true, whereas the value of 7 is entirely true of the "positive" adjective (e.g., varied) [17]. For the qualitative assessment, special questions were posed in media psychology and film analysis for both quality characteristics. These deal with perspective, impressions, cuts, as well as language, music and emotions [1, 16, 24, 26]. For interpretation, the mean values for each criterion are calculated. Afterwards the crossed values in the table are to be connected with a line from top to bottom. This allows a quick and easy visual assessment.

| | adjective | 1 | 2 | 3 | 4 | 5 | 6 | 7 | adjective |
|---|---|---|---|---|---|---|---|---|---|
| transparency | confusing | ☐ | ☐ | ☐ | ☐ | ☐ | ☐ | ☐ | clear |
| | difficult to learn | ☐ | ☐ | ☐ | ☐ | ☐ | ☐ | ☐ | easy to learn |
| | complicated | ☐ | ☐ | ☐ | ☐ | ☐ | ☐ | ☐ | easy |
| | not understandable | ☐ | ☐ | ☐ | ☐ | ☐ | ☐ | ☐ | understandable |
| efficiency | halting | ☐ | ☐ | ☐ | ☐ | ☐ | ☐ | ☐ | fluent |
| | cluttered | ☐ | ☐ | ☐ | ☐ | ☐ | ☐ | ☐ | organized |
| | slow | ☐ | ☐ | ☐ | ☐ | ☐ | ☐ | ☐ | fast |

**Fig. 4.** An extract of the UEQ, quantitative

The following figure shows an excerpt from a completed questionnaire (Fig. 5).

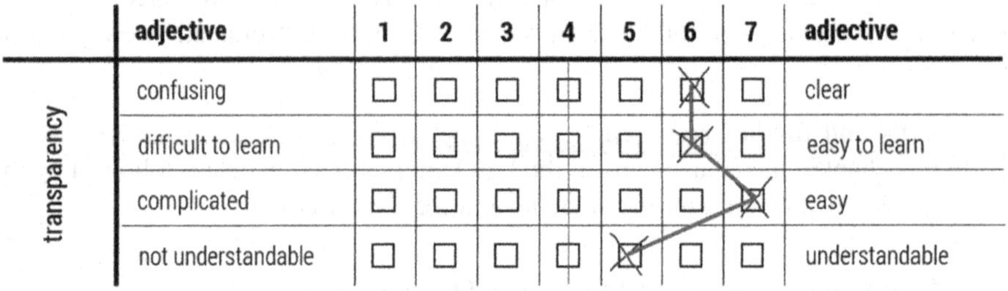


**Fig. 5.** Answered extract of Fig. 4

## 2.4 Results from the Evaluation with Students

The quantitative results of the reviewed educational videos vary depending on the user. The students of HG rated all educational videos in relation to the students of TUC fundamentally worse. This can be explained by the higher demands on design due to their discipline. From the perspective of learning success, the videos have all been judged to be effective. When comparing the HG and TUC evaluations, it should be noted that HG students used the educational videos to learn creative techniques and then used them to generate ideas. The students of the TUC, however, had produced the videos themselves and therefore judged from the perspective of designers. The qualitative evaluation also led to different results. For example, some of the students found

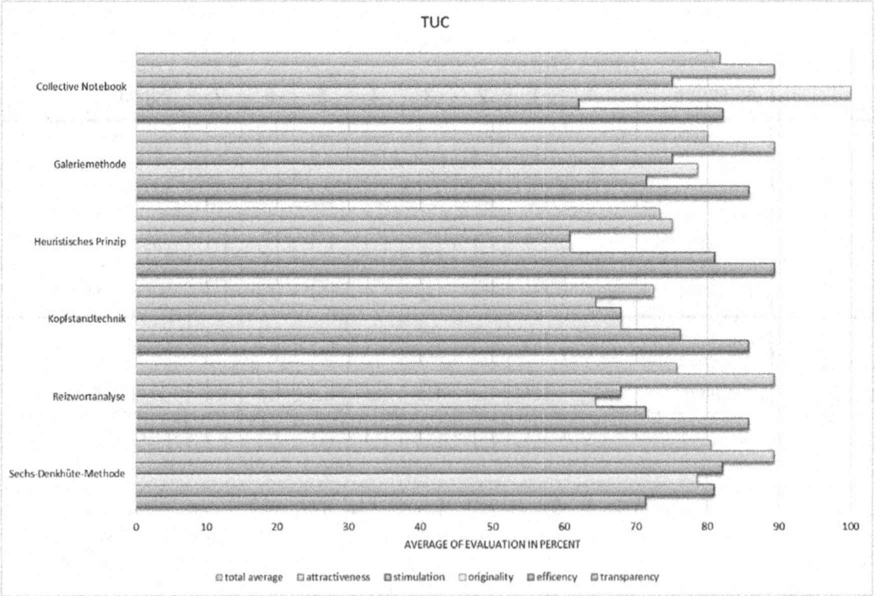


**Fig. 6.** Results of the evaluated data regarding TUC

any music disturbing, while others felt a positive effect. The notion of intelligibility of pronunciation and speed was also assessed very differently. However, more or less dialect in an educational video does not matter. A comparison of the two evaluations shows, however, that both groups have the same gradation of videos (collective notebook method as the best instructional video and the video on the heuristic principle as the worst). This proves that the UEQ works after adaptation (Figs. 6 and 7).

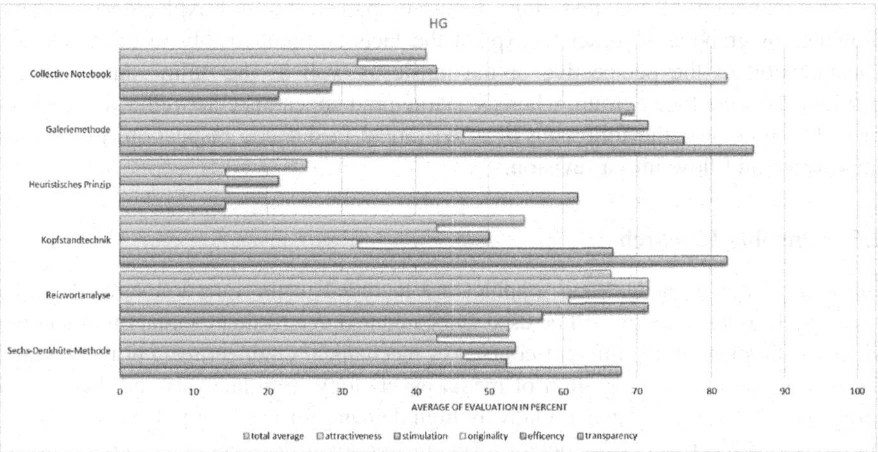


**Fig. 7.** Results of the evaluated data regarding HG

## 2.5 Results from the Evaluation with the Expert Group

The qualitative survey revealed the following points as the central aspects in the design of instructional videos: “short and concise”, “witty”, “building examples” and “clarity”. Graphics, animations, illustrations, drawings, interactive gadgets, clear forms, little text and recurring elements are proposed as measures for conveying facts. The concept of an educational video should be based on a com-mon thread. This should be divided into several units that are not overloaded with information. The educational video should be user-oriented and adapted to the knowledge of the viewer. The inclusion of examples and the transfer of theory into practice, according to respondents play an important role in the design of teaching media. The expert group was also asked about the videos from the project INNO Design. Compared to existing and on various online platforms forms retrievable videos, the project videos are more approachable, since they are partially amateurish. In addition, there is a positive difference in the clear pronunciation and the structured presentation. Suggestions for improving their own instructional videos have been to improve sound quality, interaction, and more breaks to think and understand the language.

The results of the surveyed focus group show that the criteria, elements and design types integrated in the revised UEQ clearly overlap with the proposals and their own ideas with regard to the conception of educational videos. Aspects such as a clear pronunciation, explanation of the facts with suitable examples and the tailoring of a teaching medium to specific target groups are essential. They could be verified with it. In addition to drawings, animations, graphics and lightly writ-ten text, decisive factors were mentioned for imparting factual behavior. Both, graphics and a small amount of text are not yet considered in the UEQ.

The reduction of the text must be accompanied by self-explanatory insights, drawings or graphics in order to explain the facts comprehensibly to the users. The consideration of the perspective or the music as well as the "film" cuts were not mentioned by the focus group, whereas a structured appearance as well as the division into clear units and the following of a red thread in didactic respect are given great importance and flow into a revision.

### 2.6 Ongoing Research

Currently, features on text and graphics are included in the questionnaire. With this revision, the questionnaire will be used and evaluated in a further evaluation of learning videos with students of different courses in mechanical engineering. Further research focuses on the closer integration of media psychology, film analysis and didactics. It turned out that students have a relatively high demand for learning videos. This affects both design aspects and the content level. Instructional designers cut back on design aspects such as language and cuts. For this purpose, the common denominator must be determined in future research. Furthermore, the question arises as to whether a learning video, which artistically uses language, music, graphics etc., can become a learning medium at eye level and thus increase acceptance.

## 3 Conclusion

In the context of the evaluations, the revised UEQ for educational videos showed that a video designed according to user centering with the added criteria leads to a connection between the user and the educational video if the quality features are fulfilled satisfactorily. Because of this, a positive user experience is enabled. In this context, it should be pointed out that even videos rated as poorly can fulfill a learning purpose. When these videos are reflected and discussed, they can also serve as a learning medium.

The further development of UEQ into UEQ for educational videos has made a first contribution towards a universal instrument for the use of learning videos in higher education or continuing education. It is intended as a support for the design of educational videos and their evaluation, can be used as a guide in production or take a central role in the selection of suitable videos.

## References


1. Appel M, Batinic B (2008) Medienpsychologie. Springer, Heidelberg
2. Awi.institute. https://www.youtube.com/channel/UC87L-8zhuoGLWEFvcSFe9Fg. Accessed 24 March 2018
3. Biggs JB, Tang CS-K (2011) Teaching for quality learning at university. What the student does, 4th edn. McGraw-Hill/Society for Research into Higher Education/Open University Press. Maidenhead
4. Deimann M (2016) Stärkere Individualisierung der Lehre durch neue Medien. Hochschulforum Digitalisierung. Arbeitspapier Nr. 26. Berlin
5. Deutsche Akkreditierungsstelle Technik e.V.: DATech-Prüfhandbuch. Usability-Engineering-Prozess. Leitfaden für die Evaluierung des Usability-Engineering-Prozesses bei der Herstellung und Pflege von interaktiven Systemen auf der Grundlage von ISO 13407, Version 1.4, Frankfurt/Main (2004)
6. Feldhoff A, Lohse A, Bullinger AC (2015) Studierendenaktivierung mit digitalisierter peer instruction. In: Bullinger AC (Hrsg) Mensch 2020 - transdisziplinäre Perspektiven, (S. 331–341). Verlag aw&I Wissenschaft und Praxis, Chemnitz
7. Freitas H, Oliveira M, Jenkins M, Popjoy O (1998) The focus group, a qualitative research method. ISRC Working Paper, Baltimore
8. Handke J, Kiesler N, Wiemeyer L (2013) The inverted classroom model-the 2nd German ICM-conference – proceedings. Oldenbourg, München
9. Hassenzahl M, Koller F, Burmester M (2008) Der user experience (UX) auf der Spur: Zum Einsatz von. In: Brau H, Diefenbach S, Hassenzahl M, Koller F, Peissner M, Röse K (Hrsg.) Tagungsband UP08. Fraunhofer Verlag, Stuttgart, pp 78–82. www.attrakdiff.de
10. Hevner AR, March ST, Park J, Ram S (2004) Design science in information systems research. MIS Q 28(1):75–105
11. Hinderks A, Schrepp M, Rauschenberger M, Thomaschewski J (2012) Konstruktion eines Fragebogens für jugendliche Personen zur Messung der User Experience. In: Brau H, Lehmann A, Petrovic K, Schroeder MC (ed.) Usability Professionals, pp 78–83
12. Hochschulforum Digitalisierung Homepage. https://hochschulforumdigitalisierung.de/. Accessed 25 May 2018
13. Horz H, Schulze-Vorberg L (2017) Digitalisierung in der hochschullehre. In: Digitale Gesellschaft 283, pp 57–71. Konrad Adenauer Stiftung
14. Hörold S, Ruschin D, Schubert D, Zimmermann U (2015) Der "Faktor Mensch im Fokus der Entwicklung von 3D-Technologien – Methodensammlung. TU Chemnitz, Chemnitz
15. ISO 9241-210 (2010) Ergonomics of human system interaction — Part 210: Human-centred design for interactive systems. ISO. Genf
16. Kreutzer O, Lauritz S, Mehlinger C, Moormann P (2014) Filmanalyse. Film, Fernsehen, Neue Medien. Springer, Wiesbaden
17. Laugwitz B, Schrepp M, Held T (2006) Konstruktion eines Fragebogens zur Messung der User Experience von Softwareprodukten. In: Mensch & Computer. Mensch und Computer im Strukturwandel. Oldenbourg, München, pp 125–134
18. Lorenz P (2017) Digitalisierung im deutschen Arbeitsmarkt. Eine Debattenübersicht. In: Arnold N, Köhler T (eds) Digitale Gesellschaft. 283. Konrad Adenauer Stiftung, pp 75–93
19. Nielsen J (1993) Usability engineering, 2nd edn. AP Professional, Boston
20. Morgan DL (1993) Successful focus groups: advancing the state of the art. Sage Publication, Newbury Park
21. Persike M, Friedrich J-D (2016) Lernen mit digitalen Medien aus Studierendenperspektive. Hochschulforum Digitalisierung. Arbeitspapier Nr. 17, Berlin

22. Richter M, Flückiger MD (2013) Usability engineering kompakt, 3rd edn. Springer, Heidelberg
23. Rüsseler M, Sterz J, et al (2017) Qualitätssicherung in der Lehre – Entwicklung und Analyse von Checklisten zur Beurteilung von Lehrvideos zum Erlernen praktischer Fertigkeiten. In: Zentralblatt für Chirurgie - Zeitschrift für Allgemeine, Viszeral, Thorax- und Gefäßchirurgie, 142, 1. Thieme, Stuttgart, pp 32–38
24. Stoecker D (2013) eLearning – Konzepte und Drehbuch, 2nd edn. Springer, Heidelberg
25. Süss D, Lampert C, Wijnen CW (2013) Mediendidaktik. Lehren und Lernen mit Medien. In: Medienpädagogik. Springer, Wiesbaden, pp 169–195
26. Trepte S, Reinecke L (2013) Medienpsychologie. In: Grundriss der Psychologie. 27. Kohlhammer Urban Taschenbücher. Stuttgart, pp 229–248
27. Technische Universität Chemnitz. https://www.tu-chemitz.de/mb/ArbeitsWiss/forschung/projekte/inno_design_ingenieure_und_holzgestalter_innovation_engineering_fuer. Accessed 23 May 2018
28. Vaishnavi V, Kuechler W (2008) Design science research methods and patterns: innovating information and communication technology. Auerbach Publications, Boca Raton
29. Wannemacher K, Jungermann I, Scholz J, Tercanli H, von Villiez A (2016) Digitale Lernszenarien im Hochschulbereich. Hochschulforum Digitalisierung. Arbeitspapier Nr. 15, Berlin
30. Zierer K (2017) Digitales Lernen. Möglichkeiten und Grenzen einer Digitalisierung im Bildungsbereich. In: Analysen & Argumente. 283. Konrad Adenauer Stiftung, pp 41–56

# Way-Finding and Communication Design as Strategic Systems to Improve the Well-Being of Children in Paediatric Hospitals

Laura Giraldi[1(✉)], Marta Maini[1(✉)], and Donatella Meloni[2(✉)]

[1] University of Florence, Florence, Italy
Laura.giraldi@unifi.it, marta.maini@hotmail.it
[2] St. Peter Pediatric University Hospital, AOU Sassari, Sassari, Italy
San.Pietro.donatella.meloni@aousassari.it

**Abstract.** In recent years there has been a growing need to design way-finding systems and communication more effectively, as we realize the great importance they have in everyday people's life, becoming necessary and operative guidelines tools for improving the quality of life. Sometimes these tools are used as simple signposting or mere decorations designed to embellish a specific space, as happens in place dedicated to children. Instead they are very important and strategic instruments able to improve the liveability and usability of specific environment by different kind of users. Among these ones, the children are very particular, because they feel, think and behave in a very different way beside the adults.

Generally, the hospital environment causes a high level of stress in children. Indeed, because of their young ages, they have a very limited experience and therefore they do not have the psychological and emotional state to deal with kind of stressful environment.

Way-finding systems and communication design products improving the user experience, are able not only to orient children inside unknown spaces, but also to make "familiar" a place never seen before, making the child feels at ease, avoiding or reducing the stress that this place can bring to the user.

**Keywords:** Way-finding · Communication · Children · Well-being Pediatric hospital

## 1 Introduction: Context of Reference

### 1.1 Wayfinding Systems

In our days people spend a significant part of their life outside home by living outdoor or indoor public spaces to live experiences or socialize with the others. During this kind of experience people need to "find their way". Sometimes they become disoriented in the event of there being difficulties in reaching their destination or in case of losing their sense of orientation during the journey. In such situations, way-finding systems help to keep guidance and give people the means to feel comfortable in new environments. In case of young people the problem is even more complicated because they have more difficulty in recognizing unknown or non-familiar places. Today there are


S. Bagnara et al. (Eds.): IEA 2018, AISC 826, pp. 799–810, 2019.
https://doi.org/10.1007/978-3-319-96065-4_83

just few examples of way-finding systems and communication design products, which are especially designed according to children needs and their psycho-physical and emotional characteristics.

Generally, way-finding systems dedicated to children are not easy to be found in modern cities' indoor or outdoor spaces. This is probably due to the fact that young people are always associate with an adult or live experience under the supervision of a helper. So, the guidance is demanded to the adults (parents, teachers, etc.) going along them. When children grow up and reach the age of over six/seven years, sometimes they succeed in following way-finding systems designed for adults, as, for examples in the underground big cities, such as an example Milan in Italy, because, in this case, the signs are clearly visible and recognizable, well organized, placed also on the ground (easy to see and follow). Moreover, each route is identified by graphic elements of a specific color. In spaces and along routes where way-finding systems are not installed, people follow undersigned signs they find in the environment. Indeed, this is the case, as instance, of urban scenarios, where they use for orienting shop windows, mail boxes, bus shelter, advertising signs. A similar case apply in indoor public spaces, where those elements are represented by lifts, stairs, corner, lights, signposting and so on. All of these elements realize spontaneous way-finding systems that often are not sufficient for navigation.

Unlike external environments, in interiors spaces dedicated to childhood such as schools, play-rooms and pediatric hospitals, actually it is possible to find few examples of way-finding systems. All of them are characterized by a vast wealth of colors and graphic elements, like stylized animals remembering tales and cartoons, which can be easily to recognized by children. "*In developing the wayfinding strategy and designing the sign system, the designer will have to create a family of sign types that not only addresses primary information and wayfinding needs but also recognizes secondary issues and audiences with an appropriate information hierarchy and sign-messaging protocols*" [9].

Furthermore "*symbol design is equally important to wayfinding. Symbols provide a shortcut way for large groups of people who may not share a common language to Communicate*" [9].

Generally, wayfinding systems are represented by different kind of "signs" inside connection areas, especially in the cities characterized by streets, path walks, squares, parks, gardens and playground. In the contexts, streets represent the connection areas used by people to go from one place to another, while the other urban spaces are also used for socialization and free time. In public interior environments the wayfinding systems have the same functions. The first one is practical or denotative [7] and consists in orienting people inside the space. The second "function", defined connotative [7] is related to the emotional sphere, it involves the senses of the users remembering familiar elements and actions and suggesting behaviors or relation among users.

As a consequence, we can state with a certain level of confidence that a way-finding design for little users could help them to feel comfortable even if they are in an unfamiliar space as a hospital, which is generally perceived as hostile.

If we consider children as main user of a collective interior space we can say that it is not easy to find interesting examples of places with way-finding systems designed at "children's size".

Generally speaking, the hospital environment causes a high level of stress in children. Indeed, because of their young ages, they have a very limited experience of life and therefore they do not have the psychological and emotional state to deal with kind of stressful environment. For this reason wayfinding systems and communication design products able to improve their experience as users, are able not only to orient children inside unknown spaces, but also to make "familiar" a place never seen before, making the child feels at ease, avoiding or reducing the stress that this place can bring to the user.

### 1.2 Pediatric Hospital

The hospital is a building which welcomes patients and dispenses a wide range of medical care services. To these, the medical staff and everybody who need to go in it, both patients and their relative, have be added. The pediatric hospital, from an organizational point of view, is mainly divided in three classes: hospitalization, diagnostic-therapeutic and general services.

Furthermore, there are also the spaces which host the machineries and those dedicated to didactics. Differently from a normal hospital, the pediatric have some peculiarities, first regarding the subdivision of hospitalization in relation to the patients' age. This is necessary due to the furniture, which is adapted respect to the specific measures of the children, accordingly to each age. In general, in the services class are included playing spaces, green spaces and those dedicated to didactics. All of those have a specific aim: entertain the children, distracting them from their illnesses, but also amuse the medical staff. The life in the hospital is perceived as extraneous by the routine and habits of everyday life. These are replaced by rules and strict rhythms of medical care. During the hospitalization experience can also incur relational problems, caused by a depersonalized environment.

There are indeed several studies that proved the importance of creative stimulation as a valid method to fight depression brought by disorientation and identity loss. As previously said, the hospital users are not only the patients, but also all the relatives which help them in carrying out the hospitalization or visit them. In every case, often, all the persons that are in the hospital, as main or secondary users, are requested to wait for. Consequently making enjoyable and interesting the transit and waiting places has been became of fundamental importance.

### 1.3 Children and Parents' Stress in Pediatric Hospital

Stress is a concept scientifically defined and an evincible phenomenon. In 1998, Evans and Mitchell McCoy, described the way a built environment may influence the health of persons and found which architectonic elements may lead relevant stress in the users. The characteristics of the building' spaces that may lead stress are basically four: stimulation, coherence, affordance and control.

In the hospitalization matter, the stress phenomenon assumes a relevance, because this may negatively influence the clinical outcome of the patient.

Subsequently to further studies and researches, two ways in which stress impact on patients' health were identified:

- Negatively impacting directly on health patients (physiological parameters: blood pressure, state of anxiety, depression, etc.)
- Negatively impacting on medical staff performances. In the hospital, in fact, may be found two types of stress: environmental and occupational.

According to Ulrich, Devlin, Arneill and Del Nord, the elements of the building that may represent possible stress vehicle for users are the following:

- The image itself of the hospital.
- Sensorial feelings.
- Difficulties, impossibility in control and manipulate the environment.
- Orientation difficulties.
- Physical discomfort.
- Viral risk.

**The Orientation Difficulties.** The orientation difficulties are typical of the hospital.

It is a common belief, in fact, that the hospital is somehow similar to a labyrinth, due to its endless corridors, often identical. Wayfinding problems impact not only on patients, but everybody experiences the place, i.e. the relatives and the medical staff, forced to interrupt their mansions for helping the users to find their way. Users are scared to get lost, to not find anymore the way to come back and, for that reason, they not completely enjoy whole the services offered by the structure. These problematic in the long run become, not only a relevant source of stress, but also a cost for the health company.

Differently from other hospitals, patients of pediatric hospital experience a continuous evolution and this make harder identifying precisely all possible source of stress. For this reason, is necessary looking the space with the perspective of the children, because what may appear beautiful or enjoyable for an adult, may look boring or even anxiogenic for children.

The child has a limited emotional experience to face the challenge resulting from hospitalization and, for this reason, he/she may be more vulnerable respect to different sources of stress.

### 1.4 Pediatrician's Point of View

The design and architecture of spaces dedicated to pediatric ward may have a relevant psychological impact on children and their family.

Design and architecture may improve the permanence of patients and their relatives in the hospital and, at the same time, they may have a positive impact on the activities of whole the medical staff. The latter face a stress often higher than other due to emotional implication respect to their young patients.

The child should first enter in a comfortable, lovely, relaxing and colorful place which evokes protection and peacefulness.

A child-friendly environment that arouses family like feelings, esthetically tidy and enjoyable, contributes to the rehabilitation and wellness, which also may result in a shorter permanence of the patient. All of that would lead benefit, not only to the children, but also to whole the relatives that face up the hospitalization with them.

## 2 The Aim of the Research

The aim of the present research is to propose an innovative approach in the designing for kids discipline by individuating a series of good practices for the strategic design of communication and way-finding systems to be applied in indoor spaces dedicated to children and, in particular, in several areas within pediatric hospitals. In order to improve the well-being of children and make them feel comfortable in indoor public spaces and, in particular, in the pediatric hospital interiors, it is necessary to render those spaces more friendly.

These good practices have to guide designers to design interior way-finding systems able to orient, inform, interact, entertain include all kind of children, at different ages, using one universal language of their collective imaginary. In order to improve this new design culture, the purpose of this work is to recommend a methodological referring system to be easy applied in different indoor places dedicated to children.

The suggested good practices are designed taking into account the peculiarities (skills, abilities, behaviors) of children at the different ages groups in accordance to the theory above mentioned. In other words, the purpose of this work is to propose a referring methodological system to design way-finding systems in indoor collective spaces able to communicate with all kind of children according to two different modes of interaction.

The first communication mode is designed to orientate children guiding them along interiors without losing their way and informing them about surrounding areas and services. It is based on simplified orientation and secure informative messages. These kinds of systems are usually founded on children behavior so that children use their previous experiences to find their way.

The second one stimulates the emotional sphere of children and allows them to live pleasant experience in an unknown place. It concerns the connotative meaning of signs able to entertain and reassure, suggesting familiar routine and known practices belonging to their imaginary. This kind of communication is based on the knowing of children' imaginary belonging to the experiences of their life.

Summarizing the final research's aim is to find open rules and good practices for the design of way-finding systems able to orient, to reduce stress and feel at ease a large number of children in unknown indoor collective spaces.

These goals also allow communicating with children as main users and indirectly with parents and health staff according to different levels of interactions.

## 3 Methods of Research

### 3.1 "Children Centered-Design": Multidisciplinary Approach

Way-finding systems have the function to advice or remind people about the surroundings, presenting the information at strategic points to orient and inform them. Children have a different ability of orientation in comparison with the adults and for this reason it is necessary to study their skills and behaviors at the different ages before start designing for them. Consequently, it is necessary to define a design method for way-finding systems based on and strongly taking into account the natural skills, behaviors and inclinations of children. In this way it is possible to design a system really able to guide and entertain children in a pleasant way using strategic solutions.

The methodological approach to design communication and way-finding systems in collective interiors and in particular in pediatric hospitals refers to the human centered design rules in order to inform the little users about surroundings and giving them points of view and references for finding their way in autonomy as in a familiar place.

The present work uses a multi-disciplinary and holistic approach involving different disciplines essential to study the children as main users. As a matter of fact, from the birth to adult age, young people have a continuous development of physical and psychical abilities, which are necessary to know in order to design according to children centered design approach.

The involved disciplines are mainly pedagogy, cognitive psychology and pediatrics. The contributions of each discipline represent the indispensable basic knowledge necessary to summarize all the features in the final design proposal.

Starting from pedagogy, the research refers to the educational theories of Montessori, [25] and of Loris Malaguzzi [22] whose educational theory is known as Reggio Children approach. These two Italian pedagogical theories are ones of the most followed in the world. Among their various indications, they identify the environment where children live as an essential support for the growth and development of them, to enhance their potential, resources and many intelligences. [8] Consequently, it is very important that the environments where children live - from a domestic interior to a hospital - are designed taking into account the pedagogical theories mentioned above. Consequently, the designer can use all the information to design way-finding systems for pleasant and inclusive experiences. The children's indoor environment characterized by ad hoc "elements", colors and graphics play a fundamental role, determining a scenario such as to be perceived by the child as friendly, pleasant, safe and familiar, do not allowing them, to lose their way. The scholars also identified three age groups of reference, 0–3, 3–6 and 7–11 years old. These groups are indicative of the steps of growth and related skills and abilities, as also explained in the cognitive psychology theories by Piaget [29] in the Sixties. Moreover, Piaget highlighted the importance of playing for children. It is sure enough that when children play they show their needs to communicate their own emotions. Certainly, the designers will have to take into account this information when designing for kids. Environments and products able to induce behaviors focus on the emotional aspects that, as Norman says, [27] they are the engines of psycho-physical development and learning.

The technique of playing represents a solution for the designer able to realize this contact with the little user. When children play they underline their needs to communicate and socialize and, at the same time, show their inner emotional world. In designing, it is *modus operandi* the involvement of the main and secondary users in co-working activities at the early stages of job definition.

The strategy of involving small users in co-working is necessary for highlighting the relations between the type of actions and the emotions happening in an experience inside unknown spaces.

In general, this is an important activity because it is able to make designers think in an innovative way. The co-working approach has been experimented as described below during the development of the present work.

## 4 Applied Method. Use Case: Pediatric Hospital in Sassari Italy

The research described in this paper investigated the following aspects, by relating them with the specific scenarios of the Pediatric Hospital at Sassari:

- Study of the literature: perception psychology, pedagogy, pediatrics, children skills and behavior at different ages, way-finding systems, children design, communication design, human factor design.
- Direct observation of children in connective and waiting areas of pediatric hospitals (as guests and as patient) - as explained in the use case below.
- Interviews to medical, health staff and parents for detecting problems of orientation both for them and for kids in collective areas of pediatric hospital.
- Co-working activity with children aged 7 to 12 years.
- Collection and comparison of results.

This practical activity with the children was based on the detection of their mood and needs when they are in the hospitals as guests or as patient.

The purpose of this co-working activity is also to know if and how children feel at ease and if they have difficult to orientate in public collective areas in hospitals. Moreover, the experimentation aim to identify possible real solutions interpreting children suggestions.

The results have the function to verify possible solutions and also to contribute to specify good practices to design a way-finding system able to guide little users, orienting and informing them with the use of the colors and familiar elements according to their own language and imaginary in a pediatric hospital.

Equally the method highlights the elements that are critical and do not let children feel at ease and live a pleasant experience.

### 4.1 Use Case

The experiment was set-up in Sassari (Italy), a small town in the north of Sardinia, at the St. Peter Pediatric University Hospital, which is the second most important pediatric clinic in the island and that, during summer, it becomes the point of reference also for tourists of the all northern coast.

The research has implemented three main actions which allowed to understand better children behaviors: direct observation, interviews and co-working initiatives together with children.

**First Activity Direct Observation.** It started considering the waiting areas in the ward, in emergency care units and the connection spaces among different clinics. The experimentation activities have involved children aged from about three to twelve years old visiting the hospital. Children were belonging to different nationalities. The observation showed how all the children were interacting with hospital environment and interior settings and how they were moving in the surrounding environment with and without their parents.

The method's application points out, for the most, the difficulties of all the children to be oriented and feel comfortable. We discovered that the main activity that children of all ages have in common, is playing. Besides, the observation highlighted how children, inside these spaces, used to play mainly together with well known and familiar elements, even if they were not designed specifically for playing. The observation also underlined that children were attracted mainly by multimedia and interactive elements present in the areas, even if they are not addressed to them.

Moreover, we observed that the more common behaviors of kids were inspired by different kind of space elements such those, for instance, floor's tiles inspiring the game "the floor is lava" or grouts spacing of flooring which inspire balances games.

**Second Activity Interviews.** In a second moment, frontal and questionnaire interviews were conducted both by the medical, paramedical and technical staff, and by visitors and patients' relatives to the structure. The questions asked to visitors required an evaluation of the interior hospital's and way-finding system. The results of the interviews showed that visitors found the internal signage very poor and causing a lot of difficulties in orientation within the hospital. The interviews' answers underlined that, very often, visitors were used to ask the staff the best path to reach their destination. Moreover the questions asked to the hospital's staff required an evaluation of the interior hospital's signage and wayfinding system.

The results of the interviews con-firmed that very often the hospital's staff is distracted by visitors asking information, which is usually cause of stress. This habit suggests the necessity of improving the communication and wayfinding system.

**Third Activity Co-working with Children.** The third action of co-working with children was carried out thanks to the collaboration with six schools in the city of Sassari. The workshops involved children aged 7 to 12 years. This activity, referring to the studies carried out by Filippazzi on the need for a group "vaccination" [9] to be implemented in schools to ensure that children could live better the experience inside the hospital, allowed to make the hospital known outside the hospital building, involving children and making this place more familiar to them.

During laboratories carried out in co-working, children, followed by pediatricians, pedagogues and designers, drew a series of characters that they would like to see inside a hospital, and that could "make them feel better".

The co-working allowed the researchers to highlight the thoughts, the emotions and the imaginary of the children related to the hospital environment.

## 5 Results of the Research

### 5.1 Good Practices

The present work states that collective and connective spaces, in pediatric hospitals, have to be characterized by specific way-finding systems and signs/elements designed ad hoc. These ones have to facilitate orientation and relations, to stimulate interaction and imagination and encouraging the socialization and the playing for well-being of children.

As a result, the research proposes a series of good practices, which are easily applicable to different kind of pediatric hospital interiors, both for new hospitals and for existing ones. The following steps describe the actions useful to design way-finding systems.

What is necessary to individuate:

- Collective areas (i.e., waiting room, bathroom, restaurant, play room, library, etc.) that needed to be reached by children.
- Corridors and connective spaces that are necessary to walk to go from one area to another.
- Each existing element (i.e. doors, baseboard, windows, switchboard, signage, etc.) in the above spaces, which could be necessary to re-design or taking into account for the way-finding design. This point is valid only for existing hospitals.

What is necessary to define:

- The optimized ways for easy orientation.
- All the elements (material and/or graphic) specifying the colors to be used, the shapes and the illustration according to the imaginary of children, able to improve children's perception of indoor spaces. All of these elements are designed as a "family of sign types" [10] placed in a recurrent mode along the way, according to the rules previously defined. Following this indication, it is suggested to design a coordinated image manual describing exactly the rules of the system.

What is necessary to determinate. The strategic communicative functions of each "element" used to design the way-finding systems, in particular between:

- Basic function: able to orient and to inform about the surroundings, realized by visual communication on the floor and on the walls positioned at a height not exceeding one meter and half (related to children height to be easy visible by them).

- "Emotional" function, with connotative meaning: able to reassure, engaging the senses, and entertain, realized through familiar elements and the practice of playing, remembering pleasant situations and actions of their daily life.

The connotative functions include many messages differently understood by children in relation to their ages, their cultural baggage and their knowledge. Generally, they refer to "elements" of children's imaginary. They could be iconic inspired by natural elements as animals, flowers, etc. or coming from tales, games, video games, cartoons, etc.

### 5.2 Results

Another very important practice that characterized children life is the playing as underline also Piaget [29].

It is quite sure that when kids play they show their own emotions together with their desire of communicate.

As a result of the observation of children at hospital and of the co-working activity (as described above) it is important to underline the importance on playing both in traditional way, - as for example symbolic or vertigo games - than in virtual one, as video games and app used by children, which are digital natives, with satisfaction and easiness. All the communicative elements of the way-finding system should stay on the following features:

- To be ludic: graphic or material elements suggesting or remembering familiar games-usually acted at home, at school or at playground. As an alternative, interactive elements offering familiar tools like video games to be used only in common waiting areas to entertain.
- To be defined by specific colors and simple shapes: easy to identify and recognize inside an interior, also encouraging creativity in differently aged children.
- To be sensory interactive: stimulating curiosity and pleasant emotions, and at the same time, feeling children well and at ease.

## 6 Conclusions

In this paper we have identified a set of design principles and good practices to define the hospital spaces through wayfinding systems design according children needs with the final aim to improve children hospital experience. The result of the present work highlights that collective spaces for children, as pediatric hospitals, need new wayfinding and communication systems designed ad hoc in order to render the interiors familiar for the little users feeling them at ease. Besides, the research claims that familiar elements (both material and graphic) are responsible of pleasant experience influencing the well-being of children and consequently facilitating the disappearance of stress and a speedy recovery.

Basically, children need emotional way-finding systems designed using natural stylized elements to live experiences in security and in freedom.

We hope to share the result of the present work in order to apply and to develop the proposed open rules for designing more and more hospital's interiors children-centered in the next future.

## Appendix

Author Laura Giraldi has coordinated the overall writing process of the paper and has written the following sections: 1.1 "Wayfindind Systems", 2 "The Aim of the Research", 3 "Methods of Research", 4 "Applied Method. Use Case: Pediatric Hospital in Sassary (Italy)", 5 "Results of the Research", 6 "Conclusions".

Author Marta Maini has written the following sections: 1.2 "Pediatric Hospital", 1.3 "Children and Parents' Stress in Pediatric Hospital", 4.1 "Use Case".

Author Donatella Meloni has written the following sections: 1.4 "Pediatrician's Point of View".

## References


1. Baraldi C, Maggioni G (2000) Una città con i bambini. Donzelli, Roma
2. Callois R (2000) I giochi e gli uomini. Bompiani
3. Campos Andrade C, Sloan Devlin A (2015) Stress reduction in the hospital room: applying Ulrich's theory of supportive design. J Environ Psychol 41:125e134
4. Carpuso M, Trappa MA (2005) La casa delle punture. La paura dell'ospedale nell'immaginario del bambino. Ma. Gi
5. Carpuso M (2001) Gioco e studio in ospedale. Creare e gestire un servizio ludico-educativo in un reparto pediatrico. Erickson, Trento
6. Del Nord R (2006) Lo stress ambientale nel progetto dell'ospedale pediatrico. Environmental stress prevention in children's hospital design: indirizzi tecnici e suggestioni architettoniche. Motta architettura, Milano
7. Eco U (1975) Trattato di semiotica generale. Bompiani, Milano
8. Edwards C, Gandini L, Forman G (a cura di) (2014) I cento linguaggi dei bambini. L'approccio di Reggio Emilia all'educazione dell'infanzia. Edizioni Junior
9. Filippazzi G (2004) Un ospedale a misura di bambino. Esperienze e proposte. FrancoAngeli, Milano
10. Gibson D (2009) The wayfindg handbook. Princeton Press, New York
11. Giraldi L (2014) Kidesign. Altralinea edizioni, Firenze
12. Giraldi L (2018) Il design pensato ad hoc per i bambini. In: Luisa C, Raffaella F, Benedetta SM (a cura di) Design su Misura. Atti dell'Assemblea annuale della Società Italiana di Design. Società Italiana di Design, Venezia
13. Giraldi L, Benelli E, Vita R, Patti I, Filieri J, Filippi F (2017) Designing for the next generation. Children urban design as a strategic method to improve the future in the cities. Des J 20:sup1
14. Golledge RG, Jacobson RD, Kitchin R, Blades M (2000) Cognitive maps, spatial abilities, and human wayfinding. Geograph Rev Japan 73(Ser. B), No. 2:93–104
15. Harris PL (1991) Il bambino e le emozioni. Raffaello Cortina Editore, Milano

16. Hillier B, Hanson J (1988) The social logic of space. Cambridge University Press, Cambridge
17. Hughes FP (1999) Children, play, and development, 3rd edn. Allyn and Bacon, Needham Heights
18. Jansen-Osmann P, Wiedenbauer G (2004) The representation of landmarks and routes in children and adults: a study in a virtual environment. J Environ Psychol 24:347–357
19. Kanizsa S, Dosso B (2006) La paura del lupo cattivo. Quando un bambino è in ospedale. Maltelmi, Roma
20. Lewis D, Miller C (1999) Wayfinding: effective wayfinding and signing systems guidance for healthcare facilities. Stationery Office, London
21. Lingwood J, Blades M, Farran EK, Courbois Y, Matthews D (2015) The development of wayfinding abilities in children: learning routes with and without landmarks. J Environ Psychol 41:74e80
22. Cagliari P, Castagnetti M, Giudici C, Rinaldi C, Vecchi V, Moss P (2016) Loris malaguzzi and the schools of reggio emilia. A selection of his writings and speeches. Routledge, UK, pp 1945–1993. (McGraw-Hill)
23. Monsa AK (2000) We the Peoples. United Nations, New York
24. Montessori M (1950) La scoperta del bambino. Garzanti, Milano
25. Montessori M (1949) Dall'infanzia all'adolescenza. Garzanti, Milano
26. Montessori M (1970) Educazione per un mondo nuovo. Garzanti, Milano
27. Norman DA (2004) Emotional design. Perché amiamo (o odiamo) gli oggetti della vita quotidiana. Apogeo, Milano
28. Petrillo M, Sanger S (1980) Emotional care of hospitalized children: an environmental approach, 2nd edn. JB Lippincott, Philadelphia
29. Piaget J (1967) Lo sviluppo mentale del bambino e altri studi di psicologia. Einaudi, Torino
30. Piaget J (1970) La psicologia del bambino. Einaudi, Torino
31. Vechakul J, Patel Shrimali B, Sandhu JS (2015) Human-centered design as an approach for place-based innovation in public health: a case study from Oakland, California. Matern Child Health J 19:2552–2559

# Racial Effect on the Recommended Safe Weight for Backpack Users Among Schoolchildren

Ademola James Adeyemi[1(✉)], Jafri Mohd Rohani[2], and Mat Rebi AbdulRani[2]

[1] Department of Mechanical Engineering, Waziri Umaru Federal Polytechnic, Birnin Kebbi, Nigeria
folashademola@gmail.com

[2] Department of Materials, Manufacturing and Industrial Engineering, Universiti Teknologi Malaysia, Johor Bahru, Johor, Malaysia

**Abstract.** Racial differences associated with variation in diet, climate and culture, may pose a challenge in developing a globally acceptable safe backpack weight for schoolchildren. This study therefore investigates the suitability of a multivariate backpack-related back pain model in a multiracial society such as Malaysia. Back pain related data from an average of 205 Malay, Chinese and Indian Schoolchildren were fitted into a proposed backpack-related back pain model and also into a regression model to predict safe weight from percentage body weight (PBW), body mass index and age in order to determine the level of fit. While the three racial models met the minimum requirements of the different goodness-of-fit indices, there were uncaptured significant relationships peculiar to each racial model in the proposed model. Notwithstanding, the combination of age, BMI and PBW are better predictors of back pain occurrence among the children in the three ethnic groups.

## 1 Introduction

Recent studies have highlighted the need for researchers to come up with a viable solution to the non-specific back pain problem associated with the use of backpack by school children. Although the problem also occur among adults, its effect on children raises serious concern because there is high probability that children who have back pain will also have it as an adult. Presently, there are evidences that the present recommendation of the ratio of bag weight to the body weight (PBW), which varies from 10% in Australia and Europe, to 20% in United States [1], has not met the desired expectation. This deficiency has resulted in researchers advocating for the use of multiple variables as a solution to the problem. This is because nonspecific back pain is a multifactorial problem and the present PBW is not sufficient to capture the variability among the causal factors. Hence, additional variables such as age and body mass index have been shown to have higher predictability than the PBW [2]. However, the


S. Bagnara et al. (Eds.): IEA 2018, AISC 826, pp. 811–817, 2019.
https://doi.org/10.1007/978-3-319-96065-4_84

suitability of such recommendation to fit different population or culture across the globe is another challenge. Anthropometric measures, which is a critical factor in human ability to support load varies across racial lines because of diet, climate and culture [3, 4]. Therefore, this paper is aimed at investigating the suitability of a multivariate backpack-related back pain model in recommending an appropriate safe weight among children from different racial groups.

### 1.1 The Backpack-Related Back Pain Model

The backpack-related pack pain model [2] shows how four factors, namely, anthropometry, posture, backpack volume and children's rating ability, are associated with back pain among school children. The model highlights the multiple interactions among these factors, resulting in non-specific back pain in school children. Backpack volume and posture were postulated to have direct effect on the occurrence of back pain while anthropometry has an indirect effect on back pain occurrence. Posture was also shown to play the role of a partial mediator between the volume of school materials and back pain while it serves as a full mediator between back pain and the children's anthropometry.

### 1.2 Malaysia as a Multiracial Sampling Population

Geographical, environmental and anthropometric diversities need to be considered when proffering a general safe weight for backpackers in schools. The ability to identify a well-structured sample population with unique features that is representative of the factors under consideration will go well in achieving the set out goal of developing a widely acceptable solution. Hence, Malaysia serves as a good sampling population because it is a multi-racial and multicultural society with diverse human groupings. Although these racial groups reside together, they still maintain their cultural diversity with minimal interference from one another. Malaysia consists of (Malay = 54.5%, Chinese = 24.5%, Indians = 7.3%, others = 13.7%) and other indigenous (bumiputera) tribes with distinct sociological features [5]. Malaysia also has government funded public and community based schools with multi-language delivery system but similar educational content. Children attend National (Malay-speaking), National-type Chinese and National-type-Tamil schools. The schools are guided by the same curriculum developed by the Ministry of Education. The schools serve as unique strata where different racial samples can easily be collected. The ability to stratify the sampling population into mutually exclusive strata will increase the accuracy of the estimator [6]. This strata also provides the diversity needed as educational systems including schoolchildren's behavior are associated with the societal and cultural dynamics of a country [7–9].

## 2 Methods

The study used a previously collected data from a study involving 615 schoolchildren in Johor Malaysia [10]. Analysis on racial differences was not included in the previous study because the study was aimed at validating a questionnaire and identifying significant factors associated with back pain among Malaysian schoolchildren. Details of the instruments and procedures of the data collection were well spelt out in Adeyemi et al. (2014). The data was split into three based on race. The breakdown of the population distribution is shown in Table 1.

Correlation techniques such as analysis of variance, regression analysis and structural equation modeling (SEM) were used to analyze both the subjective responses, obtained with a questionnaire, and objective anthropometric variables.

Each of the three data was fitted into the backpack weight regression model and backpack-related back pain model in Adeyemi et al. (2017) to determine their fitness level. Thereafter, structural equation model technique was also used to develop individual model for each population. The developed models were then compared with the backpack-related back pain model to identify any significant difference. The model fit was assessed based on the criteria of the goodness-of-fit indices used in [11, 2]. All statistical analysis was performed using Predictive Analysis Software (PASW) statistics 18 and Analysis of Moment Structure (AMOS) version 18.

## 3 Results

Table 1 shows the age-gender-race distribution of the participants. The sampling plan ensures good distribution based on age, gender and race, since there was no significant difference in the age-gender distribution ($p = 0.779$) and age-race distribution ($p = 0.954$).

**Table 1.** Age, gender and race distribution of the schoolchildren

| Age | Gender | | Total | Race | | | Total |
|---|---|---|---|---|---|---|---|
| | Boys | Girls | | Malay | Chinese | Indian | |
| 7 | 42 | 59 | 101 | 33 | 38 | 30 | 101 |
| 8 | 44 | 59 | 103 | 38 | 33 | 32 | 103 |
| 9 | 32 | 73 | 105 | 33 | 33 | 39 | 105 |
| 10 | 45 | 56 | 101 | 39 | 30 | 32 | 101 |
| 11 | 44 | 58 | 102 | 35 | 30 | 37 | 102 |
| 12 | 42 | 61 | 103 | 35 | 39 | 29 | 103 |
| Total | 249 | 366 | 615 | 213 | 203 | 199 | 615 |

The study shows that average PBW (mean ± SD) for Malay, Chinese and Indian schoolchildren are $17.78 \pm 6.62$, $19.17 \pm 6.38$ and $19.59 \pm 7.99$ respectively. There was also significant difference in the weight of the backpacks carried by the children from the three races $F(2,612) = 8.927$, $p < 0.0001$). Apart from the backpack, the children were also observed to carry additional school materials either in another bag or in their arms. The total load carried by the children were also significantly different among the races $F(2,612) = 111.388$, $p < 0.0001$). Subsequently, significant difference was observed in PBW among the races $F(2,612) = 3.791$, $p = 0.023$). However, there was no difference in the PBW distribution when categorized into less than 10%PBW, 10–15%PBW and > 15%PBW among the children $F(2,614) = 1.859$, $p = 0.157$). The percentage of the children that carries less than 10%PBW are 13.1,7.9 and 10.6 among Malay, Chinese and Indians respectively while 67.6, 74.9 and 69.3 percent of the children from the races carried backpack greater than 15%PBW.

The data from the three races were fitted into the proposed generalized standardized regression equation developed in Adeyemi et al. (2017).

Bag weight = 1.062PBW + 0.639BMI + 0.383Age.

Despite the significant difference in the load carried by the children from difference races, the racial data successfully fitted into the standardized regression equation as shown in Table 2.

**Table 2.** Racial Regression equation for the generalized regression model

| Var | Malay ($R^2 = 0.551$) | | | Chinese ($R^2 = 0.547$) | | | Indian ($R^2 = 0.611$) | | |
|---|---|---|---|---|---|---|---|---|---|
| | Beta | T | Sig | Beta | t | Sig | Beta | T | sig |
| BMI | 0.54 | 9.985 | 0.0001 | 0.402 | 8.153 | 0.0001 | 0.321 | 6.708 | 0.0001 |
| PBW | 0.95 | 15.579 | 0.0001 | 0.753 | 14.167 | 0.0001 | 0.787 | 15.834 | 0.0001 |
| Age | 0.524 | 9.809 | 0.0001 | 0.494 | 9.539 | 0.0001 | 0.514 | 11.012 | 0.0001 |

### 3.1 Structural Equation Models

Table 3 shows that the three racial distributions fitted into the backpack-related back pain model since they all met the minimum acceptable value for $C_{min}$ (ratio of Chi-square to degree of freedom), Goodness of Fit index (GFI), Root Mean Square Error Approximation (RMSEA), Comparative Fit Index (CFI), Non-Normed Fit Index (NNFI) and Adjusted Goodness of Fit index (AGFI). These values are found on the Malay, Chinese and Indian rows on the table. Malay2, Chinese2 and Indian2 rows show the results for the best fit for each of the three models when developed independently. The significant change column reveals that there were additional significant relationships not captured in the generalized backpack-related back pain model for the three races. The Malay and Indian models show that there was significant correlation between age and the rating ability (RATING) construct of the children. Both the Malay and the Chinese models also show that the period of experiencing pain (POP) had a direct effect on back pain rating (BAP) of the children. The Chinese model also shows that the degree of pain (PAIN) has a direct effect on the rating of back pain (BAP) by the children. The Indian model also shows that the backpack weight (BGW) has a direct effect on the neck inclination (NECK) and the neck inclination has a direct effect on the back inclination (BAK).

**Table 3.** Comparing individual racial Model with the backpack-related back pain model.

| Model | GFI | AGFI | NNFI | CFI | RMSEA | $C_{MIN}$ | AIC | BIC | IFI | Significant change |
|---|---|---|---|---|---|---|---|---|---|---|
| Malay | 0.877 | 0.82 | 0.897 | 0.92 | 0.088 | 2.639 | 331.429 | 476.366 | 0.92 | |
| Malay2 | 0.901 | 0.852 | 0.93 | 0.947 | 0.072 | 2.117 | 282.634 | 434.312 | 0.947 | (1) Age Correlates RATING<br>(2) POP has direct effect on BAP |
| Chinese | 0.861 | 0.796 | 0.867 | 0.897 | 0.103 | 3.134 | 377.464 | 519.932 | 0.898 | |
| Chinese2 | 0.887 | 0.83 | 0.904 | 0.928 | 0.087 | 2.54 | 320.582 | 472.99 | 0.929 | (1) POP has direct effect on BAP<br>(2) PAIN has direct effect on BAP<br>(3) POST has direct effect on BAP |
| Indian | 0.81 | 0.723 | 0.702 | 0.769 | 0.127 | 4.16 | 472.879 | 614.056 | 0.774 | |
| Indian2 | 0.921 | 0.879 | 0.946 | 0.96 | 0.054 | 1.577 | 234.322 | 388.632 | 0.961 | (1) Age Correlates RATING<br>(2) BGW has direct effect on NEK<br>(3) NEK has direct effect on BAK |

## 3.2 Discussion

The study reveals that racial effect plays a significant role in the factors associated with the problem of back pain among the schoolchildren using backpack. The role played by racial differences in body composition had previously been highlighted in the literature [12–14]. Personal interviews with lecturers, teachers and doctors confirm differences in pain coping abilities among the three races associated with this research. This finding on the pain behavior is significant because of the subjective nature of pain which is associated with other physical and physiological variables used in measuring effect of backpack carriage on back pain [10]. The additional relationships identified in the backpack-related back pain model testify that racial differences also affect the children's response to their backpack carriage and feelings of back pain. Significant relationship between biomechanical variables such as back and neck inclinations of the Indian children, the tallest among the three races, is an indication of the importance of anthropometry especially among taller population. Studies has previously identified stature as a predictor of back pain [15, 16].

The result in Table 2 shows that PBW, BMI and age is sufficient to predict the backpack weight for the children from the three races. Although there was significant difference in the PBW among the three races, there was not difference in the backpack behavior as about 70% of the children from the three races carried backpack load that exceed 15%PBW. This explains the high prevalence of back pain among all the children irrespective of their races. The finding shows the possibility of developing a globally acceptable multivariate backpack-back pain model, although there might be need for some of the variables to be weighted or some models might accommodate the variation in some population better than the others.

## 4 Conclusion

Although the three racial distributions fitted into the backpack-related back pain model, development of individual racial SEM model highlighted significant relationships peculiar to the different racial group. Despite these differences, age, BMI and PBW still significantly predict backpack weight in the three racial distributions, thereby justifying the importance of multivariate index to measure the backpack weight for school children. This is because multiple variables better account for the variances associated with the back pain risk factors. However, there is still need to investigate the suitability of the backpack-related back pain model for other populations since both samples used for the model development and model testing are from the same Malaysian population.

## References

1. Dockrell S, Simms C, Blake C (2013) Schoolbag weight limit: can it be defined? J Sch Health 83:368–377
2. Adeyemi AJ, Rohani JM, Abdul Rani MR (2017) Backpack-back pain complexity and the need for multifactorial safe weight recommendation. Appl Ergon 58:573–582
3. Wilson I, Desai DA (2016) Anthropometric measurements for ergonomic design of students' furniture in India. Eng Sci Technol Int J 20:232–239
4. Humphries DL, Dearden KA, Crookston BT, Woldehanna T, Penny ME, Behrman JR (2017) Household food group expenditure patterns are associated with child anthropometry at ages 5, 8 and 12 years in Ethiopia, India, Peru and Vietnam. Econ Hum Biol 26:30–41
5. Malaysia Department of Statistics (2011) Taburan_Penduduk_dan_Ciri-ciri_Asas_Demografi.pdf
6. Ott LR, Longneck M (2010) An introduction to statistical methods and data analysis. Cengage Learning, Belmont
7. Tharp RG (1989) Psychocultural variables and constants: effects on teaching and learning in schools. Am Psychol 44:349–359
8. Bacevic J (2008) Anthropological educational policy in the light of european transformations. Stud Ethnol Croat 20:37–56
9. Downey DB, Pribesh S (2004) When race matters: teachers' evaluations of students' classroom behavior. Sociol Educ 77:267–282
10. Adeyemi AJ, Rohani JM, Abdul Rani M (2014) Back pain arising from schoolbag usage among primary schoolchildren. Int J Ind Ergon 44:590–600
11. Adeyemi AJ, Rohani JM, Akanbi G, Rani MRA (2014) Anthropometric data reduction using confirmatory factor analysis. Work A J Prev Assess Rehabil 47:173–181
12. Turk Z, Vauhnik R, Micetić-Turk D (2011) Prevalence of nonspecific low back pain in schoolchildren in north-eastern Slovenia. Coll Antropol 35:1031–1035
13. Duncan JS, Duncan EK, Schofield G (2009) Accuracy of body mass index (BMI) thresholds for predicting excess body fat in girls from five ethnicities. Asia Pac J Clin Nutr 18:404–411

14. Liu A, Byrne NM, Kagawa M, Ma G, Kijboonchoo K, Nasreddine L et al (2011) Ethnic differences in body fat distribution among Asian pre-pubertal children: a cross-sectional multicenter study. BMC Pub Health 11:500
15. Adeyemi AJ, Rohani JM, Rani MA (2015) Interaction of body mass index and age in muscular activities among backpack carrying male schoolchildren. Work A J Prev Assess Rehabil 52:677–686
16. Poussa M, Heliövaara MM, Seitsamo JT, Könönen MH, Hurmerinta KA, Nissinen MJ (2005) Anthropometric measurements and growth as predictors of low-back pain: a cohort study of children followed up from the age of 11 to 22 years. Eur Spine J, 14:595–598

# Human Factors/Ergonomics Education and Certification: The Canadian Experience

Nancy Black[1(✉)] and Judy Village[2(✉)]

[1] Mechanical Engineering Department, Université de Moncton, Moncton, NB E1A 3E9, Canada
Nancy.black@umoncton.ca
[2] School of Population and Public Health, University of British Columbia, Vancouver, BC V6T 1Z3, Canada
jvillage@shaw.ca

**Abstract.** The Canadian College for the Certification of Professional Ergonomists (CCCPE) is responsible for ensuring certificants (CCPE and AE) meet minimum educational levels and at least four years work practice (for CCPE). The Association of Canadian Ergonomists – Association canadienne d'ergonomie supports the CCCPE but each educational institution determines the number and scope of ergonomics course offerings. Relevant courses exist at institutions across Canada, with the exception of the full-time programs in the Prairie provinces, but in limited number. Only four programs are designed currently to ensure graduates qualify for AE, two in Ontario in English for Kinesiology students and two in Québec in French open to various undergraduate degree holders. Nineteen post-secondary institutions offer some ergonomics courses but their number and focus varies, most frequently offered within kinesiology or movement science programs. The number of qualified, certified human factors and ergonomics practitioners is growing well, but could improve with more courses for those in the Prairie region and those with non-kinesiology backgrounds outside of Québec.

**Keywords:** Education · Certification · Canada

## 1 Introduction

### 1.1 Historical Context

Human factors/ergonomics (HF/E) in Canada is a relatively new profession. The Association of Canadian Ergonomists – Association canadienne d'ergonomie (ACE) was founded under the name of Human Factors Association of Canada (HFAC) in 1968 [1], adding the French Association canadienne d'ergonomie (ACE) in 1984 [2],

N. Black—Past President, Association of Canadian Ergonomists – Association canadienne d'ergonomie (ACE).
J. Village—President Elect, ACE; Previous Board Member Canadian College for the Certification of Professional Ergonomists.


S. Bagnara et al. (Eds.): IEA 2018, AISC 826, pp. 818–825, 2019.
https://doi.org/10.1007/978-3-319-96065-4_85

changing its name to the bilingually consistent acronym "ACE" in 2000 [2]. In 1990, a paper encouraging universities to develop major and minor programs in HF/E was presented at the Association's Annual Conference by the Committee on Professional Education [3]. At the time, there were only occasional independent courses in HF/E offered in various departments including psychology, kinesiology, systems design engineering and industrial engineering. ACE was attracting members, and HF/E specialists were working across Canada in defense, private and public sectors, as well as in education.

In 1998, the Executive Council of ACE formed the Canadian College for the Certification of Professional Ergonomists/Conseil canadien de certification des praticiens en ergonomie (CCCPE) [4], and established the first Board of the CCCPE in January 1999 [2]. According to its constitution, the CCCPE Board includes eight persons who are CCPE, including "to the extent practicable and feasible" both academics and practitioners, all geographic regions of Canada and at least two members fluent in each official language (English and French) [5]. Board members represent the range of physical and cognitive HF/E domains. CCCPE acts as a committee of ACE, with all members of CCCPE also being ACE members, and CCCPE presenting a report at the Annual General Meeting of ACE, but daily activities being largely independent.

### 1.2 Management of Certification in Canada

Certification requirement criteria were developed under two categories: Canadian Certified Professional Ergonomist (CCPE), and Associate Ergonomist (AE). Both categories have met minimum education levels, but only the CCPE also has proven sufficient practical experience (four years with one mentored or five years, generally). AE designation may be held for at most five years, anticipating that during that time, the AE would apply for CCPE designation. Recognizing the limited courses focussing on HF/E prior to 1997, for a limited duration Ergonomists could be "grandfathered" into CCPE if holding an appropriate degree and 10 years' experience. A more limited "grandfathering" was offered also until 2021 for those with 25 years' experience [6].

Members of the CCCPE Board evaluate applications in two cycles annually. Applications must include

- application form and fee
- a summary letter
- curriculum vitae
- proof of education, including course descriptions and credit allocation logs
- employment history log
- samples of work products and work product summaries.

No examination is required.

Educational requirements for both CCPE and AE categories were designed to be consistent with those of the International Ergonomics Association (IEA) [7, 8], although as of May 2018 official recognition from IEA had yet not been requested. Educational requirements cover the entire breadth of HF/E and include minimum contact hours. Initially these focussed greatly on specialised HF/E, however over time, ACE and CCCPE recognized applicants' continued challenge meeting the minimum number of educational hours, as well as laboratory and fieldwork specific to HF/E in available undergraduate programs. In response, the minimum hours were changed in 2017. The resulting current educational requirements are found in Table 1. These account for the fact that areas deemed 'foundational' (section A) may be accessed through complementary fields of study (e.g. Engineering, Kinesiology, Psychology). Other areas must be taught with specific references to HF/E (section B).

**Table 1.** Educational requirements to access AE and CCPE certification levels. All courses must be at least at the undergraduate bachelor's level. [6]

| Categories | Knowledge areas | Minimum contact hours |
|---|---|---|
| Section A: foundational for ergonomics/human factors | A1. Design concepts (other) | 20 |
| | A2. Evaluation | 75 |
| | A3. Other | 5 |
| | A4. Physical demands | 20 |
| | A5. Human performance | 20 |
| Minimum subtotal A4 + A5 | | *120* |
| Minimum total of section A | | *300* |
| Section B: Specific to ergonomics/human factors | B1. General | 5 |
| | B2. Design concepts in Ergo/HF | 10 |
| | B3. Cognitive Ergo/HF | 10 |
| | B4. Physical Ergo/HF | 10 |
| | B5. Macro Ergo/HF | 10 |
| Minimum total of section B | | *100* |
| Minimum total of sections A + B | | *500* |
| Section C: Laboratory work | Specific to Ergo/HF | 20 |
| | Foundational or specific to Ergo/HF | 80 |
| | Minimum total of section C | 100 |
| Section D: Field work | Specific to Ergo/HF | 30 |
| Minimum total of section D | | 30 |
| Minimum total of sections C + D | | *150* |
| Overall minimum total of sections A, B, C & D | | **800** |

In addition to educational requirements, when applying, CCPE applicants must demonstrate that they devoted the majority of their work time to the application, practice and/or teaching of HF/E for five years. One year less of applied work in HF/E is required if a year was overseen by a qualified mentor, recognised by CCPE. That mentor must inspect the applicant's work from time to time and provide feedback and guidance to the applicant [6].

Applicants also must include work products demonstrating HF/E professional competencies, including preliminary project definition, systematic analysis, participation in the design process, and "other" competencies. The breadth of required professional competencies are explicitly listed in the CCCPE Application kit [6], and includes "the integration of, or ability to integrate, biophysical, perceptual, cognitive and psychosocial considerations in ergonomic evaluation of existing design or recommendations for design changes in products, services or work processes". Additional "other" competencies include the preparation of materials and presentations to communication HF/E application to a range of audiences and project management. Work products include publications, reports, design specifications, while respecting confidentiality, as necessary [6]. These must demonstrate a breadth of work experience; thus they must not all be replications of similar evaluations of office or industrial workplaces.

Proof of education need not be submitted to CCPE for applicants who already hold either Certified Professional Ergonomist (CPE) or Certified Human Factors Professional (CHFP) designations administered by the Board of Certification in Professional Ergonomics (BCPE) and those holding the European Ergonomist (Eur. Erg.) administered by the Centre for Registration of European Ergonomists (CREE) [6].

## 2 Current Status of Programs Leading to Certification

In 2018, our survey of post-secondary institutions found that at least 14 institutions offer one or more courses focussing on ergonomics. The majority of these are associated with kinesiology programmes. They are available within four of the five regions recognized by ACE: BC-Y (British Columbia – Yukon); ON (Ontario); QC (Québec); Atl (Atlantic provinces – including New Brunswick, Nova Scotia, Prince Edward Island and Newfoundland and Labrador. The Prairie and Northern (PN) region, comprising the provinces of Alberta, Saskatchewan and Manitoba, is the only region with no known ergonomics educational programs, at present.

Four educational programmes in Canada have been designed specifically with access to CCPE in mind (Table 2), however programmes across Canada offer some ergonomics courses (Table 3). Note that with the exception of University of Waterloo, all programs designed to facilitate AE access are post-graduate level. Throughout this section, level of study is defined as 1 = undergraduate; 2 = master's and *Diplôme d'études supérieures spécialisées* (DESS); 3 = doctoral; 4 = certificate (typically post-undergraduate, but not equivalent to a master's degree).

**Table 2.** Summary of educational programs specifically designed for CCPE/AE certification preparation.

| Institution | Language | Region | Required background | Associated department | Level of study |
|---|---|---|---|---|---|
| Université Laval | French | QC | Bachelor's degree + 1 related course | Industrial Relations/Kinesiology/Medicine | 2 (DESS) |
| Université de Québec à Montréal | French | QC | Bachelor's degree | Life sciences & physical activity; Kinanthropologie | 2 (prof.)/ 2 DESS |
| University of Waterloo | English | ON | Kinesiology (in progress) | Kinesiology | 1, 2 |
| Fanshawe College | English | BC-Y | B. Kinesiology (completed) | School of public safety | 4 |

**Table 3.** Summary of educational programmes offering human factors and ergonomics courses in order of decreasing course offerings.

| Institution | Language | ACE Region | Department[1] | Field of study | Level of study | # semester courses |
|---|---|---|---|---|---|---|
| Fanshawe College | English | ON | Public safety | Advanced Erg. Studies | 4 (Kin. graduates) | 38 weeks inc. 10-week placement |
| Université de Québec à Montréal (UQAM) | French | QC | Life sciences & physical activity; Kinanthropologie | Erg. intervention DESS | 2 professional | 15; 7 courses + 5 intervention |
| | | | | Erg. | 2 DESS | 7 + 3 intervention |
| Simon Fraser University | English | BC-Y | Kinesiology | Biomedical physiology & Kin. Certificate in Occ. Erg. | 1 (4) | 8 |
| École polytechnique de Montréal | French | QC | Mathematics & Industrial Eng. | Cognitive | 1; 2 & 3 | 3; 8 |
| | | | | Physical | 1; 2 & 3 | 3; 6 |
| | | | | Erg. | 4 | 3–10 |
| | | | | Human computer interaction | 1 | 4 |
| Université Laval | French | QC | Industrial Relations | Erg. & Innovation | 2: DESS | 9 |
| University of Waterloo | English | ON | Kinesiology | Erg. & Injury Prevention Minor | 1 | 5 req. +5 |
| | | | Systems Design Eng. | Human Factors Elective package | 1; 2 & 3 | 1 req. + 4; 3 |
| University of Toronto | English | ON | Mechanical & Industrial Eng. | Industrial Eng. | 1; 2 & 3 | 3 req. + 4; 6 |
| University of Ontario Institute of Technology (UOIT) | English | ON | Kinesiology | Occupational Erg.; Clinical biomechanics; Erg. & HF | 1 | 3 |
| | | | Manufacturing Eng. | Industrial Erg. | 1 | 1 |
| | | | Work disability prevention | Occ. Erg. & Work Disability Prevention | 4 | 5 modules |

*(continued)*

**Table 3.** (*continued*)

| Institution | Language | ACE Region | Department[1] | Field of study | Level of study | # semester courses |
|---|---|---|---|---|---|---|
| Dalhousie University | English | ON | Industrial Eng. | Work design; ind. biomechanics | 1, 2, 3 | 2 |
| | | | Kinesiology | Erg. stream | 1, 2, 3 | 3 |
| | | | Continuing Ed. | Ergonomic program Mgment | 4 | 4 |
| University of New Brunswick | English | ON | Kinesiology | Erg. minor; Occ. biomechanics | 1, 2 | 4 |
| University of Guelph | English | ON | Human Health & Nutritional Sciences; Psychology; Eng.; Computer Science | Occ./Eng. Biomechanics; Sensation & Perception; Human computer interaction | 1 | 4 |
| Ryerson University | English | ON | Industrial Eng. | Industrial erg./psych. | 1, 2, 3 | 2 |
| | | | Occ. & Public Health | Occ. safety & health | 1 | 4 |
| University of Windsor | English | ON | Kinesiology | Occ. biomechanics; work; injury prevention | 1 | 3 (1 required) + 3 lab |
| Carleton University | English | ON | Industrial Design | B. Ind. Design M. Design | 1, 2 | 3 |
| Queen's University | English | ON | Kinesiology | Occ. biomechanics/Physical Erg. | 1 | 2 |
| Laurentian University | English | ON | Human Kinetics | Intro; methods; assessment | 1 | 3 |
| Lakehead University | English | ON | Kinesiology | Kinesiology | 1 | 1 |
| | | | | Graduate Diploma P. Kin | 4 | 2 |
| Laurier University | English | ON | Kinesiology & Physical Education | Occ. Biomechanics & Erg. | 1 | 1 |
| McMaster University | English | ON | Kinesiology | Workplace Injury Risk Assessment (occ. biomechanics) | 1 (opt.) | 1 |

[1]Eng. = Engineering; Occ. = Occupational; Kin. = Kinesiology [2] Erg. = Ergonomics

## 3 Certification in 2018

From 1999 to 2015, 237 people were designated as CCPE and another 60 as AE. In May 2018, 206 CCPE and 57 AE had active designations. Currently, the majority of applicants come from educational backgrounds in kinesiology, or human kinetics or related fields. Most certification applicants require education beyond one undergraduate program. This article summarised that while some courses are available at the undergraduate levels, most institutions provide opportunities to enhance this learning with research applications at the graduate level (both master's – either course-based or thesis-focused, and doctoral).

In the spring of 2018, CCCPE was processing approximately 13 CCPE applications, and 30 AE applications, partly as a result of Fanshawe College's certificate program that started in 2015. It is expected that changes to both educational requirements and new programs like Fanshawe's increase numbers of Certified HF/E specialists in Canada.

In 2017, ACE and CCCPE began the process to obtain IEA recognition of CCPE certification, but in the meantime, employer recognition of CCPE already supports certification within Canada.

## 4 Conclusions

Since the creation of the Canadian College for the Certification of Professional Ergonomists (CCPE) in 1998, recognition of CCPE and AE titles has grown. More work is required to provide easy access across Canada to CCCPE recognized educational breadth. In English, two programs facilitate certification by people with a kinesiology background while in French two programs are available with access to other educational backgrounds. In contrast, when the first national ergonomics association started in Canada, it was principally an engineering and psychology-based field. There is a need to ensure that human factors and ergonomics certified professionals in Canada remains representative of the variety of relevant educational backgrounds while having sufficient specialized knowledge.

**Acknowledgements.** Information included in Table 3 was obtained mostly through personal contact with representatives of each concerned academic institution. Special thanks go to (in alphabetical order): Wayne Albert (University of New Brunswick), David Andrews (University of Windsor), Anne-Kristina Arnold (Simon Fraser University), John Kozey (Dalhousie University), Nick La Delfa (University of Ontario Institute of Technology), Steven Fischer (University of Waterloo), Élise Ledoux & Nicole Vézina (Université de Québec à Montréal), Jim Potvin (McMaster, CCCPE), Jean-Marc Robert (École Polytechnique à Montréal), and Allison Stephens (Fanshawe College).

## References

1. About ACE. In: Association of Canadian Ergonomists. https://ace-ergocanada.ca/about/about-ace/index.html. Accessed 16 May 2018
2. Association of Canadian Ergonomists (2008) Communiqué: Celebrating 40 years 38(3). https://ace-ergocanada.ca/files/Communique/2008/Fall_2008_Eng_Final.pdf. Accessed 26 May 2018
3. Webb R, Stager P (1990) Conceptual and practial issues affecting professsional education in ergonomics/human factors. In: Proceedings of the 23rd annual conference of the human factors association of Canada - Comptes rendus du congrès annuel de l'Association canadienne d'ergonomie. Ottawa, ON, pp 1–5
4. Canadian College for the Certification of Professional Ergonomists (2016) History of the CCCPE. In: Canadian College for the Certification of Professional Ergonomists. https://www.cccpe.ca/about/history-of-the-cccpe.html. Accessed 16 May 2018

5. Canadian College for the Certification of Professional Ergonomists (2017) Constitution of the Canadian College for the Certification of Professional Ergonomists. https://www.cccpe.ca/files/New_CCCPE_Kit/CCCPE_Constitution_2017-07-19_EN.pdf. Accessed 16 May 2018
6. Canadian College for the Certification of Professional Ergonomists (2017) CCCPE Certification Application Kit (effective July 2017). https://www.cccpe.ca/certification/application-process.html. Accessed 28 May 2018
7. Professional Standards and Education Committee (2001) Full version of core competencies in ergonomics: units, elements, and performance criteria. IEA Website. In: International ergonomics association - resources - projects. https://iea.cc/project/index.html. Accessed 26 May 2018
8. Resources- projects: professional standards and education standing committee documents. IEA Website. In: International Ergonomics Association - Resources - Projects. https://iea.cc/project/index.html. Accessed 26 May 2018

# Musculoskeletal Disorders in Indian School Children Due to Carrying Heavy Back Packs

Deepak Sharan(✉)

Department of Orthopaedics and Rehabilitation,
RECOUP Neuromusculoskeletal Rehabilitation Centre,
312, 10th Block, Anjanapura, Bangalore 560108, KA, India
deepak.sharan@recoup.in

**Abstract.** Heavy back packs have been reported to result in pain in upper back, shoulder and neck in school children. A retrospective analysis of clinical records from a Paediatric Orthopaedic Surgeon and Rehabilitation Physician's practice in a tertiary level Neuromusculoskeletal Rehabilitation centre in India found 62 children diagnosed with severe MSD attributed to carrying overloaded backpacks. The severity of the pain was such that the child was forced to miss school for at least a day and receive physiotherapy. 37 (59.7%) of the affected children were females and 36 (58.1%) children were of age 10–15 years. Pain in the upper back (72%), neck (56%), shoulder (52%) and lower back (44%) was most common followed by forearm and wrist (24%). The mean duration of discomfort was 24 days and the mean load of the back pack was 12.5 kg. All the children were found to be carrying backpacks more than 15% of their body weights.

**Keywords:** Heavy back packs · Musculoskeletal pain · School children

## 1 Introduction

Heavy back packs have been reported to result in pain in upper back, shoulder and neck in school children. There are concerns that chronic pain in children can persist into adulthood and cause long term disability. The objective of the study was to find out the clinical presentation and risk factors for Musculoskeletal Disorders (MSD) caused by heavy backpacks among school children in India.

## 2 Methodology

A retrospective analysis of clinical records from a Paediatric Orthopaedic Surgeon and Rehabilitation Physician's practice in a tertiary level Neuromusculoskeletal Rehabilitation centre in India found 62 children diagnosed with severe MSD attributed to carrying overloaded backpacks. The severity of the pain was such that the child was forced to miss school for at least a day and receive physiotherapy. The diagnostic


S. Bagnara et al. (Eds.): IEA 2018, AISC 826, pp. 826–827, 2019.
https://doi.org/10.1007/978-3-319-96065-4_86

criteria for a MSD caused by heavy backpacks were pressure mark (redness or swelling) over neck and shoulder corresponding to the straps of the backpack, stooping posture while carrying the backpack, pain or stiffness in the neck, upper back and shoulders predominantly while carrying the backpack and absence of these symptoms during school holidays (Sharan et al. 2012).

## 3 Results

The results of the study showed that 37 (59.7%) of the affected children were females and 36 (58.1%) children were of age 10–15 years. Pain in the upper back (72%), neck (56%), shoulder (52%) and lower back (44%) was most common followed by forearm and wrist (24%). 9 (14.5%) reported burning, numbness and tingling sensation in upper back and 5 (8%) in the upper limbs. 45 (72.6%) of the children were diagnosed to have Myofascial Pain Syndrome, 19 (30.7%) Thoracic Outlet Syndrome and 7 (11.3%) Fibromyalgia Syndrome. 23 (37.1%) children were found to have generalised hypermobility of joints as diagnosed by the Beighton criteria and 41 (66.1%) were found to have an Upper Crossed Pattern described by Janda. The associated conditions included Temporomandibular Disorder, Left Sprengel Shoulder, Short Stature, Cerebral Palsy (Spastic Diplegia), Adolescent Idiopathic Scoliosis and Bilateral Patellofemoral Pain Syndrome (1 each). The mean duration of discomfort was 24 days and the mean load of the back pack was 12.5 kg. All the children were found to be carrying backpacks more than 15% of their body weights. All the 62 children had to undergo physical therapy (range 1–41 days, median 10 days). After the treatment 90% were pain free according to the VAS ratings, while 10% had pain of severity less than 3 on a scale of 0–10.

## 4 Conclusions

Limiting weight of the schoolbag to less than 10% of the body weight by photocopying books, using a lockable basket beneath the school desk to store books and carrying only loose sheets of paper in the schoolbag were found to be the most effective measures to prevent recurrence of MSD.

# Musculoskeletal Disorders in 115 Students Due to Overuse of Electronic Devices: Risk Factors and Clinical Features

Deepak Sharan(✉)

Department of Orthopedics and Rehabilitation,
RECOUP Neuromusculoskeletal Rehabilitation Centre,
312, 10th Block, Anjanapura, Bangalore 560108, KA, India
deepak.sharan@recoup.in

**Abstract.** Musculoskeletal disorders (MSD) are common in students, usually comprising of children, adolescents and young adults, ranging from 30% to 65%. The pathophysiology of MSD is multifactorial. The use of electronic devices (computer, electronic games, portable music players, electronic book readers, tablets and cell phones) has been reported as a risk factor associated with MSD. A total of 115 subjects between the age of 5 to 25 years were evaluated in the study. The evaluation and diagnosis were performed by a single orthopaedic and rehabilitation physician between the years of 2005 to 2017. 53% (n = 61) of the participants were males. The commonly used electronic devices were cell phone (80%), tablet (70%), laptop (65%), electronic games (54%), desktop computer (45%), portable music players (5%), and electronic book readers (2%). The commonest risk factors for MSD were identified as lack of rest breaks (85%), static loading (70%), hazardous body positions (60%), excessive load (30%), deficiency in design of tools/furniture or poor ergonomics (30%) and repetition (20%).

**Keywords:** Musculoskeletal disorder · Physiotherapists · Rehabilitation centre

## 1 Introduction

Musculoskeletal disorders (MSD) are common in students, usually comprising of children, adolescents and young adults, ranging from 30% to 65%. The pathophysiology of MSD is multifactorial. The use of electronic devices (computer, electronic games, portable music players, electronic book readers, tablets and cell phones) has been reported as a risk factor associated with MSD. Some of these devices are increasingly being used as a tool for education. The aim of this study was to find out the risk factors, clinical features and outcome of treatment of MSD due to the use of electronic devices among students.


S. Bagnara et al. (Eds.): IEA 2018, AISC 826, pp. 828–829, 2019.
https://doi.org/10.1007/978-3-319-96065-4_87

## 2 Methodology

A retrospective analysis was conducted among students (children, adolescents and young adults) with complaints of musculoskeletal symptoms reported after the usage of electronic devices. A total of 115 subjects between the age of 5 to 25 years were evaluated in the study. The evaluation and diagnosis were performed by a single orthopaedic and rehabilitation physician between the years of 2005 to 2017. After the diagnosis and assessment, all the patients underwent rehabilitation for 2 to 4 weeks using a sequenced protocol.

## 3 Results

The results of the study are as follows. 53% (n = 61) of the participants were male, with a median age of 16 years. The commonly used electronic devices were cell phone (80%), tablet (70%), laptop (65%), electronic games (54%), desktop computer (45%), portable music players (5%), and electronic book readers (2%). The commonest risk factors for MSD were identified as lack of rest breaks (85%), static loading (70%), hazardous body positions (60%), excessive load (30%), deficiency in design of tools/furniture or poor ergonomics (30%) and repetition (20%). Right upper limb musculoskeletal symptoms (84%) were predominant compared to bilateral (10%) and left upper limb (6%). Individuals who used the electronic device with the thumb alone reported musculoskeletal symptoms more commonly than those who used all the fingers. Myofascial Pain Syndrome (MPS) was the most commonly diagnosed clinical condition in 100% followed by Thoracic Outlet Syndrome (34%), tendinopathies of elbow, wrist or hand (22%) and Fibromyalgia Syndrome (18%). 32 (27.8%) of the subjects were found to have generalised hypermobility of joints as diagnosed by the Beighton criteria and 48 (41.7%) were found to have an Upper Crossed Pattern described by Vladimir Janda. After the rehabilitation following a sequenced protocol the VAS scale showed significant reduction in pain levels ($p < 0.01$).

## 4 Conclusions

With the increasing use of electronic devices, and lack of awareness in students, parents, academicians and medical professionals, MSD in students is likely to increase in the years to come. A coordinated multidisciplinary approach focusing on prevention, early diagnosis and appropriate rehabilitation is needed. Reducing the duration of electronic device use, training in ergonomics and body mechanics, including rest breaks, is recommended.

# Author Index





S. Bagnara et al. (Eds.): IEA 2018, AISC 826, pp. 831–834, 2019.
https://doi.org/10.1007/978-3-319-96065-4